A Companion
to Forensic
Anthropology

The *Blackwell Companions to Anthropology* offers a series of comprehensive syntheses of the traditional subdisciplines, primary subjects, and geographic areas of inquiry for the field. Taken together, the series represents both a contemporary survey of anthropology and a cutting edge guide to the emerging research and intellectual trends in the field as a whole.

A Companion to Forensic Anthropology

Edited by
Dennis C. Dirkmaat

WILEY Blackwell

Registered Office
John Wiley & Sons, Ltd, The Atrium, Southern Gate, Chichester, West Sussex, PO19 8SQ, UK

Editorial Offices
350 Main Street, Malden, MA 02148-5020, USA
9600 Garsington Road, Oxford, OX4 2DQ, UK
The Atrium, Southern Gate, Chichester, West Sussex, PO19 8SQ, UK

For details of our global editorial offices, for customer services, and for information about how
to apply for permission to reuse the copyright material in this book please see our website
at www.wiley.com/wiley-blackwell.

Library of Congress Cataloging-in-Publication Data

A companion to forensic anthropology/edited by Dennis Dirkmaat.
    p. cm. – (Blackwell companions to anthropology)
  Includes bibliographical references and index.
  ISBN 978-1-4051-9123-4 (hardback) ISBN 978-1-118-95979-4 (paperback)
1. Forensic anthropology.   I. Dirkmaat, Dennis.
  GN69.8.C659 2012
  599.9–dc23

                                        2011044947

A catalogue record for this book is available from the British Library.

Cover image: Forensic archaeologists conduct a field recovery of surface-scattered human remains from a
homicide case in Pennsylvania, USA. © Dennis Dirkmaat/Mercyhurst College; a biologist examines the
bones of Britain's oldest known TB victim, an Iron Age man from 300 BC. © Roger Bamber/Alamy; and
a partially mummified human skull from a forensic case in the US. © Dennis Dirkmaat/Mercyhurst College.

Set in 10/12.5 Galliard by SPi Publisher Services, Pondicherry, India

1   2015

Contents

List of Illustrations

List of Tables

Notes on Contributors

Bradley J. Adams, PhD, D-ABFA (University of Tennessee), is Director of the Forensic Anthropology Unit for the Office of Chief Medical Examiner in New York City. Dr Adams and his team are responsible for all forensic anthropology casework in New York City, including the ongoing recovery and identification efforts of victims of the World Trade Center attacks. Prior to accepting this position, Dr Adams worked at the Joint Prisoners of War, Missing in Action Accounting Command Central Identification Laboratory (JPAC-CIL) in Hawaii from 1997 to 2004. Dr Adams is a Diplomate of the American Board of Forensic Anthropology, a Fellow with the American Academy of Forensic Sciences, and a member of the Editorial Board of the *Journal of Forensic Sciences.*

James M. Adovasio, PhD (University of Utah), is Director of the Mercyhurst Archaeological Institute, and Provost, at Mercyhurst College, Erie, PA. He achieved world acclaim in the 1970s with his excavation of Meadowcroft Rockshelter, southwest PA, widely recognized as the earliest well-dated (*c.* 16 000 years ago) archaeological site in North America. He has also conducted multidisciplinary investigations at Mezhirich, Ukraine; Dolni Vestonice/Pavlov, Czech Republic; and Gault, TX. One of the world's foremost experts in perishable material culture (basketry, textiles, cordage) he has published more than 400 books, monographs, articles, and technical papers.

Mark O. Beary, MS (Mercyhurst College), is currently a doctoral candidate at the University of Missouri and the consulting forensic anthropologist for the Boone County Medical Examiner's Office in Columbia, MO. He received his Masters of Science at Mercyhurst College, Erie, PA. His research interests include forensic anthropology, taphonomy, archaeological theory, and bioarchaeology of the US Central Plains.

William R. Belcher, PhD, D-ABFA (University of Wisconsin-Madison), is currently Deputy Laboratory Director for the Central Identification Laboratory section of the Joint Prisoners of War, Missing in Action Accounting Command (JPAC-CIL). Dr Belcher has conducted archaeological research in northern New England and the Maritime Provinces of Canada, the North American Pacific Northwest, and Pakistan. With JPAC, he has led recovery and investigation operations throughout Southeast Asia, the UK, the Republic of Kiribati, the USA (Washington, Oregon, and California), and North Korea. Dr Belcher has published over 40 articles and chapters in professional archaeological, anthropological, and historical journals and books.

Hugh E. Berryman, PhD, D-ABFA (University of Tennessee), is a research professor in the Department of Sociology and Anthropology and Director of the Forensic Institute for Research and Education at Middle Tennessee State University. He is a faculty member of the National College of District Attorneys, and provides forensic anthropology consultation to the Joint POW/MIA Accounting Command Central Identification Laboratory (JPAC-CIL) in Hawaii, and the Office of the Tennessee State Medical Examiner. He has served on the Board of Directors for the American Board of Forensic Anthropologists, and co-chairs the Trauma and Recovery Sections of the Scientific Working Group for Forensic Anthropology (SWGANTH).

Jesper L. Boldsen, PhD (University of Aarhus, Denmark), is currently Professor of Anthropology and Head of ADBOU of the Institute of Forensic Medicine, University of Southern Denmark. His research interests revolve around population biology, epidemiology, demography, evolution, and leprosy. He is codeveloper of ADBOU (with George Milner).

Katelyn L. Bolhofner, MA (Arizona State University), is a PhD student at the Center for Bioarchaeological Research, School of Human Evolution and Social Change, Arizona State University. Katelyn works as a skeletal analyst on archaeological projects in Majorca and Cyprus, and as a bioarchaeology laboratory technician for the 4th Cataract Archaeological Project, Sudan.

David Boyer, MFS (George Washington University), is a nationally recognized authority on DNA identification in mass fatality incidents for his participation in the September 11, 2001 attacks, Hurricane Katrina, the Space Shuttle Columbia, the Tri-State Crematory incident, and other mass fatalities. From 1997 through 2006 Mr Boyer worked for the Armed Forces Institute of Pathology where he was the program manager for the Department of Defense (DoD) DNA Reference Specimen Collection Program and the Director of Operations of the DoD DNA Repository. Mr Boyer is currently working under a Department of Homeland Security grant to develop a mass fatalities response for Houston, TX, and the surrounding 13-county region.

Ciarán P. Brewster, MS (Mercyhurst College), hails from Longford, Ireland. He completed both Bachelor's and Master's degrees at Mercyhurst College, Erie, PA. During his time there, Ciarán assisted in many forensic anthropology cases. He has worked on numerous archaeological projects in the USA, Ireland, Spain, Portugal, Armenia, and Georgia. He is currently completing a PhD at University College Cork in Ireland. His research interests include Upper Paleolithic and Mesolithic societies, geometric morphometrics, and hunter–gatherer social systems.

Luis L. Cabo, MS (University of Oviedo, Spain), is the Director of the Forensic and Bioarchaeology Laboratory, and graduate student research advisor, Department of Applied Forensic Sciences, Mercyhurst College, Erie, PA. With a background in zoology and paleoanthropology, he has participated in over two dozen archaeological and paleontological field and laboratory projects in Spain, Portugal, and the USA. Since joining the Mercyhurst staff in 2003, Mr Cabo has assisted in the recovery and analysis of more than 200 forensic cases and has published numerous contributions on forensic anthropology, biostatistics, and human skeletal biology.

Erin N. Chapman, MS (Mercyhurst College), is a Research Fellow and Adjunct Faculty member in the Department of Applied Forensic Sciences at Mercyhurst College, Erie, PA. Her research interests focus on skeletal trauma analysis, forensic archaeological recovery of victims at various types of outdoor scene, and faunal analyses. Ms Chapman has assisted with human skeletal and faunal analyses for several historic and prehistoric sites in Portugal and in Asturias, Spain. She has assisted on nearly 30 forensic anthropology cases.

Christian Crowder, PhD, D-ABFA (University of Toronto), is currently the Deputy Director of Forensic Anthropology for the Office of Chief Medical Examiner (OCME) in New York City He is also an Adjunct Lecturer at Pace University, and holds a faculty position at the New York University Medical Center and an affiliation with the NYU Anthropology Department. In his present position with the OCME, Dr Crowder assists with anthropology casework in the five boroughs of New York City. He is also the site coordinator for the ongoing search and recovery of remains at the World Trade Center site. Prior to accepting the position in New York City, Dr Crowder was a forensic anthropologist at the Joint Prisoners of War, Missing in Action Accounting Command Central Identification Laboratory (JPAC-CIL) in Hawaii.

Dennis C. Dirkmaat, PhD, D-ABFA (University of Pittsburgh), is the Chair of both the undergraduate program in Applied Forensic Sciences and the Masters of Science in Anthropology (Forensic and Biological Anthropology Concentration) at Mercyhurst College, in Erie, PA, USA. Dr Dirkmaat has conducted over 400 forensic anthropology cases for nearly 40 coroners, medical examiners, and the state police in Pennsylvania, New York, Ohio, and West Virginia. He has published articles on the role of archaeology and forensic anthropology in forensic investigations, fatal-fire scenes, and mass fatalities.

Dr Dirkmaat has been a member of the US federal government's Disaster Mortuary Operational Response Team (DMORT) since its inception in the mid-1990s, having participated as a forensic anthropologist during mass fatalities in Pittsburgh (USAir Flight 427 crash in 1994), the island of Guam (KAL Flight 801 crash in 1997), and Rhode Island (Egypt Air 990 crash in 1999). In September 2001, he served as the primary scientific advisor to Somerset County Coroner Wallace Miller during the recovery and identification of victims of United Flight 93 in Shanksville, PA. Most recently, Dr Dirkmaat, consulting for the Erie County Medical Examiner's Office (Buffalo, NY), directed the recovery of victims of Continental (Colgan Air) Flight 3407 outside of Buffalo, NY, in February 2009.

Dirkmaat serves as a consultant for international companies involved in the recovery and identification of victims of plane crashes from around the world (including Kenya, Angola, and Peru) and other mass disaster events (such as the Haiti earthquake in 2010). He has completed two major grant projects for the design of national scene processing protocols for mass disasters, terrorist attacks, and fatal-fire scenes for the National Institute of Justice, US Department of Justice. He is co-chair of the Scene Search and Recovery, and Anthropology Subcommittees of the Scientific Working Group on Disaster Victim Identification (SWGDVI).

As an instructor, Dr Dirkmaat has close to two decades of experience organizing training courses and lecturing for local, state, and federal agencies in the USA, Mexico, Chile, Canada, and Spain. He has also presented over 70 lectures and papers discussing forensic investigation and anthropology at numerous regional, national, and international meetings.

Paul D. Emanovsky, PhD (University of Florida), joined the scientific staff of the Joint Prisoners of War, Missing in Action Accounting Command Central Identification Laboratory (JPAC-CIL) in 2005. Dr Emanovsky has participated in JPAC-CIL investigations and recoveries in Laos, Vietnam, North Korea, Germany, and Alaska. In 2004, Dr Emanovsky was deployed to Thailand to assist with the identification of victims just after the South Asian Tsunami. Prior to joining the JPAC-CIL, Dr Emanovsky worked on various archaeological projects as well as Physicians for Human Rights, Cyprus Mission. His research interests include length/shoe-size prediction based on the skeletal elements of the foot, and taphonomic processes affecting the decomposition and preservation of remains in a jungle environment.

Todd W. Fenton, PhD (University of Arizona), is an associate professor in the Department of Anthropology at Michigan State University, and board member of the Scientific Working Group for Forensic Anthropology (SWGANTH) and Board of Trustees for the Forensic Science Foundation. He is a past Chair of the Physical Anthropology Section of the American Academy of Forensic Sciences. His expertise includes human identification, skeletal trauma, ancestry, and forensic photo comparison. His current research is on pediatric cranial trauma. In addition to forensic case work, Dr Fenton is currently involved in bioarchaeological research in Albania and Italy.

Benjamin J. Figura, PhD (University of Binghamton), has responded in various capacities to numerous mass fatality incidents including the World Trade Center, American Airlines Flight 587, the Thailand Tsunami, and Hurricane Katrina, among others. He is currently a Forensic Anthropologist and the Director of Identification at the Office of Chief Medical Examiner in New York City where he serves as the agency's Disaster Victim Identification Coordinator and oversees the their use of the National Missing and Unidentified Persons System.

Luis Fondebrider, Lic., is cofounder and current president of the Argentine Forensic Anthropology Team (EAAF) based in Buenos Aires, Argentina. Fondebrider, as part of EAAF, has participated in over 900 cases in Argentina and abroad including Truth Commissions of Argentina, El Salvador, Haiti, Peru, and South Africa; International Criminal Tribunal for the Former Yugoslavia; Committee of Missing Persons of Cyprus; UN Secretary General Investigation Team for Democratic Republic of Congo; UN Commission of Inquiry on Darfur; and International Committee of the Red Cross (ICRC), among others. Fondebrider teaches forensic anthropology at the annual course of Legal Medicine of the Faculty of Medicine of the University of Buenos Aires.

Diane L. France, PhD, D-ABFA (University of Colorado), is the Director of the Human Identification Laboratory of Colorado and France Custom Casting. She is also an adjunct faculty member of Colorado State University. In addition to forensic anthropology, her work includes the differentiation of human and nonhuman skeletons, the location of clandestine graves and recovery of remains and associated evidence, and replication of skeletal elements for court exhibits. Dr France has assisted in hundreds of cases in the USA and internationally. She has served on the Board of Directors and as President of the American Board of Forensic Anthropology, and was the 2011 recipient of the Physical Anthropology Section's T. Dale Stewart Award for lifetime achievement in Forensic Anthropology.

Jeannette S. Fridie, MA (New York University), has been employed full-time in the Forensic Anthropology Unit at the Office of Chief Medical Examiner in New York City since 2005. She is interested in skeletal development of juveniles and bone trauma associated with dismemberment. She has recently coauthored a sharp-force trauma study funded by the US National Institute of Justice.

Heather M. Garvin, MS (Mercyhurst College), is completing her Doctoral degree in functional anatomy and evolution at Johns Hopkins School of Medicine. She was first introduced to forensic anthropology as an undergraduate volunteer at the University of Florida's C.A. Pound Human Identification Laboratory. She worked on numerous forensic anthropology cases while completing her Master's degree at Mercyhurst College, Erie, PA. She currently studies physical anthropology with special interests in modern human variation in cranial and postcranial sexual dimorphism.

Sara Getz, MS (Mercyhurst College), is currently pursuing her doctoral degree in anthropology at the Pennsylvania State University. Her research interests include the development and improvement of age-at-death estimation methods, the investigation and quantification of human skeletal variation, and the exploration of the interdisciplinary connections between biological anthropology, archaeology, the forensic sciences, and other related fields.

Desina R. Gipson, MS (Mercyhurst College), recently completed a GIS certificate program at Humboldt State University, CA. While at Mercyhurst College, Erie, PA, she was in charge of all forensic case field data collection, analysis, and presentation. She is interested in new methods in age-at-death estimation, burn trauma, morphometrics, and ancestry/population variation. She plans to focus on developing new geospatial analysis techniques for forensic site recovery and analysis for her PhD dissertation.

Jennifer Godbold, BA (New York University), graduated from the Post-Baccalaureate Certificate Program in Forensic Anthropology at Mercyhurst College, Erie, PA. She currently is employed with the New York City Medical Examiner's Office where she works with World Trade Center (WTC) families, maintains WTC remains, and works on their identification. She also works on a US National Institute of Justice grant for Missing and Unidentified Persons, performing case reviews of all New York City unidentified cases and entering them into national databases such as NamUs.

William D. Haglund, PhD, D-ABFA (University of Washington), is a forensic anthropologist specializing in forensic taphonomy and mass fatalities from war and human-rights violations. He served as Chief Medical Investigator of the King County Medical Examiner's Office, Seattle, for 14 years. From 1995 to 2009 he directed international forensic investigations, first for the United Nations and later for Physicians for Human Rights in numerous countries. Dr Haglund has coauthored and coedited a number of books and articles on forensic taphonomy, death investigation, and mass graves, including *Forensic Taphonomy*, *Advances in Forensic Taphonomy* and the *Medicolegal Death Investigator Training Manual*. He teaches medicolegal death investigation at the University of Washington.

Kristen M. Hartnett, PhD (Arizona State University), is a Forensic Anthropologist for the Office of Chief Medical Examiner (OCME) in New York City. Her job duties include forensic anthropological casework, archaeological excavation and monitoring of work at the World Trade Center, scene response, and disaster training and disaster response. In addition to duties at the OCME, Dr Hartnett has taught anthropology courses for Hunter College in New York City. Her research interests include bioarchaeology, disaster response and preparedness, analysis of blunt-force trauma, and estimation of adult age at death.

Joseph T. Hefner, PhD (University of Florida), is a forensic anthropologist with the Joint Prisoners of War, Missing in Action Accounting Command Central Identification Laboratory (JPAC-CIL) in Hawaii. Prior to his current position he worked for Statistical Research, Inc. and the Pima County Medical Examiner's Office in Tucson, Arizona, the University of Tennessee's Forensic Anthropology Center and Research Facility in Knoxville, Tennessee, and as a lecturer for Mercyhurst College, Erie, PA and the Univerity of Pittsburgh, Titusville, PA. His current research interests include nonmetric trait analysis, ancestry determination methods, quantitative data analysis, nonparametric data mining statistics, classification statistics, and geometric morphometrics.

Special Agent Michael J. Hochrein, BS (University of Pittsburgh), has served as a Federal Bureau of Investigation (FBI) Special Agent since 1988. He is currently assigned to Pittsburgh Division's Laurel Highlands Resident Agency. Prior to joining the FBI, Agent Hochrein was a field archaeologist specializing in historical archaeology.

As an FBI Evidence Response Team member, Hochrein conducts research, teaches, publishes articles, and testifies in areas of forensic archaeology, geotaphonomy, crime scene mapping, and general evidence collection. He is also an FBI-certified police instructor who has trained national police organizations in crime scene investigation techniques in the Americas, Europe, Middle East, Africa, and Southeast Asia.

R. Eric Hollinger, PhD (University of Illinois-Urbana), is a Supervisory Archaeologist in the Repatriation Office of the Smithsonian's National Museum of Natural History, Washington DC. He has excavated in the Caribbean and Ecuador but has concentrated on the Great Plains and Midwest of the USA. He is currently responsible for evaluating repatriation claims from around the country. He has published or presented on lithic, faunal, floral, human osteological, and architectural materials as well as on remote sensing, taxonomy, trade, warfare, migration, repatriation, pesticide-contaminated collections, and traditional care.

Richard L. Jantz, PhD (University of Kansas), is currently Professor Emeritus of Anthropology at the University of Tennessee-Knoxville and Director of the Forensic Anthropology Center. He developed the Forensic Anthropology Data Bank and is codeveloper of Fordisc, a popular forensic software package widely used by forensic anthropologists. His research concerns quantitative variation in human populations, especially anthropometric, osteometric, and dermatoglyphic data. He is interested in the application of forensic data to human variation topics such as secular change, and to forensic topics such as race, sex, and stature estimation from the skeleton.

Alexandra Klales, MS (Mercyhurst College), is currently a doctoral student at the University of Manitoba in biological anthropology. While completing her Master's degree at Mercyhurst she assisted on many forensic anthropology cases. Her research interests include sex and ancestry estimation of the human innominate, assessment of population affinity using geometric morphometrics, and methodological and observer error improvements in forensic anthropology.

Ericka N. L'Abbé, PhD, D-ABFA (University of Pretoria, South Africa), is a senior lecturer in the section of Physical Anthropology, Department of Anatomy, at the University of Pretoria in South Africa. Her research focuses on modern human skeletal variation, with specific application to the forensic sciences. She is the curator of the Pretoria Skeletal Collection of the Anatomy Department, and is active in research with national and international postgraduate students. She actively analyses and writes case reports for the South African Police Services and forensic pathologists.

Alicja K. Lanfear, MS (Middle Tennessee State University), is a doctoral candidate in the Department of Anthropology at the University of Tennessee, Knoxville. Alicja's doctoral research investigates the anthropometric history of Poland using the records of the *Institut für Deutsche Ostarbeit*, an anthropometric data set from World War II. Current research projects include hanging and its effect on decomposition, entomology and skeletal deposition of human cadavers, and primer-derived gunshot residue analysis on bone, research in collaboration with Dr Hugh E. Berryman, PhD, D-ABFA, that received the 2008 Ellis R. Kerley Award.

Stephen Litherland, MA (University of Birmingham, UK), is a senior forensic archaeologist at Cellmark Forensic Services leading the search, location, and recovery of human remains and other buried items. His research interests are in industrial landscape and buildings archaeology. He has worked extensively across Europe and recently assisted in confirming the presence of a series of mass graves of 250 Australian and British missing from the Battle of Fromelles (France) that took place in 1916. At present, he provides training in forensic archaeology to crime scene investigators for the National Policing Improvement Agency and at other events. During the past 12 years he has appeared as an expert witness in forensic archaeology in the UK.

Juan Luengo Azpiazu, MS (University of Oviedo, Spain), is a zoologist and physical anthropologist and President of Soluciones Antropométricas SL, a company providing anthropometric services to government agencies and private companies in sectors in which the measurement and distribution of body dimensions play an important role for the design and manufacturing of their products: clothing, shoes, automobiles, furniture, and architecture. His past research also includes areas such as kinanthropometry, the analysis of the influence of body dimensions and composition in the performance of premier adult and subadult athletes.

R. Lee Lyman, PhD (University of Washington), is Professor and Chair of Anthropology at the University of Missouri-Columbia. His research interests include late Quaternary paleomammalogy of the northwestern USA, and the history of American archaeology. He is author of *Vertebrate Taphonomy* (1994) and *Quantitative Paleozoology* (2008).

Nicholas Márquez-Grant, PhD (University of Oxford, UK), is a Forensic Anthropologist and Archaeologist at Cellmark Forensic Services (UK). He is also a Research Associate of the Institute of Human Sciences, University of Oxford. He is a specialist in human skeletal remains from archaeological sites from a variety of geographical areas in Europe, and in particular Spain. He has taught biological anthropology since 2001 at the University of Oxford. In recent years he has worked full time as a forensic anthropologist and archaeologist working for a large number of police forces in the UK in the search, recovery, location, and identification of human remains and has acted as an expert witness. He also trains crime scene investigators from a number of police forces in the UK.

Amy R. Michael, MA (Michigan State University), is a PhD student at Michigan State University in East Lansing, MI. Her research interests are primarily in the fields of skeletal biology and bioarchaeology, with a focus on dental histology and investigations of bone microstructure and issues of health and gender in prehistory using biological data. Amy works on archaeological projects in both west-central Illinois and in the Belize River Valley. Her dissertation research focuses on the analysis of human remains from a variety of cave and rockshelter contexts in Belize.

George R. Milner, PhD (Northwestern University), is currently Professor of Anthropology at the Pennsylvania State University. He specializes in human osteology and the prehistory of eastern North America, and has participated on excavations and conducted research using archaeological and skeletal museum collections in the USA, Europe, Africa, and Oceania. Recent publications include "Eastern North American population at A.D. 1500" in *American Antiquity* (Milner and Chaplin, 2010), "Archaic burial sites in the Midcontinent" in *Archaic Societies: Diversity and Complexity Across the Midcontinent* (Milner, Buikstra, and Wiant 2009), and "Advances in paleo-demography" in *Biological Anthropology of the Human Skeleton* (Milner, Wood, and Boldsen 2008).

Gregory O. Olson, MS (Mercyhurst College), completed 30 years of policing before accepting a position with the Office of the Fire Marshal for the Province of Ontario. He founded the Archaeological/Forensic Recovery Team of the York Regional Police in 1996, the only team of its kind in Canada. He is currently under part-time contract with the Canadian Military in the role of forensic anthropologist to assist in the recovery of disassociated remains in major military events. Mr Olson is also part of a three-member team to excavate historical military graves.

Stephen D. Ousley, PhD (University of Tennessee), is an assistant professor in the Departments of Anthropology/Archaeology and Applied Forensic Sciences at Mercyhurst College, Erie, PA. He served as Director of the Repatriation Osteology Laboratory in the Repatriation Office of the National Museum of Natural History at the Smithsonian Institution for 9 years. His professional activities revolve around anthropological databases, computer programming, and statistical approaches to biological anthropology, including Fordisc. His research interests focus on forensic anthropology, human growth and development, human variation, geometric morphometrics, and dermatoglyphics.

Rebecca S. Overbury, MS (Mercyhurst College), received her Master of Science degree in 2007 from Mercyhurst College in Erie, PA. Her thesis research included the study of skeletal asymmetry and how it might affect the accuracy of current aging methods employed in the practice of forensic anthropology. She is currently a medical student at the University of Tennessee Health Science Center in Memphis, TN, and will receive her medical degree in 2012 to pursue a career in internal medicine.

Nicholas V. Passalacqua, MS (Mercyhurst College), is a physical anthropologist specializing in human skeletal biology, forensic anthropology, and bioarchaeology. He is currently pursuing his PhD at Michigan State University. His research interests include age-at-death estimation, skeletal trauma and taphonomy, paleodemography, and paleopathology. Aside from his forensic anthropological work, Mr Passalacqua studies skeletal markers of medieval health and disease in Asturias, Spain.

Christopher W. Rainwater, MS (Mercyhurst College), is currently pursuing his PhD at New York University. Mr Rainwater works as a forensic anthropologist for the Office of Chief Medical Examiner in New York City engaging in active anthropological casework as well the ongoing World Trade Center recovery efforts. He also works as the Director of Photography for the agency. Mr Rainwater is currently engaged in research concerning the estimation of body mass as well as skeletal trauma, specifically sharp-force trauma and burned bone.

Julie Roberts, PhD (University of Glasgow, UK), is a senior anthropologist and archaeologist in the UK. Her professional experience includes senior forensic anthropologist with the British Forensic Team in Kosovo, exhumation and examination of British soldiers in Iraq, lead anthropologist in the London bombings, the Nimrod air crash, and British hostages murdered and buried in Lebanon. She is one of only two UK Disaster Victim Identification leaders in anthropology. She is registered with the National Policing Improvement Agency as an expert advisor in anthropology and archaeology, and has been involved in instructing crime scene investigators for the past 10 years.

Vincente Rozas, PhD (University of Oviedo, Spain), is a forest ecologist currently affiliated with the *Misión Biológica de Galicia*, of the Spanish *Consejo Superior de Investigaciones Científicas*. His research is primarily focused on the dynamics of secondary growth, population structure, and dynamics of tree species in response to environmental conditions. His research typically combines dendroecology, wood anatomy, and spatial statistics. He also develops and evaluates statistical techniques for the analysis of spatial data. Dr Rozas has assisted in the analysis of biological materials and spatial distributions from numerous archaeological and paleontological sites in Spain, Portugal, and the USA.

Norman J. Sauer, PhD, D-ABFA (Michigan State University), is Professor of Anthropology, Adjunct Professor of Criminal Justice and Radiology, and Adjunct Curator of the Museum at Michigan State University. His research has involved the anthropology of the concept of race, the relationship between trauma and the time of death, and bioarchaeology in the Midwestern USA and Latin America. He has assisted local, state, and federal agencies with hundreds of forensic anthropology cases. He has served on the Board of Directors of the American Board of Forensic Anthropology. He was the 2007 recipient of the Physical Anthropology Section's T. Dale Stewart Award for lifetime achievement in Forensic Anthropology.

John J. Schultz, PhD (University of Florida), is an Associate Professor of Anthropology at the University of Central Florida. He specializes in forensic anthropology with a primary research focus on forensic and archaeological applications of ground-penetrating radar (GPR) for grave detection, and detection of buried metallic weapons using various geophysical technologies. In addition, he is a consulting forensic anthropologist in the central Florida area for various law-enforcement agencies and the local medical examiner's office.

Natalie R. Shirley, PhD (University of Tennessee), is Assistant Professor of Anatomy at the LMU DeBusk College of Osteopathic Medicine and adjunct faculty in the University of Tennessee Anthropology Department. Her research interests include skeletal maturation in modern populations, age and sex estimation from the human skeleton, and skeletal trauma. She is an active participant in the American Academy of Forensic Sciences and received the Emerging Forensic Scientist Award (2007) for her research in skeletal maturation.

Marcella H. Sorg, PhD, D-ABFA (Ohio State University), is a Research Associate Professor at the University of Maine, jointly appointed in the Department of

Anthropology, the Margaret Chase Smith Policy Center, and the Climate Change Institute. She serves as forensic anthropologist for the Offices of the Chief Medical Examiner in Maine, New Hampshire, and Delaware. Dr Sorg has authored and coedited a number of books and articles on forensic anthropology and taphonomy, including *Bone Modification*, *Advances in Forensic Taphonomy*, *Forensic Taphonomy*, and *The Cadaver Dog Handbook*. Her current research, funded in part by the US National Institute of Justice, focuses on taphonomic approaches to remains exposed in outdoor scenes, and the estimation of postmortem interval.

Maryna Steyn, PhD (University of the Witwatersrand, South Africa), has been employed in the Department of Anatomy, University of Pretoria, South Africa, since 1988. As a specialist in human skeletal remains, she consults to the South African Police Service on decomposed and skeletonized human remains. She also conducts research on human remains from archaeological sites. She has published more than 70 papers in scientific journals, as well as several book chapters. M. Steyn is a member of the editorial board of *Forensic Science International*, and is the director of the Forensic Anthropology Research Centre, University of Pretoria.

Steven A. Symes, PhD, D-ABFA (University of Tennessee), is an Associate Professor of Anthropology at Mercyhurst College, Erie, PA. He is an expert in interpreting trauma to bone and a leading authority on saw- and knife-mark analysis, assisting federal, state, local, and non-US authorities in nearly 200 dismemberment cases and roughly 400 knife-wound cases. Dr Symes spent 16 years as forensic anthropologist for the medical examiner's office at the Regional Forensic Center for Shelby County, TN. In 2008 he was awarded the T. Dale Stewart Award for lifetime achievement by the Physical Anthropology Section of the American Academy of Forensic Sciences.

Hugh H. Tuller, MA (Louisiana State University), currently is a forensic anthropologist at the Joint Prisoners of War, Missing in Action Accounting Command Central Identification Laboratory (JPAC-CIL). He worked in the Balkans as a forensic archaeologist, first with the United Nations' (UN's) International Criminal Tribunal for the former Yugoslavia (ICTY), and later with the International Commission on Missing Persons (ICMP). In addition, he has participated in the JPAC response to the 2004 Asian Tsunami, conducted training in Colombia, and assisted in the joint Greek–Turkish recovery and identification project in Cyprus sponsored by the UN's Committee on Missing Persons.

Natalie M. Uhl, MS (University of Indianapolis), is a biological anthropologist currently pursuing a PhD at the University of Illinois at Urbana-Champaign, specializing in human skeletal variation. She received a Master of Science degree with an emphasis in forensic anthropology from the University of Indianapolis in 2007 and has assisted with casework in the USA and South Africa. Her current research interests include age-at-death estimation, statistical modeling, and brain and body-size evolution and allometry.

Traci L. Van Deest, MA (California State University, Chico), is a doctoral student in the Department of Anthropology at the University of Florida and serves as a graduate analyst for the C.A. Pound Human Identification Laboratory at the same institution. Her research interests include heat alteration of bone and human remains, cremated remains, human identification, human variation and anatomy, and skeletal biology.

Scott C. Warnasch, MA (Hunter College), is currently Lead Archaeologist for the New York City Office of Chief Medical Examiner's (OCME's) World Trade Center Human Remains Recovery Operation. Prior to this position, Mr Warnasch spent 2 years reviewing case files with the family members of World Trade Center victims. As a member of the OCME Forensic Anthropology Unit, Mr Warnasch also assists in crime scene recovery throughout the five boroughs of New York City and advises on cold cases. Mr Warnasch previously worked as an archaeologist specializing in cemetery and burial excavation and has taught archaeological field methods at field schools in California, Belize, Ecuador, and Italy.

Michael W. Warren, PhD (University of Florida), is currently an Associate Professor and Director of the C.A. Pound Human Identification Laboratory in the Department of Anthropology at the University of Florida in Gainesville. He also serves as the assistant director of the William R. Maples Center for Forensic Medicine. He received his Master's degree and PhD in anthropology at the University of Florida. Dr Warren is a Fellow and past Chair of the Physical Anthropology section of the American Academy of Forensic Sciences. He is a Diplomate and board member of American Board of Forensic Anthropology.

Ivana Wolff, MS (Mercyhurst College), is from Argentina. She serves as an Investigation Assistant in forensic missions for the Argentine Forensic Anthropology Team (EAAF) where she conducts archaeological surveys, performs extensive laboratory work, and partakes in excavations throughout Argentina. She has also spent several seasons working for the South African National Prosecutor Authority on numerous investigations and excavations. In July 2006 she was awarded the *Joven Talento Mendocino*, an honor given to those who stand out for their commitment to defending human rights, peace, and childhood.

Jamie A. Wren is an anthropology major at the University of Maine. He is a research assistant in the Regional Taphonomy project, and a forensic anthropology assistant, working with Dr Marcella Sorg. He is particularly interested in both forensic taphonomy and forensic radiology.

About This Book

In 2008, Dirkmaat et al. described how the field of forensic anthropology is experiencing what has been termed a "paradigm shift" in goals, methods, and perspectives related to five major external and internal factors. These factors include the rapidity and effectiveness of DNA to produce positive identifications, judicial decisions regarding the inclusion of science in the courtroom, the development of skeletal samples more appropriate for modern forensic populations, better biostatistics, and the incorporation of forensic archaeology.

The first step toward this shift occurred within the last 30 or so years and involved a serious reexamination of how human skeletal remains are analyzed and interpreted relative to solving modern forensic cases. New research and new perspectives in the analysis of human biological tissue resulted from this review, especially with respect to constructing biological profiles (the big four: chronological age, sex, stature, ancestry) based on skeletal evidence.

In the 1980s it was realized that a key factor causing limitations in the field was the fact that the source of most human skeletal biology research was from humans who had died close to a century earlier. Humans, in general, have undergone evolutionary changes since that time, and the appropriateness of the sample was called into question. In addition, issues were raised related to the composition of the samples, as to who was being sampled, and documentation of that sample in terms of precise information of age, ancestry, identity, and history of the individual. One way to solve this problem was to search for, and even create, more appropriate samples with remains of the more recently deceased. Various approaches to assembling new collections of human remains were tried with reasonable success.

One approach was to appeal to living populations at large to donate their corporal beings, en masse, to anthropologists who would dispose of their biological tissues in manners similar to those methods popular with the murdering denizens of modern North American society. In this case, an anthropologist would be carefully watching with notepad in hand to capture each event in the sequence of the decomposition.

Another approach was to set up shop in an excellent location by which to encounter a spectacular cross-section of American society such as the Los Angeles County Morgue. Here, the anthropologist, with Stryker saw in one hand and signed permission forms in the other, could peek over the pathologist's shoulder at the rather well-documented cross-section of American lying on the county morgue table. Small portions of their skeletal remains could be removed for analysis, e.g., bits of pelvis and clavicles; a small, unnoticed price to pay to provide one last, and lasting, benefit to society.

Still another approach was to ask forensic anthropologists to gather their measurements from their individual forensic cases and send them to a central location in eastern Tennessee. This data would be collated into one database for the use by researchers. All told, these three approaches yielded huge assets that were assembled in a relatively short time for all skeletal biologists examining modern humans to use.

Associated with these newly acquired samples of modern individuals was an insistence on analyzing trends and patterns in the distribution and configuration of skeletal attributes with more appropriate, ever more sophisticated statistical programs that could now be run on personal laptop computers rather than on CRAYS. In fact, a computer software program was developed specifically for analyzing modern forensic samples!

Armed with these new tools and samples, the analysis of human skeletal biology of modern humans took off and much good research was produced. When asked to validate the new methods used in forensic cases in preparation for courtroom challenges to their scientific validity, the discipline did not flinch and easily took on the challenge with renewed vigor and rigor and, thus, established a leadership role within biological anthropology.

A more significant challenge in general, potentially providing a hearty kick-in-the-pants, was administered by the National Academy of Science report in 2009 that questioned the scientific merit of many of the forensic science disciplines (all except DNA, of course). Dramatic federal-level oversight of the forensic sciences seems imminent, and enforcement of significant requirements through legislation and focus will likely be on institutional accreditation and personnel certification requirements of those handling evidence. A likely result will be a new wave of validation studies that will critically reexamine old standards. Both the Daubert ruling and the National Academy of Science report have already led the federal government (Federal Bureau of Investigation, National Institute of Justice, National Transportation Safety Board, and other federal agencies) to provide scientific leadership with the formation and sponsorship of Scientific Working Groups, Technical Working Groups, and sponsored research opportunities.

The result of all of this recent activity is that the field of forensic anthropology is well positioned to move forward as a strong, viable member of the forensic science team. As leaders in the field of human skeletal biology, forensic anthropologists can be held to the same expectations as other forensic specialists with scientific standards and validation.

If this were a book assembled to review and highlight recent work and current thinking with perspectives of forensic anthropology limited to 10–15 years ago, the vast majority of the book's chapters would deal with some aspect of producing ever-better human skeletal biological profiles. It would contain a large section on aging methods, chapters on how to determine sex, stature, and even ancestry, and a chapter or two on the analysis of unique skeletal features in efforts to provide positive identification. There might even be a chapter on digging up celebrities or historical figures to finally answer, "Who is buried in Grant's Tomb?" We might even attempt to identify the specifics of how the individual in Grant's Tomb lived and died. Certainly, a chapter would exist on how anthropologists provide biological profiles for each scrap of bone that enters the morgue in a mass disaster scenario. Forensic anthropologists, in general, would see the book as providing the proper focus of the field and with major topics addressed.

However, we must back up a bit. With all of the good skeletal biology research being conducted on the recently deceased, forensic anthropologists can now send law enforcement a scientifically solid report. Statistical analysis of human remains can provide more accurate posterior probabilities of sex, age, stature, and ancestry. After scanning all of those excellent Daubert-ready analytical results, the cop looks up from the scatterplots and asks... "But why are all but one of the hand bones missing?," "Why was the arm bones found 12 meters feet away from the rest of the skeleton?," and "How long do you think the remains were at the scene?" The forensic anthropologist can provide an educated guess based on years of experience of looking at stained, discolored, and even bitten bones with the realization that there really is no scientific basis for the answer provided. A better suggestion might be made to review the report of the recovery process or contextual clues as provided by law enforcement. However, this would likely reveal that the documentation of the scene was geared more toward the immediacy of the law-enforcement needs and may not be as wide-ranging and thorough as required by the forensic anthropologist. At closer examination, therefore, the educated guess based on this scene information, would not stand up in court under the unforgiving eye of Daubert. Were the cops working the case incompetent or poorly trained? Not necessarily. A review of standard law-enforcement and forensic science literature points to a glaring deficiency in what should be done at an outdoor scene. Although documentation and recovery protocols employed at indoor crime scenes are well thought out, thorough, and result in excellent, scientific reconstructions of past events, the same cannot be said of protocols applicable to forensic scenes found out of doors.

Forensic Archaeology Plays a Role

And so a small group of anthropologists/archaeologists saw the need for better recovery through the employment of archaeological methods, principles, and practices; techniques specifically tailored to modern forensic situations and scenes in the form of what was cleverly termed "forensic archaeology." And it works! It works as expected at burial sites, but also for carefully documenting evidence and remains scattered on the surface. It works for the recovery of fatal-fire scenes and even large-scale mass disaster scenes.

As a result of collecting good data at the outdoor scene through archaeology, it was realized that we now had data to reconstruct events as to what happened to the body since deposition at the outdoor scene at a level of insight already completed at the indoor scene.

FORENSIC TAPHONOMY ARRIVES

Some forward-thinking souls read the scientific literature beyond anthropological articles and tomes, and realized that the field of taphonomy worked spectacularly well in paleontology, paleoanthropology, and archaeology. They embraced it and added the F-word and forensic taphonomy was born; note, however, the late birth in the 1990s. One part of the forensic taphonomic analysis requires a thorough and detailed analysis of how individual bones are fractured, degraded, stained, sub-bleached, root-etched, discolored, and chewed upon. Another component requires the analysis of how these bones were dispersed or buried, and their location with regard to sunlight, shade and water. In other words, the contextual information of the scene is an important component of the analysis. Reasons for why bones were missing or not in their original anatomical position could now be answered scientifically. For the first time, postmortem interval estimates had a scientific basis.

A rather unexpected result of the employment of forensic taphonomy to the outdoor forensic scene was a dramatic improvement in human skeletal trauma. Although the biomechanics of breaking bones was relatively well known as a result of the engineering field, problems with recovery remained, especially issues with the timing of the trauma. Healed versus unhealed evidence is clear; a more difficult assessment is to be able to differentiate between fractures to "fresh" bone that contains the organic (collagen) component versus fractures to "dry" bone that has lost significant amounts of collagen. However, it takes a while to lose the collagen component of bone (weeks, months, years?) and if trauma is applied to bone during that period it will behave as fresh bone and it will be difficult to distinguish perimortem (as associated with the death of the individual) from an event that occurs a period of time after death. If, for example, the bone is dropped or mishandled during recovery, it is difficult to distinguish it from perimortem trauma. A well-documented recovery, conducted by individuals familiar with handling bones, provides the best scenario for maximizing information and ensures that the trauma patterns are accurately interpreted.

SHIFTING THE PARADIGM

And so…back to the point where forensic anthropology becomes much more than an attempt to provide a biological profile that will help narrow the missing person's list. The incorporation of forensic archaeology into everyday practice, principles, and methodologies has opened opportunities, new perspectives, and new roles in forensic science and medicolegal death investigation, and maybe it has even raised the field's stature in the eyes of the stern father, physical anthropology. A true paradigm shift is indicated. This book is a reflection of this new paradigm and is best exemplified by the table of Contents.

Obviously, human skeletal biology still provides the centerpiece of forensic anthropology and so chapters on new techniques and methods, as well as interpretative tools related to determinations of chronological age, sex, stature, and ancestry will be found within the pages of this book. We have included discussions of biostatistics in general, and personal identification derived from skeletal structures, including insight into highly altered remains. However, for the moment, these chapters take a back seat to some of the other aspects of the field that have gone previously unnoticed or unrealized. We have some catching up to do, especially if this book is used to supplement introductory forensic anthropology books available today that treat some of the topics in this book rather superficially. Therefore, major sections of this Companion book will focus on the forensic archaeological recovery of human remains from a wide variety of forensic scenes, as well as forensic taphonomy and human skeletal trauma. The forensic community at large has noticed these new-found skill sets of a well-trained forensic anthropologist. Forensic anthropologists are now providing significant and unique roles in day-to-day operations in medical examiner's offices, in federal government-sponsored efforts to locate and identify previously unlocated US military war dead, and at mass disaster incidents. All of these issues are more fully developed in chapters throughout this book. The perspective taken by the editor is that forensic anthropology is a vibrant discipline that is more scientific as a result of recent reevaluations, and produces unique, relevant research that is, in turn, used by the big sister discipline of physical anthropology. Forensic anthropology has become a critical partner to forensic scientists, forensic pathologists, and law enforcement.

Previewing the Book

Following a discussion of forensic anthropology in general and the important influences upon it in recent years found in Chapter 1, subsequent sections of the book will focus on major aspects of the field that are undergoing significant change and reconsideration since their first inception in the 1970s.

Part II will include several chapters on the field recovery of human biological evidence. Individual chapters will focus on the benefits of forensic archaeology in general, law-enforcement perspectives on the role of forensic archaeology, and how forensic significance is approached. Recent technological advances in the search for buried features, as well as principles and protocols in the recovery of fatal-fire scenes, mass disaster scenes, and mass graves in human-rights cases, and discussion on how more thorough analysis of the spatial distribution of evidence (including human remains) can address issues of commingled remains, will be addressed. Although this section may seem to be a bit top-heavy, this is an area of forensic anthropology that has been largely ignored and the benefits have been unrealized until recently. These chapters should provide a baseline for discussion of the role of forensic archaeology within law enforcement, the forensic sciences, and forensic anthropology.

Part III will focus on recent developments in human skeletal biology, which has historically been the bread and butter of forensic anthropological work. Rather than recite a list of individual studies related to the estimation of age, sex, stature, and ancestry, the authors were asked to discuss current trends and unique approaches and to rethink how forensic osteology is accomplished. The reader will quickly realize that those requests were well met, and very interesting and unique discussions have been produced in the chapters.

Part IV deals with a specialized aspect of forensic anthropology that is unique to the field: human skeletal trauma analysis. Primarily developed from the hands-on work conducted in the medical examiner's morgue while working alongside forensic pathologists (discussed also in Part VII of the book), the analysis of skeletal trauma has benefitted greatly from an understanding of the soft-tissue evidence derived from this interaction. Recent research in the biomechanics of skeletal trauma provides an innovative approach to the interpretation of bone fractures, which is icing on the cake.

Part V includes chapters on advances in human identification with obvious discussion of DNA collection and analysis, and how it will eventually impact forensic anthropology. We have also included interesting discussions of human cremated remains research and video superimposition efforts in producing personal identifications.

Part VI presents the topic of interpreting events surrounding the death and what occurs after the deposition of the body known as forensic taphonomy. Methodologies and practices in the collection and interpretation of field contextual information, as well as the laboratory analysis of the bones and their modifications are discussed.

Part VII highlights the history and current configuration of the roles and duties of forensic anthropologists in nonacademic settings. Specifically, the responsibilities as full-time, productive participants in the day-to-day operations of the medical examiner's office and as forensic anthropologists in the collection and analysis of the USA's previously unrecovered war dead are described.

Part VIII provides an overview of the fields of forensic anthropology and forensic archaeology outside of North America. Discussions of forensic anthropology perspectives and practices in Europe and South Africa, as well the Argentinean Human Rights Team, are a major element of this section.

The final section of the book, **Part IX**, covers a number of professional issues related to forensic anthropology. Professional ethics, the view of the field by an outside observer, as well as the role Daubert will play in the field, are presented.

The book is meant to serve as a supplement to textbooks in forensic anthropology. The few introductory textbooks available provide an overview of the field, often with an expected focus exclusively on the analysis of clean human bones. "Method" books provide details of specific studies on how to establish a human biological profile and provide positive identification from skeletal features. This book would then be used to highlight the latest thinking and perspectives in the field, advances in techniques, provide excellent up-to-date bibliographies, and current methods used by professional forensic anthropologists "in the trenches." This book could, therefore, be used in advanced undergraduate courses in forensic anthropology and graduate courses in forensic anthropology, bioarchaeology, and osteology-based physical anthropology. Hopefully, it will be of interest to professional forensic anthropologists at large.

Dennis C. Dirkmaat
Luis L. Cabo

REFERENCE

Dirkmaat, D.C., Cabo, L.L., Ousley, S.D., and Symes, S.A. (2008). New perspectives in forensic anthropology. *Yearbook of Physical Anthropology* 51: 33–52.

Acknowledgments

I wish to thank many people for their assistance, patience, diligence, and expertise during the construction of this volume. First, thanks to the authors who were asked to consider sharing their knowledge and experience from the perspective of how recent developments have significantly altered the field of forensic anthropology. From my humble perspective, they have produced excellent chapters. I hope that they will forgive me for cajoling and hounding them for their manuscripts.

My Mercyhurst colleague, Luis L. Cabo, was instrumental in early discussions of the vision for the book, and later as collaborator on more than a few chapters. He also provided a key stimulus for drawing up innovative ways to push the forensic anthropology envelope.

Next, thanks go to all the wonderful graduate students in the Masters of Anthropology program at Mercyhurst College who have, during the past few years, helped with editing, reviewing, and checking references for the book.

Four individuals must be singled out for their editing prowess: Zoey Alderman-Tuttleman, Lauren Diefenbach, Mary Kuhns, and Kathi Staaf. The difficult task of producing the index for the book fell under the competent direction of Diana Messer. Indexing could not have been completed without the assistance of the graduate students.

The National Institute of Justice funded two of our projects that now form the backbone of the chapters on documentation and recovery of mass disaster and fatal fire scenes. Thanks to project managers Danielle McLeod-Henning and Henry Maynard for their assistance and encouragement throughout all phases of those projects.

Thanks to the Erie County Medical Examiner's Office, Buffalo, NY for permitting forensic archaeologists to assist their office in the recovery of the Colgan Air crash and then allowing us to present the final details and benefits of that archaeological approach in a number of venues, culminating in the descriptions in this tome.

I would be remiss not to thank Rosalie Robertson at Wiley-Blackwell for the invitation to contribute to the impressive Companion series, her associate Julia Kirk for poking and prodding to keep the work moving, Nik Prowse for the fine-tuned copy-editing, and Leah Morin for final proofing, indexing, and production.

And, finally, thanks to my wife Karen for patiently dealing with computers, books, papers, and coffee cups strewn about the house, and specifically on the kitchen table, for the last few years.

Dennis C. Dirkmaat

CHAPTER **1**

Forensic Anthropology: Embracing the New Paradigm

Dennis C. Dirkmaat and Luis L. Cabo

Introduction: The Entity

It seems only natural that a volume devoted to a particular area of scientific expertise would start with a chapter aimed at providing some sort of general overview and definition of the field. This requirement would appear even more relevant in the present volume, as many experienced forensic anthropologists may have trouble identifying some of the areas covered in the book as part of their everyday work, or even as remotely related to the discipline that went by the name of *forensic anthropology* when they were growing their academic or professional teeth. Just a few decades ago, most practicing forensic anthropologists would likely have protested even the suggestion of including in the picture many of the subjects that are presented in this volume as well-established, integral parts of forensic anthropology.

These differences in the conception of the field go beyond the methodological discrepancies derived from the logical substitution of old with new techniques. The 1970s forensic anthropologist traveling forward in time would most likely realize right out of the time machine the relevance of gaining a better understanding of DNA analysis, for example, as the subject directly relates to victim identification, the classical goal of forensic anthropology. But forensic archaeology? Really? How does that relate to victim identification, and wouldn't it be a job for the police anyway? And what about trauma analysis? Didn't we grow up reciting every night in our bedtime prayers that

A Companion to Forensic Anthropology, First Edition. Edited by Dennis C. Dirkmaat.
© 2012 John Wiley & Sons Ltd. Published 2015 by John Wiley & Sons Ltd.

"the forensic anthropologist cannot discuss the cause and manner of death"? And they even want to look at the weapon too, as if you did not have tool-mark analysts for that.

Some modern physical anthropologists equally may be troubled by the view of the field presented in this book. Wasn't forensic anthropology supposed to be just a direct application of physical anthropology techniques and, hence, once you knew your human osteology and general physical anthropology, you were ready to take on forensic cases? What is all this nonsense about soft tissue, postmortem intervals, scene investigation, or (brace yourself) paleopathology not providing a valid foundation for trauma analysis?

It is not possible to make sense of these apparent betrayals of the sacred principles and teachings of our forensic forefathers without taking a look back at the origins and development of the discipline of forensic anthropology. As Dickensian heroes, scientific fields usually rise from humble origins and goals, often even from necessity, gaining momentum, complexity, and scope as the pages are turned, before we can come to meet the wise, mature, and successful individuals who greet us from the closing paragraphs of the novel. The character cannot be encapsulated in any single snapshot, taken at a particular moment in time, or by its current state, but we can understand it fully only as its personality unfolds during personal encounters and acquaintances, traumatic or enlightening episodes, obstacles, successes, and setbacks. In other words, scientific disciplines do not evolve from their definition, but are defined by their evolution.

As a matter of fact, if we had to look for a Fagin or an Ebenezer Scrooge in our story, it would probably be some stubborn, almost religious historical adherence to a self-inflicted, very restrictive, and initial definition of forensic anthropology: to wit, as a strictly applied laboratory field, devoted solely to aiding in victim identification. In a sense, the story that we will uncover in the remaining pages of this chapter (and we may even dare say in the remaining chapters of this book) is mainly that of the struggle to grow beyond this classic definition, climbing the conceptual ladder from the humble origins as a technical, applied field, to the heights of a fully grown scientific discipline, with interests far more diverse than just victim identification. In our view, also as with Dickens' novels, this story is fortunately one of success, even if (spoiler alert) it doesn't exactly end with our 2 m, 100 kg Tiny Tim ice skating along the streets of old London.

It could not be any other way as, to a large extent, ours is also mostly an American story, and we all know that those always end happily. Although modern forensic anthropology definitely is not just an American enterprise, given that Europe and other areas of the world have made very important contributions to the history of the field, it is in the USA where the story has presented a more linear, consistent narrative. The story of American forensic anthropology is not based only on somehow isolated individual contributions, but rather characterized by an actual continuity along a well-defined tradition of research and professional practice. It can be stated rather confidently that forensic anthropology was born and took its first and more important steps in the United States of America; maybe not necessarily as a concept, but at least as an actual professional field, with a cohesive, constant, and independent body of practitioners, rather than as an additional task for other professionals, such as forensic pathologists. Forensic anthropology, though often presented as a relatively young discipline given its formal configuration and recognition in the 1970s, has a rich history in the USA, spanning most of the twentieth century.

And here is where the Dickensian parallels end. We might talk of humble origins in relation to the initial scope of the field but, as will be discussed below, when it comes to practical terms, our hero, like Darwin in Dickens' time, was born into a quite healthy and wealthy intellectual household. The participants in the early development of the field were some of the premier physical anthropologists of the day. Much of what we know about human skeletal variation and how to determine all aspects of the biological profile (age, sex, stature, ancestry) for physical anthropological purposes arose from the initial consideration of human bones from forensic contexts by these pioneers and their direct descendants, in their efforts to solve forensic identification issues (Kerley 1978). Thus, forensic anthropology is no stunted child (neither *ignorance* nor *want*) crusted with scabs and stooped with rickets, taking arms against a sea of troubles with a stomach bloated from malnutrition. Ours would be more of a *coming-of-age* story, starring the wealthy kid who becomes a somewhat spoiled and self-centered teenager, lounges through college and, in the end, recovering from alcoholism, becomes a leader of men: once again, an American story.

Because we must admit that there was actually quite a bit of lounging after formally defining the field in the 1970s, and a period of relative stasis in which minimal research was conducted that was directly applicable to the analysis of skeletal remains in forensic contexts (Snow 1973). Few cases were referred to the forensic anthropologist, and the answers proposed to most forensically relevant questions relied on old analytical methods derived from outdated skeletal samples. Career and state were at stake, and change was required. In the late eighties İşcan even warned that "this entity can stagnate or even self-destruct if the direction of future research is not carefully planned" (İşcan 1988a: 222)

In the following sections, the history of the field will be reviewed, mostly from an American perspective. As the story unfolds, you will see the character grow and mature, shiver as outrageous fortune throws new slings and arrows in its path, mostly in the shape of legal rulings and the development of other fields, and rejoice when characters like forensic taphonomy, forensic archaeology, or trauma analysis come to the rescue. It is clear now that forensic anthropology is moving away from fulfilling İşcan's prophecies of stagnation and self-destruction. In fact, the discipline is witnessing a revitalization derived from a "new conceptual framework" (Little and Sussman 2010: 31) in philosophy, composition, and practice.

This shift transpired because of a variety of factors, but primarily resulted from: (1) a critical self-evaluation of discipline definitions and best practices; and (2) strong outside influences from DNA, federal court rulings, and Congress-mandated assessments of the forensic sciences. At one time faced with extinction because of the "threat" posed by the ability of DNA to provide quick and precise personal identifications of unknown skeletons, the field has re-emerged in the last 10 years as a robust scientific discipline, able to stand on its own because of the realization of unique strengths, perspectives, and research goals. In other words, by looking outside the (packaging) box, a stronger forensic anthropology was developed. Of course, the job is not finished completely and our hero still is to face many new challenges in the future, but it is good just to be alive.

First, A Bit of History: The Early Years

It is suggested that forensic anthropology gained notoriety and acquired a face as a scientific discipline in the late 1930s, with the publication of Wilton Krogman's series of articles in the *FBI Law Enforcement Bulletin* (Krogman 1939a, 1943; Krogman et al. 1948). Krogman can be considered as the first renowned practitioner of endeavors with police that became known as "forensic anthropology." He was a brilliant scholar, researcher, and academician who trained with the likes of Sir Arthur Keith, in Great Britain, and T. Wingate Todd, at Western Reserve University in Cleveland, Ohio (Haviland 1994; Johnston 1989). Krogman later taught at the universities of Chicago and Pennsylvania. His research work was devoted largely to child growth and development, although he cultivated many other interests in human biology during his career. Even though a recognized expert on human identification, he was not contacted very frequently by police to assist in the construction of a biological profile for the unknown human skeletal remains that were brought to his laboratory (Haviland 1994). In this laboratory-based and episodic involvement in forensic cases, his profile is very similar to most of the other "practitioners" of the day, prior to the 1970s. Before Krogman's time, the history of the field had been written mostly by the contributions of diverse "anatomists-morphologists-anthropologists" (Kerley 1978: 160), who conducted research on variation in the human skeleton, which aimed at answering questions that at times arose in forensic settings (Pearson and Bell 1919).

Although he might have attained higher celebrity status, Krogman was not alone. The aforementioned T. Wingate Todd, as well as Aleš Hrdlička, Earnest Hooton, and a few other renowned physical anthropologists of the first half of the twentieth century also provided human identification services intermittently for the police (Kerley 1978); Hrdlička perhaps more than any other physical anthropologist of the day. Working out of the Smithsonian Institution, in Washington DC, he published little on the issue of human identification but consulted with police and, especially, the Federal Bureau of Investigation (FBI) on a large number of cases, from 1936 until his death in 1943 (Ubelaker 1999). This relationship was continued by Hrdlička's hand-picked successor, T.D. Stewart, although Stewart's interest in medicolegal issues resulted in a number of important articles (Stewart 1948, 1951). Todd, on the other hand, found his forensic line of work promising enough to realize the value of constructing a significant collection of human skeletal remains, aimed at studying human variation and answering basic research questions. With this purpose, Todd started expanding a small collection that had been started by Carl A. Hamann, his predecessor at the Case Western Reserve Medical School. The results of Todd's efforts came to form the basis of the *Hamann–Todd Collection*, the largest assemblage of modern human remains in the world, comprising more than 3300 individuals. Todd and his coworkers (including Montague Cobb and Krogman) used this collection to conduct basic research in human skeletal biology, notably that including age-related changes in the cranial sutures (Todd and Lyon 1924, 1925a, 1925b) and the pubic symphysis (Todd 1920, 1921a, 1921b). These studies have served as basic references and a starting point for the work of scores of researchers in many fields of anthropology, from forensic anthropology through bioarchaeology, and even paleoanthropology. The collection is housed currently at the Cleveland Museum of Natural History, in Cleveland, Ohio, where it attracts a multitude of researchers from throughout the USA and beyond. Apart from its large sample size, the Hamann–Todd collection, under the wise supervision of Lyman M. Jellema, also may be considered one of the better-curated comparative samples of human skeletons in the world. Krogman worked in Todd's laboratory from 1931 until 1938 (İşcan 1988b; Krogman 1939b; Haviland 1994) and the forensic cases that came to the lab likely provided the stimulus for Krogman to start considering the broader applications of human skeletal biology to other disciplines, including medicolegal investigation.

As the Hamann–Todd collection was being amassed, William Terry, of Washington University, St. Louis, Missouri, was collecting unclaimed and donated bodies used in anatomy classes at their medical school and other Missouri institutions. In 1941, Mildred Trotter took over from Terry and further increased the size of the collections. In 1967, the 1728 individuals that comprised the collection at that time were sent to the Smithsonian Institution for their continued curation and availability for research (Hunt and Albanese 2005).

Wars have helped to keep 'forensic' anthropologists employed and busy during the following decades. During World War II, Charles Snow, Mildred Trotter, and Harry Shapiro assisted in the identification of US war dead and even started collecting basic biological data from these war casualties (Stewart and Trotter 1954, 1955; Trotter and Gleser 1952). T. Dale Stewart, Thomas McKern, Ellis Kerley, and Charles Warren did the same during the Korean War (McKern and Stewart 1957; Klepinger 2006) and the Vietnam War (Stewart 1970; Ubelaker 2001). This eventually led to the formation of the US federal government's Central Identification Laboratory in Hawaii (CILHI) and Thailand (in the early and mid-1970s), renamed the Joint POW/MAI Accounting Command (JPAC) in 2003. Kerley (1978) suggests that it was this work for the US Armed Forces in the 1950s that legitimatized forensic anthropology as a scientific discipline. As an example of the quality of research produced, it was at this time that new standards for determining adult stature were presented by Mildred Trotter (Trotter and Gleser 1952, 1958; Trotter 1970), finally revising the turn-of-the-century European standards. Most of the publications from these times dealt with age-related changes in a few parts of the body (McKern and Stewart 1957), although, since the vast majority of individuals involved in the conflicts were white males between the ages of 17 to 30, the sample was slightly skewed.

In times of peace the job of helping the police identify the dead in the USA remained rather sporadic, infrequent, and limited. The few anthropologists who did this work all had similar curricula vitae: essentially academicians or museum specialists who were better known for their basic research in the field of physical anthropology, whereas their forensic anthropology work was conducted on the side. In the typical scenario of the day, after collection by the police, the remains were brought back to the morgue for possible identification, hopefully through soft tissue comparisons, including tattoos and medical interventions. Dental comparison was attempted next. If this proved fruitless, often the remains were taken to artists for facial reproduction, usually via molding of clay over the cranial remains. Only as a last resort were anthropologists sought, often by calling the local university for someone familiar with human osteology, to reevaluate components of the biological profile to help narrow down the missing person list. Perhaps the age, sex, stature, or ancestry determined by the police, pathologist, dentist, or sculptor was wrong! By the time of the analysis, the remains had passed through many hands and likely had been altered in some way, either by the police during recovery or transport, the pathologist during autopsy, the dentists during the all-too-common practice of "resecting" the jaw from the cranium, or by the sculptor putting clay on bone. Unfortunately, this description still remains a fairly accurate portrayal of the current situation.

The More Recent Years

Another important turning point in the field can be attributed to Krogman and the publication of his book, *The Human Skeleton in Forensic Medicine* (1962), after which the field was visible and well presented to a much wider forensic and medicolegal audience. During the 1960s and early 1970s police began to rely more and more on physical anthropologists to provide important information for their investigations regarding skeletal remains. As a result of this interest, an increase in basic research ensued that related to the identification of the recently deceased. This research often was described in physical anthropology journals and provided better methods to determine age, sex, stature, and ancestry (e.g., Bennett 1987; Fazekas and Kosa 1978; Phenice 1969; Gilbert and McKern 1973; Giles 1970; Giles and Elliot 1964; El-Najjar and McWilliams 1978; İşcan 1989; İşcan and Kennedy 1989; Stewart 1970, 1972). Bass (1969, 1978, 1979), Kerley (1978), İşcan (1988a), Stewart (1976), and others provide a rather comprehensive list of important articles and topics in the field over the last 30 years.

Academically, a few institutions arose that focused on skeletal biology contributions to medicolegal contexts and offered training and casework experience (Ubelaker 1997). The first institution of the sort was the University of Kansas in the 1960s. Many of the key forensic anthropologists were either on the faculty, including William Bass, Kerley, and McKern, or were graduate students, including Walter Birkby, Ted Rathbun, Richard Jantz, George Gill, Judy Suchey, and Doug Ubelaker (Rhine 1998; Bass 2001). However, after a brief time the department broke up as Bass left for the University of Tennessee in 1971, Kerley for Maryland, and McKern for Simon Fraser in 1972 (Ubelaker and Hunt 1995; Rhine 1998). Birkby, one of Bass' first students, later established a program at the University of Arizona in 1983. These departments, along with the University of Florida program established by William Maples in 1968, provided the focus of forensic anthropology research and training in the USA through the 1980s and 1990s (Falsetti 1999; Maples and Browning 1994).

In addition to academia-based training centers, Clyde Snow became a training center by himself with his work with what came to be termed "human rights" in Argentina in the 1980s, on cases involving historical figures (Josef Mengele), Custer battlefield participants, mummies, victims of John Wayne Gacy, and plane crash victims (Joyce and Stover 1991). It may be suggested that he was the first individual to conduct forensic anthropology casework on a full-time basis.

By the 1970s Krogman, Snow, Kerley, and a few other physical anthropologists had been attending the American Academy of Forensic Sciences meetings in the General Section on a regular basis (Kerley 1978). More and more of their time was taken up with forensic cases, conducting research, and teaching until they decided that it was time for some recognition of this particular specialty. In 1972, the Physical Anthropology section was created within the Academy (Kerley 1978; Snow 1982). Shortly thereafter, in 1977, the American Board of Forensic Anthropology was created and certification for forensic anthropologists in North America was in place.

This core group of "forensic anthropologists" also settled on a name – forensic anthropology – for the work done with the police which attempted to provide clues to the identity of the unknown deceased found in unusual circumstances.

Apparently they were unconcerned that the Germans had originally used the term in the 1940s and 1950s after World War II to describe a field of endeavor to determine ancestry and familial relationships of kids orphaned by the war (Schwidetsky 1954; Stewart 1984).

With the construction of a new name for the medicolegal work completed by these physical anthropologists, definitions were needed. One of the key publications that served to provide a basic working outline for the discipline was Stewart's *Essentials of Forensic Anthropology* published in 1979. T. Dale Stewart was a curator at the Smithsonian who was best known for his work with the Neanderthal (*Homo neandertalensis*) remains recovered from Shanidar Cave in Iran (Stewart 1977). As described above, the common scenario was that remains were brought to Stewart in a box by the police, especially the FBI, from their primary headquarters across the street from the Smithsonian in Washington DC, for which he provided a biological profile of the unidentified individual. Nonetheless, his full-time job remained as a physical anthropologist.

Stewart's *modus operandi* with respect to forensic casework (part-time, infrequent, and after the remains had passed through many hands) formed his definition of the field, which served as a guideline to the discipline since that time:

> Forensic anthropology is that branch of physical anthropology which, for forensic purposes, deals with the identification of more or less skeletonized remains known to be, or suspected of being human. Beyond the elimination of nonhuman elements, the identification process undertakes to provide opinions regarding sex, age, race, stature, and such other characteristics of each individual involved as may lead to his or her recognition. (Stewart 1979: ix)

Other definitions by active practitioners followed, which were along the same vein:

> Forensic Anthropology encompasses the application of the physical anthropologist's specialized knowledge of human sexual, racial, age, and individual variation to problems of medical jurisprudence. (Snow 1973: 4)
> Forensic Anthropology is the specialized subdiscipline of physical anthropology that applies the techniques of osteology and skeletal identification to problems of legal and public concern. (Kerley 1978: 160)
> Forensic Anthropology is that branch of applied physical anthropology concerned with the identification of human remains in a legal context. (Reichs 1986: xv)
> Forensic Anthropology is the field of study that deals with the analysis of human skeletal remains resulting from unexplained deaths. (Byers 2002: 1)

Forensic anthropology was experiencing renewed recognition within the forensic sciences and law enforcement as a field that could provide an important and reliable role in medicolegal investigation (Bass 2006; Rathbun and Buikstra 1984; Krogman and İşcan 1986). As a result of this renewed interest, new research in human skeletal biology arose. During the 1980s and 1990s, forensic anthropology began addressing some of the more pressing issues related to modernizing the determination of a biological profile of the recently deceased: reevaluation of chronological age markers, including the pubic symphysis (Brooks and Suchey 1990; Suchey et al. 1986), cranial sutures (Meindl and Lovejoy 1985), auricular surface (Lovejoy et al. 1985), and rib ends (İşcan and Loth 1986, 1989); reconsidering assessing ancestry in modern individuals (Gill and Rhine 1990); stature estimation (Ousley 1995); and trauma (Maples 1986; Merbs 1989), to name but a few important studies. It could be argued that modern human skeletal biology experienced a renaissance in research unseen since the 1920s, because of the rise of forensic anthropology (İşcan 1988a).

By the end of the twentieth century, forensic anthropology, though now proudly with a name, definitions, and better analytical methods, still was not too dissimilar to what had been practiced throughout the previous 50 years. Forensic anthropology has been considered a subfield of physical anthropology, almost exclusively laboratory-based (Wolf 1986), and done only occasionally on an as-needed basis by academia-based consulting physical anthropologists. Still, by the turn of the new century, considering its relatively short formal history, forensic anthropology was experiencing what could be termed the "salad days," probably best exemplified by Kerley's colorfully enthusiastic endorsement of the field: "The delightful days of early summer will probably continue to disclose to the adventurous the decomposed harvest of winter's crimes, and the forensic anthropologist is still the person best trained to reconstruct the biological nature of such skeletal remains at the time of death" (Kerley 1978: 170).

CHINKS IN THE ARMOR: CONSIDERING BEST PRACTICES

As part of the reevaluation during the 1990s of forensic anthropology in general, and human skeletal biology in particular, the old reliable skeletal analytical methods and their applicability to modern forensic cases came under scrutiny. Many of the tried and true methods were developed in the first half of the twentieth century,and the samples were possibly inappropriate for comparison with modern humans. As a result, two significant developments with respect to the analysis of modern human skeletal samples derived from forensic cases occurred: (i) modern samples of human skeletal tissue and information upon which new or reevaluated analytical methods could be based were sought and (ii) better analytical statistics to interpret human skeletal variation were employed.

Seeking modern human skeletal samples

Many of the forensic anthropology or physical anthropology methods used in the 1970s and 1980s to determine biological parameters (chronological age, sex, and stature) of unknown individuals were based on studies that drew upon samples of individuals from the turn of the century – usually of lower socioeconomic status – and even from prehistoric Native American samples (Johnston 1962), consisting of individuals of unknown age. This is fine when working with historical cemeteries, individuals who died during that time period, or prehistoric Native Americans; however, it has been shown clearly that the effects of better nutrition, better health care, etc. have led to significant secular changes in many of these biological parameters (Meadows and Jantz 1995). In addition, factors of immigration, emigration, genetic mixing, hybridization, and others on the modern North American population have altered the genetic landscape rather dramatically, suggesting strongly that samples of modern humans were required to properly interpret the bones of the recently deceased from modern forensic cases (Ousley and Jantz 1998). However, the major problem with this solution is that creating large modern human skeletal collections like the Todd and

Terry Collections is rather difficult. Hamann seems to have altered the Ohio mortuary laws in order to permit the collection of unclaimed bodies, and Terry simply placed individuals from the medical dissecting room into the collections at Washington University Medical Center in St. Louis, Missouri (Hunt and Albanese 2005). In the early 1980s, Bass of the University of Tennessee, Knoxville, addressed this pressing need by starting to collect complete human skeletons of known individuals who donated their bodies to the Department of Anthropology. Initially in an attempt to address issues related to postmortem interval (Bass and Jefferson 2003) and forensic taphonomy, Bass found space on university property to place donated human remains and study decomposition patterns (see below). A residual benefit of the "Body Farm" project, as it became to be known, was an ever-growing collection of human skeletal material of known individuals (Wilson et al. 2010). The William M. Bass Donated Skeletal Collection currently contains nearly 870 individuals although the population demographics are skewed somewhat toward older white males. Later, donated forensic cases formed the basis of the William Bass Forensic Skeletal Collection, which consisted of over 100 individuals from cases conducted by the University of Tennessee Department of Anthropology as of 2009 (http://web.utk.edu/~fac/facilities).

Judy Suchey and her colleagues were pioneers in the study of the skeletal biology of modern forensic populations when she was permitted to retain as evidence, skeletal samples (clavicles, pubic symphyses, superior iliac crests) of individuals that entered the Los Angeles County Coroner's morgue in the late 1970s. Her samples totaled 1225 individuals of all shapes, sizes, and types (Suchey and Katz 1998). Some casts and photographs were made for teaching purposes.

An ongoing project involves the collection of a virtual modern human skeletal database, primarily in the form of anthropometric data, while also containing demographic and skeletal biological information. This database, termed the Forensic Anthropology Databank (FDB), was developed by Dick Jantz in the 1980s at the University of Tennessee from a National Institute of Justice grant to create a Database for Forensic Anthropology in the USA (DFAUS) and originally included information from 1523 individuals (Wilson et al. 2010). The database consists of University of Tennessee cases and data submitted from cases completed by other forensic anthropologists across the country (Jantz and Moore-Jansen 1988; Ousley and Jantz 1998). The database had information from nearly 2900 forensic cases as of 2010 (http://web.utk.edu/~fac/databank).

Longitudinal studies of human growth and development, primarily through radiographic imaging, have been conducted in the last 20 to 30 years to replace or supplement older studies (Moorrees et al. 1963; Maresh 1943) and have yielded excellent results. The most important are dental studies (Sciulli and Pfau 1994; Harris and McKee 1990) and pediatric radiographic studies detailing long-bone growth and development (Hoffman 1979). In addition, recent efforts to collect data from the vast radiographic record of forensic case individuals from medical examiner's offices around the country have proved very fruitful (Fojas 2010).

BETTER STATISTICS

Of course, with updated collections and new databases came an enhanced ability to perform quantitative analyses. Quantitative statistical analyses are far from new in physical and forensic anthropology. Most of the key statistical techniques currently employed to estimate different components of the biological profile have not only been long known and widely applied by anthropologists, but in some cases were historically first utilized to address anthropological questions. For example, least squares linear regression (LSL regression), which today is the most popular method employed to estimate parameters such as adult stature or infant age, was first utilized to assess the correlation between parental and offspring stature by Francis Galton in 1886 (Galton 1886). As a matter of fact, the Anthropometric Laboratory funded by Galton (Galton 1882), as well Galton's collaboration with Karl Pearson, played a central role in the development of modern statistics.

This lead in statistical research in the natural sciences would soon vanish, however, engulfed by the descent to the abyss of social sciences or tragic excursions into pseudoscientific crazes such as eugenics; anthropologists would subsequently contribute little to the development of new statistical methods. Still, inferential and exploratory statistical techniques would remain an important component of the toolkit of physical anthropologists, if now mostly as borrowers of methods developed by other disciplines. The practitioner would obtain a sex, ancestry, stature, or age diagnosis simply by substituting individual case measurements into the corresponding discriminant function (DF) or regression equations, or by consulting tables of cut-off values or confidence intervals for observed traits, rather than by actually performing any statistical analysis. Whenever the comparative samples from which the published estimates had been obtained were appropriate for the particular case, these methods, if conceptually destitute, were fairly effective in providing simple estimates.

The use of these methods, however, imposed severe limitations: (i) it did not allow for multigroup comparisons in classification methods, for example, when trying to assess the probability of the victim belonging to one of several ancestry groups; and (ii) it severely limited the ability of the researcher to estimate the associated probabilities in most multivariate methods. For example, published cut-off values obtained from discriminant function analysis allowed only for the assessment of whether the individual was more likely to belong to one of two groups, which means that in their simplest application – sex determination – the analyst was able to predict that the individual had a higher or lower probability of being a male or a female, based on whether a single particular measurement was larger or smaller than the provided cut-off value (see France 1998 for a discussion and numerous examples of these methods). However, the analysis could not estimate the exact probabilities associated to this diagnosis (i.e., the posterior probabilities and typicalities), which require complex calculations obtained from the raw (measurement) data. In other words, the forensic anthropologist could not distinguish a case with a 51% to 49% relative probability from a 99% to 1% case.

Off the shelf and wrong: the example from regression equations

To make matters worse, even methods requiring simple calculations, easy to perform by hand or with a pocket calculator, were typically published in many textbooks devoid of the information necessary to properly calculate these estimates. The most striking example is probably that of the prediction intervals for the LSL regression equations for stature estimation. The basic assumption in LSL regression can be written as:

$$y_i \sim N(\hat{y}_i, S_R) \qquad (1.1)$$

where

$$\hat{y}_i = a + bx_i \qquad (1.2)$$

which means that the average height of individuals with a given long-bone length is the result of the regression equation when we substitute the value measured in our individual (x_i).

In other words, if we are trying to estimate stature (y) from a long-bone measurement (x), we can read equation 1.1 as the stature (y_i) of those individuals with the same long-bone length (x_i) as our individual follows a normal distribution with a mean equal to the result obtained when we enter that length (x_i) in the regression equation, and standard deviation equal to S_R. Note that x and y refer to the generic variables, while x_i and y_i refer to the values corresponding to the exact measurement of x in our particular individual.

Since we are just dealing with a univariate normal distribution, to construct a confidence interval for our estimate we only need to add and subtract the appropriate number of standard deviations (S_R) from the solution of the regression equation. For example, a common approximation to obtain a 95% confidence interval is multiplying S_R by two (the corresponding value in a Student t distribution under the large sample approximation), and adding and subtracting the resulting figure from our discrete stature estimate.

S_R is a standard error that accounts for the three main sources of error affecting the calculation of the regression equation: (i) normal variability in stature not depending on variable x, which is accounted for by the standard deviation of the residuals (the average square difference between the values predicted by the regression equation and those observed in the real sample), usually noted as $S_{y \cdot x}$; (ii) the error associated with the calculation of the mean of the dependent variable y (i.e., the standard error of y), which determines the height in the vertical axis at which the center of the regression line will be placed; and (iii) the error in the calculation of the slope of the regression (b in equation 1.2). That is, the uncertainty linked to our estimates of where the line is to be placed, its inclination, and how close the real statures are to the regression line on average.

Note that from the explanation above it follows that S_R, and thus the breadth of our interval, will depend on the value of x that we measure in our individual. Or, simply put, that S_R must be calculated case by case, and cannot have a single constant value along the regression line. This is mostly due to the error associated with the calculation of the regression slope (b). By definition (in particular, from the definition of the regression intercept, a), the regression line must pass through the point corresponding to the means of x and y [($\bar{x}, \bar{y}$) in Cartesian coordinates]. Consequently, as we have the line anchored in this point (for large samples, with little variation from sample to sample), and there is also an error attached to the calculation of the slope, the latter will affect more severely values far from the mean of x than those which are close to it (imagine a ruler spinning around a central point: the tips of the rule cover much larger distances than do the points close to the center). The calculation of S_R is not complex, but requires the introduction of a few values, including the size of the sample employed to obtain the regression equation, as well as the sample mean and standard deviation of x, in a linear algebraic equation. The correct equation and a straightforward explanation of how to use it can be found in Giles and Klepinger (1988). Ousley (see Chapter 16 in this volume) discusses stature estimation in depth.

Most classic forensic and physical anthropology textbooks, however, usually provided *only* the value of $S_{y \cdot x}$ (i.e., the standard deviation of the residuals, which unlike S_R was assumed to be constant and independent of x), thus omitting most of the information necessary to calculate S_R, while also misguiding the reader by indicating that the corresponding 95% confidence intervals could be calculated simply by multiplying this constant $S_{y \cdot x}$ by two. This latter estimate, however, does not provide a confidence interval for the regression prediction, but for the residuals (in order to avoid confusion between both confidence intervals, it is common to refer to the confidence interval linked to S_R as the *prediction interval*). Although the difference between both estimates decreases with sample size, at the sample sizes for most classic regression equations in anthropology this difference can be in terms of centimeters in the case of very tall or very short individuals. Additionally, the confidence interval obtained from $S_{y \cdot x}$ is *always* smaller than the real intended one. The only goal of statistical analysis is to provide probability estimates, and you might think that we could at least provide the right ones.

Do it yourself: stats for the people!

Fortunately, these and other similar situations would change drastically with the collection of the samples and databases described in the previous section of this chapter, as well as with the development of personal computers. Suddenly, forensic anthropologists had in their hands both the comparative data and the tools to perform the analyses themselves, obtaining their own equations and detailed reports. Resources like the Forensic Data Bank (Jantz and Moore-Jansen 2000) allowed for the importation of standard measurements of contemporary individuals, from large samples of different groups, into conventional statistical software. This permitted comprehensive *a la carte* analyses, now with all the required correct information. Statistical analyses became part of the everyday practice of the profession, instead of a strictly research-oriented enterprise. This new emphasis and possibility for real statistical analysis, rather than the algebraic operation of existing equations, resulted in a steadily increased focus and improvement in the statistical literacy of forensic practitioners. On the other hand, the new and updated skeletal collections allowed for the refinement of already existing methods and estimates, as well as the testing of their validity and application in modern populations based on expanded sample sizes. This trend was further strengthened by court rulings that stressed the need to test and justify the validity of the methods applied in case investigations, as well as of attaching probability estimates to forensic diagnoses (see the section on court rulings below.)

The resulting expanded range of analyses and probability estimates that could now be obtained and presented in court had obvious advantages. Namely these were obtaining new, more relevant, and precise information from our set of skeletal human remains but also, as discussed above, simply getting the ones from classic analyses right. Yet, the new approach was not completely free of shortcomings. From a forensic point of view, the most important shortcoming was that the new approaches complicated the task of evaluating and presenting the results to end users (law enforcement, coroners, and medical examiners) and in court. Numerical results (and often qualitative diagnoses) depend heavily on the selection of samples, analytical methods, and even the exact computer algorithms used by different statistical packages. Analyses based on different samples, outlier-removal criteria, test method options (e.g., stepwise versus one-step selection criteria), or even different software packages will render different results, which can therefore be potentially contested in court. This forced forensic analysts to include lengthy and complex method and result justifications in their case reports, making them difficult to understand (and therefore appear as less reliable) for law enforcement, court officials, and juries.

This drawback was mostly eliminated by the development and popularization of comprehensive statistical packages like Fordisc (Jantz and Ousley 2005; Ousley and Jantz 1996). Fordisc is a statistical package that provides both the database and the statistical tools for the metric analysis of different components of the biological profile. The program offers different standard discriminant analysis methods for the assessment of sex and ancestry, as well as regression equations to obtain stature estimates, including the appropriate confidence intervals. Although the interpretation of Fordisc as a tool for strict taxonomic and systematic analysis (i.e., to infer phylogenetic group relationships, rather than to assign individuals into groups, when the problem individual belonged to one of the groups included in the analysis) has received some criticism (Elliot and Collard 2009, and references therein), its utility in forensic contexts, when properly understood and utilized, is indisputable, having become a standard in US courts and laboratories. Placing the statistical analyses most commonly utilized in forensic anthropology within a framework of standard software and common samples has enormously simplified court presentation, analysis interpretation by third parties, and analyst training. The program also serves to maintain updated comparative datasets, through the continuous addition of data from ongoing forensic cases.

Outside the black box: from naked outputs to systems and processes

Finally, the new resources and mindset also served to boost research and the introduction of new analytical methods. Most traditional analyses were based on parametric methods that could be approximated by rather simple linear algebra equations. A commonly stated goal was to allow the practitioner to obtain the estimate with paper and pencil, a goal that, sadly, was often extended to the researcher producing the method. This imposed severe constraints to the range of methods and, more importantly, perspectives from which a particular problem could be approached. For example, descriptive univariate or bivariate approaches based on raw variable scores were favored over more complex techniques requiring distribution assessments and variable transformations, or that target processes rather than variable frequencies. From a conceptual point of view, the classic line of attack limited the analytical options to basically just *black-box* approaches, in which the analyst only focuses on the inputs and outputs of the system (what enters the box and what exits it), ignoring processes (what happens inside the box).

We can probably better understand the differences between both approaches and their consequences with an example. One of the classic problems in physical and forensic anthropology is estimating the age of an individual based on a skeletal marker, which appears or changes as the individual develops or grows older. Due to the regular sequence of changes or *phases* that the marker undergoes as the individual ages, the methods in this family of age-estimation techniques are often referred to as *phase methods* (Chapter 10 in this volume provides a lengthy discussion of the different alternatives for adult age estimation, including the most widespread phase methods). Within the classic scope, the researcher would approach phase methods by first calculating the mean age of all the individuals that displayed a particular phase or trait. A confidence interval containing a particular portion of the population (typically 95%) around this mean age would then be calculated. In this way, the practitioner could say that the predicted age of an individual presenting the trait lay somewhere within a particular range with a 95% probability of being correct.

Estimate errors and, consequently, the breadth of confidence intervals, are expected to decrease as sample size increases. Thus, most of the efforts to improve classic techniques through the production of narrower (i.e., more precise) age ranges, were based on recalculating confidence intervals from new, and larger samples (which included a greater variety of more recent populations). Simply put, almost all the emphasis was placed on sampling issues, while little or no attention was placed on the nature and dynamics of the physiological processes that produced or altered the traits being utilized. As a matter of fact, if one thing has characterized forensic anthropology research during past few decades, it is that a disproportionate emphasis has been placed on sampling techniques and sample characteristics, while nearly completely ignoring organismal biology, physiology, and population structure issues. Consequently, although new useful age markers have been successfully identified and introduced since the 1970s, the results of the reappraisal of most of the classic, widely used and reliable aging methods were anything but impressive. Perhaps sample characteristics were not the only issue; perhaps the methods themselves were at fault.

As with least squares regression above, an example is probably the best way of understanding how some classic methods limited our estimates, as well as how new approaches can help us to improve both our estimates and our insight into the processes underlying them. Let's begin with an unknown individual represented by a skeleton. Imagine that we are focused on providing an age estimate based on a single trait (e.g., an ossification center) that first appears in *some* individuals at age *A*, and is present in *all* individuals at age *B*. Therefore, if the trait is absent the individual is almost certainly younger than age *B*, while the presence of the trait would indicate that our individual is most likely older than age *A*. As mentioned above, the first step in traditional phase methods would be calculating the average age of all individuals exhibiting the trait, which would be followed by constructing a confidence interval around it.

Imagine, however, that ages *A* and *B* are both less than 20 years. Also let's assume that the trait develops during the same age interval in all populations considered. This is not a far-fetched assumption, if our age markers are actually expressing normal developmental processes. Now imagine that we calculate the average age of all individuals displaying the trait in two populations with different age structures: our marker is behaving in exactly the same way in both populations, and is expressing exactly the same information and physiological changes; however, the mean age will be higher in the older population, simply because we have proportionally more individuals older than 20. In other words, the mean age will depend partly on the age range at which our marker appears, but also, very importantly, on the proportion of adults present in our sample. Therefore, the confidence intervals calculated around this parameter will be very imprecise and extremely dependent on sample characteristics, thus providing very limited information on the actual ranges for the age marker itself.

Hence, recent approaches have started focusing not on the mean age of *presence* of the trait, but on the mean age of its first *appearance*, taking advantage of more complex techniques that are based on *logit* or *probit* models, admittedly harder to estimate with paper and pencil, but that, in exchange, actually do address directly the process under study. These models are not based on the distribution of ages, but on the conditional distribution of trait presence and *absence* on age. This means that the confidence intervals are not constructed around the mean age, but around the age at which an individual taken at random from the population has a 50% chance of displaying the trait and a 50% chance of lacking it. The resulting confidence intervals will also be narrower, as they will not refer to the whole age range, but just the narrow range of ages at which all individuals either display or lack the character.

These analyses also allow for comparison and combining probabilities derived from similar analyses of other traits (see Chapter 11 in this volume). However, probably their main advantage is that they offer real insight into the physiological process under study: note how we can infer details such as when the physiological change expressed by the trait is triggered, its rate of development, and when it is completed in most of the population. Boldsen et al. (2002), Milner et al. (2008), Konigsberg et al. (2008), and Milner and Boldsen (Chapter 11 in this volume) provide excellent starting points for discussions of transition analysis, one of the most promising applications of this type of methods in a modern forensic context. Hefner and coauthors (Chapter 14 in this volume) also provide another excellent example of the new trends in the statistical analysis of discrete traits, with their assessment of ancestry, looking far more closely than in the past at trait definition, the exact distributions of the ancestry markers in different populations, and the conditional (posterior) probabilities resulting from them.

Future venues

Finally, the new approaches are also starting to benefit from enhanced data-acquisition methods, which allow one to introduce more powerful techniques like those generically known as *geometric morphometrics* (GM). Plainly put, GM is not based on the analysis of linear measurements, but rather on the exact spatial location (coordinates in an *n*-dimensional space) of the landmarks previously used to take the classic anthropometric measurements. This results in a tremendous gain of information, especially useful to define shapes. For example, in the past you could take three measurements from three anthropometric landmarks (*A*, *B*, and *C*) that form the vertices of an imaginary triangle. The classic measurements would be those of the sides of the triangle, separating each pair of landmarks. This is to say, the segments *A–B*, *B–C*, and *C–A*. What information were we missing with this approach? Basically that each landmark was part of two different measurements or, in plain Castilian, that the three points were forming a triangle. We were getting some information on dimensionality, but actually missing the most relevant information regarding the geometric shape of the object that we were examining. The information gain from using Cartesian coordinates (i.e., GM) in a multidimensional space, instead of distances between points, increases exponentially with the number of points (landmarks) that we are considering. In a nutshell, to approximate with linear dimensions the amount of information collected in a GM analysis, we would have to measure all possible distances between all pairs of landmarks, and we would be still losing some geometric information when entering them in standard statistical analyses. Collecting precise landmarks from photographs or, especially, real specimens, was a rather arduous and delicate task in the past, but modern three-dimensional digitizers allow for the collection of data as easily as we did with the distance measurements in the past. Thus, the common use of GM techniques is an emerging but clear trend in forensic anthropology. Zelditch et al. (2004) is probably the best general introduction to GM. Online resources like the Morphometrics at SUNY Stony Brook webpage (http://life.bio.sunysb.edu/morph) are also invaluable.

FINALLY, ADDING FORENSIC ARCHAEOLOGY

As described above, significant steps have been taken to address skeletal biology, and even the statistical techniques to analyze the traits, of the recently deceased since the inception of forensic anthropology. Law enforcement professionals, coroners, and medical examiners in the 1980s and 1990s began to figure out that the best avenue for analyzing unidentified skeletal remains was through a forensic anthropologist (Snow 1982). And so following the recovery, postmortem examination, and analysis performed by the forensic odontologist, or sculptor, the box of still unidentified remains would be sent to a forensic anthropology laboratory. The focus of the request was to provide a more definitive biological profile that might provide hints of identity or, at the very least, narrow down the missing-person list. However, additional questions were being asked: How long have the remains been there?, Why are some bones missing?, Why are some bones out-of-place?, Why are some of the bones broken, when did that occur, and could that relate to the death of the individual? Answers to these questions based on one's credentials could be provided; however, careful scrutiny, both in and out of court, revealed that the answers rarely had real scientific backing and had no place in the testimony for they were pure conjecture and not Daubert-worthy, which will be discussed later in this chapter.

The answers to these questions do not reside in the bones alone, but require a careful analysis of where they came from, the spatial distribution of the remains at the scene, and a careful consideration of the condition of the remains at the time of discovery. These are issues that relate to the contextual setting of the remains and cannot be obtained solely from the analysis of the bones in the box sent to the forensic anthropologist, or from the pictures of bones at the scene. And so, in order to give reliable answers, the forensic anthropologist, in turn, must retort with questions of their own that relate to the recovery and the post-recovery handling of the remains. With respect to the recovery, some of the key questions asked would be: Were the bones buried or above ground?, Were the remains exposed to the elements?, Were they on a slope or flat ground?, Could flooding from a nearby creek have disturbed the remains?, Were the remains in the shade and for roughly how long?, What types of trees are in the immediate vicinity (deciduous, coniferous)?, What is the grass cover?, What is the leaf cover?, What is the elevation of the site?, What is the soil acidity?, What animals (carnivores, scavengers, rodents) may have impacted the remains?, Where exactly were the bones relative to one another?, Where was the main concentration of remains?, and Where was the body originally located before dispersal? These are only a few of the many, many other questions essential to the work of a forensic anthropologist.

Further, other questions must also be asked by the forensic anthropologist that relate to the removal and transport of the remains from the scene, the possible effects of the postmortem examination, and by subsequent forensic analyses: What role did recovery play in altering the remains?, Were they dug out of the ground with a shovel, backhoe, or pulled out by hand?, Were some bones dropped accidentally, stepped upon, or the skull picked up incorrectly?, How were the remains handled during transport (all in one bunch in a body bag?, with other evidence placed on top of the bones?), during the postmortem examination, and during examination by other forensic specialists? All of this activity and interaction with the human bones prior to their arrival in the forensic anthropologist's laboratory is critical to the final proper forensic anthropological analysis and interpretation. Attention, therefore, shifts back to the forensic investigator and a review of how outdoor forensic scene recoveries generally are conducted by law enforcement.

As described more fully in Chapter 2, evidence-documentation protocols at indoor scenes are well constructed and yield precise notation of exact location of all evidence to other evidence and to the surrounding scene which, in turn, leads to the establishment of proper chain of custody at the outset of evidence recovery. The same meticulous protocol, however, cannot be said of the outdoor scene where remains are often hastily removed from the scene with little or no documentation of provenience. A review of the literature and training regimen of law-enforcement officials at all levels reveals that *there are no law enforcement protocols available for the recovery of human remains from outdoor contexts.* When shovels and backhoes are employed as first-line recovery tools, problems persist. Ill-conceived or incomplete recovery methods do not yield scientific answers to the aforementioned questions. The other aspect of that revelation is that simply overlaying indoor crime scene documentation and recovery protocols onto outdoor scenes will not work. As has been argued in the past (Dirkmaat and Adovasio 1997; Dirkmaat et al. 2008; and Chapter 2 in this volume), the best protocols for the recovery of outdoor scenes, therefore, do not lie within law enforcement protocols but instead lie within forensic anthropology and specifically with the forensic archaeology component of the field.

Forensic anthropology as a discipline, however, in the USA has been slow both to realize the problem and to embrace the solution. Primarily the reason for this lethargic response relates to how forensic anthropology is perceived by law enforcement and most forensic anthropologists themselves. Definitions of the field still indicate that forensic anthropology is a laboratory-based discipline focused on providing clues with respect to the identity of the victim represented by their cleaned skeletal remains. Only after the remains are recovered from the scene, examined at an autopsy *and* if other victim identification avenues (odontology, forensic sculpture) have been exhausted, are the bones then forwarded to forensic anthropologists. As is clear from a review of the vast majority of definitions and descriptions of the field provided by practitioners, forensic anthropology is considered a laboratory-based discipline. Other individuals and disciplines are relied upon to provide background information, collect the remains, document context, and construct viable scene interpretations. In the past, forensic anthropologists seemed to be perfectly content to wait in the laboratory for the remains to be brought to them.

Some forensic anthropologists with training in archaeology began to see this as a problem. Krogman, who had worked as an archaeologist early in his career, actually suggested very early on that outdoor forensic scene recovery could benefit from an archaeological perspective (Krogman 1943). Kerley also pointed out that, "one long-neglected aspect of forensic anthropology, which is of very practical interest to homicide investigators, is the application of standard archaeological techniques to the search for and recovery of homicide victims, and examination of the site of discovery of the body" (Kerley 1978: 166). Some early advocates of using archaeology in efforts to recover human remains from outdoor medicolegal contexts included Dan Morse at Florida State (Morse et al. 1983, 1984), Mark Skinner of Simon Fraser (Skinner and Lazenby 1983), Sheilagh and Richard Brooks in Nevada (Brooks and Brooks 1984), and Doug Wolf of the Kentucky state medical examiner's office (Wolf 1986). Although Bass was certainly an advocate of taking forensic anthropology into the field during the processing of outdoor scenes (Bass and Birkby 1978), as were other prac-tioners, including William Maples (Maples and Browning 1994) and Stanley Rhine (Rhine 1998), to name a few, forensic archaeology was used little and the vast major-ity of cases today still arrive at the forensic anthropology laboratory in a box after the police have collected the remains from the scene. Forensic archaeology remains a peripheral rather than an integral activity of forensic anthropology.

Forensic archaeology today

In the last few years, changes have been forthcoming. Recent research and literature have described forensic archaeology more fully as a robust discipline that does not begin and end at the buried body feature (Dirkmaat and Adovasio 1997; Dirkmaat et al. 2008; Hochrein 2002; Conner 2007; Dupras et al. 2006). Modern archaeological practices are applied to the full range of outdoor scene location, documentation, and recovery activities beginning with the search for the unlocated site. Here, shoulder-to-shoulder pedestrian searches are effective in examining 100% of the surface within search corridors. If the remains are located on the surface within the path of the searchers, they will be found! Another important role that forensic anthropologists perform uniquely during the search is the on-site, instantaneous determinations of significance of biologi-cal tissue, whether animal, human, or nonforensic (see Chapters 2 and 3 in this volume).

If forensically significant remains are discovered, forensic anthropologists will clean the surface of the scene of extraneous material, and expose and then map the remains and evidence *in situ* by hand, supplemented by electronic mapping instrumentation, such as total stations or global positioning system (GPS) units (Dirkmaat and Adovasio 1997; Dirkmaat and Cabo 2006; Dirkmaat et al. 2008; see also Chapters 2, 5, 6, and 7 of this volume). As noted by Snow, the "spatial distribution of bones, teeth, and other items recovered in surface finds can help in determining the original location and position of the body" (Snow 1982: 118).

If the remains are buried, it is a much more difficult task to find the feature and remains, especially if a few weeks or months have passed. In these cases, cadaver dogs are particularly helpful, especially in the spring when new plants are emerging (Rebmann et al. 2000). More sophisticated subsurface examination techniques, and equipment such as ground-penetrating radar (GPR), may be used in confined or well-defined areas (see Chapter 4 in this volume). Obviously, forensic archaeology is especially well suited for these recoveries. Standard excavation methods, drawn nearly unmodified from archaeology protocols, serve to provide guidelines of how to delimit backdirt piles, carefully excavate burial fill to expose the remains within, and document the stratification found within the burial feature (Dirkmaat and Cabo 2006). These methods are well defined, well refined, and well practiced (Hochrein 1997, 2002; Hochrein et al. 2000; Chapter 5 in this volume). In turn, the same guiding principles that work at the small grave feature work during the recovery of much larger grave features, such as those encountered in human-rights work (Dirkmaat et al. 2005; Tuller et al. 2008; Chapter 8 in this volume).

Processing unique outdoor forensic scenes

Recently, the forensic processing of unique outdoor scenes has benefited from forensic archaeological recoveries, in particular fatal-fire scenes and large-scale mass disaster scenes. With respect to fire scenes that contain human victims, new techniques and new technologies (described in Chapter 6), used in conjunction with standard fire investigation methods, have resulted in more efficient *in situ* documentation and recovery efforts of human remains. Damage to biological tissue, resulting from modification of the scene in order to find the body (e.g., practices such as "overhauling") is dramatically minimized. Better transport methods that further minimize damage to the fragile and friable burned remains, in conjunction with better on-site documentation of bone fracture and damage patterns, benefit subsequent forensic anthropological analyses, especially human skeletal trauma interpretation (see Chapter 17 in this volume).

Finally, archaeological recovery methods have served to make recovery efforts at plane crashes and bomb blast sites much more efficient and effective (see Chapter 7 in this volume). New recovery protocols (Dirkmaat and Hefner 2001) were developed and employed in 2000 during the recovery of the crash of Missouri governor Carnahan (Reineke and Hochrein 2008), and during the initial scene documentation efforts at the crash of United Flight 93 on September 11, 2001 in Shanksville, Pennsylvania (Dirkmaat and Miller 2003). The forensic excavation of the victims of Colgan Air Flight 3407 in Buffalo in February 2009 represents the first recovery of mass fatality victims associated with a commercial airliner crash directed and conducted by a team of trained forensic archaeologists. The search efforts, careful excavation, and screening of all excavated material resulted in the near 100% retrieval of all recoverable human tissue. Furthermore, the documentation of provenience of each set of remains (via an electronic total station) permitted a scientific explanation for the inability to recover any identifiable remains (including DNA) of one the victims (Dirkmaat et al. 2010; Chapter 7 in this volume).

PRIMARY BENEFITS OF FORENSIC ARCHAEOLOGY

In addition to ensuring the comprehensive recovery of all human remains at the outdoor scene and ensuring that nonhuman biological tissue does not enter the evidence chain, forensic archaeological principles and practices employed during the recovery of the outdoor scene provide other significant benefits for the forensic investigation: establishing chain of custody early, providing a baseline for forensic trauma analysis, and permitting the construction of scientific arguments leading to reconstruction of past events (forensic taphonomy).

Chain of custody at the outdoor scene

The National Institute of Justice has recently set out clear guidelines with regard to the handling of evidence (National Institute of Justice 2005). One of the key aspects of proper crime scene processing they note is the early establishment of chain of custody. This is defined as the chronological log of the travels of evidence from the scene to the courtroom (where, when, how, and by whom was evidence touched). The important first step in the chain is the careful and thorough documentation of evidence *in situ* (in context) at the crime scene. Protocols are in place to accomplish this task very well at the indoor scene (Saferstein 2007, 2009; James and Nordby 2003). However, when the processing of an outdoor crime scene involves picking up bones that are visible on the surface and putting them in one bag, after a few pictures are taken, then it can be stated confidently that the first step in the chain of custody is not done well and could be questioned in court. As discussed above, forensic archaeological recovery practice and the comprehensive documentation of context provides the perfect solution for establishing the first step of the chain-of-custody sequence at the outdoor crime scene. Particularly important here are the detailed plan-view maps showing the exact location and orientation of each bit of evidence.

Human skeletal trauma analysis

In the early 1990s Hugh Berryman and Steve Symes began working in the Medical Examiner's Office in Memphis, Tennessee, as forensic anthropologists (Dirkmaat et al. 2008; see also Chapter 26). This is significant because it represented a shift in perspective by a progressive-minded forensic pathologist (in this case, Dr J.T. Francisco). Forensic pathologists usually work alone during the postmortem examination and base much of their interpretations of the cause of death of the victim on analysis of the soft tissue during the short-term autopsy. If bone damage is considered it is usually only a cursory examination such as attempting to put cranial bone fragments covered in blood and tissue together during the autopsy at the morgue table in order to better understand gunshot wounds. Francisco realized that much information could be gathered from a more in-depth analysis of the cleaned, damaged bones of the victim in addition to the study of the soft tissue. This is especially true with respect to human skeletal trauma. And so, osteological samples exhibiting evidence of trauma were culled during the postmortem examinations and retained *as evidence*. Soft tissue was removed to thoroughly study the bone specimens. Macroscopic and microscopic analysis was undertaken to understand the biomechanics of the forces upon the bone, better differentiate types of trauma (blunt force versus sharp force, versus ballistics force), and document tool marks (identifying unique characteristics of tools including teeth per centimeter of serrated knives and saws) on the bones (Berryman and Symes 1998; Symes et al. 1998; Chapter 17 in this volume). As a result of these simple protocols, the study of human skeletal trauma has advanced significantly. Previously, information regarding human skeletal trauma was derived primarily from the field of paleopathology and vertebral faunal analysis (Aufderheide and Rodriguez-Martin 1998; Brothwell 1981; Ortner 2003; Potts and Shipman 1981; Shipman 1981) and perhaps many of these interpretations essentially were based on educated guesses and a whole lot of faith since the real cause of the damage was unknown. On the other hand, forensic anthropology analyses conducted in the morgue setting provide a perfect testing laboratory because of the presence of both soft and hard tissue, and, in some cases, reliable witness accounts of the sequence and timing of events, instruments used, and actors involved. Now, add experimental studies on bone material by engineers and biomechanics in the laboratory (Kroman 2010; Baumer et al. 2010; see also Chapter 19) and the field of human skeletal trauma has become a robust scientific endeavor.

It was soon realized, however, that broken bones arriving in a box after recovery, autopsy, dental examination, and clay emplacement activities were hard to interpret with respect to the *timing* of the trauma. Was the damage to the bone the result of *perimortem* (at the time of death) activities – and thus very significant with respect to determining cause and manner of death – or the result of postmortem factors related to animal activity, tumbles down cliffs, rock falls, etc? Add to these factors the potential for bone damage resulting from mishandling of bones during recovery (especially critical for fatal-fire victims, and juvenile remains), or damage occurring during transport to the morgue, during the autopsy, during handling by other forensic specialists, or during the placement of clay, and "Houston, we have a problem."

Forensic archaeology methods again aid significantly in the determination of timing of trauma because: (i) scene information (e.g., base of cliff) and other contextual information is carefully noted, (ii) standard practice requires detailed and careful notation of the condition of the remains *in situ* at the time of recovery (broken bones are carefully described and photographed prior to moving them into collection containers), and (iii) forensic anthropologists know how to handle bones without inflicting damage. As a result of employing forensic archeological methods to the recover the remains, the trauma analyst is now confident that they better understand *when* the bones were broken and they can focus on the *how* and *why* they were broken.

Forensic taphonomy

With the implementation of forensic archaeological practices and the routine collection of information related to context, another important new advance within the field of forensic anthropology could develop; the subdiscipline of forensic taphonomy.

Again, we return to the questions being asked by law enforcement along with the box of bones brought to the laboratory: Why are some of the bones missing, damaged or modified?, Was any of this bone modification done by humans at the time of death or later to hide evidence?, How long has the body been at the scene?, and Is there any evidence in the position and orientation of the body that would provide clues to cause and manner of death? The answers to these and many other questions cannot be drawn solely from the analysis of the bones pulled from the box sent by police. It requires a thorough analysis of the bones and the contextual setting from which they are derived.

Earlier in the twentieth century another scientific discipline attempted to better understand the death event and what happened to the body afterwards: geology and specifically paleontology. Questions surrounding vertebrate faunal assemblages from the standpoint of where had the animals died and what factors led to the movement, dispersion, and removal of their bone elements arose. Efremov (1940) was one of the first researchers to link bone assemblage composition (species and specific bones) with analysis of surface modification and bone chemistry changes, and with the geological site formation analysis (Lyman 1994; Gifford 1981; see also Chapters 24 and 25). He termed the field of study "taphonomy" which can be defined simply as the study of what happens to biological tissue following the death of an individual. The central concepts, methods, and techniques have been applied widely within the fields of vertebrate paleontology (Lyman 1994), fossil hominid research (Brain 1981), and bio-archaeology (Grayson 1984) to answer many different questions.

William Bass saw early on that the principles of taphonomy were applicable to the study of human remains from forensic contexts. Initially as a result of a misinterpretation of the postmortem interval on a nonforensic case (Bass 1979), he and others (Kerley 1978) realized that very little was known about human decomposition in terms of sequence of decomposition, rates of decomposition, and how that rate is affected by a variety of factors, including temperature, exposure on the surface, burial, emplacement in water, carnivore activity, and insect activity. Shortly thereafter he established the (Bass) Anthropological Research Facility, affectionately known today as the "Body Farm," at the University of Tennessee, devoted almost exclusively to the study of human tissue decomposition, in the hopes of establishing more scientific estimates of postmortem interval (Bass and Jefferson 2003).

A thorough consideration of how taphonomic principles, methods, and practices could be applied to medicolegal contexts (especially outdoor scenes) led to two tomes on the subject of what came to be more formally termed forensic taphonomy (Haglund and Sorg 1997, 2002). As currently configured, forensic taphonomy requires a careful consideration of the factors (taphonomic agents) related to: (i) the removing of soft and hard tissue from the scene, (ii) altering the spatial distribution of the remains within the scene, (iii) modifying the surface of the bones, and (iv) modifying the chemical-mineral composition of the bones. The systematic collection of data related to these issues then permits the consideration of the primary foci of a forensic taphonomic analysis: (i) whether humans, as taphonomic agents, had a role to play in modifying the remains after death and (ii) what has been the postmortem interval since emplacement of the body on the scene?

Forensic taphonomic analysis begins at the outdoor forensic scene with the careful notation of the environs of the immediate area including floral distribution (ground cover, tree cover), fauna (carnivores, scavengers, rodents), insects (flesh-eating), geology (soils, slope, water), and climate (rain, snow), among other factors. This list of potential taphonomic agents that might impact the biological tissue next is linked to the pattern of distribution of the human remains at the scene via the careful three-dimensional mapping of all of the pieces of evidence relative to one another and to the scene itself. Obviously, by far, the best way to collect data at an outdoor crime scene that allows for a taphonomic analysis of the remains is to recover it using archaeological methods, techniques, and principles (see Chapter 2 in this volume). The final piece of the puzzle in the forensic taphonomic analysis is the careful notation of bone modification patterns in the laboratory.

And so, forensic archaeology adds another dimension to the unique skills that forensic anthropologists can bring to the forensic investigation table. It will be argued here that forensic archaeology is not just a supplemental activity to be applied to the investigation of the outdoor scene, only if so inclined and if not too inconvenient, but an absolutely integral part of outdoor forensic scene processing and the subsequent proper interpretation of the activities at these scenes at a level already attained by indoor scene recoveries conducted by law enforcement.

OUTSIDE INFLUENCES

As described above, a critical review of current practices in the field of forensic anthropology in the 1980s and 1990s eventually led to much better human skeletal biology and even the initial consideration of the role of forensic archaeology. At the turn of the twenty-first century, influences from beyond the field of forensic anthropology have forced renewed evaluations and considerations of best practices. These external influences have come in the form of advances in DNA analysis; federal court rulings related to the science behind the collection, analysis, interpretations, and presentation of forensic evidence; and recent governmental oversight of the forensic sciences.

Outside influence 1: DNA

Very few practitioners would dispute the relevance of the development of DNA analysis to forensic anthropology. This importance is largely the product of a common goal of both disciplines: positive identification. The question immediately arising from this coincidence in objectives is whether both disciplines can still be compatible or whether DNA might be called upon to completely supplant forensic anthropology, rendering it irrelevant for identification purposes. Why bother assessing the biological profile of the victim if we can easily identify her through DNA?

In our view, the answer to this question will not depend so much on future developments in genetics, but rather on the ability of forensic anthropology, first, to acknowledge the question and, secondly, to diversify its goals, techniques, and intellectual framework in order to adapt to these developments.

Curiously enough, it may be the first of these conditions (i.e., the realization that DNA advances pose a real threat to the viability of forensic anthropology as originally conceived) that represents a bigger challenge. As a matter of fact, up to the present, forensic DNA analysis has been a staunch ally, rather than a rival, of forensic anthropology. The introduction and popularization of forensic DNA analysis as a routine technique dramatically increased the probability of identifying unknown individuals. Just a couple of decades ago, the possibility of positively identifying a John or Jane Doe decreased dramatically if the victim could not be matched with the short list of missing persons in the area during the initial steps of the investigation. This possibility further decreased as time went by and consultations with other neighboring or overlapping law-enforcement offices were also fruitless. Then came the introduction of genetic tests and the possibility of comparing the victim's DNA with millions of profiles contained in comprehensive databases such as Combined DNA Index System (CODIS; see Dale et al. 2006 for a clear and comprehensive description of the system) or the *ad hoc* databases compiled in human-rights or mass disaster investigations. The accuracy of DNA matches also allowed forensic practitioners to obtain positive identifications even in the absence of fingerprints or dental or medical records, through comparison with the DNA profiles of potential victim relatives.

The first consequence of these developments was stimulating investigative agencies to further their identification goals and efforts: the budget and time devoted to the identification of unmatched John Does or of the myriad fragments of human tissue present at a mass fatality scene were now much more likely to render positive results. During this time, the contribution of forensic anthropology probably became more important than at any time previously. The biological profiles provided by the anthropologist served to reduce the number of DNA comparisons to be performed in each case by limiting the list of potential matches to the subsample of individuals sharing the same sex, age range, ancestry, and/or stature of the victim. This reduction of potential matches consequently served to exponentially reduce the time requirements and cost of DNA analysis.

Within this framework, many practicing forensic anthropologists may tend to dismiss the notion that forensic DNA analysis might represent any kind of threat for the profession. If anything, the collaboration with DNA analysts has done nothing but increase their casework and enhance both their results and their importance within the forensic community. This perception, however, may be based on a flawed assumption that the relevance of reducing the list of comparisons *before* genetic analysis is derived from inherent limitations of DNA analysis and, therefore, that the relevance of pre-sorting tasks will be permanent.

As discussed more in depth by Cabo (Chapter 22 in this volume), the development of DNA analysis in recent decades has not been marked by a mere increase in the number of forensic DNA laboratories or by a superficial refinement of basic techniques and equipment. On the contrary, the evolution of DNA analysis has been characterized by a frantic development of new techniques resulting in the rapid and systematic elimination of most key obstacles hampering its development. In approximately four decades, the limiting steps of DNA analysis progressed from replication (amplification) to subcloning to sequencing and to comparison, as successive issues were resolved and each step was automated. At present, massive parallel sequencing methods allow for the sequencing of millions of bases per hour, making it potentially possible to sequence the complete genome of an individual in just a few days (Rogers and Venter 2005). In the last two decades, the time required to sequence a full human genome has been reduced by thousands of percent and the costs by hundreds of thousands percent (see Chapter 22), and the key technical limitations for forensic DNA analysis now probably lie more in the completeness of profile databases than in genetic techniques themselves.

Conversely, governments and law-enforcement agencies have also done their homework, making large investments to create the appropriate resources by which to take advantage of these developments in genetics. In the USA, the National DNA Index System (NDIS) and CODIS, of the FBI, currently contain more than 9 million profiles, and the agency is focusing now on increasing the database of missing-person profiles (www.fbi.gov/about-us/lab/codis/codis_future). If an imposing case backlog still exists, technical improvements and the investment in the modernization of forensic laboratories are already resulting in a backlog reduction, in spite of the simultaneous increase in the number of samples submitted for DNA analysis.

Thus, the near future outlined by these developments is one in which analyzing a DNA sample will be both very cheap and very quick (at present it can be argued that it is already either cheap or quick), while comprehensive databases containing profiles for the identification of virtually any missing person will be readily available. How will this affect the work of forensic anthropologists?

We can probably get an idea by looking at other competing identification techniques that are already available. We could think of fingerprints or dental records, and ask ourselves how often are bodies submitted to us just for identification purposes when any of these items are available; however, the best comparison would probably be with fresh, fully fleshed bodies. In these cases most of the components of the biological profile (save for, perhaps, the exact age range) can be directly and rather unequivocally assessed by the pathologist, and thus are not commonly submitted to the forensic anthropologist for identification purposes. With the appropriate comparative databases, DNA would have a similar quality, but could also be extracted from skeletonized or fragmentary remains.

In this scenario, the biological profile would probably still be useful as a safeguard or double-check to identify laboratory or family record errors, but its utility as a pre-sorting method, aimed at reducing the list of potential matches, would be fundamentally diminished. It would represent an analysis sometimes useful to perform *in addition* to DNA analysis, but no longer something required *before* the latter.

We could also ask what would be the parameters marking the point of no return or, in other words, the amount of time and the costs to be reached by DNA analysis in order to become a closed alternative to the forensic anthropological profile, rather than its sister technique. The answer is rather simple: to reach that point, DNA techniques must simply become cheaper and faster than forensic anthropological analyses.

Fortunately, due to the development of forensic anthropology in recent decades, the solution to the problem and the future of the discipline will not depend on progressive salary cuts and modification of anthropological report production. We do believe that the primitive, exclusive goal of helping with identification issues will lose importance with the further development of forensic DNA techniques. Still, as discussed throughout this chapter, the evolution of forensic anthropology in the last few decades has been characterized by a diversification of its goals, objectives, and techniques, paralleled by a decisive refinement and deepening of its conceptual framework and entrenchment in the natural sciences. Nowadays, forensic anthropologists do receive and deal with identified and even fully fleshed bodies in their everyday practice. They do so in tasks such as forensic archaeological recoveries and trauma analysis.

Outside influence 2: recent federal rulings related to science in the courtroom

In the last 40 years, forensic anthropologists have testified rather infrequently in court. The primary reason for this lies with the definition of the role that forensic anthropology plays in forensic investigation: forensic anthropological analysis of the skeleton can produce a biological profile that *helps* determine identity of the victim by providing circumstantial evidence of identity, which in turn helps narrow the missing-person list. Positive identification is most often provided by other disciplines (dental and DNA). And since the identity of the victim is usually not in question at the time of trial, why then would forensic anthropologists be called to testify?

Another reason for the infrequency of courtroom testimony is that the results of the forensic anthropological investigation are always turned over to the police, forensic pathologist and/or medical examiner. The primary responsibility for determining cause and manner of death belongs to the forensic pathologist (cause of death) and the coroner/medical examiner (manner of death). Law-enforcement officials generally process the crime scene and are responsible for providing reconstructions of past events that occurred at these scenes. If a case goes to trial, it is these individuals who will provide commentary and expert testimony over the whole sweep of analyses and investigations that led to the rulings.

However, with the addition of forensic archaeology at outdoor scenes and ramifications to providing unique determinations of postmortem interval, circumstances surrounding the death event and placement at the scene, and especially, skeletal trauma, testifying opportunities for forensic anthropologists will become more frequent. With the responsibility of handling evidence as one of the leads at the outdoor scene recovery, there is a need for forensic anthropologists to better understand how to properly handle evidence, establish chain of custody, and prepare court testimony as an expert witness.

As with all other forensic scientific investigators, forensic anthropologists must document, analyze, and interpret evidence according to a certain level of standards in order for it to be admissible in court. Since the 1920s, the Frye decision has provided the guideline by which trial judges determine the validity of courtroom expert witness testimony (*Frye v. United States* 1923). Frye required that the science, upon which the findings of a particular expert witness are based, must be generally accepted in the expert's particular field if it is to be presented in court (*Frye v. United States* 1923: 1014). This ruling was amended in 1975 by the Federal Rules of Evidence (FRE) 702, which read: "If scientific, technical, or other specialized knowledge will assist the trier of fact to understand the evidence or to determine a fact in issue, a witness qualified as an expert by knowledge, skill, experience, training, or education, may testify thereto in the form of an opinion or otherwise" (FRE 702, 1975). With this ruling it thus appeared that the "general acceptance" guide was replaced by the ability of testimony to assist the trier (the judge) of fact. This often led to reliance on experts who would provide the most forceful testimony and opinions. The "weight" of their expert opinion often resided on the length of their curriculum vitae and years of experience with little concern focused on the scientific basis of their interpretations.

In a 1993 ruling, *Daubert v. Merrell Dow Pharmaceuticals*, the Supreme Court attempted to alter this state of expert testimony by emphasizing that testimony related to scientific matters had to rely on the science behind the court presentation, rather than the presentation (*Daubert v. Merrell Dow Pharmaceuticals* 1993). The research and studies upon which conclusions were reached in a report should be testable, previously tested, previously presented in peer-reviewed publications that included appropriate consideration of potential error rate, and represent generally accepted scientific methods. Frye's "general acceptance" test was rejected and replaced by a new requirement that the "trial judge serve the role as gatekeeper" and "ensure that any and all scientific testimony or evidence admitted is not only relevant, but reliable" (*Daubert v. Merrell Dow Pharmaceuticals* 1993). It was hoped that focus would shift from the expert to the expert's testimony.

First, a bit of clarification
Additional clarifications, however, were needed and were presented in the form of additional court rulings, including *General Electric Co. v. Joiner* (1997) and *Kumho*

Tire Co. v. Carmichael (1999). Kuhmo, in particular, instructed that standards comparable to Daubert should apply to all expert testimony (*Kuhmo Tire Co. v. Carmichael* 1999). Even though most trial judges are not science experts and are generally limited in their ability to judge the scientific merit of much of the technical aspects of the forensic sciences, the US Supreme Court was confident that the adversarial nature of the judicial system – that is, "vigorous cross-examination, presentation of contrary evidence and careful instruction on the burden of proof" (*Daubert v. Merrell Dow Pharmaceuticals* 1993), would eventually lead to best-solutions science entering the courtroom, and suggested that the judge's inquiry remain flexible.

Finally, FRE rule 702 was amended in 2000 to read that a witness did not have to be a scientist to testify as an expert, but could also be qualified as an expert by "knowledge, skill, experience, training, or education" (FRE 702, 2000) if these findings were based on good scientific principles and best practices. The apparent wiggle room, perhaps, is necessary in many forensic sciences, including forensic anthropology. Ubelaker suggests "much of diverse anthropological analysis and interpretation must call upon experience and observation if it is indeed to maximize the information retrieved" (Ubelaker 2010: 416).

The bottom line

Most of forensic science evidence relates to crimes of violence, the vast majority of which are tried in state courts not federal courts (200 times more criminal cases are presented in state courts than federal courts). As currently constructed, Daubert applies to all federal cases and individual states adhere to their own standards (Frye, or versions of Daubert). Forensic scientists, including anthropologists, therefore, may see very little effect on their testimony in court in the near future. However, perhaps the potential for testimony to be disallowed and the case even thrown out, provides enough of a threat to raise standards across the board.

On the other hand, since Daubert attempts to get the best science into the courtroom and to diminish anecdotal accounts and years of experience arguments that focus on the status of the expert witness, it should serve as a standard-bearer and goal for all forensic scientists, including forensic anthropologists. It should not be considered a "threat" hanging over the heads of expert witnesses, but a way to level the playing field and get rid of "junk" science. With recent reevaluations of the forensic science (see the National Academy of Sciences report described below), change is afoot. The highest of standards should always be sought and Daubert provides a useable playbook. The effect of Daubert on the forensic sciences and forensic anthropology in particular is explored in greater detail in Chapter 32.

Outside influence 3: the US federal government

National Institute of Justice

As a result of the 1993 Daubert ruling, and likely a big slice of humble pie served as a result of the botched handling of forensic evidence in the O.J. Simpson case, the federal government set about to reevaluate best practices and try to improve a number of areas in the forensic sciences. To some degree the National Institute of Justice, the research, development, and evaluation arm of the US Department of Justice, has taken the lead in attempts to upgrade aspects of forensic sciences. This has been accomplished through research initiatives, funding opportunities, specialized training, and discipline-specific publications. In addition to these activities, one of the most tangible results of their activities was the formation of "working groups," assemblages of professionals from specific forensic specialties that are brought together to formulate standards for their particular discipline. Since the 1990s the National Institute of Justice has created what they call Technical Working Groups (TWGs), peer-review panels of experts (25 to 30 practioners from local, state, tribal, and federal agencies and laboratories) who were tasked with evaluating the problems and needs of participants in the forensic sciences and criminal justice community, and constructing potential solutions. Currently, 19 active TWGs are listed on the National Institute of Justice website. The direct result of many of these TWGs, are publications that present "best practices" in the form of Crime Scene Guides. They are typically widely disseminated and relatively influential. They include free publications (and pdf files) on *Fire and Arson Scene Evidence, Crime Scene Investigation, Death Investigation, Guide for Explosion and Bombing Scene Investigation*, and even *Electronic Crime Scene Investigation* (www.ojp. gov/nij/topics/law-enforcement/investigations/crime-scene/guides/twgs.htm).

FBI

In a similar vein and similarly constructed to the National Institute of Justice TWGs, the FBI created what they termed Scientific Working Groups (SWGs). There are currently 20 SWGs at various stages of development of guidelines (visit www.fbi.gov and follow links). Two of them include forensic anthropologists as key members: the Scientific Working Group for Forensic Anthropology (SWGANTH) assembled in 2008 (www.swganth.org) and the Scientific Working Group on Disaster Victim Identification (SWGDVI) created in 2010 (www.swgdvi.org). The SWGANTH includes 18 subcommittees that are attempting to provide guidelines that cover topics as diverse as ethics and conduct (see Chapter 33) to detection and recovery, as well as all of the standard forensic anthropology categories (e.g., age, sex, stature, trauma). The SWGDVI includes anthropologists at many levels of administration and subcommittees membership. Although these committees will not produce standards or guidelines to be followed under threat of retribution, they may prove relatively influential as ways to: (i) advertise the discipline and depth of "services" available to other disciplines, (ii) provide guides for new recruits to the field, (iii) provide fodder for defense lawyers upon which to cogitate and ruminate, and (iv) may serve to cause some current practitioners (unreconstructed whelps of true physical anthropologists) to pause and reconsider the quality of their efforts.

National Academy of Sciences

Probably the most influential document to effect the forensic sciences in recent years is the National Academy of Sciences' (NAS) report of the state of the forensic sciences in the USA that was published in February 2009 (NAS 2009). The appointed committee consisted of experts in many forensic disciplines and found problems in many areas, including:

1. with the *practices* within the varying forensic science disciplines themselves: "… operational principles and procedures … are not standardized or embraced, either

between or within jurisdictions. There is no uniformity in the certification of forensic practitioners, or in the accreditation of crime laboratories" (NAS 2009: 6);

2. with the underlying *principles* of individual forensic methods: "with the exception of nuclear DNA analysis, … no forensic method has been rigorously shown to have the capacity to consistently, and with a high degree of certainty, demonstrate a connection between evidence and a specific individual or source" (NAS 2009: 7);

3. with the *handling of evidence* at the crime scene and in the laboratory: "The depth, reliability, and overall quality of substantive information arising from the forensic examination of evidence available to the legal system" (NAS 2009: 6) also varied significantly according to jurisdictional levels, regions of the country, and state versus municipal versus federal agencies;

4. with the *training* of forensic specialists: "…there remains great variability in crime scene investigation practices, along with persistent concerns that the lack of standards and proper training at the crime scene can contribute to the difficulties of drawing accurate conclusions once evidence is subjected to forensic laboratory methods" (NAS 2009: 57).

One significant recommendation was the creation of an entirely new federal agency with oversight capabilities, and opportunities for funding research. The committee saw a critical need to conduct much more basic scientific research in order to: (i) "establish the scientific bases demonstrating the validity of forensic methods" (NAS 2009: 22); (ii) develop and establish "quantifiable measures of reliability and accuracy of forensic analysis" (NAS 2009: 23); (iii) develop "quantifiable measures of uncertainty in the conclusions of forensic analysis" (NAS 2009: 23); and (iv) conduct research into "human observer bias and sources of human error in forensic examinations" (NAS 2009: 24).

These findings could potentially have a significant effect on the state of forensic sciences, especially if a new federal department is created. Impacts will be felt in many areas including funding for research opportunities, especially validation of methods studies; accreditation of laboratories that handle and analyze evidence, and certification of those who handle evidence. As we have discussed above, forensic anthropology has already started improving the way that it does business: validation studies of both old and new methodologies have been conducted; much better comparative samples are used, and the statistics used to analyze the data are state-of-the-field. These efforts have improved the scientific worth of these methods and their applicability to modern forensic cases. In addition to improvements to the laboratory aspect of the field, forensic archaeology has been able to fill the investigative gap long missing in forensic investigations of outdoor scenes by dramatically improving how evidence is found, documented, and processed at outdoor scenes. The depth of interpretations of past events now rivals those drawn from indoor scenes. Forensic anthropology, it could be argued, therefore, is in good shape and would welcome any changes for the better in the forensic sciences.

SUMMARY

Forensic anthropology has gone through a number of growth spurts in the last 80 or so years. Springing from sporadic and infrequent requests in the early twentieth century for assistance to provide a biological profile from skeletal remains, a few prominent academic and museum-based physical anthropologists such as Hrdlička, Todd, Krogman, and Hooten maintained a passing interest in the medicolegal applications of human skeletal biology. With very few exceptions, professional literature on the subject was not produced.

By the mid portion of the last century, more formalized relationships with law enforcement (in this case, the FBI) arose, especially at the Smithsonian Institution with T. Dale Stewart, Larry Angel, and Doug Ubelaker, which brought more attention to the field. This interest continued to grow with the work of Clyde Snow in early human-rights work, the creation of academic programs with a strong forensic anthropology component (Kansas, Tennessee, Arizona, Florida), professional literature, and especially with the formal creation of a separate section of the American Academy of Forensic Science and Board certification in the early 1970s. Throughout this period, forensic anthropology operated primarily as a laboratory-based discipline that sprang into action when called upon by law enforcement, coroners, and medical examiners to provide a biological profile after "standard" forensic avenues to provide victim identification had been exhausted.

The classic definition of the field, as proposed by T. Dale Stewart (1979) and as understood by İşcan (1988a), thus indicated that the primary, if almost exclusive goal of forensic anthropology was aiding in the identification of human remains in forensic contexts. This goal was attained through the estimation of biological profiles (chronological age, sex, ancestry, stature, and antemortem bone modification), which served to reduce the list of potential victim identities. A quantum shift in identification possibilities occurred in the 1980s and 1990s with the amplification of DNA through the PCR, which allowed for the sequencing of DNA even from trace samples. What is more important is that it permitted researchers the ability to perform a virtually infinite number of DNA comparisons, rendering match probabilities several orders of magnitude higher than can be attained through biological profiles derived from forensic anthropological analyses.

At first glance, it may appear that DNA analysis does not necessarily imply a fundamental change from past conditions regarding the goals, functions, and perspective of forensic anthropology. After all, providing positive identification from the bones has not commonly been one of the primary court-accepted tasks of forensic anthropologists, which instead has fallen to other forensic specialists such as forensic pathologists and forensic odontologists. In addition, DNA analysis is still regarded as a relatively expensive and slow procedure, and the number of DNA samples routinely submitted for analysis overwhelms forensic laboratories. From this perspective, the classic goal of biological profile estimation from bones within forensic anthropology still remains a unique and significant role in simplifying the task of narrowing down the missing person list.

When the current trends in DNA analysis are closely examined, however, it soon becomes clear that the current state of affairs is inevitably bound to change. In the last two decades, the limiting steps of DNA analysis have rapidly shifted from DNA amplification to DNA sequencing, and thence to sample comparison and matching, resulting in a rapid decrease in DNA processing times and costs. PCR has become an almost routine procedure, available in most biomedical research and practice centers. Visual comparison from electrophoresis in agarose and polyacrylamide gels has been replaced by automated capillary electrophoresis in the modern DNA sequencers, allowing the processing and sequencing of a large number of samples simultaneously. More importantly, robust DNA databases for sample comparison have been created and made available to the forensic community, with the reference samples growing at an astounding rate.

At present, the only issues preventing routine and widespread victim identification solely based on DNA comparisons are the costs and time required for amplification, sequencing, and comparison, as well as the need to provide potential matches, currently based on samples collected *ad hoc* from the family members of the potential victims. Overcoming these limitations only requires an improvement in sequencing techniques to an extent much smaller than what has transpired during the last two decades, and the inclusion of the DNA sequences of family members of all missing persons in CODIS or equivalent databases. The question is not whether this will happen, but when. When this point is reached, if positive identification remains as the main and almost exclusive goal of forensic anthropology, forensic anthropologists (and odontologists) may become mostly superfluous in most cases, other than those involving commingled remains, where element matching will still result in a significant decrease of sampling, amplification and sequencing efforts.

Therefore, if forensic anthropology is to remain a useful, vibrant scientific discipline, it is necessary to shift the scope of the field from mere identification to a larger range of problems. Through introspection, and the pressure of outside influences, forensic anthropologists were forced to look outside of the box for answers to pressing issues. First came the call to action to produce better scientific research in forensic anthropology, starting with the identification (that is, creation) of better, more appropriate modern skeletal samples (e.g., the Bass Skeletal Collection at the University of Tennessee, the Forensic Data Bank), and computer programs (Fordisc) to analyze the date. Naturally, better research samples yielded better research.

Next came the realization that forensic archaeology could provide, for the first time, detailed information on context at outdoor scenes not provided by law enforcement. Only after scene context and notation of evidence distribution and location via detailed maps were obtained could forensic taphonomic analyses be conducted, which in turn then permitted scientific (nonconjectural) determinations of postmortem interval, past events reconstruction, and even significantly better skeletal trauma analysis.

Recent court rulings regarding the role of science in the courtroom, as well as critical review of the science behind the forensic sciences, have emphasized that the old ways of conducting forensic science, including forensic anthropology, need to be updated. In many ways, forensic anthropology has anticipated these findings and preemptively embraced scientific improvements to forensic anthropology practices. Recent considerations of "best practices," therefore, did not result in a diminution of the field, but rather a strengthening, reconfiguring it as a vibrant, robust, scientific discipline.

REFERENCES

Aufderheide, A.C. and Rodriguez-Martin, C. (1998). *The Cambridge Encyclopedia of Human Paleopathology*. Cambridge University Press, Cambridge.

Bass, W.M. (1969). Recent developments in the identification of human skeletal material. *American Journal of Physical Anthropology* 30: 459–461.

Bass, W.M. (1978). Developments in the identification of human skeletal material. *American Journal of Physical Anthropology* 51: 555–562.

Bass, W.M. (1979). Developments in the identification of human skeletal material (1968–1978). *American Journal of Physical Anthropology* 51: 555–562.

Bass, W.M. (2001). A tribute to Ellis R. Kerley: The Kansas Years. *Journal of Forensic Sciences* 46(4): 780–781.

Bass, W.M. (2006). Forensic anthropology. In W.V. Spitz and D.J. Spitz (eds), *Spitz and Fisher's Medicolegal Investigation of Death: Guidelines for the Application of Pathology to Crime Investigation*, 4th edn (pp. 240–254). Charles C. Thomas, Springfield, IL.

Bass, W.M. and Birkby, W.H. (1978). Exhumation: the method could make the difference. *FBI Law Enforcement Bulletin* 47: 6–11.

Bass, W.M. and Jefferson, J. (2003). *Death's Acres: Inside the Legendary Forensic Lab Where the Dead Do Tell Tales*. Penguin, New York.

Baumer, T.G., Passalacqua, N.V., Powell, B.J., Newberry, W.N., Fenton, T.W., and Haut, R.C. (2010). Age-dependent fracture characteristics of rigid and compliant surface impacts on the infant skull—a porcine model. *Journal of Forensic Sciences* 55(4): 993–997.

Bennett, K.A. (1987). *A Field Guide for Human Skeletal Identification*. Charles C. Thomas, Springfield, IL.

Berryman, H.E. and Symes, S.A. (1998). Recognizing gunshot and blunt cranial trauma through fracture interpretation. In K.J. Reichs (ed.), *Forensic Osteology: Advances in the Identification of Human Remains*, 2nd edn (pp. 333–344). Charles C. Thomas, Springfield, IL.

Boldsen, J.L., Milner, G.R., Konigsberg, L.W., and Wood, J.W. (2002). Transition analysis: a new method for estimating age from skeleton. In R.D. Hoppa and J.W. Vaupel (eds), *Paleodemography: Age Distribution from Skeletal Samples* (pp. 73–106). Cambridge University Press, Cambridge.

Brain, C.K. (1981). *The Hunters or the Hunted An Introduction to African Cave Taphonomy*. University of Chicago Press, Chicago, IL.

Brooks, S.T. and Brooks, R.H. (1984). Problems of burial exhumation, historical and forensic aspects. *Human Identification: Case Studies in Forensic Anthropology* (pp. 64–86). Charles C. Thomas, Springfield, IL.

Brooks, S. and Suchey, J.M. (1990). Skeletal age determination based on the os pubis: a comparison of the Acsadi-Nemeskeri and Suchey-Brooks methods. *Human Evolution* 5(3): 227–238.

Brothwell, D.R. (1981). *Digging up Bones*, 3rd edn. Cornell University Press, Ithaca, NY.

Byers, S.N. (2002). *Introduction to Forensic Anthropology*. Allyn and Bacon, Boston, MA.

Connor, M.A. (2007). *Forensic Methods: Excavation for the Archaeologist and Investigator*. AltaMira Press, Lanham, MD.

Dale, W.M., Greenspan, O., and Orokos, D. (2006). *DNA Forensics: Expanding Uses and Information Sharing*. NCJ 217992. SEARCH, The National Consortium for Justice Information and Statistics: Sacramento. http://bjs.ojp.usdoj.gov/content/pub/pdf/dnaf.pdf.

Daubert v. Merrell Dow Pharmaceuticals, 509 US 579, 113 S.Ct. 2786, 125 L.Ed. 2d 469 (US June 28, 1993) (No. 92-102).

Dirkmaat, D.C. and Adovasio. J.M. (1997). The role of archaeology in the recovery and interpretation of human remains from an outdoor forensic setting. In W.D. Haglund and M.H. Sorg (eds), *Forensic Taphonomy: The Postmortem Fate of Human Remains* (pp. 39–64). CRC Press.

Dirkmaat, D.C. and Cabo, L.L. (2006). The shallow grave as an option for disposing of the recently deceased: goals and consequences [abstract]. *Proceedings of American Academy of Forensic Sciences* 12: 299.

Dirkmaat, D.C. and Hefner, J.T. (2001). Forensic processing of the terrestrial mass fatality scene: testing new search, documentation and recovery methologies [abstract] *Proceedings of American Academy of Forensic Sciences* 7: 241–242.

Dirkmaat, D.C. and Miller, W. (2003). Scene recovery efforts in Shanksville, Pennsylvania: the role of the Coroner's Office in the processing of the crash site of United Airlines Flight 93 [abstract]. *Proceedings of American Academy of Forensic Sciences* 9: 279.

Dirkmaat, D.C., Cabo, L.L., Adovasio, J.M., Rozas, V. (2005). Mass graves, human rights and commingled: considering the benefits of forensic archaeology [abstract] *Proceedings of American Academy of Forensic Sciences* 11: 316.

Dirkmaat, D.C., Cabo, L.L., Ousley, S.D., and Symes, S.A. (2008). New perspectives in forensic anthropology. *Yearbook of Physical Anthropology* 51: 33–52.

Dirkmaat, D.C., Symes, S.A., and Cabo-Perez, L.L. (2010). Forensic archaeological recovery of the victims of the Continental Connection Flight 3407 Crash in Clarence Center, NY. [abstract]. *Proceedings of American Academy of Forensic Sciences* 16: 387.

Dupras, T.L., Schultz, J.L., Wheeler, S.M., and Williams, L.J. (2006). *Forensic Recovery of Human Remains: Archaeological Approaches.* CRC Press, Boca Raton, FL.

Efremov, I.A. (1940). Taphonomy: new branch of paleontology. *Pan American Geologist* 74: 81–93.

Elliott, M. and Collard, M. 2009. Fordisc and the determination of ancestry from cranial measurements. *Biology Letters* 5: 849–852.

El-Najjar, M.Y. and McWilliams, K.R. (1978). *Forensic Anthropology: The Structure, Morphology and Variation of Human Bone and Dentition.* Charles C. Thomas, Springfield, IL.

Falsetti, A.B. (1999). A thousand tales of dead men: the forensic anthropology cases of William R. Maples, Ph.D. *Journal of Forensic Sciences* 44 (4): 682–686.

Fazekas, G. and Kosa, F. (1978). *Forensic Fetal Osteology.* Akademiai Kiado, Budapest.

Federal Rules of Evidence (FRE), 702, Pub. L. No. 93-595, §, 1, 88 Stat. 1926. Effective January 2, 1975.

Federal Rules of Evidence (FRE), 702, Pub. L. No. 93-595, §, 1, 88 Stat. 1926. Amended April 17, 2000. Effective December 1, 2000.

Fojas, C.L. (2010). A radiographic assessment of age using distal radius epiphysis presence in a modern subadult sample. *Proceedings of the American Academy of Forensic Sciences* 16: 371–372.

France, D.L. (1998). Observational and metrical analysis of sex in the skeleton. In Reichs, K.J. (ed.), *Forensic Osteology: Advances in the Identification of Human Remains,* 2nd edn (pp. 163–186). Charles C. Thomas, Springfield, IL.

Frye v. United States, 293 F. 1013 (D.C. Cir. 1923).

Galton, F. (1882). The Anthropometric Laboratory. *Fortnightly Review* 31: 332–338.

Galton, F. (1886). Regression toward mediocrity in hereditary stature. *Journal of the Anthropological Institute* 15: 246–263.

General Electric Co. v. Joiner, 522 US 136 (1997).

Gifford, D.P. (1981). Taphonomy and paleoecology: a critical review of archaeology's sister discipline. In M.B. Schiffer (ed.), *Advances in Archaeological Method and Theory,* vol. 4 (pp. 365–438). Academic Press, New York.

Gilbert, B.M. and McKern, T.W. (1973). A method for aging the female os pubis. *American Journal of Physical Anthropology* 38: 31–38.

Giles, E. (1970). Discriminate function sexing of the human skeleton. In T.D. Stewart (ed.), *Personal Identification in Mass Disasters* (pp. 99–109). Smithsonian Institution: Washington DC.

Giles, E. and Elliot, O. (1964). Sex determination by discriminate function analysis of crania. *American Journal of Physical Anthropology* 21(1): 53–68.

Giles, E. and Klepinger, L.L. (1988). Confidence intervals for estimates based on linear regression in forensic anthropology. *Journal of Forensic Sciences* 33(5): 1218–1222.

Gill, G.W. and Rhine, S. (eds) (1990). *Skeletal Attribution of Race.* Maxwell Museum of Anthropology, Albuquerque, NM.

Grayson, D.K. (1984). *Quantitative Zooarchaeology: Topics in the Analysis of Archaeological Faunas.* Academic Press, Orlando, FL.

Haglund, W.D. and Sorg, M.H. (eds) (1997). *Forensic Taphonomy: The Postmortem Fate of Human Remains*. CRC Press, Boca Raton, FL.

Haglund, W.D. and Sorg, M.H. (eds) (2002). *Advances in Forensic Taphonomy: Method, Theory, and Archaeological Perspectives*. CRC Press, Boca Raton, FL.

Harris, E.F. and McKee, J.H. (1990). Tooth mineralization standards for Blacks and Whites from the middle southern United States. *Journal of Forensic Sciences* 34: 859–872.

Haviland, W.A. (1994). Wilton Marion Krogman. National Academy of Sciences. *Biographical Memoirs* 63: 292–321.

Hochrein, M.J. (1997). Buried crime scene evidence: the application of forensic geotaphonomy in forensic archaeology. In P.G. Stimson and C.A. Mertz (eds), *Forensic Dentistry* (pp. 83–98). CRC Press, Boca Raton, FL.

Hochrein, M.J. (2002). An autopsy of the grave: recognizing, collecting, and processing forensic geotaphonomic evidence. In W.D. Haglund and M.H. Sorg (eds), *Advances in Forensic Taphonomy: Method, Theory, and Anthropological Perspectives* (pp. 45–70). CRC Press, Boca Raton, FL.

Hochrein, M., Dirkmaat D.C., and Adovasio J. M. (2000). Beyond the grave: applied archaeology for the forensic sciences [abstract]. *Proceedings of the American Academy of Forensic Sciences* 6: 116.

Hoffman, J.M. (1979). Age estimation from diaphyseal lengths: two months to twelve years. *Journal of Forensic Sciences* 24(2): 461–469.

Hunt, D.R. and Albanese J. (2005). History and demographic composition of the Robert J. Terry Anatomical Collection. *American Journal of Physical Anthropology* 127(4): 406–417.

İşcan, M.Y. (1988a). Rise of forensic anthropology. *Yearbook of Physical Anthropology* 31: 203–230.

İşcan, M.Y. (1988b). Wilton Marion Krogman, Ph.D. (1903–1987): the end of an era. *Journal of Forensic Sciences* 33(6): 1473–1476.

İşcan, M.Y. (ed.) (1989). *Age Markers in the Human Skeleton*. Charles C. Thomas, Springfield, IL.

İşcan, M.Y. and Kennedy, K. (eds) (1989). *The Reconstruction of Life from the Skeleton*. Alan Liss, New York.

İşcan, M.Y. and Loth, S.R. (1986). Estimation of age and determination of sex from the sternal rib. In K.J. Reichs (ed.), *Forensic Osteology: Advances in the Identification of Human Remains* (pp. 68–89). Charles C. Thomas, Springfield, IL.

İşcan, M.Y. and Loth, S.R. (1989). Osteological manifestations of age in the adult. In M.Y. İşcan and K. Kennedy (eds), *The Reconstruction of Life from the Skeleton* (pp. 23–40). Alan Liss, New York.

James, S.H. and Nordby, J.J. (2003). *Forensic Science: An Introduction to Scientific and Investigative Techniques*. CRC Press, Boca Raton, FL.

Jantz, R.L. and Moore-Jansen, P.H. (1988). A database for forensic anthropology: structure, content, and analysis. *Report of Investigations No. 47*. Department of Anthropology, The University of Tennessee, Knoxville, TN.

Jantz, R.L. and Moore-Jansen, P.H. (2000). *Database for Forensic Anthropology in the United States, 1962–1991* (computer file). ICPSR version. University of Tennessee, Department of Anthropology Knoxville, TN.

Jantz, R.L. and Ousley, S.D. (2005). *FORDISC 3. Computerized Forensic Discriminant Functions, Version 3.0*. University of Tennessee, Knoxville, TN.

Johnston, F.E. (1962). Growth of long bones of infants and young children at Indian Knoll. *American Journal of Physical Anthropology* 20(3): 249–254.

Johnston, F.E. (1989). On Krogman. *American Journal of Physical Anthropology* 80: 127–128.

Joyce, C. and Stover, E. (1991). *Witnesses from the Grave: the Stories Bones Tell*. Little Brown, Boston, MA.

Kerley, E.R. (1978). Recent developments in forensic anthropology. *Yearbook of Physical Anthropology* 21: 160–173.

Klepinger, L.L. (2006). *Fundamentals of Forensic Anthropology*. John Wiley and Sons, New York.

Konigsberg, L.W., Herrmann, N.P., Wescott, D.J., and Kimmerle, E.H.

(2008). Estimation and evidence in forensic anthropology: age-at-death. *Journal of Forensic Sciences* 53(3): 541–557.

Krogman, W.M. (1939a). A guide to the identification of human skeletal material. *FBI Law Enforcement Bulletin* 8(8): 3–31.

Krogman, W.M. (1939b). Contributions of T. Wingate Todd to Anatomy and Physical Anthropology. *American Journal of Physical Anthropology* 25(2): 145–186.

Krogman, W.M. (1943). Role of the physical anthropologist in the identification of human skeletal remains. *FBI Law Enforcement Bulletin* 12(4): 17–40, 12(5): 12–28.

Krogman, W.M. (1962). *The Human Skeleton in Forensic Medicine*. Charles C. Thomas, Springfield, IL.

Krogman, W.M. and İşcan, M.Y. (1986). *The Human Skeleton in Forensic Medicine*, 2nd edn. Charles C. Thomas, Springfield, IL.

Krogman, W.M., McGregor, J., and Frost, B. (1948). A problem in human skeletal remains. *FBI Law Enforcement Bulletin* 17(6): 7–12.

Kroman, A. (2010). Rethinking bone trauma: a new biomechanical contunuum base approach [abstract]. *Proceedings of American Academy of Forensic Sciences* 16: 355–356.

Kumho Tire Co., Ltd. v. Carmichael, 526 US 137, 119 S.Ct. 1167, 143 L.Ed. 2d 238 (US March 23, 1999) (No. 97-1709).

Little, M.A. and Sussman, R.W. (2010). History of biological anthropology. In C.S. Larsen (ed.), *A Companion to Biological Anthropology* (pp. 14–38). Wiley-Blackwell, Oxford.

Lovejoy, C.O., Meindl, R.S., Prysbeck, T.R., and Mensforth, R.P. (1985). Chronological metamorphosis of the auricular surface of the ilium: a new method for the determination of adult skeletal age at death. *American Journal of Physical Anthropology* 68: 15–28.

Lyman, R.L. (1994). *Vertebrate Taphonomy*. Cambridge University Press, Cambridge.

Maples, W.R. (1986). Trauma analysis by the forensic anthropologist. In K.J. Reichs (ed.), *Forensic Osteology: Advances in the Identification of Human Remains* (pp. 218–228). Charles C. Thomas, Springfield, IL.

Maples, W.R. and Browning, M. (1994). *Dead Men Do Tell Tales*. Doubleday, New York.

Maresh, M.M. (1943). Growth of major long bones in healthy children. *American Journal of Diseases of Children* 89: 725–742.

McKern, T.W. and Stewart, T.D. (1957). *Skeletal Age Changes in Young American Males. Analyzed from the Standpoint of Age Identification*. Environmental Protection Research Division, Technical Report EP-45. Quartermaster Research and Development Center, US Army, Natick, MA.

Meadows, L. and Jantz, R.L. (1995). Allometric secular change in the long bones from the 1800's to the present. *Journal of Forensic Sciences* 40: 762–767.

Meindl, R.S., and Lovejoy, C.O. (1985). Ectocranial suture closure: a revised method for the determination of skeletal age at death based on the lateral anterior sutures. *American Journal of Physical Anthropology* 68: 57–66.

Merbs, C.F. (1989). Trauma. In M.Y. İşcan and K.A.R. Kennedy, *Reconstruction of Life From the Skeleton* (pp. 161–189). Alan R. Liss, New York.

Milner, G.R., Wood, J.W., and Boldsen, J.L. (2008). Advances in paleodemography. In M.A. Katzenberg and S.R. Saunders (eds), *Biological Anthropology of the Human Skeleton*, 2nd edn (pp. 561–600). John Wiley and Sons, New York.

Moorrees, C.F.A., Fanning, E.A., and Hunt, E.E. (1963). Formation and resorption of three deciduous teeth in children. *American Journal of Physical Anthropology* 21: 205–213.

Morse, D., Duncan, J., and Stoutamire, J. (1983). *Handbook of Forensic Archaeology and Anthropology*. Rose Printing Co, Tallahassee, FL.

Morse, D., Dailey, R.C., Stoutamire, J., and Duncan, J. (1984). Forensic arachaeology. In *Human Identification: Case Studies in Forensic Anthropology* (pp. 53–64). Charles C. Thomas, Springfield, IL.

National Academy of Sciences (NAS) (2009). *Strengthing Forensic Sciences in the United States: A Path Forward*. National Research Council of the National Academies, National Academies Press, Washington DC.

National Institute of Justice (2005). *Mass Disaster Incidents: A guide for human forensic identification*. NIJ Technical Working Group for Mass Fatality Forensic Identification, United States Department of Justice, Office of Justice Programs, Washington DC.

Ortner, D. (2003). *Identification of Pathological Conditions in Human Skeletal Remains*, 2nd edn. Academic Press, San Diego, CA.

Ousley, S.D. (1995). Should we estimate biological or forensic stature? *Journal of Forensic Sciences* 40: 768–773.

Ousley, S.D. and Jantz, R.L. (1996). *FORDISC 2.0: personal computer forensic discriminant functions*. University of Tennessee, Knoxville, TN.

Ousley, S.D. and Jantz, R.L. (1998). The Forensic Data Bank: documenting skeletal trends in the United States. In K.J. Reichs (ed), *Forensic Osteology: Advances in the Identification of Human Remains*, 2nd edn (pp. 441–458). Charles C. Thomas, Springfield, IL.

Pearson, K. and Bell, J. (1919). A study of the long bones of the English skeleton I, the femur (chapters 1 to 6). *Draper's Co. Res Mem* (Biometric Series X). Department of Applied Statistics, University College, University of London, London.

Phenice, T.W. (1969). A newly developed visual method of sexing the os pubis. *American Journal of Physical Anthropology* 30(2): 297–302.

Potts, R.B. and Shipman, P.L. (1981). Cutmarks made by stone tools on bones from Olduvai Gorge, Tanzania. *Nature* 291: 577–580.

Rathbun, T.A. and Buikstra, J.E. (1984). *Human Identification: Case Studies in Forensic Anthropology*. Charles C. Thomas, Springfield, IL.

Rebmann, A., David E., and Sorg M.H. (2000). *Cadaver Dog Handbook: Forensic Training and Tactics for the Recovery of Human Remains*. CRC Press, Boca Raton, FL.

Reichs, K. (1986). Introduction. In *Forensic Osteology: Advances in the Identification of Human Remains* (pp. xv–ccciv). Charles C. Thomas, Springfield, IL.

Reineke, G.W. and Hochrein, M.J. (2008). Pieces of the puzzzle: FBI Evidence Response Team approaches to scenes with commingles evidence. In B.J. Adams and J.E. Byrd (eds), *Recovery, Analysis, and Identification of Commingled Human Remains* (pp. 31–56). Humana Press, Totowa, NJ.

Rhine, S. (1998). *Bone Voyage: a Journey in Forensic Anthropology*. University of New Mexico Press, Albuquerque, NM.

Rogers, Y.H. and Venter, J.C. (2005). Genomics: massively parallel sequencing. *Nature* 437: 326–327.

Saferstein, R. (2007). *Criminalistics: An Introduction to Forensic Sciences*, 9th edn. Pearson/Prentice Hall, Upper Saddle River, NJ.

Saferstein, R. (2009). *Forensic Science: From the Crime Scene to the Crime Lab*. Pearson/Prentice Hall, Upper Saddle River, NJ.

Schwidetsky, I.L. (1954). Forensic anthropology in Germany. *Human Biology*. 26(1): 1–20.

Sciulli, P.W. and Pfau, R.O. (1994). A method of estimating for establishing the age of subadults. *Journal of Forensic Sciences* 39 (1) 165–176.

Shipman, P. (1981). *Life History of a fossil*. Harvard University Press, Cambridge, MA.

Skinner, M., and Lazenby, R.A. (1983). *Found! Human Remains: A Field Manual for the Recovery of the Recent Human Skeleton*. Archaeology Press, Burnaby, BC.

Snow, C.C. (1973). Forensic anthropology. In A. Redfield (ed), *Anthropology beyond the University* (pp. 4–17). Southern Anthropological Society Proceedings, No. 7. University of Georgia Press, Athens, GA.

Snow, C.C. (1982). Forensic anthropology. *Annual Review of Anthropology* 11: 97–131.

Stewart, T.D. (1948). Medicolegal aspects of the skeleton, I. Sex, age, race, and stature. *American Journal of Physical Anthropology* 6: 315–322.

Stewart, T.D. (1951). What the bones tell. *FBI Law Enforcement Bulletin* 20(2): 2–5, 19.

Stewart, T.D. (ed.) (1970). *Personal Identification in Mass Disasters*. National

Museum of Natural History, Smithsonian Institution, Washington DC.

Stewart, T.D. (1972). What the bones tell-today. *FBI Law Enforcement Bulletin* 41(2): 16–20, 30–31.

Stewart, T.D. (1976). Identification by skeletal structures. *Gradwohl's Legal Medicine*, 3rd edn (pp. 109–135). Francis E. Camps, Bristol.

Stewart, T.D. (1977). The Neanderthal skeletal remains from Shanidar Cave, Iraq: a summary of findings to date. *Proceedings of the American Philosophical Society* 121(2): 121–165.

Stewart, T.D. (1979). *Essentials of Forensic Anthropology*. Charles C. Thomas, Springfield, IL.

Stewart, T.D. (1984). Perspective on the reporting of forensic cases. In T.A. Rathbun and J.E. Buikstra (eds), *Human Identification: Case Studies in Forensic Anthropology* (pp. 15–18). Charles C. Thomas, Springfield, IL.

Stewart, T.D. and Trotter, M. (eds) (1954). *Basic Readings on the Identification of Human Skeletons: Estimation of Age*. Wenner-Gren Foundation: New York.

Stewart, T.D. and Trotter, M. (1955). Role of physical anthropology in the field of human identification. *Science* 122: 883–884.

Suchey, J.M. and Katz, D. (1998). Applications of pubic age determination in a forensic setting. In K.J. Reichs (ed), *Forensic Osteology: Advances in the Identification of Human Remains*, 2nd edn (pp. 204–236). Charles C. Thomas, Springfield, IL.

Suchey, J.M., Wisely, D.V., and Katz, D. (1986). Evaluation of the Todd and McKern-Stewart methods for aging the male os pubis. In K.J. Riech (ed), *Forensic Osteology: Advances in the Idenfification of Human Remains* (pp. 33–67). Charles C. Thomas, Springfield, IL.

Symes, S.A., Berryman, H.E., and Smith, O.C. (1998). Saw marks in bone: introduction and examination of residual kerf contour. In K.J. Reichs (ed), *Forensic Osteology: Advances in the Identification of Human Remains*, 2nd edn (pp. 389–409). Charles C. Thomas, Springfield, IL.

Todd, T.W. (1920). Age changes in the pubic bone: I. the male white pubis. *American Journal of Physical Anthropology* 3(3): 285–334.

Todd, T.W. (1921a). Age changes in the pubic bone: II the pubis of the male Negro-white hybrid; III: the pubis of the white female; IV: the pubis of the female white-Negro hybrid. *American Journal of Physical Anthropology* 4: 1–70.

Todd, T.W. (1921b). Age changes in the pubic bone: VI. The interpretation of variations in the symphysial area. *American Journal of Physical Anthropology* 4(4): 407–424.

Todd, T.W. and Lyon, D.W. (1924). Cranial suture closure; its progress and age relationships. *American Journal of Physical Anthropology* 7(3): 325–384.

Todd, T.W. and Lyon, D.W. (1925a). Cranial suture closure; its progress and age relationships, Part II. Ectocranial closure in adult males of white stock. *American Journal of Physical Anthropology* 8(1): 23–46.

Todd, T.W. and Lyon, D.W. (1925b). Cranial suture closure; its progress and age relationships, Part III. Endocranial suture closure in adult males of Negro stock. *American Journal of Physical Anthropology* 8(1): 47–71.

Trotter, M. (1970). Estimation of stature from intact long limb bones. In T.D. Stewart (ed), *Personal Identification in Mass Disasters* (pp. 71–83). National Museum of Natural History, Smithsonian Institution, Washington DC.

Trotter, M. and Gleser, G.C. (1952). Estimation of stature from long bones of American whites and negroes. *American Journal of Physical Anthropology* 10: 463–524.

Trotter, M. and Gleser, G.C. (1958). A re-evaluation of stature based on measurements taken during life and of long bones after death. *American Journal of Physical Anthropology* 16: 79–123.

Tuller, H., Hofmeister, U., and Daley S. (2008). Spatial analysis of mass grave mapping data to assist in the reassociation of disarticulated and commingled human remains. In B.J. Adams, J.E. Byrd (eds), *Recovery, Analysis, and Identification of Commingled Human Remains* (pp. 7–30). Humana Press, Totowa, NJ.

Ubelaker, D.H. (1997). Forensic anthropology. In F. Spenser (ed), *History of Physical Anthropology: An Encyclopedia*, vol. 1 (pp. 392–396). Garland Publishing, New York.

Ubelaker, D.H. (1999). Aleš Hrdlička's role in the history of forensic anthropology. *Journal of Forensic Sciences* 44(4): 724–730.

Ubelaker, D.H. (2001). Contributions of Ellis R. Kerley to Forensic Anthropology. *Journal of Forensic Sciences* 46(4): 773–776.

Ubelaker, D.H. (2010). Issues in forensic anthropology. In: C.S. Larsen (ed), *A Companion to Biological Anthropology* (pp. 412–426). Wiley-Blackwell, West Sussex.

Ubelaker, D.H. and Hunt, D.R. (1995). The Influence of William M. Bass on the development of Forensic Anthropology. *Journal of Forensic Sciences* 40: 729–734.

Wilson, R.J., Herrmann, N.P., and Meadows-Jantz, L. (2010). Evaluation of stature estimation from the Database for Forensic Anthropology. *Journal of Forensic Sciences* 55(3): 684–689.

Wolf, D.J. (1986). Forensic anthropology scene investigations. In Reichs, K.J. (ed). *Forensic Osteology: Advances in the Identification of Human Remains* (pp. 3–23). Charles C. Thomas, Springfield, IL.

Zelditch, M.L., Swiderski, D.L., Sheets, H.D., and Fink, W.L. (2004). *Geometric Morphometrics for Biologists: A Primer*. Elsevier Academic Press, New York.

PART I

Introduction and Brief History of Forensic Anthropology

PART II Recovery of Human Remains from Outdoor Contexts

Introduction to Part II

Dennis C. Dirkmaat

As emphasized throughout this book, the comprehensive documentation of context through the employment of forensic archaeology provides the key contribution to the medicolegal investigation of the outdoor death scene. With the addition of forensic archaeology to the toolbox, forensic anthropology now has something critically important to contribute beyond merely "assisting in reducing the missing-person list" through the analysis of the bones of the recently deceased. As described in Chapters 1 and 2 of this volume, not only has forensic archaeology resulted in more thorough recoveries of all types and classes of evidence at the outdoor scene, but better provenience and notation of that evidence has also allowed for a much better understanding of the scene. Further, when the scene is understood, the effects of specific taphonomic agents impacting the remains, such as those leading to scattering, removing, burying, burning, chewing, and decomposing, are better understood. This leads to better reconstructions of the sequence, trajectory, and timing of past events.

Forensic anthropologists must begin to insist that the proper scientific analysis of human remains from an outdoor scene is only achieved as a result of the complete documentation of the context of the remains. This perspective requires a change in the existing commonplace and historical definition of forensic anthropology that is focused exclusively on examining human bones in the laboratory in order to provide clues to the identity of the individual. Laboratory analysis alone, however, cannot be used to answer questions related to the death event.

Although law enforcement possesses excellent recovery and documentation protocols, and thus interpretive opportunities and abilities, for indoor crime scenes, no standard evidence-processing protocols exist for the outdoor forensic scene. Evidence is often overlooked in the field and incompletely documented, especially with respect to spatial distribution, and is usually hurriedly removed from the scene. However,

A Companion to Forensic Anthropology, First Edition. Edited by Dennis C. Dirkmaat.
© 2012 Blackwell Publishing Ltd. Published 2012 by Blackwell Publishing Ltd.

as described by Dirkmaat in Chapter 2, forensic anthropology offers the best solution to filling this glaring investigative void by providing forensic archaeological principles, methods, and practices for documenting context at a wide variety of outdoor scenes, from surface scatters to mass disaster scenes. When context is properly noted, scientific answers can be provided to questions related to postmortem interval such as why remains are not where they should be, why remains are missing, and what was the condition of the remains prior to recovery. Accurately answering these questions can provide invaluable information in a medicolegal setting.

Dirkmaat illustrates his points by focusing first on contemporary law-enforcement methods for the processing of indoor scenes, emphasizing that nothing is disturbed until context is carefully documented. Here, context refers to the notation of exact positions and locations of evidence relative to other evidence and to the scene. The product of this effort is better evaluation of forensic significance and the ability to scientifically reconstruct past events. These standards should be in place for all medicolegal death scenes. However, a search of law-enforcement protocols finds that there are no appropriate methodologies for outdoor scenes except those that refer back to forensic archaeological methods. Dirkmaat illustrates the point by comparing current law-enforcement methods for processing an outdoor scene with a forensic archaeological approach during the recovery of a set of human remains from a recent case found in a dense patch of brush. Simply attempting to answer basic questions regarding the remains and events surrounding the death following each recovery approach produces an effective selling point for forensic archaeology.

In the remaining chapters of Part II of this book, the forensic recovery of a variety of scenes and activities, including organized searches for previously unlocated forensic scenes, surface-scattered scenes, clandestine grave features, mass graves in human-rights-abuse scenarios, fatal fire scenes, and large-scale mass fatality scenes, are addressed.

SEARCH EFFORTS

One of the key aspects of an efficient and effective search for unlocated remains and scenes, primarily on the surface, is to include someone who can provide a determination of forensic significance of evidence found on site and in real time. This ability insures that nonforensic materials do not enter into the system, thus bogging it down with frivolous activities. The National Institute of Justice, in fact, cites this issue as a major problem in crime scene investigation that needs to be addressed (National Institute of Justice 1999). Simply put, this issue is resolved with the employment of forensic archaeological principles and procedures and the inclusion of the forensic anthropologist during the search.

Schultz, in Chapter 3, provides an outline of how the forensic anthropologist determines forensic significance at many different levels, including distinguishing human remains from sticks, stones, and animal bones. In the chapter, he describes how the biological evidence, context, and associated artifacts mesh together as an integrated package in the evaluative process.

The combination of proper personnel and appropriate search strategies results in an effective and efficient large-scale search effort when searching for remains that are situated on the surface. Following standard archaeological protocols, with the possible utilization of a cadaver dog, results in near 100% confidence levels that the remains will be found.

When the remains and evidence are not located on the surface, and instead are hidden under rubble, within garbage dumps or buried, the level of confidence in abilities to successfully locate them drops dramatically. In cases involving individuals potentially buried in previously undisturbed soils, searchers must rely on above-surface clues such as soil subsidence and surface cracks over the burial feature, differential growth of vegetation, and other subtle evidence. Even cadaver dogs are not as effective in these situations and a good dose of luck helps.

In some cases in which informant information is very good, or the search has been narrowed to very specific areas, sophisticated instruments that provide images of sub-surface stratigraphic profiles emphasizing anomalies or voids, or metal, beneath the surface, have been employed with success. In Chapter 4, Schultz provides a thorough review of the use of the most common subsurface viewer: ground-penetrating radar (GPR). He provides an excellent introduction to the technology, its advantages and disadvantages, what can be expected of the work, and where it works best and where it does not work well (Editor's note: it does not work after the forensic backhoe has been used to search for the burial feature).

Michael Hochrein, trained originally as an archaeologist prior to employment with the FBI, provides an interesting perspective to the investigation of a criminal "incident" involving multiple events and multiple scenes in Chapter 5. His advocacy of a multidisciplinary approach to processing each distinct crime scene encountered is well illustrated by a case he describes in the chapter. Nearly 10 distinct scenes related to one "incident" and one individual's series of criminal activities were investigated, linked, and compiled into a comprehensive reconstruction of past events. The individual sites included decomposing human remains, discarded clothing, the location of a possible suicide, and others. They were all effectively linked through a combination of documentation and mapping procedures that ranged from hand-drawn maps, to total station maps, and finally to geographic information systems (GIS). He emphasizes that the best approach to scientifically "connecting the dots" is the integration of forensic anthropology, and especially forensic archaeology, into standard law-enforcement investigative protocols when dealing with outdoor crime scenes.

Chapters 6 and 7 deal with recent developments in the documentation and recovery of rather unique outdoor crime scenes that benefit greatly from forensic archaeological perspectives and methods: fatal fire and mass fatality scenes. As discussed in Chapter 2 and further developed by Dirkmaat et al. in Chapter 6, significant gaps exist in law-enforcement protocols relative to the recovery of outdoor scenes in general. This gap widens when faced with evidence, including human remains that are significantly altered by fire. Even though fire investigators deal with fire scenes everyday, their focus is on cause, origin, and progress of the fire. If human victims are discovered in the fire rubble or thought to lie within, it is law enforcement, and not the fire investigators, who are usually tasked with processing the scene and recovering the body and associated evidence. However, since the scene is highly altered and viewed as disturbed, processing typically involves the taking of a few photographs followed by placement of recognizable tissue into a body bag. Since all of the materials within the fire scene are similarly discolored, pictures are ineffective in capturing details. Human remains are particularly harshly affected by fire and heat: soft tissue is burned away, tissue is discolored to match the surrounding matrix, and the exposed bones exhibit shrinkage, warpage, discoloration, and fracturing patterns. Not only is the tissue difficult to identify in the field, but it is also extremely friable and will be further damaged by recovery methods as well as transport in soft-sided body bags. Dirkmaat et al. discuss these issues and suggest that, again, forensic archaeological recovery methods employed by forensic anthropologists (those who are familiar with burned bone) provide the best solution. These methods result in efficient and effective location, documentation, recovery, and transport methods. Also, it is pointed out that forensic anthropologists provide an excellent sidekick to forensic pathologists working with the fatal fire victim in the morgue, for reasons described in Chapter 6.

Chapter 7 discusses the role that forensic anthropology can provide at the mass disaster scene. Certainly, forensic anthropological skills, with respect to the analysis and interpretation of fragmented, burned, and otherwise altered bones, are unique and beneficial in the disaster morgue. Forensic anthropologists have been employed to help at the initial sorting table, or triage station, as well as the anthropology station. Additionally, forensic archaeological skills provide leadership capabilities in organizing efficient searches over large areas, and developing and employing outdoor scene documentation protocols of "significant evidence" context and provenience at the large-scale, outdoor forensic scene. It is emphasized in the chapter that coroners and medical examiners are responsible for both the identification of the victims, as well as ensuring that the victims are properly recovered (issues related to scene management), tasks often not realized or appreciated by the coroner or medical examiner until the event occurs. Forensic anthropology provides a much needed helping hand. Recent research in the use of technology to further improve efficiency and effectiveness of comprehensive scene processing and victim recovery efforts are also discussed in Chapter 7.

In Chapter 8, Tuller discusses specialized forensic anthropology work conducted during the investigation of human-rights violations, especially those involving rather complicated and large-scale mass grave features. The focus, methods and personnel in charge have changed dramatically since the early days. The first investigations and recoveries of mass graves were focused on removing victims from the large burial features, establishing a positive identification, and returning the remains to families in order to provide closure. Issues were quickly raised, however, relative to whether forensic evidence was collected properly and could eventually be used in a court of law to prosecute perpetrators. The primary issue would be whether there was proper consideration of the context and association of the evidence relative to other evidence, relative to the immediate scene, and relative to the larger-scale scene. Tuller's descriptions of some of the early miss-steps suggests that there were problems. However, he describes recent work in which forensic anthropology and forensic archaeology play a more important role in the planning of the investigation, and especially with respect to the excavation and recovery of evidence at the mass grave burial feature. These features are very complicated and their recovery, analysis, and interpretation require specialized archaeological skills. He provides excellent descriptions of how this work is conducted today in stages that include grave feature location, as well as excavation and documentation procedures. Interestingly, he notes that anthropologists are also involved in the collection of background information regarding where individuals and burial locations might be found. These skills and activities in interviewing witnesses and family members mirror what is described in human-rights work conducted in Latin America (see Chapter 31 in this volume), and even, to some degree, at mass disaster scenes and the collection of antemortem data at the Family Assistance Center.

In Chapter 9, Cabo and coauthors reaffirm Tuller's assertion in Chapter 8 that the primary purpose of the investigation and excavation of mass graves is to recover evidence within context in order to link evidence, remains, and events to produce an "accurate" reconstruction of what happened in the past, very useful in attempts to prosecute perpetrators of these atrocities. Forensic archaeological principles, methods, and practices provide the best way – some may argue the only viable way – to conduct business at such a complicated outdoor forensic scene. Cabo et al. demonstrate how detailed provenience information recovered at a large archaeological burial scene, in this case a prehistoric Native American ossuary containing over 300 individuals dating to the fourteenth century, is very useful in resolving a variety of issues including sorting commingled remains and which bones likely belong to which individual, among others.

REFERENCE

National Institute of Justice (1999). *Forensic Sciences: Review of Status and Needs*. NCJ 173412. National Institute of Justice, Washington DC.

CHAPTER **2**

Documenting Context at the Outdoor Crime Scene: Why Bother?

Dennis C. Dirkmaat

INTRODUCTION

Law-enforcement protocols for processing indoor scenes are well established (DeForest et al. 1983; Fisher and Fisher 2003; Gardner 2005; Miller 2003; Saferstein 2009; Swanson et al. 2006) and provide scientifically validated, court-room-defendable reconstructions of past events. Law-enforcement protocols for processing outdoor scenes, however, are basically nonexistent. The reasons for this situation are unclear. Perhaps there is a sense that outdoor scenes contain too many variables, including the size of the scene, that do not easily allow for distinguishing forensically significant evidence from natural artifacts. Perhaps too many "agents," such as animals, rain, snow, and even gravity have conspired to modify the scene since the time of the original deposition of the body. These factors often make it seem to be nearly impossible to accurately reconstruct events surrounding the incident. Whatever the reasoning, it is clear that outdoor scenes are not processed with the same high standards as indoor scenes. Before we give up, we must make an effort to find a method or practice that attempts to produce comparable results. If the results are not found in modified indoor scene methods, then we must look elsewhere to see if other methods can produce results comparable to indoor investigations.

A Companion to Forensic Anthropology, First Edition. Edited by Dennis C. Dirkmaat.
© 2012 John Wiley & Sons Ltd. Published 2015 by John Wiley & Sons Ltd.

THE FORENSIC DEATH INVESTIGATION

In the typical investigation of a suspicious human death, a multidisciplinary approach is utilized to answer four universal questions: (i) Who is the victim?; (ii) How did they die?; (iii) What are the circumstances leading up to and surrounding the death; and (iv) Was anyone else involved and, if so, who were they? Each law-enforcement agency and forensic specialist has a specific role to play. The investigation is conducted at the crime scene as well as beyond the initial scene through interviews and background research (Swanson et al. 2006).

Data-collection episodes

A typical forensic investigation of a recently deceased individual includes three distinct episodes of data collection: at the scene, at the autopsy, and in the laboratory. The information collected at the crime scene (data-collection episode 1) regarding how the body is situated relative to other evidence is of benefit to the analysis and interpretation of the condition of the remains at the autopsy table during the postmortem examination (data-collection episode 2). The final data collection (data-collection episode 3) involves the analysis in the laboratory of the forensic evidence related to the body and the scene. The coroner or medical examiner considers data collected during all three stages in the final determination of cause, circumstances, and manner of death. Importantly, the success in each stage is strongly influenced by success in previous stages. If the initial stages are improperly documented or analyzed, subsequent stages suffer in a domino effect.

Protocols for forensic pathological examinations and most laboratory analysis of forensic evidence are well established (DeForest et al. 1983; James and Nordby 2003; Spitz and Spitz 2006; Swanson et al. 2006) and are beyond the scope of this chapter. The focus, instead, is on the processing of the crime scene, specifically the outdoor crime scene. However, before we consider outdoor scene recovery protocols and determine whether they are worth the time and effort, let us examine what is expected of a rigorous indoor scene recovery and determine whether we can apply lessons learned to the outdoor scene recovery.

CRIMINAL INVESTIGATIONS AND PROCESSING OF THE INDOOR SCENE

At indoor crime scenes, well-trained crime scene investigators conduct the documentation of evidence and reconstruction of the circumstances of death. All potential evidence in the room is left undisturbed prior to notation of its precise location and orientation relative to the body, other evidence, and to features of the room. The focus is not just on the recognition and collection of physical evidence at the scene, but on the careful notation of the precise relative position of physical evidence. The *context* and *association* of the evidence is considered just as important as the evidence itself. The working assumption is that if context is carefully noted, then very accurate, very precise, scientifically defensible reconstructions of sequences of past events can be created. Notation of context takes the form of written notes, photographs, and sketches of the scene. Likewise, *chain of custody*, the detailed, recorded path of evidence through the hands of law enforcement, forensic scientists, and the legal system, begins with the description of the evidence *in situ* at the scene (Saferstein 2007). It continues through the postmortem examination of the body, the laboratory analysis of the evidence, and eventually the presentation in court. The best example of "out-of-context" evidence is evidence that has been moved prior to full documentation at the crime scene. This transgression of proper recovery protocols could result in dire consequences for the successful litigation of the case since the potential exists for the evidence to be considered inadmissible in court.

In conclusion, we see that *much* is expected of the indoor crime scene recovery. The goal of these recoveries is not only to document the final position of physical evidence relative to other evidence and to the scene (context and association), but also to attempt to accurately reconstruct events and any modification to the scene. These scenes are handled by forensic investigators who routinely receive extensive training in the documentation and collection of physical evidence from indoor crime scene contexts. Additionally, much law enforcement and forensic science research and literature is devoted to this topic.

Following the recovery at the scene, the remains of the victim are taken to the morgue for forensic pathological examination (termed the *forensic autopsy*) with the express goals of providing an identity for the victim through unique soft-tissue characteristics such as fingerprints, tattoos, etc. (Spitz 2006) and determining a cause of death (Wright 2003a). Cause of death refers to the specific cause or sequence of events leading to the death of the individual; hundreds have been described (Spitz and Spitz 2006). Final assessment of identity and cause of death is often aided or corroborated by analyses conducted by experts in other related forensic fields, such as forensic odontology and fingerprints (for victim identity), forensic toxicology, ballistics, forensic entomology, and even forensic anthropology. These analyses have scientific validity because the context of the evidence is well documented and well known. In addition, the location, position, and orientation of the body at the scene play an important role in final determination of cause of death.

The coroner or medical examiner must then assimilate all of the information gathered by law enforcement, the forensic pathologist, and the various forensic experts in order to provide a feasible and defendable determination of victim identity and *manner of death*. The options available to the coroner/medical examiner for manner of death include *homicide, suicide, natural, accidental,* and *unknown.* Again, context of the body is paramount to proper determination of manner of death (Wright 2003b).

Finally, it is the District Attorney or prosecutor who then determines whether societal laws have been broken, which specific laws are involved, and what the appropriate course of action is for addressing these transgressions. The decision to try accused perpetrators in a court of law, however, often comes down to the "weight of the evidence"; that is, how much evidence is available and whether it has been collected, documented, and analyzed appropriately and can be used effectively in court.

In summary, the key to a properly functioning chain of investigation and prosecution is that complete information has been appropriately collected and analyzed during *all investigative levels*, beginning with the proper processing of the scene. It is clear that investigative protocols employed during the processing of the victims found at indoor crime scenes typically yield successful results.

LAW-ENFORCEMENT PERCEPTION OF THE OUTDOOR CRIME SCENE

When the crime scene is located outdoors, however, there are a vast number of extrinsic factors that can affect the crime scene and potentially lead to a sense of futility in any attempt to fully document and, ultimately, interpret the scene (Figure 2.1). These factors may include geological, environmental, plant, and animal disturbances that can alter the scene and the final disposition of evidence. Long-term exposure of the human remains to the environment and to animals inhabiting that environment can lead to the assumption that the range and depth of comparable information that is

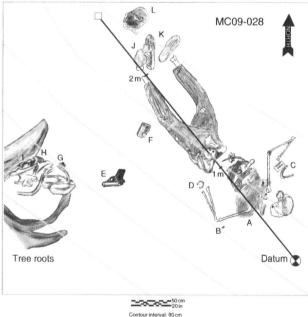

Figure 2.1   Superior view of human remains prior to recovery (top) and plan-view map of remains (bottom).

retrievable at the indoor scene was long ago swept away in the outdoor scene. Animals may have eaten biological tissues or dragged the bones away. Rain and snow may have washed away evidence or shifted it from its original position. This means that the original context of the remains at the time of emplacement has been altered. Why bother then? Significant or useful patterns to be revealed from the remains dispersal are not discernible to the naked eye. The perception is, therefore, that they are either irretrievable or not there at all. Thus, any extensive effort with regard to carefully clearing the site and carefully noting the location of evidence would likely be a waste of time and a waste of limited resources. As a result of this line of thinking, the processing of the outdoor scene often boils down to taking a series of photographs of the body or body parts, followed by the random collection within a reasonable time frame of as many of the bones as are visible on the surface. Often, this is done without knowing which skeletal elements were left behind or assuming that missing elements were removed by animals. In some cases, buried bodies are removed using shovels or a backhoe. Additionally, even though the body may have been at the scene for months or years, there is an overwhelming desire to get the remains to the morgue as quickly as possible.

In these cases, the remains are then taken to the morgue for postmortem examination, and various evidentiary materials collected at the scene are distributed to the laboratory departments as necessary. However, in this case, unlike the indoor scene, very little hard data are available for the reconstruction of the circumstances of death since little or no information regarding context was collected at the scene. Anecdotal reconstructions of what happened at the scene tend to replace defendable event reconstructions that are typically provided by thorough indoor-scene processing protocols.

CAN INDOOR RECOVERY METHODS APPLY TO OUTDOOR SCENES?

Let us assume, however, that there are actually useful and significant patterns in the seemingly disturbed jumble of bones in the outdoor context. Can we apply indoor recovery methods to outdoor scenes and expect the same results as we do of indoor scenes? A search of the law-enforcement literature (e.g., Gardner 2005; James and Nordby 2003; Saferstein 2007, 2009; Swanson et al. 2006) and training regimens reveals, however, that there are no comprehensive recovery protocols derived from the well-developed indoor methods. Aside from a few references to outdoor searches, which are outdated and ineffective, and discussion of how to process footwear and tire impressions (Bodziak 2003a, 2003b), there is no specific discussion of the outdoor scene; it is largely ignored.

WHY NOT FORENSIC ARCHAEOLOGY?

There is a discipline outside of law enforcement that actually attempts to derive vast amounts of data from outdoor scenes. That field is *archaeology*. Archaeology is the "study of the human past, the basis of which is material evidence (artifacts, ecofacts, human remains) and its context" (Stewart 2002: 2). Interestingly, the goals of crime scene processing and archaeology are identical: *reconstruct and understand past events.* This is accomplished in both disciplines through a careful consideration of the context of the evidence relative to the scene (Dirkmaat and Adovasio 1997; Fisher and Fisher 2003; Hester et al. 1997; Hurst Thomas 1998; Joukouwsky 1980; Stewart 2002).

It has been suggested that effective and efficient protocols drawn from archaeology can be effectively applied to outdoor crime scenes (Connor and Scott 2001; Dirkmaat and Adovasio 1997; Haglund 2001; Hochrein 2002; Krogman and İşcan 1986; Sigler-Eisenberg 1985; Skinner and Lazenby 1983; Stoutamire 1983) in the discipline that melds both forensic anthropology and archaeology. That discipline is *forensic archaeology.* It is time to evaluate these claims. Rather than present a list of activities that should be completed at the outdoor crime scene, it is necessary to justify the use of these methods and compare the resulting outputs with those derived from the proper processing of the indoor scene.

Contemporary Archaeological Principles

As described above, the primary goal of archaeology is to "obtain valid knowledge about the past" (Shennan 2004: 3). In practice as well as in theory, both archaeology and criminalistics share the common goal of systematically documenting, collecting, and interpreting physical evidence for the purpose of understanding the factors that affected the depositional history of that evidence (Dirkmaat and Adovasio 1997).

David Hurst Thomas has suggested that archaeology operates at three hierarchical "levels of archaeology theory": high level, middle level, and low level (Hurst Thomas 1998). The first level, termed low-level theory, can be created after basic archaeological data sets – the "material record" (Shafer 1997) – are recovered. These data are used to address questions such as: (i) *Who* was involved in creating the situation?; (ii) *What* are the material signatures of those individuals?; (iii) *How many* data sets (e.g., artifact, ecofact, paleoclimatic information) of these activities are available?; (iv) *Where*, specifically, did the event occur?; and (v) At what particular point in time (*when*) did these events transpire? Middle-level theory (Shafer 1997) attempts to address how the archaeological record was formed by the human actors, an approach championed by Binford (1983) (e.g., how did things transpire at a particular locus and time?). The link between human behavior and the resulting material culture occurs at this level. Finally, the third level (high-level theory) asks why did the event occur, searching for underlying processes and explanations. The goal is to identify generalized laws (termed *nomothetic patterns*) that can be used to explain the past (Hurst Thomas 1998; Shafer 1997).

Addressing low-level questions always begins with good, exacting archaeological excavation techniques that are used to recover evidence *in situ.* This methodology permits a thorough understanding of the relationships (spatial and temporal) between evidence, scene, and environs. James Adovasio (personal communication; Dirkmaat and Adovasio 1997) suggests that proper archaeological excavation has three primary and interdependent responsibilities:

1. *Defining site stratigraphy and stratification.* Stratification refers to the actual observed sequential layering of deposits (i.e., the pages in a book) while stratigraphy is the sum total of the processes that produces these accumulated layers

(i.e., the story). All stratification adheres to certain basic "rules" or laws, often called Steno's principles, which include the laws of superimposition, original horizontality, lateral continuity, and intersecting relationships (see Dirkmaat and Adovasio 1997). The key to understanding stratigraphy is the successful identification and differentiation of individual stratum. This requires excavation skills and attention to detail. For example, most of the individual stratum within the deposits at Meadowcroft Rockshelter, Avella, PA, were very thin and required excavation with razor blades (Carlisle and Adovasio 1982).

2. *Establishment of "context."* Context is an object's place and time in space and includes both physical and temporal coordinates. Documenting and understanding context is critical. It begins with the careful documentation of the position of any and all objects within a stratum relative to all other objects in all other strata (Joukowsky 1980).

3. *Demonstration of "association."* Association means that two or more objects entered the archaeological record at or about the same time as a consequence of the same process, event, action, or activity. Association is the most difficult concept for an archaeologist to prove. Only after context has been established via precise manual or computer-aided mapping is it possible to establish which items entered the depositional record as a result of the same process, and hence which items are associated.

These three responsibilities absolutely require good archaeological excavation methodologies, principles, and practices. Once context and association have been satisfactorily established, the archaeologist can then move on to addressing middle-level questions and, finally, generate high-level theories.

This perspective on how to extract information and answer "how, when, and where" questions from outdoor contexts that usually contain only very ephemeral evidence of past activities is of obvious value when considering how to extract information from outdoor forensic scenes with evidence of relatively recent events. Each set of questions can only be answered reliably if the questions of the previous level have been properly addressed.

In summary, comprehensive scene documentation and recovery methodologies are routinely employed by archaeologists at outdoor sites to maximize the efficient and accurate collection of a wide variety of data. The goal of the excavation and recovery efforts is not simply the exposure and collection of the artifacts, but the detailed reconstruction of past behaviors. As such, it is imperative to properly document the contextual setting of each and every artifact, as well as to document and collect more minute evidence of the environmental and climatic setting. Only then can associations of artifacts be established and the story constructed. There are lessons here for the processing of outdoor forensic scenes.

THE OUTDOOR FORENSIC SCENE

Many questions need to be answered when a body is found in an unexpected location: How and when did the body get to the scene?, Were there other individuals involved?, If so, how many?, Who were they?, Did they come back to modify or disturb the remains?, and If the person was murdered, were they killed at the scene and left there, or were they killed elsewhere? The recovery of a set of human remains at any crime scene must be undertaken with the express purpose of answering as many of these questions as possible. Again, in the same way as archaeology, this boils down to efforts to reconstruct and understand past events.

It is also clear that, in both archaeological and forensic investigation, the objects themselves recovered from these scenes (artifacts, human remains, evidence) cannot answer any of these questions without a thorough understanding of their contextual place in time and space. From a forensic perspective, evidence taken from an indoor crime scene without careful notation of *in situ* placement and position within the scene has very little prosecutorial value. It is surprising that this central tenet is misunderstood or dismissed at the outdoor scene.

It will be argued here that, in nearly every instance, comprehensive notation of context at the outdoor scene, especially in the form of carefully drawn maps of the spatial distribution of evidence, will provide significant information with respect to the original position and orientation of the body at the time of deposition, time-since-death estimates, and identification of factors altering the body since death. Such notations go a long way toward answering the important questions discussed earlier.

DEFINING CONTEXT AT THE OUTDOOR SCENE

The use of the word "context" at the outdoor crime scene relates to: (i) the primary objects of interest (i.e., physical evidence, including the body and associated items); (ii) the surrounding biotic (plants and animals), geophysical (soils, geomorphological features, and water), and climatic environs (temperature, precipitation, and humidity); and (iii) the passage of time since the evidence entered the scene (the temporal context). There must be an attempt to document the interplay of these factors at the forensic locale from the time the victim and, if relevant, the perpetrators entered the scene, to the time that the body was placed at the scene, to the time of recovery.

It is emphasized in both criminalistics and archaeology that the recovery process of artifactual or physical evidence effectively destroys context. As a result, it is imperative that comprehensive and accurate documentation of the contextual setting of each artifact be completed during the recovery. Standard documentation procedures in the archaeological investigation of a site or scene include descriptions of the geophysical, biotic, and environmental setting, as well as specifics of the associations of all of the physical evidence to the entire contextual setting. These details are documented via written notes, photographic and videographic images, and detailed maps (plan view and profile) of the spatial distribution of evidence relative to topographic and other features of the physical setting. The documentation protocols and methodologies clearly parallel those required for accurate processing of forensic scenes. Such procedures yield a wealth of information that is often crucial to the resolution of the case.

FORENSIC ARCHAEOLOGY

Archaeology, as applied to crime scenes and the processing of the recently deceased, has been termed *forensic archaeology*. Serious discussion of the components and the utility of forensic archaeology has been in play since the 1980s (Lovis1992, Morse et al. 1983,

Skinner and Lazenby 1983) and has proven beyond doubt that there is always important information to be gathered at the outdoor scene (Dirkmaat and Adovasio 1997; Dirkmaat and Cabo 2006; Dirkmaat et al. 2008; Dupras et al. 2006; Galloway et al. 2001; Komar and Buikstra 2008). Further, reconstruction of past events at outdoor crime scenes can be as thorough and far-reaching as indoor crime scenes. It is the contention and experience of this author that when efforts are made to carefully document an outdoor scene, especially with regard to mapping procedures, previously unnoticed patterns of spatial distribution of evidence will always be revealed, and significant information regarding the events at the scene will become available.

The types of evidence and information that can be obtained from employing forensic archaeological protocols include: (i) the original location and position of body at time of emplacement at the scene, (ii) the identification of taphonomic agents responsible for dispersing remains and explanation of why remains are not where they should be, (iii) the maximum of skeletal elements and evidence, and (iv) a more informed idea of how long the body has been at the site. Is this done through magic and a cast of thousands? No! Forensic archaeology, as a conceptual discipline, is based on proven contemporary archaeological principles that have been developed and used by archaeologists over the course of the past 100 years. In order to reconstruct past events in outdoor forensic settings, some modifications are required due to the nature of the evidence and legal considerations.

A Case Example

Perhaps the best way to illustrate how important context is to the final interpretation of the outdoor crime scene is to describe a recent forensic case (Figure 2.2). We will compare and contrast the outcomes of two potential scene-recovery options that law enforcement could have taken. Although details have been changed to protect the innocent, the components of the case are not atypical.

In the late spring in Pennsylvania, police were led to a scene where the body of a teenage girl had purportedly been dumped 15 summers earlier (Figure 2.2a). Wading through an overgrown patch of thick brush along a country road on private farm property, a police officer saw what looked like part of a human skull. The immediate area had also served as a dumping ground for road-kill, butchered white-tailed deer (*Odocoileus virginianus*), and other animals. Sun-bleached bones and carcasses with dried tissue were strewn all along the edge of the road. Two major options were available to the officers at this juncture.

Option 1: nonarchaeological recovery

The first option is to clear the major brush from above the body without worrying about any grass undergrowth or leaf litter covering the remains. Pictures of the bones visible on the surface around the skull are taken, after which any bones in the immediate area are collected. This bone collection likely includes animal bones as well, due to the unskilled bone collectors who are unable to differentiate between human and nonhuman elements. All evidence is then be placed in a body bag and sent to a forensic pathologist for an autopsy.

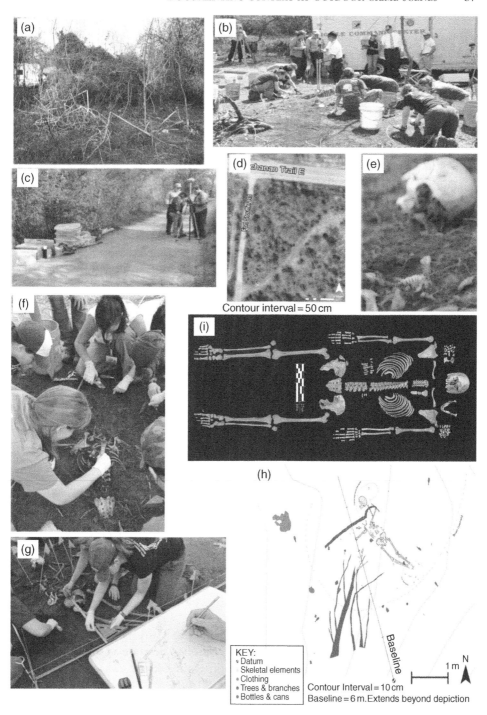

Figure 2.2 Details of case study. (a) General view of site prior to forensic archaeological recovery; (b) clearing the site of surface debris and exposing human remains; (c) taking provience data with survey-grade GPS unit; (d) geographic information system map of site; (e) close-up of skull (exhibiting trauma) and mandible; (f) final exposure of remains; (g) mapping procedure; (h) final map; (i) skeletal remains in the laboratory.

The forensic pathologist then sorts out the human bones from the animal bones (although they are not quite sure about a few of the bones, especially the broken ones) at the postmortem examination. The focus is on the skull. Many of the bones of the face are separate from the cranium and exhibit significant fractures (Figure 2.2e). Unfortunately, not all of the bones are present in the assemblage on the autopsy table. The forensic pathologist asks the police whether the bones were separate at the time of discovery, or whether they may have been broken when the skull was picked up, or during the transport in the body bag to the autopsy. No one can answer with 100% certainty.

Given the epiphyseal lines on some of the bones, the pathologist indicates that the remains are likely those of a juvenile. However, in terms of sex determination, the small size of the bones may be due either to youth or to the female sex. One important characteristic that could have been used to positively identify the individual was two distinctively chipped upper incisors noted antemortem on the suspected individual. No previous characteristic dental work had been completed on the individual. The incisors were not present in the recovered remains. Likely, they were left at the scene, given that they are single-rooted and easily fall out of their alveolar sockets during decomposition. Since possibly 15 years have elapsed from the time that the body was deposited at the site to discovery, mitochondrial DNA testing may be required, an expensive and time-consuming proposition.

However, before sending the bones away for DNA testing, the investigators determine that fine-tuning of the biological profile would be helpful. The bones are boxed up and sent to a forensic anthropologist, who may turn up some other useful information.

The box of bones arrives in the laboratory of the forensic anthropologist. The main question still regards the identity of the individual. The bones reveal that the individual was a white female, 15–20 years of age, with a stature of 162 to 172 cm (Figure 2.2i). This fits the profile of the person for whom the police were searching. However, the issue of positive identification remains, and mitochondrial DNA analysis is still required.

What about these questions…?
In addition to addressing the biological profile, other questions will be asked: Were the remains totally on the surface since the time of deposition, or might they have been partially or wholly buried?, Could the bones have been at the site for 15 years?, Did someone come back to the site and disturb the remains to hide evidence?, Does the fact that there are missing elements indicate removal by the perpetrator?, Could the body have been elsewhere for a time before being dug up and taken to this site, thus indicating a secondary burial/deposition?, and How was the body placed at the site: face up, face down, extended, or in fetal position? These are all important questions since the perpetrator provided a description of the events.

And can these questions be answered from the bones in the box?
All of these questions relate to what happened to the body during and since placement at the scene. The study of the manner in which these postdepositional factors affect the remains has been termed *forensic taphonomy* (described more fully in Chapters 1, 24, and 25 in this volume). Forensic taphonomy requires analysis of the bones *plus* evaluation of the contextual setting from which they were recovered.

In this scenario, *none* of the forensic taphonomy questions can be answered with any scientific backing; we are forced to rely on conjecture and anecdotal evidence.

Option 2: forensic archaeological recovery

Instead of hastily removing the remains with very little consideration for documenting context, law enforcement has decided to enlist the help of a forensic anthropologist or forensic archaeologist to conduct the processing of the outdoor scene. Although each scene is unique, certain basic procedures are common to every forensic archaeological outdoor scene recovery.

Step 1: clearing the scene of overlying debris unrelated to the incident

The first step of the process would be to clear all of the vegetation and debris overlying the remains (Figure 2.2b and f) in order to see the position and orientation of the bones and evidence. This includes clearing the major brush over the remains, as well as sticks and leaf litter. None of the evidence is moved at this point. Forensic anthropologists are also valuable during this process for their role in the evaluation of significance. In this case, human remains were separated from animal remains at the site and in real time. This "ability" thus increases efficiency of the process effort since extraneous evidence (in this case, animal bones) does not enter the chain of custody process.

Throughout the process three types of information are routinely collected to document context: (i) written documentation, (ii) photographic documentation, and (iii) detailed mapping of the scene and the spatial distribution of evidence.

Step 2: documenting context through written and photographic protocols

Written documentation serves as a journal of events associated with the recovery: *who is doing what, at what time, and what was found*. The second role of written documentation is to fully note the general contextual setting of the scene including topography, vegetation, amount of shade, etc. The collection of this data is best done through standardized forms so that the same information is collected for all sites. This will allow for intra- and intersite comparisons that will reveal previously hidden patterns in much the same way as crime mapping reveals patterns. Photographic protocols now include exclusively digital images, useful because of the imbedded metadata related to the image (such as camera settings, date stamp, etc.). Documentation of activities plus *in situ* images of the scene and evidence are obtained.

Step 3: documenting context through mapping protocols

The third type of context notation is through the three-dimensional mapping of the scene and surrounding area. This is done at the global level through global positioning system (GPS) units. The immediate area around the scene is documented through more sophisticated mapping instrumentation, including the electronic total station and survey-grade GPS units. Geographical information systems (GIS) software can also be used to produce a contour interval map of the scene (Figure 2.2c, d). The third level of mapping is through the production of hand-drawn maps showing the position and orientation of each element and piece of evidence found *in situ* at the scene.

The precise mapping of the scene is one of the most important parts of the recovery. The focus is to produce an accurate depiction of the scene via a plan-view map, noting where every piece of evidence is located relative to other evidence, and also to elements of the scene. This can be approached a number of different ways. If the remains are close together, a baseline or grid system can be used. If the remains are widely scattered, as in a plane crash, a total-station or survey-grade GPS can be used to piece plot each skeletal element and piece of evidence (Dirkmaat and Hefner 2001). Each evidentiary item is given a unique identifier at the scene so that the location and orientation of every bone is known and the pattern of spatial distribution of elements can be analyzed. In this particular case, a baseline mapping protocol was followed to produce the plan-view map (Figure 2.2 g, h).

Summary of the recovery
After 5 h, the scene was cleared of vegetation and all human remains (in this case, only bones) were found. The position of each piece of evidence was noted *in situ*, via a detailed plan-view map of the scene. A link between each bone and the map was established, as was the proper chain of custody.

Scene and events interpretation
The forensic archaeological recovery described above, in combination with the analysis of the bones themselves, detailing any surface modification, trauma, etc., in the laboratory, provides the only path to conducting a detailed forensic taphonomic analysis of the scene and the remains. As noted above, *forensic taphonomy cannot be completed solely on the basis of the bones.* Only with the coupling of context with bone analysis can interpretations of how the body was placed at the scene, identification of taphonomic agents modifying the remains, and production of scientific-based estimates of postmortem interval, be produced.

The vast majority of the remains were still in relative anatomical position, indicating that the individual was placed at the site shortly after death and that decomposition occurred at the scene. It became clear that the victim was placed face down on the surface at the site. The body was oriented with the head to the northwest and feet to the southeast. The legs were fully extended and the arms bent at the side of the body. There is no indication that the body had been disturbed by humans after placement to hide evidence or otherwise. There is evidence that animals were involved in moving a few bones of the hands and feet from their original position.

All of the fractured bones of the face were recovered *in situ*. Because of the forensic archaeological recovery protocols employed, no subsequent postmortem damage was inflicted on the bones after discovery due to poor excavation, collection, or transport factors. This allowed for a comprehensive trauma analysis to be conducted, which revealed multiple blunt-force blows to the face and head at the time of death.

Most of the bones were in good shape and exhibited very little surface erosion. They were stained to a consistent dark brown color. However, one element did show much more advanced degradation in the form of sun-bleaching, surface cracking, and outer-layer bone exfoliation. This one piece suggested that the postmortem interval was much longer than the 2–5 years indicated by the better-preserved bones.

The forensic osteological analysis of the bones produced the same results with respect to the biological profile as described in Option 1. However, the isolated upper incisors were recovered and led to a positive identification based on forensic odontology the day after recovery. Mitochondrial DNA identification, which may take months, was not required.

DISCUSSION

It is likely that most human remains situated in outdoor settings and discovered by other humans will be disturbed in some way. This could happen either through the curiosity or ignorance of the discoverer. However, further disturbance by law enforcement and forensic specialists prior to recovery must be prevented. In many cases the modus operandi has been the rapid removal of the remains, often crudely conducted using shovels and even heavy machinery, and accompanied by only cursory *in-situ* documentation of the crime scene. The removal of the evidentiary material from its primary depositional context with very little associated documentation severely limits the resultant analysis and interpretation of past events.

Although it is generally assumed that information relevant to reconstructing the past is very limited at the outdoor scene, it can be shown that this is quite erroneous. Applying the proper recovery method to the outdoor scene results in the emergence of a scientific, evidence-based story. Simply put, we must expect the same standard of forensic recovery of evidence and event reconstruction at the outdoor crime scene as we do at the indoor scene. However, indoor scene recovery protocols applied to outdoor scenes are not effective. Furthermore, as there are no law-enforcement protocols specific to outdoor scenes, we must turn to other disciplines. Forensic archaeology provides the solution. From effectively and efficiently locating the forensic site to the rigorous collection and documentation of all relevant evidence, standard forensic archaeological methodologies enhance and maximize the amount and quality of data retrieved at the scene. Better understanding of the context of the body certainly helps law enforcement and forensic specialists conduct their analyses and interpretations. Ultimately, the determination of cause and manner of death by the coroner or medical examiner is enhanced as well. As has been described more fully elsewhere (Dirkmaat and Adovasio 1997; Dirkmaat et al. 2008) these protocols are effective at all outdoor scenes, from surface scatters (as described above) to clandestine graves (Dirkmaat and Cabo 2006; Hochrein 1997, 2002), and from mass graves (Dirkmaat et al. 2005; Tuller et al. 2008) to fatal fire scenes (DeHaan 2008; Dirkmaat 2002), to mass disaster sites (Dirkmaat and Cabo 2009, 2010; Dirkmaat and Miller 2003; Kontanis and Sledzik 2008).

Although the law-enforcement and medicolegal communities have long recognized the value of physical anthropologists in the comprehensive analysis of human skeletal remains, especially for identification of unknown individuals (Spitz 2006), forensic anthropologists have tended to enter forensic investigations *following the removal* of human remains from the scene, or have only rarely (usually in cases involving buried remains) visited the scene during recovery (Haglund 1991; Wolf 1986).

Today, however, more sophisticated questions regarding peri- and postmortem events are asked of anthropologists, and it cannot be overstated that these questions can only be addressed when well-documented contextual information is gathered along with the physical evidence during the recovery phase of the forensic investigation. The obvious solution is, when partially decomposed human remains are suspected, to bring into play the full array of data recovery techniques in a highly integrated fashion during the investigation's incipient stages. The forensic anthropologist who is well trained in archaeological techniques is indispensable in integrating and fulfilling these objectives.

The primary physical outputs of these forensic archaeology protocols are very robust and accurate maps of the scene, detailing topographic features and patterns of the spatial distribution of physical evidence. The combination of (i) precise maps, (ii) comprehensive documentation, analysis, and interpretation of context, and (iii) well-documented recovery protocols that maximize evidence recovery and establish strong "chain-of-custody" linkage of evidence to the scene provides effective courtroom presentations of evidentiary value. More compelling reconstructions of past events and the possibility of corroborating witness testimony leads to a greater potential for eventual prosecutorial success. This methodology is also much more solidly entrenched in the scientific method, allowing for effective protocol testing and comparison, as well as more powerful and straightforward analyses of the evidence.

From the standpoint of forensic anthropology itself, it is rather surprising that the discipline is often portrayed, and even defined, as being laboratory-based. It is not uncommon for skeletal remains to be brought to the forensic anthropologist without detailed notation or documentation of evidence location within the scene, and therefore, totally out of context. It is time to pay more attention to the *contextual setting* of the human remains found at outdoor scenes. Blindly accepting a box of bones without discussing the consequences of a poorly recovered scene endorses substandard recovery principles. Limited assessment and analysis to determine a biological profile or heading to the coroner or medical examiner's office for a few hours is not a good recipe for producing a quality, scientifically valid interpretation of the recovered remains.

Without an understanding of where the remains were found and in what condition, questions regarding taphonomic issues or interpretations of bone surface modifications, especially with respect to differentiating postmortem from perimortem trauma, cannot be answered with any degree of *scientific justification*, *backing*, or *certainty*. With the eventual and inevitable implementation of the federal Daubert standards for the admission of expert witness testimony in state courts, forensic anthropology interpretations that do not consider context have the potential to be dismissed in a court of law, severely jeopardizing many cases.

On the other hand, suggesting that forensic archaeology merely involves showing up at a scene and making sure all the human bones are located is ill-founded. Archaeology and forensic archaeology operate under certain principles, methods, and practices that ensure that the context is comprehensively documented. Importantly, a detailed map of the remains identified individually, showing their spatial distribution at the scene, is the key product, the successful production of which requires training and experience. If the forensic anthropologist lacks either, an archaeologist can be enlisted to help.

Finally, when an outdoor crime scene is processed poorly (e.g., when shovels and even heavy equipment are used as the primary excavation tools), excuses related to ignorance of methods available outside of law enforcement can no longer be accepted. Important contextual details will not be noted properly or systematically at the scene and, thus, are irretrievably and forever lost. Potential reconstructions of circumstances surrounding the death event, or even the original position and orientation of the body, suffer greatly, or may not even be possible; therefore, they may not be defendable in a court of law. "In any death scene investigation, once someone has disturbed or removed anything from the scene, the context from which it came has been destroyed" (Wolf 1986:17). The destruction of context at the *indoor* scene is certainly not tolerated by the law-enforcement and judicial system. The destruction of context at the *outdoor* scene also cannot be tolerated.

REFERENCES

Binford, L.R. (1983). *Working at Archaeology*. Academic Press, New York.

Bodziak, W.J. (2003a). Forensic footwear evidence. In S.H. James and J.J. Nordby (eds), *Forensic Science: An Introduction to Scientific and Investigative Techniques* (pp. 297–311). CRC Press, Boca Raton, FL.

Bodziak, W.J. (2003b). Forensic tire impression and tire track evidence. In S.H. James and J.J. Nordby (eds), *Forensic Science: An Introduction to Scientific and Investigative Techniques* (pp. 313–326). CRC Press, Boca Raton, FL.

Carlisle, R.C. and Adovasio, J.M. (eds) (1982). *Meadowcroft: Collected Papers on the Archaeology of Meadowcroft Rockshelter of the Cross Creek Drainage*. Department of Anthropology, University of Pittsburgh, Pittsburgh, PA.

Connor, M. and Scott, D.D. (2001). Paradigms and perpetrators. *Historical Archaeology* 35(1): 1–6.

De Forest, P., Gaensslen, R.E., and Lee, H.C. (1983). *Forensic Science: An Introduction to Criminalistics*. McGraw Hill, New York.

DeHaan, J.J. (2008). Fire and bodies. In C. Schmidt and S.A. Symes (eds), *The Analysis of Burned Human Remains* (pp. 1–13). Academic Press, London.

Dirkmaat, D.C. (2002). Recovery and interpretation of the fatal fire victim: the role of forensic anthropology. In W.H. Haglund and M. Sorg (eds), *Advances in Forensic Taphonomy: Method, Theory, and Archaeological Perspectives* (pp. 451–472). CRC Press, Boca Raton, FL.

Dirkmaat, D.C. and Adovasio, J.M. (1997). The role of archaeology in the recovery and interpretation of human remains from an outdoor forensic setting. In W.D. Haglund and M.H. Sorg (eds), *Forensic Taphonomy: The Postmortem Fate of Human Remains* (pp. 39–64). CRC Press, New York.

Dirkmaat, D.C. and Cabo, L.L. (2006). The shallow grave as an option for disposing of the recently deceased: goals and consequences. *Proceedings of the American Academy of Forensic Sciences* 12: 299.

Dirkmaat, D.C. and Cabo, L.L. (2009). New mass disaster scene recovery protocols. National Institute of Justice Anthropology Grantees Focus Group, Alexandria, VA.

Dirkmaat, D.C. and Cabo, L.L. (2010). Forensic archaeological recovery of the victims of the Continental Connection Flight 3407 in Clarence Center, NY [abstract]. *Proceedings of the American Academy of Forensic Sciences* 16: 387.

Dirkmaat, D.C. and Hefner J. (2001). Forensic processing of the terrestrial mass fatality scene: testing new search, documentation and recovery methodologies. *Proceedings of the American Academy of Forensic Sciences* 7: 241.

Dirkmaat, D.C. and Miller, W. (2003). Scene recovery efforts in Shanksville, PA: the role of the coroner's office in the processing of the crash site of United Airlines Flight 93. *Proceedings of the American Academy of Forensic Sciences* 9: 279.

Dirkmaat, D.C., Cabo, L.L., Adovasio, J.M., and Rozas, V. (2005). Mass graves, human rights, and commingled: considering the benefits of forensic archaeology. *Proceedings of the American Academy of Forensic Sciences* 11: 316.

Dirkmaat, D.C., Cabo, L.L., Ousley, S.D., Symes, S.A. (2008). New perspectives in forensic anthropology. *Yearbook Physical Anthropology* 51: 33–52.

Dupras, T.L., Schultz, J.J., Wheeler, S.M., and Williams, L.T. (2006). *Forensic Recovery of Human Remains: Archaeological Approaches.* CRC Press, Boca Raton, FL.

Fisher, B.A.J. and Fisher, D. (2003). *Techniques of Crime Scene Investigation*, 7th edn. CRC Press, Boca Raton, FL.

Galloway, A., Walsh-Haney, H., and Byrd, J.H. (2001). Recovering buried bodies and surface scatter: the associated anthropological, botanical, and entomological evidence. In J.H. Byrd and J.L. Castner (eds), *Forensic Entomology: The Utility of Arthropods in Legal Investigations* (pp. 223–262). CRC Press, Boca Raton, FL.

Gardner, R.M. (2005). *Practical Crime Scene Processing and Investigation: Practical Aspects of Criminal and Forensic Investigations.* CRC Press, Boca Raton, FL.

Haglund, W.D. (1991). *Applications of Taphonomic Models to Forensic Investigations.* PhD dissertation, Department of Anthropology, University of Washington, Seattle. University Microfilms, Ann Arbor, MI.

Haglund, W.D. (2001). Archaeology and forensic death investigations. *Historical Archaeology* 35(1): 26–34.

Hester, T.R., Shafer, H.J., and Feder, K.L. (1997). *Field Methods in Archaeology*, 7th edn. Mayfield, Mountainview, CA.

Hochrein, M.J. (1997). Buried crime scene evidence: the application of forensic geotaphonomy in forensic archaeology. In P.G. Stimson and C.A. Mertz (eds), *Forensic Dentistry* (pp. 83–98). CRC Press, Boca Raton, FL.

Hochrein, M.J. (2002). An autopsy of the grave: recognizing, collecting, and processing forensic geotaphonomic evidence. In W.D. Haglund and M.H. Sorg (eds), *Advances in Forensic Taphonomy: Method, Theory, and Anthropological Perspectives* (pp. 45–70). CRC Press, Boca Raton, FL.

Hurst Thomas, D. (1998). *Archaeology*, 3rd edn. Thomsen Learning, Wadsworth, Victoria.

James S. H., and Nordby, J.J. (2003). *Forensic Science: An Introduction to Scientific and Investigative Techniques.* CRC Press, Boca Raton, FL.

Joukowsky, M. (1980). *A Complete Manual of Field Archaeology: Tools and Techniques of Fieldwork for Archaeologists.* Prentice Hall, Upper Saddle River, NJ.

Komar, D.A., and Buikstra, J.E. (2008). Crime scene investigation. In D.A. Komar and J.E. Buikstra (eds), *Forensic Anthropology: Contemporary Theory and Practice* (pp. 65–114). Oxford University Press, Oxford.

Kontanis, E.J. and Sledzik, P.S. (2008). Resolving commingling issues during the medicolegal investigation of mass fatality incidents. In B.J. Adams and J.E. Byrd (eds), *Recovery, Analysis, and Identification of Commingled Human Remains* (pp. 317–336). Humana Press, Totowa, NJ.

Krogman, W.M., and İşcan, M.Y. (1986). Crime scene investigations. In W.M. Krogman and M.Y. İşcan (eds), *The Human Skeleton in Forensic Medicine* (pp. 15–49). Charles C. Thomas, Springfield, IL.

Lovis, W.A. (1992). Forensic archaeology as mortuary anthropology. *Social Sciences and Medicine* 34(2): 113–117.

Miller, M.T. (2003). Crime scene investigation. In S.H. James and J.J. Nordby (eds), *Forensic Science: An Introduction to Scientific and Investigative Techniques* (pp. 115–135). CRC Press, Boca Raton, FL.

Morse, M., Duncan, J. and Stoutamire, J. (eds), (1983). *Handbook of Forensic Archaeology and Anthropology.* Rose Printing Co., Tallahassee, FL.

Saferstein, R. (2007). *Criminalistics: An Introduction to Forensic Science*, 9th edn. Pearson/Prentice Hall, Upper Saddle River, NJ.

Saferstein, R. (2009). *Forensic Science: From the Crime Scene to the Crime Lab.* Pearson/Prentice Hall, Upper Saddle River, NJ.

Shafer H. J. (1997). Goals of archaeological investigation. In T.R. Hester, H.J. Shafer, and K.L. Feder (eds), *Field Methods in Archaeology*, 7th edn (pp. 5–20). Mayfield, Mountainview, CA.

Shennan, S. (2004). Analytical archaeology. In J. Bintliff (ed.), *A Companion to Archaeology* (pp. 3–20). Blackwell, Malden, MA.

Sigler-Eisenberg, B. (1985). Forensic research: explaining the concept of applied archaeology. *American Antiquity* 50(3): 650–655.

Skinner, M. and Lazenby, R.H. (1983). *Found! Human Remains*. Archaeology Press, Simon Fraser University, Burnaby, BC.

Spitz, D.J. (2006). Identification of human remains. In W.V. Spitz and D.J. Spitz (eds), *Spitz and Fisher's Medicolegal Investigation of Death: Guidelines for the Application of Pathology to Crime Investigation*, 4th edn (pp. 184–239). Charles C. Thomas, Springfield, IL.

Spitz, W.V. and Spitz, D.J. (eds) (2006). *Spitz and Fisher's Medicolegal Investigation of Death: Guidelines for the Application of Pathology to Crime Investigation*, 4th edn. Charles C. Thomas, Springfield, IL.

Stewart, R.M. (2002). *Archaeology: Basic Field Methods*. Kendall/Hunt Publishing, Dubuque, IO.

Stoutamire, J. (1983). Excavation and recovery. In D. Morse, J. Duncan, and J. Stoutamire (eds), *Handbook of Forensic Archaeology and Anthropology* (pp. 20–47). Rose Printing Co., Tallahassee, FL.

Swanson, C.R., Chamelin, N.C., Territo, L., and Taylor, R.W. (2006). *Criminal Investigation*, 9th edn. McGraw Hill, Boston, MA.

Tuller, H., Hofmeister, U., and Daly, S. (2008). Spatial analysis of mass grave mapping data to assist in the reassociation of disarticulated and commingled human remains. In B.J. Adams and J.E. Byrd (eds), *Recovery, Analysis, and Identification of Commingled Human Remains* (pp. 7–29). Humana Press, Totowa, NJ.

Wolf, D.J. (1986). Forensic anthropology scene investigation. In K.J. Reichs (ed). *Forensic Osteology: Advances in the Identification of Human Remains* (pp. 3–23). Charles C. Thomas, Springfield, IL.

Wright, R.K. (2003a). The role of the forensic pathologist. In S.H. James and J.J. Nordby (eds), *Forensic Science: An Introduction to Scientific and Investigative Techniques* (pp. 15–26). CRC Press, Boca Raton, FL.

Wright, R.K. (2003b). Investigation of traumatic deaths. In S.H. James and J.J. Nordby (eds), *Forensic Science: An Introduction to Scientific and Investigative Techniques* (pp. 27–43). CRC Press, Boca Raton, FL.

CHAPTER 3

Determining the Forensic Significance of Skeletal Remains

John J. Schultz

INTRODUCTION

Determining the forensic significance, or medicolegal significance, of skeletal material is one of the primary tasks of forensic anthropologists and involves several steps. It is important to note that, prior to determining whether human skeletal remains are of forensic significance, the material should be treated as medicolegally significant until the forensic anthropologist can determine otherwise. The first step is to eliminate material that does not represent bone or teeth; this should only be an issue with fragmentary material. The second step is to distinguish human versus nonhuman bones. Once the bones are determined to be human, the next step is to assess whether the skeletal remains are of forensic significance. Finally, if the skeletal remains in question are not from a forensic context, the forensic anthropologist must then determine the context from which the nonforensic remains originated. Traditional categories of nonforensic human remains include teaching or anatomical material (including dissected material), war trophies, and archaeological material from either prehistoric or historic, including cemetery remains, contexts. Determining whether the remains are of forensic significance or of nonforensic origin is based on a number of criteria including biological affinities, taphonomic modifications, associated artifacts, and contextual material.

The purpose of this chapter is to provide an overview of the different criteria used to determine forensic significance of human skeletal remains. Further, since determining forensic significance for out-of-context skeletal material involves eliminating the classification of the material as nonforensic skeletal remains, a discussion is provided on how the forensic anthropologist is able to distinguish the nonforensic categories of human remains. When presented with out-of-context skeletal material, recognition of

A Companion to Forensic Anthropology, First Edition. Edited by Dennis C. Dirkmaat.
© 2012 John Wiley & Sons Ltd. Published 2015 by John Wiley & Sons Ltd.

teaching or anatomical material, war trophies, and archaeological material can be based solely on the combination of biological affinities and taphonomic histories of the skeletal material. When interpreting taphonomic modifications to bone, identification is based on recognizing modifications whose causal agent has already been documented (Nawrocki et al. 1997). However, recognition of one taphonomic modification generally does not provide sufficient evidence to classify skeletal remains as either forensic or nonforensic in origin. Rather, it is the suite, or co-occurrence, of taphonomic modifications comprising a taphonomic profile that provide the criteria for correctly classifying skeletal remains by origin. A taphonomic profile not only provides a detailed description of the suite of modifications for a particular type of taphonomic history, but also provides testable hypotheses for determining the origin of the modifications that pertain to the source, sequencing, and timing of the alterations to bone (Nawrocki 2009; Schultz et al. 2003).

DISTINGUISHING BONE FROM NONBONE MATERIALS

The first step in determining forensic significance is to eliminate nonbone material from purported skeletal material. This step is normally not an issue with whole bones or bone fragments exhibiting discernable morphological features. However, this step can be challenging for personnel who do not have experience distinguishing foreign materials from fragmentary, weathered, eroded, bleached, or burnt bone and tooth fragments. There are a variety of foreign materials that may be confused with bone. Forensic anthropologists who are routinely in the field recovering skeletal remains are generally experienced at distinguishing the array of nonbone materials that can be confused with skeletal material. For example, rocks, iron concretions, and dried root and plant material can be confused with bone fragments that appear dark in color due to soil staining. Bleached bone fragments may be confused with plastic pipe fragments. Lightly colored rock or shell fragments can be confused with tooth fragments. Burnt electrical wire casing and burnt car insulation can be confused with burnt bone fragments (Komar and Buikstra 2008). Additionally, when dealing with weathered and soil-stained fetal remains, bones can be confused with small sticks, and small primary and secondary ossification centers can be confused with large seeds.

When dealing with highly fragmented material or very small fragments, determining whether fragments comprise either bone or a non-osseous material can be difficult using gross methods. Ubelaker (1998) suggests using a microscope to highlight the surface features of a questionable fragment. This method is useful for eliminating materials that do not exhibit a compact surface with graininess consistent with that of the magnified surface of bone. Ubelaker et al. (2002) have also discussed how obtaining microslices and then analyzing the specimens with scanning electron microscopy and energy-dispersive X-ray spectroscopy can determine the chemical makeup of questionable objects. The chemical makeup can then be compared with the database of the FBI's Spectral Library for Identification (SLICE) to determine similarity to known samples. Furthermore, almost all nonbone materials can be easily distinguished from bone based on the calcium-to-phosphorous ratio since only a few materials have a similar ratio to that of bone.

DISTINGUISHING HUMAN BONES FROM NONHUMAN BONES

Once it has been determined that the material represents bone, the next step for the forensic anthropologist is to distinguish human bones and teeth from nonhuman bones and teeth. Since many forensic anthropologists routinely identify nonhuman bones for law-enforcement agencies, medical examiners, and coroners, all forensic anthropologists must be proficient in distinguishing human from nonhuman bones. The easiest way to distinguish human versus nonhuman is based on the morphology or shape of the bone. Overall, human bones are shaped very differently from the bones of common species that are routinely encountered. Since many nonhuman bones exhibit a different shape or architecture, it should be routine for the forensic anthropologist to distinguish human versus nonhuman, regardless of size. However, the forensic anthropologist should also be familiar with shape differences of human bones due to skeletal variation and pathology, as well as those related to developmental stages of the different bones of the skeleton. When the bones in question are of similar size to subadult bones, determination of maturity can be helpful when distinguishing between human and nonhuman bones. When small bones are located that are similar in size to, or smaller than those of, a child these bones can be easily identified as nonhuman if fused epiphyses are noted.

When an exact nonhuman species is necessary, the bones in question can be compared to either a nonhuman reference collection or a number of popular nonhuman osteology manuals (Adams et al. 2008; Elbroch 2006; France 2008; Gilbert 1990; Gilbert et al. 1996). There are also a number of general osteology overviews for distinguishing between human and common nonhuman bones (Byers 2008; Dupras et al. 2006; Komar and Buikstra 2008; Mulhern 2009; Ubelaker 1989) that are readily available.

Determining forensic significance can be very difficult with bone that is highly fragmented, particularly with small bone fragments from other large mammals. The best options for distinguishing human from nonhuman bone fragments include microscopic and bimolecular methods. While microscopic anatomy is useful for differentiating large mammalian bone fragments that can be confused with human bone fragments, it is somewhat limited because it can only reject rather than identify a bone fragment as human (Mulhern 2009). A nonhuman fragment is recognized by an overall microscopic pattern that is plexiform, or fibrolamellar, when the linear arrangement of primary osteons is oriented as rows or bands (Mulhern 2009; Mulhern and Ubelaker 2001). Since the human histological pattern can be shared with other vertebrates, Ubelaker et al. (2004) suggest that the use of protein radioimmunoassay holds promise as a biomolecular option for distinguishing human from nonhuman samples, while protein radioimmunoassay has the potential to identify species and distinguish human from nonhuman fragments. Only small quantities are required, which is an important benefit when dealing with evidence. Further, another benefit of this method includes first testing fragments to determine species, which therefore excludes nonhuman fragments prior to submitting human bone fragments for DNA analysis.

DETERMINING FORENSIC SIGNIFICANCE

The forensic anthropologist must determine forensic significance for human skeletal remains that originate from a variety of contexts. This task is easiest when dealing with skeletal remains where context has been preserved and documented, as compared with skeletal remains that are received out of context. When the forensic anthropologist is involved in active searches for forensic cases and the subsequent recoveries or excavations, it is easy to determine forensic significance, as forensic determination is based on the contextual criteria and condition of the skeletal remains. For example, when dealing with buried remains from a forensic context, contextual criteria consistent with modern skeletal remains include the association of modern clothing and jewelry, the position of the body within the grave, and a lack of mortuary artifacts. In addition, body position is important because a body placed in a forensic grave may exhibit a discordant body position inconsistent with the body position of a formal burial.

However, there are many instances when the forensic anthropologist must make the determination of forensic significance with skeletal remains that are out of context. Examples of common out-of-context skeletal material examined by forensic anthropologists may include situations in which law enforcement finds a skull used as a decoration or for a ritual in someone's home or place of business when responding to a call. Many forensic anthropologists have analyzed human skeletal remains that had been in the possession of someone until they passed away, such as archaeological material. It is not uncommon for family members to bring the remains to law enforcement for proper disposal when they are settling their estate. There are also instances in which family dogs have brought human bones to their owner's homes after finding the remains in wooded areas. The public may bring bones to law enforcement after finding them in an outdoor environment. Poor documentation may occur during instances such as these where skeletal remains are recovered without the participation of the forensic anthropologist, making it difficult to discern forensic significance based on scene criteria. For example, it is common for historic period graves to be discovered during digging associated with construction. There may be instances where no clothing remnants persist, the coffin is completely decomposed, and coffin hardware is absent. If the forensic anthropologist or archaeologist is not on site to recognize other indicators of a coffin burial, there may be no documentation of any contextual clues.

When forensic significance must be determined from human skeletal remains received out of context, forensic remains may be identified based on a number of criteria. Evidence of fresh bone quality, modern surgical procedures, and modern dental work can easily allow recognition of forensic skeletal material. Fresh bone characteristics can include the presence of soft tissues, including remnants of desiccated ligaments and tendons, as well as costal cartilage and hyaline cartilage. Other qualities that allow the recognition of modern bone without adhered soft tissue include the smell of decomposition, a heavier weight of bone, and a yellowish hue compared to dry bone due to the retention of fats, fluid, and collagen that also result in a greasy surface texture. When the bone is dry, overall quality of the bone can be excellent or good with a smooth texture that is unlike that of eroded bone from archaeological contexts.

The process for determining forensic significance also involves determining that the material does not represent one of the categories of nonforensic remains. The traditional categories the forensic anthropologist assigns to nonforensic human remains are archaeological material from either prehistoric or historic contexts, war trophies, and teaching or anatomical material (including dissected material). The criteria used to distinguish each of the nonforensic categories are discussed below.

CLASSIFICATION OF ARCHAEOLOGICAL MATERIAL

Identifying archaeological skeletal remains is based on a number of criteria. If skeletal remains are located in a buried context and properly excavated, discrimination between archaeological graves and cases of forensic interest can be based on contextual information such as grave location, grave features, and associated artifacts. However, in many instances contextual material is not available for the identification of archaeological material. In these instances, identification is based primarily on biological criteria and taphonomic modifications of the remains. The two general types of archaeological context are classified as historic, including cemetery remains, and prehistoric. In North America, the historic period refers to the period after European contact, which varies depending on the location within North America. In the USA, prehistoric skeletal remains would be classified as Native American and refer to indigenous peoples of the continental USA, as well as those of Hawaii and parts of Alaska.

Depending on the time period represented by an unmarked burial, the forensic anthropologist may also work with the state archaeologist. For example, in the State of Florida, jurisdiction and protection of burials is specified in Section 872.05 of the Florida Statutes (Florida Statutes 2010). The district medical examiner assumes jurisdiction for burials involved in a criminal investigation or for skeletal remains interred for less than 75 years. Conversely, the Division of Historical Resources may assume responsibility for an unmarked burial of an individual who has been dead 75 years or more and who is not involved in a legal investigation. Depending on the situation, the state archaeologist can recommend that the Division of Historical Resources assume jurisdiction and final disposition of the remains. Alternately, when an unmarked historic grave or graves possibly represent a recently discovered cemetery, or part of an existing cemetery, the state archaeologist may work with a local community group or church that would assume jurisdiction (Ryan Wheeler, personal communication).

Historic or prehistoric biological criteria

When an experienced forensic anthropologist encounters archaeological skeletal remains, identification of prehistoric skeletal remains is based on recognizing a suite of biological traits, including ancestral traits associated with Asian ancestry and a number of traits that are culturally influenced. Specific ancestral traits of the skull, such as the combination of shovel-shaped incisors with extreme dental attrition or wear with exposed dentine, an edge-to-edge bite or flat wear of the anterior teeth, hyper-robust muscle insertions for chewing muscles, and cranial modification, are examples of morphological traits exclusively used to identify prehistoric skeletal material and exclude historic origin. However, because of the temporal proximity of historic-period populations with those that are contemporary, the use of only ancestral traits is not helpful when attempting to discern forensic significance. Conversely, skeletal remains exhibiting evidence of a previous autopsy, older surgical procedures, and older dental work can be informative clues as to historic origin.

Contextual clues

If personnel experienced with excavation methods examine archaeological remains *in situ*, identifying archaeological material is usually fairly straightforward. Common contextual criteria that can aid with the identification of prehistoric skeletal material include proximity of burials to known prehistoric sites, such as middens; skeletons in a flexed or semiflexed position; and the association of lithic artifacts (chipped stone tools and debitage) as well as pottery and pottery sherds. Identification of historic period skeletal remains is relatively simple when skeletal remains can be analyzed *in situ*; a number of items can be documented, such as skeletal position, remnants of coffin wood or coffin stain, coffin hardware and adornment, evidence of embalming (see section below), adhered fabric from the coffin lining or pillow, and period jewelry or clothing artifacts (e.g., buttons, textiles, clasps, and pins) that either adhere to the remains or are found in association. If a general time period is desired for a historic period burial, coffin hardware and adornment may be useful as there are numerous references available that provide typologies and dates for period of use (e.g., Bell 1990; Garrow 1987; Hacker-Norton and Trinkley 1984; Kogan and Mayer 1995; Woodley 1992).

When locating and identifying unmarked historic period graves, the forensic anthropologist must consider past cultural practices indicating pioneer graves, unmarked family cemeteries, unmarked community cemeteries, unmarked graves within the documented boundaries of an existing cemetery, and unmarked graves in close proximity to an existing cemetery. Furthermore, historic graves may also be unmarked because a marker made from wood may have decomposed, or a stone headstone may have been pushed over or damaged and never replaced. Modern graves may be unmarked because the funeral home will initially only place a nonpermanent metal or plastic placard on a stake at the gravesite. It is then the responsibility of the family of the deceased, and not cemetery management, to place a permanent marker at a gravesite. In addition, many cemeteries have a section referred to as a potter's field, which is a burial area for unknown or indigent individuals; this area is usually unmarked or poorly marked and in many instances there is poor documentation of the location and number of interments. Unmarked graves can also be found just outside of the boundaries of an existing cemetery, usually in the right-of-way or under the road, due to unclear boundaries and poor documentation of unmarked graves. In addition, finding an unmarked burial in proximity to an area that contained a former cemetery serves as an indication that not all of the graves were disinterred and that burials were missed because they were unmarked.

Taphonomic modifications associated with historic or cemetery remains

When dealing with archaeological material, bones are lighter in weight, the overall quality of the bone surface may be eroded rather than smooth, and bones generally display a uniform staining ranging from tan to brown. However, historic period remains interred in a buried, wooden coffin will exhibit a suite of taphonomic modifications that are classified as examples of staining and erosion. Staining can be classified as uniform or localized. The specific uniform coloration may vary depending on the type of interment. Buried skeletal remains can display a uniform medium to rich chocolate brown coloration resulting from either tannins in the soil solution or iron oxides in the soil (Schultz et al. 2003). It is important to note that if cemetery remains are exposed on the ground surface, the exposed bone surfaces can exhibit sun-bleaching. Also, crypt burials have been reported as exhibiting a uniform orange-brown coloration that may be due to a combination of factors including embalming chemicals, lack of groundwater, and lack of direct contact with the soil (Schultz et al. 2003). Furthermore, a uniform dark coloration has been noted on remains interred in iron coffins (Owsley and Compton 1997; Schultz et al. 2003) and is most likely due to the prolonged exposure or submersion of the bones in a moist environment from either trapped decomposition fluids and/or groundwater in the iron casket. Commonly reported localized staining has been observed on historic skeletal remains in close association with copper or iron artifacts (Schultz et al. 2003). Copper or bronze oxidation from coffin hardware or jewelry can produce a localized green staining on bone as a result of metal oxidation. Close association of bone with oxidizing coffin nails or other artifacts composed of iron and tin can cause a localized orange stain that may include adhered rust.

Common taphonomic modifications resulting from erosion include cortical flaking and coffin wear. Cortical flaking is commonly observed on the shafts of long bones and is the result of continually alternating wet and dry conditions (Berryman et al. 1997; Nawrocki 1995; Schultz et al. 2003). According to Berryman et al. (1997), circumferential lamellae of long-bone shafts provide natural lines of cleavage for flaking due to differential expansion and contraction of the outer cortical surface. The outer surface becomes wet before deeper cortical bone and dries more quickly than the deeper layers. Delamination of joint surfaces can also be observed on historic remains and is characterized by flaking of the subchondral bone, which exposes the underlying cancellous bone (Schultz et al. 2003). Coffin wear (Figure 3.1, top image) can be recognized on bony projections of the dorsal aspect of the skeleton (i.e., scapular spine, posterior aspect of the occipital bone, and spinous process of the vertebra), which may exhibit excessive erosion relative to other parts of the skeleton (Berryman et al. 1997). These dorsal contact or pressure points exhibit increased wear most likely caused by constant pressure placed on the bones by the interior hard coffin surfaces. While coffin wear is most commonly observed on dorsal skeletal surfaces, it may also be observed on ventral surfaces caused by the top of the coffin collapsing on the body (Schultz et al. 2003).

Another taphonomic modification commonly observed on both historic and prehistoric period skeletal remains results from adhered roots that produce root staining and root etching. Root etching is recognized as a dendritic pattern of shallow grooves on the bone surface. This pattern results from the "dissolution by acids associated with the growth and decay of roots or fungus in direct contact with bone surfaces" (Behrensmeyer 1978: 154). Root etching is commonly observed on cemetery remains because this modification generally requires extended periods to form. Thus, root etching is generally not observed on forensic remains due to shorter postmortem intervals.

Recognizing evidence of embalming

Individuals interred in cemeteries beginning around the late 1800s may be identified through evidence associated with embalming (Figures 3.2 and 3.3). While it is common to find empty embalming fluid bottles that were placed in the coffin by the

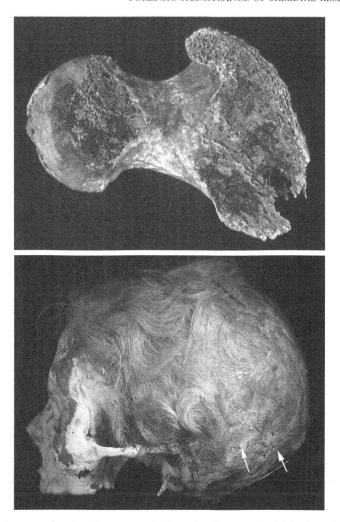

Figure 3.1   An example of coffin wear and adhered coffin wood exhibited on the dorsal aspect of a proximal right femur (top image). A historic period cranium (bottom image) looted from an aboveground crypt exhibiting evidence of embalming such as adhered hair with desiccated scalp and adhered fabric (arrows), possibly from a coffin pillow, on the posterior aspect of the cranial vault. Top image courtesy of Department of Anthropology, University of Florida.

embalmer, there may also be a number of associated embalming artifacts present (Berryman et al. 1997). Where a trocar is used to aspirate the abdomen and inject embalming fluid, the embalmer may choose to close the puncture with a trocar button (a threaded plastic cone-shaped plug) (Figure 3.2) or with a suture. While the trocar button is normally found associated with the abdominal area, the plug can also be used to seal traumatic wounds caused by punctures or gunshots (Berryman et al. 1997) and therefore may be found in other areas of the body. Berryman and colleagues (1997) also point out that a fractured cribriform plate of the ethmoid can be an indication of cranial-vault embalming. For cases of decomposition and gas formation,

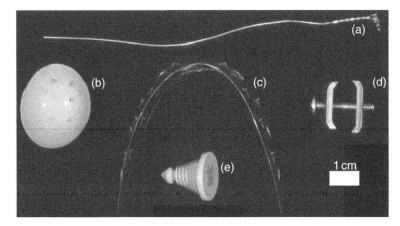

Figure 3.2 Examples of embalming artifacts that would indicate a formal interment of someone that was embalmed: (a) injector needle, (b) eye cap, (c) mouth former, (d) calvarium clamp, and (e) trocar plug.

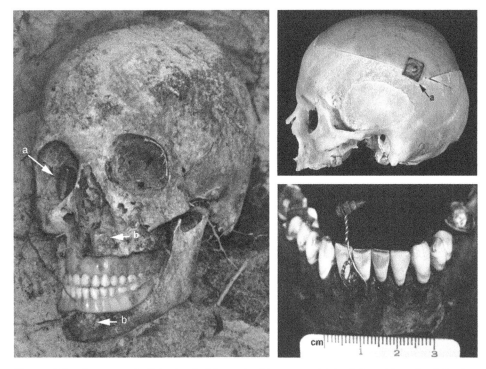

Figure 3.3 Cemetery skull *in situ* (left image) with eye cap (a) in right orbit and oxidized injectors (b) within the maxilla and mandible; note that the twisted wire from the injector needle has disintegrated. The upper right image displays a calvarium clamp (a), securing the calvarium to the lower cranial vault, and the lower right image displays the twisted injector wires, with one injector inserted in the alveolus of the mandible. Images courtesy of Department of Anthropology, University of Florida.

the embalmer can insert the trocar into a nostril and through the cribriform plate for aspiration and injection of embalming fluid. If the calvarium has been removed during autopsy, the use of calvarium clamps (Figures 3.2 and 3.3) is one option for the embalmer to secure the autopsied calvarium, or calotte, to the lower cranial vault.

Embalmers use a number of items to set the facial features (Figures 3.2 and 3.3). While some embalmers may use cotton and glue to shape the closed eye and seal the eyelids, plastic disks called eye caps are also used to shape and close the eyelid (Mayer 2000). The convex shape of the eye cap helps to maintain the convex curvature of the closed eye, and small perforations on the eye cap provide a rough surface to hold the lids closed. Eye caps are easily recognized in the orbit of the skull, allowing for iden-tification of embalmed remains (Rodgers 2005). Closing the mouth involves securing and positioning the mandible in a normal bite position. While embalmers may elect to suture the mandible in place, a widely used method for mouth closure involves using a spring-activated injector to force a grooved needle, also called an injector needle or pin, into the alveoli of the maxilla and mandible (Mayer 2000). The mandi-ble is then drawn into place and secured by twisting the wires attached to the end of the injectors. It is important to note that only the injector may be present in older cemetery remains because the wire may have completely eroded. The embalmer can also choose to use a mouth former to shape the lips; small perforations on the mouth former can provide a rough surface to hold the lips closed. Mouth formers are com-prised of a variety of materials including plastic, cardboard, or metal, and can be trimmed for a secure fit (Mayer 2000).

If the skeleton is protected from seeping groundwater, such as burials in above-ground vaults or watertight coffins, embalmed tissue may be preserved. Adhered head hair (Figure 3.1, bottom image) and eyebrows can still be found preserved and attached to the skull (Berryman et al. 1997). Desiccated brain tissue may be preserved in the cranial vault as a shriveled hard mass that may be noted as a rattling sound when the cranium is shaken. In addition, chemical analysis can be useful for detecting embalming chemicals in preserved human tissues, indicating an embalmed cemetery skeleton or historic period mummified teaching material, as well as for providing a general time period of the remains (Berryman et al. 1997; Sledzik and Micozzi 1997; Steadman 2009). For example, Johnson et al. (2000) report that a variety of chemi-cals were employed during the American Civil War, including basic preservatives such as salts of alumina, sugar of lead, zinc chloride, arsenicals, bichloride of mercury, and a host of acids, salts, and alkalis. Towards the end of the nineteenth century and into the early twentieth century, a gradual movement began to eliminate arsenic and other poisons from embalming fluid because formalin, a powerful disinfectant, was reasonably priced and readily available (Johnson et al. 2000).

CLASSIFICATION OF WAR TROPHIES

A trophy skull is a cranium or skull that has been brought home by soldiers serving in combat. Documented trophy skulls in the USA have generally represented either Japanese military personnel from World War II (Bass 1983) or military personnel from the Vietnam War (Sledzik and Ousley 1991; Taylor et al. 1984; Willey and Leach 2009). Correctly interpreting a skull as a war trophy begins with the biological affinities.

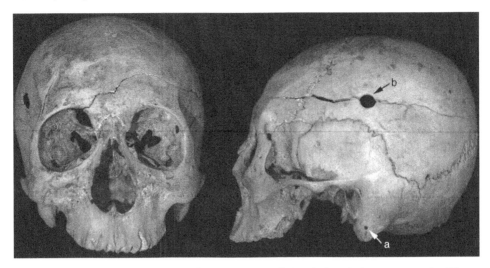

Figure 3.4 Example of a cranium (frontal and lateral views) classified as a war trophy. There are a number of handling modifications exhibited, such as drilled holes on both mastoids (a), possibly for securing a mandible; all of the mandibular teeth are missing postmortem, and there are numerous postmortem fractures, particularly to the alveolus. In addition, there is a gunshot wound (b) through the cranial vault with associated radiating fractures.

Where individuals are of Southeast Asian ancestry, Sledzik and Ousley (1991) have reported that they can be differentiated from Asian and Native American crania based on morphology because the skulls are smaller and less robust, the nasal morphology exhibits a medium width without nasal overgrowth, and the dental arcades are broad and ellipsoid. While it is generally expected that war trophies would be classified as males, caution must be taken when estimating sex, as Southeast Asian male skulls can exhibit gracile features common to female skulls (Sledzik and Ousley 1991; Willey and Leach 2009). Also, in age estimation Vietnamese samples have often been listed as adolescent and young adult (Sledzik and Ousley 1991; Taylor et al. 1984; Willey and Leach 2009). A number of documented trophy skulls have also displayed evidence of war-related perimortem trauma (Figure 3.4) such as blunt force (Sledzik and Ousley 1991; Willey and Leach 2009) and projectile trauma (Taylor et al. 1984).

According to Willey and Leach (2009), skulls classified as war trophies can display a wide range of taphonomic modifications (Figure 3.4). Handling alterations can include burnishing or a patina, and recent postmortem fractures to fragile areas such as the styloid processes, orbits, and the base of the cranium. Other, common postmortem modifications include deliberate painting, nondeliberate pigment from overspray, graffiti, and writing, including the use of crayon, and carbon burning, postmortem tooth loss, and enlargement of the foramen magnum (Bass 1983; Sledzik and Ousley 1991; Willey and Leach 2009). While documented war trophies do not exhibit evidence of being buried, skulls may be dusty as a result of being stored in a dry place for some time (Bass 1983). In addition, Willey and Leach (2009) reported drilled holes in the mandibular condyles and base of the skull for securing the mandible (Figure 3.4), and Sledzik and Ousley (1991) noted a drilled hole at the bregma, possibly for suspending the skull.

CLASSIFICATION OF TEACHING MATERIAL

The forensic anthropologist can also be involved analyzing human remains that were used as anatomical teaching material or from a variety of different contexts that include dissected material, commercially prepared mummified remains, and commercially prepared skeletal remains. Material obtained from a mortuary context and sold by a biological supply company as teaching material may also be evaluated. It is important to note that skeletal remains from an outside or buried environment do not automatically preclude commercially prepared anatomical material or dissected material used for medical purposes. The highly publicized Medical College of Georgia case presents an egregious example involving the disposing and concealment of dissected human remains that had been procured illegally by body snatching during the mid to late nineteenth century (Blakely and Harrington 1997; Harrington and Blakely 1995). If remains are discovered in a buried environment, identification of older dissected material may involve burial location, association of period artifacts, and recognition of cut and sawed bone parts consistent with dissection.

Mummified specimens

The forensic anthropologist may also be involved in identifying prepared mummified specimens. It is important to note that in the past anatomists and embalmers performed the embalming of anatomical specimens and that the preparation of these specimens was similar to the general preparation process for embalmed bodies (Tompsett 1959). As a result, similar embalming methods for prepared anatomical specimens may also be observed with remains from cemetery contexts. There are a number of characteristics that provide clues as to the origin of prepared mummified specimens (Steadman 2009). If the mummified remains exhibit evidence of dissection, identification of prepared anatomical specimens involves recognizing that the location of the dissection is consistent with medical preparation. If a bone has been sectioned with a saw, the morphology of the kerf should exhibit clean cuts that are consistent with those of a medical saw. Finally, testing the tissues for the presence of embalming chemicals may provide clues as to a temporal period that can rule out a modern specimen. According to Sledzik and Micozzi (1997), identifying anatomical specimens with preserved soft tissues can also involve sampling a variety of hardening compounds in the soft tissues that were injected to preserve and highlight blood vessels and organs.

Skeletal material

The most common commercially prepared skeletal material originates from either India or China. Prior to 1985, India provided 80% of the world's teaching skeletons until the government outlawed the export of human remains due to traders illegally obtaining bodies from sources such as grave robbers (Taylor 2000). However, Carney (2007) reported that, even 22 years after the export ban in India, the trade of skeletons continued, and that material robbed from graves was provided to vendors who were still exporting skeletal material to North America to be used as teaching specimens for

American medical programs. When teaching specimens became unavailable from India, Western countries turned primarily to China and Eastern Europe for skeletal material (Carney 2007). It was reported that skeletal material for teaching purposes during this period was obtained from Asian sources that included abandoned burial sites in China and the Cambodian killing fields (Roach 2003; Taylor 2000). In the USA, the biological supply houses that sell skeletal remains today generally indicate that the material originates from China. However, older Indian material that originated before the export ban can still be purchased online from biological supply companies. For example, The Bone Room (www.boneroom.com/skulls/adultskulls.htm) is a natural history store that also sells human skeletal material online. According to the website, The Bone Room resells a variety of human skeletal materials originating from India that were purchased from private collections, museums, estates, etc., for medical and educational purposes. More recently, The Bone Room has reported that China passed a law in 2008 outlawing the export of human skeletal remains, resulting in difficulty obtaining skeletal remains that originated from China. As a result, forensic anthropologists should now expect that skeletal material sold worldwide through online sources might originate from newer contexts that have not been reported in the forensic literature.

Correctly recognizing commercially prepared skeletal remains begins with the identification of biological affinities. While skeletons from China exhibit Asian ancestral traits, the skeletal material also consists primarily of males; this material may be more robust than the male material from India. Skulls from India will generally display a mix of Asian and European morphological traits; males usually exhibit gracile traits and may be classified as females when using discriminant function analysis. The material from India may also exhibit a black stain, most pronounced along the labial/buccal surfaces, as a result of the individuals having chewed betel nuts.

Taphonomic modifications

Common taphonomic modifications exhibited by commercially prepared teaching material can be classified as preparation modifications and easily aid in the classification of these specimens (Figure 3.5). The skulls and other skeletal materials originating from India have been commercially cleaned and are white in color, possibly by soaking in hydrogen peroxide. The teeth may be glued into the sockets and many skulls show the presence of hardware, drilled holes, and a cut calvarium. If the cranial vault was cut for access to the endocranial anatomy, there will be a horizontal cut through the superior aspect of the cranial vault to separate the calvarium, often with metal hardware then used to secure the calvarium in place. The calvarium is generally aligned to the lower aspect of the cranial vault with pins that are inserted into the diploë, and secured in place with screws and hooks. Similarly, the mandible may be secured in anatomical position to the cranium through the use of a long spring, screws, and metal wire. If the hardware has been removed, it may be possible to discern the drilled holes in the original location of the hardware. This can be useful in identifying a fragmentary skull or cranium that was a teaching specimen. A common location for a large drilled hole is at the vertex, or the most superior aspect of the cranial vault along the sagittal suture, where a hook may have been secured to facilitate hanging a completely articulated, or perhaps the axial, skeleton, on a stand. Finally, some specimens will have hand-painted major muscle origins and insertions colored blue and red with the

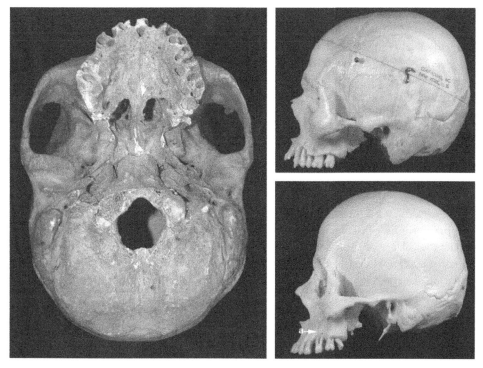

Figure 3.5 The image on the left is the inferior aspect of a cranium classified as a teaching specimen, exhibiting a number of handling modifications such as a loss of all teeth, postmortem fractures to alveolus, pterygoid plates, and the occipital condyles, and wear on the inferior aspect of the mastoid processes. The upper right image is a lateral view of a cranium classified as a teaching specimen, possibly from India, that exhibits a horizontal cut through the cranial vault, hardware to affix the calvarium to the lower cranial vault, a screw through the anterior-inferior aspect of the parietal for attaching a spring to secure the mandible, numerous teeth exhibiting postmortem fractures, a patina, and writing on the calvarium from the distributor. The lower right image is of a lateral view of a cranium classified as a teaching specimen, most likely from China, with a foreign molar (a) glued to the resorbed area of the alveolus.

names of muscles written in black ink. The quality of the Chinese material varies, and coloration can be either natural bone color or white. Skulls from China generally do not exhibit a cut calvarium with associated hardware or a spring attaching the mandible, but specimens may exhibit dental work. While the teeth are also glued into the dental alveoli, it is common to find teeth from other individuals glued into the open tooth sockets or directly to the alveoli if the tooth sockets had been resorbed due to antemortem tooth loss (Figure 3.5). Since skull pricing is generally based on the number of teeth, missing teeth may be added by anatomical preparers to increase the price at which the skulls are sold.

Teaching materials generally display a variety of common taphonomic modifications due to handling and storage (Figure 3.5). The surface of the skull may exhibit a patina or burnishing produced by the transfer of oils from the hands to the bone during extensive handling. There may be writing in ink or pencil from students labeling or tracing anatomical features. Postmortem fractures are also fairly common. While postmortem fractures can result from students accidently dropping skeletal remains, the pattern of fracturing is also commonly observed on the base of the cranium from students continually placing the cranium on hard surfaces. These postmortem fractures usually occur in areas such as the styloid processes, occipital condyles, foramen magnum, lesser and greater wings of the sphenoid, and the alveoli and tooth crowns. At the same time, taphonomic modifications from storage can include a dirty or dusty surface and flat wear on the base of the cranium in areas such as the tips of the mastoid processes.

DATING OF SKELETAL REMAINS USING ARTIFICIAL RADIOCARBON

In cases dealing with fragmentary bones or ambiguous indicators for determining the temporal origins of skeletal remains there may be issues distinguishing between archaeological remains and very recent forensic remains. According to Forbes and Nugent (2009), while there have been numerous methods investigating techniques for dating skeletal remains of forensic interest, radioisotope methods have become the most reliable for forensic anthropology.

Radiocarbon dating has been used extensively to date ancient organic material. The carbon-14 radiocarbon dating method is based on living plants and animals maintaining concentrations of atmospheric levels of this isotope. Carbon-14 concentrations gradually decay after death with a half-life of about 5730 years. Scientists measure the concentration of remaining carbon-14 in organic materials to calculate the approximate time since death. Since radiocarbon dating is most accurate for dating organic materials that are greater than 300 years old (Steadman 2009; Ubelaker 2001), it is regularly used for dating ancient archaeological material. However, while this method is not useful for providing specific dates of death for human skeletal material from recent time periods, detection of anthropometric artificial radiocarbon in skeletal remains can provide clues as to the temporal origin and forensic interest of remains that are modern.

Artificially elevated levels of carbon-14 in terrestrial organisms occurred between 1950 and 1963 as a result of atmospheric testing of thermonuclear devices (Taylor et al. 1989; Ubelaker 2001; Ubelaker et al. 2006; Ubelaker and Houck 2002). Starting in 1950, levels of artificial carbon-14 increase significantly compared to pre-1950 levels, peaking in 1963. Due to the Nuclear Ban Treaty, the reduction and cessation of thermonuclear device testing above ground after 1963 caused the artificial or "bomb" carbon-14 levels to decrease, but they still persist at 10% higher than the 1950 prebomb level (Hua 2004). The significance of testing for artificial carbon-14 in organic material allows the investigator to distinguish organic material that represents a time period prior to nuclear testing from that of a more recent post-1950 time period (Taylor et al. 1989; Ubelaker 2001; Ubelaker et al. 2006; Ubelaker and Houck 2002). In other words, comparing the artificial radiocarbon values of human remains with the "bomb" carbon-14 values can distinguish individuals who died before 1950 from those that died after 1950. The method has already been used for a number of cases to determine whether skeletal remains represent a modern forensic case or an archaeological case (Taylor et al. 1989; Ubelaker 2001; Ubelaker and Houck 2002). Wild et al. (2000) point out that it is not possible to determine an accurate date of death using artificial carbon because of the long turnover time of collagen in human bones. However, Ubelaker et al. (2006) recently reported that precision in date of death for adults is possible on the modern bomb curve by determining whether bone represents a pre-1963 or post-1963 period when using both cortical and trabecular bone samples.

CONCLUSIONS

One of the first tasks forensic anthropologists address when dealing with human skeletal remains out of context is determining forensic significance. In many instances, the taphonomic criteria may be the most useful information for making a final determination as to forensic significance and origin. If the skeletal remains in question are determined to have originated from a nonforensic context, the forensic anthropologist must then determine the origin of the skeletal remains. The nonforensic categories include war trophies, archaeological material from either prehistoric or historic time periods, and skeletal and mummified material sold for teaching, including autopsied remains. However, it is important to note that when considering only taphonomy, modifications consistent with general handling of the remains do not automatically represent teaching material since war trophies may also exhibit similar handling modifications. Therefore, when attempting to determine if the skeletal material represents a forensic or nonforensic context, multiple lines of evidence must be considered including contextual information such as artifacts and documentation associated with a grave, the taphonomic history of the remains, and the biological affinities of the skeletal remains.

Since many states do not have restrictions for purchasing skeletal remains for noneducational purposes, the general public can purchase commercially prepared skeletal remains from numerous online biological supply companies for noninstructional purposes. Further, Huxley and Finnegan (2004) have reported that skeletal material representing a variety of nonforensic categories such as archaeological and teaching, as well as possible forensic materials, are also sold worldwide to the public via internet auction sites. One particular use for skeletal remains obtained by the public involves ritual purposes. Such cases commonly involve the expertise of the forensic anthropologist determining forensic significance when this material is submitted to or discovered by law enforcement. A number of studies have reported that ritual skeletal remains, particularly skulls and crania, are commonly used as components for syncretic religions such as Palo Mayombe, and represent teaching and historic-period material (Gill et al. 2009; Mundorff 2002; Walsh-Haney et al. 2003; Wetli and Martinez 1981; Martinez and Welti 1982). Additionally, with the introduction of teaching material from a variety of different sources and of differing quality, there can be difficulty distinguishing between forensic and nonforensic material if the forensic anthropologist does not have exposure or experience with material that has been sold to the public. While forensic anthropologists have experience identifying "souvenir" teaching skulls that originated from India (Wienker et al. 1990), the commercially prepared material originating from other areas such as China may cause some confusion when determining context because the material may not exhibit the range of preparation modifications that typify the Indian material. As a result, forensic anthropologists should be familiar with the various origins involving human skeletal material that must be considered when determining forensic significance.

REFERENCES

Adams, B.J., Crabtree, P.J., and Santucci, G. (2008). *Comparative Skeletal Anatomy: A Photographic Atlas for Medical Examiners, Coroners, Forensic Anthropologists, and Archaeologists.* Humana Press, Totowa, FL.

Bass, W.M. (1983). The occurrence of Japanese trophy skulls in the United States. *Journal of Forensic Sciences* 28(3): 800–803.

Behrensmeyer, A.K. (1978). Taphonomic and ecologic information from bone weathering. *Paleobiology* 4: 150–162.

Bell, E.L. (1990). The historical archaeology of mortuary behavior: coffin hardware from Uxbridge, Massachusetts. *Historical Archaeology* 24(3): 54–78.

Berryman, H.E., Bass, W.M., Symes, S.A., and Smith, O.C. (1997). Recognition of cemetery remains in the forensic setting. In W.D. Haglund and M.H. Sorg (eds), *Forensic Taphonomy: The Postmortem Fate of Human Remains* (pp. 165–170). CRC Press, Boca Raton, FL.

Blakely, R.L. and Harrington, J.M. (1997). Grave consequences: the opportunistic procurement of cadavers at the Medical College of Georgia. In R.L. Blakely and J.M. Harrington (eds), *Bones in the Basement: Postmodern Racism in Nineteenth-Century Medical Training* (pp. 162–183). Smithsonian Institution Press, Washington DC.

Byers, S.N. (2008). *Introduction to Forensic Anthropology*, 3rd edn. Pearson Education, Boston, MA.

Carney, S. (2007). Inside India's underground trade in human remains. *Wired Magazine* 15(12), www.wired.com/medtech/health/magazine/15–12/ff_bones.

Dupras, T.L., Schultz, J.J., Wheeler, S.M., and Williams, L.J. (2006). *Forensic Recovery of Human Remains: Archaeological Approaches.* CRC Press, Boca Raton, FL.

Elbroch, M. (2006). *Animal Skulls: A Guide to North American Species.* Stackpole Books, Mechanicsburg.

Florida Statutes, 872.01-06. 2010. Offenses Concerning Dead Bodies and Graves.

Forbes, S. and Nugent, K. (2009). Dating of anthropological skeletal remains of forensic interest. In S. Blau and D.H. Ubelaker (eds), *Handbook of Forensic Anthropology and Archaeology* (pp. 164–173). Left Coast Press, Walnut Creek, CA.

France, D.L. (2008). *Human and Nonhuman Bone Identification: A Color Atlas.* CRC Press, Boca Raton, FL.

Garrow, P. (1987). A preliminary seriation of coffin hardware forms in nineteenth and twentieth century Georgia. *Early Georgia* 15(1–2). 19–45.

Gilbert, B.M. (1990). *Mammalian Osteology.* Missouri Archaeological Society, Columbia.

Gilbert, B.M., Savage, H.G., and Martin, L.D. (1996). *Avian Osteology.* Missouri Archaeological Society, Columbia.

Gill, J.R., Rainwater, C.W., and Adams, B.J. (2009). Santeria and Palo Mayombe: skulls, mercury, and artifacts. *Journal of Forensic Sciences* 54(6): 1458–1462.

Hacker-Norton, D. and Trinkley, M. (1984). *Remember Man Thou Art Dust: Coffin Hardware of the Early Twentieth Century.* Chicora Foundation Research Series 2. Chicora Foundation, Columbia.

Harrington, J.M. and Blakely, R.L. (1995). Rich man, poor man, beggar man, thief: the selectivity exercised by graverobbers at the Medical College of Georgia, 1837–1887. In S.R. Saunders and A. Herring (eds), *Grave Reflections:Portraying the Past through Cemetery Studies* (pp. 153–178). Canadian Scholars' Press, Toronto.

Hua, Q. (2004). Review of tropospheric bomb ^{14}C data for carbon cycle modeling and age calibration purposes. *Radiocarbon* 46: 1273–1298.

Huxley, A.K. and Finnegan, M. (2004). Human remains sold to the highest bidder! A snapshot of the buying and selling of human skeletal remains on eBay®, an Internet site. *Journal of Forensic Science* 49(1): 17–20.

Johnson, E.C., Johnson, G.R., and Johnson, M. (2000). The origin and history of embalming. In R.G. Mayer (ed), *Embalming: History, Theory and Practice* (pp. 457–498). McGraw-Hill, New York.

Kogan, S.L. and Mayer, R.G. (1995). Analyses of coffin hardware from unmarked burials, former Wesleyan Methodist Church Cemetery, Weston, Ontario. *North American Archaeologist* 16(2): 133–162.

Komar, D.A. and Buikstra, J.E. (2008). *Forensic Anthropology: Contemporary Theory and Practice.* Oxford University Press, New York.

Martinez, R. and Welti, C.V. (1982). Santeria: a magico-religious system of Afro-Cuban origin. *American Journal of Social Psychiatry II* 3(summer): 32–38.

Mayer, R.G. (2000). *Embalming: History, Theory and Practice.* McGraw-Hill, New York.

Mulhern, D.M. (2009). Differentiating human from nonhuman skeletal remains. In S. Blau and D.H. Ubelaker (eds), *Handbook of Forensic Anthropology and Archaeology* (pp. 153–163). Left Coast Press, Walnut Creek, CA.

Mulhern, D.M. and Ubelaker, D.H. (2001). Differences in osteon banding between human and nonhuman bone. *Journal of Forensic Sciences* 46(2): 220–222.

Mundorff, A.M. (2002). Urban anthropology: case studies from the New York City medical examiner. In D.W. Steadman (ed), *Hard Evidence: Case Studies in Forensic Anthropology* (pp. 52–62). Prentice Hall, Upper Saddle River, NJ.

Nawrocki, S.P. (1995). Taphonomic processes in historic cemeteries. In A.L. Grauer (ed), *Bodies of Evidence* (pp. 49–66). John Wiley & Sons, New York.

Nawrocki, S.P. (2009). Forensic taphonomy. In S. Blau and D.H. Ubelaker (eds), *Handbook of Forensic Anthropology and Archaeology* (pp. 284–294). Left Coast Press, Walnut Creek, CA.

Nawrocki, S.P., Pless, J.E., Hawley, D.A., and Wagner, S.A. (1997). Fluvial transport of human crania. In W.D. Haglund and M.H. Sorg (eds), *Forensic Taphonomy: The Postmortem Fate of Human Remains* (pp. 529–552). CRC Press, Boca Raton, FL.

Owsley, D.W. and Compton, B.E. (1997). Preservation in late 19th century iron coffin burials. In W.D. Haglund and M.H. Sorg (eds), *Forensic Taphonomy: The Postmortem Fate of Human Remains* (pp. 511–526). CRC Press, Boca Raton, FL.

Roach, M. (2003). *Stiff: The Curious Lives of Human Cadavers.* W.W. Norton, New York.

Rodgers, T.L. (2005). Recognition of cemetery remains in a forensic context. *Journal of Forensic Science* 50(1): 5–11.

Schultz, J.J., Williamson, M.A., Nawrocki, S.P., Falsetti, A.B., and Warren, M.W. (2003). A taphonomic profile to aid in the recognition of human remains from historic and/or cemetery contexts. *Florida Anthropologist* 56: 141–147.

Sledzik, P.S. and Ousley, S. (1991). Analysis of six Vietnamese trophy skulls. *Journal of Forensic Sciences* 36(2): 520–530.

Sledzik, P.S. and Micozzi, M.S. (1997). Autopsied, embalmed, and preserved human remains: distinguishing features in forensic and historic contexts. In W.D. Haglund and M.H. Sorg (eds), *Forensic Taphonomy: The Postmortem Fate of Human Remains* (p. 483–495). CRC Press, Boca Raton, FL.

Steadman, D.W. (2009). The pawn shop mummified head: discriminating among forensic, historic and ancient contexts. In D.W. Steadman (ed), *Hard Evidence: Case Studies in Forensic Anthropology*, 2nd edn (p. 258–270). Prentice Hall, Upper Saddle River, NJ.

Taylor, J.V., Roh, L., and Goldman, A.D. (1984). Metropolitan Forensic Anthropology Team (MFAT) case studies in identification: 2. Identification of a Vietnamese trophy skull. *Journal of Forensic Sciences* 29(4): 1253–1259.

Taylor, P. (2000). Bone dry. *Salon,* www.salon.com/health/feature/2000/08/28/skeleton.

Taylor, R.E., Suchey, J.M., Payen, L.A., and Slota, Jr, P.J. (1989). The use of radiocarbon (^{14}C) to identify human skeletal materials of forensic interest. *Journal of Forensic Sciences* 34(5): 1196–1205.

Tompsett, D.H. (1959). *Anatomical Techniques.* E. & S. Livingstone, London.

Ubelaker, D.H. (1989). *Human Skeletal Remains: Excavation, Analysis, Interpretation*, 2nd edn. Taraxacum Press, Washington DC.

Ubelaker, D.H. (1998). The evolving role of the microscope in forensic anthropology. In K.J. Reichs (ed). *Forensic Osteology:*

Advances in the Identification of Human Remains, 2nd edn (pp. 514–532). Charles C. Thomas, Springfield, IL.

Ubelaker, D.H. (2001). Artificial radiocarbon as an indicator of recent origin of organic remains in forensic cases. *Journal of Forensic Sciences* 46(6): 1285–1287.

Ubelaker, D.H. and Houck, M.M. (2002). Using radiocarbon dating and paleontological extraction techniques in the analysis of a human skull in an unusual context. *Forensic Science Communications* 4(4), www.fbi.gov/hq/lab/fsc/backissu/oct2002/ubelaker.htm.

Ubelaker, D.H., Ward, D.C., Braz, V.S., and Stewart, J. (2002). The use of SEM/EDS analysis to distinguish dental and osseous tissue from other materials. *Journal of Forensic Sciences* 47: 940–943.

Ubelaker, D.H., Lowenstein, J.M., and Hood, D.G. (2004). Use of solid-phase double-antibody radioimmunoassay to identify species from small skeletal fragments. *Journal of Forensic Sciences* 49(5): 924–929.

Ubelaker, D.H., Buchholz, B.A., and Stewart, J.E.B. (2006). Analysis of artificial radiocarbon in different skeletal and dental tissue types to evaluate date of death. *Journal of Forensic Sciences* 51(3): 484–488.

Walsh-Haney, H.A., Schultz, J.J., Falsetti, A.B., and Motte, R.W. (2003). Rituals among the Santeria: contextual clues and forensic implications. *Proceedings of the 55th Annual Meeting of the American Academy of Forensic Sciences* 9: 252.

Welti, C.V. and Martinez, R. (1981). Forensic sciences aspect of Santeria, a religious cult of African origin. *Journal of Forensic Sciences* 26(3): 506–514.

Wienker, C.W., Wood, J.E., and Diggs, C.A. (1990). Independent instances of "Souvenir" Asian skulls from the Tampa Bay area. *Journal of Forensic Sciences* 35(3): 637–643.

Wild, E.M., Arlamovsky, K.A., Golser, R., Kutschera, W., Priller, A., Puchegger, S., Rom, W., Steier, P., and Vycudilik, W. (2000). ^{14}C dating with the bomb peak: an application to forensic medicine. *Nuclear Instruments and Methods in Physics Research B* 172: 944–950.

Willey, P. and Leach, P. (2009). The skull on the lawn: trophies, taphonomy, and forensic anthropology. In D.W. Steadman (ed), *Hard Evidence: Case Studies in Forensic Anthropology*, 2nd edn (pp. 179–189). Prentice Hall, Upper Saddle River, NJ.

Woodley, P.J. (1992). The Stirrup Court Cemetery coffin hardware. *Ontario Archaeology* 53: 45–63.

4

The Application of Ground-Penetrating Radar for Forensic Grave Detection

John J. Schultz

INTRODUCTION

Searching for a buried body can be a tedious and difficult endeavor. In order to maximize the likelihood of locating a clandestine burial, a multidisciplinary approach should be incorporated that includes a variety of search methods. In particular, geophysical methods are nondestructive options that can be incorporated into the multidisciplinary protocol. While there are a variety of geophysical tools available, ground-penetrating radar (GPR) has been shown to be the most valuable geophysical tool for grave detection. For nearly 20 years throughout the USA, controlled research (France et al. 1992, 1997; Freeland et al. 2003; Schultz et al. 2006; Schultz 2007) and a growing number of successful case studies (Daniels 2004; Davenport 2001; Mellett 1992; Nobes 2000; Reynolds 2011; Schultz 2007) have documented the utility of using GPR for the detection of clandestine graves in forensic contexts. Ground-penetrating radar also is an important search option for eliminating areas falsely thought to contain a buried body so that investigations can be directed elsewhere. However, there are many instances when site conditions are not favorable for a GPR survey. Therefore, it is important to understand not only how GPR should be incorporated into the multidisciplinary search protocol but also the limitations of using this equipment for forensic grave searches.

A Companion to Forensic Anthropology, First Edition. Edited by Dennis C. Dirkmaat.
© 2012 John Wiley & Sons Ltd. Published 2015 by John Wiley & Sons Ltd.

CONTROLLED RESEARCH

Controlled geophysical research has been important for both archaeology and forensic sciences in determining the applicability of using a particular technology to locate graves, buried bodies, and buried archaeological features (France et al. 1992, 1997; Freeland et al. 2003; Isaacson et al. 1999; Schultz et al. 2006; Schultz 2007; Schurr 1997). Controlled geophysical research should consist of actually burying a body and controlling a number of variables such as time since death, body size, soil type, and depth of burial. Once these variables are documented, it is necessary to detect and monitor the burial for some period of time. Domestic pig (*Sus scrofa*) carcasses are the most common animal proxy for human bodies in controlled GPR studies because they are relatively easy to procure and are the preferred animal proxy for humans in entomologic studies (Catts and Goff 1992; Goff 1993). While initial controlled GPR studies were performed in Colorado by NecroSearch International (France et al. 1992, 1997), there have been a number of subsequent controlled GPR studies including those that have focused on regional approaches in the southeastern USA, including Florida (Schultz et al. 2006; Schultz 2008) and Tennessee (Freeland et al. 2003). Overall, the benefit of controlled research is twofold. First, this research provides the GPR operator with experience using the GPR to detect a buried body. Second, this research has been essential in determining how time since death and a particular soil or environment may influence grave detection.

PLANNING THE SEARCH

A number of issues must be considered during the planning phase of a geophysical search. Prior to performing the search, it is important to learn as much about the site characteristics and burial scenario as possible in order to plan proper subsequent search efforts. The first and foremost issue is whether the scene is suitable for a geophysical survey. After it has been determined that the scene is approropriate for a GPR survey, it is important to determine if the scene has changed since the time of disappearance. If possible, aerial images from the time of disappearance and in the present day should be requested to determine how the survey area has changed during that time span. In addition, studying aerial images prior to performing the survey can be useful for planning a grid set-up. The following questions should be asked prior to performing a GPR survey.

- When did the event occur?
- How deep was the body buried?
- Was the body wrapped in anything?
- Was anything placed in the grave, especially metallic objects?
- Was anything placed over the body to aid in concealment?
- What are the characteristics of the site including topography, vegetation, soil type, etc.?
- How has the present-day search area changed since the time at which the body was thought to have been buried?

Incorporating GPR into the Search Protocol

Search methods can be divided into two basic categories based on the presence or absence of alterations to the subsurface: nonintrusive or intrusive (Dupras et al. 2006; Hunter and Cox 2005; Killam 2004). Nonintrusive search methods consist of those methods that do not disturb the ground surface. These methods are preferable because there is no damage to the subsurface, thus preserving potential evidence and the scene. These methods may include visual, geophysical, and cadaver dog searches. Visual searches can include search lines involving personnel searching for surface indicators consistent with a clandestine grave such as vegetation changes, disturbed soil, grave depressions, soil color changes, and animal scavenging. On the other hand, intrusive search methods, or ground truthing, involve testing that disturbs the ground surface and subsurface. Caution must be exercised when utilizing these methods at a crime scene to limit damage to the potential scene, evidence, and possible grave. Ground truthing may include the use of T-bar probes or trowels, shovel shining with a flat-bladed shovel, digging with a shovel or spade, or lastly, using a forensic backhoe.

The order of search methods for a clandestine burial should begin with noninvasive methods. If no invasive testing has been performed, geophysical methods such as GPR can be used once site conditions are deemed appropriate. Once intrusive testing or digging of a site has begun, GPR cannot be used because disturbed ground will be detected with GPR and will complicate the discrimination of the clandestine burial with that of the recently disturbed areas. There are two ways in which GPR can be used. First, smaller areas that were highlighted by the visual search can be investigated. For example, if a depressed area was noted during the visual search that was consistent with the size of a clandestine burial, a small grid search could be performed with GPR to investigate the localized areas. Second, if visual searches were not useful in noting any surface characteristics consistent with a possible grave, a GPR grid search could be used to survey the entire search area to note possible locations for follow-up invasive testing. The GPR operator will then prioritize the areas that were highlighted during the survey for ground truthing.

If the pre-search information indicates that metal may have been added to a grave, a metal-detector search should be used prior to incorporating a GPR survey. The metal detector search can be important in reducing the size of the area for the GPR search or eliminating the need for a GPR survey all together. Discarded weapons, such as a pipe or a firearm, are examples of metal that may be found in a grave. In addition, metal may be placed over the body in the grave to aid in concealment. For example, Dupras et al. (2006) discussed a case study where a shallow grave was located using metal detectors because pieces of metal strips were placed over the body prior to filling the grave with the backdirt. As with any search method, a grid search should be performed when using a metal detector. The search coil should be positioned close to the ground and swung back and forth, making sure that there is overlap between contiguous search lines. If possible, it is beneficial to have someone to assist the metal-detector operator and to place nonmetallic flags at the site of each hit.

GPR EQUIPMENT

There are a number of GPR manufacturers that offer equipment for archaeological and forensic purposes. The most common and easiest-to-use technology configuration, now offered by all manufacturers, involves the mounting of all components onto a self-contained cart (Figure 4.1). The components are generally configured with a shielded monostatic antenna whereby the transmitter and receiver are contained within the housing, a control unit mounted to the top of the antenna, a separate monitor with the data-acquisition software and an internal hard drive for storing the data files, a battery, and a survey wheel located within one of the wheels of the cart. The monitor allows for real time viewing of each GPR profile as it is collected and the internal hard drive contains enough storage space to save individual files for a small to moderate-size grid survey. The data files can be downloaded for viewing on an external computer for processing or image enhancement. Several GPR systems can be operated using a laptop computer that contains the data-aquistion software and processing software rather than using a separate monitor from the manufacturer. In this

Figure 4.1   Example of a GPR unit, the MALA RAMAC X3M, configured into a cart that is being pushed over a transect line. The GPR components include the monitor (a) with an internal hard drive and the data-acquisition software, the battery (b), a 500 MHz antenna (c) with the control unit (d) mounted to the top of the antenna, and the survey wheel (e) located within the left rear wheel.

configuration, the GPR profiles are viewed directly on the laptop monitor and the data files are stored directly on the laptop. An advantage of running the GPR unit with the laptop is the ability to perform processing and filtering of GPR files directly in the field following data collection.

The GPR unit also can be configured in a number of other arrangements. One option is to mount the monitor and battery on a shoulder harness and hand-pull the antenna. In this configuration, either one person can operate the monitor and pull the antenna or two people can walk in tandem, with one operating the monitor and one hand-pulling the antenna. A third option is to connect the control unit to the antenna via a long cable. This configuration may require three people, as one person is needed to operate the control unit, a second to pull the antenna, and possibly a third person to manuever the cable. This latter option is best for smaller surveys where the cart is too large to maneuver, such as when surveying inside a residence where a body is suspected to have been buried under the cement foundation slab.

One option that must be considered before performing a GPR survey is choosing an appropriate antenna size. Antenna choice is a compromise between depth of viewing and vertical resolution. Generally a 400 or 500 MHz center frequency antenna has been shown to be an ideal antenna choice for detecting burials in sandy soils (Schultz et al. 2006; Schultz 2008) because it provides an ideal compromise between depth of pentration and the vertical resolution of subsurface features. Disadvantages of a higher-frequency antenna, such as 800 or 900 MHz, include shallower depth penetration and increased resolution or detail of subsurface objects that can make it difficult to discern the target in question. Conversely, while a lower frequency such as 250 MHz provides deeper penetration than 500 MHz, the deeper penetration is generally not needed for shallow, hand-dug burials in forensic contexts. The biggest disadvantage of a lower-frequency antenna is that less detail is highlighted in the subsurface, meaning that small targets may be missed. However, a 250 MHz antenna may be an option for soils, such as clay, that can be poorly penetrated by the GPR signal. Controlled testing prior to performing a forensic survey may be important in determining whether a 500 or a 250 MHz antenna is appropriate for the soil composition.

THE GPR PROCESS AND RESULTING IMAGERY

When using a GPR unit with all of the components mounted to a cart, the unit is operated by pushing the cart while walking at a slow pace. The GPR monitor displays the imagery data in real time and provides the highest resolution of all the land-based geophysical tools. It is this imagery viewed on a monitor that allows for an initial in-field assessment to be made when incorporating GPR as a search method. During field data acquisition, continuous electromagnetic pulses of short duration are emitted by the transmitting unit in the antenna and travel downward into the subsurface in a conical pattern (Conyers 2004). When the electromagnetic wave encounters materials of contrasting properties, such as interfaces of soil horizons that consist of different compositions, it will be reflected and scattered due to significant changes in the velocity of the radar wave. Additionally, in forensic and archaeological contexts, the electromagnetic wave is reflected and scattered when it encounters highly conductive objects (i.e., metal artifacts, metal pipes, and weapons), density differences,

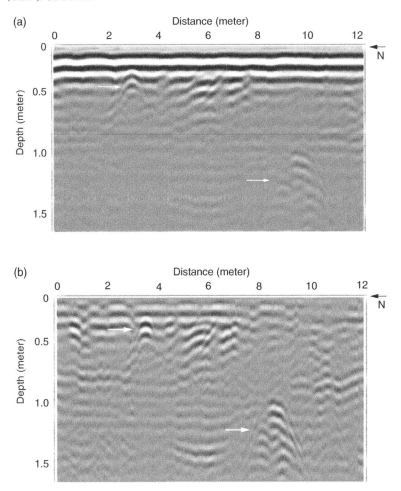

Figure 4.2 (a) Unprocessed GPR profile showing two hyperbolae (white arrows) from a shallow-buried pig carcass (left arrow) and a deep buried pig carcass (right arrow) at 4 months of burial. (b) Processed GPR profile showing increased resolution of two hyperbolae (arrows) from a shallow-buried pig carcass (left arrow) and a deep-buried pig carcass (right arrow) at 4 months of burial, compared to Figure 4.2a. GPR imagery is from a National Institute of Justice-funded research project (2008-DN-BX-K-132), and GPR profiles were imaged in REFLEXW.

changes in moisture content, and voids. The radar transmission process is repeated many times per second and the reflected data are gathered by the receiving unit in the antenna as a series of discrete waves called traces. The image or GPR profile is a composite of the reflected wave traces that are generated along the transect or grid line, and is called a GPR profile (Figure 4.2). The GPR profile is a two-dimensional picture or cross-section of the subsurface that displays time or depth along the vertical axis and length of the transect along the horizontal axis.

Time can be converted to depth by using the known relative dielectric permittivity (RDP) value for the soil being surveyed. Velocity and RDP are used interchangeably for most archaeological studies to determine depth (Conyers 2004). The RDP of a material represents how radar energy will be transmitted through a material to a certain depth. However, since soil properties may not be homogenous across a survey area, changes in moisture content can vary, resulting in different RDP values across a site. As a result, depth may only be an estimate when one value is used to convert time to depth involving a site with variable RDP values of the ground. The most accurate method to calculate depth at a site is to use the reflected-wave method (Conyers 2004). This method involves analyses of reflections at known depths on the GPR profile that allow for the calculation of average-velocity waves. For example, an accurate depth can be estimated when surveying cemeteries by running the GPR over a burial vault or coffin and then measuring the depth to the buried object by using a T-bar probe. Depth can then be estimated accurately in the field by adjusting the velocity and running the GPR over the buried item again until the depth is corrected.

An actual picture of a buried body or skeleton is not shown on the screen when a grave is detected. While the GPR profile can be difficult for untrained operators to interpret, a reflection, commonly called an anomaly, appears as a vertical series of hyperbolic (bell-shaped) curves that are the result of a buried feature, such as a grave. The buried object is located at the apex of the anomaly, and the series of hyperbolae can extend deeper than the detected buried object. The extensions, or tails, that produce the hyperbolic shape are artifacts of the wide angle of the conical radar beam that detects the buried object or features before, while, and after the antenna passes over the feature. The recognition of the hyperbola on the GPR profile is an essential skill needed by the operator to make in-field assessments. There must be enough electromagnetic contrast between the surrounding undistubed soil and the grave (which can include the backfill, a body or skeleton, or other objects) to produce a sufficiently strong amplitude of the reflected signal, resulting in a recognizable hyperbola-shaped anomaly in the GPR imagery.

The GPR reflection data can be enhanced after data acquisition is completed by using a variety of processing programs. For example, there may be horizontal lines at the top of the unprocessed profiles called noise or multiples (Figure 4.2a). In some instances, the noise can obscure shallow reflection data such as disturbed soil of a grave near the ground surface and hyperbola-shaped anomalies (Schultz et al. 2006; Schultz 2008). Removal of noise from the GPR profile and the addition of a number of other processing techniques results in increased resolution of the potentially significant hyperbola-shaped anomalies (Figure 4.2b). Note the increased resolution of the processed GPR profile of shallowly and deeply buried pig cadavers with noise removed (Figure 4.2b) compared to the unprocessed GPR profile (see Figure 4.2a), which would be similar to the imagery observed in the field during data acquisition. In addition, Conyers (2006) discusses how interference from electromagnetic sources obscures reflections in the field. When the data are processed later to remove the interfering frequencies, an archaeological feature can be discerned.

With advances in GPR processing software, numerous studies have discussed the advantages of viewing archaeological features and cemetery graves by incorporating grid data into a three-dimensional picture to view the spatial grid location of features in the subsurface (Conyers 2004; Conyers 2006; Goodman et al. 1995; Watters and Hunter 2004). A three-dimensional picture of the subsurface begins with the collection of GPR data along transects of a grid with an equal distance between each grid line (Figure 4.3a). All of the GPR profiles collected over each grid transect are then correlated and pro-

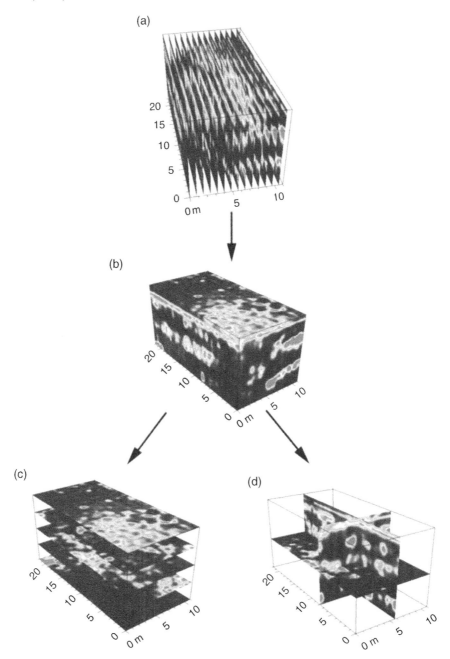

Figure 4.3 Advanced GPR processing for a grid search begins with the individual profiles (a), that are processed into a cube (b), using processing software. A variety of cuts including time slices (c) and fence diagrams (d) are commonly made through the cube to discern the spatial location, shape, and size of buried features.

cessed into a cube using GPR software that interpolates or fills in the distance between each GPR profile (Figure 4.3b). The shape and geometry of the buried features will be more accurate and smaller features will be detected when tight-transect intervals are utilized, such as 25 or 50 cm, depending on the size of the forensic target in question.

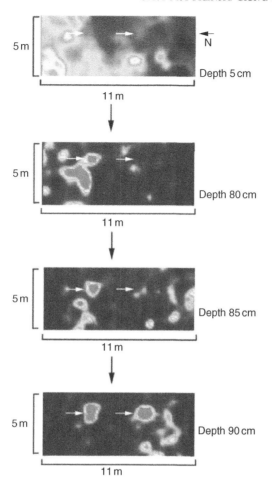

Figure 4.4 Four time slices representing progressively deeper depths (approximate depths are provided) of a 5 m × 11 m grid survey. The grid included a burial with a pig carcass (left arrow) and a control grave containing only disturbed backfill (right arrow) that were constructed with the grave floor at 100 cm. GPR time slices were processed and imaged using GPR-SLICE Open GL Volume. GPR imagery from National Institute of Justice-funded research project (2008-DN-BX-K-132).

In addition, if data are collected in *x* and *y* directions, GPR processing programs can process and integrate this data into a cube that may provide more accurate delineation and resolution of buried features than the use of data collected in one cardinal direction. There are a number of different cuts that can be made through the cube to infer the spatial location of buried features and the stratigraphy. The most common imagery presented from cuts to the cube are time slices, or horizontal slices, showing changes in amplitude through the cube at various depths (Figure 4.3c). Each time slice is akin to a plan view of a map at different depths, and the entire set of time slices for a grid survey can be used to infer the spatial distribution or location of buried features. By examining a set of time slices for a feature such as a possible grave, it is possible to determine the size, orientation, and depth of the feature (Figure 4.4). Another option for viewing the GPR imagery is to incorporate two or three cuts through the cube at different orientations to create a fence diagram (Figure 4.3d). The integration of a fence diagram allows for simultaneous visualization of a feature in more than one plane.

ADVANTAGES AND LIMITATIONS OF USING GPR FOR FORENSIC CONTEXTS

While a number of manufacturers now offer GPR models at more competitive prices, there is a growing number of law-enforcement agencies and universities that have their own in-house equipment because of the potential of GPR for the detection of archaeological features, cemetery graves, and forensic graves. GPR has become a valuable search tool for a number of important reasons (Table 4.1). First, as previously stated, the technology, which has the best resolution of all geophysical methods used on land, provides real-time imagery that allows the operator to make an in-field assessment. At the same time, this geophysical tool is a nondestructive search option that is a major advantage at a crime scene or archaeological site because there will be no damage to buried artifacts or evidence. This search option is used to pinpoint smaller or localized areas of interest at a scene or site for follow-up invasive testing, thereby reducing the amount of damage to the scene or site. Since GPR can provide valuable information about the size, depth, and position of buried features in the subsurface, it becomes possible for investigators to plan the excavation of a buried target to cause limited damage to the subsurface. The digital data files can also be stored on a hard drive to process grid data at a later time to enhance the resolution of buried features detected within the GPR imagery.

Another advantage of using GPR to search for clandestine grave detection is that it can be used to detect buried features under cement or blacktop without causing any damage. This is one of the most important advantages of using GPR because there are

Table 4.1   Advantages and limitations of using GPR for grave detection.

Advantages	Limitations
Real-time imagery with immediate results	Need level, somewhat smooth, and open terrain
No surface damage	Clutter from buried debris, rocks, and roots inhibits visibility of targets
Provides information about size, position, and depth of targets	Needs a trained operator for setting up survey, operating equipment, and software
Penetration through freshwater, ice comprised of freshwater, and snow	Expensive equipment and software
Penetration through concrete and blacktop	Slow to medium coverage speed with grid utilizing small transect interval spacing
Equipment is easy to manuver with cart	Will not work with all soils and water-saturated ground
Data files are stored on a hard drive, allowing for viewing and processing of grid data to provide additional imagery views	There must be sufficient contrast between the grave and the soil for grave detection

limited search options to use when the search must be performed over a hard surface. GPR is the preferred search option for this type of investigaton and has been used successfully in a number of cases to detect the graves of individuals buried under cement slabs for concealment. For example, Schultz (2007) confirmed the location of a grave of an individual who had been buried under a cement slab in a residential garage foundation for 15 years, and Davenport (2001) located the grave of an individual who had been buried under a patio in a residential backyard for 28 years. When performing a GPR search over cement, a plan must be formulated to test-probe any promising anomalies that are determined not to be the result of cultural features, such as buried pipes, that may run underneath cement slabs. Since a grid search can pinpoint the exact locaton and size of the buried object, invasive inspection can be performed directly over any of the anomalies that need to be tested with only localized damage to the cement slab. One option for inspecting anomalies is to cut a small section of the cement with a masonary saw and then remove the cut section. The cut section can then be easily repaired after excavation of the area is completed.

While there are numerous advantages for integrating GPR as part of the multidisciplinary protocol, there are also a number of limitations that must be considered prior to deciding the viability of GPR for a forensic survey (Table 4.1). The first issue to consider is whether the site is appropriate for a GPR survey. The site should be relatively level, somewhat smooth, and consist of open terrain. If the site contains brush and trees, the roots may cause extensive clutter, which are small discontinuities reflected in the subsurface that are not the target of the survey, making it difficult to discern the actual buried targets. In addition, if brush is cleared from the site, GPR will still reflect the stumps and roots in the subsurface. The area also should be free of buried debris, large boulders, and gravel lenses that can result in clutter. If there is extensive clutter on the radargram resulting in multiple hyperbolic anomalies, it may not be possible to discern a grave in the survey area. If there are questions as to whether the site is appropriate for a survey, digital pictures of the site can be sent to the GPR operator via e-mail for an initial assessment. Issues about the site can then be discussed before the search efforts progress.

Soil type can be a limiting factor in grave detection, possibly decreasing the effectiveness of using GPR as a search method. Soils with horizons comprised of sand have been shown to provide excellent conditions for detecting graves containing decomposing bodies with GPR (Schultz et al. 2006; Schultz 2008). Conversely, heavy clay soils and water-saturated soils are poor for GPR surveys. It may not be possible to detect recently buried remains in a clay soil with GPR due to attenuation of the radar wave (Figure 4.5a). However, disturbed soils consisting of different types of soil horizons may be detected as a disruption of the natural stratification. In addition, a sandy soil with a deep clay horizon can result in difficulty detecting a body over time if it is placed in proximity to the clay horizon because the body may appear as one of the natural undulations of the clay horizon (Schultz et al. 2006). Furthermore, the length of time the body has been buried also can influence grave detection. While the soil disturbance, or the backfill, will be detected initially with sandy soils, the disturbance from the backfill may not be detected over time, and the area around a skeleton may not provide a sufficient contrast to be detected by GPR (Schultz et al. 2006; Schultz 2008).

Items added to the grave may also contribute to grave detection. For example, items added for concealment, such as debris laid over the body before backfilling the

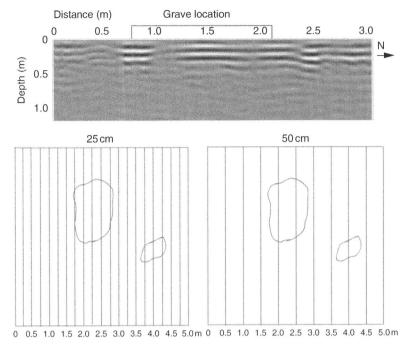

Figure 4.5 Top: note the absence of reflections in the area of a grave, approximately 60 cm deep, containing a recently buried human cadaver in dense clay soil at the Anthropological Research Facility at the University of Tennessee. GPR profile imaged in REFLEXW. Bottom: hypothetical plan view of a grid search containing an adult grave feature and a small-sized juvenile grave feature with 25 cm (left) and 50 cm (right) transect interval spacing. Note the decreased number of transects over the smaller feature using 50 cm spacing compared to 25 cm spacing that would decrease the likelihood of detecting the smaller feature.

grave, may be detected. If items such as a carpet or tarpaulin are used to wrap the body, these items may provide a contrasting area with the body that can be detected. Also, if a void forms in the grave due to soil compaction and decomposition of the body, the void space may contribute to GPR detection.

HIRING A GEOPHYSICAL CONSULTANT

A GPR unit is not normally standard field equipment for most law-enforcement agencies due to the high cost of the equipment and the limited opportunities for use. As a result, law-enforcement agencies generally need to hire an outside GPR consultant to perform a survey when needed. If a GPR consultant is sought, investigators should first inquire about the experience of the operators prior to securing their services. Considering that a forensic GPR survey involves detecting small subsurface objects, GPR operators should have specialized training in this area. However, since the equipment is expensive, it may not be possible to locate a local unit and a trained operator. Also, since the equipment requires special training for use in forensic contexts and in data collection using a transect or grid search, it is important to use an operator that has forensic search experience, including performing a grid survey. At a minimum, GPR operators should have experience detecting small archaeological features and cemetery graves. Also, if the data needs processing, training is also required to use the processing software. The best place to locate an operator with cemetery experience is through a local university or college with either an anthropology or archaeology department. A contract archaeology firm or cultural resource management firm is another option.

An environmental or survey firm may also have in-house GPR equipment. However, the expertise of a GPR operator from an environmental firm may only include interpreting large geological and soil features such as sinkholes, stratigraphic horizons, depth to bedrock, and depth to water tables. The forensic and archaeological experience of the operator should be determined prior to securing the services of someone from an environmental firm because they may not have experience using a proper grid search or surveying for small subsurface features. A clandestine burial can be missed easily if either a proper grid search is not performed with adequate transect interval spacing or the data are not interpreted correctly.

PERFORMING THE SEARCH

Two issues must be considered prior to performing a GPR survey: the size of the survey grid and whether anomalies noted during the search will be ground truthed during the day of the survey. According to Conyers (2006), areas for archaeological surveys larger than 50 m × 50 m grids can be performed in one day with a transect interval spacing of 50 cm. A forensic survey will take longer if grid transects are performed in two directions such as north to south followed by west to east, and if the transect interval spacing is 25 cm rather than 50 cm. Also, extra time will be needed if the anomalies will be ground truthed the same day that the GPR survey is performed. If only one day is scheduled to perform a search, and ground truthing will be performed in a search area larger than an acre, then a smaller-sized grid that is more manageable may have to be utilized. A half-acre or smaller area is a manageably-sized grid when data collection is performed in one direction and ground truthing is also performed the same day as the survey. If the survey takes place at a private residence, a number of small grids can be set up that encompass the backyard and front yard. If ground truthing of features is not performed the day of the survey, the data can be filtered and processed to enhance features that may not have been discerned during the initial field assessment phase using the unprocessed GPR data.

Grid transect spacing must be determined based on the size of the target. It is important to note that, as tighter grid spacing is utilized, more time will be needed to complete the survey. A general rule of thumb is to use transect spacing that is a minimum of half the size of the target in question. For example, if transect spacing of 50 cm is used for an adult burial and for a small-sized juvenile burial, there will only be adequate spacing to detect the adult burial (Figure 4.5b). There may be issues detecting the smaller-sized burial using the 50 cm spacing because it may be that no transects collect over the burial, or that only one grid transect collects over the burial. With only one grid transect over the burial, it may not be possible to detect a small grave or target. However, when the grid spacing is decreased to 25 cm, multiple transect lines will be collected over smaller targets resulting in a higher probability of detection. Also, when a buried feature is detected on contiguous transect lines, it may be possible to discern the shape and size of the target during the in-field assessment.

If there are known buried cultural features (electrical lines, pipes, septic tank, etc.) in the survey area, these buried features can be marked with nonmetallic flags prior to performing the survey to quickly exclude any anomalies that may have been produced by these features. A site survey map should be made of the area with the buried items marked as well as any other features in the area, such as trees or stumps, that may produce anomalies. When the GPR cart is pushed along each grid transect of the survey area, the GPR operator can make in-field assessments by noting the location of anomalies along each transect. The area can be marked on the grid during the survey or after completion. If an immediate assessment is desired, the operator can be followed by an assistant who places nonmetallic flags at the location of each anomaly. It is good practice to use flags with nonmetallic shafts rather than metal shafts because geophysical tools will detect the latter. After the survey is completed, it is the task of the GPR operator to prioritize the areas that need to be ground truthed. Based on site context, multiple areas can be ruled out based on features of the site. For example, anomalies near trees are probably the result of roots. In addition, anomalies that are much smaller than the target in question would be considered a low priority for invasive testing.

FUTURE CONSIDERATIONS

While the integration of GPR as part of a multidisciplinary search protocol has advantages, conditions need to be favorable for using GPR. Issues such as environmental variables, site variables, length of time the body has been buried, and items buried with or used to wrap the body will influence grave detection. At the same time, experience of the operator can influence the success of any survey. Controlled geophysical research is essential not only in learning about environmental influences, but also in providing experience to the operators that translate well during real-life searches. It may be possible to gain experience by participating in a forensic archaeology partnership with local law enforcement (Schultz 2007). This can involve performing GPR surveys during searches and collecting data over any known graves of homicide victims before the graves are excavated to gain GPR experience. The partnership can also include performing a controlled geophysical research project that involves burying either human cadavers or pig carcasses to test local environmental variables. If the intent of the controlled research project is to gain experience surveying graves containing bodies, then the test scenario should involve an actual body such as a pig carcass. Supplementing other items such as plastic bones or real bones that are dry without flesh cannot replicate the taphonomic environment of a grave containing a decomposing body.

ACKNOWLEDGMENTS

I would like to thank the personnel at the Anthropology Research Facility at the University of Tennessee for allowing GPR data collection of buried human cadavers during the summer of 2006. Also, the GPR imagery of the buried pig carcasses (Figures 4.2 and

4.4) represents preliminary data from a research project supported by award no. 2008-DN-BX-K132 awarded by the National Institute of Justice, Office of Justice Programs, US Department of Justice. The opinions, findings, and conclusions or recommendations expressed in this publication are those of the author and do not necessarily reflect those of the Department of Justice.

REFERENCES

Catts, E.P. and Goff, M.L. (1992). Forensic entomology in criminal investigations. *Annual Review of Entomology* 37: 254–272.

Conyers, L.B. (2004). *Ground-Penetrating Radar for Archaeology*. Altamira Press, Walnut Creek, CA.

Conyers, L.B. (2006). Ground-penetrating radar. In J.K. Johnson (ed.), *Remote Sensing in Archaeology: an Explicitly North American Perspective* (pp. 131–159). The University of Alabama, Tuscaloosa, AL.

Daniels, D.J. (2004). *Ground Penetrating Radar*, 2nd edn. The Institute of Electrical Engineers, London.

Davenport, G.C. (2001). Remote sensing applications in forensic investigations. *Historical Archaeology* 35: 87–100.

Dupras, T.L., Schultz, J.J., Wheeler, S.M., and Williams, L.J. (2006). *Forensic Recovery of Human Remains: Archaeological Approaches*. Taylor and Francis Group/ CRC Press, Boca Raton, FL.

France, D.L., Griffin, T.J., Swanburg, J.W., Lindemann, J.C., Davenport, G.C., Trammell, V., Armbrust, C.T., Kondrateiff, B., Nelson, A., Castellano, K., and Hopkins, D. (1992). A multidisciplinary approach to the detection of clandestine graves. *Journal of Forensic Sciences* 37: 1445–1458.

France, D.L., Griffin, T.J., Lindemann, J.C., Davenport, G.C., Trammell, V., Travis, C.T., Kondratieff, B., Nelson, A., Castellano, K., Hopkins, D., and Adair, T. (1997). NecroSearch revisited: further multidisciplinary approaches to the detection of clandestine graves. In W.D. Haglund and M.H. Sorg (eds), *Forensic Taphonomy: the Postmortem Fate of Human Remains* (pp. 497–509). CRC Press, Boca Raton, FL.

Freeland, R.S., Miller, M.L., Yoder, R.E., and Koppenjan, S.K. (2003). Forensic applications of FM-CW and pulse radar. *Journal of Environmental and Engineering Geophysics* 8: 97–103.

Goff, M.L. (1993). Estimation of postmortem interval using arthropod development and successional patterns. *Forensic Science Review* 5: 81–94.

Goodman, D., Nishimura, Y., and Rogers, J.D. (1995). GPR times slices in archaeological prospection. *Archaeological Prospection* 2: 85–89.

Hunter, J.R. and Cox, M. (2005). *Forensic Archaeology: Advances in Theory and Practice*. Routledge, London.

Isaacson, J., Hollinger, R.E., Gundrum, D., and Baird, J. (1999). A controlled test site facility in Illinois: training and research in archaeogeophysics. *Journal of Field Archaeology* 26: 227–236.

Killam, E.W. (2004). *The Detection of Human Remains*. Charles C. Thomas, Springfield, IL.

Mellett, J.S. (1992). Location of human remains with ground-penetrating radar. In P. Hanninen and S. Autio (eds), *Proceedings of the Fourth International Conference on Ground Penetrating Radar*, June 8–13, Rovaniemi, Finland. *Geological Survey of Finland, Special Paper* 16: 359–365.

Nobes, D.C. (2000). The search for "Yvonne": a case example of the delineation of a grave using near-surface geophysical methods. *Journal of Forensic Sciences* 45: 715–721.

Reynolds, J.M. (2011). *An Introduction to Applied and Environmental Geophysics*. John Wiley and Sons, New York.

Schultz, J.J. (2007). Using ground-penetrating radar to locate clandestine graves of homicide victims: forming forensic archaeology partnerships with law enforcement. *Homicide Studies* 11: 15–29.

Schultz, J.J. (2008). Sequential monitoring of burials containing small pig cadavers usingground penetrating radar. *Journal of Forensic Sciences* 53: 279–287.

Schultz, J.J., Collins, M.E., and Falsetti, A.B. (2006). Sequential monitoring of burials containing large pig cadavers using ground-penetrating radar. *Journal of Forensic Sciences* 51: 607–616.

Schurr, M.R. (1997). Using the concept of the learning curve to increase productivity of geophysical surveys. *Archaeological Prospection* 4: 69–83.

Watters, M. and Hunter, J.R. (2004). Geophysics and burials: field experience and software development. In K. Pye and D. Croft (eds), *Forensic Geosciences: Principles, Techniques, and Applications* (pp. 21–31), Special Publication 232. Geological Society of London, London.

CHAPTER 5

Crime Scene Perspective: Collecting Evidence in the Context of the Criminal Incident

Michael J. Hochrein

Introduction

Whether the primary piece of evidence in an investigation was deposited hundreds of years ago or was the result of activity in the present day, the recovery of the evidence must include associated artifacts and environmental data in order to define its spatial and temporal contexts. The goal of a crime scene investigator, as it would be a field archaeologist, should be to objectively collect all information that may serve to explain the incident under investigation rather than merely collecting the focus of the incident or one event within the incident. Crime scene investigators are more likely to be hindered by evidentiary tunnel vision than archaeologists who routinely recover artifacts in the context of a site rather than temporally limited searches which focus on finding a particular evidence item or class of items. Given that the recovery of evidence, or artifacts, must be contextual, the most difficult aspect of processing outdoor crime scenes involves a determination of the scope of the scene. That partly relies on understanding the potential of the scene's substrates. For example, the subterranean to terrestrial substrates of varied soil horizons constitute different microenvironments that affect the condition of evidence. When speaking of buried evidence, those substrates or matrices hold geotaphonomic artifacts indicative of how, when, and if the evidence may have been concealed. Surface substrates may bear impression evidence, or constitute trace evidence that has been transferred to or from the scene. Terrestrial substrates, such as vegetation foliage, or pollen deposited on various surfaces, similarly comprise evidence that may be transferred to other locations during a crime event.

A Companion to Forensic Anthropology, First Edition. Edited by Dennis C. Dirkmaat.
© 2012 John Wiley & Sons Ltd. Published 2015 by John Wiley & Sons Ltd.

An archaeologist painstakingly searches for and documents artifacts within a prehistoric or historic site. The collection of this environmental-contextual evidence requires an abundance of patience and attention to detail, but it does not necessarily require sophisticated equipment. A common piece of advice offered by seasoned crime scene investigators to new recruits is that, in spite of technological advances each year, the best tools any investigator can employ are a strong light and an open mind. Evidence that can be most easily missed at any crime scene is evidence that is contaminated or trampled under foot because investigators are unable to see it under ambient conditions. Other evidence may be dismissed as too old, too weathered, or too far away from the focus of the scene. The most salient clues at a crime scene may be subtle, but rarely are they invisible. Veteran homicide investigator Vernon Geberth (2010: 171), acknowledges as commonplace the return of investigators to scenes because, "some seemingly innocuous item was actually an important piece of evidence." The challenge of assessing the potential scope of a crime scene, recognizing and collecting both obvious and trace evidence which may have been environmentally altered, and realizing often subtle patterns in the distribution of that evidence, is amplified in outdoor settings. Investigators may fall into a trap of making determinations without collecting evidence independent of a final interpretation. For example, if called to a death scene reported as a likely suicide, there is often a temptation to let such a pronouncement influence the scope of the scene examination. Another pitfall involves procedural resignations following announcement that a confession was obtained from a suspect. False confessions are always a possibility and successful suppressions of confessions are more likely. Therefore, scene investigators must always proceed with the scene examination as if a confession was never obtained. In some instances it may be better for the crime scene investigators not to know each and every detail of witness statements in order that a more objective approach to scene processing can be taken. The author, in processing a double homicide scene, experienced an example of how such objectivity can serve to support or refute witness accounts. In that case, the victims were buried in a single clandestine grave. Once discovered, standard archaeological protocols were used to excavate the feature. Among geotaphonomic evidence encountered was an interruption, or intrusion, into the stratification (layers of soil or fill), within the grave above the bodies. During the excavation this evidence was documented in three dimensions. Following the excavation the stratigraphic intrusion was found to correspond to a statement by one suspect. That suspect told investigators that he and others attempted to relocate the bodies a few months after their burial. However, they stopped when they became overwhelmed by the smell and sight of decomposition. Had that feature been processed without adherence to archaeological protocols, including the mapping and photography of clean and plumb excavation profiles within features, this evidence may have been overlooked. Often, investigators with little or no archaeological experience process clandestine graves. As a result the fill surrounding the body in the pit feature and the undisturbed soil horizons outside the feature are treated as if they are one continuous layer, or stratum, across the site. The significance of the stratification of the feature's fill and the feature boundaries where it interfaces with undisturbed matrices are ignored. Without that recognition of disturbed versus undisturbed soils the recognition of context at its most intimate level within the crime scene is lost.

This chapter will present a typical outdoor human remains recovery in which the scope of the crime scenes, or criminal incident, rapidly exceeded the jurisdiction of any one agency. The case presented involves the collection of evidence from individual scenes relative to a criminal incident. The incident was comprised of the actions and movements of the subject and victim before, during, and after a homicide. It also demonstrates the potential importance of recognizing subtle characteristics in the condition and distribution of evidence. In presenting this case history, the author will mention basic, often overlooked, procedures that serve to preserve the scene. More sophisticated technology will also be discussed. These state-of-the-art techniques, however, do not replace the application of fundamental principles but apply them more efficiently.

No crime scene investigator is an expert in all of the types of evidence that could potentially be encountered at a scene. However, all investigators should understand how to efficiently collect, or recognize the expertise needed to collect the maximum amount of evidence from each scene. If collected in a manner that preserves contextual details of each item, every forensic specialist called to examine the case should be able to look at the evidence as if they themselves had been at the scene. Ultimately, the crime scene investigator's report should be used with ease by counsel in the presentation of the crime incident as represented by physical and testimonial evidence. In cold case investigations the contextual details should stand for decades so that subsequent investigators can virtually revisit the scene as witnesses are identified and located, subjects cooperate, and/or forensic sciences evolve.

CASE HISTORY

When an 80-year-old widowed grandmother did not show up for her usual appointments, pick up her mail, or answer her telephone over several days, her friends in a small Pennsylvania community became concerned. Their anxiety was heightened because the victim's son had recently moved in with her following his release from prison. Local police were contacted and entered the victim's house to conduct a welfare check. Neither the victim nor her son was in the residence. Although the house seemed in order at first glance, neighborhood concerns prompted police to broadcast a "bolo" (be on the lookout), for the son and his mother. Within hours the son was located in a hospital in West Virginia after having been involved in a car accident. He was interviewed by the police and admitted to murdering his mother. The son described where investigators could find her partially buried body, on public land, in a county adjacent to the county in which they lived. While investigators responded to the alleged burial site, a search warrant was obtained for the subject's/victim's residence. Although no outward signs of a struggle seemed apparent, investigators located one of the victim's denture plates on the floor in an obscure corner of a hallway. The subject had described how, in a drunken rage, he had strangled his mother during an argument over his drinking. He claimed that he did not beat her and volunteered that he had not sexually assaulted her or her corpse.

When news of the homicide was broadcast by local television affiliates and the location of the body described, a citizen came to the police and handed over personal effects identified as those of the victim's. These personal effects were found as he had hiked in a remote area of the public lands. At the time he came upon the items, the hiker did not know they were associated with a body that lay approximately 790 m away. The hiker then took crime scene investigators to that area located near yet another county line. Near the third county line, other pedestrians found the victim's clothing, which had apparently been thrown over a highway embankment.

Defining the Crime Scene

This homicide and the subject's acts, before, during, and after, comprise an incident as defined by Gardner and Bevel (2009: 37–72), in their application of event analysis to crime scene reconstruction or interpretation. The concept of considering the crime scene in the context of events leading to, comprising, and following the crime was proposed by Ludwig Benner as early as the mid-1970s with his application of multilinear events sequencing to accident and rare-event reconstructions (Benner 1975; Gardner 2009). Rynearson and Chisum (1993) also addressed the topic. Considered separately, the elderly victim's homicide and her son's actions following it (event segments), created at least seven (and potentially 10), separate scenes: (i) the victim and subject's house in which the homicide allegedly took place; (ii) the location of the victim's body; (iii) the location of the victim's personal effects as found by the hiker; (iv) the location at which the victim's clothing was dumped; (v) the subject's clothing and body; (vi) the subject's vehicle; (vii–x) the person, vehicle, and residence of the hiker who volunteered evidence he found. These ten areas, which could potentially demonstrate evidence relative to the homicide, were spread over five counties in two states.

When investigators consider contextual evidence, the crime scene can be enormous. As such it sometimes requires multijurisdictional expertise and resources. In this case the subject's initial assault on the victim took place in their home. However, she could very well have succumbed to her injuries or have again been assaulted and then killed at the location where her body was found in an adjacent county. To further complicate this scenario, we must consider whether the victim may have died in the subject's vehicle en route to the disposal site. In whose jurisdiction does the homicide occur? The manner in which each of the above scenes is processed could have a profound effect on determining the jurisdiction responsible for prosecuting the crime. If greater weight was inappropriately placed on the location at which the body was found, crucial trace and document evidence indicating intent or planning could have been lost at the house site where the victim and subject resided. Likewise, linkage of the disparate evidential sites, over 160 km apart, could have been overlooked. When the responding agency processed the scene where the victim's body was found, they realized the scope of the incident. Rather than focusing solely on one scene, assistance was sought from state and federal forensic teams. In this way, each scene comprising an event segment of the criminal incident was approached with no less focus than any other.

At the scene of the body, details such as the condition and position of the remains were documented. Insect activity and the severely decomposed condition of the victim's head relative to the rest of her nude body suggested that trauma beyond the admitted strangulation may have occurred. The autopsy confirmed that additional trauma did in fact occur. The documentation of the victim's clothing in a location separate from the body preserved evidence that indicated the subject's actions on the night of the murder constituted more than a single uncontrolled act of drunken violence. The recovered evidence produced more questions: Did a sexual assault also occur?, Would the removal of the victim's clothing, and its concealment far from the body, be within the capabilities of someone so intoxicated?, Or was this the result of multiple days of activity?, and Was it possible that one or both of the civilians who brought evidence to the police were more involved than it first appeared? These are all questions which were resolved when evidence from all the scenes were viewed in the context of each other, and testimony was collected independent of the crime scenes through witness interviews and subject interrogation. The first obligation of the crime scene investigator is not to attempt to answer all of these questions at the scene, but to objectively collect information knowing there will be a multitude of questions that need answers. Even though the other scenes did not hold what most would consider the primary focus of the investigation – the body – the physical evidence they did hold, and more so their context, were no less important. For example, the proximity of the victim's personal items to an empty bottle of alcohol and shell casings in a secluded part of the woods, nearly 790 m from the victim's body, supported the subject's admission that he contemplated suicide at one point during the incident.

The temptation for investigators to lose focus during the collection of crime scene evidence increases with knowledge of a subject's confession. The author has often been asked in the course of recovering evidence why such exacting procedures are required after a suspect has lead the investigators to where he disposed of the body, or when he has confessed his involvement in the crime. Again, experience has shown that suspect accounts are typically skewed to blame others, to fabricate defenses, or to minimize their actions. Additionally, such statements are often the subjects of suppression hearings in which a suspect's counsel may successfully prevent his client's statements from being used at trial. The product of the crime scene investigator's work will need to serve multiple experts and be able to substantiate the prosecution's case. Both the prosecutor and defender of the subject should expect the most complete description of the criminal incident through proper, thorough collection of evidence. Proper and thorough collection protocols equates with unbiased collection. Unbiased evidence collection requires that recovery protocols be applied with equal emphasis to each scene created during an incident. As in this case, multiple, associated scenes created during the incident can be linked in an exhibit with data from each scene.

DOCUMENTING THE SCENE

Regardless of the number of still photographs and amount of video footage compiled by investigators at a crime scene, two- and three-dimensional maps are essential in placing those images in context. Photographs and video images exhibit some optical distortion known as parallax, or the difference in the apparent position of objects as they are viewed along different lines of sight. Videos can also limit the observer to a tunnel-view tour of the crime scene. The use of both, in formats integrated with a two- and three-dimensional scaled diagram, can provide investigators, jurors, and jurists with references for the images they must view.

In the case presented above, larger-scale mapping had to be considered with the need for photography due to the scale and number of the widely dispersed crime scenes. Because the scenes were spread over such a wide area and not interrelated, aerial imagery, satellite imagery, and cartographic resources were utilized to depict the relationship of each scene. Aerial photographs from fixed-wing and helicopter platforms captured the scenes that occurred in Pennsylvania. These scenes included the victim's residence in relation to the site where her body was found, as well as the locations where her personal effects and clothing were found. From a more distant perspective, aerial and satellite imagery available through internet mapping sites was used to show the location of the subject's activities in West Virginia relative to the Pennsylvania sites. Global positioning systems (GPS) were used to record the approximate spatial coordinates of the distantly spaced scenes and provided the flexibility to view their relationships using US Geological Survey topographic maps, soil surveys, and satellite imagery. GPS, along with the US Global Navigation Satellite System (GNSS), is now commonly employed by law enforcement, archaeologists, surveyors, and others to record the latitude, longitude, and elevation of selected points using trilateration from a constellation of as many as 32 satellites. The position of a particular receiver positioned over a target is calculated by distance traveled over time of uniquely coded radio signals from each satellite. To improve the accuracy of location data from tens of meters to millimeters, multiple receivers are used to perform what are known as differential corrections. In order to possess the investigative and prosecutorial capabilities to "zoom in" to images of each outdoor scene from the global perspective, the spatial data-collection process began with standard photographic documentation of each site's condition upon the investigator's arrival. Photographic documentation continued throughout the processing of each scene and down to the level of individual evidence items as they were encountered. If the concept of using GNSS to collect information for interpretation and reconstruction is applied further, geographical information systems (GIS), are another tool available to crime scene investigators. By definition GIS techniques collect, archive, analyze, and demonstrate data linkages in terms of spatial relationships. In the case presented herein, GIS could have been used to relate the various event segments and ultimately interpret the various event segments of the criminal incident. Had the subject in this case been, hypothetically, a serial killer and the death of the elderly female was but one of several homicides linked to him, GIS would be a very appropriate documentary and analytical tool. As an individual homicide comprised of events spread over many miles, GIS could have been used to aid in locating evidence.

At the location of the victim's body, traditional archaeological protocols were used in the recovery of human remains. The protocols required that a three-dimensional grid consisting of x, y, and z planes of measurement be established over, and theoretically through, matrices holding the remains, burial, or depositional feature, and associated artifacts or evidence. Some of this evidence was situated in multiple matrices. Such measurements made in relation to perpendicularly oriented planes or axes are the bases for two- and three-dimensional Cartesian coordinate mapping (Figure 5.1). Connor (2007), Dirkmaat and Adovasio (1997), Dupras et al. (2006), Hochrein (1997b, 1998, 2002), and Steadman et al. (2009) provide detailed descriptions of the use of Cartesian coordinate mapping at forensically significant

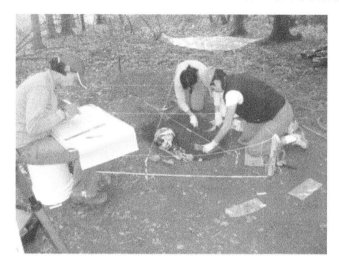

Figure 5.1 Cartesian coordinate mapping used in an outdoor context.

Figure 5.2 Polar coordinate mapping used in a subterranean context.

burial features. Two- and three-dimensional polar coordinate, or azimuth, mapping is another alternative, as may be triangulation (Hochrein 2002, 2006) (Figure 5.2).

The author recommends a combination of manual and electronic mapping techniques at most sites, especially in cases of buried evidence. In those situations the limitations of electronic mapping and scanning are their inability to map what the instruments cannot see. For example, the use of an electronic total station requires a precise set up over a control point or datum. It is usually impossible to see, and therefore map, all parts of the burial from a single vantage point. The creation of, and movement to, other subdatums around the feature is required. Other than fully extended supine positions, the configuration of the body or its parts usually requires the excavation and removal of upper portions, flexed limbs, clothing, etc., to reveal lower situated parts of the body so they can be photo-documented and mapped. In these situations it can be more efficient to use manual Cartesian or polar coordinate techniques with strings, line levels, plumb bobs, and tape measures, rather than the total station alone. Especially in deeper subterranean contexts where visual acquisition of targets is at an angle too obtuse for the declination of the surveying instrument's optics, the author prefers and recommends a combination of strategies. Burial detail is documented using manual methods; and the datum, grid lines (if applicable), and edges of the burial feature are mapped using the total station. Although prisms elevated on plumbed rods can be used in these situations, stabilization of the rods can become difficult. They could also potentially damage evidence by their insertion into the grave. These characteristics make them less desired than the collection of data using "reflectorless" or "prismless" measurement modes. Beyond the grave feature the total station can then be traversed along routes of suspected access and egress to record evidence deposited or transferred during the incident and in relation to significant landmarks. In situations where the site does not require a search warrant, or where additional evidence is encountered at later dates, the total station is an effective way of continually updating documentation of the incident as reflected by the resulting evidence. This became the nature of, and protocol used, in the referenced case. The two systems are later combined in the office or laboratory. This method of combining micromapping or individual scene mapping within the larger incident scene preserves context down to a single phalange and out to the political boundaries that determine the jurisdictions affected by each act.

The remains of the victim in the case history under consideration, although originally described as, and technically being, "buried," were in fact surface deposited within a natural depression and covered with loosened soil, leaves, and branches. In this case body position could be recorded most efficiently using the total station alone. If the body had been more deeply buried or if it had been flexed on top of itself, Cartesian coordinate mapping using a grid system would have been employed for the reasons given above. Instead, the victim's body was positioned face up, extended, and with the arms laid along side and parallel to the body. The remains were not skeletal or scavenged. The total station could also be used to quickly acquire elevations around and beneath the body to substantiate, along with sedimentological examinations, that there was no subterranean burial feature. This was an important detail in that the construction of a clandestine grave and its inherent tool marks could be used to support arguments of intent, planning, and physical activity not necessarily impeded by the subject's intoxication. The absence of typical digging implement tool marks (Hochrein 1997) supported the theory that the event was unplanned and ill-conceived. The positions of personal effects found between 6 and 790 m from the body were documented by total station. Because they were a continuation of the body-site survey, it allowed viewers of the crime scene diagram to navigate around the body and all associated evidence. Routes to and from the body, the victim's clothes, her credit cards, and other effects were included in the survey in order that investigators, forensic experts, jurists, and the jury could virtually drive or walk the alleged route taken by the suspect on the day of the body's disposal and when he revisited the clandestine burial. Other technical options available to the crime scene investigator involve the use of 360° photography and spherical videography. A complete panoramic view of a scene can be taken with a 360° camera. Spherical videography, from a mobile platform mounted on a vehicle and/or a pedestrian, records in any direction as it moves along a route and through a scene. The equipment creates a recording in which the taped video's viewer can turn and look in any direction. This capability avoids the limited, or tunnel, view of traditional video.

The outdoor event segments in the given case were related to the scene of the homicide through the application of total station mapping inside and outside of the victim's/subject's residence. By surveying the first and second floors as well as the exterior of the residence in three dimensions, data were available from which a virtual trial exhibit was constructed. As a result of the mapping and documentation of each scene in a manner which planned for their interconnectivity, the exhibit which was prepared allowed the incident to be considered as a whole rather than in piecemeal presentations of each scene.

RECOVERY

Once photo-documented and mapped, the collection of the evidence is initiated. In remote outdoor settings, like that of the case above, determining what items constitute evidence associated with the criminal incident can be problematic. In the instant case, the body was placed near an unimproved dirt road which lead to a gas well. That road, or trail, was frequented by hikers, hunters, partiers, and a murderer. Which items from which group, or entity, are forensically significant? This is when communication between the crime scene and case investigators is absolutely essential. The victim was last seen alive 3 weeks before her body was discovered. The postmortem condition of the body, given weather conditions, was consistent with her body being deposited for that length of time. One might assume that items, such as food wrappers, cans, and bottles, which appeared new and not weathered might be ruled out as relating to the victim's concealment; however, case investigators learned from the subject that he would visit his mother's remains daily until he left the state. This type of subject behavior should be considered during all investigations. It is not uncommon for murderers to revisit the scenes of a homicide, or locations of their bodies. In the given case, the subject admitted to returning to the victim's gravesite over several days before leaving the state.

During the recovery phase of processing the scene, investigators should also consider the use of outside experts. In the presented case the subject's admissions provided investigators a time line of events surrounding the victim's death. If that had not been the situation, a forensic entomologist's analyses of insect evidence may have been helpful in determining a postmortem interval. Tracking the movements of the suspect through the heavily wooded setting before and after the disposal of the body, or collecting evidence of his revisiting the site, might necessitate a man-tracker (Berryman and Michael 1998; Hull 1998, 2006, 2007). The tracker, in concert with a shoe- and tire-print expert (Bodziak 1993, 2008), may recognize subtle but significant impressions having forensic value. Given the amount of vegetation surrounding the victim's remains and intentionally used to conceal same, the assistance of a forensic botanist was also considered (Hall 1988, 1997; Bock and Norris 1995, 1997). Forensic geology also figures into the repertoire of outside expertise needed on this and other scenes comprising the incident. Even though the victim was only superficially buried, soils evidence could still have been important in interpreting the criminal incident. Soil samples collected from the burial site have the potential of being matched to soil evidence collected from the subject's clothing, tools, vehicle, etc. In similar fashion geological evidence from the vehicle could potentially link the vehicle to the disposal site (Lindemann 2000; Lombardi 1999; Murray and Tedrow 1992). The most critical aspect of any recovered environmental evidence is that its location be documented. Within the grave, Cartesian coordinates can be recorded, as with the victim's remains. The total station can most efficiently map samples collected throughout the scene in relation to the grave. Three-dimensional coordinates of known environmental samples could make the difference in establishing a suspect's direct association with the incident. Differences in soil horizons compared to trace soils on a suspect's shoes could potentially place him stepping into a clandestine grave or back-dirt halo. Without the three dimensions of sampled known soil the direct connection to the construction of the pit is less clear. Similarly, samples of vegetation should have their positions recorded in order to establish where pollen or plant fragments could have come into contact with a suspect or witness. For those agencies that do not employ or consult with one, an archaeologist with forensic experience is essential on scenes of buried evidence and a useful consultant in aboveground settings. The forensically trained archaeologist brings knowledge of excavation and mapping techniques as well as that of the geological contexts in which evidence may be encountered. More salient are the expert's insight into the geotaphonomic evidence which may reveal the peri- and postdepositional history of the victim's burial (Hochrein 1998).

CONCLUSIONS

The creation of a crime scene does not occur in a vacuum. By reconstructing the possible ways the crime was committed we must acknowledge that the boundaries far exceed the location of the victim. We must also consider that the event segments that comprise the incident may not be contemporaneous. Nor is any one scene necessarily more important than another. As Geberth (2010: 176), notes, "There is no such thing as a secondary scene. All crime scenes are primary." In recoveries from outdoor settings, in particular, investigators must avoid isolating the scenes of an incident and compartmentalizing the information which each holds. Manual, electronic, aerial, and global mediums of documentation are available which can interconnect the documentation of the event segments for consideration by investigators, experts, and juries. In the case presented above, the murder of the victim occurred at one point and place, but the actions of the murderer in relation to the homicide encompassed weeks and locations over hundreds of square kilometers. Spatially, tried and true archaeological protocols of excavation and mapping continue to be the best means of collecting evidence from both below- and aboveground settings. The temporal interpretation and representation of the criminal incident is what requires the adaptation of modern technology, such as GIS, to take the disparate groupings of evidence at each event segment within an incident and relate them in the context of subject, witness, and forensic expert statements. This same approach would also have applications in serial criminal events. Patterns of activities by the same individual(s) might be more recognizable using GIS strategies.

The crime scene investigator should take the time while examining a scene to ask himself, or herself, a question from the perspective of future investigators or jurors: "If I was presented with the investigative report to be compiled from our efforts in processing the crime scene, would I fully understand the scene as if I was there?" If a particular photograph, map, or notation is confusing to the on-scene investigator, it will be even more so for members of a jury or experts asked to consider the evidence.

In the given case, the interpreted representation, or reconstruction, of the evidence demonstrated that the subject murdered the victim in their home. He waited until dark, placed the body in his vehicle, and drove to public lands where he undressed the corpse and buried it in a natural depression covered with leaves, branches and other detritus. The subject then threw a bag containing the victim's clothing over a highway embankment. He then returned to the residential murder scene where he collected the victim's credit cards, cash savings, and a gun, then returned to the location of the victim's body. After visiting the clandestine grave over multiple days, the subject's plans were to commit suicide. During his final graveside visit the victim's son walked nearly a kilometer from his mother's remains where he drank but could not commit to taking his own life. Instead, he drove to various bars before leaving the state where he was involved in a traffic accident that resulted in hospitalization, and where he would later confess to the homicide. Ultimately the suspect was convicted on homicide charges.

Federal, state, and local law enforcement processed each of the event scenes much as Gardner and Bevel (2009) promote in their consideration of event analysis for crime scene reconstruction. This chapter argues that forensic archaeology is the discipline best suited for collecting and documenting evidence in order that such analyses can be effective. Archaeologists are rarely tasked with the search, collection, and analysis of artifacts in order to understand a single event at a particular moment. Rather, they view each excavation as preserving evidence of multiple, sequenced events at a particular location over time which must then be considered in the context of group, community, and global developments. The techniques and protocols that continue to be developed by forensic archaeologists consider this global perspective very appropriate and applicable to the reconstruction of criminal incidents over space and time.

REFERENCES

Benner, Jr, L. (1975). Accident investigations: multilinear events sequencing methods. *Journal of Safety Research* 7(2): 67–73.

Berryman, H.E. and Michael, P.R. (1998). Forensic applications of man tracking: adaptation of ancient visual tracking skills to modern crime scene investigation and law enforcement training. *Proceedings of the American Academy of Forensic Sciences*, San Francisco, CA (p. 12).

Bock, J.H. and Norris, D.O. (1995). Forensic botany: an under-utilized resource. *American Journal of Botany* 82: 105.

Bock, J.H. and Norris, D.O. (1997). Forensic botany: an under-utilized resource. *Journal of Forensic Sciences* 42(3): 364–367.

Bodziak, W.J. (1993). *Footwear Impression Evidence*. CRC Press, Boca Raton, FL.

Bodziak, W.J. (2008). *Tire Tread and Tire Track Evidence, Recovery and Forensic Examination*. Taylor Francis Group/ CRC Press, Boca Raton, FL.

Connor, M.A. (2007). *Forensic Methods: Excavation for the Archaeologist and Investigator*. AltaMira, Lanham, MD.

Dirkmaat, D.C. and Adovasio, J.M. (1997). The role of archaeology in the recovery and interpretation of human remains from an outdoor forensic setting. In W.D Haglund and M.H. Sorg (eds), *Forensic Taphonomy: The Post-Mortem Fate of Human Remains* (pp. 39–64). CRC Press, Boca Raton, FL.

Dupras, T.L., Schultz, J.J., Wheeler, S.M., and Williams, LJ. (2006). *Forensic Recovery of Human Remains: Archaeological Approaches.* Taylor Francis Group/CRC Press, Boca Raton, FL.

Gardner, R.M. (2009). A comparison of event analysis and multilinear events sequencing techniques for reconstructing unique phenomena. *Journal of the Association for Crime Scene Reconstruction* 16(1): 1–9.

Gardner, R.M. and Bevel, T. (2009). *Practical Crime Scene Analysis and Reconstruction.* Taylor Francis Group/CRC Press, Boca Raton, FL.

Geberth, V.J. (2010). *Sex-Related Homicide and Death Investigation: Practical and Clinical Perspectives*, 2nd edn. CRC Press/Taylor Francis Group, Boca Raton, FL.

Hall, D.W. (1988). Contribution of the forensic botanist in crime scene investigations. *The Prosecutor* Summer: 35–38.

Hall, D.W. (1997). Forensic botany. In W.D. Haglund and M.H. Sorg (eds), *Forensic Taphonomy: The Post-Mortem Fate of Human Remains* (pp. 353–366). CRC Press, Boca Raton, FL.

Hochrein, M.J. (1997a). The dirty dozen: the recognition and collection of tool marks in the forensic geotaphonomic record. *Journal of Forensic Identification* 47(2): 171–198.

Hochrein, M.J. (1997b). Buried crime scene evidence: the application of geotaphonomy in forensic archaeology. In P. Stimson and C. Mertz (eds), *Forensic Dentistry* (pp. 83–99). CRC Press, Boca Raton, FL.

Hochrein, M.J. (1998). An autopsy of the grave: the preservation of forensic geotaphonomic evidence. *Proceedings of the 50th Annual Meeting of the American Academy of Forensic Sciences*, San Francisco, CA.

Hochrein, M.J. (2002). Polar coordinate mapping and forensic archaeology within confined spaces. *Journal of Forensic Identification* 52(6): 733–749.

Hochrein, M.J. (2006). Crime scene mapping: train the trainer. Unpublished manuscript.

Hull, M. (1998). Tracking as an enforcement and investigation tool. Unpublished manuscript.

Hull, M. (2006). *V.I.T.A.L.E.: Visually Imperative Tracking Applications for Law Enforcement*, 3rd edn. Faber, Virginia.

Hull, M. (2007). Combining tracking with technology: use of helicopters in tracking operations. www.swatdigest.com/archives/07SEPT/HelosTrack.pdf.

Lindemann, J.W. (2000). Forensic geology. *The Professional Geologist* 37: 4–7.

Lombardi, G. (1999). The contribution of forensic geology and other trace evidence analysis to the investigation of the killing of Italian Prime Minister Aldo Moro. *Journal of Forensic Sciences* 44(3): 634–642.

Murray, R.C. and Tedrow, J.C.F. (1992). *Forensic Geology.* Prentice Hall, Englewood Cliffs, NJ.

Rynearson, J.M. and Chisum, W.J. (1993). *Evidence and Crime Scene Reconstruction*, 3rd edn. National Crime Investigation and Training, Redding, CA.

Steadman, D.W., Basler, W., Hochrein, M.J., Klein, D., and Goodin, J.C. (2009). Domestic homicide investigations: United States. In S. Blau and D. Ubelaker (eds), *Handbook of Forensic Anthropology and Archaeology* (pp. 351–362). Left Coast Press, Walnut Creek, CA.

CHAPTER 6

The Role of Forensic Anthropology in the Recovery and Interpretation of the Fatal-Fire Victim

Dennis C. Dirkmaat, Gregory O. Olson, Alexandra R. Klales, and Sara Getz

Fire scenes that include human victims in the rubble, hereafter termed the *fatal-fire scene*, provide some of the most difficult investigative challenges for fire responders, investigators, forensic experts, and law-enforcement agents. Fatal-fire scenes are often much more complex than other outdoor crime scenes not only because the body and individual skeletal elements are significantly modified by fire, but also because the entire surrounding contextual environment is likewise dramatically modified (Figure 6.1). Thermal modifications result in a homogeneous coloration of the human remains and surrounding matrix, making it difficult to distinguish the body and burned skeletal elements from the surrounding substrate debris. These circumstances increase the chance of missing some of the burned and fragmented remains during the typical forensic processing of the scene. If not detected, skeletal remains may be trampled upon during the recovery process or left behind at the scene. Added to these considerations is the fact that fire suppression and extinguishing efforts by firefighters often involve spraying the scene with water under high pressure, which typically results in dispersed and further fragmented biological remains. All of these issues adversely influence subsequent investigative efforts.

Given the real and perceived effects of fire on human remains, it is not surprising that fire is a common method for attempting to conceal evidence of criminal activity

A Companion to Forensic Anthropology, First Edition. Edited by Dennis C. Dirkmaat.
© 2012 John Wiley & Sons Ltd. Published 2015 by John Wiley & Sons Ltd.

Figure 6.1   General view of a fire scene. Note that coloration of fire-altered debris is homogeneous.

inflicted on human victims (DeHaan 2008). Common misconceptions exist that fire will destroy (i) the body, (ii) features such as facial features and fingerprints that permit victim identification, and (iii) evidence related to the circumstances surrounding the death. Although these perceptions are not entirely true, modification of soft and hard tissues by fire, often coupled with poor recovery and transportation methods, does significantly impede analyses by other forensic specialists, including forensic pathologists. As with the processing of outdoor scenes in general (see Chapter 2 in this volume), it is necessary to reevaluate how these scenes can be processed better from the standpoint of evidence collection and the subsequent analysis and interpretation of burned and heat-altered biological tissues. Also, as presented in Chapter 2, it will be argued that forensic anthropology, employing forensic archaeological methods, provides dramatic improvements to current practices in both of these areas.

   This chapter considers recent advances made by forensic anthropology with respect to the recovery and analysis of fatal-fire victims, including documentation of contextual data, laboratory analyses of these remains, and the analysis of skeletal trauma. Careful consideration is given to the complexity of fatal-fire scenes at a number of different levels and how these factors require modified recovery protocols and analytical methods as compared to typical outdoor crime scenes.

EFFECT OF FIRE ON THE SCENE AND VICTIM

Modification of the forensic scene

Most house fires burn at temperatures of over 900 °C, a temperature influenced by the material of the structure, items contained within the structure, and the duration of the fire (DeHaan 2002, 2008; Icove and DeHaan 2009; Newman 2004). Although fire alters nearly every element within the scene, it does so in a relatively predictable manner. Highly flammable household items such as wood, cloth, and paper ignite quickly and are often reduced to ash. Plastic, rubber, glass, and even some metals will melt into liquid molten form and eventually congeal into rounded, though irregular, masses and beaded shapes. With respect to the structure itself, fire alters the structural integrity of the building and will often lead to collapsed floors (Hobson 1992; Stauffer et al. 2008), which results in material forming thick layers of stratified debris on the lowest solid floor. Gravity and Steno's laws still hold true to form and most of the material will fall into relatively horizontal layers such that the levels of the structure will be duplicated in cross-section from top to bottom (roof to basement). The recognition of these effects of compaction and stratigraphic layering effect common to structure fires can be beneficial to the forensic investigation and the interpretation of the scene. Objects can often be stratigraphically identified relative to the floor of the structure from which it originated and may even remain associated with other objects (bodies, evidence, weapons, etc.) from that floor or room.

Modification of the fatal-fire victim

In addition to the structure at a fatal-fire scene, human victims found within the structure are also modified significantly. Contrary to conventional wisdom, fire does not completely consume the soft tissue of the body unless the remains are exposed to temperatures in excess of 2500 °C for over 12–18 h, as found in the crematory retort (Symes et al. 2008). During the burning process, biological tissues exhibit a combination of heat-induced bending of limbs at joints, destruction and removal of soft tissue that exposes hard tissue, and changes in the chemical and structural properties of bone such that it becomes broken, warped, discolored, and fragile. A detailed discussion of these alterations can be found in a number of pathology, anthropology, and law enforcement publications (e.g., Bass 1984; Bohnert et al. 1998; Hegler 1984; Redsicker and O'Connor 1996; Richards 1977; Symes et al. 2008). A summary of the heat-induced changes commonly observed in the human body at the fire scene are presented below.

Stage 1: initial soft-tissue modification

Heat-induced water loss in muscle tissue results in muscle contraction (Crow and Glassman 1996; Eckert et al. 1988; Mayne-Correia 1997). The larger, more powerful muscle masses in the limbs will win the tug of war between flexor and extensor muscle groups and the limb will contract or pull towards the direction of the largest muscle; usually the flexor muscles (Bass 1984; Symes et al. 2008). Hence, the limbs and components of the limbs will bend in a predictable pattern that forms the body into what has been termed the "pugilistic" pose, because the upper body posture resembles a defensive boxing posture. If the muscles cannot react naturally to heat and are somehow obstructed (e.g., lying facedown, or against a wall), or if the bone/muscle relationship has been compromised (e.g., previously fractured bone), the full-blown pugilistic pose may not be attained (for details, see Symes et al. 2008). At a fatal-fire scene the observed deviation from this *normal* pattern provides cause for suspicion.

Stage 2: extensive soft-tissue modification and bone exposure
Later during the heating process, fatty tissues may ignite and the soft tissue will burn (DeHaan 2008; DeHaan and Nurbakhsh 2001). The rate of soft tissue loss may be related to a number of factors, including fat content. Subsequent exposure of bone is especially dependent upon muscle tissue thickness overlying the bone. Bone surfaces covered with a thicker layer of muscle tissue will be protected longer from the effects of heat than bone that is covered by thinner layers of muscle tissue. For example, the area around the knee is basically bone, tendons, and skin and will be modified by heat and fire much earlier than most of the shaft of the femur, which is protected by thick musculature. Therefore, to some degree, heat-inducted bone exposure patterns are somewhat predictable on the basis of anatomy alone (Symes et al. 2008).

Stage 3: bone modification
As muscle tissue is lost, bone is exposed to heat and fire directly. Significant changes in the structure, composition, size, shape, and color of bone ensue (Devlin and Herrmann 2008; Thompson 2004). These changes are variable and are often related to the intensity and duration of heat and direct flames.

BONE STRUCTURE AND COMPOSITION MODIFICATION

The response of bone to heat is influenced by its structural make-up. Bone is a two-part composite substance consisting of both organic and inorganic components. The organic component is composed primarily of type I collagen which gives bone its flexibility and elasticity. The inorganic component is largely hydroxyapatite, which gives bone its hardness, rigidity, and strength (Shipman et al. 1985). Living bone tissue also includes water, blood vessels, nerve tissue, fats, and other tissue. All of these characteristics of skeletal material, as well as the thickness of the cortical bone, have an affect on patterns of thermal alteration.

The reaction of bone to heat follows a sequential progression as outlined by Mayne-Correia (1997). These stages are defined by observed structural and compositional changes to the bone. Dehydration (loss of water) is the initial stage of modification, which results in both shrinkage, or reduction, in overall size and dimensions of the bone, and changes in shape, described as "warpage." Size and dimension changes are likely related to loss of the organic component of bone (described below), as well as modifications to fat content, marrow, and blood (DeHaan and Nurbakhsh 2001). Some of the literature suggests that bone can be reduced in overall size by as much as 20%, but it is likely much less (Devlin et al. 2006) and depends on the temperature, the type of bone, and the changes occurring in the crystal matrix of the bone itself (Thompson 2005). Shrinkage in turn can also lead to changes in the overall shape and in some cases can significantly warp the bone to the point at which it is difficult to differentiate between human and animal remains. Continued exposure to heat and fire results in loss of organic component and even an alteration of the mineral components of bone, which leaves the bone brittle and friable (see Mayne-Correia 1997 for details; Herrmann 1977; Thompson 2005).

Bone color alteration

In addition to structural and composition changes resulting from exposure to heat and fire, bone will exhibit color changes. Color changes have been suggested to roughly correlate with fire exposure, temperature gradients, and duration (Devlin et al. 2006; McCutcheon 1992; Walker et al. 2008), although no consensus exists. Shipman et al. (1984) found that color gradation begins with yellows, then moves to reds and purples, followed by the more recognizable "neutral hues" of burnt bone, including black, grays, and white. Other authors have noted browns and, less commonly, pinks, purples, and green (Dunlap 1978). Interestingly, Walker et al. (2008) suggest that color provides a good indicator of collagen (and, likely, DNA) content. Extended exposure to heat and fire typically results in bone that is charred or calcined. Charred refers to osseous and periosteal portions of bone that become carbonized and blackened during thermal alteration (Herrmann 1977). Charred bone usually does not exhibit shrinkage or warpage and often will retain diagnostic features, even though the organic component of the bone is modified extensively, if not destroyed completely. Bone that has been further exposed to heat and flame will eventually become colored white and is considered calcined. At this stage all of its organic matrix and moisture has been destroyed (Mayne-Correia 1997).

In summary, when dealing with fatal-fire scenes, law enforcement, coroners/medical examiners, and other investigative personnel tasked with the processing of the scene are faced with a daunting task. The initial problem is that the scene is altered dramatically. Material has been burned away, structures have collapsed and compressed, and materials within the structure have turned into a homogeneous mess. Another problem is that the associated human victims also are altered significantly (see Baker-Bontrager and Nawrocki 2008; Ubelaker 2009). The biological tissue of humans changes dramatically when exposed to heat and fire. Soft tissue is generally differentially burned away, which exposes bone that is then reduced in size, warped, and discolored in a variety of colors. These heat-induced bone modifications make locating human victim remains within a scene difficult. When found, these remains are brittle and easily damaged during recovery and transport from the scene. It is no wonder it often seems easier to quickly remove the body from the scene and take it to the morgue for the postmortem examination.

COMMON PRACTICES IN THE INVESTIGATION OF FIRES

The investigation of fires is usually performed by specially trained fire investigators, who typically operate as a specialty division within law-enforcement agencies. In addition, other individuals from fire departments, forensic specialists, and insurance companies may become involved (Hine 2004; Hobson 1992; Roblee and McKechnie 1981; Stauffer et al. 2008). These individuals are tasked with reconstructing events surrounding the fire including the origin, progress, and temperature ranges of the fire. In the past, individual agencies have been responsible for implementing their own investigative protocols for fire recoveries; however, within the last two decades the need for standardized methods has been recognized.

In the 1990s, the Technical Committee in Fire Investigations section of the National Fire Protection Association (NFPA) was established as a way to standardize and improve the fire investigation process while enhancing the quality of event information derived from the process (NFPA 2008). The committee addressed and strongly supported the application of the scientific method to fire investigation. In this regard, the NFPA 921 Guide for Fire and Explosion Investigations code was specifically established in 1992 to provide "recommendations for the safe and systematic investigation or analysis of fire and explosion incidents" (NFPA 2008: section 1.2.1). It provides guidelines for best practices that focus on the collection and interpretation of empirical data through a series of analytical steps that result in professional recovery and investigative techniques. Protocols developed since the inception of NFPA 921 have drawn on this document as the standard for scene processing (NFPA 2008). Additionally, in 2000 the National Institute of Justice published the *Fire and Arson Scene Evidence: A Guide for Public Safety Personnel* to familiarize those involved in fire investigations with the basic protocols for physical evidence collection and preservation, with the exception of human remains (National Institute of Justice 2000). With effective guidelines established and a strong emphasis placed on the systematic scientific approach to the processing of fire scenes, fire investigations have improved dramatically within the last 20 years (DeHaan 2002).

COMMON FATAL-FIRE VICTIM RECOVERY PROTOCOLS EMPLOYED TODAY

Documenting and recovering the human remains from a fatal-fire scene typically falls under the jurisdiction of law enforcement, and not that of fire investigators. Despite the importance and difficulty of the task, well-constructed procedures for the location, recovery, and processing of human remains from fatal-fire scenes are virtually absent from the fire investigation and law-enforcement literature and protocols. The police are, therefore, ill-prepared to deal with forensic scenes that involve heat and fire-altered environs and victims. On the other hand, fire investigators are trained and experienced with complex fire scenes, but have not been trained to document and investigate human victims of forensic crime scenes. As a result of this lack of recovery protocols, fatal-fire scenes are often processed haphazardly, without standard operating procedures (or SOPs), and, usually, by an individual with no assistance and little necessary training. Once human remains have been located, victim recovery protocols may then consist of merely contacting medical or mortuary professionals to remove the body from the scene with little or no documentation of body position or context. This lack of rigorous documentation fails to capture the contextual evidence of the scene, which provides the key to reconstructing and interpreting past events, especially as related to original body location, position, and orientation, postmortem interval, and timing of trauma. Even more disconcerting is that in spite of presenting a more complex scenario, the presence of a human victim at a fire scene often results in modification, or even disregard, of typical fire investigation protocols (Lentini 2006; Olson 2006). It is not unusual for fire investigators, and their efforts to determine cause and manner of the fire, to be relegated to a secondary position when it is suspected that a human victim is located within the fire scene. Priority is instead given to the rapid location and removal of the victim from the scene in order to return the

Figure 6.2   General view of a fire scene showing overhauling efforts with heavy machinery.

remains to the family as quickly as possible. This often is done at the expense of rigorous fire investigation protocols, which are postponed or even prevented, and the final interpretation of the fire scene, therefore, is compromised (Olson 2006).

If the suspected victim is not found early or easily in the search process, a common practice by fire fighters is to rake through the debris of the fatal-fire crime scene until human remains are found. This can lead to the displacement, dispersal, and damage to fragile bone tissue, which translates into the destruction of evidence in order to locate the human remains as quickly as possible. Another common "recovery" method involves the use of heavy equipment to remove fire-altered debris from within the structure. In this method, debris is picked up by the grapple pincers of the demolition equipment or the bucket of the backhoe and placed outside of the structure in a process known as "overhauling" (Figure 6.2). Generally, these techniques are employed to reduce the risk of flare-ups and secondary fires (Roblee and McKechnie 1981); however, in this case, the use of heavy equipment is to find and remove the victim from the fire scene as quickly as possible. Confounding these problems are subsequent transport methods that include placing the body in flexible body bags, which make the recently recovered remains highly susceptible to further fragmentation and postmortem damage even before they get to the examination table of the forensic pathologist.

PROBLEMS WITH THESE APPROACHES

Three major problems are incurred as a result of these common fatal-fire victim recovery practices: (i) the body is significantly disturbed from its original location, position, and orientation (i.e., it is no longer *in situ*) and now is out of context,

(ii) further damage, aside from the fire, is inflicted on the body, and (iii) biological remains, primarily skeletal material, are almost always left behind. Although these problems may be recognized by fire investigators and law enforcement (Olson 2009), the implied logic for using these methods is that the destructive forces associated with fire have altered the scene and body to such a degree that significant contextual data are irretrievable and efforts to document the scene would be a waste of time and resources.

The rapid removal of the body from the scene leaves the forensic pathologist with an incomplete body at autopsy that is significantly modified by the fire, recovery process, and transport to the morgue. This results in a much-reduced ability both to analyze perimortem trauma by forensic specialists and to determine the cause and manner of death by the coroner or medical examiner. In turn, law enforcement is left with a body out of context, significantly negatively influencing their ability to associate evidence, and, ultimately, to reconstruct past events.

Investigators of fatal-fire scenes must reconstruct the circumstances surrounding the death of the victim by utilizing evidence, such as the context of the body at the scene, in conjunction with the determined cause of death. In nonburned cases, this evidence is often primarily derived from the body location and positioning of the body at the scene and the identification of perimortem trauma. Both of these assessments are more difficult if the scene and the victim are subjected to fire.

Unlike most other forensic cases where the investigator is dealing with recent remains that are in relatively good condition (complete skeletons or at least complete bones), as described above, significant modification of skeletal material occurs at fatal-fire scenes which influences dramatically the manner of the investigation. This skeletal modification can make even the identification of specific bones difficult. Burning may also obscure individual skeletal features and the fire-induced fracturing of bone may complicate trauma analysis. If an explosion is involved, there will be additional fragmentation and horizontal dispersement or displacement of bones. In addition to possibly missing elements during recovery, if the position and orientation of the remains are not carefully documented at the time of recovery and *in situ*, especially graphically via plan-view maps, this information will be lost. Due to all materials at a fire-altered scene seemingly having essentially the same coloration, the location of individual skeletal elements or evidence cannot be reconstructed from photographs (Figure 6.3).

Forensic Archaeology at the Fire Scene

Just as the application of forensic archaeological practices have revolutionized the forensic processing of all outdoor scenes and the comprehensive documentation of context (Dirkmaat and Adovasio 1997; Dirkmaat et al. 2008; see also Chapter 2), so too they will benefit the processing of fatal-fire scenes. It is argued here that these scenes, in particular, require forensic archaeological documentation and excavation methodologies, principles, and practices, due to the significant fire modification of the scene and the biological tissue and the difficulty of identifying such modified remains at the scene. The need to maximize the recovery of the often fractured and fragmented skeletal elements is especially important in the proper interpretation of

Figure 6.3   Human remains discovered at fatal-fire scene. Note that it is difficult to distinguish human tissue and bone from other fire altered debris (insulation, wood), and to determine the exact position and orientation of the body from a photograph.

trauma, which is often a contentious and very important issue in these situations. The proper documentation of the position and orientation of the body at the scene and the proper interpretation of evidence of skeletal trauma are critical components in the final determination of cause and manner of death.

Determination of forensic significance at the scene

In the first phase of the investigation of a fatal-fire scene, the forensic anthropologist can be of most use in the initial search for remains. Due to alteration by heat and fire and smoke, non-osseous materials (e.g., wood, glass, leather, plastics, insulation, etc.) may possess characteristics that might be mistaken as bone by nonexperts. More importantly, critical skeletal elements that may indicate the number of individuals present at the scene, potential trauma to the victim(s), or individualizing characteristics that may help in victim identification may easily be damaged, destroyed, or missed in a search by personnel unfamiliar with the effects of fire on these materials. Once biological tissue is found it is also often necessary to sort out the remains of household pets and vermin from human remains. This is easily accomplished by a trained forensic anthropologist, even when the remains have been highly fragmented and severely altered.

Forensic archaeological recovery

In addition to possessing skills in locating and identifying human remains, the forensic anthropologist plays a large role in the documentation of the search and recovery process, as well as the interpretation of contextual relationships. Mayne-Correia and Beattie (2002) provide a critical review of fatal-fire recovery techniques, stressing the need for improvement. An approach based on high-resolution archaeological recovery

Franklin Center, PA Fire
Excavated 11/07/09 – 11/08/09

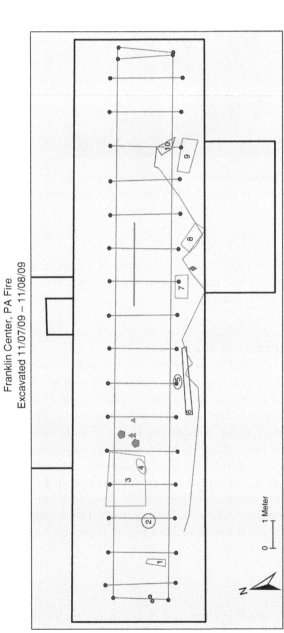

0   1 Meter

80°14'22.927"W, 41°54'12.076"N

Aerial view of house, on Crane Rd., Franklin Center, PA
Green dot = Reference point where driveway meets road

Prepared by the Department of Applied Forensic
Sciences Mercyhurst College – ERIE, PA

Figure 6.4 Final report map of a fire scene, georeferenced and utilizing GIS resources.

methodologies (Dirkmaat and Adovasio 1997; Lovis 1992; Morse et al. 1976; Sigler-Eisenberg 1985), suitably modified to fit the characteristics of fatal-fire scenes, works best (Dirkmaat 2002). From this point of view, the core of fatal-fire scene investigation should include both the classic investigation methods for determining the cause and origin of the fire, and comprehensive forensic archaeological methods for locating, recovering, and documenting the human remains and associated artifacts within the fire scene. It is clear that very little information can be gleaned through casual documentation and low-resolution recovery methods such as random scene searches, minimal photography, and rapid collection of the remains. Instead, exacting forensic archaeological recovery methods – the benefits of which are well documented in all manner of outdoor forensic scenes (Dirkmaat and Adovasio 1997 and references therein) – will significantly assist in locating all potentially significant physical evidence *in situ*, while noting the precise location and position of the body relative to the scene and to all other important physical evidence (Dirkmaat 2002).

Dirkmaat and Adovasio (1997) outlined key components of archaeological methodologies to be employed during the recovery of all outdoor forensic death scenes while emphasizing the concepts of "context" and "association," including at fire scenes (Dirkmaat 2002). Scene processing involves extensive notes, videographic and digital photographic documentation, and the generation of plan-view and profile maps (Dirkmaat 2002).

BRIDGEVILLE FATAL-FIRE RECOVERY PROTOCOLS

Recent research conducted as part of a US National Institute of Justice grant (Dirkmaat and Symes 2009) and first used at a mock fire scene in Bridgeville, PA, USA, has led to further improvements in archaeologically-based recovery protocols specific to fatal-fire scenes. Once the fire has been extinguished and fire fighters have deemed the scene safe, the recovery can commence. The primary participants in the process are the forensic anthropologists and the fire investigator. As with all outdoor forensic scenes (see Chapter 2), written and photographic documentation continues throughout the process. In addition, the overall scene (structure, roads, topography, etc.) is mapped with a total station and global positioning system (GPS), or a survey-grade GPS. Georeferencing the scene – documenting its precise location on the Earth in a defined coordinate system – permits the incorporation of geospatial data analysis, through geographical information system (GIS) programs, into subsequent analysis and preparation of the final report (Figure 6.4).

If the remains have not been located, but are suspected within the structure, the first phase (Phase 1) of the recovery will involve a large-scale scene search. Searchers line up shoulder to shoulder in tight formation (Figure 6.5a). They start at one end of the site or structure and progress in a straight line. If the scene is large, the search area can be subdivided into either search corridors (usually 6–9 m in width; Figure 6.5b), or if rooms are still present, into grids composed of individual rooms. Included in the search team are anthropologists who are familiar and experienced with the identification of burned biological tissue, and the fire investigator who is searching the scene for evidence of the fire origin and course. It is also important to have someone who can determine the forensic significance of the biological material

Figure 6.5 Bridgeville fatal-fire scene recovery protocols. (a) Large-scale search through the debris; (b) establishing a grid or corridor system; (c) rapid excavation in vicinity of remains; (d) hand screening of excavated fire-altered debris on tarps; (e) production of a hand-drawn plan-view map; (f) careful, fine excavation in immediate vicinity of victim remains.

encountered, primarily distinguishing human from animal remains. As the search progresses, searchers can move overlying debris to view material hidden by collapsed debris. It is best to move the overlying debris to a location behind the search line. It is also useful at this time to note the stratigraphic profile (roof to lowest structure levels) of the debris pile. This process can proceed fairly rapidly. Properly trained and experienced cadaver dogs and their handlers can also be incorporated into the search at this time.

When remains are found, the large-scale upright pedestrian search ends, and Phase 2 begins. A more exacting, though still relatively rapid, hands-and-knees search begins at a distance 1.5–3 m from the immediate area surrounding the remains (Figure 6.5c). The goal in this phase of the excavation is to find the edge of the immediate crime scene, i.e., body and associated evidence. In this way, evidence directly associated with the remains is not disturbed. Again, searchers line up in tight formation. Debris is rapidly excavated in this stage by hand or trowel. A modified "cake-cutting" method is used in the excavation of sediments. As employed at all archaeological sites, the trowel is held perpendicularly to the debris pile and used to cut down through the sediments, much in the same way a cake is cut. The debris pile is excavated from the top of the debris matrix straight down through all layers to the base level, starting from the outside of the burnt debris pile progressing inwards. Excavated matrix is removed from the scene by buckets and placed on a tarp outside of the fire scene where rapid sorting can be done by hand to make sure that evidence has not been missed by the excavators (Figure 6.5d). If potentially significant evidence is found during excavation or found on the tarp, the excavation progresses into a more detailed, more exacting, and slower-moving excavation mode (Phase 3; Figure 6.5f). Once significant material is encountered, debris is removed completely from atop the evidence using a "top-down" excavation method (Joukowsky 1980) to fully expose the remains and evidence. All excavated debris in this phase of the recovery is screened off-site through 6 mm mesh screens. The goal of this phase of the recovery is to fully expose the remains and associated evidence in order to document the precise position and orientation of the body, and to more fully recover all of the biological tissue. Hand-drawn plan-view maps showing the spatial distribution of evidence are constructed (Figure 6.5e) and linked to the total station data and the produced GPS maps (see Figure 6.4).

In cases in which the remains are significantly modified by fire (cremated), it is beneficial to collect all of the debris in the immediate vicinity of the body into containers for further sieving with finer geological sieves in the laboratory in order to recover all elements and especially dental remains (Dirkmaat 1991, 1999, 2002).

The final stage of the recovery (Phase 4) involves the removal of remains and preparation for transport. The hands, feet, head, and other fragile areas should be wrapped in heavy-duty plastic wrap or aluminum foil to prevent further fragmentation. The human biological tissues should be wrapped in clean white sheets and placed into body bags that contain a solid, flat, and broad underlayment, such as plywood. This procedure will limit damage to the fragile human remains during transport from the scene.

AT THE POSTMORTEM EXAMINATION

The forensic anthropologist also has an important, though often unrealized, role to play during the postmortem examination of the human remains, including the ability to separate human from animal remains and determine biological profile from fragmented and even heat-altered bones. Many of these interpretations can be done through the examination of radiographs (see Chapter 7). The forensic anthropologist also has experience with interpreting the effects of heat and fire on human bones and differentiating postmortem fractures due to fire from perimortem fractures. When cremains are involved, no one is better at drawing information from tiny fragments of bones than the forensic anthropologist (Warren 2008; Warren and Schultz 2002; see also Chapter 20).

LABORATORY ANALYSIS OF THE FATAL-FIRE VICTIM

In cases where identification of the victim cannot be determined at autopsy, or questions remain regarding trauma or circumstances of death, the forensic anthropologist can be called upon to carefully remove the soft tissue from the remains in order to analyze the hard tissue and conduct a comprehensive forensic anthropological analysis. This analysis will include an inventory of the remains, determination of a biological profile, interpretation of skeletal trauma, and interpretation of forensic taphonomy of the events regarding the victim associated with the fire scene.

Skeletal inventories

A skeletal inventory is a list and written description of each skeletal element found and the physical condition of that element. During the inventory process, individualizing characteristics of the skeleton, antemortem trauma, evidence of disease or medical intervention, and dental work are also noted. In addition to the inventory, a "homunculus" illustration is often created. A homunculus, or "little man," is a visual representation of the inventory of the elements found and their associated modifications, including trauma. This figure allows for the easy identification in the patterns of presence or absence of skeletal elements, limbs, or body segments, and in the areas of animal activity or trauma. It is particularly useful for illustrating and identifying normal versus abnormal burn patterns on the victim's body in fatal-fire cases (see below). The remains undergo photo-documentation at this time to preserve a record of the condition of the remains as they entered the forensic anthropology laboratory.

The biological profile and victim identification

In many forensic cases, if no soft-tissue markers of personal identification are available, a positive identification of a victim is sought through an antemortem/postmortem comparison of dental work by a trained forensic odontologist. At a forensic fire scene, these common methods of identification may not be available due to the modification of soft tissue, teeth, and organic material in the skeleton by extreme heat and fire. In these cases, before automatically going to DNA analysis, the forensic anthropologist can contribute significantly to the identification of a victim through the creation of a biological profile. The biological profile is a description of the individual based on skeletal characteristics and includes an estimate of the age, sex, ancestry, and stature of the indivdual, as well as a description of any potentially individualizing characteristics (e.g., healed previous trauma, prior surgeries, dental work, medical devices, etc.), trauma, or indicators of disease. The biological profile provides a starting point in the identification process because it can corroborate other presumptive identifications such as victim location (in the missing person's house or car), and clothing and wallet contents associated with the victim. However, physical and chemical alterations to the bone resulting from fire, discussed above, can destroy identifying skeletal characteristics used for the assessment of the biological profile of an individual (Maples 1986; Rhine and Curran 1990; Sauer 1998). Because forensic anthropologists are familiar with the wide range of skeletal variation, including unique features, found in the gross examination of bones and radiographs, they can compare ante- and postmortem records and radiographs (if such records exist for an individual) and possibly provide positive identification. Since human bodies in house fires usually are not completely consumed, if no unique skeletal features are found, definitive scientific proof of identity can then be obtained from DNA drawn from remaining tissues.

PATTERNS OF THERMAL DAMAGE AND TRAUMA ANALYSIS

The final interpretation of cause and manner of death and the circumstances surrounding the death of the fatal-fire victim resides in the analysis of the (i) position and orientation of the victim in the structure (discussed above), (ii) burn patterns on the victim, and (iii) patterns of skeletal trauma.

Burn patterns and burn progress

The typical reaction of the body as it is burned is to assume the pugilistic posture. As the limbs flex and the muscles shrink, specific portions of the body (e.g., the inner elbow surface) are shielded from the fire. This differential tissue shielding produces characteristic patterns of burning of the body, as described above. These "normal" burn patterns can then be distinguished from "abnormal" burn patterns, often aided by the use of the homunculus. Deviations from the expected pattern may indicate that a portion of the body was restrained or was unable to flex at the time of the fire and denote that further investigation is needed (Symes et al. 2008).

Symes et al. (2008) suggest that color gradients on a single bone can be used as an indicator of the progression of thermal damage (Figure 6.6). Earlier, Symes et al. (1996) attempted to simplify heat alteration patterns on bone by focusing on two major macroscopic changes observed on bone surfaces: one in which bone structure is modified by heat without resulting color changes, and another in which bone surface

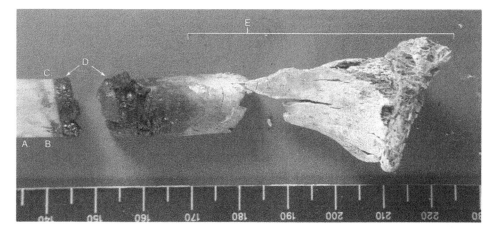

Figure 6.6 Burn patterns on a single bone (a human ulna). Note: A, unburned bone; B, heat line; C, translucent border; D, charred; E, calcined. The bone was exposed to heat/flame longest at the distal end and least on the shaft.

color changes are noted. Within the category of bone modified by heat, though exhibiting no color changes, two distinct patterns are defined (Symes et al. 1996). One is termed a "translucent border" and is described as an area of bone, translucent in appearance, in which heat has begun the process of modifying the chemical structure of the bone through dehydration and removal of collagen. It is always found in the intermediate area between unburned bone protected by muscle tissue and the area of exposed bone that has been either charred or calcined. On occasion, a thin opaque "heat line" may be found between the edge of the translucent border and the charred bone. These stages of heat alteration effects on bone can be used to interpret the direction of burning of osseous material. As the heat makes contact with the muscle tissue, the tissue will retract from the bone. The exposed bone is then altered by heat to form the translucent border configuration. After longer periods of exposure, a heat line may appear, followed by various color gradients, including charring and calcination as the bone loses moisture, collagen, and structural integrity (Symes et al. 2008).

Fracture patterns: postmortem fire damage versus perimortem trauma

The interpretation of skeletal trauma in fire victims begins with the ability to differentiate between perimortem and postmortem alterations to bones. This is not an easy task and is an area of interpretation that has been addressed recently in the literature and in primary research. Particularly critical in assessment of perimortem trauma is the ability to recognize and identify fragmentary human remains, thus permitting re-association and reconstruction of skeletal elements (Krogman and İşcan 1986; Ubelaker 1989). The physical changes to bone resulting from exposure to heat and fire (dehydration, fractures, patina, splintering, and delamination of the bone's surface) can mimic or mask perimortem skeletal trauma. Common types of fracture in burned bone are described as longitudinal, step, transverse, and curved transverse fractures (see Symes et al. 2008). Once again, certain predictable patterns do occur in that specific types of fracture are found consistently in specific areas of the body. An understanding of each of these fracture types and patterns is key to being able to differentiate normal from abnormal burning patterns, as well as to sort perimortem trauma from postmortem thermal damage resulting from subsequent recovery, transport, or analysis factors. Symes et al. (2008) provide an exhaustive description of each of the typical fracture patterns. Research has shown that although thermal alteration causes significant changes in bone, evidence of perimortem trauma can still be recognized and separated from postmortem thermal fractures and damage (Bohnert et al. 1997, 2002), although it most often requires the aid of the microscope (Herrmann and Bennett 1999; Pope and Smith 2004).

Tool-mark analysis

One of the key components of trauma analysis is determining not only the fracturing pattern resulting from traumatic impact, but also characteristics of the instrument or tool that inflicted the damage. Tool-mark analysis in blunt-force trauma scenarios has proven to be relatively successful in unburned bone with respect to documenting a variety of instruments, such as hammers and other tools, through both macroscopic

Figure 6.7 *Sus scrofa* remains in situ at fire scene, showing: (left) trauma and cut-mark evidence even after significant fire; and (right) the fragility of the evidence, and that associations of bones will be disturbed by recovery and transport efforts.

and microscopic analysis (Symes et al. 2006). In addition, evidence of saws and knives used in sharp-force trauma have been satisfactorily documented, even to the level of general tool class, power versus hand, and teeth per centimeter (Symes et al. 2006), which can aid as a reference in future trauma analysis involving saw or knife markings. Recently, research has been conducted on burned bones in order to determine whether fire alters or even destroys evidence of tool marks on bones (Devlin et al. 2006; Emanovsky et al. 2002; Herrmann and Bennett 1999; Pope and Smith 2004; Symes et al. 1996, 1999, 2005). In nearly all cases, evidence remains (Figure 6.7), with the caveat that the bone morphology must be well documented *in situ*, prior to transport to the analysis site.

FORENSIC TAPHONOMIC ANALYSIS

The final phase of the full forensic anthropological analysis is the forensic taphonomic analysis leading to scientific interpretation of what has happened to the body since death. In most outdoor scenes, this focuses on the determination of postmortem interval and the reason why elements are missing or are not in their proper anatomical position (i.e., what taphonomic agents have come into contact with the bones). In the case of the fatal fire, the primary taphonomic agents are fire and humans (if the remains are disturbed by fire suppression, recovery, and transport efforts). In most cases, the postmortem interval is often not significantly different from the time of the onset of the fire, even in cases of homicide. In other cases (e.g., extensive decomposition followed by fire), abnormal burn patterns on the tissue, and especially the bones, may provide important clues.

DISCUSSION

The fatal-fire scene is one of the most complicated forensic scenes encountered by law enforcement, the fire services, and the coroner or medical examiner's office. Not only are nearly all of the materials within the fire scene altered (burned, melted, moved) by fire, and have become homogeneous in coloration, but the scene generally has collapsed upon itself and the associated materials have compressed. Firefighter efforts to extinguish the fire often lead to further modification of the scene. Fire investigators now possess specialized skills and protocols to sort through these chaotic situations to interpret the cause, origin, and progress of the fire (Federal Emergency Management Agency 1993; Hine 2004; Hobson 1992; NFPA 2008); however, when we add a human victim to the scene and the possibility that the victim may have been intentionally placed in the burned structure to hide evidence of foul play, it is noted that neither the fire investigator nor law-enforcement officials have been trained to deal with the forensic recovery of heat-altered human remains in situations with burnt structures.

Similar to the scene itself, the body has been significantly modified. Color changes exhibited by human soft and hard tissues mirror color schemes of other items at the scene, thus making even the process of finding the remains in the fire rubble difficult. Heat alteration and exposure to fire will also remove soft tissue, thus exposing bone which then is itself altered in terms of loss of organic component or destruction and mineral alteration, leading to size changes, warpage, fragmentation, brittleness, and disassociation from the main concentration of human tissue. When the remains are moved inadvertently during search efforts, recovered improperly, or carelessly transported further fragmentation is assured.

The goal of any proper forensic recovery of human remains from a forensic setting, whether indoors, outdoors, or at a fire setting, is to collect evidence that will allow for the most accurate reconstruction of past events at the scene. This process requires that the scene be carefully documented *as it was found*. Nothing is moved, and context is comprehensively noted. If any evidence is moved, it is out of context and its evidentiary value is compromised.

Law enforcement has created, researched, practiced, and put into effect excellent protocols for the processing of indoor crime scenes; however, as discussed in other chapters in this book (Chapters 2 and 7), protocols for the processing of outdoor scenes (surface scatters and buried bodies) are generally lacking from law-enforcement perspectives. They are found instead within the field of forensic anthropology, and, specifically, within the subfield of forensic archaeology. Forensic archaeology has also developed the best protocols for the processing of fatal-fire scenes by incorporating principles, methods, and practices from contemporary archaeology.

When human victims are suspected to lie within the burned structure, but cannot be discovered after a preliminary search of the premises, it is no longer appropriate to use heavy machinery as a first line of recovery until the body turns up. These activities will always result in the alteration the body such that original position and orientation of the body are unrecoverable and context is irrevocably destroyed. Instead, there are forensic archaeological methods for locating the human remains and associated evidence within the site that are efficient (they do not significantly increase scene-processing times) and effective (bodies will be found and documented without destroying context).

The forensic archaeological search through the rubble of a structure fire is systematic, thorough, and relatively rapid until the vicinity of the body is reached or forensically relevant evidence is found. The search then becomes a more detailed excavation on hands and knees, often with hand trowels, brushes, and dustpans. Debris is removed from above the victim until all parts of the body are fully exposed and the remains can be picked straight up from the pile. In this manner, no parts of the body are trapped beneath rubble such that pulling on the remains results in disassociated parts. Full documentation of the body *in situ* is completed, including photographs, notes, and, importantly, the hand-drawn plan-view map of the body relative to the structure and associated evidence (and even to fire cause and origin-related evidence identified by the fire investigator). At fire scenes, the similar coloration of all elements of the scene greatly diminishes or even negates the value of photographs as a documentation tool. It is particularly important, therefore, that *hand-drawn plan-view maps clearly depicting features and their relative positions, to scale, be completed.*

While scene recovery and documentation of evidence continues, the larger scale of the scene will be mapped with total stations or survey-grade GPS units, thus establishing context at all levels, from the body, to the structure, to the area, and to the globe. The involvement of the forensic anthropologist ensures the maximal recovery of all heat-altered human elements, notation of the spatial distribution of remains and evidence, and establishment of context prior to removal and transport from the scene to the postmortem examination.

The forensic anthropologist provides important supplementary services for the forensic pathologist at the autopsy with respect to the analysis of heat-altered human skeletal remains. This includes the differentiation of human versus animal remains, biological profile from radiographs, preliminary analysis of fracture patterns, and separation of perimortem trauma from postmortem fire effects.

In addition to being able to identify nearly all of the fragmented, shrunk, and warped human bones associated with the fatal-fire victim resulting from exposure to fire, the bones can be reassembled in the laboratory. The forensic anthropologist is familiar with the normal patterns of burning of the human body in fires. The forensic anthropologist is also knowledgeable about the fracturing patterns of human bone and the characteristics resulting from various types of perimortem trauma (blunt, sharp, and ballistics force), as well as differentiating them from fire- or heat-induced trauma. The laboratory information, in conjunction with scene information (position and orientation of the body within the scene as represented at the scene), permits forensic taphonomic interpretations of what transpired at the scene during and subsequent to the death event, ensuring better determinations of the cause and manner of death.

Conclusions

Fire investigators know what they are doing at fire scenes. They possess the training and protocols to scientifically investigate fires for their cause and origin. Documentation and recovery protocols specific to the forensic processing of human victims within fire scenes, however, are not found in law-enforcement, fire-fighting, or fire-investigation training, research, or literature. They are found, instead, in forensic archaeology. Although human remains found at the forensic fire scene are often drastically altered by intense heat, flame, and attempts at fire suppression, the analysis and interpretation of these remains is still possible. In order for best results, it is necessary that: (i) all surviving skeletal elements be recovered from the scene, (ii) contextual and taphonomic information relating to each element and the scene be documented, (iii) remains be properly recovered, packaged, and transported with as little additional damage as possible, (iv) normal versus abnormal patterns of thermal modification (soft-tissue alteration, size, composition and color changes in bone, fracture patterns) be understood, and (v) perimortem trauma be carefully differentiated from heat-induced trauma via macroscopic and microscopic means. These are, uniquely, components of the forensic anthropological recovery and analysis of burned human remains in fatal-fire scenarios.

In summary, the forensic anthropologist trained in contemporary forensic archaeological methods is particularly well suited to work at fatal-fire scenes. He or she can conduct effective searches for possible human remains, assess the forensic significance of located items on site and in real-time, properly and comprehensively document context of the evidence and the remains, assist in the identification of the victim(s), and assess possible indicators of trauma, all while adhering to the rigorous evidentiary and legal standards required for the investigation and prosecution of a modern forensic case. Fire investigation is a complicated effort that requires a multidisciplinary approach. Forensic anthropology must be one of the key disciplines utilized early in that effort.

REFERENCES

Baker-Bontrager, A. and Nawrocki, S.P. (2008). A taphonomic analysis of human remains from the Fox Hollow Farm serial homicide site. In C.W. Schmidt and S.A. Symes (eds), *The Analysis of Burned Human Remains* (pp. 211–226). Academic Press, London.

Bass, W.M. (1984). Is it possible to consume a body completely in a fire? In T.A. Rathbun and J.E. Buikstra (eds), *Human Identification: Case Studies in Forensic Anthropology* (pp. 159–167). Charles C. Thomas, Springfield, IL.

Bohnert, M., Rost, T., Faller-Marquardt, M., Ropohl, D., and Pollak, S. (1997). Fractures of the base of the skull in charred bodies – post-mortem heat injuries or signs of mechanical traumatisation? *Forensic Science International* 87(1): 55–62.

Bohnert, M., Rost, T., and Pollak, S. (1998). The degree of destruction in human bodies in relation to the duration of the fire. *Forensic Science International* 95: 11–21.

Bohnert, M., Schmidt, U., Perdekamp, M.G., and Pollak, S. (2002). Diagnosis of a captive-bolt injury in a skull extremely destroyed by fire. *Forensic Science International* 127(3): 192–197.

Crow, D.M. and Glassman, R.M. (1996). Standardization model for describing the extent of burn injury to human remains. *Journal of Forensic Science* 41(1): 152–154.

DeHaan, J.D. (2002). *Kirk's Fire Investigation*, 5th edn. Brady Publishing, Upper Saddle River, NJ.

DeHaan, J.D. (2008). Fire and bodies. In C.W. Schmidt and S.A. Symes (eds), *The Analysis of Burned Human Remains* (pp. 1–13). Academic Press, London.

DeHaan, J.D. and Nurbakhsh, S. (2001). Sustained combustion of an animal carcass and its implications for the consumption of human bodies in fires. *Journal of Forensic Sciences* 46(5): 1076–1081.

Devlin, J.B. and Herrmann, N.P. (2008). Bone color as an interpretive tool of the depositional history of archaeological remains (pp. 109–128). In C.W. Schmidt and S.A. Symes (eds), *The Analysis of Burned Human Remains*. Academic Press, London.

Devlin, J.B., Kroman, A.M., Symes, S.A., and Herrmann, N.P. (2006). Heat intensity vs. exposure duration. Part I: macroscopic influence on burned bone [abstract]. *Proceedings of the 58th Annual Meeting of the American Academy of Forensic Sciences*, February 20–25, Seattle, pp. 310–311.

Dirkmaat, D.C. (1991). Applications of forensic anthropology: the recovery and analysis of cremated remains [abstract]. *Proceedings of the 43rd Annual Meeting of the American Academy of Forensic Sciences*, February 18–23, Anaheim, p. 154.

Dirkmaat, D.C. (1999). Recovery and identification of the fire victim [abstract]. *Proceedings of the 51st Annual Meeting of the American Academy of Forensic Sciences*, February 15–20, Orlando, pp. 205–206.

Dirkmaat, D.C. (2002). Recovery and interpretation of the fatal fire victim: the role of forensic anthropology. In W.H. Haglund and M. Sorg (eds), *Advances in Forensic Taphonomy: Method, Theory, and Archaeological Perspectives* (pp. 415–472). CRC Press, Boca Raton, FL.

Dirkmaat, D.C. and Adovasio, J.M. (1997). The role of archaeology in the recovery and interpretation of human remains. In W.H. Haglund and M. Sorg (eds), *Forensic Taphonomy: The Postmortem Fate of Human Remains* (pp. 39–64). CRC Press, Boca Raton, FL.

Dirkmaat, D.C. and Symes, S.A. (2009). New protocols for the recovery and interpretation of burned human remains. NIJ Forensic Anthropology R&D Focus Group Meeting, Arlington, VA.

Dirkmaat, D.C., Cabo, L.L., Symes, S.A., and Ousley, S. (2008). New perspectives in forensic anthropology. *Yearbook of Physical Anthropology* 51: 33–52.

Dunlop, J. (1978). Traffic light discoloration in cremated remains. *Medicine, Science, and the Law* 18(3): 163–173.

Eckert, W.G., James, S., and Katchis, S. (1988). Investigation of cremations and severely burned bodies. *American Journal For Medicine and Pathology* 2(4): 347–357.

Emanovsky, P., Hefner, J.T., and Dirkmaat, D.C. (2002). Can sharp force trauma to bone be recognized after fire modification? An experiment using *Odocoileus virginianus* (white-tailed deer) ribs [abstract]. *Proceedings of the 54th Annual Meeting of the American Academy of Forensic Sciences*, February 11–16, Atlanta, pp. 214–215.

Federal Emergency Management Agency (1993). *Basic Tools and Resources for Fire Investigators: A Handbook (FA-127)*. United States Fire Administration, Washington DC.

Hegler, R. (1984). Burned remains. In T.A. Rathbun and J.E. Buikstra (eds), *Human Identification: Case Studies in Forensic Anthropology* (pp. 148–158). Charles C. Thomas, Springfield, IL.

Herrmann, B. (1977). On histological investigations of cremated human remains. *Journal of Human Evolution* 6: 101–103.

Herrmann, N.P. and Bennett, J.L. (1999). The differentiation of traumatic and heat-related fractures in burned bone. *Journal of Forensic Science* 44(3): 461–469.

Hine, G.A. (2004). *An Introduction for Chemists found in Analysis and Interpretation of Fire Scene Evidence.* (pp. 38–74). CRC Press, Boca Raton, FL.

Hobson, C.B. (1992). *Fire Investigation: A New Concept.* Charles C. Thomas, Springfield, IL.

Icove, D.J. and DeHaan, J.D. (2009). *Forensic Fire Scene Reconstruction.* Prentice Hall, Upper Saddle River, NJ.

Joukowsky, M. (1980). *A Complete Manual of Field Archaeology.* Prentice-Hall, Englewood Cliffs, NJ.

Krogman, W. and İşcan, Y. (1986). *The Human Skeleton in Forensic Medicine.* Charles C. Thomas, Springfield, IL.

Lentini, J.J. (2006). *Scientific Protocols for Fire Investigation.* CRC Press, Boca Raton, FL.

Lovis, W. (1992). Forensic archaeology as mortuary anthropology. *Society of Science and Medicine* 34(2): 113–117.

Maples, W.R. (1986). Trauma analysis by the forensic anthropologist. In K.J. Reichs and C.C. Thomas (eds), *Forensic Osteology: Advances in Forensic Anthropology* (pp. 218–228). Charles C. Thomas, Springfield, IL.

Mayne-Correia, P.M. (1997). Fire modification of bone: a review of the literature. In W.H. Haglund and M. Sorg (eds), *Forensic Taphonomy: The Postmortem Fate of Human Remains* (pp. 275–293). CRC Press, Boca Raton, FL.

Mayne-Correia, P.M. and Beattie, O. (2002). A critical look at methods for recovering, evaluating, and interpreting cremated human remains. In W.H. Haglund and M. Sorg (eds), *Advances in Forensic Taphonomy: Method, Theory, and Archaeological Perspectives* (pp. 435–450). CRC Press, Boca Raton, FL.

Morse, D., Crusoe, D., and Smith, H.G. (1976). Forensic archaeology. *Journal of Forensic Science* 21(2): 323–332.

National Fire Protection Association (NFPA) (2008). *NFPA 921: Guide for Fire and Explosion Investigations.* National Fire Protection Association; Quincy, MA.

National Institute of Justice (2000). *Fire and Arson Scene Evidence: A Guide for Public Safety Personnel.* NCJ 181584. National Institute of Justice, Washington, DC.

Newman, R. (2004). *ASTM Approach to Fire Debris Analysis Found in Analysis and Interpretation of Fire Scene Evidence* (pp. 165–191). CRC Press, Boca Raton, FL.

Olson, G.O. (2006). Science vs. sympathy. Paper presented at the Northeast Forensic Anthropology Association annual meeting, Erie, PA.

Olson, G.O. (2009). The recovery of human remains from a fatal fire setting using archaeological methodology [abstract]. *Proceedings of the 61st Annual Meeting of the American Academy of Forensic Sciences*, February 16–21, Denver, pp. 365–366.

Pope, E. and Smith, O. (2004). Identification of traumatic injury in burned cranial bone: an experimental approach. *Journal of Forensic Science* 49(3): 431–440.

Redsicker, D.R. and O'Connor, J.J. (1996). *Practical Fire and Arson Investigation*, 2nd edn. CRC Press, Boca Raton, FL.

Rhine, J.S. and Curran, B. (1990). Multiple gunshot wounds to the head: an anthropological view. *Journal of Forensic Science* 35(5): 1236–1245.

Richards, N.F. (1977). Fire investigation – destruction of corpses. *Medicine, Science, and the Law* 17(2): 79–82.

Roblee, C.L. and McKechnie, A.J. (1981). *Investigation of Fire.* Prentice Hall: Englewood Cliffs, NJ.

Sauer, N.J. (1998). The timing of injuries and manner of death: distinguishing among antemortem, perimortem and postmortem injuries. In K.J. Reichs (ed), *Forensic Osteology: Advances in the Identification of Human Remains*, 2nd edn (pp. 321–332). Charles C. Thomas, Springfield, IL.

Shipman, P., Foster, G., and Showninger, M. (1984). Burnt bones and teeth: an experimental study of color, morphology, crystal structure, and shrinkage. *Journal of Archaeological Science* 11: 307–325.

Shipman, P., Walker, A., and Bichell, D. (1985). *The Human Skeleton.* Harvard University Press, Cambridge, MA.

Sigler-Eisenberg, B. (1985). Forensic research: expanding the concept of applied archaeology. *American Journal of Antiquity* 50: 650–655.

Stauffer, E., Dolan A., and Newman R. (2008). *Fire Debris Analysis.* Academic Press, London.

Symes, S.A., Berryman, H.E., Smith, O.C., Peters, C.E., Rockhold, L.A., and Haun, S.J. (1996). Bones: bullets, burns, bludgeons, blunderers, and why? [abstract]. *Proceedings of the 48th Annual Meeting of the American Academy of Forensic Sciences*, February 19–24, Nashville, pp. 10–11.

Symes, S.A., Smith, O.C., Berryman, H.E., and Pope, E. (1999). Patterned thermal destruction of human remains. Conference Sponsored by the Smithsonian Institution and Central Identification Laboratory, Hawaii.

Symes, S.A., Kroman, A.M., Rainwater, C.W., and Piper, A.L. (2005). Bone biomechanical considerations in perimortem vs. postmortem thermal bone fractures: fracture analyses on victims of suspicious fire scenes [abstract]. *Proceedings of the American Association of Physical Anthropologists* 40: 202–203.

Symes, S.A., Rainwater, C.W., Myster, S.M.T., and Chapman, E.N. (2006). Knife and saw toolmark analysis in bone: research designed for the examination and interpretation of criminal mutilation and dismemberment. Workshop presented to the Bureau of Criminal Apprehension, Minnesota State Crime Laboratory, St. Paul, MN.

Symes, S.A., Rainwater, C.W., Chapman, E.N., Gipson, D.R., and Piper, A.L. (2008). Patterned thermal destruction of human remains in a forensic setting. In C.W. Schmidt and S.A. Symes (eds), *The Analysis of Burned Human Remains* (pp. 15–54). Academic Press, London.

Thompson, T.J. (2004). Recent advances in the study of burned bone and their implications for forensic anthropology. *Forensic Science International* 146S: 203–205.

Thompson, T.J. (2005). Heat-induced dimensional changes in bone and their consequences for forensic anthropology. *Journal of Forensic Science* 50(5): 1008–1015.

Ubelaker, D.H. (1989). *Human Skeletal Remains: Excavation, Analysis, and Interpretation*, 2nd edn. Taraxacum Press, Washington DC.

Ubelaker, D.H. (2009). The forensic evaluation of burned skeletal remains: a synthesis. *Forensic Science International* 183(1–3): 1–5.

Walker, P.L., Miller, K.W.P., and Richman, R. (2008). Time, temperature, and oxygen availability: an experimental study of the effects of environmental conditions on the color and organic content of cremated bones. In C.W. Schmidt and S.A. Symes (eds), *The Analysis of Burned Human Remains* (pp. 129–135). Academic Press, London.

Warren, M. (2008). Detection of Commingling in Cremated Human Remains. In B.J. Adams and J.E. Byrd (eds), *Recovery, Analysis, and Identification of Commingled Human Remains* (pp. 185–197). Humana Press, Totowa, NJ.

Warren, M.W. and Schultz, J.J. (2002). Post-cremation taphonomy and artifact preservation. *Journal of Forensic Science* 47(3): 656–659.

CHAPTER **7**

Forensic Anthropology at the Mass Fatality Incident (Commercial Airliner) Crash Scene

Dennis C. Dirkmaat

One of the most complicated forensic scenes to be encountered by the forensic and law-enforcement communities involves the crash of a commercial airliner (Figure 7.1). Given the fact that many different agencies are involved in the response and subsequent investigation of a crash, planning, preparation, and training are critical to achieving success in handling the incident. Since the events of September 2001, most jurisdictions, including the federal government, have expended much time, effort, and money in preparing for a wide range of disaster scenarios that include plane crashes (World Health Organization 2005; Virginia Department of Health 2005; NIJ 2005; National Association of Medical Examiners 2010; City of London 2010). The US government's Federal Emergency Management Administration (or FEMA) has provided a set of guidelines, but not a response plan, on how to manage a disaster of a scale that lies just beyond the capacity of the local jurisdiction or community. The National Incident Management System (NIMS) was established with the expressed goal of creating a positive response to a disaster that seamlessly integrates multiple agencies, several levels of government, the private sector, and various specialties and disciplines into a "collaborative incident management" structure (Federal Emergency Management Administration 2008: 3).

Many different agencies at the state and local level have constructed Mass Fatality Incident (MFI) Response Plans specific to a community, structure, or public place such as an airport. In most pre-incident planning and training the focus has been on

A Companion to Forensic Anthropology, First Edition. Edited by Dennis C. Dirkmaat.

Figure 7.1   General view of the Colgan Air crash site, February 2009.

the initial response to the incident. The response plans include who is to be notified and in what order, how to find and remove survivors, extinguish fires, and establish security (World Health Organization 2009). After the smoke clears, however, what to do with the deceased victims is often overlooked or it is presumed that the coroner/ medical examiner's office has constructed a unique plan on their own, which may or may not be the case.

RESPONSIBILITY FOR THE INVESTIGATION

Following the crash of a commercial airliner in the USA, a variety of local and state agencies will typically respond. However, three principal agencies will hold the ultimate responsibility of the investigation of the crash and the identification of the victims: the National Transportation Safety Board (NTSB), the Federal Bureau of Investigation (FBI), and the coroner/medical examiner's (C/ME) office in whose jurisdiction the plane crashed. These three primary agencies have disparate investigative objectives, expertise, and – consequently – diverse responsibilities. The key component for the success of the operation is that all of the actors involved understand and acknowledge these different fields of expertise and associated roles and duties, not so much in terms of authority, but in terms of accountability.

The NTSB

The NTSB is an independent federal agency that responds to all major transportation accidents and incidents involving US carriers (NTSB 2002). The primary tasks of the agency involve determining the probable cause of transportation accidents, promoting safety through studies and recommendations, and providing assistance to victims and families of disaster events. They will serve as the lead agency in noncriminal cases involving primarily, but not exclusively, malfunctioning vehicle components and errors in pilot judgment. When an accident occurs that falls under NTSB guidelines for investigation, investigators and specialists will be sent to the scene to collect evidence relative to the cause of the crash. They are responsible for investigating all aspects of the crashed vehicle at the crash site. They are *not* responsible for victim recovery or victim identification, although investigators often serve an important advisory role to the C/ME and other incident responders. Typically, in these situations, the FBI will provide support to the NTSB and local authorities (FBI 2009; see www. azdhs.gov/phs/edc/edrp/MassFatConf/8a-Grosof-Sledzik-slides.pdf).

The FBI

If it is determined that criminal intent is involved in the incident, the FBI will lead in the investigation as per their mission, which is "to protect and defend the United States against terrorist and foreign intelligence threats and to enforce the criminal laws of the United States" (FBI 2009). The FBI will be in charge of documenting and collecting evidence and reconstructing events leading to the crash. This task is usually completed through the deployment and employment of agents organized into Evidence Response Teams (ERTs), specially trained in crime scene documentation (Reinecke and Hochrein 2008).

The C/ME's office

The third critical member of the response team is the C/ME's office. While the criteria for establishing whether the FBI or NTSB will provide the lead investigative role and the roles and duties of each agency are well established, the same cannot be said regarding the C/ME involvement at the MFI. This problem was addressed when the Technical Working Group for Mass Fatality Forensic Identification (TWGMFFI) sponsored by the US National Institute of Justice (NIJ 2005) created a guide for human identification in which they indicated that the C/ME was "in charge of the documentation, disposition, and certification of all remains as well as morgue operations" (NIJ 2005: 15). Thus, victim recovery and victim identification are identified as primary goals of mass disaster scene management and, importantly, fall under the jurisdiction of the C/ME. Unfortunately, however, for many C/ME jurisdictions, especially smaller coroner offices, these scene-management responsibilities and even victim identification duties at an MFI lie far beyond their capabilities in terms of personnel, equipment, training, and knowledge base. For other larger C/ME jurisdictions it is often unclear how to specifically address these responsibilities as there is no MFI plan in place or it is assumed that the federal government will "take over" both scene-management and victim identification tasks.

In everyday practice, the C/ME is in charge of victim remains at a crime scene while law enforcement is in charge of investigating the circumstances surrounding the death. In some jurisdictions, the C/ME office will send representatives to the scene solely to collect the remains for transport to the county morgue. Law enforcement will be relied upon to collect evidence associated with the remains and to provide an interpretation of events. In other jurisdictions, the C/ME will send their own investigators to document specific aspects of the scene and the body. In conjunction with the law-enforcement investigation, the C/ME will then provide an interpretation of the death event, especially with respect to the final determination of the manner of death. Given the scale of the MFI in terms of the size of scene, number of victims, and intense external focus on the event, significant modifications, or even reconfigurations to these everyday protocols, are required when managing the scene.

In a noncriminal MFI plane crash, the C/ME faces incredible pressure from a variety of sources to remove victim remains from the crash site, make identifications, and return the remains to the families of victims as quickly as possible. It is assumed by the public, politicians, families of victims, and even other investigative agencies that since the manner of death of the victims is rather clear (i.e., accidental), the goal of the scene recovery should be to remove the victims from the scene and take them to the morgue within a few days in order to begin the process of victim identification to ensure a rapid return of the remains to the families. Those in charge of scene management, therefore, have typically believed that rapidity of scene recovery is paramount. Due to this focus, very little consideration has been given to properly documenting and recovering victims or to the negative ramifications of a poor scene recovery as discussed below.

In those cases involving criminal intent, it is understood that evidence of the criminal act must also be obtained and the pressure to remove victims as quickly as possible is diminished slightly. In these cases, the cause and manner of death of some of the individuals on the plane becomes more critical and evidence of non-crash-related perimortem trauma may be sought.

MFI Management Responsibility 1: Victim Identification

Historically, attention of the C/ME has been fixed almost exclusively on morgue operations and victim identification. As part of this responsibility, many important decisions need to be made very early in the process including but not restricted to: (i) where victim identification work is to be accomplished: an existing morgue or a separate temporary mass fatality morgue site? And are the requisite supplies and equipment available?; (ii) who will make those identifications: existing personnel or outside experts?; (iii) what is an acceptable form of victim identification: DNA only, dental, fingerprints, unique clothing or jewelry, or a combination of these?; and (iv) who will pay for all of the identification work, especially if DNA is desired?

Asking for assistance

One important early decision is whether to ask for assistance in the identification process. Although victim identification is a task that is completed everyday by coroners and medical examiners, the scale of the incident relative to numbers and fragmentation of victims in a MFI may overwhelm the resources (facilities, equipment, and personnel) of a given office. The C/ME then could request help from local assets, regional assets, or the federal government. These requested assets may include equipment and supplies for the disaster morgue and/or qualified personnel to provide the victim identifications. The C/ME could turn to previously established regional networks of agencies such as the Florida Emergency Mortuary Operations Response

System (FEMORS) in Florida (Bedore et al. 2006), which could provide resources such as equipment, autopsy supplies, and even a complete temporary morgue unit, which would be readily available and likely already on pallets in preparation. A more difficult problem to solve at a local or regional level is where to find trained, professional personnel to provide reliable, scientific, and accurate victim identifications. Since the mid-1990s, this problem has been reasonably solved in the USA by the creation of the federal government's Disaster Mortuary Operational Response Team (DMORT) (Stimson and Woolridge 2007).

The 10 regionalized DMORTs comprise professional and certified experts from the forensic disciplines of anthropology, pathology, odontology, fingerprints, and DNA (National Disaster Medical System 2005; NTSB 2006; Saul et al. 2002; Sledzik and Kauffman 2007) who are federalized (temporary federal employees) when deployed for a specific incident. These forensic specialists are supported by photographers (usually law-enforcement officers), radiographic professionals, logistics experts, and mortuary personnel (Sledzik et al. 2003).

The typical procedure in a MFI is for the local C/ME to assess the situation, quickly realize that help is required, and then request assistance from the state or federal government through appropriate government channels. The request passes through the administrative hierarchy very rapidly and, if approved, the disaster victim identification (DVI) team is assembled on site within a day or two. If morgue equipment is needed, DMORT also maintains three self-contained, fully stocked portable morgue (or DPMU) units that can be shipped or driven to the site within a short time (NTSB 2006).

General MFI morgue protocols

The temporary morgue is often established in a building or structure separate from the permanent C/ME morgue to limit interference with day-to-day C/ME operations. The details of the configuration of the temporary morgue are presented elsewhere (London et al. 2003; DMORT 2006; US Army/Department of Justice 2005). Protocols for the documentation and analysis of materials specific to each station in the MFI morgue are well constructed and also are described in detail elsewhere (London et al. 2003; NIJ 2005; NTSB 2006; Saul and Saul 2003; Saul et al. 2002; Sledzik 1996; Sledzik et al. 2003). To summarize, the three most important modifications to the disaster morgue in the last 20 years have been: (i) creation of the triage station, (ii) creation of the DNA-collection station, and (iii) recognition of the role that anthropology plays in the process.

The primary goal of the disaster morgue operation is to efficiently "process" the human biological tissue derived from the crash site. Processing in this situation specifically means to thoroughly describe the remains and attempt a positive identification. The first step in the morgue process is to sort through the materials sent in from the field at the *triage station* so that only potentially identifiable human remains enter the morgue and all other debris (plane parts, disassociated personal effects, and common tissue) are stored elsewhere (Kontanis et al. 2003; Kontanis and Sledzik 2008). Each discrete piece of human tissue is given a unique morgue number and then fully documented via photography, radiography, and written descriptions at the *photography*, *radiography*, *anthropology*, and *pathology stations*.

While this documentation process commences, some identification efforts are being completed. Tissue with dermal ridges, and all dental remains are sent to the *fingerprints station* and the *dental station*, respectively. Lastly, a DNA sample is taken at the *DNA station* and the remains, once identified, are moved into storage and out of the morgue (Sledzik 2009).

Role of anthropology in the MFI morgue

Anthropologists now play an important role at both the triage and anthropology stations. At the triage station, the unique skills possessed by forensic anthropologists with respect to assessing fragments of human tissue permit determination of the potential of a specimen to yield a positive identification (Kontanis et al. 2003; Kontanis and Sledzik 2008; Mundorrf 2008).

At the anthropology station the general procedure includes creating a description of the bone element present for each discrete biological specimen (London et al. 2003; see also DMORT website, www.dmort3.org). Much of this can be completed at the anthropology table through analysis of the radiograph associated with the item including information on bone, bone portion, and side represented. If enough of the bone(s) are available, assessment of nonmetric traits (e.g., configuration of the pubic bone, the greater sciatic notch, and cranial features) might yield estimates of sex and probable ancestry. In some cases measurements might be taken from the radiographs (Frazee et al. 2009) and entered into the computer program Fordisc (Jantz and Ousley 2005) for probability statements regarding sex and stature. With respect to determining chronological age, efforts are often made to remove tissue through manual excision of tissue, boiling, or even microwave cooking, in order to examine articular surfaces, such as the auricular surface, pubic symphysis, and the sternal end of the clavicle or rib. At the very least, general age categories (e.g., juvenile, young adult, older adult) might be provided. These determinations potentially could be used to narrow down the list of passengers and, in conjunction with location of both the remains within the site and the associated personal effects, be used to associate tissue with a particular person. After remains are thoroughly documented, anthropologists are involved in attempts to provide positive identification through radiographic comparison of unique bone features, healed trauma, or prosthetic devices which may require extraction from tissue and bone.

Overview of MFI morgue operations

In summary, the disaster morgue victim identification system has proven to work very well in numerous cases in the USA and abroad (McGivney 2002; Fulginiti et al. 2006; Saul and Saul 1999; Sledzik et al. 2009; Sledzik and Rodriguez 2002). The protocols have served as a guide for mass casualty incidents on smaller scales at the local level. Interpol's DVI protocols (Interpol 2009; DeValck 2007) are similarly well constructed, generally effective (Bassendale 2010; Tyrell and Kontanis 2006; Tyrell et al. 2006), and have been used in numerous incidents, including the 2004 Thailand tsunami (Black 2009; Ciaccio 2006; Haines 2006) and Haiti earthquake disaster in 2010. It must be noted, however, that the role of the specially trained forensic anthropologist is nearly nonexistent in the Interpol model (Ciaccio and Haig 2007). In both the

USA and the Interpol disaster morgue configurations, the human biological tissue courses through the morgue in a "high-throughput" logical manner from initial triage to DNA sample collection. Every piece is documented via picture, radiograph, and written description. The disaster morgue operation associated with the identification of the victims of the United Flight 93 crash in Shanksville, PA, on September 11, 2001 (Sledzik et al. 2003) provides a telling example of the potential efficiency of the process. The materials collected in the field were brought to the morgue for processing at midday and end of the day, each day. When field operations were terminated by late morning of September 27, the last of the biological tissues was processed in the morgue that afternoon.

But Changes Are Afoot…

Within the last 10 years, a fundamental change in how human victims of crime are identified has had tremendous repercussions throughout the forensic science community. Since DNA was first used to uniquely identify human biological tissue in forensic cases in the 1980s, improvements in speed and accuracy have made DNA ubiquitous in forensic scientific identification (see Chapter 1 in this volume).

As a result of improvements in the analytic technology of DNA and the efficiency of the process to create positive identifications, the MFI morgue has experienced major reevaluations of standard operating procedures and philosophical underpinnings related to: (i) what is now defined as "identifiable" human remains, and (ii) a shift in focus from identifying victims to identifying all of the remains of the victims.

Defining "identifiable" human remains

Prior to the late 1990s, very little of the fragmented human biological tissue recovered during a MFI – especially a plane crash resulting in highly fragmented, widely scattered human remains – could be identified to a particular individual. Dental remains, skin from fingers exhibiting dermal ridges, healed broken bones, radiographic images of unique skeletal features, and prosthetic devices (which, as an assemblage, might be termed "unique biological identifiers") comprised much of the "identifiable" collection of human remains (Kontanis and Sledzik 2008). The vast majority of the recovered human remains were labeled as "unidentifiable" and termed "common tissue." In some cases, at the discretion of the C/ME, unique personal effects (e.g., jewelry, clothing, shoes) directly associated with human tissue that did not possess a unique biological identifier were used to provide presumptive identification. Opportunities for mismatches and even misidentifications were present.

The standardized collection and analysis of DNA has completely changed the definition of "identifiable." Identifications based on more complete remains and tissues exhibiting unique biological identifiers are now supplemented with identifications based on small fragments of bones and tissue, which with a high degree of probability yield unique DNA profiles. As a result of the standardized employment of DNA analyses in the MFI, the disparity in the quantity of positively identified human remains to common tissue has reversed. Common tissue is now a relatively small component of the human assemblage at the MFI.

Change in victim identification philosophy

Prior to DNA sequencing of the vast majority of tissues, the primary focus of the identification process at the disaster morgue was to positively identify at least one biological item of each individual listed on the plane manifest. This identification provided proof that the individual was on board the plane, thus satisfying families of victims and insurance companies. Focus in the morgue was directed toward high-probability human tissue candidates such as dental remains, fingerprints, and prosthetic devices, as described above. Much of this material was found in the more complete and larger remains recovered early in the field recovery process.

The recent ability of DNA laboratories to obtain a unique profile from even a minute scrap of human tissue and increasingly more rapidly led to a significant shift in the overall goal of victim identification in the mass disaster morgue. Families of victims now insist on more comprehensive efforts to not only provide a positive identification for each victim, but to identify and return all human tissue of the victim. With the reevaluation of what is now "identifiable," DNA samples are to be taken and tested from nearly all of the human tissue recovered. Morgues run until all tissue collected from the field, no matter how small, has been evaluated and identifications attempted (Kontanis and Sledzik 2008).

How does this affect the role of the anthropologist in the disaster morgue?

As described above, the primary role of the anthropologist at the MFI has always been to provide a description of the biological tissue recovered from the scene through the evaluation of the bone associated with the tissue. On rare occasions, positive identifications were attempted on prosthetic devices, radiographic comparisons of unique features, and the occasional ear lobe (Saul and Saul 1999). Much time and effort was often spent in the morgue attempting to clean tissue from bones to see joint and articular surfaces primarily for age determination. Reevaluation of those efforts has resulted in the realization that those efforts are unnecessary. Age estimates based on these skeletal biological attributes, especially in adults (the vast majority of typical passengers), produce only broad age estimates (typically, a range of 10–15 years), which is of limited value, particularly since DNA from those bones will provide a positive, rather than a presumptive, identification.

Therefore, current roles and duties of the anthropologist at the disaster morgue involve: (i) sorting material at the triage station (anthropologists still provide the best evaluators), and (ii) at the anthropology station analyzing biological tissue. The "analysis" conducted at the anthropology station, however, has evolved into documenting the remains primarily through the analysis of the radiographic image of the specimen with limited or no cleaning of that specimen. Bone, portion, and side represented are noted for each specimen. Assessment of sex is based on size and robusticity of the specimen and sexually dimorphic characters. A general categorization of age (juvenile, young adult, middle aged, and older adult) is provided and derived from epiphyseal-diaphyseal fusion patterns and condition of the joint surfaces based primarily on osteophytic activity. Antemortem trauma and unique skeletal features are also noted. Victim remains pass through the station quite efficiently.

MFI Management Responsibility 2: Victim Recovery

In addition to victim identification, the other major responsibility of the C/ME at the MFI is the management of the scene (NIJ 2005), an activity that focuses on human victim tissue location, recovery, and transport, as well as recovery of victim personal effects. As described above, the pressure to remove remains as quickly as possible from the scene and to return those remains to families of the victims is very intense. In the past, victim remains were removed rapidly from the scene by means and methods not clearly defined in the literature. However, this haste led to other critical issues such as an incomplete recovery of remains, further fragmentation, and commingling of the tissue. These issues have significant consequences relative to morgue operations, including victim identification (misidentification in some cases), duration of the victim identification efforts, attempted association of commingled remains, and the magnitude and cost of DNA analytical efforts (Kontanis and Sledzik 2008).

Too often it is assumed, or hoped, by the C/ME that one or both of the other major responding agencies (namely the NTSB or FBI) automatically will handle victim recovery activities, even though it is outside of their official mandates. In other cases, the C/ME may ignore these scene-management responsibilities or plead ignorance of these responsibilities and pass them off to other agencies such as local law enforcement, fire fighters, or the FBI.

However, the NIJ strongly suggests a different method other than hastily processing MFI scenes to fit predetermined time frames. They provide important directives, including: (i) "carefully document every piece of physical evidence recovered from the scene" (NIJ 2005: 9); (ii) "diagram/describe in writing items of evidence and their relationship to the remains" (NIJ 2005: 9); (iii) "maintain the chain of custody throughout the recovery process" (NIJ 2005: 9); and (iv) "document the collection of evidence by recording its location at the scene and time of collection" (NIJ 2005: 10). In order to manage a mass disaster scene properly and efficiently, it is necessary for the C/ME to look beyond the familiar, in which the primary focus is almost exclusively on the techniques available for victim identification, and consider how best to recover the human victims contained within the fatality scene. The problem is that MFI scenes are very complex, involving a commingled tangle of highly fragmented human remains, personal effects, vehicle parts, and damaged vegetation, all of which are widely dispersed in large outdoor scenes. This scenario is not one typically encountered in day-to-day C/ME work and represents alien territory for most C/ME offices. As discussed in Chapter 2 these scenes represent alien territory for most law-enforcement agencies as well.

The directive to thoroughly recover and document all human remains at the MFI requires a comprehensive regimen of efficient and effective scene search methods as well as body documentation and recovery protocols that include detailed notation of the position and location of human remains within the context of the scene. With a few recent exceptions (NIJ 2005; FEMORS 2004) these protocols are not found in the MFI response plans, literature, or training of coroners, medical examiners, and their investigators. Nor are there large-scale scene recovery protocols to be found in law-enforcement literature or training (see Chapter 2). Forensic archaeology, however, has constructed effective protocols for these situations.

CHANGES IN THE FIELD RECOVERY PHILOSOPHY

Historically, MFI scene recoveries involved random searches over a few days in which the largest items were flagged, usually along paths of least resistance. Nothing was removed during these searches. After a few days, the search would stop and recovery teams would collect the flagged material without noting the provenience or exact location in three dimensions. Once this process was completed, the search was renewed for the largest pieces of human tissue not initially discovered. This meant that human remains were entering the disaster morgue only intermittently, resulting in intense periods of morgue documentation and identification efforts followed by stretches of inactivity. Thus, these scene recovery efforts were: (i) random, in terms of search patterns as searchers choose their own path through the debris field, (ii) inefficient, as the search and subsequent recovery efforts were separate and distinct activities, impacting morgue activities, and (iii) without notation of the specific location of the found item, resulting in the chain of custody being broken at the scene.

The previously held argument for not carefully noting provenience was that the scene of an air crash did not represent a crime scene and all of the deceased individuals died as a result of catastrophic impact with the ground, and that the cause and manner of death was, therefore, pretty clear. It was deemed unnecessary to carefully note the location of each bit of tissue since that would only increase the time that the remains are on the scene, and seemingly would not produce any additional significant information. After a couple of weeks of scene processing and with noticeable diminishing returns on locating unrecovered remains, especially large pieces, a decision would have had to be made to cease the recovery operation. It would have been suggested that any pieces of tissue that would be found in subsequent searches could not be identified to a particular individual; that is, all human remains found after that point are, essentially, unidentifiable common tissue. The most common final solution was to consider the debris field as hallowed ground (often as a cemetery) within which lie the commingled remains of the victims. A layer of virgin soil would cover the scene, memorial statues erected, and everyone would be in agreement that the right thing had been done.

Reconsiderations of these lines of thinking in the early 2000s resulted from two new influences: (i) DNA was being used routinely to identify victim remains, as described above, and (ii) there was now an increased likelihood that criminal intent, such as terroristic activity, could lead to MFIs.

DNA, again!

DNA testing has instituted a dramatic change in MFI morgue goals, philosophy, and abilities, as described above. An even more significant shift has occurred in the goals and philosophies of the field recovery of MFI victims. With the commonplace utilization of DNA sequencing for victim identification in day-to-day forensic investigations, it was realized that DNA was of great benefit to the identification of the often highly fragmented and commingled plane crash victims. This meant that nearly all of the tissue recovered from the scene of a MFI, no matter how fragmented or small, could potentially be identified as belonging to a particular individual. Therefore, the arguments that: (i) crash site searches could be completed within a few weeks, since only unidentifiable, common tissue would be found in the latter stages of the search, and (ii) even though a lot of tissue remained on the site, to save time and effort, burying them in a common grave was a reasonable solution, were no longer acceptable. The effort to find every piece of human tissue was now required, no matter how long it took to complete the recovery. An arbitrary time frame could no longer be placed on the duration of the search at a MFI. Any recovery protocol used required that the location of biological tissue and even personal effects had to be enhanced, systematized, and comprehensive. Very little, if any, tissue could be left behind.

Criminal intent and terror attacks

The September 2001 terrorist attacks at three separate sites in the USA led to the full realization that airplanes could be used as weapons of mass destruction. This meant that the resulting crash site could no longer be considered just a debris field containing the remains and personal effects of unfortunate victims of a catastrophic accident; the debris field must be considered a crime scene (NIJ 2005). As a crime scene, it is important to establish a chain of custody of the evidence, consider its evidentiary value in court, and, as is the case at all indoor and outdoor crime scenes, thoroughly document context. Minimally, this means, that the *in situ*, undisturbed provenience of the evidence is to be noted at the time of discovery. The best way to do this is through better search strategies, and more precise and efficient large-scale mapping procedures. Protocols to accomplish these goals are found exclusively in the discipline of forensic archaeology.

Forensic archaeology and MFI scene

Forensic archaeology, as a scientific discipline, is concerned with the proper location, documentation, and recovery of physical evidence at a wide variety of outdoor scenes, ranging from surface scatters to fatal-fire scenes (see Chapter 2). The key aspect of the field is that the comprehensive contextual notation of the evidence allows for a scientifically defensible reconstruction of past events. As described elsewhere in detail, protocols for obtaining this goal have been tested and applied to outdoor forensic scenes for at least the last 25 years (Dirkmaat 1993, 1998, 2001; Dirkmaat and Adovasio 1997; Dirkmaat and Cabo 2006).

Typically, the recovery of an outdoor scene involves a series of sequential steps, first from photographic and written notation of the scene prior to recovery and then to detailed mapping of the spatial relationships of the evidence to other evidence, to the scene itself, and even to the regional and global scale. These outdoor scene protocols result in a very efficient, effective, and comprehensive recovery. Reconstructions of past events – created by determining time since death, the original orientation of the body, effects of natural taphonomic agents, and the amount of human intervention – are all arrived at in a scientific manner and do not rely on anecdotal information accepted because of the presenter's credentials (Dirkmaat 2010b).

WELDON SPRING PROTOCOLS

In the late 1990s, a modified forensic archaeology recovery protocol was constructed that was specific and applicable to a large-scale outdoor scene involving both diversity and a multitude of evidence (Dirkmaat et al. 1995; Dirkmaat and Hefner 2001; Reinecke and Hochrein 2008). The new protocol involved systematic straight-line pedestrian searches through the crash site by "search teams" whereby significant evidence (human remains, personal effects, and identifiable plane parts) were flagged and left in place. Following behind the search team was a "provenience team" that used an electronic total station to piece-plot the precise location of each flagged piece of evidence. Next, the "photography team" documented the evidence *in situ* via digital images. Each picture contained a scale and a north arrow next to the evidence. After this was completed, the remains were collected into receptacles by the "collection team" and taken off the site to the "intake station" and eventually brought to the MFI morgue. The application of these large-scale scene-processing recovery protocols results in a steady and thorough progress through the search area flagging and documenting the precise location of all significant evidence. A chain of custody is established early and the scene is cleared rapidly and systematically from the back end of the search line (concurrently during the search) and prepared properly for transport to the morgue.

These protocols were tested in an exercise conducted as part of a weeklong professional training short course sponsored by the FBI in 2000 in Weldon Spring, MO, USA. The Weldon Spring Protocols, as they came to be named, were demonstrated to be quantifiably better at locating evidence, precisely noting their spatial location in three dimensions, and completing the documentation and recovery efforts in a timely manner (Dirkmaat and Hefner 2001).

Eleven days after the completion of the Weldon Spring Protocol exercises, the FBI, the Missouri Highway Patrol, and local law enforcement utilized the methods successfully during the recovery of a fatal three-person DC-10 plane crash involving the governor of Missouri. The scene was heavily wooded and steep-sloped, requiring modifications to the protocols including dividing the scene into a series of parallel search corridors and the employment of two total stations, one for the human remains and the other for plane parts. Again, the protocols were successful in producing an efficient and effective recovery operation in the course of only 3 days (Reinecke and Hochrein 2008).

In addition to these experiences during training exercises and real cases, seven tests of the Weldon Spring Protocols have been conducted, via mock scenes involving car detonations and animal models, in attempts to increase the efficiency of the process (Figure 7.2a). The NIJ-sponsored research (Dirkmaat and Cabo 2009) provided three key additional modifications to the protocols that resulted in significant improvements in the process. The first modification was the addition of bar-coding technology (Figure 7.2c) to track evidence from the time the provenience team notes the location of the evidence to the removal of the evidence from the scene and notation at the intake station. This results in a greatly enhanced documentation of the chain of custody. The second important modification was the replacement of the electronic total station with a survey-grade global positioning system (GPS) unit within the US

(a)   (b)   (c)

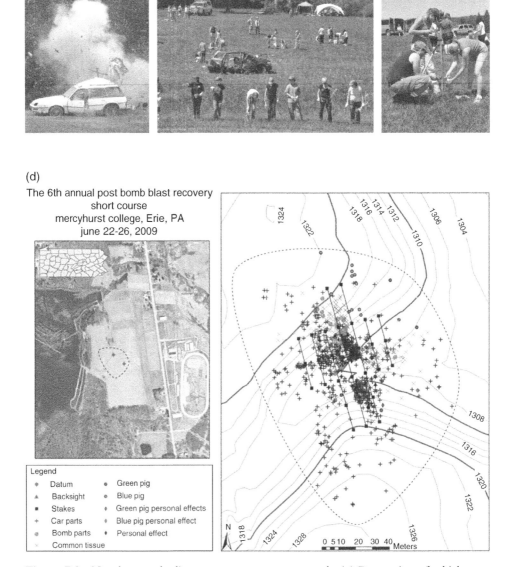

(d)

The 6th annual post bomb blast recovery
short course
mercyhurst college, Erie, PA
june 22-26, 2009

Legend

✦	Datum	●	Green pig
▲	Backsight	●	Blue pig
■	Stakes	✦	Green pig personal effects
+	Car parts	✦	Blue pig personal effect
●	Bomb parts	✦	Personal effect
×	Common tissue		

Figure 7.2   New large-scale disaster scene recovery protocols. (a) Detonation of vehicle;
(b) Weldon Spring Protocols in action, including straight-line pedestrian search team
(foreground), provenience team, photography team, collection team, and intake station
(white canopy); (c) close-up of provenience team showing evidence bar-coding operation;
(d) final geographical information system (GIS)-enhanced presentation of the mock disaster
scene recovery operation.

Global Navigation Satellite System (GNSS), resulting in an over 300% reduction in
provenience collection time (Dirkmaat and Cabo 2009). The third modification was the
addition of a second photography team and a second collection team. All scene-recovery
teams now operate in close proximity, which enhances on-site communications between

the groups. Future development of the Weldon Spring Protocols will increase on-scene, real-time spatial analysis, ensure more secure data storage, provide a better link between scene information and data collected during morgue operations, and provide protocols for geographical information systems (GIS) analysis at the site in real-time (Figure 7.2d).

With these technological and logistical improvements, the Weldon Spring Protocols provide the best scene-documentation and -recovery methods and techniques currently available for processing large-scale scenes containing a vast amount of evidence (including human remains and personal effects) that are widely scattered (Figure 7.2b). They are applicable to plane crashes, bomb incidents, and other large-scale forensic scenes (Dirkmaat and Cabo 2009).

The Continental (Colgan Air) Flight 3407 Crash Site

On February 12, 2009, at approximately 10:15 pm, Continental Flight 3407 operated by Colgan Air en route to Buffalo International Airport in Buffalo, NY, from Newark, NJ, crashed into a two-story house in the Buffalo suburb of Clarence Center (NTSB 2010). A total of 50 individuals perished including 49 on the plane and one individual in the house. The local fire department spent much of the night extinguishing the fire associated with the crash. However, one persistent fire continued within the house rubble, linked to the natural gas line coming into the house. The fire burned into the next morning.

That evening, the Erie County Medical Examiner's Office, Buffalo, NY, requested forensic anthropologists from Mercyhurst College, Erie, PA, to assist in the planning of the recovery and victim identification process. This case represents one of the first MFI recovery efforts conducted by forensic anthropologists trained in forensic archaeology.

The initial observation of the site revealed a completely destroyed house with plane rubble overlaying it, seemingly confined to a single house lot (Figure 7.3a). The first step of the recovery process focused on the search of the neighborhood (including lawns, rooftops, and gutters) surrounding the house. Nothing related to the plane crash was found in this search. Next, a straight-line pedestrian search of the immediate perimeter area (including the street, driveways, and lawns of houses across and adjacent to the site) was conducted and yielded very little human tissue. Following these searches, the area adjacent to the crash site could be used to set up tents, debris containers, and heavy equipment necessary in the next steps of the recovery process. The focus of the recovery was then directed exclusively on the debris field within the house lot. A total station was set up to record spatial location points of the human tissue and physical evidence discovered before their collection.

A closer inspection of the debris pile revealed a dense concentration of burnt, heat-altered plane wreckage, personal effects, and human remains, all encased by a thick layer of ice from the fire-extinguishing attempts in temperatures below −7 °C (Figure 7.3a). The plane descended in a near-vertical drop directly onto the house, with the underside of the plane nearly parallel to the ground (NTSB 2010). Though exhibiting heat damage, many of the victims were relatively well preserved, some with cranial structures still intact. Many were still in their seats, strapped in with seat belts.

Figure 7.3 Recovery of Continental (Colgan Air) Flight 3407 crash scene. (a) General view of crash site the morning after the incident; (b) forensic archaeological excavation in progress; (c) collecting provenience data via the total station; (d) final view of crash site (house foundation) following excavation and removal of all debris of the aircraft and house; (e) screening station on site; (f) sorting efforts in the Erie County Medical Examiner's Office, Buffalo, NY, weeks after recovery was completed.

The exception to this pattern was in the area where the gas line entering the house had burned for 10–12 h. Victim remains here consisted of charred tissue and black to calcined bone fragments.

This concentrated, stratified debris pile is quite different than what has been found at other plane crashes which have impacted the ground at great speeds and at steep angles (e.g., the USAir Flight 427 crash near Pittsburgh, PA, in 1994 and the United Flight 93 crash near Shanksville, PA, in 2001), resulting in widely dispersed evidence and debris (Dirkmaat and Miller 2003). The Weldon Spring Protocols would not be particularly applicable in the scene processing of Flight 3407. A better recovery strategy was quickly constructed which mirrored more of an archaeological site excavation of a thick layer of matrix. "Excavators" lined up in a semi-circle around the front end of the fuselage and progressed from the back of the plane, outside to inside of the debris field, until human remains were found (Figure 7.3b). As plane and house debris were taken from the debris piles or removed from around the victims, they were inspected briefly for possible significance by NTSB officials, and then placed behind the excavator. A second line of workers then took all of the excavated debris and emptied it into dumpsters located off site. With this method, debris removal proceeded very efficiently and the area directly surrounding the main concentration of the plane wreckage was kept free of rubble.

The primary goal of the excavation process was to carefully remove all of the debris overlying the victim so that the remains could be picked straight up and removed from the debris pile. No portion of the body, leg, or foot, should be trapped under debris requiring pulling and tugging to extract the limb or body part, which would result in unnecessary further fragmentation and disassociation of the tissue.

When an individual body was exposed, the next step was to note the provenience of the remains (Figure 7.3c). It was decided early in the recovery process that only one total station point of each victim, no matter how complete, described by the term "center mass," would be taken. In the future, multiple points (primarily main articular joints) will be taken in order to better show the position and orientation of the body within the debris field. This is now standard procedure in the excavation of mass graves and has been shown to result in significant improvements in pattern recognition and event reconstruction (Tuller et al. 2008).

Following collection of the provenience information, the remains were photographed. If the body was somewhat intact, the head of the victim was wrapped in heavy-duty kitchen wrap in order to keep all of the bones and tissue together. If not for this procedure, there would be an increased likelihood of fragmentation and separation in the body bag. This is a serious problem for the disaster morgue, since each discrete piece of tissue must potentially be considered a separate and distinct individual (Kontanis and Sledzik 2008). A unique morgue number must then be given to each piece and, importantly, a discrete DNA sample must be taken. In the last stages of the Colgan Air crash site recovery, the body was extracted from the debris field, placed into a body bag, and transported to the medical examiner's office.

On the third day of the recovery project, the area of the plane and house in which the fire from the gas line had burned for 10–12 h was encountered. The human remains in this area were altered significantly by heat and were comprised of tissue burned black and calcined bone. The recovery process here involved collecting all of the debris from the burned area into receptacles without attempting to separate bone from ash and other debris. Screening of this debris, if needed, and examination could be completed in the morgue under better lighting conditions.

While the excavation commenced, a total of 10 hand screens were built of wood and 6 mm wire-mesh fabric. The screens were set up over tarps and large bins to collect screened matrix. Two individuals were assigned to each screen: at least one with osteological experience dealing with fragmented and burned human bones (Figure 7.3e). All of the burned ashes, dirt, and fire debris recovered from the fuselage debris pile were carefully screened for evidence of human remains.

After five days, the site was cleared of all debris (Figure 7.3d). Considering that (i) exhaustive searches of areas around the scene and in the neighborhood were conducted to find additional biological tissue; (ii) no human remains were found more than 30 m away from the crash site; (iii) the crash site consisted of a very confined concentration of crash debris; (iv) a comprehensive forensic archaeological excavation of the site was conducted by trained, professional forensic archaeologists, and (v) all of the ash and smaller debris was screened through 6 mm mesh screens, it is clear that as close to 100% of the recoverable human remains (soft tissue and bone) was recovered in this operation.

A very important benefit was derived from utilizing forensic archaeological protocols in this case, the specifics of which are now termed the Clarence Center Protocols (Dirkmaat 2009, 2010a; Dirkmaat and Cabo 2009). Following field recovery, morgue operations, and identification efforts including the first round of DNA sample analysis, one individual listed on the plane manifest remained undiscovered or unidentified. A second search of the fragmented and commingled burned remains recovered from the area of the house and fuselage which burned for 10–12 h yielded both dental fragments and viable DNA samples (Figure 7.3f); however, it did not provide identification for the 50th victim. Two additional searches of the human tissue assemblage at Erie County Medical Examiner's Office led to the submission of bone samples that were increasingly more heat-altered. The last submission of primarily calcined bone fragments yielded very few DNA profiles. At this point, the plane manifest showing the original seat assignments for all passengers was compared to the total station map showing the location of all identified remains. The seat assignment for the missing person was right in the middle of the long-burning intense fire.

The most likely explanation for not finding any identifiable remains was that all potential DNA of that individual had been destroyed in the fire. Due to the extensive recovery protocol using forensic archaeological techniques, the medical examiner was able to be confident in providing evidence to family members that: (i) no remains were left behind since multiple search efforts were undertaken, (ii) maximum efforts were taken to provide an identification (four separate searches through the human remains assemblage), and (iii) a scientific explanation could be provided for the lack of identifiable remains (Dirkmaat et al. 2010).

Conclusions

One of the primary concerns of the families of victims, the public, media, and politicians in the recovery of victims of a large commercial airline crash is how rapidly human remains can be extracted from the wreckage, identified, and returned to the families. Disaster identification teams such as the US government's DMORT and Interpol's DVI groups, enhanced morgue protocols, and progression in DNA processing now result in efficient and relatively rapid identifications. However, advances in the recovery of victims have lagged far behind. Seemingly it still is acceptable to employ heavy equipment in cases in which the plane and victims are relatively intact, pull bodies from debris piles, disregard provenience information, and use non-archaeologically trained personnel to process these scenes. Inefficiencies and a focus on the rapidity of scene processing come at a cost. Further fragmentation and disassociation of remains due to poor recovery methods result in longer processing and identification efforts in the disaster morgue and the DNA laboratory. A fractured skull in the field, if not wrapped, can lead to a fragmented skull in the morgue requiring additional morgue numbers, increased documentation efforts by forensic experts, and additional DNA samples needed of testing. Pulling a leg from the pile may result in the now-unassociated foot being left behind, again resulting in increased time, effort, and resources for the medical examiner.

At a large-scale widely scattered scene, random searches and a lack of provenience information result in an inefficient and ineffective search, which ultimately leads to an inability to reconstruct crash dynamics scientifically. By applying forensic archaeological procedures and practices, the large-scale scene can be broken down into manageable parts, searches can cover 100% of the surface, provenience and a chain of custody is established early, and a proper reconstruction of crash dynamics is possible.

The trade-off in using archaeological methods to process a mass disaster scene is that the scene has to be processed completely in a reasonable time frame. The recovery of the 49 victims of Flight 3407 was completed within parts of 5 days and under 30 h of on-site excavation time. Every set of victim remains was exposed carefully and entirely so that the extraction process did not include pulling at entrapped body parts. A precise notation of location was obtained for all remains, which significantly aided the chain of custody and allowed for a scientific explanation for the lack of a positive identification for one passenger on the plane. As a *direct result* of the employment of forensic archaeological principles and practices, the processing of the mass fatality scene becomes both *very efficient* and *effective*.

REFERENCES

Bassendale, M. (2010). Disaster victim identification after mass fatality events: lessons learned and recommendations for disaster response planning [abstract]. *Proceedings of the American Academy of Forensic Sciences* 16: 207–208.

Bedore, L.R., Goldberger, B.A., and Lonesk, K.M. (2006). FEMORS- The state of Florida's Mass Fatality Response Team [abstract]. *Proceedings of the American Academy of Forensic Sciences* 12: 164–165.

Black, S.K. (2009). Disaster anthropology: the 2004 Asian Tsunami. In S. Blau and D.H. Ubelaker (eds), *Handbook of Forensic Anthropology and Archaeology* (pp. 397–406). Left Coast Press, Walnut Creek, CA.

Ciaccio, F.A. (2006). Thailand disaster-tsunami 2004 (an international response)

[abstract]. *Proceedings of the American Academy of Forensic Sciences* 12: 164.

Ciaccio, F.A. and Haig, N. (2007). An argument for the increased involvement of forensic anthropologists in mass fatality incidents in the United States, United Kingdom, and Europe [abstract]. *Proceedings of the American Academy of Forensic Sciences* 13: 362.

City of London (2010). *London Mass Fatality Plan: Version 3.* London Resilience, Government Office for London, London.

DeValck, E. (2007). Disaster victim identification in a global community: issues and challenges [abstract]. *Proceedings of the American Academy of Forensic Sciences* 13: 249c.

Dirkmaat, D.C. (1993). The role of archaeology in the forensic investigation:

'Start spreading the news.' [abstract]. *Proceedings of the American Academy of Forensic Sciences*, 154.

Dirkmaat, D.C. (1998). Reconsidering the scope of forensic anthropology: Data collection methodologies prior to the laboratory [abstract]. *Proceedings of the American Academy of Forensic Sciences* 4: 204.

Dirkmaat, D.C. (2001). Crime scene archaeology: better ways to 'recover physical evidence at the death scene' [abstract]. *Proceedings of the American Academy of Forensic Sciences* 7: 241–242.

Dirkmaat, D.C. (2009). Forensic archaeological documentation and recovery of the victims of the Continental Connection Flight 3407 crash in Clarence, NY. *The 43rd Annual Meeting of the National Association of Medical Examiners*; San Francisco, CA.

Dirkmaat, D.C. (2010a). Enhancing scene processing protocols to improve victim identification and field detection of human remains in mass fatality scenes. *The NIJ Conference; 2010*; Arlington, VA.

Dirkmaat, D.C. (2010b). *Forensic Anthropology: Thinking Outside-the-Box (of Bones)*. Presented at Syracuse University Dialogues in Forensic Science: Looking to the Future of Forensic Anthropology; Blue Mountain Lake, NY.

Dirkmaat, D.C. and Adovasio, J.M. (1997). The role of archaeology in the recovery and interpretation of human remains from an outdoor forensic setting. In W.D. Haglund, and M.H. Sorg (eds), *Forensic Taphonomy: The Postmortem Fate of Human Remains* (pp. 59–64),CRC Press, Boca Raton, FL.

Dirkmaat, D.C. and Cabo, L.L. (2006). The shallow grave as an option for disposing of the recently deceased: goals and consequences [abstract]. *Proceedings of the American Academy of Forensic Sciences* 12: 299.

Dirkmaat, D.C. and Cabo, L.L. (2009). *New Mass Disaster Scene Recovery Protocols*. National Institute of Justice Anthropology Grantees Focus Group, Alexandria, VA.

Dirkmaat, D.C. and Hefner, J.T. (2001). Forensic processing of the terrestrial mass fatality scene: testing new search, documentation and recovery methodologies [abstract]. *Proceedings of the American Academy of Forensic Sciences* 7: 241–242.

Dirkmaat, D.C. and Miller, W. (2003). Scene recovery efforts in Shanksville, Pennsylvania: the role of the Coroner's Office in the processing of the crash site of United Airlines Flight 93 [abstract]. *Proceedings of the American Academy of Forensic Sciences* 9: 279.

Dirkmaat, D.C., Quinn, A., and Adovasio, J.M. (1995). New methodologies for search and recovery. *Disaster Management News* December: 1–2.

Dirkmaat, D.C., Symes, S.A., and Cabo, L.L. (2010). Forensic archaeological recovery of the victims of the Continental Connection Flight 3407 in Clarence Center, NY. [abstract]. *Proceedings of the American Academy of Forensic Sciences* 16: 387.

Disaster Mortuary Operational Response Team (DMORT) (2006). *Flight 93 Morgue Protocols*. www.dmort.org/FilesforDownload/Protocol_Flight_93.pdf.

Federal Bureau of Investigation (FBI) (2009). *Quick Facts*. www.fbi.gov/about-us/quick-facts.

Federal Emergency Management Administration (2008). *National Incident Management System*, P-501. US Department of Homeland Security, Washington DC.

FEMORS (2004). *FEMOR Field Operations Guide*, 3rd edn. Florida Emergency Mortuary Operations Response System, www.femors.org.

Frazee, K., Stull, K., and Ousley, S. (2009). The value of radiographic standardization in a medico-legal setting. *The Annual Meeting of the National Association of Medical Examiners*, San Francisco, CA.

Fulginiti, L.C., Warren, M.W., Hefner, J.T., Bedore, L.R., Byrd, J.H., Stefan, V., and Dirkmaat, D.C. (2006). Anthropology responds to Hurricane Katrina [abstract]. *Proceedings of the American Academy of Forensic Sciences* 12: 305.

Haines, R.S. (2006). Interpol- its role in mass fatality incidents such as the December 26, 2004 tsunami in Southeast Asia [abstract]. *Proceedings of the American Academy of Forensic Sciences* 12: 203.

Interpol (2009). *Disaster Victim Identification Guide*. www.interpol.int/INTERPOL-expertise/Forensics/DVI.

Jantz, R.L. and Ousley, S.D. (2005). *FORDISC 3. Computerized Forensic Discriminant Functions, Version 3.0.* University of Tennessee, Knoxville, TN.

Kontanis, E.J. and Sledzik, P.S. (2008). Resolving commingling issues during the medicolegal investigation of mass fatality incidents. In B.J. Adams and J.E. Byrd (eds), *Recovery, Analysis, and Identification of Commingled Human Remains* (pp. 317–336). Humana Press, Totowa, NJ.

Kontanis, E.J., Ciaccio, F.A., Dirkmaat, D.C., Falsetti, A.B., and Jumbelic, M.J. (2003). The triage station: recent advances in mass fatality incident morgue operatons [abstract]. *Proceeding of the American Academy of Forensic Sciences* 9: 130–131.

London, M.R., Mulhern, D.M., Barbian, L.T., Sledzik, Dirkmaat, D.C., Fulginiti, L., Hefner, J.T., and Sauer, N.J. (2003). Roles of the biological anthropologist in the response to the crash of United Airlines Flight 93 [Abstract]. *Proceedings of the American Academy of Forensic Sciences* 9: 279–280.

McGivney, J. (2002). Dental identification in the Egypt Air Disaster [abstract]. *Proceedings of the American Academy of Forensic Sciences* 8: 141.

Mundorff, A.Z. (2008). Anthropologist-directed triage: three distinct mass fatality events involving fragmentation of human remains. In. B.J. Adams and J.E. Byrd (eds), *Recovery, Analysis, and Identification of Commingled Human Remains* (pp. 123–144). Humana Press, Totowa, NJ.

National Association of Medical Examiners (2010). *National Association of Medical Examiners Mass Fatality Plan 2010.* http://thename.org/index.php?option=com_docman&task=cat_view&gid=38&Itemid=26.

National Disaster Medical System (2005). *What is a Disaster Mortuary Operational Response Team (DMORT)?* US Department of Homeland Security, National Disaster Medical System Section, Washington DC. http://oep-ndms.dhhs.gov/dmort.html.

National Institute of Justice (NIJ) (2005). *Mass Fatality Incidents: A Guide for Human Forensic Identification.* NIJ Technical Working Group for Mass Fatality Forensic Identification, United States Department of Justice, Office of Justice Programs, Washington DC.

National Transportation Safety Board (NTSB) (2002). *What is the National Transportation Safety Board?* Publication reference NTSB/SPC-05/02. National Transportation Safety Board, Washington DC.

National Transportation Safety Board (NTSB) (2006). *Standard Operating Procedures for National Transportation Safety Board Activations.* Disaster Mortuary Operational Response Team. Publication reference NTSB/DMORT SOP. National Transportation Safety Board, Washington DC.

National Transportation Safety Board (NTSB) (2010). *Loss of Control on Approach.* Colgan Air, Inc. Operating as Continental Connection Flight 3407 Bombadier DHC-8-400, N200 WQ Clarence Center, New York February 12, 2009. PB 2010-910401. National Transportation Safety Board, Washington DC.

Reineke, G.W. and Hochrein, M.J. (2008). Pieces of the puzzle: F.B.I. Evidence Response Team approaches to scenes with commingled evidence. In B.J. Adams and J.E. Byrd (eds), *Recovery, Analysis, and Identification of Commingled Human Remains* (pp. 31–56). Humana Press, Totowa, NJ.

Saul, F.P. and Saul, J.M. (1999). The evolving role of the forensic anthropologist: As seen in the identification of the victims of the Comair 7232 (Michigan) and the KAL (Guam) air crashes. [abstract]. *Proceedings of the American Academy of Forensic Sciences* 5: 222.

Saul, F.P. and Saul, J.M. (2003). Planes, trains, and fireworks: the evolving role of the forensic anthropologist in mass fatality incidents. In D.W. Stedman (ed), *Hard Evidence: Case Studies in Forensic Anthropology* (pp. 266–277). Prentice Hall, Upper Saddle River, NJ.

Saul, F.P., Sledzik, P.S., Ciaccio, F.A., Shank, R.L., McGivney, J., Messner, P., McGowan, D.E., de Jong, J.L., Saul,

J.M., Warnick, A.J., Kenney, J.P., Sibert, R.W., Smith, B.C., and Jordan, R.B. (2002). The Disaster Mortuary Operational Response Team (DMORT) model for managing mass fatality incidents (MFIs). *Proceedings of the American Academy of Forensic Sciences* 8: 12–13.

Sledzik, P.S. (1996). Federal resources in mass disaster response. *Cultural Resource Management* 19(10): 19–20.

Sledzik, P.S. (2009). Forensic anthropology in disaster response. In S. Blau and D.H. Ubelaku (eds), *Handbook of Forensic Anthropology and Archaeology* (pp. 374–387). Left Coast Press, Walnut Creek, CA.

Sledzik, P.S. and Rodriguez, W.C. (2002). Damnum fatale: the taphonomic fate of human remains in mass disasters. In W.D. Haglund and M.H. Sorg (eds), *Advances in Forensic Taphonomy: Methods, Theories and Archaeological Perspectives* (pp. 321–330). CRC Press, Boca Raton, FL.

Sledzik, P.S. and Kauffman, P.J. (2007). The mass fatality incident morgue: a laboratory for disaster victim identification. In M.W. Warren, H.A. Walsh-Haney, and L.E. Freas (eds), *The Forensic Anthropology Laboratory* (pp. 97–116). CRC Press, Boca Raton, FL.

Sledzik, P.S., Miller, W., Dirkmaat, D.C., deJong, J.L., Kauffman, P.J., Boyer, D.A., and Hellman, F.N. (2003). Victim identification following the crash of United Airlines Flight 93 [abstract]. *Proceedings of the American Academy of Forensic Sciences* 9: 195–196.

Sledzik, P.S., Dirkmaat, D.C., Mann, R.W., Holland, T.D., Mundorff, A.Z., Adams, B.J., Crowder, C.M., and DePaulo, F. (2009). Disaster victim recovery and identification: forensic anthropology in the aftermath of September 11. In D.W. Steadman (ed), *Hard Evidence: Case Studies in Forensic Anthropology*, 2nd edn (pp. 289–302). Prentice Hall, Upper Saddle River, NJ.

Stimson, P.G. and Woolridge, E.D. (2007). A brief discussion of the formation of the national Disaster Mortuary Team (DMORT) [abstract]. *Proceedings of the American Academy of Forensic Sciences* 13: 248–249.

Tuller, H., Hofmeister, U., and Daly, S. (2008). Spatial analysis of mass grave mapping data to assist in the reassociation of disarticulated and commingled human remains. In B.J. Adams and J.E. Byrd (eds), *Recovery, Analysis, and Identification of Commingled Human Remains* (pp. 7–29). Humana Press, Totawa, NJ.

Tyrell, A.J. and Kontanis, E.J. (2006). Restructuring data collection strategies and investigation priorities in the resolution of mass fatality incidents [abstract]. *Proceedings of the American Academy of Forensic Sciences* 12: 165.

Tyrell, A.J., Benedix, D.C., Dunn, K.N., Emanovsky, D.D., Gleisner, M.R., and Kontaris, E.F. (2006). Death and diplomacy: multinational forensic responses to mass fatality incidents [abstract]. *Proceedings of the American Academy of Forensic Sciences* 12: 160–161.

US Army/Department of Justice (2005). *Mass Fatality Management for Incidents Involving Weapons of Mass Destruction.* US Army Research Development and Engineering Command Military Improved Response Program and Department of Justice, Washington DC.

Virginia Department of Heath. (2005). *Guidelines for Reporting and Managing Mass Fatality Events with the Virginia Medical Examiner System.* Virginia Department of Health, www.vdh.virginia.gov/medexam/documents/mass_guidelines.pdf.

World Health Organization (2005). *Management of Dead Bodies in Disaster Situations. Disaster Manuals and Guidelines Series*, no. 5. Publication reference NCJ 199758. World Health Organization, Geneva.

World Health Organization (2009). *Management of Dead Bodies after Disasters: A Field Manual for First Responders.* Pan American Health Organization, World Health Organization, International Federation of Red Cross and Red Crescent Societies. Publication reference NLM WA 840. World Health Organization, Geneva.

8 Mass Graves and Human Rights: Latest Developments, Methods, and Lessons Learned

Hugh H. Tuller

The investigation of human-rights abuses is founded in international (and in some cases national) law, and as such must involve a judicial body and the appropriate experts in which to conduct such procedures. When the focus of an investigation involves collecting evidence and/or the identification of the victims from buried contexts, the appropriate experts include forensic anthropologists and archaeologists. This chapter will explore archaeological methods and techniques regularly utilized in the excavation and recovery of human remains from mass-grave contexts and discuss ongoing developments in the field.

The field of archaeology has provided numerous examples of well-documented mass-grave excavations, from Native American ossuaries in the Americas to post-battle medieval and World War I mass graves in Europe (Fiorato et al. 2000; Saunders 2007; Ubelaker 1999). These excavations are fine examples of the application of archaeological principles, methods, and techniques to mass graves. The subsequent reports and publications associated with these archaeologically led mass-grave excavations are a credit to the field, and should serve as a foundation for the investigation of human-rights abuses. Indeed, most of the "latest developments and methods" to be discussed in this chapter are not necessarily derived from experiences and research conducted while investigating human-rights abuses; rather these methods have filtered over from standard archaeological practices and been applied to contemporary mass graves. It is unfortunate that such filtration must take place, and that the use of

A Companion to Forensic Anthropology, First Edition. Edited by Dennis C. Dirkmaat.
© 2012 John Wiley & Sons Ltd. Published 2015 by John Wiley & Sons Ltd.

archaeology at crime scenes with buried remains has not automatically taken place. This brings us to the principal development regarding mass-grave/human-rights investigations: the subtle, albeit slow, acceptance of sound archaeological practices at contemporary mass-grave sites.

ARCHAEOLOGY AND THE FORENSIC ANTHROPOLOGIST

Anyone can dig up a body. No special archaeological skills are needed. One may develop a particular method to get the job done more efficiently (e.g., gaining experience in spoil management), but this does not make one an archaeologist. Similarly, many nonbiological anthropologists have taught themselves to recognize the bones of the body and even the basic difference between perimortem and postmortem trauma, but that does not make them forensic anthropologists. Like all scientific fields, archaeology has its paradigms, theoretical bases, and sets of methods and techniques. An archaeological excavation should aim to answer questions first by posing a hypothesis (or several) and then excavating the site in a manner suitable for testing the hypothesis. Unfortunately, what is regularly asked at contemporary mass-grave sites is simply "Are there bodies buried here?" and if so, "How many bodies, and how big/deep is the grave?" While these are important questions, they do not necessarily lead to a controlled excavation and can result in the loss of information, otherwise known as evidence. Excavating a complex mass grave in a manner that maximizes the collection of evidence, recording opportunities, and assisting in the identification process requires the application of true archaeology. Unfortunately, this is a skill that many in our field lack.

Owsley (2001) pointed out that the forensic anthropologist would be shrewd to include an archaeologist on the crime scene investigation team, noting that their training and experience in locating burials, excavation planning and logistics, skills in evidence recovery and documentation, site mapping, and report-writing abilities would be vital to a recovery operation. This would seem especially true if the forensic anthropologist does not possess a solid background or experience in archaeology. Yet archaeologists have not always been included in the excavation of mass graves. Instead, it is the forensic anthropologist who regularly leads these excavations (Steadman and Haglund 2005). On one hand, this is logical considering the background and experience forensic anthropologists are expected to have in human osteology and archaeology, as well as their regular interaction with police cases. However, not all forensic anthropologists are as familiar with archaeology as they should be (Sigler-Eisenberg 1985; Skinner et al. 2003). The educational background of a forensic anthropologist is foremost based in skeletal biology, and may not necessarily include training and experience in archaeology. Indeed, many university departments simply do not adequately instruct their students in recovery techniques (Galloway and Simmons 1997). Nevertheless, graduates from these programs have been engaged in the investigation of mass graves and, consequently, archaeological methods have not been consistently applied.

Connor and Scott (2001) describe a sadly amusing story about forensic anthropology's lack of knowledge of archaeological principles that was highlighted during a 1993 Physicians for Human Rights (PHR) investigation in Croatia. PHR was investigating a series of graves, presumably marked or otherwise visible to the team. Connor and Scott, working as the excavation team's archaeologists, established a site datum, set up a total station, and began digitally surveying the site and plotting the grave features in standard archaeological practice. A stake-and-string grid system was not created over the site as the total station would be far more accurate at creating a map than any string grid system. In addition, the archaeologists recognized that the appropriate method of excavation would be to treat each grave as a separate feature instead of excavating the graves by arbitrary grid units. This line of logic is second nature to archaeologists, who are familiar with excavations of historic and prehistoric cemeteries and mass graves. It was assumed that the forensic anthropologists on the PHR team were likewise experienced and familiar with these archaeological principles. However, during the excavation, Connor and Scott learned that the forensic anthropologists were concerned that a grid system had not been established and that without one provenience data were being lost. To appease the forensic anthropologists, the archaeologists set up a baseline string from which they never actually measured a point (although it did apparently serve as a suitable trip line for several team members). Connor and Scott (2001) point out that, while archaeological techniques (the use of grid systems to control provenience and help make maps) had transferred over to forensic anthropology, the *principles* behind the techniques (when and why to use a certain mapping technique as opposed to another) had not.

In the following years, excavating graves as features and using total stations to assist in mapping became standard and accepted practice at mass-grave sites in the Balkans with field teams from the International Criminal Tribunal for the former Yugoslavia (ICTY). Unfortunately, the author has a second chapter to add to the Connor and Scott story. In 2002, a full 9 years after the introduction of the modern survey techniques to the PHR team in Croatia, and 1 year after the Connor and Scott publication, the author found himself introducing a total station to an excavation team consisting of a combination of local Bosnian Commission authorities and International Commission on Missing Persons (ICMP) personnel. The ICMP was assisting the Bosnian Commission team in the excavation of a mass grave by supplying two forensic anthropologists and, for the first time, two archaeologists. While the ICTY teams operating in the Balkans took care to record their excavations to the best of their ability, the local Bosnian teams focused mainly on body recovery. Recording the crime scene was a secondary consideration at best, and this was done in a rudimentary manner. To the author's knowledge, in the 2 years he worked with the Bosnian Commission team, they never produced a single comprehensive map of the graves they excavated. The ICMP anthropologists, who had worked side by side for years with the Bosnian Commission team, lent their talents to the recovery of remains and sorting out commingling issues within the grave, but not to mapping, recording body positions, excavation planning, or other standard archaeological practices.

After setting up the total station, the surrounding area and the outline of the mass-grave feature were surveyed. The initial surveying procedures were mostly ignored by the Bosnian Commission team, but the total station became an issue when the first set of remains was about to be lifted. The standard practice with ICTY excavations was to record 14 points on a complete body from which a digital "stickman" would be mapped (Hanson et al. 2000). The Bosnian team, unclear about the reasoning for taking all these points, wanted an explanation of the total station's capability and assurances that the surveying would not interrupt the flow of work. Although the author explained the total station and how it could dramatically assist in recording the site and bodies, and how it would not significantly affect the rhythm of work, the leaders of the Bosnian Commission team wanted further reassurance from an ICMP forensic anthropologist at the site with whom they had worked for years and whom they trusted. This particular forensic anthropologist had worked with PHR in the Balkans in the past, and indicated to the author that they were familiar with the archaeological survey equipment. However, when questioned by the Bosnian leaders, the forensic anthropologist expressed the opinion that using the total station to record the body points would take too long and would not be a useful endeavor. The author had to argue with the Bosnian Commission team leaders to be allowed to record at least a single point at the center of each body for basic mapping purposes. A few days later, the author demonstrated to the investigative judge, who was ultimately in charge of the excavation, that the 14 points needed could be recorded in 3 to 4 min, approximately the same amount of time that a single point was being recorded on the body by the Bosnian Commission team using tape measures. With this demonstration, the judge agreed to allow the total station to be fully implemented in subsequent mass-grave excavations.

In many ways, this experience reflected the situation Connor and Scott faced 9 years prior. A forensic anthropologist, through the lack of knowledge of accepted archaeological techniques, was the catalyst for preventing the full employment of the most powerful tool the team had to record evidence provenience at a mass-grave site. While not all archaeologists are familiar with the operation of total stations and the associated mapping software (they are rather expensive pieces of equipment that may not be available to an archaeologist or university department for training purposes), all archaeologists are familiar with the concept of accurate surveying and understand how such equipment would be an asset at a complex excavation. The ICMP forensic anthropologist, employed as an international expert to assist the Bosnian Commission teams in excavating and identifying remains, had no knowledge of the Connor and Scott (2001) publication, nor that Conner and Scott had highlighted forensic anthropologists' ignorance of excavation strategy and total station surveying techniques to underscore the need for our field to seek more experience in archaeology in order to participate in these complex sites. While this need among forensic anthropologists to accept and incorporate archaeological principles to crime scene contexts has long been recognized (Bass and Birkby 1978; Dirkmaat and Adovasio 1997; Morse et al. 1976; Skinner 1987; Skinner and Lazenby 1983), the implementation has lagged.

One possible reason why archaeological principles are not utilized at some mass-grave investigations may be the belief that a full archaeological excavation is not warranted in forensic cases. Hoshower (1998) makes a good case as to why, in certain cases, namely during humanitarian missions, the forensic anthropologist may not need to strictly adhere to archaeological protocol. She argues that forensic excavations often are not conducted in the "blind," but that the investigator is regularly provided with specific information on the case (e.g., location of the burial, size of the grave, and how the remains were purportedly buried). This information allows the forensic anthropologist to forgo strict procedures designed to tease out the unknown events of the past in the manner that standard archaeology does. However, Hoshower's message should not be misconstrued as an excuse to be slack on their archaeological training, but a call for flexibility in forensic excavations with an understanding of when to correctly apply archaeological methods and when they can be bypassed. The only way an anthropologist is to know when and how to use such archaeological methods is to be thoroughly familiar with them. While the forensic anthropologist may have received some training in archaeology at the graduate level, they would be wise to recognize that the field of archaeology is not stagnant, that just as the field of physical anthropology is constantly advancing, so too do the methods, techniques, and paradigms used by archaeologists. For a forensic anthropologist, following the advances in archaeology is just as important as following the advances in skeletal biology.

As suggested above, perhaps the most significant reason for the absence of archaeology at some mass-grave investigations is likely to be the manner in which forensic anthropologists are trained. Often the archaeological abilities of an individual forensic anthropologist depend upon their personal experiences and interests, and not on the program in which they study (Haglund 2001). A student interested in archaeology, or one who understands that a solid base in archaeological principles would be an asset as a forensic anthropologist, may take advantage of elective courses in archaeology, field schools, and job opportunities in archaeology during their career in graduate school, while another student may focus on other aspects of the discipline such as anatomy or biology. Sledzik et al. (2007) have called for a review of how forensic anthropology is taught, and have recommended, among numerous other things, graduate-level training in archaeology. It is increasingly apparent that an understanding of archaeological principles is a necessary component of forensic anthropology training, and that as a whole the field is lagging. Currently, the Scientific Working Group for Forensic Anthropology (SWGANTH), cosponsored by the Joint POW/MIA Accounting Command's Central Identification Laboratory (JPAC-CIL) and the Federal Bureau of Investigation (FBI), is debating the future direction of forensic anthropology in the USA, including educational requirements and qualifications. The guidelines that come out of this endeavor will likely have a dramatic affect on the field of forensic anthropology.

One also needs to keep in mind that human-rights investigations are usually international affairs involving experts from many different countries. As such, the training and experience of forensic anthropologists and archaeologists can vary widely not only between universities, but also between countries. For example, in the USA, a forensic anthropologist is someone who most likely studied anthropology as a four-field approach (physical, cultural, linguistic, and archaeological), and likely focused on the physical anthropology and archaeology aspects of their training. In contrast, many European and Scandinavian nations have an education system where the study of forensic anthropology and forensic archaeology are separate fields, taught in separate departments (Juhl and Olsen 2006). There, physical anthropologists are trained in skeletal biology but receive little training in the principles of archaeology. While the aforementioned SWGANTH recommendations will hopefully add needed motivation for educators to include rigorous training in archaeology for forensic anthropology students in the USA, one can only hope that other nations with forensic anthropology programs will likewise consider how they teach the subject when it comes to the recovery of remains.

While this discussion has so far emphasized the failures of forensic anthropology to fully utilize archaeology, the author does not wish to give a false impression of the state of the field. There is good news. The recommendations through the years for the forensic anthropologist to embrace archaeology have not fallen on deaf ears. While many forensic anthropology programs still fail to emphasize archaeological training, individual instructors experienced in archaeological principles are passing this knowledge on to their students, encouraging them to seek out additional training, and many students are taking it upon themselves to learn standard archaeology. As a result, over time archaeological thought has slowly gained acceptance at mass-grave excavations. In most cases, basic archaeological recording principles at least are expected to be conducted at a grave. In better cases, with the flexibility Hoshower (1998) emphasized, archaeology is taking a more prominent role in mass-grave excavations.

It took years for forensic anthropology as a field to convince police departments to include anthropologists on their scene-of-crime teams, or, even better, to allow them to direct the recovery of surface scatters and buried remains. The application of archaeology to mass-grave investigations is likewise taking time, but the effort is paying off. Those who apply archaeology in mass-grave contexts have been demonstrating its usefulness in interesting ways. The remainder of this chapter illustrates some examples of how archaeology has contributed to the investigation of human-rights abuses beyond merely mapping features and body positions.

LOCATING MASS GRAVES

Most perpetrators of human-rights violation, being well aware that they could be prosecuted if their activities are discovered, frequently attempt to conceal their crimes. Burying their victims and then trying to conceal the grave is a regularly used method to avoid detection. Forensic anthropologists are often in the forefront of trying to find these graves, and over time have developed a fairly standard system of investigation. Aside from the broadly recognized search methods (e.g., vegetation changes, presence of depressions, probing, surface scraping, etc.), this section will explore the uses of some nonintrusive methods that have been tried, and will examine how the traditional methods are utilized in a human-rights context.

Witnesses
An often-unappreciated aspect of locating a mass grave is the forensic anthropologist's role in interviewing potential witnesses. In most standard forensic cases, the forensic anthropologist is contacted by the police to assist in locating a grave. Any witnesses will already have been interviewed by the police. The forensic anthropologist rarely comes in contact with a witness and usually gets information secondhand from the police. This is not necessarily the case in international investigations of mass graves.

Unlike single murder/burial cases, the making of a mass grave often generates a number of witnesses who are willing to talk about the event. The creation of such graves takes a certain amount of planning and usually involves a number of people to carry it out. These individuals may not have taken part in the killing of the victims, and once an investigation begins are more than willing to explain their involvement. There may be other witnesses in the area that saw the grave being made, or noticed later that a large area of ground was disturbed when there was no reason it should have been. In rare cases, individuals have survived graveside executions to later become witnesses. In any case, these witnesses are often on hand during international investigations and available for interview by the forensic anthropologist.

Like the cultural anthropologist, the forensic anthropologist working in human-rights investigations will likely come in contact with these witnesses and must be ready and able to conduct proper interviews. Skills are required to tease out the details of the past events in a manner that does not intimidate the witness and that will lead to solid information that can assist in the location and excavation of mass graves. This also is an aspect of international human-rights work not addressed in a forensic anthropologist's formal education, but should be considered an important part of the job. Witnesses are always the best source of information on the location of a concealed grave. A good interview of a firsthand witness will provide far better information than any archaeological method and should not be taken for granted. Countless graves have been located through witness interviews where other methods had failed. It would be wise for the forensic anthropologist to sharpen his/her interview skills in order to gain this information.

Remote-sensing and pattern analysis

A number of remote-sensing search methods have been employed in locating contemporary mass graves. While not necessarily new technology, the adaptation of aerial photographs for locating contemporary graves has been successful in the Balkans and Iraq. Constantly updated intelligence aerial photos have demonstrated the change of landscapes from undisturbed settings to areas where obvious machine activity had occurred. In some cases, bodies and construction machinery have been photographed at sites. Satellite/aerial imagery of landscapes has also been employed to suggest locations where large graves could have been constructed, and may also be used to exclude potential areas. Recently, the use of spectral analysis, which measures changes in the composition of the ground surface, has also been used in an effort to locate mass graves in Iraq and Bosnia and Herzegovina. While such technology can be a powerful tool, it is not available in most human-rights investigations. These methods are usually costly and require close cooperation with a military intelligence source.

While most investigations will not have the ability to obtain remote-sensing data, they will over time compile data on mass-grave locations through their own excavation activities. These spatial data in turn may be of assistance in locating additional grave sites and possibly in helping to identify victims. In archaeology, we learn that people operate in predictable patterns and that, through careful observation, we can identify these patterns and make certain predictions regarding this behavior. Congram (2010) has been exploring the factors that influence where perpetrators decide to dig a grave. Congram's primary work has been in the context of the Spanish Civil War, but he also has experience in the Balkans. His preliminary findings suggest that graves are typically located 1–10 km from the place where the victims were detained (the point of origin); that the graves are almost always within 100 m of a principal road; that these locations are usually not visible from the point of origin; and that pre-existing features such as wells, mines, and ravines are regularly utilized. Through such pattern analysis, a greater understanding of the factors that influence the location of these types of graves could result in more effective searches.

An interesting aspect of this line of research is its potential to assist not only in the predictive patterning of grave locations, but also to assist in identifying victims after a grave is located. According to the research, victims within a grave likely originated nearby (1–10 km). If 15 individuals went missing from a certain village and 15 or so bodies are recovered from a mass grave 6 km away, this could be a logical argument of presumptive identification of the group. In such a situation, an investigation team may consider focusing their initial identification efforts with the village, effectively narrowing their pool of potential candidates to a manageable number. All available antemortem data for the missing from the village and DNA reference samples from surviving family members could be collected and compared to the remains. This information may be particularly important when funding, time, and personnel are limited.

Geophysics

Over the years, the quickest and most reliable methods to confirm or deny the presence of a grave have also been found to be the simplest: physically examining the ground with probes, hand tools, and heavy machinery. This fact is often lost on both the general public and many members of the forensic community where geophysics (e.g., ground-penetrating radar, resistivity, and magnetometry) and cadaver dogs have captured the imagination. In reality, the use of these methods in a human-rights context is rare because of the expense (unless their service was offered for free) and time it takes to organize the activity and receive results. Most organizations investigating human-rights abuses do not have geophysical experts and equipment on staff. In the time it takes to organize an outside expert or team to participate in an investigation, conduct their tests, and produce results, the targeted location could have been physically examined with 100% accuracy using probes or excavation equipment.

The lack of efficiency associated with these methods is compounded by the fact that they are less reliable than physical examination of suspected mass-grave sites. Proponents tend to tout positive results (or the potential for positive results) while ignoring failures. While geophysics and cadaver dogs have been used in the search for mass graves, little has been published on these endeavors. This is likely due to the fact that most attempts to find graves have failed. Independent evaluation of the ability of these methods to locate buried remains concluded that results can be confusing and should be viewed with caution (Buck 2003; Emanovsky 2004). Owsley (1995), who also evaluated a number of geophysical methods, agrees that careful observation and probing will likely produce better results.

Additionally, all anomalies identified through geophysics will have to be physically examined to determine their relevance. If, as a first step, the targeted area is physically checked, the presence or absence of a grave will immediately be confirmed. This is much quicker than first running a geophysics analysis, then physically checking the identified anomalies. In addition, the absence of anomalies in the data does not necessarily mean the absence of a grave. The geophysical examination may simply have failed to detect anything. The area will still need to be checked physically.

While geophysics has been inconsistent in locating graves, an avenue that may prove to be a time saving and useful endeavor is its application to mass-grave excavation strategy. In 2002, the author was engaged in the excavation of a series of large clandestine, mass graves outside of Belgrade, Serbia. A very short deadline of 4 weeks was given to complete the excavation of the last grave in the series before the end of the year and the onset of winter. The authorities feared that the political climate might change over the winter months and the window of opportunity to excavate this grave would close. The problem for the excavation team was that every other grave in the series had taken much longer than 4 weeks to complete, and it was likely that this deadline would not be met. A way to accelerate the excavation process was sought.

While the size and shape of the grave feature were identified by removal of the top several centimeters of soil, the excavation team had no way of knowing what was beneath. In order to get a better understanding of the subsurface structure of the grave, geophysical testing using electronic resistivity was conducted over the feature. The results provided the team with a glimpse of the size and general shape of the buried body mass and, more importantly, the knowledge of how much overburden could be removed before encountering bodies. A construction excavation machine was then employed to remove the overburden to a depth slightly above the body mass. It is estimated that the use of this machine saved at least a week of excavation time. In this case, the electronic resistivity data resulted in the safer removal of sterile overburden than would have occurred if the team attempted to do this without advance knowledge of the body mass depth, and enabled the team to complete the project in the time allotted (Tuller and Sterenberg 2005).

This example demonstrates that, while geophysics may have limitations when searching for mass graves, such methods do hold potential as useful tools when planning an excavation. In any case, it is hoped that further research will streamline geophysical examination of sites, make it more accurate and useful for the human-rights investigator, and demonstrate additional uses for this technology.

Excavation

While the focus of many excavations has been the recovery of remains, more emphasis has recently been placed on understanding the formation process of a grave and on the collection of evidence for cause of death (Skinner et al. 2003). As mentioned, people act in predictable ways, leaving behind patterns of their activity. Careful excavation and recording of the formation process of a grave can reveal these past activities. The ICMP, which has been using forensic archaeologists since 2002 to assist in mass-grave excavations, began referring to this documentation as "telling the story of a mass grave." The use of this terminology by an organization whose mandate is the identification of victims (and not the prosecution of perpetrators) underscores the subtle shift from simple body recovery to an awareness that the excavation process can contribute to their organization's goals. Part of this shift was the realization that family members of victims seek not only closure through the return of the remains of their loved ones, but an understanding of what happened to them as well.

To understand what happened at a site before, during, and after the creation of a mass grave, a series of questions must be answered. What was the land used for prior to the creation of the grave? How was the grave created? When was the grave dug in relationship to the killing of the victims? Were the victims killed elsewhere and then transported to the grave, or were they killed next to/within the grave? How were the victims transported to the grave? How were they deposited in the grave? Were the remains placed in the grave according to any detectable pattern? What, if any, are the relationships between the bodies and the nonbiological evidence in the grave? How was the grave filled? What other activities have occurred within or near the grave? Has any modification to the grave taken place since it was filled? Was there any attempt to conceal the grave after it was filled? These are just some of the questions that a careful excavation may seek to answer. Specifically, those conducting the excavation must be mindful of any patterns that are manifested. The simple uncovering and removal of remains as they are exposed will not suffice in answering these questions.

Excavation methodology

The actual excavation should, if possible, be conducted in a manner that reveals the order in which the grave was constructed. This usually means uncovering the evidence in the reverse order from which it was placed in the grave. Two methods of excavation are commonly employed to expose remains: the pedestal method and the stratigraphic method. In most cases, the stratigraphic method will be found to be the more appropriate technique (Tuller and Đurić 2006). This method views the whole grave as an archaeological feature where, health and safety concerns withstanding, the walls of the grave adjacent to the body mass are preserved, thereby maintaining the grave contents *in situ*. Removal of the walls, particularly in older graves where the remains have skeletonized, will cause the body mass to erode with skeletal elements and other evidence shifting out of position. This will cause the loss of provenience and disarticulation of skeletal elements. Maintaining the skeletal elements in articulation should be one of the main goals of a mass-grave excavation. Komar and Potter (2007) demonstrated that a significant link exists between identification rates and the percentage amount of recovered remains. A complete body is more likely to be identified than a body missing elements. The stratigraphic method better retains the bodies in their original positions and facilitates excavation of the grave in reverse order from how it was created.

   The other method of excavation, pedestaling the body mass, removes the grave walls, allowing access to the evidence from a variety of angles. Haglund (2001) reports the successful employment of the pedestal method to expose remains, but acknowledges the removal of the grave walls is a loss of evidence of the grave's creation. Although the stratigraphic method demonstrated a greater ability to maintain body part articulation (Tuller and Đurić 2006), the pedestal method should not be discounted. Unskeletonized remains will not pose as much of a disarticulation problem through erosion as do skeletonized remains. In certain politically charged or potentially dangerous international contexts where a quick excavation is deemed most suitable and the grave is of recent origin, the pedestal method may be more appropriate. While information will be lost, safety of the excavation team takes precedence.

Depositional events within a grave

Another aspect of archaeology that has recently been applied to the excavation of mass graves is the documentation and understanding of the depositional events that made the graves. Separate depositional events in a grave will manifest themselves as distinct strata. One of the prime concepts of archaeology (and geology) is the law of superposition, which states that any stratigraphic unit underlying another must be older than the overlying unit. Defining stratigraphy in a grave is the key to understanding the chronological sequences of events that led to its formation.

Dump trucks depositing their loads of human remains and other evidence in a grave will create separate stratigraphic units. The unit on the bottom of the grave was placed there first, the second deposit of bodies on top of the first, and so on. A good excavation plan will reveal the number of stratigraphic units and thus the sequence of their creation. The division between these units is sometimes clear, while at other times it can be very subtle. Bulks of soil between deposits of remains may be present. The orientation of the bodies may help identify separate deposits. One deposit may consist mostly of male victims while another is made up of female victims. Clothing or lack thereof may be inconsistent between deposits. Perhaps one deposit is of primary origin, while another is determined to have been removed (robbed) from a previous grave and deposited in this new location.

To form a clear view of what has occurred in the grave, all the sediment above the deposits of remains must be removed. Observations are noted, photographs taken, and mapping of the exposed deposit(s) is conducted. If individual remains and unassociated evidence are removed from the grave before first exposing the entire deposit, their context within the deposit and the grave as a whole will be lost. If this occurs, the activity becomes less of an excavation and more of a simple body-recovery operation. Understanding the relationship between separate deposits and the remains contained within not only reveals the formation process, but also has the potential to assist in the reassociation of disarticulated remains and in the identification process.

Disarticulation of remains is a common aspect of a mass grave. This is particularly true of graves where the remains have skeletonized and sites were remains removed from one grave and deposited into a different grave (what Jessee and Skinner 2005 refer to as secondary inhumation sites). Attempts at reassociating disarticulated remains are usually conducted during autopsy. If a body is without a limb, a right leg for example, all the disarticulated right legs from the grave may be examined to see if one can be matched to the body. This process can be time-consuming if the grave contains a large number of disarticulated remains. However, the time needed could be reduced if the deposit from which the remains were recovered was examined separately (Tuller et al. 2005).

The remains within a specific deposit represent a collection of individuals associated both temporally and spatially. At some point, these individuals were brought together and eventually ended up deposited in a collective grave. While remains within a deposit share an association, relationships *between* deposits may be nonexistent. When a body is recovered missing an element, such as the exemplary right leg, the chances are strong that the disarticulated element was recovered from the same deposit as the body. Thus, when disarticulation occurs, one should first look for possible matches within the same deposit before expanding the search to other deposits (Tuller et al. 2008). For instance, let's say eight disarticulated right legs are recovered from a mass grave made of five deposits. A significant amount of time could be spent examining those eight legs and running tests on each to determine whether they may match a certain body. However, if only two of those eight legs were recovered from the same deposit as the body missing the limb (the other six legs coming from other deposits), then the examination should first focus on these two limbs. By recording the provenience of the body and the legs, the forensic anthropologist in the field can save the laboratory a lot of time and effort.

In addition to sorting commingled remains, recording depositional events may also assist in the identification process. As mentioned, victims may have been transported from different locations to the grave to form separate deposits. For example, perpetrators may have visited several villages in an area and killed a number of victims at each location. Trucks could have been dispatched to each village to gather up the bodies and transport them to a location where a mass grave was dug. Each deposit of remains within the grave would then be representative of a geographical location from where the victims came. During genocide or "ethnic cleansing" operations, entire families are sometimes killed, and survivors spread out over the globe as refugees. This situation greatly inhibits DNA-led identification projects. While forensic anthropology may produce biological profiles of the victims, the bodies will remain unidentified without family reference samples or antemortem data for comparison.

If, however, if a number of remains from a certain deposit within a grave are identified and a background check of these individuals reveals that they came from a specific village, it is likely that the remaining unidentified individuals in the same deposit came from the same village (Tuller et al. 2005). This then provides an opportunity to generate additional antemortem data and possibly DNA family reference samples from the village. A review of the people reported missing from the specific village can be compared to the list of those bodies that have been identified. The remaining names on the missing persons list may very well be among the unidentified bodies of that specific deposit. Efforts can then be focused on finding surviving relatives and recording antemortem data of the reported missing at that village. In this manner, the recording of depositional events could assist in the identification process by providing an avenue for further data collection.

EVIDENCE RECORDING

Both archaeologists and crime scene investigators regularly use the overlapping documentation of note taking, photography, and mapping to guarantee that each article of evidence gets recorded. In a mass grave with hundreds of bodies and thousands of pieces of evidence, documentation becomes very complicated very quickly. To assist in the documentation deluge, investigators at mass graves have embraced *pro forma* body-recovery forms and electronic theodolites (total stations) when available.

Body forms

Pro forma body-recovery forms seek to standardize a minimal required amount of information from a set of remains before removal from the grave. With multiple sets of remains being removed from a grave simultaneously, it would be impossible for a single individual to record all the relevant information about each set of remains in a notebook. With a crew of investigators excavating a grave, it is common for more than one set of remains to be in the process of being removed simultaneously. Likewise, it would be impossible to transfer the notes from a single notebook to the autopsy room for review by a pathologist or forensic anthropologist in a comprehensive manner. Usually, multiple autopsies are carried out at once in these investigations. *Pro forma* body-recovery forms allow individuals working in the grave to fill in the necessary information on separate forms, which are then transferred to the mortuary operations and filed. In the mortuary, these separate forms can be viewed by the pathologist or forensic anthropologist to assist in their examination. Some of the basic information regularly recorded on body recovery forms includes:

- unique evidence code;
- basic inventory of remains (e.g., body, missing left arm);
- position of the body in grave;
- associated material evidence;
- basic inventory of clothes;
- possible evidence associations;
- sketch map of remains and associations.

Additional information sections are included to allow the recovery team to add any further documentation they feel is warranted. For example, someone might note a bullet round mixed with the fragments of a shattered humeral head. If the round is simply placed in the body bag along with the remains without recording where it was located, those working in the mortuary will be unaware of the context of the round. By noting the round's provenience, the association with the shattered humeral head is preserved. The pathologist and/or forensic anthropologist in the mortuary will then be able to include that information in their analysis of the trauma.

Surveying

Total stations are now a regular feature at mass-grave investigations. These electronic theodolites allow for a much quicker and more accurate survey of the site features and contents than do traditional tape measure and compass, grid system, or standard theodolites. Unlike tape measures and grid systems, a total station can easily measure long distances and does not interfere with other site activities (no tape measures or strings to impede the work of others). With computer software, a variety of maps can be created to illustrate the layout of the site or to demonstrate the spatial relationships between different types of evidence (Hanson 2003; Wright et al. 2005). This can be done in two or three dimensions and displayed as a plan view, sectional views, or on a computer screen as a three-dimensional rotating image. If desired, the specific distribution and association of evidence within a grave or a particular deposit of evidence within the grave can be illustrated.

Data collected by a total station can also display features that are eventually destroyed during the excavation process. When deep graves are encountered or when the surrounding soil matrix is unstable, the grave walls must be stepped back for safety purposes. Unstable walls can collapse onto workers in the grave. In such cases, much of the original wall surrounding the grave is removed during the excavation. This means that final photographs demonstrating the size/depth of the grave are compromised because of the absence of the grave walls. However, recording the grave walls with a total station prior to their removal preserves them in an electronic format. Computer software can take these data and display the entire grave, walls intact, in three dimensions.

One of the more interesting uses of the total station at mass graves has been developments with recording exposed remains. Many international organizations base their total station survey procedures on protocols developed with ICTY as described by Hanson et al. (2000). Typically, 14 to 15 points are recorded on a complete body. These points include the head, all major limb joints, and a central pelvis point. Sometimes the organization may include a central mass point of the body for mapping purposes. Computer software is then used to connect these points, creating a stick figure representing the body. Disarticulated body parts are also recorded in this manner. For example, a complete but unassociated arm will have three points taken on it: the shoulder (humeral head), elbow, and wrist.

The stick figures created by the computer software can be viewed in two dimensions on a printed map or three-dimensionally on a computer screen in a variety of ways, just as with all mapping data. If desired, remains from different deposits within the grave can be color-coded to better define them, bodies missing elements can be displayed along with possible matching elements, associations between individual bodies and evidence (e.g., identification media or shell cases) can be demonstrated, or any number of combinations can be shown. While such illustrations are helpful in demonstrating past events at a site, manipulation of the data behind the mapping can actually assist in reassociation of disarticulated remains.

Disarticulation of remains may occur through intentional and/or natural activity. Secondary inhumations and older graves where the remains have had a longer time to decompose will likely be more commingled than newly created primary graves. Each of the points taken on a body is recorded in three-dimensional (x, y, and z) coordinates. The distances between these points can then be computed to develop a rank order of possible matching disarticulated remains (Tuller et al. 2008). For example, suppose a body within a grave is missing its lower right arm (ulna, radius, and hand). An x, y, and z coordinate point was recorded on the distal end of the right humerus (the right elbow point), but, because the rest of the arm was absent, no point was recorded at the right wrist. However, all the disarticulated remains within the grave also had a series of coordinate points taken on them, including each of the disarticulated right lower arms. A computer program is able to calculate the distance from nearest to farthest between the right distal humeral point on the body to all the proximal right lower arm ends in the grave. The principle is that the matching disarticulated limb is more likely to be the closest physically to the body. In this manner, the excavation team is able to provide the mortuary team with a list of possible reassociations from most to least likely. While not definitive proof that a particular limb matches with a particular body, the list does give the autopsy team a starting point for their reassociation efforts. The usefulness of point coordinate spatial analysis was demonstrated with a large mass grave where DNA-reassociated matches were used as a control (Tuller et al. 2008). With large, complicated graves containing multiple deposits of human remains, the spatial calculations can even be computed from each separate deposit, rather than from the entire grave (disarticulated remains are more likely to be found in the same deposit as the bodies).

LINKING EVIDENCE

In the former Yugoslavia, perpetrators trying to muddle and hide evidence employed a tactic of digging up mass graves and reburying the contents in other locations. While this tactic does indeed make investigating cases more difficult, no site can ever be fully cleansed of evidence. Careful examination of the evidence from multiple sites may reveal information that links them. Proving that links between sites exist demonstrates that perpetrators were acting in an organized manner, likely on specific orders from higher authorities. Such evidence could be used to refute claims that perpetrators acted spontaneously and independently from their superiors.

An excellent example of tying crimes scenes together was demonstrated with the ICTY investigations in Bosnia and Herzegovina. Over the course of several years, numerous execution sites, primary graves, disturbed graves, and secondary inhumation sites were examined. These investigations provided physical evidence showing the relationship between these different types of crime scene (Manning 2000). A primary inhumation site is defined as an undisturbed grave in which individuals are buried after being killed. A disturbed grave (or robbed grave) is one that has been opened and the remains are removed for the purpose of transporting them to another, more secret location, creating a secondary inhumation. The ICTY discovered that a number of secondary inhumation sites were created from human remains originating from more than one primary inhumation site, and that a number of remains from such primary sites had been divided up among several secondary sites. Several types of recovered evidence led to this conclusion. Spent shell cases recovered from execution and grave sites were examined for unique ejector marks; embossed nicks formed on the shell case by the ejection mechanism of a weapon immediately after firing its round. When identical ejector marks are found on different shell cases, they can be shown to have been fired from the same weapon. A number of execution sites were linked to primary inhumation sites, and in some cases primary sites to secondary inhumation sites, based on identical ejector marks. Soil and pollen samples, bottle glass shards and bottle labels, and ligatures and blindfolds were also used to link these different types of crime scenes. These items were inadvertently mixed in with the remains as they were dug up from a primary inhumation site and transported to secondary locations. This evidence was recovered at disturbed primary inhumation sites and found to match evidence recovered from excavated secondary locations (Manning 2000).

In addition to evidence that the perpetrator may have added to a grave (shell cases, ligatures, etc.), the victims may also have had possessions that can help the investigation. The remains may have artifacts associated with them that can identify them as belonging to a particular ethnic group. In Bosnia and Herzegovina, a grave containing a number of Muslim-related artifacts (e.g., Muslim prayer beads, Korans) indicates that the remains are likely those of Bosnian Muslims, while Orthodox Christian-related artifacts (e.g., Orthodox cross, Bibles) can indicate Serbian victims. Roman Catholic artifacts would represent Croatian affiliation. In Iraq, clothing style has been used to indicate from where the victims come. Anthropologists, with their training and general interest in the differences and commonalities between people, are likely to be the ones to point out this form of evidence.

Concluding Remarks

In this short chapter I have attempted to provide an understanding of what is necessary to meet the challenges of mass-grave excavation, as well as to highlight some aspects of work conducted at these unique types of sites. The most important requirement to conduct an excavation of these complex graves is a solid background in archaeology. Although the field of forensic anthropology recognizes this and has been slowly moving to address deficiencies in the manner in which forensic students are taught, it is ultimately the responsibility of the individual forensic anthropologist to ensure that he/she has the skills to do the job properly. Other topics in this chapter have tried to demonstrate the interesting intersection between archaeology and the forensic investigation of these graves. While some activities would be familiar to any standard archaeologist, such as excavating in a manner that reveals the sequence of past events at a grave, others have been adapted to assist in the investigation in surprising ways (e.g., using survey data to help sort commingled remains). As new ideas and methods are developed in this field, it is important that we keep abreast of these changes.

Over the past few decades, interest in the investigation of mass graves within a human-rights context has steadily increased. Governments and the general worldwide public are becoming more aware of the capabilities of forensic science to collect evidence from these crime scenes and to identify victims. While in the past the forensic scientist had to convince authorities of the validity of such investigations, today it seems the mere accusation of a human-rights violation is accompanied by calls for an international forensic response. Politics aside, it is likely that these types of investigation will continue to increase. As forensic anthropologists are usually in the lead when it comes to excavating mass graves, we need to be prepared to meet this challenge.

REFERENCES

Bass, W.M. and Birkby, W.H. (1978). Exhumation: the method could make the difference. *FBI Law Enforcement Bulletin* 47: 6–11.

Buck, S.C. (2003). Searching for graves using geophysical technology: field tests with ground penetrating radar, magnetometry, and electrical resistivity. *Journal of Forensic Science* 48(1): 5–11.

Congram, D. (2010). Spatial patterning of clandestine graves in the investigation of large scale human rights violations: the example of the Spanish Civil War rearguard repression. *Proceedings of the American Academy of Forensic Sciences*, Seattle, WA.

Connor, M. and Scott, D.D. (2001). Paradigms and perpetrators. *Historic Archaeology* 35(1): 1–6.

Dirkmaat, D.C. and Adovasio, J.M. (1997). The role of archaeology in the recovery and interpretation of human remains from an outdoor forensic setting. In W.W. Haglund and M.M. Sorg (eds), *Forensic Taphonomy: The Postmortem Fate of Human Remains* (pp. 39–64). CRC Press. Boca Raton, FL.

Emanovsky, P.D. (2004). *Preliminary results on the use of cadaver dogs to locate Vietnam War-era human remains.* AAFS Oral Presentation, Annual Meeting, Dallas, TX.

Fiorato, V., Boylston, A., and Knüsel, C. (eds) (2000). *Blood Red Roses: The Archaeology of a Mass Grave from the Battle of Towton AD 1461.* Oxbow Books, Oxford.

Galloway, A. and Simmons, T. (1997). Education in forensic anthropology: appraisal and outlook. *Journal of Forensic Science* 42(5): 796–801.

Haglund, W.D. (2001). Archaeology and forensic death investigations. *Historic Archaeology* 35(1): 26–34.

Hanson, I. (2003). Advances in surveying and presenting evidence from mass graves, clandestine graves, and surface scatters. *Proceedings of the American Academy of Forensic Sciences*, Chicago, IL.

Hanson, I., Sterenberg, J., and Wessling, R. (2000). *Survey procedures: ICTY Forensic Field Team Bosnia: 2000 field season.* Unpublished survey protocol, ICTY Bosnia and Herzegovina.

Hoshower, L.M. (1998). Forensic archaeology and the need for flexible excavation strategies: a case study. *Journal of Forensic Science* 43(1): 53–56.

Jessee, E. and Skinner, M. (2005). A typology of mass grave-related sites. *Forensic Science International* 152: 55–59.

Juhl, K. and Olsen, O.E. (2006). Societal safety, archaeology and the investigation of contemporary mass graves. *Journal of Genocide Research* 8(4): 411–435.

Komar, D.A. and Potter, W.E. (2007). Percentage of body recovered and its effect on identification rates and cause and manner of death determination. *Journal of Forensic Science* 52(3): 528–531.

Manning, D. (2000). *Srebrenica Investigation: Summary of Forensic Evidence – Execution points and Mass Graves.* Forensic summary report to the United Nations International Criminal Tribunal for the Former Yugoslavia. UN ICTY Evidence Report, May 16, 2000. UN ICTY, The Hague.

Morse, D., Crusoe, D., and Smith, H.G. (1976). Forensic archaeology. *Journal of Forensic Science* 21(2): 323–332.

Owsley, D.W. (1995). Techniques for locating burials, with emphasis on the probe. *Journal of Forensic Science* 40(5): 735–740.

Owsley, D.W. (2001). Why the forensic anthropologist needs the archaeologist. *Historical Archaeology* 35(1): 35–38.

Saunders, N.J. (2007). *Killing Time: Archaeology and the First World War.* Stutton Publishing, Pheonix Mill.

Sigler-Eisenberg, B. (1985). Forensic research: expanding the concept of applied archaeology. *American Antiquity* 50(3): 650–655.

Skinner, M.F. (1987). Planning the archaeological recovery from recent mass graves. *Forensic Science International* 34(4): 267–287.

Skinner, M. and Lazenby, R.A. (1983). *Found! Human Remains: A Field Manual for the Recovery of the Recent Human Skeleton.* Archaeology Press, Simon Fraser University, Burnaby, BC.

Skinner, M., Alempijevic, D., and Djuric-Srejic, M. (2003). Guidelines for international forensic bio-archaeology monitors of mass grave exhumations. *Forensic Science International* 134: 81–92.

Sledzik, P.S., Fenton, T.W., Warren, M.W., Byrd, J.E., Crowder, C., Drawdy, S.M., Dirkmaat, D.C., Galloway, A., Finnegan, M., Fulginiti, L.C. et al. (2007). *The Fourth Era of Forensic Anthropology: Examining the Future of the Discipline.* Proceedings of the American Academy of Forensic Sciences, San Antonio, Texas.

Steadman, D.W. and Haglund, W.D. (2005). The scope of anthropological contributions to human rights investigations. *Journal of Forensic Science* 50(1): 23–30.

Tuller, H. and Sterenberg, J. (2005). Not for the passive: the active application of electronic resistivity in the excavation of a mass grave. *Proceedings of the American Academy of Forensic Sciences*, New Orleans, LA.

Tuller, H. and Đurić, M. (2006). Keeping the pieces together: a comparison of mass grave excavation methodology. *Journal of Forensic Science International* 156: 192–200.

Tuller, H., Hofmeister, U., and Daley, S. (2005). The importance of body deposition recording in event reconstruction and the re-association and identification of commingled remains. *Proceedings of the American Academy of Forensic Sciences*, New Orleans, LA.

Tuller, H., Hofmeister, U., and Daley, S. (2008). Spatial analysis of mass grave mapping data to assist in the reassociation of disarticulated and commingled human remains. In B.J. Adams and J.E. Byrd (eds), *Recover, Analysis, and Identification of Commingled Human Remains* (pp. 7–29). Humana Press, Totowa, NJ.

Ubelaker, D.H. (1999). *Human Skeletal Remains: Excavation, Analysis, Interpretation*, 3rd edn. Taraxacum Press, Washington DC.

Wright, R., Hanson, I., and Sterenberg J. (2005). The archaeology of mass graves. In J. Hunter and M. Cox (eds), *Forensic Archaeology: Advances in Theory and Practice* (pp. 137–158). Routledge, New York.

CHAPTER **9**

Archaeology, Mass Graves, and Resolving Commingling Issues through Spatial Analysis

Luis L. Cabo, Dennis C. Dirkmaat, James M. Adovasio, and Vicente C. Rozas

INTRODUCTION

The widespread investigation of human-rights violations and abuses throughout the world during the past two decades has generated a renewed interest in the recovery and investigation of human remains from many different depositional contexts, especially multiple victim burial features or mass graves (Haglund 2002; Schmitt 2002; Skinner et al. 2003 and references therein). These features are characterized by their complexity, usually containing multiple commingled individuals (Haglund 2002; Haglund et al. 2001; Hunter et al. 2001; Schmitt 2002; Skinner 1987). Commingling dramatically affects efforts to understand even simple issues such as quantification (i.e., the number of individuals interred in the feature).

Typically, commingling issues are dealt with in the laboratory. Human skeletal variation at various levels, including those related to sex, chronological age, ancestry, stature, pathology, and idiosyncratic factors, can be utilized to help sort remains into various categories (Adams and Konigsberg 2004; Ubelaker 2002). However, Tuller

A Companion to Forensic Anthropology, First Edition. Edited by Dennis C. Dirkmaat.
© 2012 John Wiley & Sons Ltd. Published 2015 by John Wiley & Sons Ltd.

et al. (2005, 2008) have recently demonstrated the importance of careful archaeological provenience techniques to solve commingling issues, reporting matching success rates close to 100% in a real mass-grave scenario, through the application of spatial analysis of point data (nearest-neighbor techniques). This chapter discusses the policy and practice implications for human-rights investigations (HRIs) of these findings, and presents evidence supporting the applicability of the spatial distributional premises assumed by Tuller et al. (2008), even in secondary burials that have undergone extreme postdepositional disturbance.

The empirical demonstration by Tuller et al. (2008) of the link between proper spatial data recordation and the solution of commingling problems serves to abridge the debate on the necessity of the application of careful archaeological methodologies in HRIs. While in conventional outdoor forensic settings involving human remains (i.e., forensic recovery as practiced in North America) the importance of well-recognized traditional archaeological methodologies designed to maximize the identification, documentation, and collection of relevant information is well established (Dirkmaat and Adovasio 1997; Hochrein 1997; Skinner 1987), these techniques are often dismissed in HRIs based on an alleged trade-off between victim-identification and event-reconstruction needs, supposedly derived from time, personnel, and monetary constraints.

A plethora of severe constraints do affect HRIs. Political, religious, cultural, weather, and security issues, lack of facilities, inadequate logistical support, limited financing, equipment, and supplies are just a few (Haglund 2002; Schmitt 2002). Most of these factors are beyond the control of forensic professionals. However, other constraints do clearly fall within the area of responsibility of forensic investigators: primarily derived from the rapid expansion in the number of mass-grave excavations conducted throughout the world, a large number of these projects are being excavated by inexperienced or even inappropriate individuals (Skinner et al. 2003: 82). For example, Skinner et al. (2002) describe how a number of sites are dug by local laborers, or "autopsy assistants supported by pathologists and various Members of the State Commission." Anthropologists may be involved solely as monitors in these projects. In some instances, the features are exposed by bulldozer or backhoes. Understandably, there may be some reluctance on the part of archaeologists to "utilize untrained human power or a backhoe because of the risk of damaging evidence" (Schmitt 2002). However, in our view, even under these extreme circumstances the primary concern must still be that significant forensic evidence is missed and destroyed. Therefore, we must take issue with statements that these problems have to be "balanced against the risk of not undertaking or completing excavation" (Schmitt 2002: 280).

Tied to this "utilitarian" view may be some misconceptions regarding the nature of human rights and the interpretation of the significance of complexity in forensic settings.

Human rights and forensic protocols

HRI is a relatively new and still evolving discipline, but was clearly defined within the forensic sciences from its origin, counting among its primary goals to "collect, preserve, and objectively interpret physical evidence" (Doretti and Snow 2002: 309) leading to the prosecution of perpetrators. Still, determination of personal identity and return of the remains of victims to families as soon as possible is sometimes perceived as the primary purpose of HRIs (e.g., Williams and Crews 2003). This orientation seems to clear the perceived trade-off between victim identification and event reconstruction: forensic investigation can (and sometimes *must*) be sacrificed when necessary to speed victim identification and restitution of the remains to relatives. This results in operations not primarily concerned with the recovery and analysis of forensic evidence, operations often referred to as *humanitarian efforts* (Steadman and Haglund 2005).

As argued elsewhere (Dirkmaat et al. 2005) this view stands in stark contrast with the letter and spirit of the modern human-rights concept. The primary goal of the United Nations (UN) Universal Declaration of Human Rights [UN General Assembly Resolution 217 A (III) of December 10, 1948] was to translate what had basically become an abstract statement of good will (namely the central doctrines and principles originally presented in the Declaration of the Rights of Man and of the Citizen, approved by the National Assembly of France on August 26, 1789) into an effective body of law. In the words of one of the drafters of the Declaration, Stepháne Hessel, "I felt that we had to move fast so as not to succumb to the hypocrisy of victors promoting allegiance to values that no one had the intention of enforcing faithfully" (Hessel 2011: 28).

In other words, the UN declaration was entirely aimed at converting human rights into *legal rights*. As such, judicial inquiry, court accountability, and reparation are not mere ingredients, but constitute the backbone of the modern human-rights concept. Articles 6–12 of the UN resolution, in addition to presenting rights related to guarding against illegal detention and guaranteeing a fair trial for the suspects (again, already stated in 1789), also explicitly develop the right of victims to be heard in court, as well as to defend oneself against attacks upon personal honor or reputation. The right to an effective remedy by the competent national tribunals for acts violating the fundamental rights of any individual is granted by Article 8 of the Universal Declaration of Human Rights. That is to say, making human-rights violations and abuse accountable in court is not simply a way of enforcing human rights, but one of the basic human rights itself.

The corollary of these considerations is that any investigation of human-rights violations or abuses *must* be primarily conducted in such a way as to allow for the effective presentation of the case in a court of law, keeping no lesser standards of proof than those accepted in democratic national judicial systems to guarantee the rights of victims, plaintiffs, and defendants. From the strict human-rights purview, any intentional destruction or negligent recovery of significant evidence will, in fact, constitute a new violation of basic human rights. In the opinion of the authors, this fact should also be balanced against the risk of not undertaking or completing excavation when the investigator confronts the dilemma of accepting or rejecting substandard resources or work conditions.

Complexity and information retrieval

A second misconception that may lead to the dismissal of archaeological scene processing comes from what seems to be a paradoxical perception of complexity as information loss. Complex features such as mass graves may be perceived as indecipherable conundrums with little to be gained from carefully documenting them. The trade-off in this case would be between recovery effort and information gain. Identification and documentation of even gross stratigraphic profiles or basic plots of the positioning of human remains and other evidence may be easily recognized as useful tools to detect major depositional events or to identify some articulated individuals, but the utility of collecting precise spatial coordinates or soil descriptions beyond that point may not be as evident in what may appear to be a chaotic accumulation of human remains, randomly patterned by a multitude of confounding factors.

However, complexity cannot be equated to information loss, but rather represents a wealth of information that offers increased inferential and evidentiary possibilities, requiring more refined analytical methodologies. Mass graves with large numbers of commingled individuals do not represent either unsolvable problems or unique scientific situations. Professional archaeologists have been excavating similarly complex features for nearly 100 years in the form of prehistoric ossuaries, and solid archaeological techniques and methodologies to process large burial features containing many individuals are fully developed and in place. Briefly, these techniques are based on – and their application requires training and expertise in – (i) the delineation and interpretation of stratigraphy and stratification, (ii) a thorough and comprehensive understanding of the concept of context, and (iii) the establishment of any associations between recovered materials. The key to understanding stratigraphy is the successful identification of individual strata and their interfaces following Steno's principles, which include the laws of superposition, original horizontality, lateral continuity, and intersecting relationships (Dirkmaat and Adovasio 1997: 45). Very precise three-dimensional mapping protocols, aimed at establishing precise and clearly defined contextual relationships and associations of evidence to the depositional environments, are absolutely critical to subsequent interpretations. This level of expertise can only be found in the works of professional archaeologists or forensic archaeologists.

Similar to physical evidence, contextual relationships and associations of evidence are not less important in HRIs than in conventional forensic cases. On the contrary, the identification and clear delimitation of serial depositional patterns becomes particularly relevant in human-rights scenes, as the legal definitions of the offenses under investigation heavily rely on the presence of consistently repeated patterns of action across seemingly discrete episodes of mass killings. Legally, the *Elements of Crimes* adopted by the International Penal Court, following the structure of the corresponding provisions of articles 6, 7, and 8 of the Rome Statute (International Penal Court, Official Records, ICC-ASP/1/3, September 2002), focus on the conduct, consequences, and circumstances associated with each crime. The criteria used to categorize an action as genocide, crimes against humanity, or war crimes, require determinations such as whether the conduct (i) took place in the context of a manifest pattern of similar conduct and (ii) was part of a systematic attack directed against an entire or a particular group of the civilian population. Further, the level of knowledge of the suspect about these processes is also considered.

Consequently, contextual information, including stratigraphic data and precise spatial coordinates of physical evidence, constitutes court-relevant evidence by itself. Optimal field recovery and recordation of these data are of crucial importance, as contextual information is irreparably destroyed during scene processing, due to the very nature of the recovery process (Dirkmaat et al. 2005). This is particularly true at outdoor scenes in which excavation is involved. As a result, the loss of relevant information not only handicaps current attempts to make those responsible accountable for their crimes in court, and the victims remedied, but also any future efforts to fulfill the ethical and legal obligations imposed by the Universal Declaration of Human Rights.

Linking victim identification and recovery methods

The legal, ethical, and scientific considerations described above should suffice to support the systematic application of comprehensive forensic and archaeological methods in HRIs of mass-grave scenarios, but speeding victim identification to the detriment of case investigation when only substandard resources are available may still be perceived as a valid alternative by nontrained individuals, especially when strong social or political pressures are present. Within this framework, the results described by Tuller et al. (2008) represent a turning point to settling this argument. Their analyses are only possible when detailed archaeological protocols, including precise spatial provenience of all evidence, are applied. This fact, combined with the optimal victim-identification rates obtained, demonstrate that comprehensive archaeological processing is not only important for event reconstruction and prosecutorial purposes, but also crucial for proper victim identification, thus dissolving the alleged trade-off between both factors. In addition to obstructing legal efforts to make the perpetrators of human-rights violations accountable in court, so-called humanitarian efforts that disregard contextual information do not actually speed the release of victims to their families, but can actually hinder it by impeding or delaying positive victim identification.

It may still be argued that the applicability of spatial analysis to commingling issues would be highly dependent on the degree of alteration shown by the burial feature, and therefore that spatial data recordation would not be necessary in all cases. Clear articulation patterns and discrete stratigraphic events were manifest in the mass grave studied by Tuller et al. (2005, 2008). Under these conditions, the basic assumption underlying the entire analytical process – that spatial proximity is related to previous element articulation – is clear. However, in most complex burial features, such as secondary or disturbed mass graves, articulation may be completely lost at the time of excavation. This may lead the investigator to assume that spatial analysis (and therefore "costly" field recordation of spatial data) would be superfluous in these cases. As no articulation pattern is observed during excavation, two articulating elements from the same individual can be found at any distance within the burial feature. If we multiply this effect by the number of individuals interred in the feature, a totally random distance patterning of previously articulated elements might be expected.

This chapter presents an example demonstrating that while the observation of articulation or association patterns during excavation directly proves their existence, a lack of immediately observable such patterns cannot be interpreted as proof of their absence. Furthermore, it also shows how the detection and analysis of these patterns through spatial analysis techniques can provide very useful information, for example regarding matters such as the average and maximum distances at which two originally articulating skeletal elements can be expected to be found at the feature. Orton Quarry, a highly disturbed, partially destroyed, Late Prehistoric ossuary in northwestern

Pennsylvania, USA, excavated in the early 1990s, is employed to demonstrate the persistence of detectable and quantifiable regular spatial (proximity or association) patterns between pairs of articulated skeletal elements, even in highly disturbed secondary burials, with no articulation or association detected at the time of excavation. The requisite of correlation between original articulation and distance within the feature is therefore still present in these features, demonstrating that methodologies akin to the ones developed by Tuller et al. (2005, 2008) are potentially applicable even under these extreme circumstances.

It has been argued elsewhere (Dirkmaat and Adovasio 1997) that methodological standards routinely employed in the recovery of contextual data associated with archaeological sites can be directly applied to modern scenes in what can be called forensic archaeology. It can be added that these sites, typically free from most of the legal, time, and material constrictions frequent in forensic cases, also represent the best alternative to develop and test techniques and methodologies aimed primarily at forensic investigation.

In this respect, Orton Quarry is a particularly illustrative example for different reasons. First, it presents an extreme degree of spatial alteration, not often encountered in mass-grave scenarios. In addition to being a secondary burial (disarticulation, therefore, suspected to have occurred prior to or during deposition), and having been partially destroyed, intentional association of some elements is evidenced by the presence of bone clusters or bundles. As a result, simple detection of spatial association patterns between pairs of articulated bones would not suffice to demonstrate that spatial association patterns are related to anatomical association of these elements. Mass graves resulting from human-rights violations, where the amount of soft tissue and degree of articulation are typically much higher due to the time scale, are expected to represent much simpler analytical problems.

Secondly, the site was excavated before any prior application in archaeology of the types of techniques employed in the analysis of its spatial patterning, and without predicting such application. Still, the careful provenience techniques applied to process the site allow for comprehensive analysis of its spatial patterning more than a decade after excavation. This serves to illustrate how (i) while the presence of patterns observable during excavation may not require further proof, the absence of observable patterning does not imply the inexistence of spatial patterning and, similarly, (ii) unawareness at the time of excavation of the existence of analytical techniques with the capability of extracting relevant information does not imply the inexistence of such techniques. Therefore, comprehensive documentation of contextual evidence in multiple burials with commingled remains is *not an option that can be decided at the time of excavation*, but must be performed by default in all cases.

METHODS

The Orton Quarry Ossuary

The Orton Quarry site (36ER243) is a Late Prehistoric ossuary situated in northwestern Pennsylvania along the coast of Lake Erie (Figure 9.1). In March 1991, heavy-equipment operators working at the commercial gravel pit observed bones protruding from the working face of the pit. Work was halted and Mercyhurst

(a)

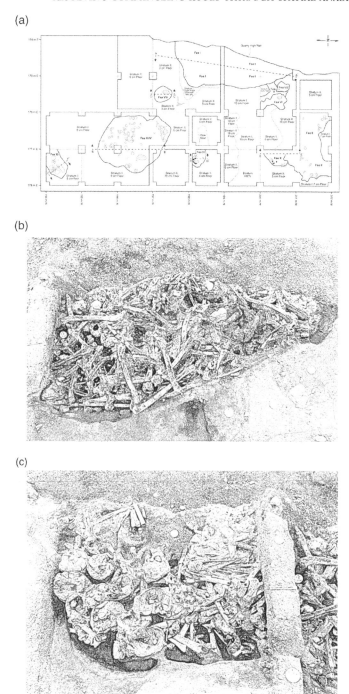

(b)

(c)

Figure 9.1   (a) Plan view map of the Orton Quarry site excavation in Pennsylvania, USA; (b) superior view of northern half of the ossuary feature; (c) superior view of southern half of the ossuary feature.

Archaeological Institute, Mercyhurst College, Erie, PA, was subsequently contracted to assess the extent of the site and begin stabilization, exploration, and recovery operations.

Initially, 10 m-wide transects were established over the terrace above and east of the gravel pit. These transects were stripped of vegetation and examined for the presence of surface artifacts. All artifacts identified in the transects were pin-flagged, mapped with a total station, and collected. Concurrent with this operation, a single north–south transect of shovel test probes was excavated across the terrace in the vicinity of the ossuary feature to assess the site's stratigraphy. The boundaries of the site east of the quarry wall were then established through the excavation of randomly selected 1 m × 1 m units, and the limits of the ossuary pit proper were determined through the excavation of 23 contiguous 1 m × 1 m units. The ossuary and 13 associated cultural features were then excavated using high-resolution archaeological methods, described in Quinn et al. (2000).

The density of bones, the severe constraints of inclement weather and the threat of vandalism required that an efficient but highly precise proveniencing technique be used in the excavation of the Orton assemblage. Following the horizontal exposure of a large number of bones (Figure 9.1a), the ossuary was photographed using a medium-format (6 cm × 7 cm) camera and the resultant high-resolution black-and-white exposures were printed on 28 cm × 35 cm paper. Tracing paper was superimposed over the photograph and used to outline each individual bone, the three-dimensional grid (Cartesian) coordinates of which were then noted on provenience tags. Each bone was lightly coated with preservative (15% Acrysol), removed from the matrix individually, and secured on supportive platforms with associated provenience information.

The 14 cultural features identified at the Orton Quarry site include the ossuary, backdirt associated with the aboriginal excavation of the ossuary, refuse pits ($n=2$), firepits ($n=5$), mixed-use fire/refuse pits ($n=2$), a probable bell-shaped storage pit, and a conical pit with an associated posthole (Figure 9.1b). The largest and most significant feature at the site is the ossuary (field designation F5). Unfortunately, approximately two-thirds of the original ossuary was destroyed as a consequence of the gravel-mining activity, leaving only the eastern third intact. This undisturbed portion measures approximately 3.5 m north–south, 1 m east–west, and only 0.5 m in depth. The original configuration of the feature was apparently circular to ovoid in plan and basin-shaped in profile. The extant walls are steep-sided to slightly "belled" and clearly have been incised with digging sticks, the parallel impressions from which were still extant on the edges of the feature. The base of the pit is essentially flat and slopes slightly to the west, where the center of the original undisturbed ossuary was located. The bone bed ranges 10–23 cm in thickness.

The entire extant surface of the ossuary was exposed to reveal three distinct burial patterns. The first pattern is a mass of bone with no readily apparent consistent or regular orientation. The elements consist of crania, disassociated mandibles, and heavier long bones such as femora, tibiae, humeri, and ulnae. This pattern occurs mainly in the northern half of the feature. The second pattern is represented by six highly compacted long-bone groups which are interpreted to be the remnants of wrapped bone bundles. These compact groups occur in the center and along the western edge of the extant portion of the feature with their long axes oriented north–south. The final interment pattern is represented by a discrete group of disarticulated bones in the southeastern portion of the ossuary. This grouping consists of layers of crania, postcranial elements, and long bones. Initial exposure of the bone bed in this portion of the site identified 14 inverted crania. Continued excavation revealed a layer of horizontally oriented long bones beneath the concentration of skulls. Other postcranial elements (e.g., scapulae, pelves, or sterna) were occasionally positioned directly beneath the inverted skulls. Beneath this layer of postcranial bones was a second layer of skulls.

The concentration of human remains in the southeastern portion of the ossuary feature (i.e., that exhibiting the third interment style) appears to lie outside of the regular, circular perimeter of the ossuary. Consequently, this concentration may represent a discrete and somewhat temporally removed burial episode (Figure 9.1c). Although artifact density in the entire feature is extremely low, no artifacts whatsoever were recovered in the southeastern portion of the ossuary. Furthermore, while evidence of a possible thermal event is indicated in the uppermost levels of the northern and central portions of the ossuary, evidence for burning is absent in the southeastern portion of the feature.

The minimum number of individuals (MNI) represented by the remains in the intact portion of the ossuary was calculated from the mastoid-petrous region of the temporal bone. The MNI for this portion of the ossuary is 77. Since we estimate that two-thirds of the ossuary was destroyed by quarrying, this figure can be extrapolated to a minimum number around 230 individuals for the entire ossuary feature. The age profile for the Orton Quarry population was derived from examination of epiphyseal union of long bones and degree of cranial suture closure. The great majority (86%) of individuals are between the ages of 20 and 50; those individuals less than 20 years old (7.3%) and greater than 50 years old (1.2%) are under-represented. Sex attribution of the cranial remains was established by examination of the supraorbital tori and mastoid and supramastoid regions, and indicates a sex division of 57.5% male and 42.5% female.

Although infrequent, cut marks are present on 6.5% (86 of 1323 skeletal elements) of cranial and postcranial elements. None of the cut marks show evidence of healing, and they may be attributable to preparation for secondary inhumation or to perimortem trauma.

Statistical methods

As explained above, the burial feature in Orton Quarry showed what appeared to be a marked secondary (i.e., postmortem) rearrangement of the human remains. There were at least 77 individuals represented, with an extremely high density of bone elements. Three distinct burial patterns were initially observed. The main feature consisted of large numbers of apparently randomly distributed commingled human bones. The other burial patterns showed rather obvious aggregated spatial distribution: six dense clusters of long bones ("bone bundles"), embedded and intermingled in the main burial feature, primarily in the central area of the grave feature, and one unique, highly ordered accumulation of predominately layered crania and long bones, clustered on the southeastern edge of the ossuary.

The bone clusters/aggregates clearly represent short-term, temporally distinct episodes that can be subjected to further analysis (e.g., studying cut-mark patterns and radiocarbon dating of the clusters); however, the main component of the ossuary is more difficult to interpret. At first glance it can be described as "a mass of chaotically positioned bone with no consistent or regular orientation" (Quinn et al. 2000: 8). If this is the case, bones were likely "thrown" into the burial pit fully disarticulated. This implies that the skeletal remains were "processed" and all soft tissue was removed prior to emplacement, or that the remains were interred in their primary burial locations for periods of time sufficient to completely decompose and remove soft tissue.

On the other hand, if there is some patterning to the emplacement of osseous remains (such as the inclusion of partially articulated individuals or body parts), a very different interpretation of cultural (burial) patterns may be invoked.

The hypothesis of random patterning seems to be supported by the presence of cut marks in some of the skeletal elements. Cut marks are present in 6.5% of all specimens, including 59 limb and 27 nonlimb (mostly scapulae and clavicles) elements. This strongly suggests the presence of a cultural pattern of defleshing and dismemberment before secondary burial, which would result in the deposition of nonarticulated or associated body parts. Still, the elements showing cut marks represent a relatively low proportion of the total number of diagnostic elements, which may be interpreted as indicating relatively long periods of primary interment and/or curation (long enough to allow for most of the bodies to become completely skeletonized before secondary burial, thus not requiring defleshing) prior to secondary deposition. A second interpretation of the relatively low frequency of cut marks might be that there was an alteration in interment patterns over time, and preburial defleshing was not practiced continuously.

In the past it may have been difficult to determine which of the alternative hypotheses (random emplacement versus patterned emplacement) provides the best solution. However, with ever-sophisticated statistical methods, "hidden" patterns can be rather easily revealed, although with the caveat that the basic data are still required. When considering spatial distribution, basic data in the form of two- or three-dimensional locations of each item considered are necessary.

In the present analysis the patterning of spatial associations between different pairs of skeletal elements was used to test whether the pattern seen in the "chaotic mass of bones" was in fact chaotic. The primary tool to be used is the bivariate Ripley's K-function analysis.

Ripley's K-function

If a set of points are randomly distributed on a plane following, for example, a Poisson distribution with λ density, the expected number of points within a circle of radius t is $\lambda \pi t^2$. The deviation from this random pattern can be quantified through the Ripley's K-function, $K(t)$ (Besag 1977; Dixon 2002; Ripley 1981; Upton and Fingleton 1985).

In its bivariate extension, Ripley's K-function $K_{12}(t)$ quantifies the interaction between two series of points (1 and 2) distributed on a plane (Lotwick and Silverman 1982) by computing the number of type 2 points located at a distance equal to or smaller than t from each focal type 1 point:

$$\hat{K}_{12}(t) = n_1^{-1} n_2^{-1} A \sum_{i=1}^{n_1} \sum_{j=1}^{n_2} w_{ij}^{-1} I_{ij}(t); \, i \neq j \qquad (9.1)$$

Where n_1 and n_2 are the total numbers of type 1 and type 2 points, respectively; A is the total area, expressed in m²; and w_{ij} is the edge effect correction factor.

The term *edge effect* refers to the smaller number of neighboring points present at the edges of the study plot when compared to its central areas. This is primarily corrected through modifications of the geometry of the relocation area. Different edge effect correction factors are available for rectangular, circular, and irregular surfaces (Diggle 1983; Getis and Franklin 1987; Goreaud and Pélissier 1999; Haase 1995; for a comprehensive comparison of some of the main methods see Yamada and Rogerson 2003).

Finally:

$$I_{ij}(t) = \begin{cases} 1 & if\ d_{ij} \leq t \\ 0 & otherwise \end{cases} \tag{9.2}$$

Where d_{ij} is the distance between the points i and j.

Therefore, contrary to the traditional closest-neighbor statistics, which are based on the mean distances between points, $K_{12}(t)$ is a second-order statistic based on the variation of such distances, combining both point counts and distance measurements (for a comprehensive comparison of Ripley's K-function with other spatial analysis techniques see Fortin et al. 2002).

Given that often $K_{12}(t) \neq K_{21}(t)$, the spatial association between two series of points is calculated as a linear combination of both bivariate statistics (Diggle and Milne 1983; Upton and Fingleton 1985):

$$\hat{K}_{12}(t) = \frac{n_2 K_{12}(t) + n_1 K_{21}(t)}{n_1 + n_2} \tag{9.3}$$

The estimated value of $\hat{K}_{12}(t)$ is then usually transformed into a new function, $L_{12}(t)$:

$$\hat{L}_{12}(t) = \sqrt{\hat{K}_{12}(t) / \pi} \tag{9.4}$$

$L_{12}(t)$ is intended to linearize and stabilize the variance, since $K_{12}(t)$ for a Poisson process is proportional to πt^2.

Hypothesis testing

The hypothesis of spatial independence (i.e., random patterning; see Figure 9.2 for an illustration of the main types of spatial relationships referred to in this chapter) is tested through Monte Carlo simulation of random circular or toroidal relocations of the type 2 points (the shape of the relocation area plays a role in the correction of the edge effect; basically, the toroidal geometry is the result of conjoining opposite edges of the study area). Confidence intervals (*envelopes*) for the null hypothesis of complete spatial randomness (CSR) at each distance t are based on the values of $K_{12}(t)$ obtained in each of the Monte Carlo iterations (Figure 9.3; see Manly 1997 for a comprehensive explanation of serial Monte Carlo methods). Departures of $K_{12}(t)$ above the upper confidence limit indicate the presence of an aggregative pattern (either

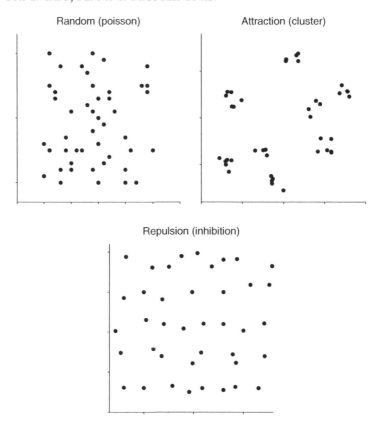

Figure 9.2 Example of the three main possible types of spatial relationship between sets of points on a surface.

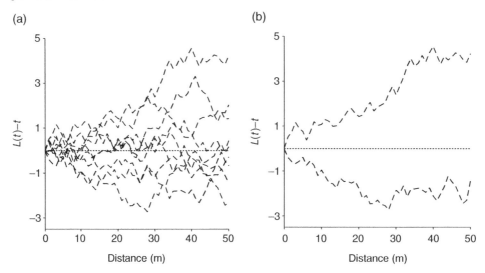

Figure 9.3 Schematic representation of the method employed to obtain the confidence envelopes for the complete spatial randomness (CSR) hypothesis. (a) $L(t) - t$ functions obtained from 10 Monte Carlo simulations. (b) Confidence intervals defined by the extreme values observed in (a).

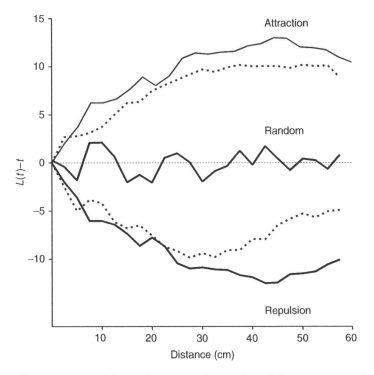

Figure 9.4   Interpretation of the random, attraction, and repulsion patterns in $L(t) - t$ plots.

clustering or association between type 1 and type 2 points), while departures below the lower confidence limit are indicative of repulsive (regular) patterning (Figure 9.4).

As the Monte Carlo-produced samples are independent between values of t, the number of confidence intervals to be estimated does not impose a restriction on the number of values of t to be considered. Thus, spatial independence can be tested at distance intervals (e.g., each 10 cm, 50 cm, etc.) chosen on the basis of research objectives, rather than on statistical considerations. The only restriction to the distances to be tested is that the maximum value of t must be equal to or smaller than half the length of the shorter side of the area under study. This is to grant that the maximum radius of the relocation area does not intercept more than two of the borders of the study area.

The capability to test the hypothesis of CSR at different distances is particularly important in Ripley's K-function analysis, as the spatial pattern observed at each distance not only indicates the presence and sign of a departure from spatial independence, but provides measurements of the *scale* at which the two types of elements are associated. Scale will be indicated by the critical values of $K_{12}(t)$; that is, those at which the $K_{12}(t)$ of the sample departs from the confidence intervals. For instance, the clustering of two particular types of skeletal element in just one area of the burial feature would be exhibited as an attraction pattern in small scales, as opposed to a regular or repulsive pattern when larger distances are considered. It is also possible to calculate the intensity of the association as a function of spatial lags. Put simply, it would reflect cluster densities at different areas. Originally described for Ripley's K-function by Getis and Franklin (1987), intensity estimates require additional calculations, and are not described in this chapter.

Within this framework, the independence from one another of individual Monte Carlo tests allows us to use the same subset of data (in the present example, a particular type of skeletal elements, such as humeri, ulnae, etc.) to test the spatial pattern at all the selected distances and, in general, different though related hypotheses, without forcing a reduction of the fixed level (α) (i.e., Bonferroni's correction). This characteristic, combined with the various analytical alternatives available to configure the statistic (e.g., edge effect correction factors or intervals of t), allows for easier modeling and design of test batteries and, as a result, allows for the testing of more complex hypotheses.

Finally, another important advantage of Ripley's K-function is its ability to produce simple and meaningful graphs that explain and visualize the main aspects of the observed pattern, namely the presence, sign, intensity, and scale of departures from spatial independence. There are three basic ways of representing graphically (through bivariate plots) the results of Ripley's K-function analyses, depending on which statistic is plotted against the distance t: $K_{12}(t)$, $L(t)$, or $L(t) - t$. The latter, $L(t) - t$, is probably the most informative, and the one employed in the present work. The main reason for subtracting t from $L(t)$ is that under CSR $K_{12}(t) = \pi t^2$ which, following equation 9.4, implies that $L(t) = t$, and therefore $L(t) - t = 0$. Spatial aggregation, therefore, translates into positive values of $L(t) - t$, while repulsion will result in negative values. This transformation thus allows for a more immediate interpretation of the sign of the pattern, both logically and visually (Figure 9.4).

Analysis of the Orton Quarry assemblage

In the present study, the statistical analyses of spatial patterning of the human remains of the Orton Quarry burial were performed using Richard Duncan's Spatial Analysis Program (He and Duncan 2000), following Getis and Franklin's (1987) method for edge correction. Only the bones in both the "chaotic" area and bone bundle areas were analyzed for spatial patterning. The area of highly ordered skulls and long bones at the southeastern portion of the grave was excluded from the analysis as, containing almost exclusively skulls and mandibles, its aggregated nature was evident and, especially, it was in spatial discontinuity with the "chaotic" and "bundles" area.

Two-tailed 95% confidence intervals for the null hypothesis were obtained after a series of 100 Monte Carlo iterations. This should be the minimal (and theoretically, sufficient) number of simulations required to estimate confidence intervals at this α level (see Manly 1997 for a comprehensive review of the matter). It has been shown that 1000 randomizations are almost certain to give the same result as the full distribution, except in rather borderline cases where p values are very close to 0.05 (Manly 1997; Marriot 1979). With the modern software and computers available today, the number of iterations to be performed is no longer a matter of concern, and so 1000 simulations are advisable for future studies (Besag and Clifford 1989; Besag and Diggle 1977; Manly 1997; Marriott 1979), and 5000 if the α level is fixed at 0.01 (Manly 1997). According to the dimensions of the shorter side of the grave (see above, in the section on Hypothesis testing), Ripley's K-function was estimated at intervals of 2.5 cm, to a maximum $t = 57.5$ cm.

The general assumption for all analyses is that if there were some degree of skeletal articulation (i.e., intervening tissue) at the time of deposition, conjoining skeletal elements would show association patterns at some scale. The expected outcomes under the null hypothesis of *no* spatial association between conjoining elements are, in all analyses: (i) association patterns at short distances (t), reflecting the presence of the bone "bundles" and (ii) random or even repulsion patterns at greater distances.

If association patterns are merely attributable to the presence of bone "bundles," association should appear at similar distance (t) ranges for all the pairs of conjoining long bones represented in these "bundles." On the other hand, if some of the conjoining elements were at least partially articulated in the "chaotic" area, different attraction patterns should be evident across distances, depending on the stability of the articulation between the different pairs of skeletal elements. Distinguishing between association patterns due to the presence of the bundles, versus those only explainable by partial articulation or anatomical association of the bones at the moment of deposition, required the comparison of the patterns displayed by different types of bones with stronger or weaker anatomical articulation.

Spatial relationships between three different pairs of skeletal elements (chosen because they had large sample sizes and were ubiquitous across the entire surface of the burial feature) were considered: (i) distal femora ($n=53$) versus proximal tibiae and fibulae ($n=49$), (ii) distal humeri ($n=46$) versus proximal radii ($n=37$), and (iii) distal humeri ($n=46$) versus proximal ulnae ($n=39$). The spatial relationship between distal femora and the pooled sample of proximal tibiae and fibulae was first studied to test for the presence of spatial association. Even if a consistent attraction pattern between the knee joint bones were found, it could still be due simply to an ordered deposition of the bones, rather than to retained partial articulation. To test this possibility, the spatial relationships between three other bones with very different soft-tissue decay rates and articulation patterns were also studied.

The upper limb elements were selected as the humero-ulnar joint is much more powerful and mechanically stable than the humero-radial articulation. Following this assumption, the formal hypothesis for analyses (ii) and (iii) is that if articulation was present at deposition in the burial feature, distal humeri would be expected to show a much stronger (consistent) pattern of association to proximal ulnae than to proximal radii, across all distances. In other words, association patterns paralleling those found in the lower limb would indicate that spatial association could still be explained by the presence of the bone bundles, while a departure from this pattern in the upper limb, with the pair humerus-ulna showing a consistently higher pattern of association than the pair humerus-radius, could only be explained by the presence of articulation or anatomical association patterns at deposition (and thus at the feature).

RESULTS

Distal femora showed a clear pattern of association with proximal tibiae and fibulae with critical $L_{12}(t)$ (intersections with the upper confidence limit) at 5.0 and 27.5 cm (Figure 9.5a). The line of observed $L_{12}(t)$ values for distal humeri and proximal radii (Figure 9.5b) runs very close to the upper confidence limit, intersecting it on nine occasions ($t=10.0$, 12.5, 17.5, 20.0, 27.5, 32.5, 37.5, 52.5, and 57.5 cm). Distal humeri and proximal ulnae show a more consistent aggregation pattern (Figure 9.5c). $L_{12}(t)$ departs from the upper confidence limit at 12.5 cm, and does not intersect the limit back into the acceptance area until $t=47.5$ cm.

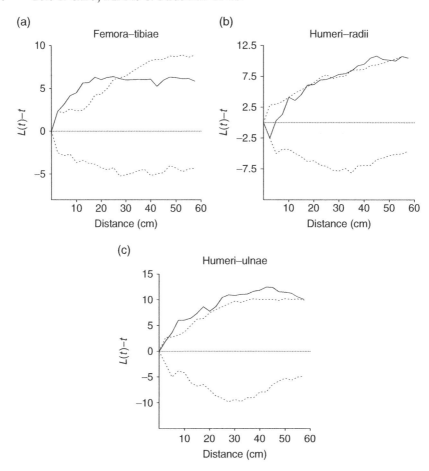

Figure 9.5 Results obtained for the spatial relationships between (a) distal femora and pooled proximal fibulae and tibiae, (b) distal humeri and proximal radii, and (c) distal humeri and proximal ulnae (see text for explanation).

The spatial association at small to intermediate scales of femora and tibiae/fibulae pairs could be due to the presence of the bone bundles, not necessarily implying articulation or anatomical association of the skeletal elements at the time of emplacement in the burial feature. However, the association patterns exhibited by the upper limb pairs stands in stark contrast with the null hypothesis of "no articulation" of skeletal elements at the time of deposition into the burial feature. The association pattern between humeri and ulnae is evident at all distances, which cannot be explained by the bundles alone, especially when the results for lower limb elements are considered. The differences with the weaker pattern of association shown by the humeri-radii pair, which intersects the upper envelope limit at all distance ranges, precisely matches the predictions made under the alternative hypothesis of "articulation," and can be explained by the stronger musculoskeletal connections at the humero-ulnar joint, when compared to the humero-radial joint. Further, the combined results of all three analyses are inconsistent with the hypothesis of nonarticulation at the time of body disposal, and strongly indicate that association patterns between conjoining skeletal elements are still present, detectable, and quantifiable in the sample.

These results effectively demonstrate the capability of spatial analysis techniques – specifically Ripley's K-function – to detect and quantify skeletal articulation patterns even in highly disturbed sets of commingled human remains, even when merely considering a bidimensional Cartesian space (although the method, in its current configuration, needs further testing in deeper graves). It must again be emphasized that this is possible only when the proper recordation and analysis techniques are applied during recovery efforts.

These findings are relevant for at least four aspects of HRIs and, in general, of any forensic scenario where commingled multiple victims are present:

1. the ability to distinguish between primary and secondary burials, based on the degree of alteration of association patterns; criminal attempts to further conceal evidence can be documented and objectively proven;
2. individualization of interment episodes, detected by different patterns of association in different areas of the feature;
3. approximate estimation of sequence and temporal lapse between episodes; older episodes are expected to show more relaxed spatial association patterns than more recent ones. On the other hand, the duration of time lapses between depositional episodes will also be reflected in the articulation patterns, which in turn are based on the degree of decomposition attained by soft tissues;
4. improvements in victim individualization and identification; spatial analysis of point data using closest-neighbor techniques has already proven to be a highly reliable method for matching skeletal elements in commingled cases, with success rates close to 100% (Tuller et al. 2008). However, for secondary burials in which postdepositional disturbances may have taken place it may seem that the spatial patterns are far too complex to render any useful information.

DISCUSSION

The results presented in this chapter show that, even after intense postburial alteration, spatial associations between articulated elements still persist, and it is possible to estimate the spatial scale at which this association is present and detectable. Therefore, the proximity criterion underlying the closest-neighbor analysis presented by Tuller et al. (2008) still applies to these types of crime scenes. Analysis of the intensity of association through Ripley's K-related techniques will most likely allow for estimation of probability areas where this proximity criterion can be applied (the closest neighbor would be substituted by all potentially matching bones within a given distance, to complement the morphological similarity criterion).

It is important to stress at this point that the method described here is but one of many possible hypothesis-testing designs, probably representing the simpler of such designs available for this type of spatial analysis. Employing a higher number of Monte Carlo iterations, utilizing more advanced edge-detection methods or using a greater variety of intensity indices, and even dividing the feature surface into subareas would allow for far more complex and powerful analyses. The range of analytical techniques available and the number of specific problems that they are able to confront are very vast, and are growing by the day. Two new alternative three-dimensional methods – specifically designed for situations in which, as in mass graves, the entire area of the feature under study is occupied by different subsets of elements – are being developed and tested by the authors. The flexibility of these simulation methods allows for the development of new methodologies in an almost case-specific manner. In addition to purely statistical techniques, the range of potential designs based on different sample parameters and alternative hypotheses is almost endless. For example, in the present case other skeletal elements can be considered, the "bundles" removed, or more precise anatomical points be used.

More importantly, the results and observed patterns can be compared between different sites, and the influence of relevant factors modeled and assessed. For example, distinct clusters of skeletal remains can be detected. This will allow for identification of specific depositional episodes, while the degree of disarticulation can be quantified through comparison with other graves, potentially allowing for estimation of postdepositional time intervals (based on the degree of perturbation showed by the articulation patterns). In this way, common patterns among graves can be detected and precisely described, even if they are not immediately evident.

Although much more powerful analyses and software are currently available, the software utilized in the study serves to illustrate another aspect of the importance of exhaustive scene data recovery. Duncan's Spatial Analysis Program was one of the first PC programs available for Ripley's K analysis. It was still being tested while the backhoes uncovered the first human remains at Orton Quarry, and was published later in 1991. At that time, Ripley's K-function analysis had been barely used beyond the boundaries of system ecology and, to our knowledge, it had never been previously applied to archaeological or any other anthropological context. Still, the availability of high-quality spatial and contextual data, comprehensively documented through systematic and carefully designed archaeological excavation, allowed for the sophisticated analysis of the Orton Quarry remains more than a decade after its excavation.

The lesson here, then, is that at first glance the exhaustive recordation of spatial and contextual data may erroneously appear to be overkill and unnecessary (or presented as particularly cost-inefficient). But, given the rapid development of analytical techniques, and their potential for solving a wide variety of important issues (many still unrealized), it is foolhardy, short-sighted, and irresponsible to take the easy way out and not fully document these ephemeral features using state-of-the-field contextual documentation techniques.

The Orton Quarry example illustrates how the types of information potentially retrievable from well-excavated ossuary features (and by default, mass graves) are many and varied. They include: (i) the size and shape of the burial feature itself, (ii) the instrument used to excavate the feature (whether shovel or backhoe) through tool mark evidence on both the walls in the floor of the feature, (iii) unique/discrete episodes of excavation or infilling, (iv) sequences of individual body/body-part placement, (v) primary (complete bodies) versus secondary (decomposing partially to fully disarticulated remains) deposition, and (vi) evidence of attempts to hide evidence by removing dirt or adding more dirt to the top of the grave the feature, or even partial removal of bodies (Dirkmaat and Adovasio 1997; Hochrein 2002; Skinner et al. 2002).

The analysis also illustrates how documentation strategies cannot be designed based on immediate short-term objectives alone, but must be aimed at retrieving and safeguarding the highest possible amount and quality of information, foreseeing that other relevant analyses, new, more sophisticated or simply unknown to us at the moment, may be available in the future. The Orton Quarry site was excavated in the early 1990s, prior to the availability of many of the rather sophisticated statistical analytical tools discussed here. And still, we are able to apply the full suite of analyses on the data collected. We must reiterate that the forensic archaeological excavation of a large-scale, multiple-individual burial feature is not merely removing dirt from around interred remains and evidence (i.e., exposure of bodies or "exhumation"). It is the reconstruction of the complete history of that feature from initial digging of the hole to professional excavation, and it requires advanced skills in archaeological excavation and interpretation.

As mentioned in the introduction, the concern is that with the rapid expansion in the number of mass-grave excavations that are being concurrently conducted throughout the world, a large number of these projects are being excavated by inexperienced or even inappropriate individuals (Skinner et al. 2003: 82) under a plethora of rather severe constraints. When perceived or real constraints are placed on forensic investigations there is a tendency to cut corners with respect to the proper collection of forensic evidence inherent in these scenes. The perception is that if a detailed forensic archaeological recovery of such features is attempted, the sheer size of the feature, the large number of victims interred, and apparent complexity of the feature will result in tremendous overruns in both time and resources. This is often couched in terms of delaying the return of victims to families.

These arguments can be countered at two levels. First, if we are conducting "forensic" recoveries, the primary goal should be on the collection of forensic evidence that can someday be used in a court of law. As forensic scientists we should not be swayed from that objective. Obviously, the humanitarian consideration of judiciously identifying victims in order to return remains quickly to families is not an ill-considered goal. However, approaching the scene from a scientific/forensic perspective allows for both the proper, contextually based recovery of forensically significant evidence and potentially enhances (and certainly does not detract from) the ability to identify victims. An archaeological excavation is obviously a destructive process and once completed we cannot go back and extract new data if context is not considered at the outset of the recovery.

The primary intent of human-rights activities, as originally documented in published doctrines, is to collect evidence that leads to successful prosecution of the guilty. This requires that all efforts must be made to apply current methods in the forensic sciences in the collection and interpretation of evidence in a court of law. When dealing with the excavation of mass graves with the expressed goal of obtaining forensic scientific evidence, only methodologies and expertise derived from traditional archaeology/forensic archaeology are acceptable. As illustrated by the discussion of the Orton Quarry ossuary feature, the new, more sophisticated data-acquisition techniques and instrumentation [global positioning system (GPS), geographical information systems (GIS), total station surveying instruments, hand-held scanners, etc.] are not merely introduced in the field as an attempt to ease the field work of the professionals, but to respond to a parallel, even more dramatic improvement in the arena of analytical techniques, which require more precise and larger amounts of data, offering in exchange the possibility of answering a much larger range of relevant questions, in a far more powerful way than traditional methods.

Conversely, objectivity in the collection, analysis, and interpretation of evidence is tantamount in the forensic sciences. There is no talk on the stand or in published descriptions of forensic science principles, methods, and techniques focusing exclusively on the needs of the victim's families. When one side or another is taken, objectivity and potential bias come under scrutiny and the effort is rejected as a scientific pursuit. When the goal is the acceptable presentation in a court of law there exists no difference between a lone burial feature in the middle of a cornfield in Iowa and a large mass grave in Bosnia.

In sum, the excavation of mass graves, now a common occurrence in the investigation of human rights, is a very complex undertaking. Merely exposing and removing bodies from burial contexts – properly termed "disinterment" or "exhumation" – should never be confused with proper forensic archaeological excavations. Without the documentation of detailed contextual associations of all evidence (human remains and personal effects) with geophysical, environmental, and archaeological evidence, the expansive reconstruction of past events is not possible; moreover, such reconstructions will remain anecdotal and hence inadmissible in court. On the other hand, proper forensic archaeological recovery does not necessarily entail dramatic time and resource overruns. When done properly, a well-thought-out excavation protocol results in maximal data collection within a reasonable time period. The recovery becomes both temporally and cost efficient and effective, even under difficult circumstances, as illustrated by the superb work of many HRI teams in countries like Argentina or Guatemala. Hence, constraints which are significant enough to compromise the application of forensic science to body and evidence recovery may indicate that the project should not be started until the commissioning agency can provide better resources, or the situation becomes stable and favorable enough to proceed with the investigation. The classification as a humanitarian effort, with the resulting dramatic decrease in its utility for prosecutorial purposes, should be an exceptional last-resort measure, derived from truly extraordinary circumstances, rather than the norm.

REFERENCES

Adams, B.J. and Konigsberg, L.W. (2004). Estimation of the most likely number of individuals from commingled human skeletal remains. *American Journal of Physical Anthropology* 125(2): 138–151.

Besag, J. (1977). Contribution to the discussion of Dr. Ripley's paper. *Journal of the Royal Statistical Society Series B (Statistial Methodology)* 39: 193–195.

Besag, J. and Diggle, P.J. (1977). Simple Monte Carlo tests for spatial pattern. *Applied Statistics* 26: 327–333.

Besag, J. and Clifford, P. (1989). Generalized Monte Carlo significance tests. *Biometrika* 76: 633–642.

Diggle, P.J. (1983). *Statistical Analysis of Spatial Point Patterns.* Academic Press, London.

Diggle, P.J. and Milne, R.K. (1983). Bivariate Cox processes: some models for bivariate spatial point patterns. *Journal of the Royal Statistical Society Series B (Statistial Methodology)* 45: 11–21.

Dirkmaat, D.C. and Adovasio, J.M. (1997). The role of archaeology in the recovery and interpretation of human remains from an outdoor forensic setting. In W.D. Haglund and M.H. Sorg (eds), *Forensic Taphonomy: The Postmortem Fate of Human Remains* (pp. 39–64). CRC Press, Boca Raton, FL.

Dirkmaat, D.C., Cabo, L.L., Adovasio, J.M., and Rozas, V.C. (2005). Mass graves, human rights and commingled remains: considering the benefits of forensic archaeology. Paper presented at the 57th annual meeting of the American Academy of Forensic Sciences, New Orleans.

Dixon, P.M. (2002). Ripley's K function. In A.H. El-Shaaraui and W.W. Piergorsch (eds), *The Encyclopedia of Environmetrics* (pp. 1796–1803). Wiley, New York.

Doretti, M. and Snow, C.C. (2002). Forensic anthropology and human rights: the Argentine experience. In D.W. Steadman (ed.), *Hard Evidence: Case Studies in Forensic Anthropology* (pp. 290–310). Prentice Hall, Upper Saddle River, NJ.

Fortin, M.J., Dale, M., and Ver Hoef, J. (2002). Spatial analysis in ecology. In A.H. El-Shaaraui and W.W. Piergorsch (eds), *The Encyclopedia of Environmetrics* (pp. 2051–2058). Wiley, New York.

Getis, A. and Franklin, J. (1987). Second-order neighborhood analysis of mapped point patterns. *Ecology* 68: 473–477.

Goreaud, F. and Pélissier, R. (1999). On explicit formulas of edge effect correction for Ripley's K-function. *Journal of Vegetation Science* 10: 433–438.

Haase, P. (1995). Spatial pattern analysis in ecology based on Ripley's K-function: Introduction and methods of edge correction. *Journal of Vegetation Science* 6: 575–582.

Haglund, W.D. (2002). Recent mass graves, an introduction. In W.D. Haglund and M.H. Sorg (eds), *Advances in Forensic Taphonomy: Method, Theory, and Archaeological Perspectives* (pp. 243–261). CRC Press, Boca Raton, FL.

Haglund, W.D., Connor, M.A., and Scott, D.D. (2001). The archaeology of contemporary mass graves. *Historical Archaeology* 35: 57–69.

He, F. and Duncan, R.P. (2000). Density-dependent effects on tree survival in an old-growth Douglas fir forest. *Journal of Ecology* 88: 676–688.

Hessel, S. (2011). *Time for Outrage!* Quartet Books, London.

Hochrein, M.M. (1997). Buried crime scene evidence: the application of forensic geotaphonomy in forensic archaeology. In P.G. Stimson and C.A. Mertz (eds), *Forensic Dentistry* (pp. 83–99). CRC Press, Boca Raton, FL.

Hochrein, M.J. (2002). An autopsy of the grave: recognizing, collecting, and preserving forensic geotaphonomic evidence. In W.D. Haglund and M.H. Sorg (eds), *Advances in Forensic Taphonomy: Method, Theory, and Archaeological Perspectives* (pp. 45–70). CRC Press, Boca Raton, FL.

Hunter, J.R., Brickley, M.B., Bourgeois, J., Bouts, W., Bourguignon, L., Hubrecht, F., De Winne, J., Van Haaster, H., Hakbijl, T., De Jong, H., Smits, L., Van Wijngaarden, L.H., and Luschen, M. (2001). Forensic archaeology, forensic anthropology and human rights in Europe. *Science & Justice: Journal of the Forensic Science Society* 41: 173–178.

Lotwick, H.W. and Silverman, B.W. (1982). Methods for analysing spatial processes of several types of points. *Journal of the Royal Statistical Society Series B (Statistial Methodology)* 44: 406–413.

Manly, B.J. (1997). *Randomization, Bootstrap and Monte Carlo Methods in Biology*, 2nd edn. Chapman & Hall, London.

Marriott, F.H.C. (1979). Barnard's Monte Carlo tests: how many simulations? *Applied Statistics* 28: 75–77.

Quinn, A.G., Adovasio, J.M., Pedler, C.L., Pedler, D.C., Dirkmaat, D.C., and Buyce, M.R. (2000). The Orton Quarry Site (36er243) and the late prehistory of the Lake Erie Plain. Paper presented at the symposium *Current Archaeological Research in Pennsylvania and Related Areas, 65th Annual Meeting of the Society for American Archaeology*, April 5–9, Philadelphia, PA. http://mai.mercyhurst.edu/.

Ripley, B.D. (1981). *Spatial Statistics*. Wiley, New York.

Schmitt, S. (2002). Mass graves and the collection of forensic evidence: genocide, war crimes, and crimes against humanity. In W.D. Haglund and M.H. Sorg (eds), *Advances in Forensic Taphonomy: Method,*

Theory, and Archaeological Perspectives (pp. 277–292). CRC Press, Boca Raton, FL.

Skinner, M. (1987). Planning the archaeological recovery of evidence from recent mass graves. *Forensic Science International* 34(60): 267–287.

Skinner, M., York, H.P., and Connor, M.A. (2002). Postburial disturbance of graves in Bosnia-Herzegovina. In W.D. Haglund and M.H. Sorg (eds), *Advances in Forensic Taphonomy: Method, Theory, and Archaeological Perspectives* (pp. 293–308). CRC Press, Boca Raton, FL.

Skinner, M., Alempijevic, D., and Djuric-Srejic, M. (2003). Guidelines for international forensic bio-archaeology monitors of mass grave exhumations. *Forensic Science International* 134: 81–92.

Steadman, D.W. and Haglund, W.D. (2005). The scope of anthropological contributions to human rights investigations. *Journal of Forensic Sciences* 50: 23–30.

Tuller, H.H., Cropper, C., Yazedjian, L., Sterenberg, J., and Davoren, J.M. (2005). New tools for the processing of human remains from mass graves: spatial analysis and skeletal inventory computer programs developed for an inter-disciplinary approach to the re-association of commingled, disarticulated and incomplete human remains. Paper presented at the 57th annual meeting of the American Academy of Forensic Sciences, New Orleans.

Tuller, H.H., Hofmeister, U., and Daley, S. (2008). Spatial analysis of mass grave mapping data to assist in the reassociation of disarticulated and commingled human remains. In B.J. Adams and J.E. Byrd (eds), *Recovery, Analysis, and Identification of Commingled Human Remains* (pp. 7–29). Humana Press, Totowa, NJ.

Ubelaker, D.H. (2002). Approaches to the study of commingling in human skeletal biology. In W.D. Haglund and M.H. Sorg (eds), *Advances in Forensic Taphonomy: Method, Theory, and Archaeological Perspectives* (pp. 331–351). CRC Press, Boca Raton, FL.

Upton, G.J.G. and Fingleton, B. (1985). *Spatial Data Analysis by Example, Vol. 1. Point Pattern and Quantitative Data.* Wiley, Chichester.

Williams, E.D. and Crews, J.D. (2003). From dust to dust: ethical and practical issues involved in the location, exhumation, and identification of bodies from mass graves. *Croatian Medical Journal* 44(3): 251–258.

Yamada, I. and Rogerson, P.A. (2003). An empirical comparison of edge effect correction methods applied to *K*-function analysis. *Geographical Analysis* 35: 97–109.

PART III Developments in Forensic Osteology

Introduction to Part III

Luis L. Cabo

The seven chapters in this part focus on the assessment of the key elements that define the basic biological profile of the individual: age, ancestry, sex, and stature. As discussed in Chapter 1, the study of these four areas constituted the very core of forensic anthropology as originally defined and classically approached for most of the twentieth century. By assessing and identifying these four elements the forensic anthropologist was able to reduce the list of potential matches, providing an important service to aid in victim identification. Forensically oriented research in this area also provided maybe the most important contributions of forensic anthropology to its parent field, physical anthropology, in the form of extensive comparative skeletal collections and reliable methods for the assessment of all the components of the biological profile, continuously refined, tested, and filtered through the reality checks of everyday forensic practice under the demanding conditions of the medicolegal system.

Forensic research on the different components of the biological profile also came to define the scope and personality of forensic anthropology within physical anthropology: while paleoanthropology and other related subfields of physical anthropology sought to infer populational characteristics parting from the individual, forensic anthropology bought the roundtrip ticket, summoning the populational data back to infer the biological profile of isolated individuals.

As also discussed in Chapter 1, the history of the discipline during the last few decades can be understood in terms of a gradual but decisive broadening in scope and objectives, departing from the original sole goal of aiding in victim identification through the assessment of the biological profile. However, the addition of new tasks and goals must not be interpreted as a weakening of this original objective. Forensic anthropologists may now process scenes using forensic archaeological techniques, or help in reconstructing the events surrounding death, through forensic taphonomy and trauma analysis, but the assessment of the biological profile of the victim is still

A Companion to Forensic Anthropology, First Edition. Edited by Dennis C. Dirkmaat.
© 2012 Blackwell Publishing Ltd. Published 2012 by Blackwell Publishing Ltd.

their bread and butter, and the necessary starting point for most other analyses. In a diversified environment in which narrow specialization in some complex areas is an increasingly necessary trend, what unites and defines one and all forensic anthropologists is their common expertise in forensic osteology.

This section of the book reflects the dual nature of forensic osteology as a centenary field with a wealth of traditional methods to know and master, but also as a vibrant one booming within a rapidly growing discipline, immersed in an extremely exciting process of expansion, maturation, and conceptual redefinition. In this way, the assessment of age at death and biological sex, the two subjects with perhaps the longest tradition, are treated each in two separate chapters, intimately overlapping with the discussion of ancestry, stature, and the statistical methods and resources discussed in the remaining chapters.

Chapter 10 offers a critical review of the most widespread classic and novel methods for age assessment, providing an extended bibliography with the key references on the subject, as well as important advice and hints for their review and evaluation. The chapter can be understood as a compass to navigate the massive body of literature on the subject developed during more than one century, but rather than just enumerating a series of methods, the authors also use them to articulate a discussion on the general problems and constraints of age assessment at large, as well as the most challenging and promising research questions ahead.

Once the reader has become acquainted with the main age-estimation techniques and literature, in Chapter 11, Milner and Boldsen further develop the questions posed in the previous chapter, taking a closer look at the biological and statistical underpinnings of age estimation. In that chapter, the authors review key conceptual and methodological issues rarely discussed in the classic literature, such as the requisites to ask from truly reliable skeletal age markers, the importance of the reference samples employed to define them and infer their correlation with age, or the estimation procedures and statistical methods applied to obtain the corresponding age intervals.

Chapters 12 and 13 follow a similar approach in relation to the assessment of biological sex. In Chapter 12, Garvin presents a critical review of the most widespread sex-estimation methods, posing key methodological questions that are addressed in depth in Chapter 13. Special attention is paid to subjects like allometric scaling, biomechanics, growth and development, and the basics of sexual selection, all intimately related to sexual dimorphism and sex determination, and rarely discussed from an integrative perspective in the anthropological literature. As with age determination, the twin chapters combine to provide more than a set of methodological recipes, serving to supply important tools and insight for the critical evaluation, utilization, and development of sex-assessment methods. As with the chapters devoted to age assessment, these two chapters also offer a thoughtful glimpse into the most promising research areas for the future development of the field, both from the practical and conceptual points of view.

In Chapter 14, Hefner, Ousley, and Dirkmaat adopt a similar highly conceptual approach to discuss the estimation of ancestry. The authors utilize a historical perspective to delineate and explain the coevolution of both the conceptual framework and the estimation methodologies associated with ancestry assessment. In their discussion, the authors pay special attention to issues such as the comparison between metric and nonmetric techniques, or between different statistical methods for the analysis of ancestry markers, arming the reader with a good understanding and decision-making tools to evaluate and select different alternatives. They close their exposition with a practical example that puts the theoretical considerations to work, offering a template for the application of the conceptual and practical considerations discussed in the paper to the reader's own analyses. The chapter focuses more on the analysis of nonmetric traits, as metric analyses are discussed in depth in Chapter 15.

As the proverbial rug, Chapter 15 ties the sex- and ancestry-devoted chapters together with a comprehensive examination of metric methods for the assessment of sex and ancestry. Ousley and Jantz focus on analyses from linear measurements through their acclaimed software Fordisc. However, the chapter is far more than a quick reference guide or a brief version of the program's manual, discussing in depth discriminant function analysis (DFA), the family of statistical techniques chiefly utilized in Fordisc. The authors review and provide extremely useful insight and advice regarding issues like sample requirements, best practice and usage cautions, or the correct interpretation of each statistic in DFA outputs. The section on Fordisc usage does not limit itself either to supplying the reader with optimal guidelines and advice to use and interpret the program correctly, but also stops to discuss conceptual and methodological issues such as the nature and main components of human craniometric variation. Finally, Chapter 16 examines the most straightforward component of the biological profile – stature estimation – taking in this case a more practical approach.

In sum, taken as a whole, this section provides the reader with a set of invaluable conceptual and practical tools to approach and review critically the literature on age, sex, ancestry, and stature assessment, understand the biological and methodological underpinnings behind each component of the biological profile, and apply the new knowledge to successfully confront both real case investigations and future research on the subject. Departing from the usual cookbook approach, the authors stop often to consider and explain general biological and statistical issues affecting the analysis and interpretation of the individual components of the profile, and of the biological profile as a whole. The result is a set of tightly interrelated chapters that can be read at different levels and are relevant to the novice reader seeking to get acquainted with forensic osteology for the first time, the practicing professional always in need of keeping updated with the current literature and trends in the field, and the researcher embarked on the development of new methods. We believe that the section comes to provide a very rare resource within the forensic anthropological literature, examining the issue with a depth and breadth very rarely observed in previous volumes. The practical advice and, especially the conceptual insights offered by the authors, will solve many doubts and save more than one headache to both the student and practitioner, and we are sure that readers will find themselves coming back to the chapters in this part of the book very often, long after the first reading.

CHAPTER **10** Developments in Forensic Anthropology: Age-at-Death Estimation

Heather M. Garvin, Nicholas V. Passalacqua, Natalie M. Uhl, Desina R. Gipson, Rebecca S. Overbury, and Luis L. Cabo

ADULT SKELETAL AGE ESTIMATION

Age estimation is an important step in constructing a biological profile from human skeletal remains. The goal of the forensic anthropologist is to assist medicolegal officials with identification by presenting a probable age range of the deceased. In adults, this is typically done by examining various skeletal traits which have been shown to degenerate with age in a predictable manner. Using trait characteristics, forensic anthropologists strive to provide as narrow an estimated age range as possible, but, as you will see, human variation in degenerative traits and variation in the rate of the aging process necessitates somewhat broader age-range estimates.

Chronological versus biological age

One major source of lack of precision in age estimates is disassociation of chronological and biological age (see Nawrocki 2010 for a full discussion). Chronological age is strictly defined by time: how many calendrical years, months, and days have passed since birth. Without a known birth date exact chronological age cannot be determined.

A Companion to Forensic Anthropology, First Edition. Edited by Dennis C. Dirkmaat.
© 2012 John Wiley & Sons Ltd. Published 2015 by John Wiley & Sons Ltd.

Biological age, however, refers to the physiological state of an individual, which is reflected in skeletal remains. Because a general correlation exists between biological and chronological age, forensic anthropologists use the biological age estimate (from the remains) to predict chronological age (recorded in missing persons files). Biological age, however, is dependent on genetic and environmental factors, and consequently activity, health, and nutrition may all influence biological age by altering the aging rate of various tissues (including the skeleton). Because these influences may vary between individuals, at any given chronological age, individuals within a population may display various biological ages (İşcan 1989). Furthermore, as an individual's chronological age increases, so does the accumulation of these extrinsic factors resulting in greater variation in biological age and hence broader age estimates.

Precision versus accuracy

Faced with human variation in the aging process and the pressure from law officials for narrow estimates, forensic anthropologists are constantly compromising between precision and accuracy. The narrower, or more precise, the age estimate given, the more helpful it can be to law enforcement when eliminating possible identities. However, as you narrow an age estimate, you increase the probability of accidentally eliminating the true identity of the individual. In contrast, a broader age range is more likely to include the true age, but may not be as helpful when attempting to narrow a missing persons list.

Forensic anthropologists typically approach this dilemma using statistical data presented by the methods to determine confidence or prediction intervals. These intervals allow the forensic anthropologist to say with a certain degree of confidence (typically 95%), that the given age range will encompass the true age of the deceased. Most skeletal aging methods are developed by recording observed skeletal trait characteristics and categorizing certain characteristics into phases, which are accompanied by various statistical age descriptors. Among the statistical data reported by methods are: observed age ranges, mean ages and standard errors or standard deviations, confidence intervals, and prediction intervals.

Currently there are no standards regarding which statistical information aging methods should report and which information should be used for age-estimation procedures. However, forensic anthropologists should be cognizant of which statistics are reported and their subtle differences in order to correctly evaluate methods and interpret age-estimation results.

Juvenile versus adult

Although this chapter focuses on adult skeletal aging, a brief description of juvenile aging techniques is warranted. Because growth and development are programmed more strictly by evolution and genetics than adult degenerative processes (Crews 1993; Zwaan 1999), skeletal growth characteristics such as long-bone lengths, epiphyseal fusion, and dental eruption provide a more precise and accurate indication of age than most adult skeletal traits. In particular, dental development has been shown to be less environmentally sensitive, thereby resulting in the most reliable indicators of juvenile chronological age (Scheuer and Black 2000).

While more recent studies exist, traditional sources consulted by forensic anthropologists include: Fazekas and Kósa (1978) for fetal diaphyseal lengths, Stewart (1979) for lengths of diaphyses at birth, Johnston (1962), Ubelaker (1978) or Hoffman (1979) for infant and juvenile long-bone lengths, Moorrees et al. (1963a, 1963b) for dental calcification, Ubelaker (1978) or Schour and Massler (1941) for dental eruption sequences, and Stewart (1979) and Scheuer and Black (2000) for epiphyseal fusions. Many of these studies are based on samples differing in ancestry and health status, and therefore researchers are encouraged to use the most sample-appropriate method, and should always be aware of possible sample biases. Scheuer and Black (2000) provide a detailed account of developmental osteology which is an essential resource when working with infant and juvenile remains. It describes the developmental stages for each skeletal element from birth through adulthood, and presents general male and female age ranges at each developmental stage when possible.

The complete fusion of all long-bone epiphyses, the eruption of the third molars, and fusion of the spheno-occipital synchondrosis (basilar suture) are all used as markers of adulthood. With a few exceptions, such as medial clavicle fusion (see below), growth and development are then complete, and further age-related changes are the result of skeletal maintenance and degeneration.

Traditional Adult Skeletal Aging Methods

While the skeletal system undergoes numerous transformations with age, adult skeletal aging methods have traditionally focused on four main regions: the pubic symphysis, auricular surface, sternal rib ends, and cranial sutures. Studies describing predictable age-related changes in these skeletal traits have been documented since the early 1920s. Since then, methods using these traits have been evaluated and reevaluated on various samples, and many of the original methods continue to be among the most popular adult aging techniques.

Pubic symphysis

The pubic symphysis has long been regarded as the most reliable skeletal indicator of age in adults (Meindl et al. 1985). Todd was one of the first to begin documenting age-related changes in the pubic bones of European males in 1920, and published similar studies on European females and African-American males and females in 1921 (Todd 1920, 1921). He described 10 morphological phases with associated age ranges. The general trait progression described by Todd begins with billowing of the pubic symphysis in young adults. These billows begin to fill in on the dorsal margin and ossific nodules appear on the superior and inferior surfaces. As the billows continue to fill in across the rest of the symphyseal surface and the ossific nodules spread, dorsal and ventral margins become better defined, forming the symphyseal rim in middle-aged adults. Finally, as age progresses, the pubic symphysis is characterized by more degenerative features such as lipping, erosion, and breakdown of the symphyseal rim (Figure 10.1).

While subsequent pubic symphysis aging methods and resultant age ranges vary slightly, all methods are based on these general trait characteristics and sequence of

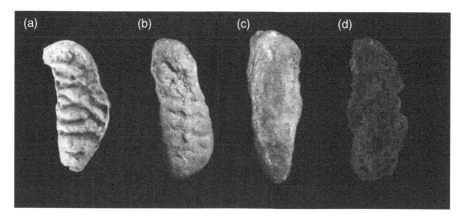

Figure 10.1   Age-related changes to the pubic symphysis. Age progression of the pubic symphysis from young to old (left to right). Young adults display a characteristic billowy surface (a). With age, ossific nodules form and the billows begin to disappear (b). This is accompanied by the formation of a dorsal plateau and a ventral rampart. Eventually the billows disappear, and the symphyseal rim is complete (c). Erratic ossification and irregular lipping and margins are characteristic of older individuals (d). Refer to Brooks and Suchey (1990) for detailed descriptions, illustrations, and accompanying age ranges. Photographs taken by Heather Garvin.

transformation. For example, McKern and Stewart (1957) used Todd's descriptions and a sample of Korean War dead to develop a three-component system of pubic symphysis age estimation. Their method involved scoring the dorsal plateau, ventral rampart, and symphyseal rim development, each on a scale of 0 to 5. The total score was then compared to a chart with given age ranges (±2 standard deviations). They argued that a component system was less restricting than a phase system since each of the trait characteristics could be scored independently. It should be noted, however, that their method was developed on war dead (primarily young white males) and consequently has a tendency to under-age older individuals. Gilbert and McKern (1973) presented a similar three-component system method applicable to females. A recent survey suggests that a good portion of forensic anthropologists still use these three methods for age determination (Garvin and Passalacqua 2011).

In a large study of male pubic symphyses Katz and Suchey (1986) pointed out considerable problems with the samples and techniques of both Todd (1920, 1921) and McKern and Stewart (1957) (see also Suchey et al. 1986). The Suchey–Katz sample was a large, documented, modern sample of 739 males autopsied at the Department of Chief Medical Examiner-Coroner, County of Los Angeles. Ages at death ranged from 14 to 92 years, and contained individuals from a diverse background (birthplaces in 32 countries). The pubic symphyses were first scored according to the Todd (1921) and McKern and Stewart (1957) methods. The observed ranges were found to be much wider than those reported in the original studies. Their results supported earlier studies (Brooks 1955; Meindl et al. 1985) that found that the Todd system systematically over-ages individuals and that neither the Todd nor the McKern and Stewart system can account for the sum total of human variation, especially in older age phases. Furthermore Katz and Suchey (1986) rejected the three-component approach of McKern and Stewart (1957) asserting that the three components do not vary independently and that an approach focusing on the entire pattern of morphological change (i.e., Todd's phase method) is easier to use. In light of these conclusions, they proposed a modified Todd method, where the 10 phases have been reduced to six phases.

In 1990, Brooks and Suchey added 273 female pubic symphyses to the original male sample of 739 individuals to refine the morphological descriptions of the six phases and present sex-specific age ranges. They reported that previous research justified the need for a separate set of standards for females because of shape and pregnancy-related changes in the pelvis. Because of their large sample, detailed phase descriptions, and availability of corresponding casts, the Suchey–Brooks method is the most widely used method today for aging the pubic symphysis (Garvin and Passalacqua 2011). Brooks and Suchey (1990) argue that their method is appropriate for use in a wide array of contexts because the Los Angeles sample was derived from individuals born throughout North America, Europe, South America, and Asia, and includes diverse socioeconomic backgrounds. However, they warn that for forensic work the large amount of actual variation observed in phases III–VI must be kept in mind and that multiple age indicators should be employed whenever possible. Recently the Brooks and Suchey method was revised with a contemporary morgue sample, altering the age ranges slightly and adding a seventh phase dealing with more advanced age morphologies (Hartnett 2010).

Auricular surface

Lovejoy et al. (1985a) first proposed using the auricular surface of the ilium as an indicator of age at death because they noted a high correlation between skeletal age indicators and morphological changes of the auricular surface. Since their publication, this method has been widely used, but it has only recently been subjected to scientific scrutiny. The auricular surface is of great importance because it is more durable than other skeletal elements used for aging, and its morphology does not appear to be affected by sex or ancestry (Osborne et al. 2004). As with pubic symphyseal studies (Meindl et al. 1985), Lovejoy et al. (1985a) stress the importance of understanding human aging as a process. They recommend eight phases with corresponding age ranges based on observations. The first seven ranges are narrow, each only encompassing 5 years, while the final phase encompasses all individuals of 60 years and older. The age-related progression of characteristics starts with a young billow surface that gradually reduces to transverse striae. Around middle age, surface granularity increases followed by densification and appearance of micro- and macroporosity. Finally, as in the pubic symphysis, old age is characterized by overall breakdown, lipping, and irregularities (Figure 10.2).

In 2002, Buckberry and Chamberlain undertook a validation study of the auricular surface method. In an attempt to accommodate more individual variation and make the method easier to apply, they developed a component system similar to the McKern and Stewart (1957) pubic symphysis method. Using the descriptions in Lovejoy et al. (1985), Buckberry and Chamberlain (2002) devised progressive scores for transverse organization, surface texture, microporosity, macroporosity, and morphological changes in the apex and retroauricular areas. After a preliminary test of these indicators, the retroauricular area was determined to be a poor individual indicator of age

(a)

(b)

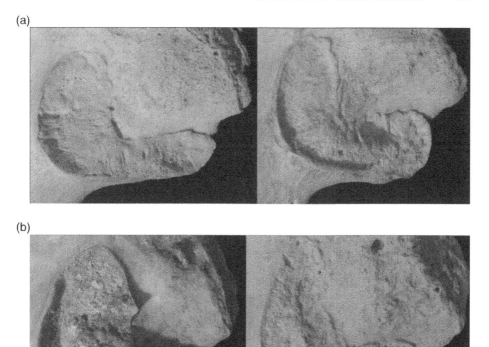

Figure 10.2 Age-related changes to the auricular surface. Examples of the auricular surface in younger (a) and older (b) adult individuals. In young adults the auricular surface is finely granular with marked transversely organized billows (a). With age, the transverse organization disappears, and the texture becomes more coarsely granular with the appearance of islands of dense bone (b). Eventually all granulation is lost, micro- and macroporosity increases, and irregular changes to the apex and retroauricular areas become evident. Photographs taken by Nicholas Passalacqua.

and was excluded from the method. Each of the remaining components, however, demonstrated a high correlation with true age and was therefore retained in the analyses. Following their method, each of the components are scored independently and then summed to create a composite score. The authors provide a table which groups these composite scores into seven auricular surface stages, and provide corresponding age ranges and statistics for each stage. Blind tests conducted on the sample from the Christ Church Spitalfields Collection (at the Natural History Museum of London) suggest low intraobserver errors. Although Buckberry and Chamberlain suggest that their method performed well on the Christ Church Spitalfields Collection, further testing on more diverse modern populations with a better representation of younger individuals is necessary. The Buckberry and Chamberlain method was recently tested on a sample of nineteenth–twentieth-century US blacks and whites from the Terry and Huntington Collections housed at the Smithsonian Institution National Museum of Natural History in Washington DC, USA (Mulhern and Jones 2004). The results suggested that the method was applicable to both black and white males and females, although they found that the original Lovejoy et al. (1985a) method more accurately aged younger individuals (20–49 years of age).

In 2004, Osborne et al. tested the Lovejoy et al. (1985a) method on modern samples from the Terry Collection, and on the Bass Donated and Forensic Collections (at the University of Tennessee), and found that only 33% of the individuals were correctly aged using the original 5-year age ranges. The results of the analysis indicate that the 5-year age ranges originally published by Lovejoy et al. (1985a) are much too narrow to capture the entire range of human variation in the auricular surface. Very young and very old individuals (phases I–III and phases VII–VIII) were over-aged while middle-age individuals (phases IV–VI) were under-aged. In an attempt to increase accuracy, each age range was expanded to include the phase before and after it, providing age ranges of 15 years or more. Accuracy, however, only increased slightly, and still remained relatively low (59% correct). In an attempt to increase accuracy and provide statistical descriptors, Osborne et al. (2004) calculated new mean ages and 95% prediction intervals for each of the eight Lovejoy et al. (1985a) phases. Furthermore, they found that the prediction intervals for the first and last phases completely overlapped, and therefore combined them, resulting in a modified six-phase approach. While accuracy significantly increased (with percentage correct in the upper 90 percentiles), the prediction intervals span, on average, 43.5 years (assuming a lower limit of 18 years).

A more recent study by Igarashi et al. (2005) used a large Japanese sample ($n=700$) to develop a new auricular method using a binary classification system. In their method 13 surface and texture variables are scored for presence/absence. The total number of present features was summed and mean ages and standard deviations calculated. Based on these age ranges, composite groups were created, but the probability densities for these groups overlapped too much to facilitate aging. The authors then presented a different approach, using the binary scores in a multiple regression analysis with dummy variables to obtain an estimated age. Based on whether a trait is scored as present or absent, a certain value can be added or subtracted, and the total sum of the 13 values is matched with an estimated age range. As with any new method, however, further evaluation on different samples and additional statistical analyses regarding accuracy, standard deviations, and intra- and interobserver errors are necessary for validation.

Cranial sutures

Scientists have been aware of the connection between the extent of cranial suture closure and age at death since Vesalius first published on the subject in 1542 (Masset 1989; Todd and Lyon 1924). Over the next four centuries a few scientists, including Welcker, Ferraz de Macedo, and Frédéric, continued to explore this relationship (Masset 1989).

In 1924, Todd and Lyon published an extensive study of endocranial suture closure in Euro-American males, which was followed by an ectocranial suture study in 1925 (Todd and Lyon 1924, 1925). They initiated this research stating "the whole question of the relation of suture union to age remains an intricate and unsolved problem" (Todd and Lyon 1924: 329). Their study included over 300 crania of known ages ranging from 18 to 84 years of age. Based on broad observations of closure, the authors attempted to apply suture closure to individual skulls as an indicator of age at death and subsequently caution that the "results in individual cases leave much to be desired" (Todd and Lyon 1924: 379). Todd and Lyon conclude that cranial suture closure is not a very reliable aging indicator; however, they do feel that suture closure is valuable for individual cases when used in conjunction with other parts of the skeleton.

Meindl and Lovejoy (1985) published a new method for age estimation from cranial suture closure which involved observing 1 cm lengths of specific sites on sutures (for repeatability) and a four-point scoring system: 0=no observable closure, 1=minimal closure (about 1–50%), 2=significant closure (about 50–99%), 3=complete obliteration (Figure 10.3). Using the previous literature as a guide, the authors narrowed down 10 specific sites for observation and limited their study to ectocranial sutures to increase the practicality. These 10 sites were divided into the "vault system" and the "lateral-anterior system," and modal patterns were investigated in each. The lateral-anterior system was found to be more regular. Similar to the pubic symphyseal aging techniques of McKern and Stewart (1957) and Gilbert and McKern (1973), the authors produce a table of composite scores and a mean age and standard deviation for each score. However, as with other aging techniques, the standard deviations are rather large, as are the observed ranges. The authors note that "the relationship between degree of closure and age is therefore only general" (Meindl and Lovejoy 1985: 62).

In 1998, Nawrocki expanded on the work of Meindl and Lovejoy (1985) and developed a method of scoring 27 cranial landmarks along ectocranial, endocranial, and palatal sutures. Nawrocki (1998) used 100 crania from the Terry Collection, including 50 females (25 European-American and 25 African-American) and 50 males (24 European-American and 26 African-American), ranging in age from 21 to 85 years with a mean age of 53.71 years. Sutures on the vault were scored in 1 cm segments on a four-point scale, in accordance with Meindl and Lovejoy's method. Palatal sutures were observed along their entire length and are scored on the same scale.

Nawrocki's results show a moderate correlation between age and the summed cranial suture score for an individual. Group-specific equations were developed that improve the correlation (with r^2 values as high as 0.86 for African-American females) and reduce the standard errors. The outcome produced two general equations, and six ancestry and sex-specific equations for the entire cranium. They also presented two additional general equations and five group-specific equations based only on the calotte.

Zambrano (2005) reevaluated and tested Nawrocki's methods and found that the general "All Groups" equation out-performed the ancestry- and sex-specific equations, based on the percentage of individuals whose actual age fell within the ±2 standard error intervals. Further, Zambrano tested for secular trends and found that although Nawrocki's equations were developed on the nineteenth–twentieth-century Terry Collection, they are applicable to modern, forensic casework. Overall, however, because of the broad age intervals, most forensic anthropologists report relying on cranial suture methods only when other postcranial elements are not available or to determine a general age group (young versus old) (Garvin and Passalacqua 2011). In an interesting examination of cranial suture closure recently conducted by Kroman

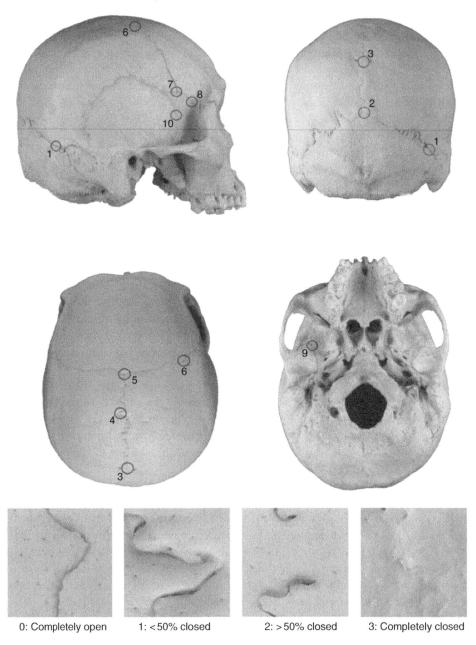

| 0: Completely open | 1: <50% closed | 2: >50% closed | 3: Completely closed |

Figure 10.3   Top: regions scored by Meindl and Lovejoy (1985) in their cranial suture age-estimation method. Circled are the ten 1 cm areas on the vault and lateral-anterior aspects of the crania described by Meindl and Lovejoy (1985). Using their method, the sutures in each of these regions is scored from 0 to 3 based on the degree of sutural closure (see lower panel) and the compiled composite score is then associated with estimated age ranges. Bottom: Meindl and Lovejoy (1985) ectocranial suture closure scoring criteria. Each centimeter region is scored based on the degree of closure, ranked from 0 to 3 as shown. These 1 cm regions are scored independent of changes occurring in any other portion of the sutures. Photographs taken by Dominique Semeraro and Nicholas Passalacqua.

and Thompson (2009), they suggest that cranial suture closure is actually more closely correlated to somatic dysfunction (e.g., sacroiliac fusion, ankylosing spondylitis, severe scoliosis) than advancing age.

Sternal rib ends

Osteological changes to the sternal rib ends have also been shown to be useful in adult age estimation. Currently, most anthropologists employ the technique described by İşcan et al. (1984a, 1984b). Expanding on the work of Kerley (1970) and Ubelaker (1978), İşcan et al. (1984a) first described changes in three components of the right fourth rib morphology: pit depth, pit shape, and rim wall, creating a component system like many of the other original aging methods. Later that year, however, they converted their component system into a phase method, similar to the Suchey–Brooks methods for the pubic symphysis. This new phase method was developed from right fourth rib of 118 modern white males, and described the same aging characteristics as the original component system. It is this phase method which forensic anthropologists continue to use and may even refer to as "the rib-end method." Eight phases were developed based on age-related changes, including the formation, depth, and shape of a pit, the configuration of the walls and rim around the pit, and the overall texture and quality of the bone. With chronological age, the sternal rib ends proceed from a flat and billow articular surface, to deep V-shaped and then wide U-shaped pit. The rib rims, which begin as rounded, become scalloped or wavy and eventually more and more irregular with thin sharp edges. The overall bone quality of the rib also deteriorates with age, most notably by decreasing in density (Figures 10.4a and b).

Investigation into population differences in the metamorphosis of the sternal extremity of the rib continued for several years, primarily by İşcan and colleagues (İşcan 1991; İşcan et al. 1985, 1987). These studies found significant differences in the timing of morphological changes for different sexes, ancestries, and even occupations. İşcan et al. (1987) published new statistics and photos for African-American males and females. While these studies are widely used and cited, it should be noted that the sample sizes are much reduced. For example, the African-American female sample includes 14 individuals. While population differences may exist, methods based on such small sample sizes cannot be used with high confidence. Russell et al. (1993) found when testing the phase rib-end method on the Hamann–Todd Collection (at the Cleveland Museum of Natural History) that not only was the overall method accurate and reliable, but use of the white standards was successful in estimating the age of black individuals.

One possible advantage to the method of İşcan et al. is that changes to the rib ends can be observed using medical imaging. Dedouit et al. (2008) successfully applied the İşcan et al. phase method to two- and three-dimensional computed tomography images. If this can be further applied to radiography, a few simple X-rays taken at the medical examiner's office could produce immediate preliminary age estimates prior to bone processing, on fresh bodies, or in scenarios when skeletal maceration is not possible.

On a similar note, radiographic analysis of costal cartilage ossification has also been suggested as a forensic aging technique. Many of the rib-end osteophytic changes described in the İşcan et al. method are related to the calcification or ossification of

(a)

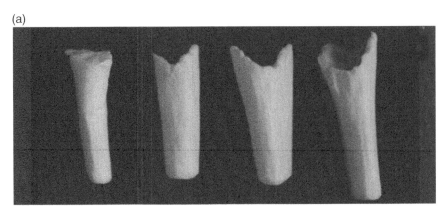

(b)

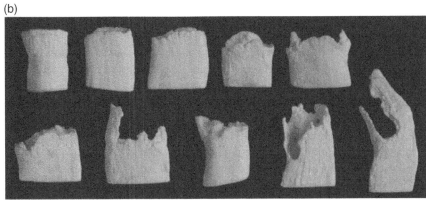

(c)

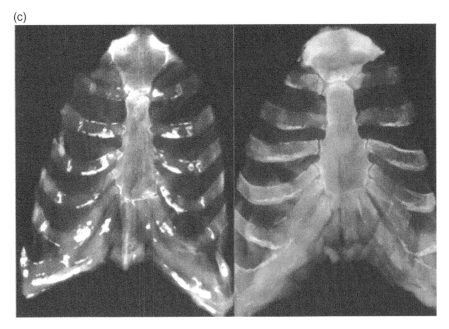

Figure 10.4 (a) Age progression of the sternal rib pit formation. Superior views of four sternal rib end casts (France casting), illustrating the progression of pit formation with age (youngest to oldest) as described by Iscan et al. (1984a, 1984b, 1985, 1987). While the sternal rib end originally displays billows and no pit, an amorphous indentation will occur with age.

the costal cartilage, creating at times long, "craggy," bony extensions. Sensibly then, radiographic analysis of the overall cartilage ossification patterns could be useful in adult age estimation.

The use of costal cartilage ossification patterns for age determination has been explored by numerous researchers (Barchilon et al. 1996; Barres et al. 1989; Eichelberger and Roma 1954; King 1939; McCormick 1980; McCormick and Stewart 1983; Semine and Damon 1975). Although anthropologists are likely familiar with McCormick's work on costal cartilage ossification, it is seldom used in the field. In a recent study by Garvin (2010), general age progression trends in the anterior chest plate (costal cartilage and sternum) were described. Sternal body, manubrium-sternum, and xiphoid-sternum fusion were used to describe minimum age estimates. For example, fusion of the manubrium and xiphoid indicated an individual was at least 25 years of age. Costal cartilage ossification was typically first observed at the manubrium notch, followed by peristernal ossification (at the sternum costal notches). Centrichondral ossification was more variable, and displayed sex-specific patterns, but indicated an individual was at least 30 years of age (Figure 10.4c).

Garvin (2010) developed a new costal cartilage age-estimation method using the presence/absence of eight chest-plate characteristics to create a composite score. Descriptive statistics were provided for the composite scores, including standard deviations and observed age ranges. Garvin found that because the method utilized presence/absence scores, it was equally applicable to males and females despite sexual differences in ossification patterns. Comparable to most other adult aging methods, standard deviations in ages ranged from 3.1 years in the early stages of ossification up to 16.9 years in the later stages. While the use of cartilage ossification for forensic skeletal aging needs further validation, cartilage ossification sequences may prove useful in the medical examiner setting and may reveal further information on sternal rib-end aging.

OTHER METHODS

While the traditional four skeletal traits described above are the most commonly used and evaluated in adult skeletal aging, there are a multitude of other variables and methods which forensic anthropologists may use in conjunction with the conventional methods.

Figure 10.4   (*cont'd*) This is followed by a V-shaped pit, and then a narrow U-shaped pit which continues to widen with age. (b) Age progression of sternal rib end margins. Anterior views of sternal rib-end casts (France casting), illustrating the progression of rim margins with age (top left, youngest; bottom right, oldest) as described by Iscan et al. (1984a, 1984b, 1985, 1987). Note the increased irregularities, thinning of the rim walls and osteophytic activity with age. (c) Costal cartilage ossification with age. Examples of a female (left) and male (right) chest plate radiographs taken at autopsy displaying degrees of costal cartilage ossification consistent with older age. Note the sex-specific patterns of ossification with the female ossification characterized by dense ossific globules and the male displaying linear ossific extensions from the superior and inferior rib margins. Photographs taken by Heather Garvin.

Medial clavicle epiphyseal fusion

Because the epiphysis of the medial clavicle is the last to fuse, it holds potential in aging young adults. Fusion typically begins at puberty but is not completed until the late 20s or early 30s. Langley-Shirley and Jantz (2010) provide a concise historical account of medial clavicle aging methods and present a new method, scoring the degree of fusion of the medial epiphysis, and supply descriptive statistics and results from a Bayesian approach.

Maxillary suture closure

Mann et al. (1987) first proposed a method of estimating skeletal age from the fusion of maxillary sutures on a small sample of 36 individuals. In 1991 Mann and colleagues evaluated and revised the method on a larger sample ($n = 186$; Mann et al. 1991). Their method involves scoring the various maxillary sutures as open, partial fusion, or complete fusion and then comparing those scores with tables presenting minimum age at start of fusion and minimum age at complete fusion. Ginter (2005) tested Mann et al.'s revised method on a large sample of South African individuals ($n = 155$) and documented an accuracy rate of 83% correct, determining that the method is useful when used in conjunction with other methods.

Tooth-root translucency

Lamendin et al.'s (1992) method using tooth-root translucency is another popular adult aging method. It involves viewing a single rooted tooth under a light (or on a light box), and taking two measurements: the distance from the cement-enamel junction to the line of soft-tissue attachment and the height of transparency from the apex of the tooth root. These measurements are then placed into a formula to calculate an age estimate. Lamendin et al. reported error estimates of approximately 8 years. Prince and Ubelaker (2002) tested the Lamendin method on the Terry Collection and, after coming up with similar error estimates, developed sex- and ancestry-specific equations.

Dental cementum annulations

Dental cementum annulations have recently been proposed as an accurate skeletal aging method in humans. The method involves cross-sectioning a tooth and counting the number of alternating dark and light bands in the cementum. Wittwer-Backofen et al. (2004) provide a detailed description of band formation and procedures involved. While the method is destructive and requires specific equipment, Wittwer-Backofen et al. report error estimates of less than 2.5 years. Further evaluation, however, is required to determine how accuracy may vary across samples or in remains exposed to various environmental factors.

Bone histology

Histomorphometric analysis of cortical bone has many of the same downfalls as the dental cementum annulations. It is destructive, requires special equipment and training, and is relatively time-consuming. On the other hand, it is objective and can be applied to fragmentary or burned remains (Bradtmiller and Buikstra 1984). Kerley (1965) was the first to describe a histological method for estimation of age from cortical bone. The method involved counting the number of osteons, osteon fragments, and non-Haversian canals and inserting the counts into bone-specific regression formulae to obtain an age estimate. Kerley and Ubelaker (1978) provided revised formulae and method specifications. Since then, numerous researchers have evaluated the use of bone histology in aging, with somewhat mixed results (e.g., see Lynnerup et al. 1998; Ericksen 1991; Stout 1988; Stout and Gehlert 1980; Stout and Paine 1992).

Osteoarthritis

Although the formation of osteoarthritic characteristics (e.g., ostephytes and lipping) are certainly related to age, because of the degree of variation their use in skeletal age estimation remains limited. Stewart (1958) published an aging method using degree of vertebral osteoarthritis. Snodgrass (2004) used Stewart's five-stage classification system to further evaluate patterns of osteoarthritis with age. While Snodgrass confirmed a significant correlation between age and degree of osteophytic activity he admits a high degree of variation, suggesting the use of arthritis in general age estimation or for determining lower and upper age boundaries.

THINGS TO CONSIDER WHEN AGING A SKELETON

Sex, ancestry, and age

Because intrinsic and extrinsic factors both influence age, it is important to control for these factors whenever possible. This includes using sex- and ancestry-specific study information when available. The closer you can match your study sample to your case, the more accurately you should be able to estimate age. This, of course, is assuming that the sex- and ancestry-specific information provided is based on well-developed studies of large sample sizes, which is not always the case. For example, in the sternal rib study by İşcan et al. (1987), the statistics for black females are based on an inadequate total sample size of 14. It is up to the forensic anthropologist to take note of such details and make an educated decision on which methodological standards to apply. On the other hand, however, Konigsberg et al. (2008) caution that while there are minor age-trait variations between populations, these differences are not necessarily significant, and that more emphasis should be placed on obtaining larger reference samples in order to better understand the age-related changes than on focusing on interpopulation variations.

It has also been found that certain skeletal aging methods perform better for certain age groups. If a certain method has been found to consistently under-age older individuals or over-age younger individuals, this should be taken into consideration. Given the general age range of the individual, certain methods may be more applicable and accurate than others. Two articles, by Cunha et al. (2009) and Ritz-Timme et al. (2000), address some of these concerns by evaluating numerous skeletal aging methodologies (both traditional and more recent) and providing recommended approaches dependent on the specific scenarios (condition of remains, elements present, general age, sex, and race). Interestingly, they do not agree on all recommendations. For example, Ritz-Timme et al. (2000) suggests using the rib ends to estimate ages under 40 years, while Cunha et al. (2009) suggest that the ribs are most reliable for ages over 60 years. It is therefore clear that further evaluations and possible revisions of adult skeletal aging methods should be conducted. However, the overall idea of forming approach recommendations and summarizing which techniques are most reliable under specific situations could help standardize forensic skeletal aging techniques.

Asymmetry

It is important to note that the morphological timelines by which osteological markers are known to progress are not always symmetrically stable. Developmental and degenerative rates may vary across skeletal elements and between left and right sides. Biomechanical forces can vary across traits and sides, influencing the expression of age characteristics. The progression of age-associated markers is also influenced by both the length of the maturation period (Halgrimsson 1995; Kobyliansky and Livshits 1989) and environmental factors (Albert and Greene 1999). Therefore, with a prolonged biological maturation and a gamut of highly variable environmental influences, human development has ample opportunity for the accumulation of asymmetry. Regardless of the causal factors, biological asymmetry could possibly interfere with the accuracy of aging of skeletal remains.

A recent study utilizing the Suchey–Brooks method as a model for discrete phase-based aging methods found asymmetrical aging characteristics in 63% of the left and right symphyseal faces from a modern population of 140 white males (Overbury et al. 2009). Almost 75% of this asymmetry was great enough to cause conflicting Suchey–Brooks phases. This conflict creates discrepancies when both right and left elements are present and can produce inaccuracies when an individual is represented unilaterally. There are presently no standards regarding the handling of asymmetric traits. Some argue that the elemental side used should be the same as what was used in the study (if documented). Others prefer an average of the two sides (Garvin and Passalacqua 2011). Overbury et al. (2009), however, found that when applying the Suchey–Brooks method the morphologically older element of an asymmetrically phased individual was the most accurate (increasing accuracy rates from 78 to 91%). Similar future studies may not only help increase the accuracy rates of aging methods, but may help anthropologists better understand the aging process.

Multifactorial approaches

Just as left and right sides may vary, different skeletal traits may be under different influences and hence reflect different biological ages. Consequently, it is a commonly held notion that multiple indicators of age at death used together are more precise than single indicators. However, there are currently no standards regarding how to combine information from multiple methods. Some common practices include: using the overlap of age ranges provided by the studies, using the entire range of all the studies, or combining the lowest range of the method providing the oldest age and the highest range of the method providing the lowest age [in a recent survey (Garvin and Passalacqua 2011) this last technique was described by participants as a technique presented by Kerley]. Others may prefer to use the age ranges presented by methods they feel are most reliable, disregarding other inconsistent estimates. Still others will use a combination of techniques to provide both a more conservative broad estimate and narrower "most likely" age range to officials. In reality, however, none of these techniques are statistically valid, given that different methods are developed on different samples, under different assumptions, and may even present different statistical information. True multifactorial methods devised from numerous traits and methods utilizing transition analysis can resolve these statistical dilemmas, but remain relatively underused in the field.

With the exception of transition analysis (see below), most attempts at combining methods revolve around regression approaches (e.g., Aykroyd et al. 1999; Martrille et al. 2007; Uhl 2008a). The methods combined, however, usually only include the conventional traits (cranial sutures, pubic symphysis, auricular surface, and sternal rib ends) and disregard other aging methods. Furthermore, publication of such multifactorial approaches is rare. Lovejoy et al. (1985b) provide a multifactorial summary aging technique, but it was developed for estimating age distributions in archaeological populations and is not applicable to individual forensic age estimates (Kemkes-Grottenthaler 2002). Samworth and Gowland (2007) present a method for using single or combined age estimates from the Suchey–Brooks pubic symphysis and/or the Lovejoy et al. (1985a) auricular surface methods. Passalacqua (2010) evaluated the effectiveness of these look-up tables and found them to outperform the original single methods, although final results are still slightly below ideal.

Transition analysis

Age-at-death estimation poses many challenges for osteologists because the very nature of aging markers creates statistical problems that should be addressed rather than ignored. One alternative way of approaching age estimation is using transition analysis, so termed because the analysis relies upon the estimated age of transition between adjacent stages of an age phase or age trait. For a classification scheme to be valid, morphological change must progress with age along a consistent sequence of distinguishable phases, where no phase is skipped or revisited. Transition analysis can be used for any aging indicator that is arranged in a series of discrete stages. This is a valuable technique because most of the commonly used age methods use discrete phases and the transition analysis eliminates some of the statistical issues inherent in discrete data.

In traditional age-at-death estimation, osteologists think of an individual's age as dependent on an aging indicator. For example, one might see a pubic symphysis that is a Suchey–Brooks stage IV and report that age at death was in the late 30s (Brooks and Suchey 1990). The first step in transition analysis is to essentially invert that process. Using a sample of known-age individuals scored for an age marker, an ordinal probit model generates the probability of being in a certain morphological state conditional on age. The parameters from this analysis can be converted, via maximum likelihood analysis, to the mean and standard deviation for a distribution of the age at transition from one stage to the next (Boldsen et al. 2002). Konigsberg created a

Fortran program that uses a probability density function for calculating mean age at transition for single skeletal traits. More information and free download of the program are available at http://konig.la.utk.edu/nphases2.htm, and access to other related computer program scripts may be found at https://netfiles.uiuc.edu/lylek/www/.

While transition analysis is fairly straightforward for a single skeletal trait, osteologists agree that it is ideal to use more than one aging indicator when estimating age at death (e.g., Brooks 1955; Lovejoy et al. 1985; Uhl 2007). However, this becomes statistically complicated for a number of reasons. Beside the complexity from adding more parameters, the assumption of independence cannot be made for multiple aging indicators on the same skeleton. If those indicators are all varying with age it is likely that they are correlated with each other and this will affect the statistical inferences. An easy way to correct this problem in age-at-death estimation is to condition all of the indicators on age because it is unlikely that these indicators are correlated in any way other than the information they provide about age at death. This conditioning allows osteologists to calculate an age-at-death estimate from several skeletal indicators at once.

Boldsen et al. (2002) developed the ADBOU computer program, which leads the user through data collection on several skeletal traits (pubic symphysis, auricular surface, and cranial sutures) and then uses transition analysis to estimate age at death. While it seems that a computer program may make skeletal analysis simpler for the less experienced user, there is some manual input required. For example, the output includes two maximum likelihood estimates: one that is calculated with a uniform prior and one that is calculated with an informed prior. If the program user indicates that the unknown remains are "archaeological" the reference population for the informed prior is a seventeenth century Danish cemetery population. If the remains are deemed "forensic" the informed prior is from 1996 US homicide data. The efficacy of the informed priors provided has been questioned because this program has shown only limited success with modern Americans (Bethard 2005) and modern South Africans (Uhl 2008). Thus, while transition analysis does offer solutions to problems that have traditionally plagued age-estimation techniques (e.g., age-mimicry, inaccurate representation of estimation uncertainty, open-ended age-intervals), it is important to note that this method, like any aging method, works only as well as the associated reference samples and scoring systems allow.

Despite issues with the informed priors, ADBOU does have several advantages. Many multifactorial methods use techniques that cannot handle missing data (e.g., multiple regression), but ADBOU can calculate a maximum likelihood estimate with the smallest amount of data. However, if the remains are complete ADBOU collects a large amount of data. There are 19 components from three skeletal elements that are recorded. Some have questioned whether it is advantageous to break skeletal scoring into several components rather than just scoring elements as one morphological unit (Passalacqua and Uhl 2009). Another advantage of ADBOU is that the output includes the multifactorial likelihood estimate as well as a maximum likelihood estimate for each separate skeletal element. These maximum likelihood estimates are point estimates, but 95% confidence intervals are also included.

Overall, the ADBOU program gives anthropologists a wonderful resource for statistically combining age-at-death estimates from multiple skeletal estimates. Some information may be lost with the use of component scoring and uniform priors; however, the maximum likelihood estimates ultimately stand on solid statistical ground.

FUTURE DIRECTIONS

It has been stated that, despite recurrent attempts to quantify age at death, "age determination is ultimately an art, not a precise science" (Maples 1989: 323). Not only is this statement incorrect, but detrimental to the discipline of forensic anthropology in the face of the Daubert challenge (*Daubert v. Merrell Dow Pharmaceuticals* 1993; Christensen 2004). The utmost goal of the forensic anthropologist is to perform, within the purview of the scientific method, skeletal analysis and identification, including challenging parameters such as age-at-death estimation.

With the progression of the field of forensic anthropology and the need to continuously update and validate methods, traditional approaches to age-at-death estimation may be challenged. The development of such probabilistic statistical procedures as transition analysis (e.g., Boldsen et al. 2002) are in fact not only promising, but groundbreaking. However, transition analysis methods may not necessarily outperform other regression attempts at multifactorial aging (Uhl 2008b), and it is likely that we will see continued advancements in both these areas. The future of age-at-death estimation lies in the hands of researchers and it is through the investigation of unexplored anatomical areas and new statistical procedures that age-at-death methods will meet not only judicial requirements, but mathematical and practical needs as well.

REFERENCES

Albert, A.M. and Greene, D.L. (1999). Bilateral asymmetry in skeletal growth and maturation as an indicator of environmental stress. *American Journal of Physical Anthropology* 110: 341–349.

Aykroyd, R.G., Lucy, D., Pollard, A.M., and Roberts, C.A. (1999). Nasty, brutish, but not necessarily short: a reconsidering of the statistical methods used to calculate age at death from adult human skeletal and dental indicators. *American Antiquity* 61(1): 55–70.

Barchilon, V., Hershkovitz, I., Rothschild, B., Wish-Baratz, S., Latimer, B., Jellema, L., Hallel, T., and Arensburg, B. (1996). Factors affecting the rate and pattern of the first costal cartilage ossification. *American Journal of Forensic Medicine and Pathology* 17: 239–247.

Barres, D.R., Durigon, M., and Paraire, F. (1989). Age estimation from quantification of features of "chest plate" x-rays. *Journal of Forensic Sciences* 34: 28–33.

Bethard, J. (2005). *A Test of the Transition Analysis Method for Estimation of Age-at-Death in Adult Human Skeletal Remains.* MA thesis, University of Tennessee, Knoxville, TN.

Boldsen, J.L., Milner, G.R., Konigsberg, L.W., and Wood, J.W. (2002). Transition analysis: a new method for estimating age from skeletons. In R.D. Hoppa and J.W. Vaupel (eds), *Paleodemography: Age Distributions from Skeletal Samples* (p. 73–106). Cambridge University Press, Cambridge.

Bradtmiller, B. and Buikstra, J.E. (1984). Effects of burning on human bone microstructure: a preliminary study. *Journal of Forensic Sciences* 29: 535–540.

Brooks, S.T. (1955). Skeletal age at death: the reliability of cranial and pubic age indicators. *American Journal of Physical Anthropology* 13: 567–589.

Brooks, S. and Suchey, J.M. (1990). Skeletal age determination based on the os pubis: a comparison of the Ascadi-Nemeskeri and Suchey-Brooks methods. *Journal of Human Evolution* 5: 227–238.

Buckberry, J.L. and Chamberlain, A.T. (2002). Age estimation from the auricular surface of the ilium: a revised method. *American Journal of Physical Anthropology* 119: 231–239.

Christensen, A.M. (2004). The impact of Daubert: implications for testimony and research in forensic anthropology (and the use of frontal sinuses in personal identification) *Journal of Forensic Sciences* 49: 427–430.

Crews, D.E. (1993). Biological anthropology and human aging: some current directions in human aging research. *Annual Review of Anthropology* 22: 395–423.

Cunha, E., Baccino, E., Martrille, L., Ramsthaler, F., Prieto, J., Schuliar, Y., Lynnerup, N., and Cattaneo, C. (2009). The problem of aging human remains and living individuals: a review. *Forensic Science International* 193: 1–13.

Daubert v. Merrell Dow Pharmaceuticals, 509 US 579, 113 S.Ct. 2786, 125 L.Ed. 2d 469 (US June 28, 1993) (No. 92–102).

Dedouit, F., Bindel, S., Gainza, D., Blanc, A., Joffre, F., Rouge, D., and Telmon, N. (2008). Application of the Iscan method to two- and three-dimensional imaging of the sternal end of the right fourth rib. *Journal of Forensic Sciences* 53: 288–295.

Eichelberger, L. and Roma, M. (1954). Effects of age on the histochemical characterization of costal cartilage. *American Journal of Physiology* 178: 296–304.

Ericksen, M.F. (1991). Histologic examination of age at death using the anterior cortex of the femur. *American Journal of Physical Anthropology* 84: 171–179.

Fazekas, G and Kósa, F. (1978). *Forensic Fetal Osteology*. Akadémiai Kiadó, Budapest.

Garvin, H.M. (2010). Limitations of cartilage ossification as an indicator of age at death. In K. Latham and M. Finnegan (eds), *Age Estimation of the Human Skeleton* (pp. 118–133). Charles C. Thomas, Springfield, IL.

Garvin, H.M. and Passalacqua, N.V. (2011). Current practices by forensic anthropologists in adult skeletal age estimation.

Journal of Forensic Sciences doi 10.1111/j.1556-4029.2011.01979.x.

Gilbert, B.M. and McKern, T.W. (1973). A method for aging the female os pubis. *American Journal of Physical Anthropology* 38: 31–38.

Ginter, J.K. (2005). A test of the effectiveness of the revised maxillary suture obliteration method in estimating adult age at death. *Journal of Forensic Sciences* 50: 1303–1309.

Halgrimsson, B. (1995). Interspecific variation in fluctuating asymmetry among primates. *American Journal of Physical Anthropology* 20: 104.

Hartnett, K.M. (2010). Analysis of age-at-death estimation using data from a new, modern autopsy sample - part I: pubic bone. *Journal of Forensic Sciences* 55: 1145–1151.

Hoffman, J.M. (1979). Age estimations from diaphyseal lengths: two months to twelve years. *Journal of Forensic Sciences* 24: 461–469.

Igarashi, Y., Uesu, K., Wakebe, T., and Kanazawa, E. (2005). New method for estimation of adult skeletal age at death from the morphology of the auricular surface of the ilium. *American Journal of Physical Anthropology* 128: 324–339.

İşcan, M.Y. (1989). Research strategies in age estimation: the multiregional approach. In M.Y. İşcan (ed.), *Age Markers in the Human Skeleton* (pp. 325–339). Charles C. Thomas, Springfield, IL.

İşcan, M.Y. (1991). The aging process in the rib: an analysis of sex- and ancestry-related morphological variation. *American Journal of Human Biology* 3: 617–623.

İşcan, M.Y., Loth, S.R., and Wright, R.K. (1984a). Metamorphosis at the sternal rib end: a new method to estimate age at death in white males. *American Journal of Physical Anthropology* 65: 147–156.

İşcan, M.Y., Loth, S.R., and Wright, R.K. (1984b). Age estimation from the rib by phase analysis: white males. *Journal of Forensic Sciences* 29: 1094–1104.

İşcan, M.Y., Loth, S.R., and Wright, R.K. (1985). Age estimation from the rib by phase analysis: white females. *Journal of Forensic Sciences* 30: 853–863.

İşcan, M.Y., Loth, S.R., and Wright, R.K. (1987). Racial variation in the sternal extremity of the rib and its effect on age determination. *Journal of Forensic Sciences* 32(2): 452–466.

Johnston, F.E. (1962). Growth of the long bones of infants and children at Indian Knoll. *American Journal of Physical Anthropology* 20: 249–254.

Katz, D. and Suchey, J.M. (1986). Age determination of the male os pubis. *American Journal of Physical Anthropology* 69: 427–435.

Kemkes-Grottenthaler, A. (2002). Aging through the ages: historical perspectives on age indicator methods. In R.D. Hoppa and J.W., Vaupel (eds), *Paleodemography: Age Distributions from Skeletal Samples* (pp. 48–72). Cambridge University Press, Cambridge.

Kerley, E.R. (1965). The microscopic determination of age in human bone. *American Journal of Physical Anthropology* 23: 149–163.

Kerley, E.R. (1970). Estimation of skeletal age: after about 30 years. In T.D. Stewart (ed.), *Personal Identification in Mass Disasters* (pp. 57–70). National Museum of Natural History, Washington DC.

Kerley, E.R. and Ubelaker, D.H. (1978). Revisions in the microscopic method of estimating age at death in human cortical bone. *American Journal of Physical Anthropology* 49: 545–546.

King, J.B. (1939). Calcification of the costal cartilages. *British Journal of Radiology* 12: 2–12.

Kobyliansky, E. and Livshits, G. (1989). Age-dependent changes in morphometric and biochemical traits. *Annals of Human Biology* 16(3): 237–247.

Konigsberg, L., Herrmann, N.P., Wescott, D.J. and Kimmerle, E.H. (2008). Estimation and evidence in forensic anthropology: age-at-death. *Journal of Forensic Sciences* 53(3): 541–557.

Kroman, A.M. and Thompson, G.A. (2009). Cranial suture closure as a reflection of somatic dysfunction: lessons from osteopathic medicine applied to physical anthropology. *Proceedings American Academy of Forensic Sciences Annual Meeting*, Denver, CO, pp. 326–237.

Lamendin, H., Baccino, E., Humbert, J.F., Tavernier, J.C., Nossintchouk, R.M., and Zerille, A. (1992). A simple technique for age estimation in adult corpses: the two criteria dental method. *Journal of Forensic Sciences* 37: 1373–1379.

Langley-Shirley, N. and Jantz, R.L. (2010). A Bayesian approach to age estimation in modern Americans from the clavicle. *Journal of Forensic Sciences* 55(3): 571–583.

Lovejoy, C.O., Meindl, R.S., Pryzbeck, T.R., and Mensforth, R.P. (1985a). Chronological metamorphosis of the auricular surface of the ilium: a new method for the determination of adult skeletal age at death. *American Journal of Physical Anthropology* 68: 15–28.

Lovejoy, C.O., Meindl, R.S., Mensforth, R.P., and Barton, T.J. (1985b). Multifactorial determination of skeletal age at death: a method and blind tests of its accuracy. *American Journal of Physical Anthropology* 68: 1–14.

Lynnerup, N., Thomsen, J.L., and Frolich, B. (1998). Intra- and inter-observer variation in histological criteria used in age at death determination based on femoral cortical bone. *Forensic Science International* 91: 219–230.

Mann, R.W., Symes, S.A., and Bass, W.M. (1987). Maxillary suture obliteration: aging the human skeleton based on intact or fragmentary maxillae. *Journal of Forensic Sciences* 32: 148–157.

Mann, R.W., Jantz, R.L., Bass, W.M., and Willey, P.S. (1991). Maxillary suture obliteration: a visual method for estimating skeletal age. *Journal of Forensic Sciences* 36: 781–791.

Maples, W.R. (1989). The practical application of age-estimation techniques. In M.Y. İşcan (ed.), *Age Markers in the Human Skeleton* (pp. 319–324). Charles C. Thomas, Springfield, IL.

Martrille, L., Ubelaker, D.H., Cattaneo, C., Seguret, F., Tremblay, M., and Baccino, E. (2007). Comparison of four skeletal methods for the estimation of age at death on White and Black adults. *Journal of Forensic Sciences* 52(2) 302–307.

Masset, C. (1989). Age estimation on the basis of cranial sutures. In İşcan, M.Y. (ed.), *Age Markers in the Human Skeleton* (pp. 71–104). Charles C. Thomas, Springfield, IL.

McCormick, W.F. (1980). Mineralization of the costal cartilages as an indicator of age: preliminary observations. *Journal of Forensic Sciences* 25: 736–741.

McCormick, W.F. and Stewart, J.H. (1983). Ossification patterns of costal cartilages as an indicator of sex. *Archives of Pathology and Laboratory Medicine* 107: 206–210.

McKern, T.W. and Stewart, T.D. (1957). *Skeletal Age Changes in Young American Males Analysed from the Standpoint of Age Identification.* Technical Report EP-45. Quartermaster Research and Development Command, Natick, MA.

Meindl, R.S., Lovejoy, C.O., Mensforth, R.P., and Walker, R.A. (1985). A revised method of age determination using the os pubis, with a review and tests of accuracy of other current methods of pubic symphyseal aging. *American Journal of Physical Anthropology* 68: 29–45.

Meindl, R.S. and Lovejoy, C.O. (1985). Ectocranial suture closure: a revised method for the determination of skeletal age at death based on the lateral anterior sutures. *American Journal of Physical Anthropology* 68: 57–66.

Moorrees, C.F.A., Fanning, E.A., and Hunt, E.E. (1963a). Formation and restoration of three deciduous teeth in children. *American Journal of Physical Anthropology* 21: 205–213.

Moorrees, C.F.A., Fanning, E.A., and Hunt, E.E. (1963b). Age variation of formation stages for ten permanent teeth. *Journal of Dental Research* 42(6): 1490–1502.

Mulhern, D.M. and Jones, E.B. (2004). Test of revised method of age estimation from the auricular surface of the ilium. *American Journal of Physical Anthropology* 126: 61–65.

Nawrocki, S.P. (1998). Regression formulae for the estimation of age from cranial suture closure. In K.J. Reichs (ed.), *Forensic Osteology: Advances in the Identification of Human Remains*, 2nd edn (pp. 276–292). Charles C. Thomas, Springfield, IL.

Nawrocki, S.P. (2010). The nature and sources of error in the estimation of age at death from the skeleton. In K. Latham and M. Finnegan (eds), *Age Estimation of the Human Skeleton* (pp. 79–101). Charles C. Thomas, Springfield, IL.

Osborne, D.L., Simmons, T.L., and Nawrocki, S.P. (2004). Reconsidering the auricular surface as an indicator of age at death. *Journal of Forensic Sciences* 49: 905–911.

Overbury, R.S., Cabo, L.L., Dirkmaat, D.C., and Symes, S.A. (2009). Asymmetry of the os pubis: implications for the Suchey-Brooks method. *American Journal of Physical Anthropology* 139(2): 261–268.

Passalacqua, N.V. (2010). The utility of the Samworth and Gowland age-at-death "look-up" tables in forensic anthropology. *Journal of Forensic Sciences* 55(2): 482–487.

Passalacqua, N.V. and Uhl, N.M. (2009). Phase versus component systems in age-at-death estimation I: the methodology and usage of component systems. In *Proceedings of the American Association of Physical Anthropologists 78th Annual Meeting*, Chicago, IL, p. 289.

Prince, D.A. and Ubelaker, D.H. (2002). Application of Lamendin's adult dental aging technique to a diverse skeletal sample. *Journal of Forensic Sciences* 47: 107–116.

Ritz-Timme, S., Cattaneo, C., Collins, M.J., Waite, E.R., Shutz, H.W., Kaatsch, H.J., and Borrman HIM. (2000). Age estimation: the state of the art in relation to the specific demands of forensic practice. *International Journal of Legal Medicine* 113: 129–136.

Russell, K.F., Simpson, S.W., Genovese, J., Kinkel, M.D., Meindl, R.S., and Lovejoy, C.O. (1993). Independent test of the fourth rib aging technique. *American Journal of Physical Anthropology* 92: 53–62.

Samworth, R. and Gowland, R. (2007). Estimation of adult skeletal age-at-death: statistical assumptions and applications. *International Journal of Osteoarchaeology* 17: 174–188.

Scheuer, L. and Black, S. (2000). *Developmental Juvenile Osteology.* Academic Press, San Diego, CA.

Schour, I. and Massler, M. (1941). The development of the human dentition.

Journal of American Dental Association 28: 1153–1160.

Semine, A.A. and Damon, A. (1975). Costochondral ossification and aging in five populations. *Human Biology* 47(1): 101–116.

Snodgrass, J.J. (2004). Sex differences and aging of the vertebral column. *Journal of Forensic Sciences* 49(3): 458–463.

Stewart, T.D. (1958). The rate of development of vertebral osteoarthritis in American whites and its significance in skeletal age identification. *The Leech* 28: 144–151.

Stewart, T.D. (1979). *Essentials of Forensic Anthropology*. Charles C. Thomas, Springfield, IL.

Stout, S.D. (1988). The use of histomorphology to estimate age. *Journal of Forensic Sciences* 33: 121–125.

Stout, S.D. and Gehlert, S.J. (1980). The relative accuracy and reliability of histological aging methods. *Forensic Science International* 15: 181–190.

Stout, S.D. and Paine, R.R. (1992). Brief Communication: histological age estimation using the rib and clavicle. *American Journal of Physical Anthropology* 87: 111–115.

Suchey, J.M., Wiseley, D.V., and Katz, D. (1986). Evaluation of the Todd and McKern-Stewart methods for aging the male os pubis. In K.J. Reichs (eds), *Forensic Osteology – Advances in the Identification of Human Remains* (pp. 33–67). Charles C. Thomas, Springfield, IL.

Todd, T.W. (1920). Age changes in the pubic bone I: the male White pubis. *American Journal of Physical Anthropology* 3: 285–334.

Todd, T.W. (1921). Age changes in the pubic bone II-IV: the pubis of the male Negro-White hybrid, the pubis of the White female, the pubis of the female Negro-White hybrid. *American Journal of Physical Anthropology* 4: 1–70.

Todd, T.W. and Lyon, D.W. (1924). Endocranial suture closure, its progress and age relationship: part I. Adult males of the white stock. *American Journal of Physical Anthropology* 7: 325–384.

Todd, T.W. and Lyon, D.W. (1925). Cranial suture closure, its progress and age relationship: part, I.I. Ectocranial suture closure in adult males of the white stock. *American Journal of Physical Anthropology* 8: 23–45.

Ubelaker, D.H. (1978). *Human Skeletal Remains: Excavation, Analysis, Interpretation*. Aldine Publishing Company, Chicago, IL.

Uhl, N.M. (2007). *Multifactorial Determination of Age at Death from the Human Skeleton*. Thesis, University of Indianapolis, Indianapolis, IN.

Uhl, N.M. (2008a). Multifactorial determination of age-at-death from the human skeleton. In *Proceedings of the American Academy of Forensic Sciences 60th Annual Meeting*, vol. XIV, Washington DC, pp. 341–342.

Uhl, N.M. (2008b). ADBOU age-at-death estimation in South Africa. In *Proceedings of the American Association of Physical Anthropologists 77th Annual Meeting*, Columbus, OH, p. 211.

Wittwer-Backofen, U., Gampe, J., and Vaupel, J.W. (2004). Tooth cementum annulation for age estimation: results from a large known-age validation study. *American Journal of Physical Anthropology* 123: 119–129.

Zambrano, C.J. (2005). *Evaluation of Regression Equations used to Estimate Age at Death from Cranial Suture Closure*. MS thesis, University of Indianapolis, Indianapolis, IN.

Zwaan, B.J. (1999). The evolutionary genetics of aging and longevity. *Heredity* 82: 589–597.

CHAPTER 11

Skeletal Age Estimation: Where We Are and Where We Should Go

George R. Milner
and Jesper L. Boldsen

Estimating age is a critical part of establishing the identity of a deceased person, yet despite about a century of work current methods based on bones are far from ideal. A number of age-estimation procedures are covered elsewhere in this volume (Chapter 10). In this chapter, we skate lightly over existing practices, highlighting several issues and ongoing work that point toward promising new research directions. Our focus is on easily observable aspects of the skeleton since they are commonly used to estimate age, and they remain the quickest, cheapest, and arguably the best means of doing so.

Although this volume focuses on forensic anthropology, it is impossible to disentangle the development of methods for medicolegal purposes from those intended for archaeological skeletons because the means of estimating age in these sister disciplines are largely the same. The reasons for examining skeletons, however, are quite different. Forensic anthropologists are interested in age as a step toward the identification of specific people. Their archaeological colleagues, in contrast, are concerned with both individual skeletons (but not personal identification) and population age-at-death distributions. On the forensic side, much of what drives current work are court-inspired demands not only for improved methods, but for better characterizations of the error associated with old and new procedures when applied to a wide range of populations (Dirkmaat et al. 2008). Such concerns, common to all aspects of forensic-related work (National Academy of Sciences 2009), are a sharp spur encouraging the development of new procedures and the evaluation of how well they and earlier methods perform. Among bioarchaeologists, however, there is a

A Companion to Forensic Anthropology, First Edition. Edited by Dennis C. Dirkmaat.
© 2012 John Wiley & Sons Ltd. Published 2015 by John Wiley & Sons Ltd.

decidedly mixed response to a frequently voiced concern over the effectiveness of standard methods, even though problems with paleodemographic findings have been recognized for at least three decades (Bocquet-Appel and Masset 1982; Milner et al. 2008).[1]

Forensic osteologists have one big advantage over their archaeological colleagues: they can check their conclusions when individuals are positively identified. Archaeological osteologists rarely enjoy that luxury, unless they happen to study a known historical figure or a modern skeletal collection as part of their research. The weaknesses of existing age-estimation methods, therefore, are not as immediately apparent as they might be if results could be routinely compared to known-age skeletons.[2] Archaeological osteologists, however, have one thing in their favor. When looking at large groups of skeletons, as opposed to individual cases examined sequentially, it is possible to detect unusual age-at-death distributions attributable to biased age-estimation methods. Such observations in the early 1980s initiated much of an ongoing debate over adult age-estimate accuracy, and it, in turn, has prompted the development of much-needed new methods (Bocquet-Appel and Masset 1982; Howell 1982).

At the outset, it is best to explain why *age estimation* instead of *age determination* is used, even though both appear in the osteological literature. All osteologists realize that an age assigned to a skeleton is only an approximation of that individual's true age. That is because what can be observed is biological age, as inferred from various skeletal indicators, but what one wants is its chronological counterpart. The correspondence between the two is imperfect, although it is better for juveniles than adults. The ages of the former are based on tightly constrained bone and tooth development, whereas those of the latter are largely derived from irregularly occurring degenerative processes, especially from the late 20s onward. In adults, the skeletal changes that occur with advancing age are no doubt responsive to both genetics and life-history events, including disease and trauma experience. Estimation, as opposed to determination, underscores the uncertainty inherent in inferring age from bones and teeth.

AGE INDICATORS

The requirements for a good age indicator are deceptively simple, as there are only three of them. First, the characteristics being examined must display progressive, unidirectional change with advancing age. In a sequence of young to old stages, there can be no reversal to an earlier morphological state once a later, or older-looking, one has been reached. That is straightforward for measurements of bones or teeth as longer implies older, but it demands more attention when defining a sequence of stages that characterize changes in complex anatomical structures. Second, it should be possible to reliably classify (categorical data) or measure (continuous data) the age-informative morphological features. That is, low observer error is necessary to ensure consistent, hence valid, results. Third, observable alterations in the anatomical features of interest should take place at roughly the same time in all people, perhaps with skeletons divided by sex or some other readily distinguishable characteristic, such as geographical origin. Of great practical significance is whether a particular procedure can be applied to a skeletal sample other than the one it was originally based on. That is, it should be possible to use a method on more than a single population. If that cannot be done, the procedure has only limited practical importance. Observation error and variation among individuals and populations – the second and third age-indicator criteria – are why there is uncertainty in all age estimates.

For juveniles, osteologists focus on tooth formation, classified as a series of stages or measured in terms of length; tooth eruption; epiphysis appearance and, especially, union with the rest of the bone; and the dimensions of skeletal elements, often major limb bones. Tooth formation and eruption have long been regarded as more accurate for estimating age than bone development and size (e.g., Lewis and Garn 1960; Ubelaker 1987). Bone dimensions, in particular, are susceptible to growth faltering attributable to poor health, which contributes to variation in body size among people of the same age.

The most common anatomical structures used to estimate the age of adult skeletons are the pubic symphysis, the iliac portion of the sacroiliac joint (auricular surface), the sternal ends of ribs (especially the fourth rib), and cranial sutures. Age-related change is recorded in two ways. Entire anatomical structures can be ordered from young to old according to their overall appearance (Brooks and Suchey 1990; İşcan et al. 1984, 1985, 1987; Katz and Suchey 1986; Kimmerle et al. 2008; Lovejoy et al. 1985a; Meindl et al. 1985; Oettlé and Steyn 2000; Todd 1920, 1921). Alternatively, these structures can be divided into several individually scored parts that are then combined in some fashion to yield age estimates (Boldsen et al. 2002; Buckberry and Chamberlain 2002; Chen et al. 2008; DiGangi et al. 2009; Hanihara and Suzuki 1978; Igarashi et al. 2005; McKern and Stewart 1957).

For adults, in contrast to juveniles, finding skeletal structures that fulfill the three age-indicator criteria have long bedeviled efforts to develop methods that yield accurate and precise estimates. For example, if one were only to consider the first two criteria for an effective age indicator, the cranial sutures would be counted as being among the best skeletal features available. With increasing age, open sutures eventually become partly filled and then completely obliterated. So the requirement for unidirectional change is met, and observer error should be low because it is easy to classify the changes that take place. Yet, for the past half century, sutures have been widely, and rightly, criticized as age indicators because they fail miserably when it comes to the third criterion (McKern and Stewart 1957; Powers 1962; Singer 1953; Stewart 1952).[3] A little information about age can be gleaned from suture closure during the first two decades, or so, of adulthood, but from that point onward little can be gained by looking at sutures, regardless of how they are scored or analyzed (Lynnerup and Jacobsen 2003; Milner 2010; Perizonius 1984).

The reason the pubic symphysis, sacroiliac joint, sternal rib ends, and cranial sutures have attracted so much attention is simple: they display enough change with increasing age to allow the definition of multiple stages. Whether they yield much in the way of useful information is another question entirely, as is widely recognized for cranial sutures.

Scant attention has been directed toward other parts of the skeleton, largely because they individually contribute little to age assessments. They include, among others, rough areas where tendons and muscle attach, the lipping of joint margins, and the characteristics of vertebral body borders that range from rounded to angular and lipped. These features have two strikes against them. They only indicate an individual

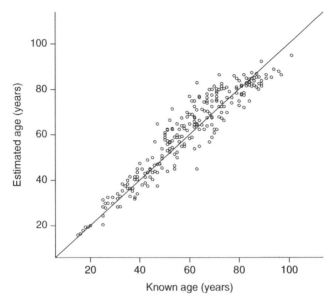

Figure 11.1 Actual ages of 254 skeletons, both males ($n=131$) and females ($n=123$) from the Bass Collection ($n=241$) and Mercyhurst forensic case load ($n=13$) compared to midpoints of experience-based age ranges. The identity line is indicated to highlight the dispersion around a perfect match between estimated and actual ages.

was "young" or "old" for a particular trait, and many of them are presumably affected by life experience, such as a heavy workload.

There are, however, several good reasons why these largely overlooked skeletal features deserve more attention. First, they collectively contribute much to the accuracy and precision of age estimates, as indicated by experience-based assessments of known-age skeletons where such traits figure prominently (Milner 2010). In fact, they are often more important than the standard suite of anatomical structures used for age estimation. Figure 11.1 shows actual versus estimated (range midpoints) ages for 254 skeletons, both males ($n=131$) and females ($n=123$), from the Bass Donated Collection ($n=241$) and Mercyhurst forensic cases ($n=13$). The correspondence between estimates and reported ages is far greater than one would achieve with any standard age-estimation method. Examinations of the Portuguese Coimbra Collection by both authors and Svenja Weise have yielded broadly similar results. Second, the assessments can be made quickly with no special equipment, making them suitable for situations where speed and the capacity to work in less than ideal circumstances are required. Third, a quantitative procedure is being developed that uses expert assessments to generate a population age-at-death distribution before the ages of individual skeletons are estimated (Weise et al. 2009). This procedure conforms to requirements outlined in Hoppa and Vaupel (2002) for obtaining the unbiased estimates essential for paleodemographic analyses. Fourth, the skeletal characteristics can be accommodated within existing quantitative methods, along the lines of what has been termed transition analysis, which hold promise for forensic applications where the emphasis is on individual skeletons, not large groups of them (Boldsen et al. 2002; DiGangi et al. 2009; Ríos et al. 2008).

It might be said that since these skeletal characteristics individually contribute little to the assessment of age they are not worth much relative to anatomical structures that can be divided into multiple age-progressive stages. That objection, whatever its merits, must be weighed against the fact that no method based on the pubic symphysis, sacroiliac joint, ribs, or cranial sutures works well for the entirety of adulthood. In fact, existing methods commonly, if not universally, produce biased age estimates. Even recently developed transition analysis, which focuses on the pubis, ilium, and cranium, yields results that are far from ideal (Milner 2010). In contrast, experience-based assessments that rely heavily on a loose and ill-defined group of age-informative markers distributed throughout the skeleton perform rather well in blind tests, as indicated by Figure 11.1. It is important to note that we are not advocating simply guessing ages: the tools are available, or are being developed, to handle these traits in a quantitatively appropriate fashion.

REFERENCE SAMPLES

Data used for age-estimation come from two sources: modern (living) populations and skeletal collections. It is widely recognized that what is visible in radiographs, computed tomography scans, or other imaging technology is not the same as what can be observed or measured on dry bones. For example, radiographic and computed tomography scan data compiled from several studies of the union of the medial clavicle epiphysis are similar, but are quite different from what can be seen in direct examinations of bones (Meijerman et al. 2007). So the source of information can have a big effect on the ages assigned to completely skeletonized remains.

Unfortunately, juveniles are poorly represented in known-age skeletal collections, which means it is especially important to see what can be done with data derived from studies of living people. Tooth crown and root formation are a notable example of where information from living subjects is of great value. A wide range of populations from different countries and socioeconomic positions have been examined in clinical settings and surveys of specific groups of people, providing a good feel for the range of variation in tooth development that exists around the world. Forensic and archaeological skeletons can be radiographed when teeth are still embedded in their sockets and the roots are likely to be incomplete, and loose teeth can be assessed by eye for the amount of crown and root formation. Eruption, of course, can be directly observed in both living subjects and skeletons, so it is of great use as there are many studies of tooth emergence in different populations.

Known-age skeletal collections consist mostly of adults, but there are still relatively few large and well-documented samples, and they do not represent cross-sections of the national populations from which they were derived (e.g., Komar and Grivas 2008; Usher 2002). For example, the frequently used Terry Collection, originally from St. Louis, Missouri, in the USA, is heavily weighted toward members of low socioeconomic groups (Hunt and Albanese 2005).[4] Particularly for forensic purposes, there is a need for additional sources of comparative data on modern populations, such as the University of Tennessee's Bass Donated Collection and the Forensic Data Bank (Dirkmaat et al. 2008; Ousley and Jantz 1998). Even when the country of origin and ancestry remain the same, measurable and analytically important differences in skeletal dimensions occur in the skeletons of people who died over the past century or so (Jantz 2001; Jantz and Jantz 1999, 2000; Ousley and Jantz 1998; Meadows and Jantz 1995). The extent to which age-informative markers have changed over such a short period of time, as opposed to what has happened to craniofacial and limb dimensions, has yet to be determined.

It is commonly, and correctly, believed that it is best to get the closest match possible between the study skeletons and reference collections when choosing the most appropriate procedures to use. In other words, under ideal circumstances forensic investigations and paleodemographic analyses should be guided by a "regional" approach, following the advice of Ubelaker (2006: 3, 9; 2008: 607), among others. In this context, regional refers to more than the word's usual meaning as it covers any readily classifiable characteristic of a group of skeletons that might affect how observable skeletal features are related to chronological age, including geographical origin, ancestry, time period, sex, and socioeconomic position.

Another reason for having many comparative collections is that it is essential to use skeletons that are as dissimilar as possible to the original reference sample when conducting validation studies. One would like to know if a particular method is broadly applicable, or whether it is limited to the population that gave rise to the reference sample. So if a particular method was developed on, say, the Terry Collection, one would not be inclined to choose the Hamann–Todd Collection when designing a validation study. Both heavily used skeletal samples are similar because they are from the USA and originated in much the same way, being assembled by anatomists, at roughly the same time. It would be far better to evaluate the method on one of the other collections that exist, such as the Coimbra skeletons from Portugal (Cunha and Wasterlain 2007).

As there are many studies of tooth development, especially eruption, in living populations, the dentition is perhaps the best candidate for putting a truly regional approach into effect. Yet despite the undeniable importance of such an approach – and its value is not being criticized here – the most satisfactory studies for age-estimation purposes are not always those that match the skeletal samples most closely in terms of geographical origin or ancestry. That is because the original research design, including analytical methods as well as sample size and composition, has a lot to do with the appropriateness of published materials when it comes to estimating the ages of skeletons from forensic or archaeological contexts.

An alternative way to proceed would be to base estimates on as large and diverse an array of populations as possible by combining the results of numerous well-documented studies. One such meta-analysis, designed for rapid age assessments in surveys of modern children, is based on simple counts of erupted deciduous teeth (Townsend and Hammel 1990). By pursuing such a strategy, some precision is undoubtedly lost (age intervals are wider than in individual studies), and estimates are likely to deviate a bit from true ages when applied to any given population, much like what happened in well-nourished Finnish children (Nyström et al. 2000). The loss of precision, however, is likely balanced by a gain in accuracy as skeletons, in general, are more likely to fall within their proper age intervals. More importantly, a perfect match between a reference sample and the skeletons being examined – that is, complete correspondence in geographical origin, ancestry, general living conditions, and the like – is an elusive goal. It is one that will rarely be achieved for forensic skeletons, and is downright impossible for archaeological ones.

ESTIMATION PROCEDURES

Requirements

Various procedures, including widely used ones, are not equal in terms of their capacity to estimate age accurately, precisely, reliably, and economically. All four are important, regardless of whether an osteologist's primary concern is with forensic or archaeological skeletons.

Accuracy refers to the correspondence between estimated and true ages. It measures the degree to which point estimates, generally single years, match actual ages.[5] Age intervals, however, are usually used as there is not a perfect correspondence between biological and chronological age; that is, individuals vary in rates of aging as inferred from the appearance of bones and teeth. So when ranges are used, accuracy refers to the extent to which intervals encompass true ages. Precision pertains to age interval length. For adults, intervals span years, if not decades, whereas for infants they can be as tight as a few months.

Accuracy and precision are related to one another: when one goes up, the other goes down. A long interval is more likely to encompass a person's true age than a short one, and it can be a realistic appraisal of the relationship between a particular skeletal characteristic used for estimation purposes and actual age. Yet lengthening intervals, if pushed too far, devolves into an all but useless exercise. This point is not as contrived as it might at first seem because age intervals, including the upper ones in the heavily used Suchey–Brooks pubic symphysis procedure (Brooks and Suchey 1990; Katz and Suchey 1986), can span many decades, up to much of adulthood.

Procedures should also be reliable in terms of classification or measurement error, which should be low. Researchers must be able to replicate their own observations, and others should be able to produce the same results, as long as they are adequately trained. Familiarity with the human skeleton and adequate instruction are important, as there is no such thing as a cookbook approach to age estimation. One must possess knowledge of what is being examined in order to classify highly variable skeletal structures correctly. A good classification, however, does not guarantee satisfactory results (that is, estimates closely approximating real age). As noted previously, it is easy to classify cranial suture stages correctly (with little error), but because of the considerable inter-individual variation in closure rates, the resulting age estimates are poor.

Finally, procedures should have practical value in real-world settings. That is, it must be possible to produce estimates quickly and cheaply enough to meet research needs. Osteologists often work in field, museum, or morgue settings with limited or no access to specialized equipment. Forensic anthropologists can find themselves under pressure to produce assessments in a timely fashion. Their archaeological colleagues require large samples for paleodemographic purposes, so they likewise face time constraints as they cannot become bogged down when dealing with individual skeletons.

Problems

Existing age-estimation methods typically involve classifying bones and assigning ages based on the particular stage, as defined in the reference sample, that best conforms to the specimen being examined. There are several problems with doing so.

The composition of the reference sample used to develop an estimation procedure influences the ages assigned to forensic or archaeological skeletons; that is, it contributes to biased estimates. Bocquet-Appel and Masset (1982) first brought this issue to the attention of osteologists, and it has been much discussed since then (e.g., Hoppa and Vaupel 2002; Konigsberg and Frankenberg 1992, 1994). The problem can be most easily visualized when the original sample is heavily weighted toward a certain part of the age distribution, such as the McKern and Stewart (1957) Korean War sample where no soldier was over 50 years old, and most of them were much younger than that.

To obtain unbiased estimates, one must know the age distribution of the population before estimating the ages-at-death of individual skeletons (Hoppa and Vaupel 2002). The issue boils down to the influence of prior distributions, in Bayesian terms, on the ages being estimated. The difficulty, of course, is that a reasonable approximation of the population age distribution is precisely what paleodemographers hope to obtain after they look at a bunch of skeletons. For forensic anthropologists, this problem has been partly addressed by the inclusion of the ages at death of American homicide victims in a computerized version of transition analysis (Boldsen et al. 2002).

The good news about the effects of different prior distributions is that transition analysis, when used on Bass Collection skeletons, yielded maximum likelihood age estimates that usually did not differ by more than 5 years, and discrepancies were trivial in early adulthood (Milner 2010). This finding should not encourage complacency about the effect of prior distributions on age estimates. Yet difficulties raised by inappropriate prior distributions are dwarfed by problems stemming from the poor accuracy and precision of conventional skeletal indicators of age, specifically the pubic symphysis, sacroiliac joint, and cranial sutures. To return to an earlier point, much-needed improvements in analytical methods must be accompanied by a larger battery of systematically examined age-informative skeletal characteristics.

Osteologists also face the problem of how best to combine information to produce a composite age estimate. Often contradictory evidence is balanced by placing the greatest weight on skeletal structures that are believed to work best, such as the pubic symphysis in preference to the cranial sutures. There are also several formal means of combining information about age, with the multifactorial and complex procedures being the most widely used (Lovejoy et al. 1985b; Workshop of European Anthropologists 1980). These procedures, however, do not automatically solve the problem of producing better assessments of age since they are based on methods that yield biased estimates in the first place. Combining likelihood curves for various anatomical structures to obtain overall age estimates (maximum likelihood estimates and confidence intervals), as is done in transition analysis, is an analytically better approach (Boldsen et al. 2002). Yet the transition analysis estimates are still limited by the skeletal structures they are based on (Milner 2010). The cranial sutures scored in the present version of transition analysis, for example, do not provide much, if any, useful information beyond early adulthood. The advantage with transition analysis is that likelihood curves for cranial sutures tend to be so flat that they usually contribute little to final estimates, especially when the pubic symphysis and sacroiliac joint curves are relatively peaked so they exert a much greater effect on the composite estimate.

Conventional age-estimation methods do not provide a sure means of determining how certain one is about age when assigning a skeleton to a particular stage. First, some published age intervals are not accompanied by measures of central tendency and dispersion calculated from the original reference sample. They include Todd's (1920, 1921) 10 pubic symphysis stages and subsequent work where several categories were collapsed (Meindl et al. 1985). Second, the age intervals associated with various morphological stages are affected by the age composition of the reference sample. So even when the ages of individuals associated with a particular stage are summarized by means and standard deviations, as is done for the six Suchey–Brooks pubic symphysis stages (Brooks and Suchey 1990; Katz and Suchey 1986), those figures are not unbiased estimates of the distribution of the stages in the population as a whole. Once again, that problem can be visualized most easily with the McKern and Stewart (1957) Korean War sample where individuals in the first two decades of adulthood dominated the sample. A better approach is to generate likelihood curves based on the transition from one stage to the next (Boldsen et al. 2002). Third, the process of allocating skeletons to a fixed set of stages involves an indefensible assumption: all skeletons assigned to a particular morphological category are equally likely to fall within the associated age range. Common experience shows that age-informative markers for some people are reasonably consistent, whereas in other individuals very different estimates can be obtained from different parts of single anatomical structures as well as the skeleton as a whole. Thus, it stands to reason that more confidence can be placed in estimates of age for some skeletons than in others. Instead of using fixed age intervals, it is better to tailor those intervals to the specific mix of age-informative characteristics that are present in individual skeletons, as is done in transition analysis (Boldsen et al. 2002). In that procedure, confidence intervals for people of exactly the same age will usually be different. They will be identical only when the same age-informative skeletal traits are present, and these anatomical features are scored in exactly the same way.

Biased age estimates, as mentioned above, are perhaps easiest to detect in archaeological studies because much of that work involves large samples, not case-by-case studies of individual skeletons. It has long been recognized that most paleodemographic results show a marked tendency for an excessive number of deaths during early adulthood. So seemingly too few individuals survived beyond 50 years or so (Aykroyd et al. 1999; Howell 1982; Milner et al. 1989; Paine 1989). Over the past several decades a debate has sputtered along over whether age-at-death distributions generated from archaeological samples can be taken at face value as reflecting past reality, or whether something else is going on, most importantly a problem with the age estimates (Aykroyd et al. 1999; Bocquet-Appel and Masset 1982; Howell 1982; Lovejoy et al. 1977; Meindl and Russell 1998; Mensforth 1990). It would seem prudent to start with the assumption that most archaeological age-at-death distributions, except for those from samples that originated under unusual circumstances such as mass disasters, should not depart terribly far from the general characteristics of modern populations, including those documented in ethnographic studies (Chamberlain 2000; Howell 1976, 1982). If true, then the common archaeological pattern of having only a few individuals over 50 years and correspondingly large numbers of younger adults, especially those in their 30s and 40s, is partly a result of procedures that underestimate the ages of old people. Unfortunately, the controversy over whether the age-at-death distributions are partly a result of poor age-estimation methods has had little effect on the practice of bioarchaeology outside a few loosely defined research groups.

Another limitation of conventional age-estimation procedures is not being able to do much, if anything, with the skeletons of old people. It would be quite useful to be able to say something about the deaths that occur in the upper part of the lifespan. That is obviously true for forensic work that focuses on modern populations where many people live to advanced years. It is also important for archaeological investigations if one suspects, as we do, that the sparse record of individuals who survived beyond 50 years in the distant past is partly a result of inadequate age-estimation methods.

While the use of an open-ended terminal interval such as 50+ years nicely accommodates the limitations of conventional methods, there are promising signs that future developments will permit osteologists to estimate age in the upper part of the lifespan. For example, estimates for two premodern skeletons of known individuals – one from early modern Denmark and the other from a prehistoric Maya site – show that the present version of transition analysis can provide information about the ages of old people in samples much different than the reference samples (Terry and Coimbra) used when developing the procedure (Boldsen et al. 2002; Buikstra et al. 2006).

Open-ended terminal intervals accommodate the often-repeated statement that variation among individuals in skeletal age indicators eventually becomes so great that one can only say a person was old when he or she died. A recent study of Bass Collection skeletons, however, indicates that it is not entirely true that variation in age-related skeletal markers simply accumulates with advancing years (Milner 2010). Confidence intervals generated with transition analysis decreased in the oldest estimated ages relative to those associated with middle age. What is taking place is not entirely clear, although selective mortality likely has an effect on observable skeletal variation through the preferential removal of people who were the most frail at each age. Different rates of change in age-informative bony traits produce variation in the appearance of the skeletons of people who are the same age, but that variation is simultaneously diminished by selective mortality. In old age, the balance is noticeably tilted toward the latter, so estimates improve as far as confidence interval lengths are concerned. What is surprising is not what is happening, but the fact that it can be observed in a suite of traits focused narrowly on the skull and pelvis. If these findings can be reproduced with other skeletal characteristics and the interpretation is correct, it is likely that middle-aged adults will always prove to be the most difficult group to age accurately and precisely, not the oldest people.

FUTURE DIRECTIONS

The sheer number of articles on age estimation published in recent years might lead one to believe we are doing it better today than several decades ago. There is some truth in that statement. But the sad fact of the matter is that progress – as measured by the efficacy of commonly used procedures – has not been all that great, and gains have been agonizingly slow in coming.

From an osteological perspective, there are at least two ways to improve age estimates. The first involves analyses of information that already exists (e.g., Townsend and Hammel 1990). To repeat much of this work would be impossibly expensive and time consuming, particularly for the great amount of information available on tooth development and eruption. The second requires thoroughly evaluating the skeleton for any feature that undergoes age-progressive change. The wide-ranging work of McKern and Stewart (1957) serves as a model of this kind of approach, although the analytical tools available at that time were primitive in comparison to those of today. The means of making simultaneous use of many traits for forensic purposes, all of which individually yield little information, now exist, such as something along the lines of an elaborated version of transition analysis (Boldsen et al. 2002).

If osteologists shift to a much larger array of age indicators, it will be necessary to collect data to establish ages of transition from one stage to the next, generally simply "young" versus "old" for each trait. Estimates of classification error are also required, as are assessments of the degree to which various traits are correlated with one another and how they might vary among different populations. So much work is required before it will be possible to make the most effective use of the numerous age-informative features that seem to be distributed widely throughout the skeleton. Yet despite the length of the road that lies ahead of us, we can be reasonably confident that moving in this direction will result in marked improvements in accuracy and precision, to judge from experience-based age estimates that rest heavily on informal assessments of a broad array of skeletal characteristics (Milner 2010; Weise et al. 2009). In short, it is time to shift from looking at a little to a lot when it comes to estimating age, especially for adults.

The osteological work on a wide array of skeletal characteristics must be accompanied by the development of quantitatively rigorous procedures that accommodate problems with age estimation that are covered in detail in the paleodemographic literature (e.g., Hoppa and Vaupel 2002). The osteological and analytical components are equally important: one cannot result in significant improvements without the further development of the other. Should simultaneous progress be made on both, it is possible to be optimistic about the future of skeletal age estimation for both forensic and archaeological purposes.

ACKNOWLEDGMENTS

We would like to thank Lee Meadows Jantz who arranged access to the Bass Donated Collection at the University of Tennessee, and Steven Symes and Stephen Ousley who did the same for forensic cases at Mercyhurst College. Many colleagues have contributed over the years to our thinking about skeletal age estimation, but in particular we would like to thank Jutta Gampe and Svenja Weise at the Max Planck Institute for Demographic Research in Rostock, Germany.

NOTES

1   There are, of course, other differences. For example, the consequences of making mistakes in medicolegal investigations are much worse than they are for researchers who deal with old bones from archaeological excavations. Quite simply, analyses of skeletons from forensic contexts should facilitate, not impede or even mislead, investigations. In contrast, getting an ancient population's age distribution wrong will unnecessarily muddy conclusions about life in the distant past, but there will be little "real-world" cost other than professional embarrassment.

2   For convenience, and following common practice, we write known-age skeletons when, in fact, it is often more accurate to substitute *reported* for *known*. Frequently used reference samples, such as the Terry and Hamann–Todd Collections, contain many individuals whose ages were estimated or self reported, not documented through birth records or other such documents.

3   A notable exception is the spheno-occipital synchondrosis, which generally closes during the teenage years.

4   That they lived hard lives is evident from the large number of broken facial bones, many presumably from domestic abuse or barroom brawls, that one runs across when examining the Terry Collection skeletons. One of us (GRM) who spent many years doing archaeological work in the area can attest to the hazards of living there!

5   A year is considered a point estimate, in keeping with usual practice, although it too is an interval because it spans 12 months.

REFERENCES

Aykroyd, R.G., Lucy, D., Pollard, M.A., and Roberts, C.A. (1999). Nasty, brutish, but not necessarily short: a reconsideration of the statistical methods used to calculate age at death from adult human skeletal and dental age indicators. *American Antiquity* 64: 55–70.

Bocquet-Appel, J.P. and Masset, C. (1982). Farewell to paleodemography. *Journal of Human Evolution* 12: 353–360.

Boldsen, J.L., Milner, G.R., Konigsberg, L.W., and Wood, J.W. (2002). Transition analysis: a new method for estimating age from skeletons. In R.D. Hoppa and J.L. Vaupel (eds), *Paleodemography: Age Distributions from Skeletal Samples* (pp. 73–106). Cambridge University Press, Cambridge.

Brooks, S. and Suchey, J.M. (1990). Skeletal age determination based on the os pubis: a comparison of the Acsadi-Nemeskeri and Suchey-Brooks methods. *Human Evolution* 5: 227–238.

Buckberry, J.L. and Chamberlain, A.T. (2002). Age estimation from the auricular surface of the ilium: a revised method. *American Journal of Physical Anthropology* 119: 231–239.

Buikstra, J.E., Milner, G.R., and Boldsen, J.L. (2006). Janaab' Pakal: the age-at-death controversy re-visited. In V. Tiesler and A. Cucina (eds), *Janaab' Pakal of Palenque: Reconstructing the Life and Death of a Maya Ruler* (pp. 48–59). University of Arizona Press, Tucson, AZ.

Chamberlain, A. (2000). Problems and prospects in palaeodemography. In M. Cox and S. Mays (eds), *Human Osteology in Archaeology and Forensic Science* (pp. 101–115). Greenwich Medical Media, London.

Chen, X., Zhongyao, Z., and Luyang, T. (2008). Determination of male age at death in Chinese Han population: using quantitative variables statistical analysis from pubic bones. *Forensic Science International* 175: 36–43.

Cunha, E. and Wasterlain, S. (2007). The Coimbra Identified Osteological Collection. In G. Grupe and J. Peters (eds), *Skeletal Series and Their Socio-Economic Context* (pp. 23–33), Documenta Archaeobiologia 5. Verlag Marie Leidorf, Rahden, Germany.

DiGangi, E.A., Bethard, J.D., Kimmerle, E.H., and Konigsberg, L.W. (2009). A new method for estimating age-at-death from the first rib. *American Journal of Physical Anthropology* 138: 164–176.

Dirkmaat, D.C., Cabo L.L., Ousley, S.D., and Symes, S.A. (2008). New perspectives in forensic anthropology. *Yearbook of Physical Anthropology* 51: 33–52.

Hanihara, K. and Suzuki, T. (1978). Estimation of age from the pubic symphysis by means of multiple regression analysis. *American Journal of Physical Anthropology* 48: 233–240.

Hoppa, R.D. and Vaupel, J.W. (eds). (2002). *Paleodemography: Age Distributions from*

Skeletal Samples. Cambridge University Press, Cambridge.

Howell, N. (1976). Toward a uniformitarian theory of human paleodemography. In R.H. Ward and K.M. Weiss (eds), *The Demographic Evolution of Human Populations* (pp. 25–40). Academic Press, New York.

Howell, N. (1982). Village composition implied by a paleodemographic life table: the Libben site. *American Journal of Physical Anthropology* 59: 263–269.

Hunt, D.E. and Albanese, J. (2005). History and demographic composition of the Robert J. Terry Anatomical Collection. *American Journal of Physical Anthropology* 127: 406–417.

Igarashi, Y., Uesu, K., Wakebe, T., and Kanazawa, E. (2005). New method for estimation of adult skeletal age at death from the morphology of the auricular surface of the ilium. *American Journal of Physical Anthropology* 128: 324–339.

İşcan, M.Y., Loth, S.R., and Wright, R.K. (1984). Estimation from the rib by phase analysis: white males. *Journal of Forensic Sciences* 29: 1094–1104.

İşcan, M.Y., Loth, S.R., and Wright, R.K. (1985).Estimation from the rib by phase analysis: white females. *Journal of Forensic Sciences* 30: 853–863.

İşcan, M.Y., Loth, S.R., and Wright, R.K. (1987). Racial variation in the sternal extremity of the rib and its effect on age determination. *Journal of Forensic Sciences* 32: 452–466.

Jantz, R.L., (2001). Cranial change in Americans: 1850–1975. *Journal of Forensic Sciences* 46: 784–787.

Jantz, L.M. and Jantz, R.L. (1999). Secular change in long bone length and proportion in the United States, 1800–1970. *American Journal of Physical Anthropology* 110: 57–67.

Jantz, R.L. and Jantz, L.M. (2000). Secular change in craniofacial morphology. *American Journal of Human Biology* 12: 327–338.

Katz, D. and Suchey, J.M. (1986). Age determination of the male os pubis. *American Journal of Physical Anthropology* 69: 427–435.

Kimmerle, E.H., Konigsberg, L.W., Jantz, R.L., and Baraybar, J.P. (2008). Analysis of age-at-death estimation through the use of pubic symphyseal data. *Journal of Forensic Sciences* 53: 558–568.

Komar, D.A. and Grivas, C. (2008). Manufactured populations: what do contemporary reference skeletal collections represent? A comparative study using the Maxwell Museum Documented Collection. *American Journal of Physical Anthropology* 137: 224–233.

Konigsberg, L.W. and Frankenberg, S.R. (1992). Estimation of age structure in anthropological demography. *American Journal of Physical Anthropology* 89: 235–256.

Konigsberg, L.W. and Frankenberg, S.R. (1994). Paleodemography: "not quite dead." *Evolutionary Anthropology* 3: 92–105.

Lewis, A.B. and Garn, S.M. (1960). The relationship between tooth formation and other maturational factors. *The Angle Orthodontist* 30: 70–77.

Lovejoy, C.O., Meindl, R.S., Pryzbeck T. R., Barton, T.J., Heiple, K.G., and Kotting, D. (1977). Paleodemography of the Libben site, Ottawa County, Ohio. *Science* 198: 291–293.

Lovejoy, C.O., Meindl, R.S., Pryzbeck T. R., and Mensforth, R.P. (1985a). Chronological metamorphosis of the auricular surface of the ilium: a new method for the determination of adult skeletal age at death. *American Journal of Physical Anthropology* 68: 15–28.

Lovejoy, C.O., Meindl, R.S., Mensforth, R.P., and Barton, T.J. (1985b). Multifactorial determination of skeletal age at death: a method and blind tests of its accuracy. *American Journal of Physical Anthropology* 68: 1–14.

Lynnerup, N. and Jacobsen, J.C.B. (2003). Age and fractal dimensions of human sagittal and coronal sutures. *American Journal of Physical Anthropology* 121: 332–336.

McKern, T.W. and Stewart, T.D. (1957). *Skeletal Age Changes in Young American Males Analysed from the Standpoint of Age Identification.* Technical Report EP-45. Quartermaster Research and Development Command, Natick, MA.

Meadows, L. and Jantz, R.L. (1995). Allometric secular change in the long bones from the 1800s to the present. *Journal of Forensic Sciences* 40: 762–767.

Meijerman, L., Maat, G.J.R., Schulz, R., and Schmeling, A. (2007). Variables affecting the probability of complete fusion of the medial clavicular epiphysis. *International Journal of Legal Medicine* 121: 463–468.

Meindl, R.S. and Russell, K.F. (1998). Recent advances in method and theory in paleodemography. *Annual Review of Anthropology* 27: 375–399.

Meindl, R.S., Lovejoy, C.O., Mensforth, R.P., and Walker, R.A. (1985). A revised method of age determination using the os pubis, with a review and tests of accuracy of other current methods of pubic symphyseal aging. *American Journal of Physical Anthropology* 68: 29–38.

Mensforth, R.P. (1990). Paleodemography of the Carlston Annis (Bt-5) Late Archaic skeletal population. *American Journal of Physical Anthropology* 82: 81–99.

Milner, G.R. (2010). Transition analysis and subjective assessments of age in adult skeletons. In J.L. Boldsen and P. Tarp (eds), *ADBOU 1992–2009: Forskiningsresultater* (pp. 15–27). ADBOU, Syddansk Universitet, Odense.

Milner, G.R., Humpf, D.H., and Harpending, H.C. (1989). Pattern matching of age-at-death distributions in paleodemographic analysis. *American Journal of Physical Anthropology* 80: 49–58.

Milner, G.R., Wood, J.W., and Boldsen, J.L. (2008). Advances in paleodemography. In M.A. Katzenberg and S.R. Saunders (eds), *Biological Anthropology of the Human Skeleton*, 2nd edn (pp. 561–600). Wiley-Liss, New York.

National Academy of Sciences (2009). *Strengthening Forensic Science in the United States: A Path Forward*. National Academies Press, Washington DC.

Nyström, M., Peck, L., Kleemola-Kujala, E., Evälahti, M., and Kataja, M. (2000). Age estimation in small children: reference values based on counts of deciduous teeth in finns. *Forensic Science International* 110: 179–188.

Oettlé, A.C. and Steyn, M. (2000). Age estimation from sternal ends of ribs by phase analysis in South African Blacks. *Journal of Forensic Sciences* 45: 1071–1079.

Ousley, S.D. and Jantz, R.L. (1998). The Forensic Data Bank: documenting skeletal trends in the United States. In K.J. Reichs (ed.), *Forensic Osteology: Advances in the Identification of Human Remains*, 2nd edn (pp. 441–458). Charles C. Thomas, Springfield, IL.

Paine, R. (1989). Model life table fitting by maximum likelihood estimation: a procedure to reconstruct paleodemographic characteristics from skeletal age distributions. *American Journal of Physical Anthropology* 79: 51–61.

Perizonius, W.R.K. (1984). Closing and non-closing sutures in 256 crania of known age and sex from Amsterdam (A.D. 1883–1909). *Journal of Human Evolution* 13: 201–216.

Powers, R. (1962). The disparity between known age and age as estimated by cranial suture closure. *Man* 62: 52–54.

Ríos, L., Weisensee, K., and Rissech, C. (2008). Sacral fusion as an aid in age estimation. *Forensic Science International* 180: 111.e1–111.e7.

Singer, R. (1953). Estimation of age from cranial suture closure. *Journal of Forensic Medicine* 1: 52–59.

Stewart, T.D. (1952). *Hrdlička's Practical Anthropometry*, 4th edn. Wistar Institute of Anatomy and Biology, Philadelphia, PA.

Todd, T.W. (1920). Age changes in the pubic bone. I. The male white pubis. *American Journal of Physical Anthropology* 3: 285–334.

Todd, T.W. (1921). Age changes in the pubic bone. II, III, IV. *American Journal of Physical Anthropology* 4: 1–70.

Townsend, N. and Hammel, E.A. (1990). Age estimation from the number of teeth erupted in young children: an aid to demographic surveys. *Demography* 27: 165–174.

Ubelaker, D.H. (1987). Estimating age at death from immature human skeletons: an overview. *Journal of Forensic Sciences* 32: 1254–1263.

Ubelaker, D.H. (2006). Introduction to forensic anthropology. In A. Schmitt, E. Cunha, and J. Pinheiro (eds), *Forensic*

Anthropology and Medicine: Complementary Sciences from Recovery to Cause of Death (pp. 3–12). Humana, Totowa, NJ.

Ubelaker, D.H. (2008). Issues in the global applications of methodology in forensic anthropology. *Journal of Forensic Sciences* 53: 606–607.

Usher, B.M. (2002). Reference samples: the first step in linking biology and age in the human skeleton. In R.D. Hoppa and J.W. Vaupel (eds), *Paleodemography: Age Distributions from Skeletal Samples* (pp. 29–47). Cambridge University Press, Cambridge.

Weise, S., Boldsen, J.L., Gampe, J., and Milner, G.R. (2009). Calibrated expert inference and the construction of unbiased paleodemographic mortality profiles. *American Journal of Physical Anthropology Supplement* 48: 269 (abstract).

Workshop of European Anthropologists (1980). Recommendations for age and sex diagnoses of skeletons. *Journal of Human Evolution* 9: 517–549.

CHAPTER 12 Adult Sex Determination: Methods and Application

Heather M. Garvin

There are a number of methods that have been traditionally used by physical anthropologists to determine sex in adults. These can be primarily broken down into metric methods distinguishing mostly size differences between the sexes, and nonmetric methods evaluating pelvic and cranial traits generally characteristic of males and females. This brief chapter is by no means aimed at providing a comprehensive review of all available methods and their inner workings. Its purpose is merely summarizing the more widespread traditional sexing methods, providing some key references, visual examples, and information regarding published accuracy rates. Detailed information on these methods can be easily obtained through the provided references, or in most forensic anthropological texts (e.g., Buikstra and Ubelaker 1994; Bass 2005; Stewart 1979; Krogman and İşcan 1986). A quick literature search will also reveal a number of "nontraditional" sexing techniques such as the use of metatarsal or cervical vertebrae measurements (e.g., Robling and Ubelaker 1997; Wescott 2000), or more advanced techniques such as geometric morphometric analyses of crania and pelves (e.g., Gonzalez et al. 2011; Bytheway and Ross 2010; Kimmerle et al. 2008). Instead, this chapter acts more as a prelude, providing some general information regarding popular sexing methods, and necessary background to the subsequent chapter which focuses primarily on some theoretical issues in human skeletal sex determination, a subject much less publicized.

A Companion to Forensic Anthropology, First Edition. Edited by Dennis C. Dirkmaat.
© 2012 John Wiley & Sons Ltd. Published 2015 by John Wiley & Sons Ltd.

METRIC SEX DETERMINATION

As a universal rule, within each population males are on average larger than females (Chapter 13 in this volume discusses this issue in detail). For this reason, forensic anthropologists have been able to use measurements of body size (bone lengths, articular surface dimensions, and cranial size) to help determine sex in unknown remains. These measurements can be used in univariate analyses, where only a single measurement is used for sex determination, or in multivariate analyses where a number of measurements are used in conjunction, typically through discriminant function analysis or, less frequently, logistic regression, to estimate sex.

Stewart (1979) provided univariate plots for sex determination using the femoral and humeral heads. Sectioning points that have been provided suggest that femoral heads less than 42.5 mm are female and more than 47.5 are male, with 42.5–43.5 mm representing probable females, 46.5–47.5 mm probable males. For the humerus Stewart supports a sectioning point of 45 mm, with less than 43 mm "almost certainly" female, and more than 47 mm "almost certainly" male. Many anthropologists still use such sectioning points. Both Stewart (1979) and Krogman and İşcan (1986) provide a review of other historical univariate studies, which provide slight variations in the reported sectioning points.

More recently, Spradley and Jantz (2011) used a large sample from the Forensic Data Bank to investigate the accuracy of numerous postcranial metric measurements in sex determination. In regards to univariate analyses, they present percentage correct classification rates of 86% in blacks and 88% in whites for femoral head diameters, and 86% in blacks and 83% in whites for humeral head diameters. They actually found femoral epicondylar breadth, proximal tibial epicondylar breadth, and scapula height to be the best individual measurements for sex estimation in American blacks (with 89, 88, and 87% correct classification rates, respectively). In American whites, the top three univariate measurements were proximal tibial epicondylar breadth (90%), scapula height (89%), and femoral head diameter (88%).

When multivariate analyses were used, the humerus was the best sex indicator for American whites (93.84% correct classification based on maximum length, epicondylar breadth, maximum vertical head diameter, and maximum diameter at midshaft), and the radius for American blacks (94.34% correct classification based on maximum length, sagittal diameter at midshaft, and transverse diameter at midshaft). Numerous other elements also presented classification rates above 90% and are all presented in tables along with classification functions (see Spradley and Jantz 2011).

Note that the first sentence of this section stated "*within each population* males are *on average* larger than females." The phrase "within each population" is necessary because population differences do exist. An average Asian male may provide body size measurements that are in the range of American white females. On the other hand, the average Asian male is guaranteed to be larger than the average Asian female. Also note the use of the terms *average* male and *average* female. Within each population there is some overlap in size between males and females (and in some populations there is more overlap than in others). There are some females who are larger than some males and vice versa, hence the reason why classification rates are not 100% (see Chapter 13 for an in-depth discussion regarding interpretation of classification rates).

Fordisc 3 (Ousley and Jantz 2005) helps alleviate some of these issues, or at least take them into account (Chapter 15 discusses the program in detail). It permits not only the input of numerous postcranial and cranial metrics for multivariate analyses, but also allows the researcher to identify which populations should be incorporated in the analysis. So, for example, if you have already determined that the individual is of African-American ancestry, it will compare the measurements from the unknown individual with known male and female measurements of African-American individuals only. It also reports statistical results, such as posterior probabilities and typicalities, which can help interpret the degree of confidence in the classification. So for these reasons, combined with its ease of use, Fordisc is among the most popular methods used by forensic anthropologists for metric sex determination.

Cranial metrics have also been used for sex estimation (Giles and Elliot 1963). Although a multivariate analysis using numerous cranial dimension will undeniably capture some shape changes in the skull and does include such features as the glabella and the mastoid (see the nonmetric section below), size remains the primary variable. Spradley and Jantz (2011) found that multivariate analyses of cranial metrics produced correct classification rates between 90 and 91%, which are still good, although lower than their values for many of the postcranial remains. Fordisc 3 (Ousley and Jantz 2005) can combine cranial and postcranial metrics into a multivariate analysis of sex, and also provides the opportunity to partially correct for size, and perform analyses based on "shape." But as in the postcrania, use of the correct group comparisons are important in receiving accurate results (e.g. see Guyomarc'h and Bruzek 2011).

Nonmetric Sex Determination

Pelvis

There is an overall consensus that the pelvis is the most reliable skeletal element for sex estimation in humans (Phenice 1969; Spradley and Jantz 2011; Krogman and İşcan 1986; Stewart 1979). This is because there is a difference in biological function between the male and female pelvis. In both males and females it gives support to internal organs and provides a necessary weight-bearing joint for locomotion. Because *Homo sapiens* are bipedal, selection pressures have lead to the evolution of a relatively narrow pelvis to facilitate shifting the body weight over the stance leg during locomotion (thereby increasing efficiency). Females, however, have an added selective pressure: childbirth. If the female pelvis is too narrow, both her and her child's life are at risk. These sex differences in pelvic functional adaptation have resulted in a number of morphological differences in the male and female pelvis, which can be used reliably for sex estimation (Figure 12.1).

One of the most popular methods for nonmetric pelvic sex determination is presented by Phenice (1969). Phenice looked at three particular pelvic traits: presence of a ventral arc, subpubic concavity, and medial aspect of the ischiopubic ramus. Applying these three traits to black and white individuals from the Terry Collection, he reported an accuracy rate of 96% (although see Ubelaker and Volk 2002; Lovell 1989; Kelley 1979; McLaughlin and Bruce 1990; Sutherland and Suchey 1991 for reports of similar and dissimilar accuracy rates, as well as discussion regarding intra- and interobserver rates, and application to other samples). Phenice did point out that,

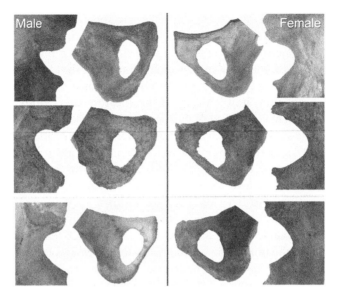

Figure 12.1   Examples of male and female pelvic traits. Examples of male (left) and female (right) greater sciatic notch and pubic bone morphology are provided. The top two individuals represent classic male and female morphologies. Males typically display a narrow notch (described by some as "J-shaped"), and females a broader, more obtuse notch (more "L-shaped") (see Walker 2005). Females also display a more concave subpubic concavity (inferior border of the ischiopubic ramus), and generally have a more "stretched out" appearance of the pubis. The bottom male and female individuals represent examples where signals may be mixed. This particular male displays a broader greater sciatic notch and more concave ischiopubic ramus than would be expected for a male. Although the bottom example of the female maintains a relatively broad greater sciatic notch, the pubic bone appears more ambiguous. All examples are from documented samples with known sex.

at times, one or two of the traits may be ambiguous, but that "there is almost always one of the criteria which is obviously indicative of male or female" (Phenice 1969: 299). Furthermore, he notes that the ischiopubic ramus is the least reliable of the three traits, suggesting that reliance should only be placed on this trait in the absence of the other two criteria.

There are a number of other sexually diagnostic pelvic and sacral traits that Phenice did not analyze, and Rogers and Saunders (1994) summarize and evaluate a number of these pelvic traits. They provide accuracy and intraobserver rates for 17 different morphological pelvic traits, as well as multivariate results when traits were combined. They found that certain trait combinations, such as obturator foramen shape and ventral arc presence, when used in conjunction, provide correct classification rates as high as 98%. Individual traits, such as sacrum or obturator foramen shape, could reach accuracy levels of 94% when observed independently. They also indicated which traits were more unreliable, producing intraobserver error rates above 10%. In concordance with Phenice's words of caution, the ischiopubic ramus was among the four traits determined unreliable.

Another popular method is presented by Bruzek (2002). This study focused on five main characters: the preauricular sulcus, the greater sciatic notch, the composite arch, the inferior pelvis, and the ischiopubic proportions. Bruzek reported correct classification rates of 95% based on two European samples. The method is slightly more complex than Phenice's method, requiring the observation of numerous traits within each character. However, Bruzek does provide detailed descriptions of the various male, female, and indeterminate forms which may be observed, discusses their functional importance, and provides detailed instructions for sex assessment, as well as a number of historical references for each trait: all valuable tools to the physical anthropologist.

A quick literature search will reveal numerous other articles testing many of these methods/traits on various samples, revising methods, and presenting metric techniques in an attempt to objectively capture many of these morphological traits. Regardless of the preferred technique, however, the pelvis remains the most favored element for adult sex determination.

Crania

Nonmetric traits in the skull have also been traditionally used for sex determination (Figure 12.2). Although Krogman (1955) presented 13 cranial traits indicative of sex, five particular traits have received most of the attention: the nuchal crest, mastoid process, supraorbital margin, supraorbital ridge/glabella, and mental eminence. While these traits were studied and presented as early as Broca (1875) and Acsádi and Nemeskéri (1970), the scoring system presented in the popular Standards (Buikstra and Ubelaker 1994) is the most commonly employed. In this method, each trait is presented as a series of five line drawings, illustrating a gradation of the traits from very "feminine" (1) to very "masculine" (5). The illustrations and accompanied descriptions are used to score each cranial trait, and based on the overall pattern of the scores they determine sex. While visually scoring features does introduce a degree of subjectivity, and interobserver scores are not ideal (Walker 2008; Walrath et al. 2004; Williams and Rogers 2006), this method has been used by forensic anthropologists for decades with reported high degrees of accuracy.

Walker (2008) applied this ordinal scale method to samples of modern European-Americans, African-Americans, and English (total $n=304$), and a sample of Native Americans ($n=156$). He reported that percentage correct classification rates for univariate analyses of the traditional traits ranged from 69 to 83% in modern samples, and only up to 70% in prehistoric North American samples. The mastoid process and glabellar region presented the highest accuracy in both samples. Using multivariate analysis he obtained accuracy rates up to 89% for the modern samples.

Rogers (2005) and Williams and Rogers (2006) pooled these five traditional characteristics with 12 additional ones, to obtain a total of 17 cranial and mandibular sex markers. Instead of using the ordinal scales described above, these studies used a binomial method, scoring each feature as displaying either male or female morphology. Rogers (2005) reports percentage correct values and intraobserver errors for each trait. Applying all 17 traits to a sample of nineteenth-century Canadian cemetery remains ($n=46$), she obtained a correct rate of 89.1%. Of the five traditional traits discussed above, the supraorbital ridge ranked the highest. Williams and Rogers (2006) applied this same binomial method to a sample of modern American whites from the Bass Donated Skeletal Collection ($n=50$). The only modification they made was to divide the individual components of the mandible into four separate scoring

(a)

(b)

(c)

(d)

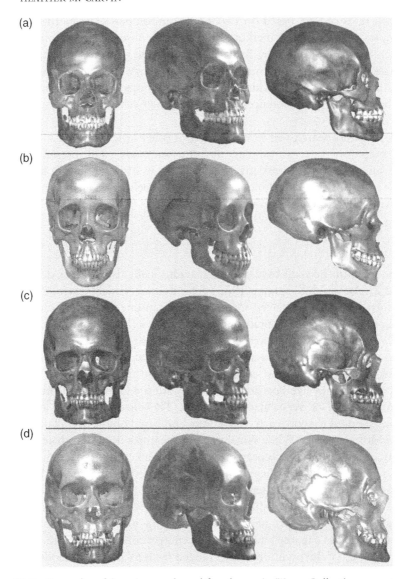

Figure 12.2 Examples of American male and female crania (Terry Collection, courtesy of Physical Anthropology Division, Smithsonian Institution): (a) black male (specimen no. 509), (b) black female (specimen no. 1500), (c) white male (specimen no. 380), (d) white female (specimen no. 37R). Generally, males (a, c) are characterized by more "robust" crania, displaying more prominent muscle markings, "squared" chins with more prominent mental eminences, more pronounced supraorbital ridges, larger mastoid processes, and greater gonial eversion. However, variation does occur. For example, the chin morphology in individual (a) is characteristically male, while the white male's (c) chin morphology is not much different than the white female's (d). On the other hand, the white male's (c) supraorbital ridge is much more pronounced and dimorphic than the black male's (a). Note that these specimens were chosen to illustrate individual variation, and do not represent group average morphologies.

traits. They found that by combining these 20 traits they obtained a percentage correct classification between 92 and 96%. Again, percentage correct classification and intraobserver errors were documented per trait. As in Rogers (2005) the surpaorbital ridge was ranked most reliable (86–90% accurate with 6% intraobserver error). However, the rankings of the remaining traits differed between the two studies. For example, Williams and Rogers (2006) ranked the mastoid process (82% accuracy and 8.0% intraobserver error) with the supraoribital ridges in the most reliable group, while Rogers (2005) ranked the mastoid among the group of third most reliable traits (44.7% accuracy and 0.0% intraobserver error). While the different results between the two studies could be the result of interobserver error, they may also reflect variation in cranial trait expression between the two samples (modern European-American versus nineteeth-century Canadian).

Walker (2008) compared his ordinal cranial scores between his ancestrally diverse samples (European-American, African-American, English, and Native American). He found that the degree and pattern of expression of these traits, as well as their sexual dimorphism, varied across populations. For example, he found that the modern English sample was more gracile overall than the African- or European-American samples. European-Americans had more pronounced glabellar regions, while African-American individuals displayed more prominent mental eminences. The Native American sample in his study displayed more overall robust crania, but the sexual dimorphism (difference between males and females) was comparatively low.

Given such population variations, you can see how applying the traditional scoring method using a universal set of illustrations and standards would be problematic. Population-specific standards should be applied, but are currently not available. Anthropologists have yet to quantify the variation in these traits within or across populations. Furthermore, whereas factors such as hormones, sexual selection, body size, muscle mechanics, age, health, and nutrition have all been suggested to influence these trait morphologies, empirical evidence remains limited. Therefore, it is suggested that when applying nonmetric cranial sexing methods the researcher should be familiar with the study population, adjusting the scoring method to the patterns and degree of expression exhibited by that specific population.

CONCLUSION: WHAT TO USE?

Now you're thinking "Thanks for all the info, but you didn't tell me which methods/elements I should use." You're right. That's because there is no absolute answer, only recommendations and preferences that will vary depending on the source. One truth that is accepted by physical anthropologists is that using multiple indicators will produce better results than any single trait used alone. It is also a general consensus that the pelvis provides more reliable sex indicators than any other skeletal element. Beyond those statements, however, preferred sex determination methods will vary according to the anthropologist's personal preference and experience, as well as to which skeletal elements are available for analysis.

Spradley and Jantz (2011) argue that their 94% correct classification rate using postcranial metrics exceeds even the best models using the cranium. And this appears to be true, for both cranial metrics and nonmetrics, with the exception of some accuracy rates that Williams and Rogers (2006) reported when using certain multivariate trait combinations. Many argue that metric methods are superior to nonmetric methods because they are objective, resulting in lower intra- and interobserver rates regardless of an individual's experience. Cranial and postcranial metrics can also be more easily subjected to statistical analyses, facilitating within- and between-population comparisons, multivariate, probabilistic, and error analyses. On the other hand, metric analyses typically require intact skeletal elements and rely on size, which is known to vary greatly between populations.

Visual assessment of nonmetric analyses can be carried out rapidly, without the need of special equipment, and can be applied to fragmentary remains (although see Konigsberg and Hens 1998 for a critique of nonmetric cranial methods). More importantly, it is clear that sex differences certainly do exist in pelvic and cranial/mandibular shape even if size is removed. Variation in documented accuracy rates of these traits likely reflect population variations in trait patterns/expression, and method subjectivity (with researcher experience, both with the method and with that specific sample being a particularly influential variable). Traditional metric methods are unable to adequately capture these morphological traits, but continuing technological advances promise future quantitative methods that would eliminate the majority of these issues.

Until then, the debate between metric and nonmetric methods will continue. Regardless of your preference, it remains crucial that population variation is considered in analyses. With the increasing medicolegal and judicial pressures (see Chapter 1), it also remains vital that the forensic anthropologist be familiar with and ready to report the published error rates for the methods they choose, and to continue evaluating the reliability of methods (old and new) on a variety of samples.

REFERENCES

Acsádi, G. and Nemeskéri, J. (1970). *History of Human Life Span and Mortality.* Akadémiai Kiadó, Budapest.

Bass, W.M. (2005). *Human Osteology: a Laboratory and Field Manual*, 5th edn. Missouri Archaeological Society, Columbia, MO.

Broca, P. (1875). Instructions craniologiques et craniométriques de la Société d'Anthropologie de Paris. *Bulletin de la Societe Anthropologie*16: 534–536.

Bruzek, J. (2002). A method for visual determination of sex, using the human hip bone. *American Journal of Physical Anthropology* 117: 157–168.

Buikstra, J.E. and Ubelaker, D.H. (1994). Standards for data collection from human skeletal remains. *Proceedings of a seminar at the Field Museum of Natural History*, Arkansas Archeological Survey.

Bytheway, J.A. and Ross, A.H. (2010). A geometric morphometric approach to sex determination of the human adult os coxa. *Journal of Forensic Sciences* 55(4): 859–864.

Giles, E. and Elliot, O. (1963). Sex determination by discriminant function analysis of crania. *American Journal of Physical Anthropology* 21: 129–135.

Gonzalez, P.N., Bernal, V., and Perez, S.I. (2011). Analysis of sexual dimorphism of craniofacial traits using geometric morphometric techniques. *International Journal of Osteoarchaeology* 21: 89–91.

Guyomarc'h, P. and Bruzek, J. (2011). Accuracy and reliability in sex determination from skulls: a comparison of Fordisc 3.0 and the discriminant function analysis. *Forensic Science International* 208: 180. e1–180.e6.

Kelley, M.A. (1979). Parturition and pelvic changes. *American Journal of Physical Anthropology* 51: 541–546.

Kimmerle, E.H., Ross, A., and Slice, D. (2008). Sexual dimorphism in America: geometric morphometric analysis of the craniofacial region. *Journal of Forensic Sciences* 53(1): 54–57.

Konigsberg, L. and Hens, S.M. (1998). Use of ordinal categorical variables in skeletal assessment of sex from the cranium. *American Journal of Physical Anthropology* 107: 97–112.

Krogman, W.M. (1955). The skeleton in the forensic medicine. *Postgrad Med* 17(2): A48–A62.

Krogman, W.M. and İşcan, M.Y. (1986). *The Human Skeleton in Forensic Medicine.* Charles C. Thomas, Springfield, IL.

Lovell, N.C. (1989). Test of Phenice's technique for determining sex from the os pubis. *American Journal of Physical Anthropology* 79: 117–120.

McLaughlin, S.M. and Bruce, M.F. (1990). The accuracy of sex identification in European skeletal remains using the Phenice criteria. *Journal of Forensic Sciences* 35: 1384–1392.

Ousley, S.D. and Jantz, R.L. (2005). *FORDISC 3.0: Personal Computer Forensic Discriminant Functions.* University of Tennessee, Knoxville, TN.

Phenice, T.W. (1969). A newly developed visual method of sexing the os pubis. *American Journal of Physical Anthropology* 30: 297–301.

Robling, A.G. and Ubelaker, D.H. (1997). Sex estimation from the metatarsals. *Journal of Forensic Sciences* 42(6): 1062–1069.

Rogers, T.L. (2005). Determining the sex of human remains through cranial morphology. *Journal of Forensic Sciences* 50(3): 1–8.

Rogers, T. and Saunders, S. (1994). Accuracy of sex determination using morphological traits of the human pelvis. *Journal of Forensic Sciences* 39: 1047–1056.

Spradley, M.K. and Jantz, R.L. (2011). Sex estimation in forensic anthropology: skull versus postcranial elements. *Journal of Forensic Sciences* 56(2): 289–296.

Stewart, T.D. (1979). *Essentials of Forensic Anthropology.* Charles C. Thomas, Springfield, IL.

Sutherland, L.D. and Suchey, J.M. (1991). Use of the ventral arc in sex determination. *Journal of Forensic Sciences* 36: 501–511.

Ubelaker, D.H. and Volk, C.G. (2002). A test of the Phenice method for the estimation of sex. *Journal of Forensic Sciences* 47: 19–24.

Walker, P.L. (2005). Greater sciatic notch morphology: sex, age, and population differences. *American Journal of Physical Anthropology* 127: 385–391.

Walker, P.L. (2008). Sexing skulls using discriminant function analysis of visually assessed traits. *American Journal of Physical Anthropology* 136: 39–50.

Walrath, D.E., Turner, P., and Bruzek, J. (2004). Reliability test of the visual assessment of cranial traits for sex determination. *American Journal of Physical Anthropology* 125: 132–137.

Wescott, D.J. (2000). Sex variation in the second cervical vertebra. *Journal of Forensic Sciences* 45(2): 462–466.

Williams, B.A. and T. Rogers. (2006). Evaluating the accuracy and precision of cranial morphological traits for sex determination. *Journal of Forensic Sciences* 51: 729–735.

CHAPTER 13 Sexual Dimorphism: Interpreting Sex Markers

Luis L. Cabo, Ciarán P. Brewster, and Juan Luengo Azpiazu

INTRODUCTION: EASY SEX?

Sex is a priori the most incontrovertible component of the biological profile. First, unlike in the remaining components of the profile, the alternatives in a sex diagnosis are limited to two clear-cut options: *male* or *female*, without any continua or gray areas between them. Although humans simultaneously displaying masculine and feminine organs and secondary sexual traits do exist (see Kim et al. 2002 and references therein for a brief review of the subject), their frequency is too low to be relevant in forensic or paleontological settings (well, at least until we get one of those cases, which may be tomorrow morning).

Hence, when attempting to determine the sex of our set of human remains, we should theoretically benefit from a certain freedom from some of the woes associated with other classification issues. For example, unlike in ancestry determination, we do not have to worry about "mixed" individuals or overlapping groups and subgroups, which may or may not be represented in our reference samples. Neither are there concerns regarding whether the sources are placing the victim into an objective biological group, or just categorizing her based on a cultural construct without much basis in biology.

Sex assessment also compares favorably with other seemingly equally objective components of the biological profile, such as stature or age. No worries about whether the person reported correctly the stature in her driving license, if she experienced a particularly dramatic growth spurt after getting her wheels, or stooped before the last

A Companion to Forensic Anthropology, First Edition. Edited by Dennis C. Dirkmaat.
© 2012 John Wiley & Sons Ltd. Published 2015 by John Wiley & Sons Ltd.

renewal; whether she had a hard or easy life, making her body age at a faster or slower pace, or if she simply took a chance to add some years to her birth date at some point, maybe seeking to appear more attractive or less convictable. In the case of sex, the antemortem records and family reports are much more straightforward and less exposed to subjective interpretations. Released from these considerations and meta-physical enquiries, the question in sex determination remains very simple: was it a "he" or a "she"?

It appears that solving this dichotomy in a set of skeletal remains should be easy enough as the sex of a human being also remains constant throughout life, being eas-ily assessed in the fleshed individual. Masculine and feminine sexual organs can already be distinguished in the fetus from week 12 after fertilization (Gilbert-Barness and Debich-Spicer 2004: 12). From this age, the simplest of physical examinations (including just playing doctors) can serve to determine the sex of a fully fleshed per-son with little margin of error, even in the most ambiguous cases. As we will discuss in later sections, when dealing with living adults we can even do without the medical profession (real or pretend), in addition to keeping the subject's pants on, as we all are natural-born sex assessment experts. Beyond the external genitalia, multiple conspicu-ous external cues guide us to accurate sex identification in living individuals. Our evolutionary fitness is extremely dependent on both displaying these traits and being extremely proficient at recognizing and interpreting them. These innate sexing capa-bilities remain active and proficient long after other basic instincts have deserted us, such as in those occasions when our senses are impaired by factors like alcohol consumption, a dim light, and the right Patsy Cline song playing in the jukebox (regarding the musical selection, we had already established that we had been drinking).

Dealing with such an unambiguous, permanent, and easily recognizable dichoto-mous biological trait, it would appear that, compared to the remaining components of the biological profile, skeletal sex assessment should be very easy, shouldn't it? Well, in case the paragraphs above had awakened a new vocation, inspiring you to leave everything in pursuit of a career as a professional human skeletal sexer, brace yourself for the bad news: It is not that easy.

PROBLEMS AND QUESTIONS: NAUGHTY BITS AND NOISY SEX

The way in which we described biological sex above is mostly a matter of soft tissues. Soft tissues in the gonads secrete hormones (*gonadal sex steroids*), which in turn promote the development of other soft tissues, many of which will become those secondary sexual traits of which we became so fond and appreciative during our teen years. Therefore, it is from all these overgrown soft tissues (textually, soft tissues on steroids) that determining sex is so clear cut, simply in a presence/absence basis.

As we will discuss during the rest of this chapter, sex hormones also affect (and effect) the development of our skeleton, but they do so in more systemic and indirect ways than in the case of target soft tissues. By *systemic* we simply mean that, unlike in some target soft tissues (e.g., those in the gonadal glands or secondary sexual traits), sexual hormones do not promote the formation of any sexually specific bone structure in humans. They do so in other species, for example with the formation of baculi (singular *baculum*, or *os penis*) in many mammals, but skeletal sexual dimorphism in humans will be more related to systemic effects on all bone tissues, resulting in general differences in overall body size and body proportions. Sex steroids do act directly on the bone tissue. However, the effects of this direct histological action pale in comparison to their *indirect* effect on sexual dimorphism, through the induction of sex-related divergences in developmental timing and biomechanical stresses. These last two factors are in turn intimately related to sexual differences in overall body size, proportions, and composition.

In spite of being more subtly and systemically expressed than in soft tissues, sexual differences in the human skeleton do exist and can be utilized to estimate the sex of a deceased individual. There is a wealth of detailed how-to articles dealing with skeletal sex determination in the literature, with many of the most relevant of them reviewed in Chapter 12 in this volume. In addition, almost any reputable manual on physical or forensic anthropology (and many seedy ones) provide comprehensive lists of skeletal sex markers, with abundant indications on how they are expressed in males and females. If you check the corresponding sections in references like Bass (2005), Buikstra and Ubelaker (1994), Krogman and İşcan (1986), or White and Folkens (2005) you will probably find out that most of the information provided is fairly similar and repetitive. There is good reason for it: these classic skeletal sex markers work pretty well in most cases. To wit, at least in those ones where we confront a decently "gracile" female or a reasonably "robust" male.

Thus, when asked to write a chapter on sex determination, the authors soon realized that they could contribute little to the classic how-to approach, other than compiling and reheating the old trait menu for the *n*th time in recent history. If anything, many of the virtues and defects distinguishing the classic references are related to the quality of their graphic depictions, with regard to their ability in encapsulating the subtle sexual differences in trait morphology and development. But it happens to be the case that none of the authors is a particularly gifted artist. Indeed, we collectively have the artistic ability of a 5 year old (and one who would have been very happy of getting rid of it). So, we will instead refer you to Chapter 12 for a brief but insightful review of the main sexing methods, and devote this chapter to solving some issues that may arise when trying to apply or combine them.

Because the fact is that, rather than our ability to remember and identify all those widely publicized classic skeletal sex markers, what tormented us in all those sleepless nights of restless pondering as human osteology students, and still occasionally haunt us today in our professional practice (although we do sleep better, thank you), were those cases in which the diagnostic traits seemed to contradict each other. If, let's say, the mastoid process, occipital protuberance, and general frontal morphology adamantly insist on our individual being a male, but the mental protuberance, suprameatal crest, and glabellar area seem to cry out loud that what is on the laboratory table is a female skull, how should we go about it? Is there any way in which we could rank these traits from more to less reliable? Or maybe we should go for a raw-numbers approach: if seven traits say "he" and just three traits say "she," should we go for the male diagnosis?

Metric methods served to alleviate our suffering, helping us to see at least some of the forest beyond the traits. As we will explain below, multivariate methods like discriminant function analysis (DFA) allow us to simultaneously compare a large number of dimensions of our individual with those of hundreds of other persons of known sex. This helps us decide not only whether our individual is more likely to be a male or a female, but also to assess precisely how much likely is it to be so. This appeared to be just what we had been praying for: several traits but only a true single diagnosis, with precise probability assessments as an extra bonus. Fantastic software packages like Fordisc (Ousley and Jantz 2005; see also Chapter 15 in this volume) include not only a comparative dataset from a growing number of different populations, but also the statistical tools to analyze it, having simplified and made these metric analyses widely (and maybe even wildly) accessible to forensic practitioners in the last decades.

However, although these multivariate methods have provided us with some very powerful statistical garlic and holy water to battle the fiend, the drapery has not been completely drawn to let all sunshine in. The beast will keep tapping on our window more frequently than decency would advise and, in those fearful nights of unfathomable darkness, it is typically still pretty difficult to reach and find the third intercostal space where to stick our statistical stake. The master of deceit will keep making our probabilities dance like candle lights in a draft every time we add or subtract a variable or ancestry group. Our individual will keep mutating from male to female depending on whether we include, let's say, jaw measurements or Vietnamese males in our analysis. The results from cranial and postcranial data (which must be analyzed separately in Fordisc) may also stubbornly contradict each other, claiming that our individual was a female head trapped on a male body, or vice versa. Finally, in the climax of this pandemonium, our metric and morphological analyses can both render apparently clear, but completely contradictory, results.

Uunlike the clean-cut diagnoses provided by their soft-tissue counterparts, skeletal sex markers tend to provide noisy and confusing results, no matter whether we are looking at discrete traits or at metric variables. This noisy nature of sex estimates, plagued by ambiguous trait distinctions and confusing figures, sets us at risk of providing the wrong diagnosis. But in the forensic arena our problems do not end here.

MORE PROBLEMS: NO MEANS NO... BUT YOU HAVE TO EXPLAIN WHY

Conflicting results are a nuisance in paleoanthropological studies, but have an even deeper impact in forensic analyses, serving to illustrate some of the fundamental differences between both fields. The human paleontologist can usually rely on experience to weigh the consistency of each diagnostic alternative. Baccino and colleagues (1999) examined the reliability of different metric and nonmetric age-estimation techniques, depending on the experience of the analyst. Their results showed that (i) a comprehensive approach, based on the application of several methods, worked better than the examination of any single age marker alone, and (ii) morphological (qualitative) methods worked better for experienced professionals, while metric analyses rendered better results for more inexperienced analysts (Baccino et al. 1999).

We believe that both principles also apply to sex determination. The first premise (which we can summarize as *never trust an isolated trait*) is sufficiently demonstrated by the aforementioned common attainment of contradictory results from different anatomical areas and skeletal sex markers (although we will later discuss the superiority as sex markers of whole anatomical regions such as the pubic symphyseal area).

As for the success of qualitative methods depending largely on the experience of the observer, the interpretation of nonmetric sex markers is always comparative. We already mentioned that humans do not display any bones or osseous structures appearing exclusively in one of the sexes. Hence, human nonmetric sex markers are almost invariably defined as elements being "larger" or "more marked" in one of the sexes. Consequently, it will be easier to decide whether a mastoid process is relatively large or small if we have seen a hundred, than if we have confronted just 10 of them in our career. Similarly, in the absence of well-established "reliability rankings" for the different methods and markers, we will have a better impression of which ones we should trust best once that we have observed their respective success and failure rates in a large number of cases and situations.

The comparative nature of qualitative sex markers and the reliance for their success on personal experience take a toll on the possibility of objectively ranking traits as universally better to worse sex predictors, as their relative success rates (e.g., their percentage of correct classification) will also depend largely on the observer. One anthropologist may swear by the sex markers in the temporal bone, while a second one will place his undivided affection on facial traits; and very likely both of them will be right: the better trait is the one that you know and understand better and, consequently, which works for you. It is based on these personal experiences and abilities, in the very distinct overall impression of the individual being a male or a female, regardless of what some individual traits may say, that even moderately experienced paleoanthropologists can feel very comfortable and safe providing a sex diagnosis from apparently contradictory evidence; occasionally, even being right about it (but see Kruger and Dunning 1999).

However, being confident in their intuitions does not suffice for forensic anthropologists. Unlike that of the paleoanthropologist, the path of the righteous forensic anthropologist is beset on all sides by the requirements of the legal system and, in the USA, the tyranny of the *Daubert standard*. The latter is a rule of evidence regulating the admissibility in US federal courts of expert witness testimony on scientific matters. The standard derives from the US Supreme Court decision in the case *Daubert v. Merrell Dow Pharmaceuticals* (1993). Dirkmaat et al. (2008) and Chapter 1 in this volume present lengthier discussions of the Daubert standard and its implications for forensic anthropology, while Huber (1993) provides a very entertaining (if also rather depressing) in-depth discussion of the Daubert v. Merrell Dow case itself. However, without getting into these depths, the main consequence of the Daubert ruling for our purpose is that it shifts the spotlight from the expert to the expert's report, substituting appeals to authority by scientific criteria as the only acceptable means to justify expert diagnoses and conclusions. In its attempt to adapt the forensic standards and protocols to mainstream scientific criteria, the implications and meaning of the Daubert standard transcend those of a mere legal rule, relevant at a particular territory or court jurisdiction, coming rather to provide a set of universal principles and values for the forensic sciences at large. If Moses had been a forensic scientist, Yahweh would have handed him the Daubert standard at Mount Sinai, while snatching the magic rod from him.

How does this affect our problem? If as scientists we should already be very wary of founding our diagnoses on intuition (the specter of bad science feeds on overcredited hunches and the decaying corpses of spoiled brainchildren), the Daubert standard makes a forensic case of such practice itself. The *bêtes noires* of forensic reports are vagueness and contradictions. If they are present in your report, and you fail to clearly establish why you opted for the result of a particular analysis over a second contradictory one, your opponent on the other side of the stand will brandish it in court as the most prized trophy: the incarnation of reasonable doubt, when not proof of your worthlessness as an expert witness. After Daubert, referring to your past experience or credentials as a justification for your election is simply not an option. If you want to be safe, you will need to provide very convincing objective arguments and reasoning to support your sex diagnoses.

From a practical point of view, this means first being able to explain why some traits or analyses can render a "male" diagnosis in a female individual or vice versa, and secondly being able to assess whether, how, and to what extent the factors usually promoting such confusion are affecting our particular case. The objective of this chapter is providing the most common answers to the first question, and some general principles that will be useful to answer the second in the most frequent scenarios.

Most of our reasoning and explanations from this point will be articulated around metric methods and variables. This approach was not selected (only) to torment anthropology students who, as we are sadly well aware, are as likely to be intensely enticed by the joys and thrills of biometry as dentists are to recommend sugar-rich chewing gum. It is not intended either to argue in favor of any alleged intrinsic superiority of metric over morphological methods. On the contrary, our higher focus on metric traits simply spans from two very convenient characteristics of quantitative traits: (i) they are easier to describe, analyze, and display in simple univariate and bivariate plots and (ii) the vast majority of qualitative traits employed in sex determination are just ranked classifications of quantitative traits. When we say "larger than" or "more marked," we are just deciding to look at broad intervals within the continuous size distribution of the trait, rather than at its exact dimensions, just as when we describe our shirt size as "S" or "L" instead of providing our exact neck, sleeve, and chest dimensions. For that reason, many of the problems in interpreting qualitative traits emerge from the distribution of their metric dimensions, and are better and more easily approached from a metric perspective.

We also favored examples based on postcranial rather than cranial elements when explaining general principles. While these principles apply to both anatomical regions, relationships between bone size, bone morphology and factors such as growth, development, overall body size, or body mass are easier to grasp through examples based on beam-shaped, weight-bearing structures, such as long bones.

Given the introductory nature of the chapter, rather than limiting the bibliography to the mandatory citations of the works used to reference our points, we opted for explicitly pointing the reader to some monographs and journal articles that we find especially relevant or useful when approaching the study of sexual dimorphism for the first time, through a Recommended Readings section. In in-text citations, review and general articles with expanded bibliographies were preferred to highly specialized studies, even when the latter were the original source of the information in the review article. This somehow unconventional practice was adopted with the goal of providing a bibliography section representing a selection of accessible readings, which can be realistically covered during a class semester by the (usually lone) anthropology student wishing to delve deeper into the subject. In particular, we paid especial attention to introducing key references and comprehensive reviews from the general biological literature, with which professional or aspiring anthropologists may thus be less familiar.

Finally, this chapter is about sex and death, those being the key constituents of any literary bombshell and, in our experience, the only two subjects than can capture a college student's attention during a 6pm 2-h class session. We hope that you will excuse us when you think that we are going a pinch too far trying to take advantage of this fact, in an attempt to attain the impossible: that our students would *actually* read it, without dropping like flies prey to either boredom or despair (after all, we already have our exams carefully designed for the latter).

SEXUAL SIZE DIMORPHISM: SIZE DOES MATTER

We have established that, from a strictly biological point of view, sex is a discrete variable, which for any given individual can only take one of two values. However, the diagnostic skeletal traits that we employ to assess the sex of an individual show continuous distributions that overlap between sexes. Whether we are defining a particular qualitative trait as "larger" or "marked," or trying to quantify these characteristics through metric measurements, we generally will not be confronting sharp crevices separating the sexes, but gentle slopes gradually sliding toward areas of higher or smaller relative probabilities for each sex group. It is like asking the waiter if the fish *du jour* is cod or tuna, and getting the dish's price as the only answer.

The key question then becomes whether the discrepancy between the binomial (male or female) nature of the variable that we are trying to infer, and the continuous nature of our estimates, is simply due to measurement error or random noise, like the static in an old TV not allowing us to perceive clearly the images.

If there were no correlation between dish price and contents, and cod gives us a rash, we should probably go for the salad. However, if such a correlation exists, we would only need the appropriate information to help us decide whether, when ordering the dish, we would be on our way to tuna paradise or to the closest dermatologist. The first and probably most important clue to solve our inquiry would come from the prices of other dishes in the menu. Is this a cheap or an expensive restaurant? Fresh tuna may be more expensive than cod, but cod at an expensive restaurant may be pricier than tuna at a cheaper one.

Is there any general factor, equivalent to restaurant expensiveness in our example, which could allow us to make similar inferences respect to skeletal sex markers? The answer is, of course, yes. Figure 13.1 displays the distribution of four different variables in a sample of "white" American men and women from the Forensic Anthropology Data Bank (FDB; Jantz and Moore-Jansen 2000). The female group is the smallest one in all the variables, no matter whether they refer to overall body size (i.e., stature and body weight) or individual bone dimensions (i.e., the mediolateral breadth of the distal epiphysis and the total length of the humerus). This reveals a simple but fundamental principle that, as we will discuss, is essential to understand and interpret sex assessments: men are on average taller and heavier than women and, consequently, their bones are also longer and larger (we will later explain what we may mean by "larger").

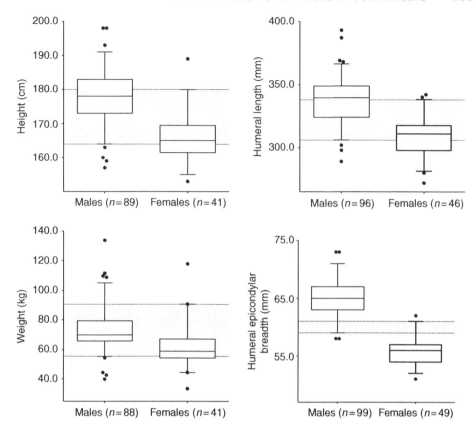

Figure 13.1 Distribution of two overall body-size estimates (left) and two bone measurements (right). The gray areas highlight the overlap between the upper 5% tail of the female distribution and the lower 5% tail of the male one, indicating the degree of overlap between the two sex groups. Note how the epicondylar breadth of the humerus (lower right quadrant) shows a much smaller overlap than the remaining variables.

Sexual differences in overall size explain such a large portion of sexual dimorphism in most species that, unless a specific factor such as differences in coloration or ornamentation is specified, it is usually understood that by "sexual dimorphism" we are referring to sexual *size* dimorphism (SSD). Apart from its quantitative importance (i.e., the large portion of total variance of overall sexual dimorphism typically explained by SSD), this relevance is also related to the enormous influence of body size in almost any other significant physiological or ecological property of an organism. The size of an organism influences key biological processes from the molecular to the populational level, as diverse as metabolic rates and the timing of physiological processes, life history, the range of organisms that it can prey on or be a prey to, energy requirements and expenditure, biomechanical configurations and constraints, the size and number of its potential offspring (e.g., the number of eggs that a female can lay and their average size at oviposition), or stable group and population sizes.

Hence, whether we are talking about the physiological, reproductive, ecological, or behavioral implications or consequences of sexual dimorphism at large, SSD always arises as a key element to model and understand the former. SSD can reduce competition between males and females (the resulting differences in energy requirements or prey selection serving to shift apart the niches or even the whole spatial distribution of the two sex groups), condition group structure and the number of potential mates, and affect success when competing with other individuals for these mates or to achieve a higher status, or the number and quality of the offspring.

In other words, it is impossible to properly approach, understand, or analyze sexual differences without taking into account their physiological, ecological, and evolutionary implications, keeping in mind that body size plays an essential role in all of them and, consequently, is the primary suspect to explain both most of those differences and the biological processes driving them. Of course, the same is true for anatomical questions, such as the ones that we will confront when assessing skeletal sex differences.

With this information, it will not come as a surprise that most of the skeletal traits employed to assess sex in humans are body size markers, to a smaller or larger extent. Just as most dishes will be pricier at an expensive restaurant, most skeletal measurements are higher in larger individuals, and men are on average larger than women. However, an examination and good understanding of what SSD entitles and how it arises in humans will allow us to go much further than that simple distinction in our efforts to assess the sex of skeletal remains.

SSD in the human skeleton is the result of gonadal steroid action, being only fully apparent in adults. Figure 13.2 illustrates the conditional distribution of body mass on body height in a sample of 12 year olds from Spain. The children were enrolled in a swimming training program, with little distortion in their weight figures caused by variation in body fat (i.e., body composition). At this age, significant sexual differences are detected neither in stature nor in weight. More importantly, when body mass is expressed as the cubic root of weight, in order to linearize the relationship, both boys and girls can be fitted into a single regression line describing the relationship between weight and height, following the general equation:

$$y = a + bx \tag{13.1}$$

Where y is the cubic root of body weight, x is body height (stature), b is the slope of the relationship, and a the intercept of the regression line with the vertical (y) axis. We will get very familiar with these simple but extremely useful and explanatory linear relationships during the rest of the chapter.

Both sexes sharing the same values of a and b in equation 13.1 (whose exact values for our reduced sample are not important here) means that, *for any given height*, boys and girls have on average the same body mass (i.e., the distribution of weight for both boys and girls of a given height will be normally distributed around the solution of equation 13.1). Conversely, skeletal mass (and thus the growth of the skeleton) is mostly fixed around ages 18 to 20, although some studies suggest that it can keep increasing well into the third decade of life in some individuals (Slemenda et al. 1994). This means that little or no skeletal mass increase is experienced after the growth spurt. In other words, no significant sexual differences are detected before the growth

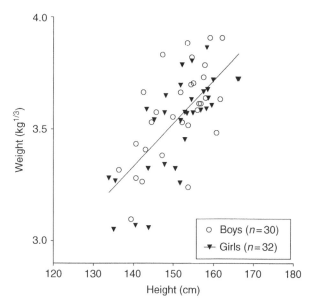

Figure 13.2 Conditional distribution of body mass on height for a sample of 12-year-old Spanish children. Body mass is expressed as the cubic root of weight, to linearize the relationship. The two sex groups share the same regression line (static allometry), evidencing the absence of sexual dimorphism at this age.

spurt, and little or no skeletal growth is experienced after the growth spurt. Therefore, the significant differences between adult males and females illustrated in Figure 13.1 develop after puberty onset and are basically fixed by the end of the growth spurt.

Note that in the last phrase of the previous paragraph we wrote "after puberty onset" and not "from puberty" or "as a result of the growth spurt." The reason for this choice of words is that a large proportion of SSD in humans does not develop as a result of males and females growing at very different rates during their respective growth spurts, but rather as a consequence of the different timings of puberty onset in males and females. This is to say, in those periods when one group is undergoing the growth spurt and the other is not.

Both before and during the growth spurt, skeletal growth (measured in terms of bone mass and density) is mostly just correlated with increases in height and body mass, with little sexual differences in this relationship, which are in any case much smaller than those related to the level of physical activity of the individual, independent of sex (Slemenda et al. 1994). Although body mass may be the most relevant estimate of overall body size from the physiological and ecological points of view in most species, it can be argued that the former only makes sense when compared with body length or, in humans, stature. Putting it simply, while body mass will depend largely on body length, the opposite is not necessarily true. If you increase your stature you will increase your body mass, due to the added tissue but, as most of us have experienced at some point in life, you do not become taller when you gain weight. This is to say, weight depends on height, rather than the opposite. Thus, to review how overall body size develops in humans, it appears sensible starting by taking a look at how adult stature is attained.

After some initial cat-and-mice play, in which girls catch up with boys (females display smaller average weights at birth, and lower growth rates during the first 6–8 months of life), after around age 5 both sexes show basically the same height increase rates until approximately age 9 (Veldhuis et al. 2005). Similar to Figure 13.2, the two sexes following the same growth line means that, even if slight average differences may exist, it will be virtually impossible to distinguish the two sexes based on body size during this period. Given that secondary sexual traits have still not developed at these ages either, precise and accurate anthropological sex assessment is basically only possible in adults, at least with the techniques currently available.

At around age 9, girls begin their pubertal growth spurt, but boys will keep gaining height at the slower prepubertal rate for another 2 years, until they begin their own growth spurt at around age 11 (Veldhuis et al. 2005). As a result, in Western countries girls around 14 are taller than boys of the same age.

Overall, girls gain height at the high pubertal rate between around 9 to 14.5 years, while boys will do the same from 11 to 17 years. During this pubertal growth period, skeletal mass basically doubles, with boys and girls attaining maximum height increase velocities of 8.3 and 9.5 cm per year, respectively (Riggs et al. 2002; Veldhuis et al. 2005). After the zenithal ages, both sexes keep gaining height at a much smaller rate (around 1 cm per year, with boys growing very slightly faster than girls) until they are around 20 years of age, with again some small differences in favor of the male group. At the end of the process men are around 13 cm taller than women. In our relatively small FDB sample, we actually observed a difference in height between 8.5 and 14.3 cm (95% confidence interval; $P < 0.001$). But what is more relevant for our purpose is that most of this difference is not accounted for by the sex-specific discrepancies in growth rates during the respective growth spurts, but by the 8–11 extra centimeters gained by males during the 2 extra years of growth at prepubertal rates (Veldhuis et al. 2005). This is to say, the bulk of the sexual size differences in height (and thus in overall body size when considered as a function of body length) are not related to the direct differential action of sex steroids on bone during puberty, but rather to timing differences in the onset of puberty and, consequently, to prepubertal "asexual" growth rates and patterns, mediated primarily by growth hormone rather than directly by the gonadal sex steroids.

From a practical point of view this means that many sexual differences in body size are to a certain extent just that, raw metric differences, not necessarily linked to disparities in how estrogens or androgens influence the development of specific bones during puberty. Tall females will share many metrical (and rank qualitative characteristics) with short or average men, and vice versa.

The problem is further complicated by the correlation between height and body mass, as markers related to muscle insertions will also reflect the increased stresses (and thus higher bone deposition) derived from a larger body size. Longer bones and distances between muscle insertions mean longer lever arms that, paired with the higher mass supported by the system, result in higher torques and stresses placed on its skeletal components. Once again, males will display larger muscle attachments and impressions, but tall or "robust" females may also display them.

This correlation of overall size with many sex markers explains much of the noise and uncertainty associated with the sex diagnoses obtained from these traits. A second

Table 13.1 Pearson correlations between different body dimensions, bone measurements, and the discriminant scores obtained from the latter for sex classification. Note the high correlation of all measurements with body height, and the larger correlation of epiphyseal measurements with body weight.

	Height	Weight$^{1/3}$	HXLN	BEPB	FXLN	FHDD
Weight$^{1/3}$	0.44					
Humeral length (HXLN)	0.84	0.27				
Humeral biepicondylar breadth (BEPB)	0.65	0.40	0.62			
Femoral length (FXLN)	0.87	0.27	0.89	0.58		
Femoral head diameter (FHDD)	0.68	0.32	0.72	0.77	0.67	
Discriminant scores	0.70	0.38	0.71	0.95	0.66	0.93

All correlations significant at the 0.01 level (two-tailed).

look at Figure 13.1 will help us to realize that, while females are on average smaller than males for all the variables considered, the size ranges of both sexes show important overlaps, represented in the figure by the gray areas running between the upper 95th percentile of the female distributions, and the lower 5th percentile of the male distributions. In particular, the distributions of stature, weight, and humeral length show very large overlaps between sexes, with the gray areas going as far or even beyond the mean values of the opposite sex. This means that a large proportion of females will show values close to the mean of males for all those three variables (i.e., for body size), and vice versa. Furthermore, if we must also consider populations with different average body sizes, males of the shorter population will be even closer in their SSD markers to females of the taller population.

However, something different seems to happen with the breadth of the distal epiphysis of the humerus (*epicondylar breadth*), where the overlap between both groups is just around 2 mm (Figure 13.1). Very little females will have epicondylar breadths as large as even the smallest males, and vice versa. Is it possible that this trait may not be merely expressing overall body size, therefore constituting a true sexual difference not related to SSD? But, if that is the case, how can we explain the high correlations displayed in Table 13.1 between epicondylar breadth and the three remaining variables? Humeral epicondylar breadth is actually the variable of the lot displaying a higher correlation with body mass, which is the noisiest of the three variables in Figure 13.1. If so much of its variability appears to be explained by body mass, why does this variable show so little noise and sex overlap itself? The answer to this question will provide us with a key tool and criterion to interpret our sex markers, as well as for solving and explaining apparent contradictions between the sex diagnoses obtained from different traits or anatomical areas.

I TELL YOU, IT'S A WHOLE DIFFERENT SEX!

The header of this section is a quote from Billy Wilder's 1959 movie *Some Like it Hot*. In Wilder's masterpiece, Jack Lemmon and Tony Curtis play two speakeasy musicians who, after accidentally witnessing Saint Valentine's Day massacre in 1929 Chicago, must hide from the killers by posing as women travelling with an all-female music band. In the referred scene, Lemmon and Curtis test for the first time their female attires, clumsily struggling with their high heels as they walk toward the train where they are to join the band, on a tour to Florida. Curtis is trying to wave off Lemmon's concerns about the possibility of actually passing for women, despite their elaborate costumes and efforts to look feminine. As they approach the train, Marilyn Monroe (the female lead in the movie) walks pass them, coming to support Lemmon's pessimistic view and triggering his despaired comment: "Look how she moves! It's like Jell-O on springs. Must have some sort of built-in motor or something. I tell you, it's a whole different sex!"

Billy Wilder's simple observation, expressed through Lemmon's character, set the basis not only for the remaining of his script, but for the development of a complete movie comedy genre involving men pretending to be women. The success of such a simple plot is entirely based on the disparity between the perception of the viewer and that of the characters inside the movie. What makes it so funny when Osgood Fielding III (Joe E. Brown) asks Lemmon's character to marry him is not that we *know* from the plot that Lemmon is actually a man, but that we can clearly *see* it, while Osgood seems to be completely blind to that fact.

The plot would not work if the audience (all of us) were not extremely good at distinguishing men from women. No surprise, as that is a pretty relevant skill from an evolutionary point of view. The question is, what is it in the male and female physiques that we can recognize so rapidly and intuitively? The previous section may have led you to suspect that it might be just a raw difference in body size, resulting only from 2 extra years of prepubertal growth in human males. However, this does not seem to make much sense. We do not tend to take tall women for men, or short men for women. Sigourney Weaver is around 1.82 m tall and John Hurt just 1.75, but you could distinguish them pretty well in Ridley Scott's 1979 *Alien*, even in the scenes when both of them were wearing the same uniform (make it two movies at the video store). It could all be just about the soft tissues, but then, how can you explain that the hypothetical confusion is not any more frequent in the arctic winter of Erie, Pennsylvania, where no sane human being would adventure outdoors with less than seven layers of thick clothing, a long scarf, and a rabbit-fur bomber hat?

No, the distribution of soft tissues and certain minor details, such as facial hair, certainly help in milder climates and at close distances but, once again, Billy Wilder got it right (write it down and get yourself a t-shirt with the motto: Billy Wilder was always right). Women have a different body frame than men, resulting in biomechanical differences in force directions, lever arms, and torques. Marilyn Monroe did have a different skeletal shape than Jack Lemmon, and definitively moved in a different way. You could tell them apart even in the distance. It is the recognition of these differences in body shape, and their biomechanical consequences, that allows us to determine so intuitively and easily that the person trying to keep a polar bear at bay at the corner with Pine Avenue, two blocks from your place, is a woman, even when there aren't any clear size references around in the white infinity of the arctic tundra (either the bear is huge, or she is very short), and she is wearing what appears to be some kind of blue Michelin Man costume.

Once again, where do these differences come from, if all our variables seem to be so correlated with body height, and body height appears to be expressing just raw SSD? The first component of the answer to this question comes from body mass: while a portion of the sexual differences in body mass are explained by those in height, differences in hormonal action during the growth spurt and sexual maturation themselves play a much more important role in this case; in particular, regarding the development of muscle mass.

As happened with height, girls are also lighter than boys at birth, and keep gaining weight at a slightly slower pace during the first year of life. By year two both sexes are gaining weight at the same rate, and at age 7 boys are just around 1 kg heavier than girls, a difference and pattern that will be maintained until puberty (see Figure 13.2). A large proportion of this increase in body mass is related to linear growth. As an example, height explained 42% of the variation in weight in our 12-year-old sample ($P < 0.001$; Figure 13.2).

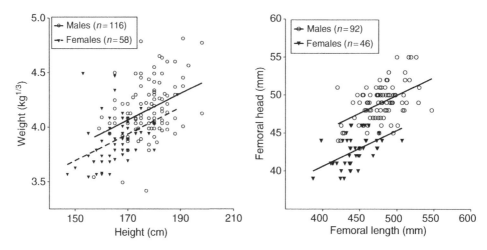

Figure 13.3 Conditional distributions of body mass on height (left) and femoral head diameter on the total length of the femur (right) for a "white" adult sample from the Forensic Data Bank. Note how the bivariate relationship between bone measurements parallels that between the overall body dimensions. In both cases the two groups share the same slope, but have different intercepts, so that for any given value of the independent variable males are on average larger than females by a constant figure (the difference between the y intercepts of both lines). Compare this situation with the one observed in Figure 13.2.

Still, unlike in the case of body height, androgens do induce much higher increases in both skeletal mass and overall body mass in males. Figure 13.3 depicts a panorama of our FDB adult sample that is very different from the one displayed by the 12-year-old children in Figure 13.2. As with children, linearity tests confirm that, when the cubic root of body weight is utilized to linearize the relationship, both sexes fit into straight lines following the general form of equation 13.1. However, unlike in children, an analysis of covariance (ANCOVA) reveals that the adult men and women in the FDB sample must be fitted into two separate regression lines, sharing the same slope (b in equation 13.1 above), but different intercepts with the vertical axis (a in equation 13.1 above; $P=0.005$).

Both sexes sharing the same value for the slope (b) means that an increase in one unit of height results in an identical increase in average weight in both men and women (note that b is the solution to the first derivative of equation 13.1, and thus represents the instant rate of change of y as we increase x). The average difference in weight between a man of height h and a man of height $h+1$ is identical to that between two women with heights h and $h+1$. Thus differences in height explain approximately the same amount of differences in body weight within each sex: around 14% of variance explained in females and 15% in males in our sample. However, these figures are much smaller than the 42% variance explained observed in children, indicating very important changes in body composition (i.e., in the relative overall distribution and proportions of the different body tissues), as compared to the childhood pattern.

But the difference in intercepts (a) is an even more important factor, meaning that, for any given height, adult men are heavier than adult women by a constant mass (the difference between the intercepts of the two lines). In our sample, a man is on average around 1.4 kg heavier than a woman of the same height.

So, if independent from height, where do those extra 1.4 kg come from? The answer is basically bone and muscle but, as already hinted above, the development of muscle tissue plays a more important role. Androgens (and in particular testosterone) directly induce higher rates of bone formation through different molecular and cellular mechanisms (Riggs et al. 2002; Veldhuis et al. 2005). Androgens also inhibit fat accumulation (Riggs et al. 2002; Singh et al. 2003), which is higher in females, so that the difference in muscle and bone mass is much higher than the roughly 1.5 kg suggested by our analysis. However, the indirect action of androgens on bone, through the induction of muscle development, is probably much more important and of particular relevance for sex assessment in humans. Androgens induce muscle formation by stimulating both muscle protein synthesis and muscle cell formation (MacLean and Handelsman 2009; Singh et al. 2003, 2009).

The influence of muscle mass on bone growth is also twofold. First, the extra muscle mass results in male bone diaphyses supporting higher compressive forces. The ability of a beam to support a particular compressive load depends primarily on its total cross-sectional area (Biewener 2003; Ruff 2002 and references therein). Thus male long bones will have relatively larger transverse diameters (and therefore section areas) than those of females, due to their proportionally larger body mass. This effect will not affect just weight-bearing bones, as bones in the upper body will also experience increased compression and bending forces due to the increased muscle volumes and tensions associated with the male physique (Henneberg et al. 2001; O'Brien et al. 2009, 2010). These increased forces will also promote enlarged bone remodeling and deposition in the areas of muscle and tendon insertion (Robling 2006).

Whenever the load is maintained perfectly balanced, with the normal force aligning with the beam axis, compression forces will be the only ones present, and the length of the beam will play no role in its ability to withstand the load. However, this would mean that the beam is perfectly rigid, still, and straight, whereas bone tissue is a viscoelastic material, the main function of the skeletal system is movement, and vertebrate long bones are far from straight beams. Thus, as soon as we unbalance the load, or change the direction of the bone axis away from a straight line, bending forces will appear. Bending forces are the result of the appearance of compression forces on one side of the bone section and proportional tension forces on the opposite side, as the load force is displaced toward the compression side. Think about an archery bow: compression in the bow shaft will appear on the string side, while the opposite surface of the shaft experiences tension forces.

In order to avoid bending to the point of fracture, the shape of the cross-section must be modified to respond to the tension and compression forces acting on opposite sides of the bone. Bone tissue is much stronger in compression than in tension (Chapter 17 in this volume discusses more in depth the reaction of bone tissue to tension and compression forces, in relation to bone fracture), so bending will be reduced by altering the shape of the section through the accumulation of more bone tissue in the compression areas (note that by reducing deformation due to compression on the string side of a bow, we are also reducing tension on the opposite surface), increasing a mechanical property known as *second moment of area* (Biewener 2003; Robling 2006; Ruff 2002). A related parameter often employed in biomechanical analyses of bone is the *polar section modulus*, which measures resistance to both bending and torsion (Ruff 2005).

The greatest tension and compression stresses (and deformations under stress, or *strains*) occur at the surfaces in the plane of bending (this is to say, at the outermost surfaces of the bone), resulting in a gradient from the maximum tension to the maximum compression surfaces. Hence, tension and compression forces will balance each other in the central area of the section, resulting in a neutral plane in which the net bending force is zero (Biewener 2003: 7–9). This means that the materials in the central plane basically do not contribute to the capacity of the structure to resist bending forces, other than by adding their mass to the load forces, and lighter, hollow beams will be more efficient at withstanding bending forces. As a consequence, increases in both total cross-sectional areas and specific compression areas will be more mechanically efficient when based on increases in the thickness of the outer walls of the structure or, in osseous structures, of cortical bone. While estrogens act to conserve bone mass by reducing bone turnover (Prior 1990; Riggs et al. 2002), androgens increase periosteal apposition of bone, resulting in males displaying thicker bones, although the exact mechanism is still poorly known, and it is not completely clear the relative contribution to this phenomenon of direct and indirect hormonal action (Riggs et al. 2002).

Therefore, not only are bone diameters enlarged, but the cross-sectional shape of male diaphyses is also modified to withstand their absolute and relative larger muscle (and bone) mass. Although expressing body mass to a lesser degree than diaphyseal breadths (Ruff 2002 and references therein), articular and epiphyseal dimensions of long bones display relationships with total length dimensions paralleling those between body mass and body length (Figure 13.3).

In plain Castilian, SSD may constitute the key factor in human sexual dimorphism, but its differential expression in different tissues and dimensions results in differences that go beyond males being simply larger than females. As a combined result of both the developmental path followed to achieve adult size, and the biomechanical consequences derived from the body composition developed during this path, "size-free" shape differences, expressed as differences in the bivariate relationships between different dimensions (such as the ones in Figure 13.3), will be also present.

TECHNIQUE IS (ALMOST) EVERYTHING

The utility of these shape differences to discriminate between the male and female skeleton will not depend just on their existence, but on our ability to detect them, which in our context means being able to strip our variables from the influence of raw body size. This ability will in turn depend primarily on two factors: (i) their magnitude, compared to the effect of raw size, and (ii) the resolving power of our analytical techniques, depending on whether they can detect differences at the order of magnitude in (i).

We already have all the elements to evaluate (i) in relation to body mass and height, which are, as we saw, the main components of SSD in humans. According to Ruff (2002), the average sexual dimorphism in body mass in modern human populations is around 15%. Due to the differences in body composition mentioned above, this can translate to postpubertal boys displaying an average bone mass as high as 25% more than postpubertal girls (Riggs et al. 2002). As explained above, the most optimistic estimates of the difference in body height would be around 10% (the average difference in stature between adult males and females in our sample was between 8 and 14 cm, for male and female median heights of 165 and 178 cm respectively). This seems a huge difference, which should allow us to distinguish very clearly between males and females, just by comparing their weight and height.

Our FDB sample fits these estimates very well, with a mean difference in body mass of around 12 kg, which is close to 20% of the average body mass of the women and around 15% of that of the men in the sample, figures even higher than the average obtained by Ruff (2002). However, we also saw that, when we controlled for body length, males and females of identical heights were only 1.4 kg apart on average. In other words, the body mass difference not explained by height represents in our sample just around 12% of the total differences in body mass (1.4 of 12 kg). Using the most conservative figures, the 15% sexual difference in body mass would be composed of a 13.2% due just to differences in height, for just 1.8% truly representing body shape and composition differences between sexes. We told you that raw SSD was important.

Thus, true size-free shape differences are much smaller than size differences. Does this mean that they will be so comparatively small that we will be left again with the only possibility of distinguishing just average or petite females from average or large males? Is there no hope in satisfactorily solving the conundrum posed by doubtful individuals falling in the middle of both size distributions? Figure 13.3 taught us that males are *proportionally* heavier than females, as well as that the same relationship could be observed when comparing transverse or articular measurements (more influenced by body mass) with total bone lengths (highly correlated with stature). But, more importantly, it revealed that differences in shape and proportions can be detected by comparing different variables. In the case of the kids in Figure 13.2, both sexes fitting the same regression line meant that prepubertal boys and girls of identical heights were also expected to have the same weight. Body shape and composition had therefore to be the same. However, the separate lines in Figure 13.3 indicated that men were heavier than women not only in absolute but also and decisively in relative terms. Two spheres of the same radius are supposed to weigh the same. If they don't, it means that either one or both of them are not really spheres, they have different densities (i.e., "body compositions"), or both.

Multivariate techniques, such as the already mentioned DFA, try to take advantage of this fact, by combining and comparing different measurements of the individual. Chapter 15 describes in depth both the technique and the current standard tool to perform the analysis in forensic settings, the computer program Fordisc. However, let's take a quick look at it here to try answering question (ii) above, by examining whether DFA goes beyond detecting raw size differences, as we would expect. With this purpose, we ran a DFA utilizing two epiphyseal breadths and their corresponding bone lengths from our FDB sample of "white" individuals; the four variables of the human femur and humerus in Table 13.1.

Table 13.2 displays the number of individuals correctly and incorrectly classified within their real sex group, in a Punnet square fashion. Even with such a small number of variables to choose from, the analysis rendered a very impressive 95% rate of correctly classified individuals. We used a stepwise analysis, which lets the analysis

Table 13.2 Percentages of correct classification for the discriminant function using two humeral and two femoral variables. Note how all incorrectly classified individuals are males (see text for details).

	Classified as		Total
	Female	Male	
Females	41 (100%)	0 (0%)	41
Males	6 (7%)	83 (93%)	89

95% correct classification (cross-validated).

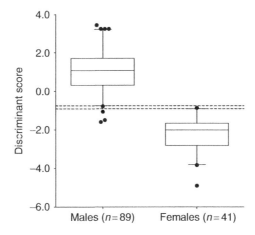

Figure 13.4 Group distribution of the discriminant scores obtained from two femoral and two humeral measurements. Note how the overlap between the two sex groups is now much smaller than for any of the raw variables in Figure 13.1.

select those variables showing a higher separation between the two groups (minimizing the ratio of the variance within groups divided by the variance between groups, a parameter known as Wilks' lambda, or λ), to create a new set of variables: the linear discriminant functions.

In a nutshell, the goal of the analysis is combining the original variables to create new ones showing the smallest overlap between the two groups (the gray areas in Figure 13.1). In our analysis, only one of such discriminant functions was created, and any potential additional ones from our bone measurements did not serve to improve the classification. Figure 13.4 displays the distribution of this new variable in each of the sex groups. Note how the gray area between the groups is actually much thinner than in any of our previous variables, and thus appears to be fulfilling its intended function. Very few females will show discriminant function scores within the range of variation of males, and vice versa. What is this new variable? Is it expressing differences in raw SSD or in size-free shape? The answer is *both*.

Only the two epiphyseal variables entered the analysis, which means that the two bone lengths would not be adding any new information to that contained in the femoral head diameter and the epicondylar breadth of the humerus. This looks good,

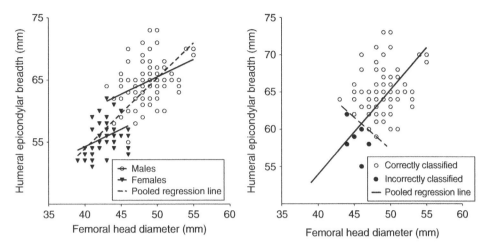

Figure 13.5 Left: conditional distribution of the two variables entering the discriminant function analysis. Both variables show a relationship similar to those in Figure 13.3, with the two sex groups sharing the same slope but displaying different intercepts. The segmented line represents the regression line obtained when pooling both groups together. Note the almost circular shape of the two dots of points, indicating a weak correlation between the two variables. Right: position along the pooled regression line of the six males incorrectly classified as females in the discriminant function analysis. Note how all six incorrectly classified individuals fall below a particular value (dashed line) along the pooled regression line.

as we saw that long-bone lengths are highly correlated with stature, and the latter is basically just a measure of raw size, rather than sexual shape differences. Figure 13.5 displays the relationship between the two variables. Similar to the pairs of variables in Figure 13.3, males and females follow different regression lines, with the same slope but different intercepts ($P < 0.001$). A surprising observation is that the clouds of points representing each group are somewhat circular, rather than the long ellipses that we would expect from a very strong correlation between the two variables (0.77 in Table 13.1). Table 13.3 reveals that many of the strong correlations in Table 13.1 disappear when we calculate them individually for each sex, instead of pooling males and females together.

Remember our question above, about how was it possible that the epicondylar breadth of the humerus showed so little overlap between sexes if it was so strongly correlated with height and weight, which in turn show huge overlaps? Here you have your answer. As a matter of fact, the variables are not strongly correlated. The high correlation observed in the pooled sample is just expressing that the two groups are very different in that variable. It is not that an increase in either height or weight translates into an increase of a similar magnitude in epicondylar breadth, but that almost every time that we have an individual with a broad elbow articulation it will be a man, and thus larger due to SSD, and individuals with small epicondylar breadths will tend to be women, who also happen to be smaller. The largely size-free difference is "fooling" the correlation to provide a result suggesting that the two variables vary together, when what is actually happening is that they display very different values in each sex.

A second look at Figure 13.5 will help us to understand how this affects our discriminant functions. The dashed line crossing through both groups is the pooled

Table 13.3 Pearson correlations between the variables in Table 13.1, now calculated separately for each sex group. Note the much smaller correlations of the biepicondylar breadth of the humerus (BEPB) with all variables, as compared to the high values observed when the two groups were pooled. In particular, it is very weakly correlated with height and weight in males, and uncorrelated in the female group (probably an effect of the smaller sample size of the latter), indicating that this variable is actually very weakly influenced by overall body size and, thus, by SSD. Slightly smaller within-group correlations are also observed for the diameter of the femoral head, although to a much lesser extent than in BEPB, while bone lengths maintain high correlations with body height, indicating that the latter are stronger raw SSD markers and thus more influenced by the overall body size of the victim.

	Height	Weight$^{1/3}$	HXLN	BEPB	FXLN	FHDD
Males						
Weight$^{1/3}$	0.33	0.30	0.76	0.44	0.80	0.48
Humeral length (HXLN)	0.75	—	—	0.21	—	—
Humeral biepicondylar breadth (BEPB)	—	—	**HXLN**	0.36	0.85	0.56
Femoral length (FXLN)	0.88	—	—	**BEPB**	0.30	0.41
Femoral head diameter (FHDD)	0.40	—	0.78	0.36	**FXLN**	0.50
			0.37	0.38	0.41	**FHDD**
Females						

All correlations significant at the 0.05 level (two-tailed).

regression line, obtained when performing our regression analysis with all males and females together, without distinguishing between groups. Our initial Pearson's correlations in Table 13.1 "thought" that the relationship between femoral head diameter and humeral epicondylar breadth followed that line, so that an increase of one unit in femoral head diameter would result in an increase in humeral epicondylar breadth equal to the slope of the line. We now know that not to be the case, with the real relationships between the variables being much weaker and following one of the two different solid lines, depending on the group. However, the diagram on the right in Figure 13.5 shows the placement along that "fake" pooled regression line of the six male individuals that were incorrectly classified by our discriminant function. All of them fall below a particular perfectly perpendicular section of the line (short, dashed line). Which indicates that the discriminant function is basically finding a similar line and using the alignment of each individual across that line (the projection on the line) as the new variable in which the distance between the two groups will be the largest possible.

Do not think for a second that we do not know what you are thinking: It is (hopefully) great to understand how DFA works from a geometric point of view, but how is it of any help to interpret our DFA results, not to mention morphological analyses based on discrete traits? Well, for starters, our example shows that there *are* morphological traits that are sharply different between the two sexes and which, unlike traits being more correlated with height or body mass, will be good sex markers independently of the body size of our problem individual. Humeral epicondylar breadth is one of them, and in fact relatively (only relatively) homogeneous across populations. Secondly, it shows that DFA can be used to detect them and, finally, it will serve to explain some of the main potential problems and contradictions arising during metric analysis.

The key idea is that the degree of overlap between our groups along the discriminant functions will depend to a large extent on the slope of the relationships between groups, which in turn is related to (although not determined by) how strongly correlated they are within each group: correlation between versus correlation within groups. Imagine that the situation is the one in Figure 13.2. The two groups share a same regression line, so that most of the variability in both groups occurs in the direction of the line. The correlation between and within groups is the same. Even if there were differences in height and (consequently) weight between the two sexes, the two clouds of points would show an important overlap, making it difficult to distinguish tall girls from short boys.

Now imagine rotating 45° counterclockwise the two regression lines and point clouds in Figure 13.5. Now both the between and within correlations are high, following step lines, even when they are different. When there would still be two different lines and clouds, the upper end of one line and the lower end of the other would now be very close to each other, so that the overlap between both distributions would be much larger.

Finally, imagine that both regression lines were completely horizontal, indicating that both variables are completely independent from each other (high between-group correlation, but no within-group one). The separation between the two clouds of points (and thus their overlap) would be the maximum possible (unless one of the variables actually became smaller as the other grows, which is almost never the case with bone measurements, although it is a common relationship between some physical and physiological variables across species, such as heartbeat rate over body mass).

In other words, the less correlated the variables, the larger the separation between their conditional sex distributions. This is because the more correlated with body size (measured as either body length or body mass) two variables are, the more correlated they will be among themselves, and DFA will tend to select variables being less correlated among themselves (within group), and thus with overall body size.

In conclusion, answering question (ii) above, DFA does detect and utilize sexual differences beyond overall body size (i.e., raw SSD). Still, we also saw that the only way in which we can tell whether a variable is expressing just body size, or a sexual difference in size-free shape, is by checking the linear relationship (*sensu stricto*) between that variable and a body size marker. Thus, the discriminant functions will also select and be influenced by overall body size markers, to utilize that comparison, especially when a single discriminant function is extracted.

In reality, as you may already know from Chapter 15 in this volume, DFA does not operate by calculating the pooled regression line, but rather the direction of the line coming closer to the central points (group centroids) of the male and female groups. However, that line is very similar to what we termed "between-group correlation" above. As a consequence, when only two groups and two variables are considered, the smaller the correlation between the variables is, the closer the regression line and the axis of the discriminant function will be, becoming virtually identical when both variables are independent (as the regression line would them be just joining the two group centroids). In our example, the projections of the points on the regression line show a Pearson correlation of 0.70 with the discriminant scores (the values shown by each individual in the discriminant function).

This is relevant because it implies that, when large size differences between groups are present, DFA will tend to select variables showing high pooled ("between-group") correlations but weak within-group ones, just like the first component extracted in principal component analysis (PCA). This usually translates into extremely high correlations between the first discriminant function and the first principal component extracted, which will therefore be expressing the same (Wolff 2010). When the analysis is performed from linear measurements, this first component (PC1) is clasically considered to be expressing *allometric size* (Bookstein 1989; Klingenberg 1996), which is usually understood as raw size plus that fraction of variability in body shape that can be explained by allometry, or those changes in proportions related exclusively to the size of the individual (all the bivariate relationships that we have seen so far can be read as static allometric patterns, particularly as the independent variables are body size markers). Allometric size is sometimes interpreted as expressing just size-related differences within the same allometric line, such as the ones among the children in Figure 13.2, in which differences in the dependent variable (weight, in that case) are explained exclusively by differences in the independent one (height), rather than in group origin. However, we have already seen that our discriminant function, and thus the first principal component in a PCA, can be also expressing shape differences as the ones in Figures 13.3 and 13.5. Across groups the intercept differences can still be explained by those between the male and female average body sizes: a female of height *h* does not look like a male of the same height, but the female group as a whole

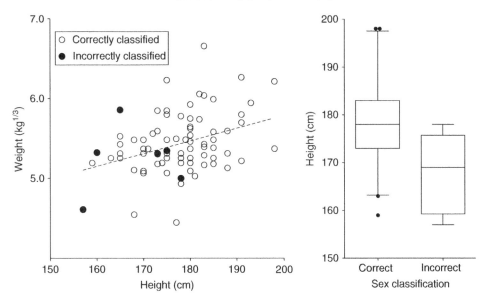

Figure 13.6   Height distribution (right) and conditional distribution of weight on height (left) of the incorrectly and correctly identified males in the discriminant function analysis. All incorrectly classified individuals are in the lower tail of the male height distribution, indicating that the overall body size of the indiviual keeps playing a role in the obtained diagnoses. Note how all misclassified individuals are either very short, or show a low weight for their particular height (i.e., low robusticity). However, other gracile individuals around the latter are correctly classified, indicating that classification is not just based on the gracility of the individual.

would probably look like the male one if it were on average as large as the latter (Gould 1971 is the classic reference for the correct interpretation of allometric intercepts). Shape differences are present, but from an evolutionary point of view they are still the result of just SSD.

The bottom line is that the discriminant functions obtained in forensic anthropological analyses are more often than not expressing allometric size. Even when they capture sexual shape differences, size and size markers will keep playing an important role in DFA. This explains why all six incorrectly classified individuals above are males from the lower half of the distribution of height (Figure 13.6). The good news is that DFA reduces very significantly the influence of the raw body size of our individual in our sex assessments, by weighting body size markers with variables being less influenced by raw SSD and, in general, is very effective in producing reliable sex diagnoses. Note also how body mass seems to be less of an issue, largely due to the larger sexual differences in muscle mass described above, now weighted against other body size markers (femoral head diameter is highly correlated both with body length and body mass), although the results might have been different if we had included shaft diameters, which are, as explained above, much more related to muscle mass.

The bad news is that body size keeps playing a very important role and posing a potential problem for our sex assessments from DFA, if less frequently than in the examination of isolated traits.

A second look at Table 13.2 also reveals that the possibility of wrongly classifying a male as a female or vice versa is not the same, depending on the variables employed. This means that we can be more comfortable classifying individuals as males than as females, or the other way around, depending on the analysis and trait that we are examining. We will come back to this issue in our final section.

THERE IS MORE THAN ONE POSTURE

In the previous sections we assumed that sexual shape differences such as those observed in Figures 13.3 and 13.5 were mostly the result of the absolute and relative higher average muscle mass of adult males, largely dismissing differences in height as merely expressing raw SSD. Conversely, while height is indeed mostly a size marker, we also saw that the differences in height arose from the different timings of puberty onset and completion in males and females. Taking into account that childhood and pubertal growth rates and patterns differ, these timing differences imply that during around 2 years human males and females grow apart in shape by following markedly different allometric patterns. The presence of allometric patterns entails that the overall shape of the individual changes as body size increases. The children in the upper end of the allometric line in Figure 13.2 are not just absolutely heavier than the children in the lower end of the line, but they are also heavier in relative terms, as body weight increases faster than body height. Therefore, boys have a different average body shape than girls when they enter their respective pubertal growth spurts. At the starting point males do not display the body proportions of a generic child, but those of a child who kept growing for an extra 2 years. Consequently the differences in developmental timing largely responsible for SSD will also have a deep impact on general body proportions that, in turn, will have important biomechanical consequences. Figure 13.7 displays the relationships between body weight and the total breadth of the hip (from left to right greater trochanters of the femur) in a group of adult male and female Spanish swimmers followed and monitored by one of the authors (JLA) along a string of official competitions and training exercises. Once again, males and females share the same slope, but have different intercepts ($P < 0.001$), with men weighing on average close to 4 kg more than women for any given bitrochanteric breadth. The sexual difference in body mass in this much fitter and more trim sample of high-ranked athletes was 17 kg, approximately 40% more than in the forensic population from the FDB (12 kg). Men averaged 77 kg and women 60 kg, which would make 17 kg represent a weight difference of around 28%, much closer to the difference up to 25% in skeletal mass proposed by Riggs et al. (2002). This confirms that the differences in body composition (fat mass) partially mask those in muscle and skeletal mass, with SSD being higher in the latter than the approximately 15% difference observed in total body mass.

If we have a sample with an approximately 40% higher sexual dimorphism in overall body mass than the FDB sample, the 3.8 kg sexual difference in body mass for any given bitrochanteric breadth is 270% that obtained when weight is compared instead to total body height (1.4 kg). The most interesting aspect of this striking difference is its biomechanical implications, as a large proportion of this extra weight rests on the structure

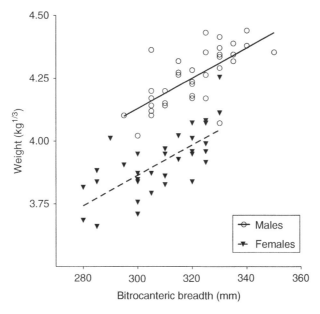

Figure 13.7 Conditional distribution of body mass on the total breadth of the hip (bitrochanteric breadth) in a sample of Spanish athletes. The relationship follows the same type of relationship as the variables in Figures 13.3 and 13.5, with both sex groups displaying the same slope and diverging intercepts. The male pelvic lever system supports a much larger load than the female one for any given bitrochanteric breadth. Males manage to take the extra load while maintaining the same biomechanical safety factors as females by displaying a different pelvic and femoral morphology that reduces stress levels in the different components of the lever, conferring them a biomechanical advantage.

defined by trochanteric breadth, which is further accentuated by men having relatively longer upper bodies than both women and prepubertal children (Brewster 2006).

If we think in biomechanical terms about the relationship described in Figure 13.7, we can picture the system as an inverted capital T, in which the lower horizontal line is trochanteric breadth and the vertical line is the load force of the upper body mass, transmitted through the vertebral column. Trochanteric breadth would hence be approximately indicating the moment arm of the complete lever supporting all the weight of the upper body (the whole structure is actually a system composed of different levers, but note that, independently of how the resultant forces are distributed across the system, it is the structure as a whole supporting the load. We can draw a parallel with the bone sections described above: even when just the bone walls support a vertical load it is the total cross-sectional area of the bone determining its ability to support it). The torque or bending force supported by the whole system is proportional to the product of the moment arm by the mass of the load, which means that, for any given trochanteric breadth, an increase Δw in mass results in an increase in the bending force withstood by the system proportional to Δw times the moment of the arm. If we conservatively assumed that just one half of the difference in weight independent of trochanteric breadth corresponds to the upper body, Δw would be equal to 2.

Based on this model, the male pelvis and femur would appear to show a mechanical disadvantage compared to the female ones, the male articulation being subject to proportionally higher bending forces. This is a similar problem to the one faced by species increasing their body size across taxa. Think about sheep and cows, for example. They are very closely related, being classified within the family Bovidae and sharing the same body plan, but cows (*Bos taurus*) have evolved a much larger body size than sheep (*Ovis aries*). We have seen that a way of increasing the strength of the bone in response to higher loads is to increase bone wall thickness and cross-sectional area. However, larger animals show relatively smaller cross-sectional areas and wall thicknesses than one would expect from their differences in weight from their smaller counterparts. What is more surprising is that they do so while maintaining the same safety factors (the ratio between the bending force necessary to fracture the bone and the peak forces experienced during high-speed running or jumping), which are maintained at a factor of between two and four independent of body size (i.e., their bones have a strength two to four times higher than the peak stresses they will normally experience; Biewener 1989). How they manage this? The answer is that they modify their posture.

Sheep have a more croaching posture than cows. By having their limbs more vertically oriented, the load in cows is oriented more parallel to the vertical axis of the bone, thus reducing load imbalance and bending. The same is promoted by placing their limbs closer to each other, thus reducing moment of arm and torque. Torque (and with it bending force) does not depend just on the product of moment arm by load, but also on the sine of the angle formed by the resultant force and the lever. Those angles can be reduced by orienting the muscles more parallel to the bone axis (thus the force being exerted at minimal angles), which will reduce the stress placed on both bone and muscle (Biewener 1989, 1990, 2003). Just as with the distance at which the limbs are placed from each other, moment arms can also be reduced by reducing bone (and consequently limb) lengths (Christiansen 1999).

The human male pelvic girdle displays the whole list of posture modifications given above. The pelvis is relatively narrower than in females, bringing the articulations of the femoral heads closer to each other. The ilea are more vertically oriented than in females, placing the glutei minima more parallel to the axis of both ilia and femora. The femoral neck is also relatively shorter, with the major axis of the femur at a smaller angle with the central plane and, as we mentioned above, males also have relatively shorter legs than females and children. In sum, SSD in body mass, which resulted in larger cross-sectional areas and modified section geometries of long bones, translates in turn into a complete remodeling of the pelvic area.

As explained in Chapter 12, the second determinant of pelvic morphology is parturition. Females require a larger pelvis inlet for the birth canal. The ventral area of the pelvic girdle (the symphyseal area) is less influenced by upper body mass, being loaded laterally, almost like a bow, by resultant compression forces exerted by the femoral heads on the pelvic acetabuli. Either simply due to the general female pelvic pattern, with broader sections and a smaller vertical development than in males (see above), or to the presence of lower biomechanical constraints related to body mass, leaving more room for the morphology of this area to adapt to parturition, the pubic symphyseal area displays dramatic sexual differences that may be evolutionarily related to SSD, but not correlated with overall body size at the individual level. I tell you, it's a whole different sex.

The problem of whether some sexual differences reflect selective pressures affecting the male or female group is a classic one in evolutionary biology. Did human males develop narrower and more vertical pelvises to be able to increase their body size (or maintain the SSD of heavier species of hominids)? Or did female morphology develop in response to reproductive pressures? A classic approach to answer these questions is comparing the adult and juvenile patterns (Braña 1996 and Reyes-Gavilán et al. 1997 are particularly elegant and meaningful examples of this type of analysis). In this case, the female pattern seems to be closer to the juvenile one, which is also true regarding other general body proportions, and elements such as a strong correlation between lower body length and trochanteric breadth, which is absent in the male group (Brewster 2006). The male pelvis of other fossil hominids is also relatively shorter and broader than in modern human males, although the sexual differences in the pubic symphyseal area are very similar to those in modern humans (Arsuaga et al. 1999). The question is not irrelevant for our subject, as the relative utility for sex determination of the weight-bearing dorsal structures of the pelvis and its ventral area largely depends on its answer (are any of them more correlated with body size at the individual level?), coming to illustrate the relevance of evolutionary issues for forensic estimation techniques. However, the answer to this question is still far from clear.

Whatever the evolutionary explanation, the fact is that the male pelvis has to withstand a heavier load, while the female pelvis must leave room for the birth canal, resulting in derivations of the classic method proposed by Phenice (1969) or the more recent one by Bruzek (2002) being possibly the most reliable methods to assess sex in skeletal human remains. Sex estimates based on pelvis morphology are actually the gold standard when testing or developing other sexing techniques from archaeological materials of unknown sex. But note how its effectiveness spans mainly from the existence and definition of a whole morphological pattern, rather than from the separate consideration of isolated traits. If the subpubic concavity is broader, the obturator foramina will be more triangular, the ischiopubic rami thinner mediolaterally, and the symphyseal bodies shorter. We may err over- or underestimating how marked one of these traits is, but it will be much harder to miss or misinterpret all of them when considered as a whole.

As discussed in the previous chapter, and above with the example of the epicondylar breadth of the humerus, while the morphology of the pelvis is an example of an area where sexual dimorphism is weakly or not correlated with the size of the individual, it is not the only one, and postural changes are not the only mechanism to originate this type of trait.

Wrapping It Up

The main message conveyed throughout this chapter is that sexual differences in the human skeleton are basically expressing size dimorphism (SSD). In this way, the main source of error, as well as of the conflicting observations that may arise when attempting to assess the sex of a set of skeletal human remains, will be related to the victim showing an overall body size atypical of its group. In most cases in which we are plagued by conflicting results or puzzled by the inconsistent morphology of our individual we will be confronting either a "big" female or a "small" male. However, we have also learnt that SSD is expressed at least in four general, different ways, which we can use to our advantage in some cases, by comparing our trait with body size estimates, as follows.

Raw SSD

Both sexes follow the same static allometric relationship, just as the children in Figure 13.2. In this case, even if one of the sexes may be larger on average, individuals of the same body size will have the same appearance independently of their sex. Thus, in this type of trait we will not gain anything by comparing the trait score (metric or qualitative) with the corresponding body size marker, as in these cases SSD and simple allometric size are one and the same thing. It is like looking at an isolated trait influenced by body size, as we did in Figure 13.1. This results in the trait having a much lower value as a sex marker, in particular when more than one population is considered, as discussed in the previous chapter. For example, if different populations with dissimilar average body sizes share the same allometric line, the position of an individual along that line will not depend on its gender, but rather on its population of origin, our estimate assessing ancestry instead of sex.

Traits expressing body length (such as maximum lengths of limb bones or the skull), or bone-deposition rates resulting from direct responses to biomechanical stresses, such as wall thicknesses and, specially, muscle insertions, will be more prone to show these relationships. Their ability to express either height or raw body mass makes them relatively less reliable as sex markers.

Intercept shifts

Both sexes follow different static allometries, usually with the same slope but different intercepts, as the pairs of traits in Figures 13.3 and 13.5. From an evolutionary point of view, these traits are still just expressing SSD. Following Gould's (1971) classic interpretation, these allometric relationships do not represent a change in body plan or posture, but rather the evolutionary response to the nonadaptive effects of allometry, once the increase in size has taken place. From this perspective, we can see the changing levels of androgens in pubertal males, and their effect on height, muscle, and bone-mass development, as the adaptive response to their larger body size. Growing larger gives males an evolutionary advantage, but larger skeletons are also subject to proportionally greater stresses than small ones. Thus they must further increase their muscle and bone section areas to compensate for their relatively larger mechanical loads.

However, unlike in traits expressing raw SSD, this results in absolute differences in body proportions. The long bones of a tall woman are subject to more stresses than those of a short one, due to both the increased body mass and the longer distances (lever arms) between her muscle insertions, which result in larger compression and bending forces. Therefore her shaft diameters and, to a lesser extent, articular dimensions may be in the higher end of the female distribution for those parameters. However, when we compare those dimensions with those of males with similar bone lengths (or height, if you prefer), they will likely appear in the lower end or even outside the male distribution for that particular body length. In other words, by comparing these traits with the size of the individual (specially expressed as body length), we will be able to assess her relative gracility or robusticity and compare it with those of the male and female groups.

Multivariate analyses such as DFA take advantage of this type of difference, being very effective in detecting and incorporating them into the analysis. We can do the same thing when interpreting qualitative traits, by contrasting them with body size markers, like bone lengths. However, the ability of the analysis to detect these relationships will depend on the variables available, and raw size differences will more often than not still play an important role in DFA, simply because body size is the best parameter to distinguish average males from average females.

Size-independent spandrels

The different developmental paths taken to generate SSD (namely, the timing of puberty onset and completion, or even that of the prenatal hormone surge) may also result in markedly different morphologies in anatomical traits that are either weakly or not correlated with biomechanical stresses derived from overall body size or muscle mass. The epicondylar breadth of the distal epiphysis of the humerus, discussed above, may be the result of this type of process. The morphology of the glabelar area of the frontal, or that of the mandibular symphysis, discussed in the previous chapter, may represent other examples, the latter having been suggested to display slight sexual differences in newborns (Loth and Henneberg 2001). Further research on the exact developmental paths followed by these and other traits, as well as on the biomechanical stresses to which they are subject, is required to confirm or reject this possibility. Due to their low dependence on overall body size and, likely, their direct relationship with sexual hormone surges, these traits can represent very reliable sex markers although, as discussed in Chapter 12, they are also subject to individual variation.

Postural changes

Last but not least, modifying cross-sectional areas and geometries may not be enough or the most efficient alternative to compensate for the increased loads derived from SSD. When the increase in body mass is affecting complex lever systems, such as the conformation of the human pelvis and proximal femora, where a large proportion of the total body mass is loading the system in the maximum bending direction (i.e., perpendicular to the major axis of the system), it may result in marked postural modifications to reduce the resultant forces in the different components of the lever system. In addition, as discussed above and in the previous chapter, differing selective pressures such as parturition and sexual selection may be affecting these and other anatomical structures, resulting in further diverging force loads and postural morphologies.

These changes are not shaped by the specific dimensions or forces arising at the individual level during development, but rather represent evolutionary responses to the different average sizes and selective pressures affecting each sex group. As a result, these skeletal areas will offer true morphological differences largely independent of the body size of the individual and, consequently, optimal sex markers. In particular this will occur when the morphology of the structure is interpreted as a whole. As in the previous trait category, further research is needed to assess the exact developmental mechanisms molding the sexual dimorphism of these structures, and the relative contribution of individual body size to some of its components, in order to improve their interpretation.

Being able to assign and interpret our sex markers within one of these general categories will be essential for our previously stated main goals of (i) explaining why some traits or analyses can be rendering a "male" diagnosis in a female individual or vice versa, and (ii) assessing whether, how and to what extent the factors usually promoting such confusion are affecting our particular case. In the next section we describe some practices and rules of thumb, taking advantage of this approach, which have been particularly useful to the authors when confronting difficult or confusing cases.

DOING IT: SOME PRACTICAL ADVICE

Maybe the most common mistake and temptation when embarking not only on skeletal sex assessment, but on any physical anthropology analysis, is reading the alternative methods as technical recipes aimed at producing a discrete result or figure. The problem confronted by physical anthropologists is very different from that of a chemist trying to figure out whether a particular substance is really absent from a sample or if the instrumental technique being used is not precise enough to detect the substance in low concentrations. Sex assessment techniques do not detect some hidden sex particle, but rather try to infer the sex of the victim indirectly, from other biological parameters (e.g., body length, muscular development, and body posture), which are related but not completely determined by genetic sex and hormonal action.

Thus, any method or result will require comparison and interpretation from the anthropologist. This demands assessing and understanding the evolutionary, biomechanical, and developmental factors influencing the sex markers entered in the analysis, as well as the technical or statistical underpinnings of the techniques employed. These are five basic principles and rules of thumb that work well for us when trying to attain these goals:

Size is part of the equation

All sex assessments should probably include an evaluation of the overall body size of the individual. As we have seen in this chapter, this will be the main source of error and confusion in sex assessments. On the other hand, with SSD being the main component of skeletal sexual dimorphism in humans, all of our sex analyses are, to a certain extent, body size assessments. Hence, we would rather make this assessment in an explicit manner, distinct from other metric dimensions. Stature estimates (see Chapter 16) are typically going to be part of our report, and can serve to this purpose: Is our individual tall or short within the sex group to which we are assigning it? In case it falls in the gray areas between the distributions of both sex groups, we may want to check and review the results rendered by some of our sex markers.

Know what each trait is expressing

Height is a component of body size, but we have seen that body mass is also very important, and one cannot be fully interpreted without looking at the other. Products of body length by body mass, or of long-bone length by pelvic breadth, can also provide meaningful estimates of "body frame size" (Ruff 2002 and references therein), basically modeling the human body as either a cylinder or a rectangle.

But we also must take into account that many of our sex markers are so just because they are primarily expressing muscle or body mass. Thus, it is necessary to interpret them as such, and mechanically comparing them with method charts and diagrams will not be enough. When learning each method from the literature in this and the previous chapter, do not limit yourself to the technical and descriptive details. Review what muscles attach to the structure, and the anatomy, movement, and, consequently, bending and compression forces to which the particular area is subject. Is it related to other traits? A marked nuchal line may seem to be additional evidence in favor of a male diagnosis when we also observed a large mastoid and occipital eminence. But the three structures are intimately related from the functional and anatomical points of view, being insertion points for the powerful neck muscles rotating and stabilizing the skull. Therefore the three traits may just be telling the same story: the individual had a muscular neck.

Try to place each trait at least into one of the broad categories described above and, in the case of metric analyses, check in the outputs which variables have a higher weight in the analysis, and in which ones your individual is closer or farther away from the mean of each group. This is specially important when comparing the victim with different populations of varying overall body sizes. We have learned that a result based primarily on diaphyseal diameters is in essence just telling us that the individual was heavy, while one based on bone lengths that it was tall. Could this explain discrepancies in the results obtained from different anatomical areas, such as the cranium and the postcranial skeleton? Also, is the result based on a consistent pattern common to all variables, or is it just one or two of them placing our individual away from the male or female groups?

When doing and publishing research, include information regarding all these aspects, such as the function and range of movement of the bone, muscle attachments, embryological origin, and relationships with other areas. A brief review of pathological conditions can be particularly useful to understand and describe many of these factors.

Knowing what each trait is expressing will also help to weed out some traits that are not adding any new information, but may confound your results by doing so in different scales. For example, all the information contained in bone circumferences (e.g., Safont et al. 2000) can be inferred from the corresponding diaphyseal diameters, but the opposite is not true. Bone sections are not perfectly circular, and the radii (and therefore shape) of most other geometric shapes cannot be inferred directly from their perimeters. Differential calculus was originally created precisely to try approximating this problem in the case of ellipses. On the other hand, as we have learnt, the relevant information that both epiphyseal diameters and perimeters convey is related to the cross-sectional area of the bone, which is correlated with the body mass of the individual. The cross-sectional area depends directly on the diameters, its relationship with perimeters taking much more complex forms. Perimeters increase faster than radii (in the case of circumferences, 2π times faster), and can only be related to cross-sectional areas (which increase proportionally to the square of the diameters) as a function of the diameters. Therefore, if you have the diameters you do not need the perimeters at all. The latter will provide just the same information as bone diameters, only in a complex, uncertain, and inflated form, extremely difficult to model in relation to the remaining linear measurements in the analysis.

Other measurements to avoid are simple ratios, obtained by dividing one variable by another, as the allometric relationships between biological variables rarely take a form allowing for this transformation (see Packard and Boardman 1999 for an extremely clear and meaningful discussion of this problem). Angles are also basically undercover ratios, and besides are measured in a scale where "−1" is actually bigger than "358." Thus avoid combining all these complex variables with simple linear measurements. They are completely different animals.

Combine metric and nonmetric analyses

As we have seen, the results from a DFA can provide very useful information on aspects such as the overall body size of the individual, its robusticity, or the set of variables placing it away from or closer to a particular group. This can give you very important clues about how to interpret related nonmetric traits in your individual. In this sense, the fact of metric and nonmetric analyses sometimes rendering divergent results is not a disadvantage, but an important benefit that will often help us to better characterize our individual from a biological point of view.

Resources like Fordisc can also help in evaluating the relative value of some nonmetric traits. Simply run a univariate analysis utilizing its metric counterpart and you will get a good estimate of its relative reliability in the shape of a percentage of correct classification. The same is true for combinations of traits, in this case useful not only in examining the percentage correct classification, but also the number of discriminant functions extracted. If your particular combination of traits only results in a single discriminant function, you already know that allometric size will be playing a very important role in placing your individual closer to one of the groups, while major shape differences are not present in the form of additional discriminant roots.

Following the previous recommendation, once that you are more familiar with what some sex markers are expressing, you can also use metric analysis to assess bivariate relationships similar to the ones reviewed in this chapter. For example, a comparison through Fordisc of the epicondylar breadth of the humerus with its length will give you a good estimate of the robusticity in this variable, in terms that you can already justify, understand, and evaluate.

Some methods work better for one sex than the other

In our metric example above, we saw how the discriminant function classified all females correctly, but was not equally successful with the male group. This is common for many methods, sometimes caused by the different sample sizes of the male and female samples utilized to develop the method (smaller samples result in broader confidence intervals, thus reducing the chance of misclassifying individuals from that group), but can also be due to the trait being simply better at characterizing one of the sexes.

For example, in our experience a very acute angle in the sciatic notch, similar to the category labeled as *M* in Bruzek (2002: see Fig. 3 on p. 161), is a very reliable indication of the individual being male, whereas wider angles do not necessarily imply that the victim was a female. In this way, it is useful examining in detail the percentage correct tables obtained from metric analysis or provided in the original publications describing a particular method, looking not only for the overall figures but also for the ones corresponding to each sex. In medicine, the proportion of individuals receiving a particular diagnosis that actually suffer from the diagnosed disease (i.e., actually belonging to the group to which they were assigned) is called the *positive predictive value* of the test, and is an extremely important parameter for evaluating clinical tests. Unfortunately, the equivalent parameter is often not reported in forensic anthropological analysis, when it represents a very useful tool to evaluate not only a particular technique, but the reliability of each alternative diagnosis within that technique. If trait A is rendering a *male* diagnosis, and trait B a *female* one, but we know that trait B happens to err more often by diagnosing males as females, our election and report explanations are made much simpler.

Express your doubts

Hopefully, these general recommendations will make it happen less often, but sometimes you simply do not have a clear diagnosis. When we confront an atypical individual or contradictory results from different sex markers we may be tempted to focus our report on those traits supporting our intuitive overall impression. However, as discussed in relation to the Daubert standard, this is rather dangerous when facing a court of law and not particularly ethical or professional. Report all your results and justify your elections and diagnoses rather than ignoring those that do not fit. More often than not your overall impression will be right, but even in these cases you will be serving the case and victim better when truly acting as a scientist. Furthermore, producing a detailed and thorough report requires us to avoid the temptation to cut corners or examine the evidence less carefully, and is a good habit to keep us on our toes, helping us to keep growing as professionals.

WHAT'S NEXT?

This chapter may have made you more aware of how many questions remain unanswered in relation to human sexual dimorphism and sex determination from skeletal elements. Most of the ones posed and discussed in the chapter are related to two main subjects: (i) the exact developmental paths explaining trait morphology, including the evolutionary processes that molded these paths, and (ii) the best methods to identify, describe, and analyze both potential sex markers and their relationship with other traits and general biological characteristics of the individual and the sex group to which it belongs. After a period in which most efforts were aimed at identifying new sex markers and refining existing methods, most current research is slowly turning to address mainly question (ii), and we believe that this will in turn promote a growing focus on understanding what biological processes we really are examining when confronting a new or already recognized sex marker (i.e., question i).

From a practical point of view, the most interesting new development in the field is an increased effort to look at qualitative traits in a more objective way. For example, statistical models (mainly logit and probit models) are being increasingly introduced in the analysis of morphological traits, trying to obtain posterior probabilities similar to those provided by DFA for metric traits (Konigsberg and Hens 1998 is probably the seminal work within this approach). Even more important, new techniques, such as geometric morphometrics (GM), combine both approaches by incorporating in the analysis the geometric location of the landmarks utilized to obtain our metric variables in classical studies. In this way, the size and shape components of a particular skeletal element can be more efficiently defined, separated, and analyzed. The main forensic development in this sense is the development of tools like the software 3D ID (Slice and Ross 2009), which utilizes three-dimensional cranial landmarks rather than linear measurements in statistical analyses otherwise basically identical to the ones in Fordisc. You can find a few key references in the Recommended Readings section below to introduce you to GM. However, always keep in mind that GM represents only a new way of recording our data, incorporating geometric information to the purely metric one. The biological principles, scaling techniques, and statistical analyses applied once we have recorded and transformed our landmark data are basically the same than the ones described in this chapter. GM is a powerful tool, not a magic wand to circumvent biological problems and questions. Remember: technique is *almost* everything.

RECOMMENDED READINGS

Andersson (1994) is the basic reference for sexual selection and dimorphism, and an important lecture when approaching any issue related to sexual dimorphism.

Klingenberg (1996) is probably the best reference for getting acquainted with allometry and heterochrony (a concept to which, for simplicity, we loosely referred as "developmental timing differences" in this text), particularly from the geometric and statistical points of view, covering all key mathematical elements of the subject in a succinct but extremely, clear, deep, and insightful review. Brown and West (2000) offers a comprehensive and extremely clear review of the problem from a biological perspective, being a mandatory lecture before approaching any issue in which allometric scaling may play a role, including sexual dimorphism (seriously, we mean the "mandatory" part: read it).

Peters (1986) and Calder (1996) are classic references and, if slightly outdated in some subjects, are, with Brown and West (2000), still two of the best introductory texts to the biological importance of body size and its relationship with a wide range of key physiological and ecological processes. The former also contains an excellent, very accessible mathematical primer, covering the basis of the main statistical techniques involved in the analysis of body size in a very straightforward and painless fashion (Peters 1986: 10–23).

Damuth and MacFadden (2005) provides a more comprehensive review of the techniques for the estimation and analysis of body size from osteological materials of a wide variety of mammalian taxa, including humans and other primates. In the same line, Ruff (2002) examines the issue from an anthropological perspective, including its relationship with sexual dimorphism, human evolution, temporal trends, and biomechanics, in a 22-page review that, if we include its 100-plus references, probably contains everything that you always wanted to know about overall size and shape variation in modern humans; at least up to 2002, that is, but its insight and logical framework are still completely up to date. The article also points to all the must-read classical literature on shaft biomechanics, in which Dr Ruff himself is a key contributor.

Veldhuis et al. (2005) is an extremely good review of both sexual developmental differences in height and body mass, and of the main effects of sexual hormones on bone, providing more than 500 references. Once you are done with the former, Riggs et al. (2002) is probably the best next step to delve deeper into the endocrinology of bone formation and maintenance, including those aspects related to sexual dimorphism. Robling et al. (2006) examines everything concerning bone remodeling, wrapping up the key endocrine, molecular, and biomechanical aspects of the process better than any other short reference with which we are acquainted.

Biewener (2003) is the basic biomechanical reference, in particular regarding postural aspects, by possibly the leading author in the field. Alexander (2006) is another useful and longer reference, on the same line and by another key contributor. Ruff (2003) examines in depth the correlation between humeral and femoral strength and different measurements of body size and muscle mass in humans, in a longitudinal study following individuals of both sexes from birth to late pubescence. Ruff (2005) is a review article approaching the subject from a paleoanthropological point of view, also offering very relevant insight into methodological issues.

Zelditch et al. (2004), also known as the *Green Book*, is the best and most accessible practical introduction to GM and the analysis of shape. Hammer and Harper (2005) combines an introduction to both linear and GM methods, illustrating it with applications to different biological issues, most of them very relevant to physical anthropology, being a reference that you will want to have in your library. Slice (2010) introduces the subject from a purely anthropological perspective, with an updated bibliography. Hood (2000) provides a short introduction to the application of GM to the study of SSD, linking it with the traditional conceptual approaches to the analysis of the problem from linear dimensions, followed in most of our chapter.

REFERENCES

Alexander, R.M. (2006). *Principles of Animal Locomotion*. Princeton University Press, Princeton, NJ.

Andersson, M.B. (1994). *Sexual Selection*. Princeton University Press, Princeton, NJ.

Arsuaga, J.L., Lorenzo, C., Carretero, J.M., Gracia, A., Martínez, I., García, N., Bermúdez de Castro, J.M., and Carbonell, E. (1999). A complete human pelvis from the middle Pleistocene of Spain. *Nature* 399: 255–258.

Baccino, E., Ubelaker, D.H., Hayek, L.A., and Zerilli, A. (1999). evaluation of seven methods of estimating age at death from mature human skeletal remains. *Journal of Forensic Sciences* 44: 931–936.

Bass, W.M. (2005). *Human Osteology: A Laboratory and Field Manual.* Missouri Archaeological Society.

Biewener, A.A. (1989). Scaling body support in mammals: limb posture and muscle mechanics. *Science* 245: 45–48.

Biewener, A.A. (1990). Biomechanics of mammalian terrestrial locomotion. *Science* 250: 1097–1103.

Biewener, A.A. (2003). *Animal Locomotion.* Oxford University Press, Oxford.

Bookstein, F.L. (1989). "Size and shape": a comment on semantics. *Systematic Zoology* 38(2): 173–180.

Braña, F. (1996). Sexual dimorphism in lacertid lizards: male head increase vs. female abdomen increase? *Oikos* 75: 511–523.

Brewster, C.P. (2006). *Sexual Differences in Allometric Relationships in the Human Pelvis.* MS thesis, Department of Applied Forensic Sciences, Mercyhurst College, Erie, PA.

Brown, J.H. and West, G.B. (2000). *Scaling in Biology.* Oxford University Press, Oxford.

Bruzek, J. (2002). A method for visual determination of sex, using the human hip bone. *American Journal of Physical Anthropology* 117: 157–168.

Buikstra, J.E. and Ubelaker, D.H. (1994). Standards for data collection from human skeletal remains. *Proceedings of a seminar at the Field Museum of Natural History*, Arkansas Archeological Survey.

Calder, III, W.A. (1996). *Size, Function and Life History.* Dover Publications, Mineola, NY.

Christiansen, P. (1999). Scaling of the limb long bones to body mass in terrestrial mammals. *Journal of Morphology* 239: 167–190.

Damuth, J. and MacFadden, B.J. (2005). *Body Size in Mammalian Paleobiology: Estimation and Biological Implications.* Cambridge University Press, Cambridge.

Daubert v. Merrell Dow Pharmaceuticals, 509 US 579, 113 S.Ct. 2786, 125 L.Ed. 2d 469 (US June 28, 1993) (No. 92-102).

Dirkmaat, D.C., Cabo, L.L., Ousley, S.D., and Symes, S.A. (2008). New perspectives in forensic anthropology. *Yearbook Physical Anthropology* 51: 33–52.

Gilbert-Barness, E. and Debich-Spicer, D. (2004). *Embryo and Fetal Pathology.* Cambridge University Press, Cambridge.

Gould, S.J. (1971). Geometric similarity in allometric growth: a contribution to the problem of scaling in the evolution of size. *American Naturalist* 105(942): 113–136.

Hammer, Ø. and Harper, D.A.T. (2005). *Paleontological Data Analysis.* Wiley-Blackwell, Oxford.

Henneberg, M., Brush, G., and Harrison, G.A. (2001). Growth of specific muscle strength between 6 and 18 years in contrasting socioeconomic conditions. *American Journal of Physical Anthropology* 115: 62–70.

Huber, P.W. (1993). *Galileo's Revenge: Junk Science in the Courtroom.* Basic Books, New York.

Hood, C.S. (2000). Geometric morphometric approaches to the study of sexual size dimorphism in mammals. *Hystrix* 11(1): 77–90.

Jantz, R.J. and Moore-Jansen, P.H. (2000). *Database for Forensic Anthropology in the United States, 1962–1991* (computer file). ICPSR version. Department of Anthropology, University of Tennessee, Knoxville, TN.

Kim, K.R., Kwon, Y., Joung, J.Y., Kim, K.S., Ayala, A.G., and Ro, J.Y. (2002). True hermaphroditism and mixed gonadal dysgenesis in young children: a clinicopathologic study of 10 cases. *Modern Pathology* 15(10): 1013–1019.

Klingenberg, C.P. (1996). Multivariate allometry. In L.F. Marcus, M. Corti, A. Loy, G.J.P. Naylor, and D.E. Slice (eds), *Advances in Morphometrics* (pp. 23–49). NATO ASI Series A: Natural Sciences, vol. 284. Plenum Press, New York.

Konigsberg, L.W. and Hens, S.M. (1998). Use of ordinal categorical variables in skeletal assessment of sex from the cranium. *American Journal of Physical Anthropology* 107: 97–112.

Krogman, W.M. and İşcan, M.Y. (1986). *The Human Skeleton in Forensic Medicine.* Charles C. Thomas, Springfield, IL.

Kruger, J. and Dunning, D. (1999). Unskilled and unaware of it: how difficulties in recognizing one's own incompetence lead to inflated self-assessments. *Journal of Personality and Social Psychology* 77(6): 1121–1134.

Loth, S.R. and Henneberg, M. (2001). Sexually dimorphic mandibular morphology in the first few years of life. *American Journal of Physical Anthropology* 115: 179–186.

MacLean, H.E. and Handelsman, D.J. (2009). Unraveling androgen action in muscle: genetic tools probing cellular mechanisms. *Endocrinology* 150(8): 3437–3439.

O'Brien, T.D., Reeves, N.D., Baltzopoulos, V., Jones, D.A., and Maganaris, C.N. (2009). Strong relationships exist between muscle volume, joint power and whole-body external mechanical power in adults and children. *Experimental Physiology* 94(6): 731–738.

O'Brien, T.D., Reeves, N.D., Baltzopoulos, V., Jones, D.A., and Maganaris, C.N. (2010). In vivo measurements of muscle specific tension in adults and children. *Experimental Physiology* 95(1): 202–210.

Ousley, S.D. and Jantz, R.L. (2005). *FORDISC 3.0: Personal Computer Forensic Discriminant Functions.* University of Tennessee, Knoxville, TN.

Packard, G.C. and Boardman, T.J. (1999). The use of percentages and size-specific indices to normalize physiological data for variation in body size: wasted time, wasted effort? *Comparative Biochemistry and Physiology A* 122: 37–44.

Peters, R.H. (1986). *The Ecological Implications of Body Size.* Cambridge University Press, Cambridge.

Phenice, T.W. (1969). A newly developed visual method of sexing the os pubis. *American Journal of Physical Anthropology* 30: 297–301.

Prior, J.C. (1990). Progesterone as a bone-trophic hormone. *Endocrine Review* 11(2): 386–398.

Reyes-Gavilán, F.G., Ojanguren, A.F., and Braña, F. (1997). The ontogenetic development of body segments and sexual dimorphism in brown trout (*Salmo trutta* L.). *Canadian Journal of Zoology* 75: 651–655.

Riggs, B.L., Khosla, S., and Melton, III, L.J. (2002). Sex steroids and the construction and conservation of the adult skeleton. *Endocrine Review* 23(3): 279–302.

Robling, A.G., Castillo, A.B., and Turner, C.H. (2006). Biomechanical and molecular regulation of bone remodeling. *Annual Review of Biomedical Engineering* 8: 455–498.

Ruff, C. (2002). Variation in human body size and shape. *Annual Review of Anthropology* 31: 211–232.

Ruff, C. (2003). Growth in bone strength, body size, and muscle size in a juvenile longitudinal sample. *Bone* 33: 317–329.

Ruff, C. (2005). Mechanical determinants of bone form: insights from skeletal remains. *Journal of Musculoskeletal and Neuronal Interactions* 5(3): 202–212.

Safont, S., Malgosa, A., and Subirà, M.E. (2000). Sex assessment on the basis of long bone circumference. *American Journal of Physical Anthropology* 113: 317–328.

Singh, R., Artaza, J.N., Taylor, W.E., Gonzalez-Cadavid, N.F., and Bhasin, S. (2003). Androgens stimulate myogenic differentiation and inhibit adipogenesis in C3H 10T1/2 pluripotent cells through an androgen receptor-mediated pathway. *Endocrinology* 144(11): 5081–5081.

Singh, R., Bhasin, S., Braga, M., Artaza, J.N., Pervin, S., Taylor, W.E., Krishnan, V., Sinha, S.K., Rajavashisth, T.B., and Jasuja, R. (2009). Regulation of myogenic differentiation by androgens: cross talk between androgen receptor/b-catenin and follistatin/transforming growth factor-b signaling pathways. *Endocrinology* 150(3): 1259–1268.

Slemenda, C.W., Reister, T.K., Hui, S.L., Miller, J.Z., Christian, J.C., and Johnston, Jr, C.C. (1994). Influences on skeletal mineralization in children and adolescents: evidence for varying effects of sexual maturation and physical activity. *Journal of Pediatrics* 125: 201–207.

Slice, D.E. (ed.) (2010). *Modern Morphometrics in Physical Anthropology.* Kluwer Academic/Plenum Publisher, Dordrecht.

Slice, D.E. and Ross, A. (2009). *3D-ID: Geometric Morphometric Classification of Crania for Forensic Scientists.* Software available at www.3d-id.org/.

Veldhuis, J.D., Roemmich, J.N., Richmond, E.J., Rogol, A.D., Lovejoy, J.C., Sheffield-Moore, M., Mauras, N., and Bowers, C.Y. (2005). Endocrine control of body composition in infancy, childhood and puberty. *Endocrine Review* 26(1): 114–146.

White, T.D. and Folkens, P.A. (2005). *The Human Bone Manual*. Elsevier Academic Press, Amsterdam.

Wolff, I. (2010). *Sexual Dimorphism in an Argentinean Population. What Are We Classifying?* MS thesis, Department of Applied Forensic Sciences, Mercyhurst College, Erie, PA.

Zelditch, M.L., Swiderski, D.L., Sheets, H.D., and Fink, W.L. (2004). *Geometric Morphometrics for Biologists: A Primer.* Elsevier Academic Press, Amsterdam.

CHAPTER **14**

Morphoscopic Traits and the Assessment of Ancestry

Joseph T. Hefner, Stephen D. Ousley, and Dennis C. Dirkmaat

Assessing ancestry from discrete traits is not an easy undertaking. These assessments are not easy because traditional approaches to ancestry assessment rely more on observer experience than on any understanding of the distribution of traits among modern humans. This traditional approach, used in most forensic anthropology laboratories around the USA, however, is unscientific. The approach is also rooted in typology and outdated race studies, and needs to be abandoned. This chapter presents a brief history of the study of ancestry assessment, addresses the difficulties of the analysis of categorical data, and suggests a more scientific approach based on an understanding of the variation, distribution, and frequencies of common cranial traits.

ANATOMY, SYSTEMATICS, AND THE SEARCH FOR ORDER

The modern anthropological view on race was founded on the efforts of eighteenth- and nineteenth-century European scholars and naturalists attempting to classify the natural world: from the inorganic to the organic, from the smallest of organisms to the largest (Comas 1961; Montagu 1974; Shanklin 1994). Classification of humans into discrete biological packages was first attempted by the father of modern taxonomy, Carolus Linnaeus (1707–1778). Linnaeus, in the tenth edition of *Systema Naturae* (1759), distinguished four groups of human subspecies: "*Homo sapiens africanus,*" "*H. s. americanus,*" "*H. s. asiaticus,*" and "*H. s. europaeus,*" based primarily on shared anatomical characteristics, though some perceived differences in behavior

A Companion to Forensic Anthropology, First Edition. Edited by Dennis C. Dirkmaat.
© 2012 John Wiley & Sons Ltd. Published 2015 by John Wiley & Sons Ltd.

and social interaction crept into his classifications (Brace 2005). In doing so, Linnaeus effectively established the comparative anatomical approach to classification still pursued by today's scientists.

Johann Blumenbach (1752–1840), a German anatomist and naturalist, did not agree with Linnaeus. Blumenbach could not see how the natural world and its inhabitants could be pigeon-holed into discrete categories through the use of "arbitrary and artificial framework of medieval logic" (Brace 2005). Blumenbach argued progress of humans from one "variety" to another. Differences between these varieties resulted from differing climates, nutrition, and modes of life, which had an effect on the *nisus formativus*, or vital force, of humans (Count 1950). Blumenbach did not see the varieties of humans as fixed (emplaced, as he saw it, at the time of the creation), but rather, as degenerations (from Latin the *degeneris*, or "removed from one's origin") from an original, perfect form (in his case, humans from the Caucasus area). Differences between the varieties of humans could be attributed to population migrations and environmental shifts. These new environments caused soft tissue and skeletal changes, which, after a period of time, would become heritable, in a Lamarckian sense (Brace 2005). Blumenbach's monogenist classification of humans included five varieties – Caucasian, Mongolian, Ethiopian, American, and Malayan – but the varieties were not arranged in a specific order or hierarchy.

As scientists and scholars became increasingly cynical of the biblical account of creation (Harris 1968), the belief in a single origin for all humans (i.e., monogenism) also faded into relative obscurity. The role of religion in the interpretation of natural phenomenon was losing ground to a new breed of scholars who looked to the scientific method for answers (Wolpoff and Caspari 1997). At the turn of the nineteenth century, scholars were gradually becoming more aware of the extent of human diversity and variation, but they continued to explain this diversity using the typological approach of Linnaeus and Blumenbach, which focused on a small number of phenotypic traits, like hair form and skin color, to assign an individual to a particular race.

TYPOLOGY, POLYGENISM, AND THE AMERICAN SCHOOL

In the USA, the cultural mixture of displaced indigenous Americans, Chinese laborers, European immigrants, and African slaves led to the creation of an uneasy stewing cauldron of entangled diverse cultures. It is of little surprise, therefore, that the polygenist movement was first championed in the USA in the late nineteenth century (Gould 1996).

Polygenism held that human races were "separate biological species" (Gould 1996), each descended from its own Adam and Eve. The polygenist explanation of human variation was used as justification for inequality: if American Indians, Africans, or the Irish represent another separate and distinct race of humans than oneself, then equal rights do not need to apply to them (Brace 2005; Gould 1996). The most prominent polygenist in the USA was Samuel G. Morton (1799–1851), anatomist, physician, and craniologist. As a physician, he worked on the diagnosis and treatment of tuberculosis (Morton 1834a), and his research in paleontology (Morton 1834b) and anthropology (Morton 1839) essentially established each of these fields as a scientific discipline in the USA (Brace 2005; Harris 1968). However, Morton's work in craniology represents his most recognized contribution.

Morton defended the polygenist explanation for the differences in human groups by using 13 cranial measurements collected from 250 individuals representing "Caucasian, Malayan, American, Negroid, and Mongolian" groups (Harris 1968). Morton borrowed his five groupings directly from Blumenbach, but, unlike Blumenbach, Morton's races were unrelated to each other and had come about not through climactic adaptation, but through an act of divinity: "each Race was adapted from the beginning to its peculiar local destination" (Morton 1839, quoted in Brace 2005).

While Morton's analytical abilities are debated, his conclusions concerning human intellect and cranial capacity have had a major impact on American anthropology and, unfortunately, racist research, including Rushton (1999). Brace (2005) suggests that American anthropology after Morton, but prior to the modern era, is composed of researchers who were attempting to hold on to the last scraps of slavery in the USA. Not only did these anthropologists support slavery, but some, like Josiah Nott, actively fought to protect the southern bastion of inequality (Brace 2005).

As the nineteenth century was nearing its end, America was recovering from the Civil War and American anthropology was reeling from misconceptions about race (Gould 1996). However, two anthropologists with very different views on the scientific validity of race were emerging and would have a major impact on the modern views of race in the USA: Franz Boas at Columbia University and Earnest A. Hooton at Harvard University (Spencer 1981; Washburn 1983; Wolpoff and Caspari 1997).

One of Boas' greatest contributions to biological anthropology was the classic study *Changes in Bodily Form of Descendants of Immigrants* (1910), in which Boas tested the influence of environment on body form (Relethford 2004). Morton believed that the form of the skull was constant in each race and that measurements and calculated indices would allow one to draw conclusions about the race as a whole. Adamantly opposed to this concept of human variation, Boas, under a grant from the USA Immigration Commission, collected anthropometric, craniometric, and pedigree data from nearly 18 000 individuals representing seven populations. From those data he concluded that the environment significantly influenced cranial morphology, even within a single generation (Boas 1910).

Reanalysis of Boas' original data by several research cohorts has led to very different interpretations of the same dataset. Gravlee and colleagues (2003) feel they confirm Boas' interpretation of cranial plasticity by modeling the effect of age, sex, birthplace, and immigrant group on cephalic index using analysis of covariance. However, Sparks and Jantz (2002) argue that the underlying pattern of the populations is not affected significantly by plasticity and that a principal coordinate analysis (PCO) separates Boas' groups into east–west environmental grades, suggesting gene flow as a partial explanation for the patterns observed in the Boas dataset. Relethford's (2004) analysis suggests that cranial plasticity is evident; however, the magnitude of the changes brought on by cranial plasticity is not enough to negate the underlying patterns of the population relationships.

Boas' influence on American biological anthropology was matched only by E.A. Hooton's influence at Harvard University. Hooton, however, provided an entirely different approach to ancestry and race. While Hooton's influence on forensic anthropology has been largely overshadowed by his eccentric work in criminal anthropology and the eugenics movement (Stewart 1979), the methods of analysis he initiated, particularly his work with nonmetric traits, are still very much in use today (Brues 1990; Rhine 1990).

Early in his career at Harvard, Hooton was looking for combinations of metric and nonmetric traits to define races. He wanted traits that were not affected by environmental factors, because he felt that adaptive traits would not provide information on population-level common descent (Hooton 1926). He would later, however, decide that his "insistence ... on nonadaptive characters in human taxonomy [was] impractical and erroneous" (Hooton 1946). Even still, he would always retain a polygenist outlook on human variation, with a particular interest in typological categories and biometric methods (Wolpoff and Caspari 1997).

The typological approach to human variation was, of course, completely contrary to the emerging school of thought at Columbia University. At the end of the Holocaust attitudes about race and the study of race were changing, largely due to the efforts of Boas and his students (Wolpoff and Caspari 1997). While Boas stressed the importance of human variation, Hooton and his students, particularly Carleton Coon, wanted to demonstrate the existence of distinct biological races (Comas 1961) through the collection of massive amounts of anthropometric data. Hooton himself collected data on thousands of civilians and criminals for his Lombroso-like work in criminal anthropology, and a large amount of data for the US Air Force and the Quartermaster Corps related to the design of their equipment, including helmets, seat belts, and safety harnesses. He also personally collected craniometric and nonmetric data on all of the skeletal material recovered from Pecos Pueblo (Hooton 1930).

Hooton could also be considered a biological determinist, as he included various aspects of behavior along with biological traits in his definition of each race, even to the point of looking for an association between various criminal acts and race (Hooton 1939, 1946). Although Lombroso believed individual criminals exhibited inferior physiological differences that could be phenotypically detected (Lombroso 1896, translated 2006), Hooton (1946) took this thinking a step further and suggested that some *races* may be more likely to commit a specific criminal act than others. He published these findings in *The American Criminal: An Anthropological Study* (1939). In that volume, Hooton analyzed data from nearly 14 000 criminals and 3000 civilians. He found that in 19 of his 33 measurements criminals were significantly different from civilians. Like Lombroso, Hooton offered morphological criteria for distinguishing criminals (e.g., low foreheads, high pinched nasal roots, excess of nasal deflections), but, unlike Lombroso, Hooton also provided frequency data on the relationship between race and criminal acts.

Hooton (1926) identified three biologically discrete human races: White, Negroid, and Mongoloid. Each of these primary races was further subdivided into secondary races. For example, Hooton divided the Negroid race into African Negro, Niolitic Negro, and Negrito. In *The Indians of Pecos Pueblo* (1930), Hooton introduces a third racial category: pseudo-types. Using the skeletal material recovered from the Pecos site, Hooton isolated eight "morphological types," first by morphology and later using metric analysis (Woodbury 1932). As an example, individuals with morphological features (i.e., nonmetric traits) that were more commonly found in Africans were classified Pseudo-Negroids. Hooton (1930) did not see a genetic relationship between these Native Americans and African populations. Instead, he saw the discordant nature of these morphologies as evidence of heterogeneity among the Pecos Pueblo population.

Hooton's theoretical approach to race was adopted by many of his students (Brace 2005). For instance, Carleton Coon's dissertation explored the adaptive significance of racial features and "typical racial forms" (Wolpoff and Caspari 1997) and Harry Shapiro's research centered on "hybrid vigor" in a "racially mixed" population on the Pitcairn Islands in the Southern Pacific Ocean (Spencer 1981). Coon's view on race was founded on the assumption that there are a fixed number of races corresponding to divisions that occurred prior to the evolution of *Homo sapiens* (Coon 1962). In Coon's opinion, the fossil record provided direct evidence of the five categories of humans: Australoids, Mongoloids, Caucasoids, Congoids, and Capoids. He believed the five races had each evolved at different rates to their present *H. sapiens* form and hence differences existed in the level of their cultural development, as well. This view of the evolution of humans and human variation is today considered inconsistent with both the fossil record (Stringer and McKie 1996) and modern genetic evidence (Sykes 2002).

Stanley Garn was another Hooton student who studied the classification of human variation, but in a completely different manner than Coon. Garn noted that groups of people in the same geographical area resemble each other more closely than they do inhabitants of other geographical areas. Based on this observation, Garn divided humans into nine all-encompassing geographical races: Amerindian, Asiatic, Australian, Melanesian, Micronesian, Polynesian, Indian, African, and European. The criterion for group membership was geography alone, not skin color, or head form, or nasal breadth. As Garn (1961) saw it, gene flow takes place more often within a geographical area than between geographical areas. Garn (1961) also divided geographical races, into "local races." According to Garn, local races are either (i) distinct, isolated groups (remnants of larger units; i.e., the Ainu of Japan), or (ii) large local races with greater levels of gene flow (i.e., Eastern Europe). He recognized the large amount of variation in these local races, and divided them into arbitrary smaller units which he called microraces (Garn 1961). The actual definition of a microrace is unclear, however. Garn (1961) states that "precise boundaries [between microraces] cannot be drawn," and, in fact, members of one microrace may express phenotypes more similar to another. This line of thinking is not unlike the pseudo-races used by Hooton at Pecos Pueblo, but Garn (1961) recognized the role geography plays in gene flow and the dynamic nature of human variation more so than did Hooton.

The impact of Hooton's students, particularly Coon and Garn, on the modern views of race in the anthropological community cannot be understated. But it was condemnation and criticism to Coon's approach that most affected anthropological studies (Montagu 1963). In fact, Marks (1995) suggests that the general reaction to Coon's work in the anthropological community precipitated the change from studies of human races to the study of human variation. The switch to studies centering on human variation, however, was delayed within forensic anthropology.

This delay may be attributed to another of Hooton's students, Alice M. Brues. Brues is best known for her work in human variation and genetics (Brues 1946, 1959). Her book, *People and Races*, published in 1977 (Brues 1977), exemplifies the Hootonian approach to race: broad assertions concerning human variability within races, but few empirically supported models of the distribution of traits or character-istics. Her interest was in aspects of morphology and evolution, but also included the application of physical anthropological techniques to criminal investigation

(Brues 1958), an interest she passed on to many of her students, including Stanley Rhine. Rhine's typological approach to race is best exemplified by the influential publication *The Skeletal Attribution of Race* (1990), in which morphological trait lists (i.e., cranial nonmetric traits) are presented for each race (Mongoloid, Caucasoid, Negroid) in a very Hootonian manner. This approach to the determination of ancestry in human skeletal remains is still common among the majority of active forensic anthropologists (Gill and Rhine 1990), although recently metric analysis of human skeletal morphology in the assessment of ancestry has assumed a more prominent role (Dirkmaat et al. 2008).

NONMETRIC VERSUS METRIC ANALYSIS OF HUMAN SKELETAL TRAITS

Metric and nonmetric analyses have been utilized alternatively in efforts to categorize and distinguish human groups. Metric analyses are characterized by standard definitions of measurements, collected using precise instruments, and analyzed statistically. Statistical methods used to analyze these data vary from indices and ratios to univariate and multivariate analyses, but in all cases the most common cranial measurements employed conform to the standards established at the *Frankfurter Verständigung* (Frankfurt Convention) of 1882 (Hursh 1976). While the early efforts by Giles and others are noted (Giles and Elliot 1963; Howells 1970, 1973, 1989, 1995; Olivier 1969), today the computer program Fordisc 3 is the most common tool for analyzing craniometric data in forensic contexts. Metric approaches to the determination of ancestry are discussed in depth in Jantz and Ousley (2005) and in Chapter 15 in this volume, and thus we will focus mainly on nonmetric methods.

Although nonmetric traits have never been subjected to the same level of scientific standardization as metric analysis, they are still often preferred over metric methods because nonmetric observations can be obtained from fragmented assemblages, they are relatively easy to collect, and "best of all: *they work*" (Buikstra 1974, her emphasis). For instance, nonmetric traits have been used to show familial inheritance (Cheverud 1981; Laughlin and Jørgensen 1956) and biological distance (Laughlin and Jørgensen 1956). In fact, nonmetric traits are most commonly used in taxonomic analysis at large, through cladistic methods.

Hooton preferred using nonmetric traits to categorize human groups because "they [nonmetric traits] are capable of classification according to presence or absence, [and] grade of development and form, if the observer is *experienced* and is able to maintain a consistent standard for his morphological appraisals" (Hooton 1926, emphasis added). However, scoring individual traits in a consistent manner is often difficult (i.e., high inter- and intraobserver error), even for the expert. Hooton noted that "even veteran anthropologists have difficulty in maintaining consistency in these subjective ratings and still greater difficulty in equating their standards with those of equally experienced observers" (Hooton 1946), an observation that led Hooton to push for standardization.

Hooton made an effort to standardize observations with the so-called Harvard list (Brues 1990), a series of data-collection sheets that include both metric and nonmetric observations (Figure 14.1). The Harvard list, along with Hooton's illustrations of the various nonmetric observations (Figure 14.2), represent one of the earliest

Code A

PEABODY MUSEUM OF HARVARD UNIVERSITY — CRANIAL OBSERVATIONS AND INDICES

Catalogue No. ....Sex.............................

Area ....Special Locality ....Tribe............................

Observer ....Date.............................

1 Description	2 Age	3 Deformation	4 Form	5 Frontal Region	6 Frontal Region
1 Cranium	1 Infant (x–3)	1 Occipital	1 Ellipsoid	Brow Ridges	Height
2 Calvarium	2 Child (4–6)	2 Right Occipital	2 Ovoid	1 Median	1 Very low
3 Calvaria	3 Child (7–12)	3 Left Occipital	3 Spheroid	2 Divided	2 Low
4 Calva	4 Adolescent (13–17)	4 Lambdoid	4 Pentagonoid	3 Continuous	3 Medium
Condition	5 Subadult (18–20)	5 Fronto-occipital	5 Rhomboid	Brow Ridges Size	4 High
5 Poor	6 Young Adult (21–35)	6 Other.............	6 Sphenoid	4 Trace	5 Very High
6 Fair	7 Middle-aged Adult	....	7 Brisoid	5 Small	Slope
7 Good	(36–55)	Degree Deformation	Sex	6 Medium	6 None, Bulging
Sex Criteria	8 Old Adult (56–75)	7 Trace	10 Male	7 Large	7 Slight
8 Uncertain	9 Very Old (76–x)	8 Small	11 Female	8 Very Large	8 Medium
9 Certain	Weight	9 Medium	12 Doubtful	Glabella	9 Pronounced
Muscularity	10 Light	10 Pronounced		9 Small	10 Very Pronounced
10 Small	11 Medium	Cause Deformation		10 Medium	Metopism
11 Medium	12 Heavy	11 Artificial		11 Large	11 Traces
12 Large		12 Pathological		12 Very Large	12 Complete

7 Frontal Region	8 Parietal Region	9 Temporal Region	10 Occipital Region	11 Lambdoid Flattening	12 Serration
Postorbital Constriction	Sagittal Elevation	Fullness	Curve	1 None	Coronal
1 Small	1 Small	1 Flat	1 None	2 Small	1 ?
2 Medium	2 Medium	2 Small	2 Small	3 Medium	2 Simple
3 Large	3 Large	3 Medium	3 Medium	4 Pronounced	3 Submedium
Bosses	4 Very Large	4 Large	4 Pronounced	Transverse Suture	4 Medium
4 Small	Postcoronal Depression	Mastoids	Inion	5 Absent	5 Pronounced
5 Medium	5 Small	5 Small	5 None	6 Present	6 Very Pronounced
6 Large	6 Medium	6 Medium	6 Small	Serration	Serration
Median Crest	7 Large	7 Large	7 Medium	Lambdoid	Sagittal
7 Small	Bosses	Supramastoid Crest	8 Large	7 ?	7 ?
8 Medium	8 Small, Medium	8 Small	Torus	8 Simple	8 Simple
9 Large	9 Large	9 Medium	9 Absent	9 Submedium	9 Submedium
Breadth	Foramina	10 Large	10 Small	10 Medium	10 Medium
10 Small	10 None	Sphenoid Depression	11 Medium	11 Pronounced	11 Pronounced
11 Medium	11 Small, Medium	11 Small	12 Large	12 Very Pronounced	12 Very Pronounced
12 Large	12 Large	12 Medium, Large	Shape of Torus		
			4, 8 Ridge		
			6, 9 Mound		

13 External Occlusion	14 External Occlusion	15 Pterion Form	16 Condyles Elevation	17 Pharyngeal Fossa	18 Tympanic Plate
Coronal	Lambdoid	Right	1 Small	1 None, Submedium	1 Thin
1 Open	1 Open	1 H	2 Medium	2 Medium	2 Medium
2 Beginning	2 Beginning	2 K	3 Large	3 Large	3 Thick
3 Medium	3 Medium	3 X	Basion	Lacerate Foramina	4 Very Thick
4 Advanced	4 Advanced	4 Retourné	4 Low	4 Small	Auditory Meatus
5 Complete	Wormian Bones	5 Epicteric	5 Medium	5 Medium	5 Round
External Occlusion	Lambdoid	Pterion Form	6 High	6 Large	6 Oval
Sagittal	5 None	Left	Styloids	Glenoid Fossa Depth	7 Ellipse
6 Open	6 Few (1–3)	6 H	7 Small	7 Small	8 Slit
7 Beginning	7 Medium (4–6)	7 K	8 Medium	8 Medium	Petrous Depression
8 Medium	8 Many (7–x)	8 X	9 Large	9 Large	9 Absent
9 Advanced	Wormian Bones	9 Retourné	Pharyngeal Tubercle	Postglenoid Process	10 Small
10 Complete	Others	10 Epicteric	10 Absent, Submedium	10 Small	11 Medium
Os Incae	9 Temporo-occipital	Median Occipital Fossa	11 Medium	11 Medium	12 Large
11 Single	10 Other	11 Small	12 Large	12 Large	
12 Multipartite		12 Medium, Large			

19 External Pterygoid	20 Orbits Shape	21 Infra-orbital Suture	22 Malars Size	23 Zygomatic Process	24 Nasal Root Breadth
Plate	1 Oblong	Right	1 Small	Thickness	1 Very Small
1 Small	2 Rhomboid	1 None	2 Medium	1 Small	2 Small
2 Medium	3 Square	2 Facial	3 Large	2 Medium	3 Medium
3 Large	4 Ellipse	3 Orbital	4 Very Large	3 Pronounced	4 Large, Very Large
Internal Pterygoid Plate	5 Round	Infra-orbital Suture	Malars Lateral Pro-	Nasion Depression	Nasal Bridge Height
4 Small	Orbits Inclination	Left	jection	4 Absent	5 Very Low
5 Medium	6 None	4 None	5 Small	5 Small	6 Low
6 Large	7 Small	5 Facial	6 Medium	6 Medium	7 Medium
Pterygo-basal Foramina	8 Medium	6 Orbital	7 Large	7 Deep	8 High
Right	9 Pronounced	Suborbital Fossa	Malars Anterior Pro-	Nasal Root Height	9 Very High
7 Absent	Lacrimo-ethmoid	7 Absent	jection	8 Very Low	Nasal Bridge Breadth
8 Indicated	Articulation	8 Slight	8 Small	9 Low	10 Small
9 Complete	10 Absent	9 Medium	9 Medium	10 Medium	11 Medium
Pterygo-basal Foramina	11 Small	10 Deep	10 Large	11 High	12 Large
Left	12 Medium, Large	Os Japonica	Marginal Process	12 Very High	
10 Absent		11 Absent	11 Absent, Submedium		
11 Indicated		12 Present	12 Medium, Large		
12 Complete					

25 Nasal Profile	26 Subnasal Grooves	27 Total Prognathism	28 Palate Shape	29 Palatine Torus Size	30 Mandible Size
1 Straight	1 Absent	1 Absent	1 Parabolic	1 Absent	1 Small
2 Concave	2 Small	2 Slight	2 Hyperbolic	2 Small	2 Medium
3 Concavo-convex	3 Medium	3 Medium	3 Elliptical	3 Medium	3 Large
4 Convex	4 Pronounced	4 Pronounced	4 Small U	4 Large	4 Very Large
Nasal Sills	Mid-facial Prognathism	Alveolar Border	5 Large U	Palatine Transverse	Chin Form
5 Absent	5 Absent	Absorption	Palate Height	Suture	5 Median
6 Dull	6 Slight	5 None	6 Low	Direction	6 Bilateral
7 Medium	7 Medium	6 Slight	7 Medium	5 Transverse	Chin Projection
8 Sharp	8 Pronounced	7 Medium	8 High	6 Anterior	7 Negative
Nasal Spine	Alveolar Prognathism	8 Pronounced	9 Very High	7 Posterior	8 Neutral
9 Absent	9 Absent	Alveolar Border	Palatine Torus Form	Postnasal Spine	9 Small
10 Small	10 Slight	Preservation	10 Ridge	8 Absent	10 Medium
11 Medium	11 Medium	9 Poor	11 Mound	9 Small	11 Large
12 Large	12 Pronounced	10 Fair	12 Lump	10 Medium	
		11 Good		11 Large	
		12 Perfect			

Figure 14.1 Example of the Harvard list. Courtesy of the Hooton Archives, Peabody Museum of Archaeology and Ethnology, Harvard University.

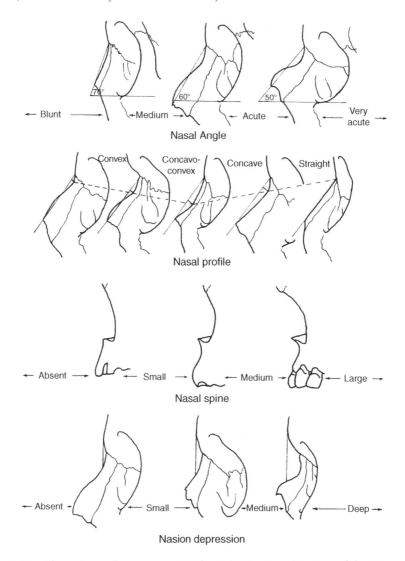

Figure 14.2   Illustrations of nonmetric traits by E.A. Hooton. Courtesy of the Hooton Archives, Peabody Museum of Archaeology and Ethnology, Harvard University.

attempts to provide a scoring system for cranial nonmetric traits. In *A Handbook of Anthropometry* (1960), Ashley Montagu, a Boas student from Columbia University, includes one version of the Harvard list which he attributes to J. Lawrence Angel, a Hooton student and later a curator in the Physical Anthropology Department at the Smithsonian Institution.

CONSIDERING NONMETRIC TRAITS

Traditional cranial nonmetric, or discrete, traits ("epigenetic variants" following Hauser and DeStafano 1989) are best defined – following Buikstra and Ubelaker

(1996) – as "dichotomous, discontinuous, epigenetic traits; nonpathological variations of skeletal tissues that can be better classified as present or absent (or as a point on a morphological gradient; e.g., small to large) rather than quantified by a measurement." The five major categories of epigenetic variants in the cranium are (i) extra-sutural bone (e.g., Inca bone); (ii) proliferative ossifications (e.g., pterygo-alar *bridging*); (iii) ossification failure (e.g., septal aperture); (iv) suture variation (e.g., metopic suture); and (v) foramina variation (e.g., zygomaticofacial foramen number) (Buikstra and Ubelaker 1996). Although the role played by the genome and the environment in the inheritance of cranial nonmetric traits is poorly understood, these traits are routinely used in biological distance studies as a measure of relatedness within and between populations (see Sjøvold 1977, 1984, 1986) and as a proxy for identifying familial relationships within cemeteries (Pilloud 2009).

Ousley and Hefner (2005) used the term "macromorphoscopic" trait to describe the cranial nonmetric traits used in forensic anthropological research, but they are not the same as the discrete traits described above. Macromorphoscopic traits are quasicontinuous variables of the cranium that can be reflected as soft-tissue differences in the living. In a sense, these traits are similar to Brues' (1958) second class of traits, which "due to the contour of bone in areas where it closely follows the surface are apparent in both skeleton and living" (Brues 1958). Later, Hefner (2009) simplified Ousley and Hefner's term to *morphoscopic* traits, while maintaining the original characterization of the variables. With a nod to Earnest Hooton, a scoring system for a series of quasicontinuous nonmetric cranial traits was developed by Hefner (2007, 2009) and proved useful for assessing ancestry. These traits are subdivided into five classes: (i) assessing bone shape (e.g., nasal bone structure); (ii) bony feature morphology (e.g., inferior nasal aperture morphology); (iii) suture shape (e.g., zygomaticomaxillary suture); (iv) presence/absence data (e.g., post-bregmatic depression); and (v) feature prominence/protrusion (e.g., anterior nasal spine) (Hefner 2009).

Unlike traditional analyses based on cranial nonmetric traits used to study frequencies of expression of the features in populations, morphoscopic traits are used to assess (or estimate, predict, determine, etc.) the ancestry of a single individual for the purpose of identification. Table 14.1 presents the morphoscopic traits more commonly employed to assess ancestry, as identified by Hefner (2009) and drawn predominately from trait lists found in Rhine (1990). Most introductory forensic anthropology textbooks also provide trait lists; however, standardization of trait descriptions and analysis of the distribution of these traits within large reference samples have only recently been completed (Hefner 2009).

NONMETRIC TRAIT ANALYSIS IN FORENSIC ANTHROPOLOGY

With respect to the morphoscopic approach to assessing ancestry, in the most-referenced study, Rhine (1990) provided a list of 45 nonmetric cranial traits for four ancestry groups, compiled by members of a regional forensic anthropology conference. While Rhine does not provide a clear methodological approach for using his trait lists (in fact, he cautions against using his lists without further testing and evaluation; Rhine 1990), his trait lists are widely cited in reports and literature.

Table 14.1   Traditional morphoscopic traits used to assess ancestry.

Anterior nasal spine
Inferior nasal aperture
Interorbital breadth
Malar tubercle
Nasal aperture width
Nasal bone contour
Nasal overgrowth
Postbregmatic depression
Supranasal suture
Transverse palatine suture
Zygomaticomaxillary suture

See Hefner (2009) for a complete description of these traits.

Slight variations in traits related to cranial form, such as those presented by Rhine (1990), are difficult to define and measure on an interval scale and a discussion of the variation associated with those traits (e.g., character states) has generally been avoided. However, most of the skeletal traits listed in Rhine (1990) are expressed – in real life – in a quasicontinuous manner (multiple expressions of a trait). Attempting to describe these traits merely in a presence/absence configuration adversely skews the results; unfortunately, these trait lists have been used by forensic anthropologists rather indiscriminately for 20 years.

Rhine (1990) examined 45 nonmetric cranial traits in four groups ("whites," $n=53$; "blacks," $n=7$; "Hispanics," $n=15$; and Amerindians, $n=12$). Based on his analysis of the frequency distribution of this dataset, he concluded that nonmetric traits are useful for assessing ancestry. Yet, as reflected above, his sample sizes were less than ideal to draw such a general conclusion (Rhine 1990), and his observations were based on rather paltry results offering little empirical support (Rhine 1990). Although Rhine (1990) clearly admits that his samples, particularly American "blacks" ($n=7$), are very small, his list of "expected" trait values (Rhine 1990: 13) can be found in most forensic anthropology textbooks (see Burns 1999; Byers 2002; Klepinger 2006). In other words, even with small sample sizes and less-than-ideal results, the Rhine (1990) study continues to be used as an almost exclusive reference for race attribution from nonmetric traits.

Perhaps a closer look at the some of the frequency distributions for "expected" trait values (*sensu* Rhine) is required. For instance, in many textbooks the presence of a postbregmatic depression (PBD) is an *expected* trait for American blacks. However, only 33% of the Rhine sample, or one individual ("American black," $n=3$, for this trait), actually displays a PBD. Rhine is puzzled by this finding as well ("These … features are commonly encountered on the Blacks of the Terry Collection"; Rhine 1990), which may explain why he included PBD as an "American black" trait. PBD is not an isolated case. In fact, of the 45 traits that Rhine reported, 19 did not match the "expected race" value (Rhine 1990). In other words, over 40% of the expected values in the Rhine trait lists were not the observed values for his sample.

When using trait lists, it seems that an expert-level experience with human variation, rather than the analysis of the traits and their association with specific ancestral groups, is the key factor for providing an accurate ancestry assessment (Gill 1998; Rhine 1990). Yet, research on the judgment process has concluded that expert judgments may not be as accurate as the expert would like to believe, and judgment theory increasingly indicates that the expert's evaluation of their own performance is often at odds with reality (Hefner et al. 2007; and see Kruger and Dunning 1999). The frank expert should ask, "Am I doing any better than flipping pennies?" when experience alone is used to make a judgment (Meehl 1986).

The ability to accurately predict unknown events from visual information is subject to "certain systematic flaws; perhaps the most prominent … is simple overconfidence" (Meehl 1986). Researchers from fields like psychometrics and medicine have identified several problems inherent in the human judgment process, which, if correct, greatly diminish the value of the current analytical approaches to nonmetric trait analysis (Hefner et al. 2007). Psychological experiments have shown that experts tend to rely on very little information to make a prediction of an unknown event, in part because feedback is often not available until long after a judgment has been made (Dror et al. 2005; Hastie and Dawes 2001).

The same is true in forensic anthropological analyses; only after an ancestry assessment is made and a positive identification is established does the forensic anthropologist learn of the accuracy of the judgment. Information obtained after the fact leads to adjustments of the relative importance of each trait used to assess ancestry (Hefner et al. 2007). This leads to post hoc trait selection for ancestry dependent on the cranial Gestalt, which Hefner and Ousley (2006) have suggested has no empirical basis. In fact, this Hootonian approach to ancestry determination presents a paradox: although traits have been historically associated with certain groups, the actual trait frequencies in modern populations may not be as high as previously suggested. These differences in trait frequencies are likely a reflection of the lack of standard observations, poor critical evaluation, and the inadequate sample sizes of earlier studies (Hefner 2003).

Rhine (1990) described the nonmetric method of ancestry determination thusly: "as much art as science … contribut[ing] to the mistrust of some who feel that to quantify is to do science, while to evaluate is to produce a system wholly experiential, and thus unverifiable by retest" (Rhine 1990). Therefore, it must be concluded that the current approach to ancestry determination from nonmetric traits is unscientific and, thus, must be abandoned.

MORPHOSCOPIC TRAITS AND ANCESTRY ASSESSMENT IN A STATISTICAL FRAMEWORK

When assessing ancestry from an unknown set of skeletal remains, especially the skull, rather than relying on typological trait lists that supposedly typify the skull of an individual derived from a specific ancestral group, *sensu* Rhine (1990), a better approach involves focusing on individual traits (*characters*) and the variable expression of those traits (*character states*) within a given population.

No single known trait is found exclusively in only one population. For example, shovel-shaped incisors have been used to indicate Asian ancestry. That trait occurs in 70–85% of Asians worldwide (not 100%) (Scott and Turner 1997). Shovel-shaped incisors are also found in most other populations, although in much lower frequencies (3–10% of Europeans; 8–11% of Africans; Scott and Turner 1997). By studying the frequency of expression of individual traits, combining them into suites of significant traits, and then analyzing them within a statistical framework, we might begin to see patterns emerge that will allow us to make scientific and statistically valid assessments of ancestry based on nonmetric traits. These assessments, however, require (i) significant reference data (large sample sizes from multiple and diverse skeletal populations), (ii) better and standardized protocols for trait and character state recordation and coding, and (iii) rigorous classification statistics appropriate for categorical data analysis.

To that end, Hefner (2003, 2007, 2009) collected data on the expression of a large number of morphoscopic traits from multiple skeletal populations (Table 14.2). Following Hooton's lead, Hefner (2009) provided a series of simple, direct illustrations of character states for 11 traits (Figure 14.3). He also provided concise written

Table 14.2 Demographic composition of Hefner's (2003, 2007, 2009) skeletal populations.

Sample	Female (n)	Male (n)	Total (n)	Age
American black	80	140	220	19th century to modern
American white	96	89	185	20th century to modern
Amerindian	230	134	364	Pre- and protohistoric
Asian	35	63	98	Modern
Hispanic	6	33	39	Modern
Total	447	459	906	

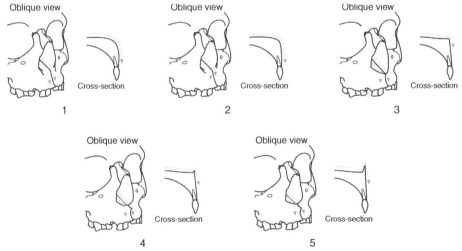

Definition. The most inferior portion of the nasal aperture, just lateral to the anterior nasal spine, which, when combined with the lateral alae, constitutes the transition from the nasal floor to the vertical portion of the maxillae, superior to the anterior dentition.

Guttering (1): A pronounced sloping of the nasal floor beginning within the nasal cavity and terminating on the vertical surface of the maxilla, producing a smooth transition. This morphology is can be differentiated from INA 2 (see below) by nothing the more posterior origin and greater slope of INA 1.

Partial Guttering (2): A moderate sloping of the nasal aperture beginning more anteriorly than in INA 1, and with more angulation at the exit of the nasal opening.

Straight (3): A straight transition from nasal floor to the vertical maxilla with no intervening projection of bone. This morphology is usually angled sharply, although more blunted forms have been observed.

Partial Sill (4): Any superior projection of the anterior nasal floor, creating a weak (but present) vertical ridge of bone traversing the inferior nasal border (partial sill).

Nasal Sill (5): A pronounced ridge (sill) obstructing the nasal floor-to-maxilla transition.

Figure 14.3 Example of trait illustrations and character states: inferior nasal aperture (INA) morphology.

Table 14.3   Frequency distribution of inferior nasal aperture (INA) morphology.

INA	American black (n=180)		American white (n=184)		Amerindian (n=283)		Asian (n=75)		Hispanic (n=37)	
	n	*%*	*n*	*%*	*n*	*%*	*n*	*%*	*n*	*%*
1	64	35.6	1	0.54	11	3.89	9	12.00	1	2.70
2	59	32.8	6	3.26	66	23.32	13	17.33	6	16.22
3	33	18.3	41	22.28	**159**	**56.18**	48	**64.00**	20	**54.05**
4	18	10.0	76	**41.30**	46	16.25	3	4.00	10	27.03
5	6	3.3	60	32.61	1	0.35	2	2.67	0	0.00

The most common INA morphology for each group is shown in bold. For a description of each INA type (1–5), see Figure 14.3.

descriptions for each character and character state along with frequency tables by population (see Table 14.3). The complete list of traits (characters and character states) and populations are fully described elsewhere (Hefner 2009). Utilizing the statistical methods described below, he was able to document intergroup variation (without making assumptions about so-called racial groups) in a clearer way, and to develop empirically supported forensic anthropological techniques for assessing ancestry with high success rates.

Following data-collection efforts and construction of frequency tables, the distribution of a trait within and between groups can be considered. As an example, Table 14.3 presents the distribution of the character states of the inferior nasal aperture (INA) among a sample of American black, American White, Asian, and Amerindian crania. Note that all of the character states of INA are expressed in each group, but in different frequencies. This distribution does not suggest admixture (which might be a common explanation by many forensic anthropologists), or that the sample is composed of a large number of idiosyncratic individuals. Rather, the distribution of each character state within ancestral groups illustrates the true nature of morphoscopic traits and human variation. Importantly, these distributions document that a relationship does exist between populations and trait expression (Hefner 2007), which is best illustrated by a correspondence map of the INA morphology as shown in Figure 14.4. Note how this trait clearly displays much more within-group variation than a simple trait list would suggest. Hefner (2009) presents frequency data for another 10 traits, always revealing a similar pattern: in every instance the distributions within and between populations suggest much more variation (multiple character states) than one would expect from looking at a trait list that includes only one variant of each feature.

Once we make our observations of each individual trait, how do we combine these observations and construct a scientifically valid assessment of ancestry? The answer lies in the myriad of robust statistical techniques and methods that are currently available. Rao (1989) has suggested that the advantage of a statistical analysis of complex data is that the inductive reasoning inherent in the human thought processes "can be made precise by specifying the amount of uncertainty involved in the conclusions drawn" (Rao 1989). As outlined by Hefner (2007, 2012), a number

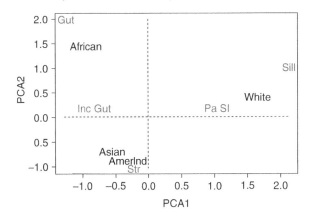

Figure 14.4   A two-dimensional plot of the INA frequency table. Note: the Hispanic sample is not included in this analysis. AmerInd, Amerindian; Inc Gut, incipient guttering; PCA, principal component analysis; Pa SI, partial sill; Str, straight.

of statistical approaches have proven useful to analyze morphoscopic data. Two of these methods are summarized below.

k-nearest neighbor

The nearest neighbor nonparametric classification method is an intuitive statistic, in part because nearest neighbor analysis proceeds quite similarly to the human thought process: in simple terms, those objects that are more similar to one another should be more closely related. The *k*-nearest neighbor methods classify objects on the basis of a training set of individuals of known group membership plotted (as vectors) in a multidimensional feature space. An unknown individual is assigned to the group characterizing most of the training sample individuals surrounding it (i.e., its nearest neighbors) in the multidimensional space. During a learning phase, the training set is used to define the coordinate system of this space, storing its vectors and class labels. For a classification, the unknown individual is also plotted as a vector within this coordinate system (based on observed trait values) and the distance to each of the stored vectors (known individuals) from the training set are computed using some distance measure (e.g., city block, Euclidean). The *k* closest individuals in the training set are selected and the unknown individual is classified with the most frequent group within these nearest neighbors in the training set. Choosing the optimal number of neighbors for a classification depends principally on the dataset; generally, larger values of *k* blur boundaries between the groups, but also reduce any noise inherent in a dataset. There are a variety of statistical packages that offer nearest neighbor analysis. For instance, the *k*-nearest neighbor analyses in R version 2.14.0 are easy to use (and R is a free program; see http://cran.r-project.org/). The R code for the *k*-nearest neighbor statistic used herein can be found in the Appendix to this chapter.

A two-group *k*-nearest neighbor analysis of American "blacks" and "whites" illustrates the usefulness of morphoscopic trait data in a nonparametric model. A random 20% of the dataset was drawn to serve as an initial training set. Next, 12

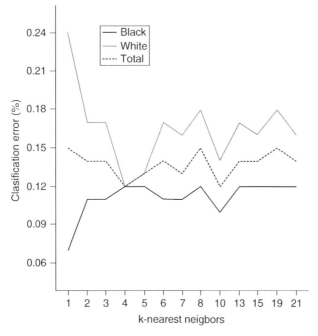

Figure 14.5 Comparison of the error rates among *k*-nearest neighbors.

Table 14.4 Classification matrix from a two-group *k*-nearest neighbor analysis.

Group	American black	American white	Percentage correct
American black	121	17	87.68
American white	14	100	87.72
Total	135	117	87.7

analyses were performed using an increasing number of nearest neighbors, including 1–7, 10, 13, 15, 19, and 21 neighbors. Figure 14.5 compares the error rates (misclassification percentage) for each number of *k* nearest neighbors. Based on this plot, *k* was set at four neighbors for the remaining analyses. As Figure 14.5 illustrates, the number of *k* nearest neighbors affects the overall classification rate, but perhaps more importantly, the plot also confirms that setting *k*=4 reduced the error over the entire model to nearly equal values for all groups. In a two-group analysis with four neighbors, *k*-nearest neighbor correctly classified approximately 88% of the individuals (Table 14.4).

Canonical analysis of principal coordinates

Compare the *k*-nearest neighbor method to a canonical analysis of principal coordinates (CAP). Legendre and Legendre (1998) suggested that a canonical discriminant analysis using the transformed values of the principal coordinates was possible. Anderson and Willis (2003) implemented the method for numerical ecology studies, which led to the development of CAP, a DOS-based computer program by Anderson (2004). A CAP analysis offers several promising qualities to biological anthropologists and forensic anthropologists, alike. First, the method is highly flexible. Several distance and similarity/dissimilarity measures can be selected for an analysis. These include, but are not limited to: Euclidean distance, chi-square distance, Bray–Curtis dissimilarity, Manhattan block distance, and the Canberra distance. Second, the method accounts for correlations of the data. By accounting for the correlative structure in the response matrix, significant patterns in the dataset can be highlighted, particularly in reference to a priori hypotheses, like group membership. The complex relationship of morphoscopic traits, environment, genetics, and head form can be explored more thoroughly if the interrelationships of the data are properly understood.

The final advantage of the CAP method is that unknown individuals can be assigned to a group through a "generalized discriminant analysis based on distances" (Anderson and Robinson 2003). This is particularly useful to forensic anthropologists who seek to identify the ancestry of an unknown skeleton using categorical variables from the cranium. Although traditional discriminant function analysis produces excellent results using categorical variables (Berg 2008; Ousley and Hefner 2005), the underlying assumptions of multivariate normality and equal variance/covariance are not met using categorical data. Krzanowski (2002) reviews the use of categorical data in multivariate discriminant analysis and acknowledges that underlying assumptions will be violated, but he also points out that the method may be appropriate for descriptive purposes and as an exploratory method. However, the CAP method circumvents this issue by transforming data values from categorical variables into principal coordinate scores; in a sense, the data are changed from categorical responses to continuous variables (Anderson and Willis 2003).

The CAP method applies a principal coordinate analysis (PCO) (Anderson and Willis 2003) using any one of several distance measures. The PCO is used to explain a multidimensional dataset through fewer dimensions, although the resulting dimensions are still linear combinations of the original variables. Once the PCOs are obtained, an appropriate number of axes are selected to maximize the classification rate. The final step in the CAP method is a canonical discriminant analysis on the first m axes of the PCO, followed by a leave-one-out cross-validation procedure. Assumptions of CAP deal strictly with the selection of a distance/dissimilarity measure. The measure giving the highest classification rate is assumed to be the best-fitting model, regardless of any underlying assumptions.

A two-group CAP analysis produces classification rates similar to those observed in the k-nearest neighbor analysis (Table 14.5). For example, using five variables and three PCOs – which explains over half of the total observed variation – the CAP method correctly classified 87% of the sample. A three-group classification is also promising (Table 14.6) using the chi-square distance measure (with $m = 12$). The CAP method correctly classified 75% of the sample using a leave-one-out cross-validation procedure.

There are numerous statistical methods that can be applied to morphoscopic trait data. The only limitations beyond statistical assumptions are available reference datasets and the sample sizes within those datasets.

Table 14.5 Classification rates (cross-validated) for the two-group CAP analysis for eight distance measures.

Distance measure	Number of variables	m	Percentage correct American black	American white	Overall
Chi square	5	3	81	93	87
Chi square (metric)	13	10	88	80	84
Euclidean	5	3	83	84	84
Bray–Curtis	5	3	83	84	84
Manhattan	5	3	82	84	83
Orloci's	5	3	80	84	82
Canberra	5	3	73	88	81
Jaccard	5	3	63	85	74

Table 14.6 Classification rates (cross-validated) of a three-way CAP analysis.

Group	American black	American white	Amerindian	Percentage correct
American black	146	14	24	79.35
American white	24	130	26	72.22
Amerindian	43	57	297	74.81

$m = 12$.

Assessing Ancestry from an Unknown Skull

Abandoning the traditional, experience-based method of ancestry assessments is not a call to completely abandon morphoscopic trait analysis. In fact, in a statistical model morphoscopic traits perform as well as a metric analysis. The following section guides the reader through a typical analysis of morphoscopic traits, and presents reporting strategies following data analysis.

The first step of the analysis requires the observer to select one of several character states that best matches the configuration exhibited by the unknown specimen (see example below). This is completed for each observable trait. Following this, an appropriate statistic (e.g., CAP, *k*-nearest neighbor, discriminant function analysis) and suitable reference groups (e.g., American black, American white, and Hispanic) are selected. Once the statistical analysis is conducted on the unknown specimen, the probability of group membership for the unknown specimen, *and* the overall error rate (misclassification rate or classification accuracy) of the model, is reported along with the assigned group membership.

This approach should seem familiar, as it is the same reporting strategy used in metric analyses, particularly those that incorporate Fordisc as an analytical tool. The computer program Macromorphoscopics was designed specifically for the collection of morphoscopic trait data and is available from the authors (Hefner and Ousley 2005). Figure 14.6 shows a screen shot of the program in action.

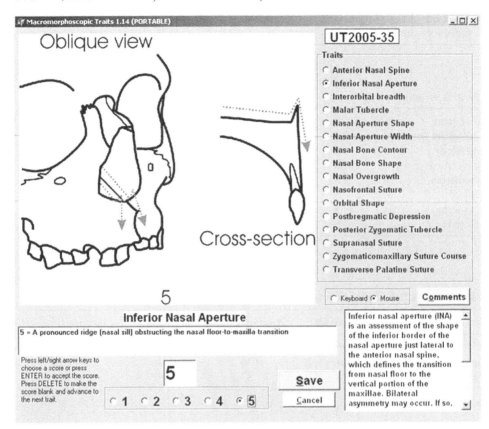

Figure 14.6   Screen capture of the computer program Macromorphoscopics.

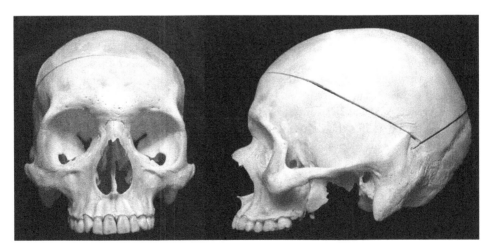

Figure 14.7   A cranium recovered along the USA/Mexico border in the desert Southwest.

An isolated cranium (Figure 14.7) was recently recovered along the USA/Mexico border. The circumstantial evidence recovered at the scene suggests that the individual was likely Hispanic (Anderson 2008; Birkby et al. 2008). The results of the analysis of morphoscopic traits are consistent with this evidence, but empirical support of these observations is needed. Hefner's (2009) illustrations and definitions were used to make the following observations from the photographs: (i) the anterior nasal spine is well developed and markedly protrudes from the face (ANS=3); (ii) the INA is consistent with the straight morphology (INA=3); (iii) interorbital breadth is intermediate (INA=2); (iv) nasal aperture width is intermediate (NAW=2); (v) the nasal bones exhibit steep lateral walls, with an accompanying broad surface plateau (NBC=2); (vi) nasal overgrowth is pronounced (NO=1); and (vii) no PBD is observed (PBD=0). These visual observations and the circumstantial evidence all certainly suggest that this individual is Hispanic. Using the CAP statistical method outlined above, the unknown specimen is placed in the Hispanic category (posterior probability=0.879) in a three-group analysis that correctly classified 89% of the sample (cross-validated).

By reporting analytical findings in this manner, the error rate (how often the model will be wrong) and the strength of the classification (posterior probability) add statistical credibility to the ancestry assessment and offer empirical support for the assessment. In other words, Stewart's (1979) "indefinable something" no longer needs to be used to address confidence levels or error rates, and the "years of experience" of the observer is no longer the appropriate response to the question, "Well how do you know the skull is Hispanic?"

Conclusions

Assessing ancestry on the basis of comparison with typological trait lists that are suggested to represent the morphotyopes of each major racial category is based almost entirely on reliance on the experience of the observer and on trait distributions that have never been empirically supported. This leads to a priori conclusions drawn from some overall impression of the cranial Gestalt. The conventional approach to morphoscopic trait analysis is, therefore, not very different from the early typological efforts of Hooton and his students. Hefner (2009), in fact, demonstrated that when using only five of the "best" traits [INA, interorbital breadth (IOB), NAW, nasal bone structure (NBS), and PBD], only 17% of American blacks and 51% of American whites exhibited the expected traits listed by Rhine (1990). This suggests that either these traits do not work or the true extent of variation in these traits has never been explored properly. The suggestion by some researchers (Rhine 1990) that idiosyncratic variation and racial admixture is a possible cause for some of the low observed frequencies of these traits is considered by these authors to be unfounded and typological.

The research described in this chapter has clearly established that the "typical" morphoscopic traits of African-, Asian-, and European-derived groups commonly used by forensic anthropologists to assess ancestry in unknown skeletal remains are not found at the expected frequencies that would warrant confidence in "accurate" assessments. This traditional, visual approach, which focuses on extreme trait expressions (i.e., a single character state) for these morphoscopic traits, is not very reliable and has no scientific basis. However, discarding the utility of nonmetric traits in ancestry determination is not correct either. By studying the variable expression of these morphoscopic traits (character states) in large skeletal samples from multiple populations, frequencies of trait expression can be calculated. In combination with modern classification statistics for discerning significant patterns of the interaction of suites of morphoscopic traits, analyses of nonmetric traits show great promise. Indications are that this is particularly true for discriminating between American blacks and whites, although more research is needed and larger, more representative samples are necessary.

All of the statistical methods presented herein work better than the traditional approach of visual estimation and guesswork. Each of these methods highlights some important aspect of the multivariate data while enabling the practicing forensic anthropologist to assess the ancestry of a cranium using morphoscopic traits. In this chapter, the morphoscopic approach to ancestry determination using traits originally presented by Hooton and Rhine is discussed within a statistical and scientific framework, a framework which provides the replicability and reliability needed to meet the Daubert challenge (*Daubert v. Merrell Dow Pharmaceuticals* 1993; see Christensen 2004).

In summary, morphoscopic traits work well if considered within a solid statistical framework, which includes consideration of variation and frequency of trait expression, and do not rely solely on the experience of the observer or comparisons to typological trait lists.

APPENDIX

k-nearest neighbor

```
###k-nearest neighbor
###Script adapted from R code on CRAN website
knn (train, test, cl, k=1, l=0, prob=FALSE, use.all=TRUE)
{
train <- as.matrix(train)
if (is.null(dim(test)))
dim(test) <- c(1, length(test))
test <- as.matrix(test)
if (any(is.na(train)) || any(is.na(test)) || any(is.na(cl)))
stop("no missing values are allowed")
p <- ncol(train)
ntr <- nrow(train)
if (length(cl) !=ntr)
stop("'train' and 'class' have different lengths")
if (ntr<k) {
warning(gettextf("k=%d exceeds number %d of patterns",
k, ntr), domain=NA)
k <- ntr
}
if (k<1)
stop(gettextf("k=%d must be at least 1", k), domain=NA)
nte <- nrow(test)
if (ncol(test) !=p)
stop("dims of 'test' and 'train differ")
clf <- as.factor(cl)
```

```
nc <- max(unclass(clf))
Z <- .C(VR_knn, as.integer(k), as.integer(l), as.integer(ntr),
as.integer(nte), as.integer(p), as.double(train), as.integer(unclass(clf)),
as.double(test), res=integer(nte), pr=double(nte),
integer(nc+1), as.integer(nc), as.integer(FALSE), as.integer(use.all))

res <- factor(Z$res, levels=seq_along(levels(clf)), labels=levels(clf))

if (prob)
attr(res, "prob") <- Z$pr
res
}

#now load training set and test set
#
train <- read.table("***.txt", header=TRUE, sep="", na.strings="NA", dec=".",
strip.white=TRUE)
test <- read.table("***.txt", header=TRUE, sep="", na.strings="NA", dec=".",
strip.white=TRUE)
test <-as.matrix(test)
train <-as.matrix(train)
#cl is a matrix of the factors for your training set
#replace GroupNo with name of factors and n with number of individuals
cl <- factor(c(rep("Group1",n), rep("Group 2",n)))
#here you can select the number of neighbors to use for the classification (k), as well
#as various output options
knn(train, test, cl, k=21, prob=T, use.all=T)
#this final line returns the class prediction for the test set
attributes(.Last.value)
```

REFERENCES

Anderson, B. (2008). Identifying the dead: methods utilized by the Pima County (Arizona) Office of the Medical Examiner for undocumented border crossers: 2001–2006. *Journal of Forensic Sciences* 53(1): 8–15.

Anderson, M.J. (2004). *CAP: A Computer Program*. Auckland.

Anderson, M.J. and Robinson, G.R. (2003). Generalised discriminant analysis based on distances. *Australian and New Zealand Journal of Statistics* 45(3): 301–318.

Anderson, M.J. and Willis, T.J. (2003). Canonical analysis of principal coordinates: a useful method of constrained ordination for ecology. *Ecology* 84: 511–525.

Berg, G.E. (2008). *Biological Affinity and Sex Determination Using Morphometric and Morphoscopic Variables from the Human Mandible*. Dissertation, University of Tennessee, Knoxville, TN.

Birkby, W., Fenton, T., and Anderson, B. (2008). Identifying Southwest Hispanics using nonmetric traits and the cultural profile. *Journal of Forensic Sciences* 53(1): 29–33.

Boas, F. (1910). *Changes in Bodily Form of Descendants of Immigrants*, United States Immigration Commission (p. 113). Government Printing Office, Washington DC.

Brace, C.L. (2005). *Race is a Four-Letter Word*. Oxford University Press, New York.

Brues, A.M. (1946). A genetic analysis of human eye color. *American Journal of Physical Anthropology* 21:3, 429–32.

Brues, A.M. (1958). Identification of skeletal remains. *Journal of Criminal Law, Criminology, and Police Science* 48(5): 551–563.

Brues, A.M. (1959). The Spearman and the Archer. *American Anthropologist* 61: 458–469.

Brues, A.M. (1977). *People and Races.* Macmillan Publishing, New York.

Brues, A.M. (1990). The once and future diagnosis of race. In G.W. Gill and S. Rhine (eds), *Skeletal Attribution of Race: Methods for Forensic Anthropology* (p. 1–8). Maxwell Museum of Anthropology, Albuquerque, NM.

Buikstra, J. (1974). Non-metric traits: the control of environmental noise. Paper presented before the *American Association of Physical Anthropologists,* Amherst, MA.

Buikstra, J. and Ubelaker, D. (1995). Standards for data collection from human skeletal remains. *Proceedings of a Seminar at the Field Museum of Natural History.* Arkansas Archeological Survey Research Series 44, Arkansas.

Burns, K. (1999). *Forensic Anthropology Training Manual.* Prentice Hall, Upper Saddle River, NJ.

Byers, S.N. (2002). *Introduction to Forensic Anthropology,* 2nd edn. Allen and Bacon, New York.

Cheverud, J. (1981). Phenotypic, genetic, and environmental morphological integration in the cranium. *Evolution* 36: 499–516.

Christensen, A. (2004). The impact of Daubert: implications for testimony and research in forensic anthropology (and the use of frontal sinuses in personal identification). *Journal of Forensic Sciences* 49: 1–4.

Comas, R. (1961). 'Scientific' racism again? *Current Anthropology* 2(4): 310–340.

Coon, C.S. (1962). *The Origins of Races.* Knopf, New York.

Count, E.W. (1950). *This is Race.* Henry Schuman, New York.

Daubert v. Merrell Dow Pharmaceuticals, 509 US 579, 113 S.Ct. 2786, 125 L.Ed. 2d 469 (US June 28, 1993) (No. 92-102).

Dirkmaat, D.C., Cabo, L.L., Ousley, S.D., and Symes, S.A. (2008). New perspectives in forensic anthropology. *Yearbook of Physical Anthropology.* 51: 33–52.

Dror, I.E., Charlton, D., and Peron, A.E. (2005). Contextual information renders experts vulnerable to making erroneous identifications. *Forensic Science International* 156: 74–78.

Garn, S. (1961). *Human Races.* Charles C. Thomas, Springfield, IL.

Giles, E. and Elliot, O. (1963). Sex determination by discriminate function analysis of crania. *American Journal of Physical Anthropology* 21:53–68.

Gill, G.W. (1998). Craniofacial criteria in the skeletal attribution of race. In K.J. Reichs (ed.), *Forensic Osteology: Advances in the Identification of Human Remains,* 2nd edn. Charles C. Thomas, Springfield, IL.

Gill, G.W. and Rhine, S. (eds) (1990). *Skeletal Attribution of Race: Methods for Forensic Anthropology.* Maxwell Museum of Anthropology, Albuquerque, NM.

Gould, S.J. (1996). *The Mismeasure of Man.* New York: W.W. Norton & Co.

Gravlee, C.C., Bernard, H.R., and Leonard, W.R. (2003). Heredity, environment, and cranial form: a re-analysis of Boas's immigrant data. *American Anthropologist* 105: 125–138.

Harris, M. (1968). *Rise of Anthropological Theory.* Thomas Y. Crowell Company, New York.

Hastie, R. and Dawes, R.M. (2001). *Rational Choice in an Uncertain World: The Psychology of Judgment and Decision Making,* 2nd edn. Sage Publications Newbury Park, CA.

Hauser, G. and De Stefano, D.F. (1989). *Epigenetic Variants of the Human Skull.* E. Schweizerbart'sche Verlagsbuchhandlung, Stuttgart.

Hefner, J.T. (2003). *Assessing Nonmetric Cranial Traits Currently used in the Forensic Determination of Ancestry.* Master's thesis, University of Florida, Gainesville, FL.

Hefner, J.T. (2007). *The Statistical Determination of Ancestry Using Cranial Nonmetric Traits.* Unpublished PhD dissertation, University of Florida, Gainesville, FL.

Hefner, J.T. (2009). Nonmetric cranial traits: new approaches for the determination of ancestry. *Journal of Forensic Sciences* 54(5): 985–995.

Hefner, J.T. (2012) Cranial morphoscopic traits and the assessment of American black, American white, and Hispanic. In G.E. Berg and S.C. Ta'ala (eds), *Biological Affinity in Forensic Identification of Human Skeletal Remains: Beyond Black and White*. Taylor and Francis, Boca Raton, FL (in press).

Hefner, J.T. and Ousley, S.D. (2005). *Morphoscopics* [computer program]. Beta version. Hefner and Ousley, Kaneohe, HI.

Hefner, J.T. and Ousley, S.D. (2006). Morphoscopic traits and the statistical determination of ancestry II. *Proceedings of the 58th Annual Meeting of the American Academy of Forensic Science*, Seattle, WA.

Hefner, J.T., Emanovsky, P.D., Byrd, J., and Ousley, S.D. (2007). The value of experience, education, and methods in ancestry prediction. *Proceedings of the 59th Annual Meeting of the American Academy of Forensic Sciences*, San Antonio, TX.

Hooton, E.A. (1926). *Lecture Notes of EA Hooton*. Peabody Museum Archives, Cambridge, MA.

Hooton, E.A. (1930). *The Indians of Pecos Pueblo*. Yale University Press, New Haven, CT.

Hooton, E.A. (1939). *The American Criminal: an Anthropological Study*. Harvard University Press, Cambridge, MA.

Hooton, E.A. (1946). *Lecture Notes of EA Hooton*. Unpublished manuscripts of the Peabody Museum Archives of E.A. Hooton, Cambridge, MA.

Howells, W.W. (1970). Multivariate analysis for the identification of race from the crania. In T.D. Stewart (eds.), *Personal Identification in Mass Disasters* (pp. 111–112). Smithsonian Institution Press, Washington DC.

Howells, W.W. (1973). *Cranial Variation in Man: A Study by Multivariate Analysis of Patterns of Difference Among Recent Human Populations*. Papers of the Peabody Museum, vol. 67. Peabody Museum of Archeology and Ethnology, Harvard University, Cambridge, MA.

Howells, W.W. (1989). *Skull Shapes and the Map: Craniometric Analyses in the Dispersion of Modern Homo*. Papers of the Peabody Museum of Archaeology and Ethnology, Harvard University, vol. 79. The Museum, Cambridge, MA.

Howells, W.W. (1995). *Who's Who in Skulls: Ethnic Identification of Crania from Measurements*. Papers of the Peabody Museum of Archaeology and Ethnology, Harvard University, vol. 82. The Museum, Cambridge, MA.

Hursh, T.M. (1976). Multivariate analysis of allometry in crania. *Yearbook of Physical Anthropology* 18: 111–120.

Jantz, R.L. and Ousley, S.D. (2005). *FORDISC 3.0*. University of Tennessee, Knoxville, TN.

Klepinger, L. (2006). *Fundamentals of Forensic Anthropology*. John Wiley and Sons, Hoboken, NJ.

Kruger J. and Dunning, D. (1999). Unskilled and unaware of it: how difficulties in recognizing one's own incompetence lead to inflated self-assessments. *Journal of Personality and Social Psychology* 77(6): 1121–1134.

Krzanowski, W.J. (2002). *Principles of Multivariate Analysis: A User's Perspective*. Oxford University Press, Oxford.

Laughlin, W.S. and Jørgensen, J.B. (1956). Isolate variation in Greenlandic Eskimo crania. *Acta Genetica et Statistica Medica* 6: 3–12.

Legendre, P. and Legendre, L. (1998). *Numerical Ecology*, 2nd edn. Elsevier, Amsterdam.

Linnaeus, C.V. (1759). *Systema Naturae*. Theodorum Haak, Lugduni Batavorum.

Lombroso, C. (1876). *L'Uomo Delinquente*. Hoepli, Milan (translation *Criminal Man* (2006), a new translation with introduction and notes by M. Gibson and N. Rafter, Duke University Press, Durham, NC).

Marks, J. (1995). *Human Biodiversity: Genes, Race, and History*. Aldine de Gruyter, New York.

Meehl, P. (1986). *Clinical Versus Statistical Prediction: A Theoretical Analysis and Review of the Literature*. University of Minnesota Press, Minneapolis, MN.

Montagu, A. (1960). *A Handbook of Anthropometry*. Charles C. Thomas, Springfield, IL.

Montagu, A. (1963). *The Concept of Race*. Collier-Macmillan, London.

Montagu, A. (1974). *Man's Most Dangerous Myth: The Fallacy of Race*, 5th edn. Oxford University Press, Oxford.

Morton, S.G. (1834a). *Illustrations of Pulmonary Tuberculosis.* Key and Biddle, Philadelphia, PA.

Morton, S.G. (1834b). *Synopsis of the Organic Remains of the Cretaceous Group of the United States.* Key and Biddle, Philadelphia, PA.

Morton, S.G. (1839). *Crania Americana: A Comparative View of Skulls of Various Aboriginal Nations of North and South America.* Simpkin, Marshall and Co, London.

Olivier, G. (1969). *Practical Anthropology.* Charles C. Thomas, Springfield, IL.

Ousley, S.D. and Hefner, J.T. (2005). The Statistical Determination of Ancestry. *Proceedings of the 56th Annual meeting of the American Academy of Forensic Sciences,* Seattle, WA.

Pilloud, M.A. (2009). *Community Structure at Neolithic Çatalhöyük: Biological Distance Analysis of Household, Neighborhood, and Settlement.* PhD dissertation, Ohio State University, Columbus, OH.

Rao, C.R. (1989). *Statistics and Truth: Putting Chance to Work,* 2nd edn. World Scientific, London.

Relethford, J.H. (2004). Boas and beyond: migration and craniometric variation. *American Journal of Human Biology* 16: 379–386.

Rhine, S. (1990). Nonmetric skull racing. In G.W. Gill and S. Rhine (eds), *Skeletal Attribution of Race: Methods for Forensic Anthropology* (pp. 7–20). Maxwell Museum of Anthropology, Albuquerque, NM.

Rushton, J.P. (1999). *Race, Evolution, and Behavior: A Life-History Perspective,* 2nd special abridged edn. Charles Darwin Research Institute, Port Huron, MI.

Scott, G.G. and Turner, II, C.G. (1997). The Anthropology of Modern Human Teeth: Dental *Morphology and Its Variation in Recent Human Populations.* Cambridge University Press, Cambridge.

Shanklin, E. (1994). *Anthropology & Race.* Wadsworth Publishing Company, Belmont, CA.

Sjøvold, T. (1977). Nonmetrical divergence between skeletal populations: the theoretical foundation and biological importance of C.A.B. Smith's mean measure of divergence. *Ossa* 4(suppl. 1): 1–133.

Sjøvold, T. (1984). A report on the heritability of some cranial measurements and nonmetric traits. In G.N. van Vark and W.W. Howells (eds), *Multivariate Statistical Methods in Physical Anthropology* (pp. 223–246). D. Reidel, Dordrecht.

Sjøvold, T. (1986). Infrapopulation distances and genetics of nonmetrical traits. In B. Hansel and B. Hermann (eds.) *Innovative Trends in der Prähistorischen Anthropologie: Mitteilungen der Berliner Geselhchaft für Anthropologie, Ethnologie und Urgeschichte* (pp. 81–93). Verlag Marie Leidorf, Berlin.

Sparks, C.S. and Jantz, R.L. (2003). A reassessment of human cranial plasticity: Boas revisited. *Proceedings of the National Academy of Sciences* 99:14636–14639.

Spencer, F. (1981). Charter members of the American Association of Physical Anthropologists. *American Journal of Physical Anthropology* 56(4): 531–535.

Stewart, T.D. (1979). *Essentials of Forensic Anthropology.* Charles C. Thomas, Springfield, IL.

Stringer, C. and McKie, R. (1996). *African Exodus: The Origins of Modern Humanity.* Henry Holt and Company, New York.

Sykes, B. (2002). *The Seven Daughters of Eve: The Science that Unravels our Genetic History.* W.W. Norton and Co, New York.

Washburn, S. (1983). Evolution of Teacher. *Annual Review Anthropology* 12: 1–24.

Wolpoff, M.H. and Caspari, R. (1997). *Race and Human Evolution: A Fatal Attraction.* Simon and Schuster, New York.

Woodbury, G. (1932). North America: the Indians of Pecos Pueblo. *American Anthropology* 34(1): 142–143.

CHAPTER 15

Fordisc 3 and Statistical Methods for Estimating Sex and Ancestry

Stephen D. Ousley and Richard L. Jantz

INTRODUCTION

As part of the biological profile, estimating sex and ancestry have long been tasks for the forensic anthropologist. Sex and ancestry in forensic anthropology are *estimated* because they are determined or defined by other means. For example, male sex is *determined* by the presence of a Y chromosome. If you doubt that sex is estimated, imagine telling a widow after positive identification of her husband's remains, "Mrs. Wilson, I'm sorry to inform you that based on certain pelvic features, we have determined that your husband was a female." Nongenetic sex indicators show variability and overlap between the sexes, and certain morphological observations, whether measured or categorical, provide better information as to the sex of the decedent than others. Discriminant function analysis (DFA) is a statistical method of incorporating observations into a mathematical formula to evaluate biological aspects of unknown individuals. Whereas nonmetric observations can be used in DFA, measurements have most often been analyzed because computationally simpler statistical methods can be employed. Fordisc (Jantz and Ousley 1993, 2005; Ousley and Jantz 1996) is a popular tool for analyzing remains using skeletal measurements and DFA. While Fordisc is a recent development, it uses statistical methods first published over 70 years ago (Fisher 1936; Mahalanobis 1936). This chapter is a brief overview of estimating sex and ancestry using DFA and especially using Fordisc. Further information about using Fordisc can be found in Jantz and Ousley (2012).

A Companion to Forensic Anthropology, First Edition. Edited by Dennis C. Dirkmaat.
© 2012 John Wiley & Sons Ltd. Published 2015 by John Wiley & Sons Ltd.

DFA has proven to be quite useful for estimating sex and ancestry, and the use of DFA has shown three distinct phases. The first phase is characterized by the calculation and publication of DFA formulas. Thieme and Schull (1957) provided one of the earliest DFA applications to sex estimation, and Giles and Elliott followed with sex (1962) and ancestry (1963) DFA using cranial measurements. Giles and Elliott's ancestry functions classified an individual into one of three groups: American whites, American blacks, and American Indians. These publications enabled anthropologists to estimate sex by collecting measurements and calculating DFA scores by hand to derive a classification; using Giles and Elliott (1962), practitioners could also produce a two-dimensional plot of an individual's scores by hand, and the estimated ancestry was based on where the individual was plotted (Figure 15.1). The second phase in the use of DFA occurred with the rise of the personal computer, when users could automatically calculate discriminant function scores by entering measurements and formulas into a spreadsheet or other software program. If the measurements were entered correctly and the software was programmed properly, DFA scores could be calculated without error. However, a weakness of using a published DFA formula was that all measurements had to be present in the remains at hand in order to use them. DFA could not be used on incomplete remains with measurements missing. The third phase started in 1993, when Fordisc enabled the generation of made-to-order discriminant functions based on all measurements that could be taken. Most forensic anthropologists are somewhat familiar with the capabilities and limitations of DFA and Fordisc, but few know enough of the statistical assumptions and guidelines to use DFA and Fordisc most effectively.

DISCRIMINANT FUNCTION ANALYSIS

DFA is part of a family of multivariate statistical classification methods that sort individuals into groups using various criteria, and include procedures such as cluster analysis. DFA analyzes specific reference groups with known membership in discrete categories such as ancestry, language, sex, tribe, or ancestry, and provides a basis for the classification of new individuals with unknown group membership. In order to understand DFA, it is essential to understand multivariate statistics. A good first read is Kachigan (1991). There are several excellent articles and book chapters that use relatively little mathematics (Afifi and Clark 1997; Albrecht 1980; Huberty 1994; Huberty and Olejnik 2006; Manly 2005; Tabachnick and Fidell 2001). There are also very good resources that use more mathematics (Krzanowski 2000; Legendre and Legendre 1998). Johnson and Wichern (2007) provide many exercises and worked-through examples.

The most commonly used method of DFA is linear DFA (LDFA). In LDFA, a factor (or numerical weight) is calculated for each measurement or observation that, when summed, maximizes mean differences among groups. The factors are the linear discriminant function and can then be used to classify additional individuals. The sum of the factors multiplied by the measurements in a two-group LDFA is known as the discriminant function score, and the difference between the two groups is the Mahalanobis distance or D-square (Mahalanobis 1936). The Mahalanobis distance is a multivariate measurement because it takes into account the univariate variation in all

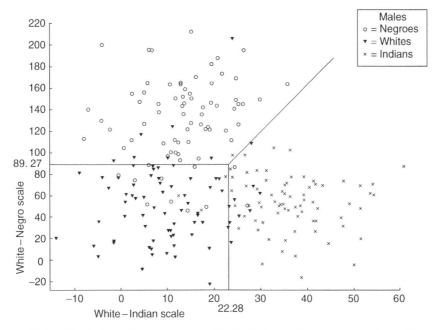

Figure 15.1   Discriminant function graph of individuals in reference groups from Giles and Elliott (1962). A new individual's scores for each axis were calculated from its measurements and then plotted. Individuals who plotted in the upper left area would be classified as black, in the lower left area as white, and in the right area as American Indian. Originally published by Callaghan and Company.

measurements, the relationships among measurements within groups, and the mean measurements in each group, to objectively represent differences among groups. Using the Mahalanobis distance, differences among groups are expressed in terms of the average variation within groups, similar to comparing univariate means in two groups by using their standard errors. If there are more than two groups, more than one discriminant function score can be calculated, and more scores are calculated on additional discriminant axes, which maximize the among-group differences remaining after the first discriminant axis; this procedure is known as canonical variates analysis (CVA). Each additional axis explains a smaller proportion of the total differences among group means. The differences on all axes are summed for the Mahalanobis distance, and all axes are uncorrelated with each other. Because more than one axis is involved in group comparisons in CVA, and the axes are uncorrelated with each other, the group canonical scores have a more or less elliptical distribution and the mean group scores are called centroids (Figure 15.2). Classification of an unknown individual is based on overall similarity and is simple: the discriminant function score of the unknown individual in LDFA is compared to the centroids for each reference group; the unknown is classified in the group to which it shows the smallest Mahalanobis distance, which is the most similar group in a multivariate sense. Once again, an unknown is simply classified in the reference group to which it is most similar. While the LDFA classification rule is quite simple, based on similarity, there are a number of requirements for the LDFA to produce functions with the highest classification accuracy.

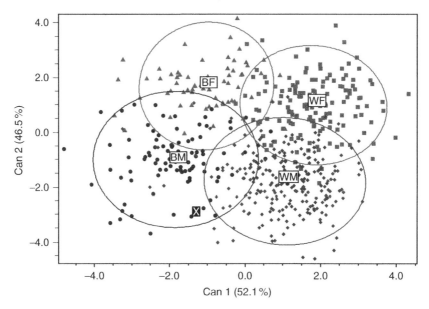

Figure 15.2 CVA in Fordisc 3.0. Reference group individuals and centroids are plotted by their scores on canonical axes one and two. The position of the classified individual is shown by the X.

Requirements
Sample size

As in all statistical analyses, the paramount concern is sample size. Sufficient sample sizes are extremely important – both the total sample size and the sample size of each group in an analysis – because other requirements often depend on having large enough samples. Sufficient sample sizes are especially important in multivariate analyses such as LDFA for accurate calculations of parameters such as means, variances, and covariances, and these are necessarily related to the number of measurements to be analyzed. A crucial question is how many measurements can be analyzed given a specific sample size. In analyzing m measurements, there are $(m*(m+1))/2+m$ parameters estimated. Some of the suggested sample sizes relate to total sample size (N), while some are applied to the minimum sample size among all groups (n). One basic criterion is that n must be larger than the number of measurements (m), although Tabachnick and Fidell (2001) suggest that a sample size of at least $4\,m$ produces more reliable parameter estimates, and we certainly want reliable estimates of group parameters, but there are no hard and fast rules. Huberty (1994) suggests that n must be larger than $3\,m$, and we use and recommend Huberty's (1994) standard because it should yield stable estimates *that are less subject to sampling variation* while allowing enough measurements for effective classification. In all analyses using Fordisc, the first results to be consulted are the sample sizes of the reference groups, to see if they meet Huberty's rule of thumb. Because many of us will analyze relatively complete crania, we calculate the target, or maximum, number of measurements as $n/3$ because in many initial Fordisc analyses some group sample sizes will be small and the number of measurements will need to be limited.

Sample size is especially important in DFA in order to avoid overfitting the data, known as the curse of dimensionality. Generally, using more measurements will more likely bring out differences among groups than fewer measurements, and increase classification accuracy. But after a certain point, as more and more measurements are used, classification accuracy tends to plateau and then decrease. This phenomenon is likely due to the presence of redundant information, resulting in the introduction of greater statistical "noise," and poor estimates of group parameters. DFA uses a basic model of within- and among-group variation, and as more and more measurements are added, the model becomes more complex. For instance, a curvilinear distribution with 10 points can be fit perfectly with a model using 10 parameters and a ninth-degree polynomial, but a simpler model with some error is a better model that will be more appropriate for future predictions. Also, as more measurements are added and the distances among individuals increase, a group sample in multivariate space becomes more and more sparse and more diffuse. Using far too many measurements relative to a group's sample size can inflate the F typicality probability of a classified individual (see below) to nearly 1.0 even when the individual is highly atypical.

Given sufficient sample sizes, to optimally estimate parameters and thus optimize classification performance, LDFA involves two further requirements: (i) the data are "multivariate normal;" and (ii) the variation within all groups is the same. Neither of these assumptions is likely to be strictly true when working with finite samples, and tests for multivariate normality and equal variance-covariance matrices can be misleading. However, these requirements should be critically evaluated and, if violated, there are several classification methods that may perform better with certain kinds of data. However, the performance of LDFA, and the simple classification rule based on overall similarity, can work remarkably well even when the requirements are not met. Quadratic DFA (QDFA) is an approach that can accommodate differences in within-group variation, but generally shows poorer performance due to overfitting.

Multivariate normality

Univariate and multivariate normality of the data are important requirements for the correct estimation of parameters and probabilities using LDFA. This requirement for measurements can be misleading, because it is far more important that the discriminant function scores are more or less normally distributed. Because skeletal measurements are very often more or less normally distributed, and are summed in linear functions, the discriminant function scores become normalized (due to the central limit theorem), and multivariate normality is simply assumed.

Outliers

A far more important concern than multivariate normality per se is the presence of outliers in reference samples. Outliers are individuals that are so unusual that they stick out from the total sample or from other members of their particular group. Outliers can be due to recording errors, encoding errors, measurement errors, pathological conditions, or any combination of these reasons. Outliers can drastically affect estimated multivariate parameters such as group means and variation within samples. Removing outliers, especially in multivariate analyses, is always recommended, and when sample sizes are large enough they can be removed with few consequences. The presence of one outlier can mask the presence of other outliers, so any indications for outliers must be reexamined after the removal of any outlier. Discovering univariate outliers is straightforward using the familiar ideas of the normal distribution and a standardized distance (the standard deviation) to the mean and calculating a p value. Discovering multivariate outliers is quite similar, and employs the Mahalanobis distance to calculate a p value. Methods for detecting outliers are incorporated into Fordisc and can be seen by clicking on Individual Scores on the Options page, analyzing, and then looking at p values on the Extended Results page. Outliers can often be detected graphically as members of a group that are far from the group centroid in Fordisc's canonical plots; right-click on the outlier to get its ID. Outliers can be removed one at a time by typing the ID into the Exclude IDs box, one per line, on the Options page. The excluded individuals will be listed on the Basic Results page.

Equal variation within groups

In univariate analyses, the standard deviation, a measure of within-group variation, is used to quantify differences among groups. Likewise, in multivariate analyses, the pooled (average) within-group variance-covariance matrix (VCM) is used to calculate the Mahalanobis distance among groups and from group centroids to the classified individual. Therefore, as in univariate analyses, the within-group variation must be relatively similar so they can be compared using a common measure of variability. The variance is the standard deviation squared; it is always positive. The covariance, which can be negative, is the variation in one measurement that can be accounted for by variation in another measurement. The covariance of two measurements is the product of their standard deviations times their correlation coefficient. The pooled VCM is calculated with every analysis and will change as groups and measurements change. The pooled VCM is always shown on the Extended Results page in Fordisc, and each group's VCM will also be displayed by clicking on Group VCMs on the Options page. Measurements with large (positive or negative) covariances are more highly correlated to each other: Figure 15.3 shows Fordisc output, including the pooled within-group VCM and correlation matrix from American black and American white males and females. Glabello-occipital length (GOL; maximum cranial length) shows a very high correlation to nasion-occipital length (NOL) because the anterior landmarks of glabella and nasion are very similarly located, and the posterior landmark, determined by spreading calipers, is virtually identical. The correlations and covariances of GOL with the other measurements are much smaller, but positive. Cranial measurements generally show positive correlations between 0.25 and 0.75 with each other; the few negative correlations are not often significantly different from zero.

Figure 15.3 also shows other important information about the VCM provided in Fordisc. The determinant of the VCM is calculated as the overall variance minus the overall covariance, representing the amount of independent variation in the data. The natural log of the determinant is provided, and generally, if it is greater than zero, it means that there is enough variation to provide the best matrix calculations, but a negative value, especially with high magnitude, usually occurs when there are too many measurements being analyzed for the sample size, known as overfitting the data (see below). Wilks' lambda is a ratio of within-group variation to among-group variation, and lower values indicate greater separation among groups. Figure 15.3 also shows the p value for a test of equal within-group variability (Kullback 1959, cited in Legendre and Legendre 1998). There are many tests for equal within-group variability,

Pooled within-class variance-covariance matrix

	GOL	NOL	OBH	XCB
GOL	51.4186	49.7227	1.2570	8.0660
NOL	49.7227	49.9824	1.5451	7.6956
OBH	1.2570	1.5451	4.0522	1.1079
XCB	8.0660	7.6956	1.1079	32.3358

Natural log of determinant = 9.3880
Wilks' Lambda = 0.5482
VCVM homogeneity test (kullback) ChiSq = 33.24 with 30 df: p = 0.31

Pooled within-class correlation matrix

	GOL	NOL	OBH	XCB
GOL	1.0000	0.9808	0.0871	0.1978
NOL	0.9808	1.0000	0.1086	0.1914
OBH	0.0871	0.1086	1.0000	0.0968
XCB	0.1978	0.1914	0.0968	1.0000

Figure 15.3   Fordisc output showing the pooled within-group VCM and correlation matrix for American white and black males and females.

and each test has its strengths and weaknesses. In this case we would fail to reject the null hypothesis that each group's VCM is the same, in other words, there is no indication that the group VCMs differ significantly. The p value from this test should be treated very carefully, because it will vary depending on the number of groups and measurements, and only p values less than approximately 0.000001 (1×10^{-6}) should be addressed. If groups show very different levels of variability based on the p value, other statistical procedures, including QDFA, logistic regression, or nonparametric methods, may be necessary, although they may not produce the most accurate results.

Criteria for discriminant functions

If the essential requirements are met, there is a virtually unlimited number of discriminant functions that can be calculated depending on which groups and which measurements are used. Therefore, criteria for judging discriminant functions are necessary.

Classification accuracy

The most important criterion in DFA is that of classification accuracy when applied to the reference groups. Classification accuracy is a reasonable estimate of how the DFA will perform on unknown cases and is a direct estimate of validity. First and foremost, classification functions must demonstrate classification accuracy better than the random, or prior, probabilities, in order to be useful. A classification accuracy of 50% for sex does not aid in estimating sex if the prior probabilities are even for male and female, but 50% classification accuracy is much better than random in DFA of ancestry using four groups with equal prior probabilities ($1/4 = 25\%$). Likewise, a classification function for handedness that is 90% accurate is not better than the prior probability that 90% of a random sample of humans will be right-handed.

As it turns out, classification accuracy using craniometrics from human groups is quite often much greater than random. In fact, Ousley et al. (2009) demonstrated that using three or more measurements classifies the 28 male groups from around the world in the Howells craniometric database (see below) at a statistically significant rate greater than random ($1/28 = 3.6\%$), despite claims to the contrary (Williams et al. 2005). Further, when using 24 of the best measurements, classification accuracy was 75%, and, importantly, 89% of the individuals from the reference groups were classified into their own group or into a group from the same region, illustrating the strong geographic patterning in craniometric variation. The craniometric results echo those from molecular genetics (Bamshad et al. 2003; Rosenberg et al. 2002).

As mentioned, generally speaking, the more measurements used in an analysis, the better the classification accuracy: there are simply more measurements that can differ among groups. As long as DFA requirements are met, a DFA with higher classification accuracy for the reference groups is better than a DFA with lower classification accuracy for the reference groups, and should be more valid when classifying unknown remains. However, due to the curse of dimensionality, the forensic anthropologist is faced with a Goldilocks dilemma: using many measurements, which encompass more morphological variation, more often produces good separation and classification of many groups, but using too many variables produces overfitting and lower classification rates.

The consequences of overfitting can best be illustrated by using the most often recommended way of estimating classification accuracy, called leave-one-out cross-validation (Lachenbruch and Mickey 1968). Leave-one-out cross-validation avoids the optimistic bias of higher classification accuracy inherent in resubstitution, in which each member of every reference group is classified into the reference groups, including the group of which he or she is a member, as was standard in Fordisc 2 (Ousley and Jantz 1996). In leave-one-out cross-validation, one individual in the reference groups is removed from his or her reference group, the discriminant function is calculated using everyone else, and that individual is then classified into the most similar reference group using that function (Figure 15.4). A record is kept of the group into which the individual was classified; rarely are all individuals classified correctly using the simple classification procedure of LDFA. That individual is then added back into his or her group and another individual is removed from his or her group and classified, until all individuals are so classified. Thus, in cross-validation, each individual is classified using functions calculated from everyone else in the reference groups. When all have been classified in this manner, the number of correctly classified individuals divided by the total sample number is the overall cross-validated classification accuracy. There are other methods of estimating classification accuracy, notably k-fold cross-validation and various resampling methods (Efron and Tibshirani 1993, 1997). Fordisc will incorporate additional methods of cross-validation in future versions because they provide confidence intervals for classification accuracy.

Figure 15.5 illustrates the consequences of trying to use too many measurements and overfitting the data as well as the importance of estimating classification accuracy using leave-one-out cross-validation. In a DFA of 200 Americans, comprising 50 black females, 50 white females, 50 black males, and 50 white males, 231 measurements were used; the estimated correct classification rate was 100% using resubstitution. The cross-validated accuracy was only 40%, however, meaning that the function was so fragile that removing one individual caused errors. The natural log of the determinant of the VCM in this case was −812, indicating a great deal of data redundancy. In contrast, when 23 measurements were used with the same sample, the cross-validated

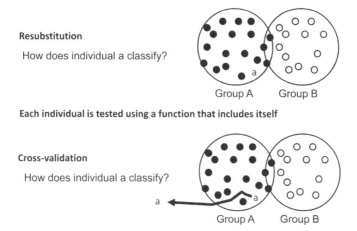

Resubstitution

How does individual a classify?

Group A   Group B

Each individual is tested using a function that includes itself

Cross-validation

How does individual a classify?

Group A   Group B

Each individual is tested using a function that includes all other individuals

Figure 15.4   Estimating classification accuracy using resubstitution and cross-validation.

accuracy was higher, at 84%; this rate is typical using these groups. Additionally, the natural log of the determinant of the VCM was 49, indicating that there was enough variation in the measurements given the sample size.

It is often the case with complete remains that many measurements can be taken, but sample sizes are often limited. While keeping in mind the target number of measurements, there are statistical procedures that can aid in finding a certain number of measurements that best separate groups. Forward stepwise selection of measurements, choosing the best measurement combination by sequentially adding one measurement at a time, is most often used in DFA to find the measurements that show the greatest among-group differences. The group differences expressed in these variable combinations can be estimated using multivariate statistics such as Wilks' lambda, which measures among-group separation of centroids, or assessed directly through classification accuracies, though the latter is computationally more intensive. Fordisc users can choose from these two methods of estimating classification accuracy when using stepwise selection of measurements. Depending on the variables and group separation, a minimum number of eight measurements may be reasonable, but fewer measurements may at times be necessary and justifiable. However, two or three stepwise-selected measurements, no matter how apparently accurately they perform, should be avoided in favor of using more measurements.

While the overall classification rate is the primary indicator of DFA validity, classification rates for each group are also important. If there are wide disparities in correct classification percentages when classifying two groups, such as 90 and 75%, then the level of variation within groups is different, and, thus, a requirement for an optimized DFA is not being met. In such cases, the group with the lowest classification rate has greater within-group variation than the other groups, and other classification methods may produce a higher correct classification rate. Disparity in correct classification percentages is a practical indicator of different amounts of within-group variation and can occur despite having an acceptable p value from a statistical test for the equality of VCMs. When classifying into more than two groups, disparities

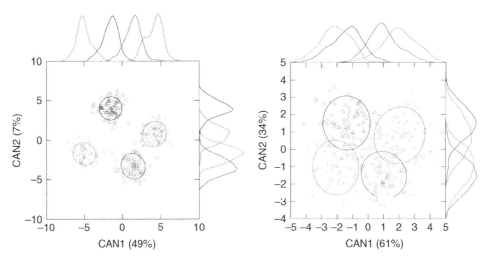

Figure 15.5 CVA showing overfitting due to too many measurements (left) and more realistic results using fewer measurements (right).

in correct classification percentages can indicate different levels of within-group variation, but they can also indicate that a group is morphologically intermediate. For example, when Fordisc's Hispanic samples are compared to American whites, American blacks, and Americans Indians, Hispanics will often show a lower correct classification rate simply because they are not as divergent as the other groups. Groups showing intermediate variation should be detectable in the Mahalanobis distance matrix that Fordisc provides in the extended results, and especially in Fordisc's canonical variates plot.

Appropriate and representative samples
In DFA, as in statistics as a whole, samples should be random and representative; in the case of forensic DFA, having appropriate groups relates to sample validity. Data from modern Americans show that standards derived from the nineteenth century are not appropriate, or – at a minimum – do not provide the most valid reference data, for the assessment of twentieth-century groups. Thieme and Schull (1957) and Giles and Elliott (1962, 1963) used American black and white samples from the Terry Collection and an American Indian sample from the Archaic site of Indian Knoll, and the discriminant functions showed lower accuracy when applied to modern forensic cases (Ayers and Jantz 1990; Ousley and Jantz 1997). Also, postcranial samples from the Hamann–Todd Collection used by İşcan and Cotton (1990) perform very poorly when applied to modern cases, largely due to secular increases in stature in American blacks and whites (Ousley and Jantz 1997). Most importantly, we know that the earlier methods are inadequate only because of the Forensic Anthropology Data Bank (FDB) at the University of Tennessee.

The FDB was started in 1986 with a grant from the National Institute of Justice and is the premier source for morphological data from modern Americans. The FDB contains extensive demographic information for many cases, including place of birth, medical history, occupation, stature, and weight. The skeletal information for individuals includes cranial and postcranial metrics, suture closure information, various aging criteria scores, nonmetric cranial information, perimortem trauma, congenital traits, and dental observations. A significant part of the FDB is made up of Terry Collection individuals born after 1900, who are much more similar to other individuals in the FDB born in the twentieth century. This data source is exhausted, however, because all of the Terry Collection individuals born after 1900 are now in the FDB. Data from modern forensic cases and donations at the University of Tennessee help forensic anthropologists keep up with the changing populations of the USA. Currently, the FDB has over 3400 cases, and almost 2300 of these are from positively identified individuals. The FDB provides the modern data in Fordisc for comparisons, and is most appropriate for recent forensic cases. For older remains, Fordisc also has nineteenth-century samples and they can be compared with twentieth-century samples and others in the Howells database.

Evaluating DFA Results from an Unknown Individual

Along with reviewing DFA requirements and evaluating DFA criteria, when an unknown individual is classified using DFA there are a number of results to examine and concerns to be addressed. There are many DFA results that are far more important than the classification of the unknown. In fact, the classification of the unknown, while important to the anthropologist, is the least important result, all things considered; it is the other results that inform us how likely the classification is correct and alert us to possible errors.

Data errors

Data errors comprise the gamut of mistakes that can happen between collecting measurements and analyzing them. Following Howells (1973), there are numerous opportunities for errors to come about: misunderstanding the measurement definition; using an uncalibrated instrument; misreading a measurement; writing or typing a wrong measurement value in a data table; and typing a wrong measurement value in an analysis. In our experience, data errors are the usual cause of unusual results when running DFA. Naturally, entering incorrect data for an individual being classified will produce incorrect results, even if the classification is eventually found to be correct. Because data errors are so common, when using Fordisc an initial check based on minimum and maximum values for each measurement is run before the analysis of the unknown begins. After analysis, Fordisc compares each measurement of the unknown with the mean measurements of each reference group. If the unknown's measurements are within plus or minus one standard deviation of all means, there is no special notation in the results. If the unknown's measurement is between one and two standard deviations above or below each group mean, then the measurement is noted with + or – respectively. Likewise, differences of between two and three standard deviations are designated ++ or –– and differences greater than three standard deviations are designated +++ or –––. Having several measurements with one or two plus signs generally means that the unknown is large. Having a mixture of pluses and minuses suggests that there are shape differences between the unknown and the reference groups, especially if there are several signs per measurement; this may suggest that the unknown comes from a group not in the analysis. There is also a multivariate statistic known as the typicality probability that may indicate that an unknown does not belong to any of the groups.

Typicality probabilities

Typicality probabilities (tps) represent the probability that the classified individual belongs to each group based on the Mahalanobis distance to each group, and range from nearly 0 to nearly 1. Because the pooled within-group VCMs of the groups analyzed are used to calculate distances to groups, tps will change for an individual depending on which groups and which measurements are chosen. An individual's tp of 0.33 for a group means that 33% of the total sample from that group would be expected to be as far or farther away from that group's centroid, or, in other words, more divergent from the group's mean morphology. The p value is similar to the familiar p value in hypothesis testing, and corresponds to the probability that the classified individual is acceptable as a member of each group. As a rule of thumb, tps above 0.05 are acceptable. In practice, tps below 0.05 are suspicious, and all measurements should be double-checked; a tp below 0.01 for a group indicates doubtful probability of membership in that group. Also, tps are most important for the most similar group, the group into which the unknown was classified, and a tp less than 0.01 for the most similar group would be a very questionable classification; the unknown would be a very unusual member of that group, if it is indeed a member of that group. When all tps are low for all groups, the unknown should not be classified; double-check for measurement errors and unusual morphology due to pathological or taphonomic conditions; also, try to identify the most unusual measurements that contribute to the low tp, and avoid using them in further analyses.

Fordisc calculates tps in three ways, using the F distribution, the chi-square distribution, and ranked distances. Each calculation has advantages and disadvantages depending on the number of measurements used, the number of groups analyzed, and the sample sizes of each group. Using the F distribution takes into account the Mahalanobis distance and sample size, because there is greater uncertainty about group sample parameters with small sample sizes. tps can often be relatively high for many groups in an analysis if the group sample sizes are small, even if the groups are quite different from the classified individual. Also, as the number of variables approaches a group's sample size, the F ratio tp becomes more and more inflated. Thus, there is practical justification for also examining the tps using the chi-square distribution, which are based on Mahalanobis distance alone. The chi-square tps assume infinite sample sizes, and are therefore almost always much lower than the tps based on the F distribution. Chi-square tps also tend to call more individuals atypical than F tps. Chi-square tps may be justified biologically if one believes that a larger sample would not change group parameters significantly. Fordisc also provides tps based on ranked Mahalanobis distances to each group's centroid. The classified individual is treated as if it is a member of each group, and the rank of the individual determines the p value. For example, if the individual is ranked 90th out of a total sample of 100 (including the unknown), then the p value is 0.10. The ranked probabilities are empirical rather than theoretical and require adequate sample sizes for consistency. Large and especially homogeneous samples may have group members that are not very far from the group mean (in comparison to other groups) and could produce a very low ranked tp for an unknown. All three probabilities converge as the group sample sizes get larger.

Posterior probabilities

As long as the DFA requirements have been met, there are no indications of data errors, and the tps are acceptable, another statistic, the posterior probabilities (pps), which are directly related to the classification, are evaluated. pps are the probability of membership in each group for the classified individual based on the relative distances to each group, as opposed to the absolute distances to each group as in calculating tps (Tatsuoka 1971). The pps for each group in an analysis range from nearly 0 to nearly 1 and they sum to 1 (100%). The most similar group to the classified individual will have the highest pp, and very high pps, such as those above 0.90, are considered "stronger" classifications because they seem to show overwhelming similarity to one group as opposed to all other groups. However, there is no required cutoff in pps to classify an unknown individual, as maintained by Elliott and Collard (2009), who misunderstood information presented at a Fordisc workshop and who do not sufficiently understand multivariate statistics (Siegel and Ousley 2011).

The major assumption in evaluating pps is that the unknown actually belongs to one of the reference groups, because pps are calculated using only the groups selected for analysis. pps can vary a great deal in different analyses depending on which groups are included in each analysis. Because DFA will classify using any set of measurements, even those from another species, both tp and pp must be evaluated in DFA. Figure 15.6 (left) presents a relatively common situation in DFA, with good separation between two groups and an unknown that is appreciably closer to one group's centroid than to the other, which is reflected in the pp. Figure 15.6 (right) illustrates a different situation in which the classified individual is "unusual"; that is, far from both group centroids. Because both distances are large, the tps of the classified individual for each group are very low, but the pp for the closest group will be very high. With this kind of result, measurement error is the most likely cause, and measurements should be double-checked. If the measurements are accurate, it is possible that the classified individual comes from a group not included in the analysis, and more groups should be analyzed. If a classified individual is very different from all groups, the remains should be carefully examined for pathological conditions such as premature craniosynostosis that can affect morphology. The reference samples in forensic DFA are more or less normal individuals, with no obvious morphological disturbances; a person with affected morphology cannot be expected to classify correctly.

FORDISC AND HUMAN VARIATION

Choosing reference groups

As mentioned, one assumption of all classification functions is that the individual being classified belongs to one of the reference groups, or could be considered a member of that group as opposed to the other reference groups. More precisely, DFA will indicate to which reference group the classified individual is most similar. Fordisc will indicate the

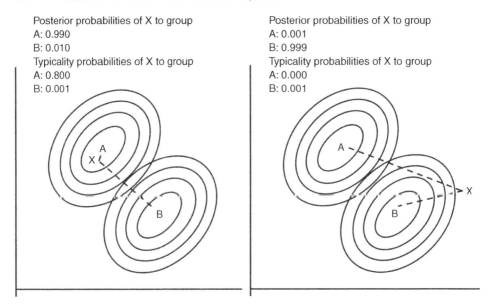

Figure 15.6   Canonical variates plot illustrating typicality and posterior probabilities with a "normal" individual (left), and with an "unusual" individual (right).

morphologically most similar group, and in one sense Fordisc, and DFA, is never wrong: As long as the requirements are met and other statistical results are acceptable, Fordisc will always indicate the most similar group based on the groups and measurements used. Also important, as mentioned, is how accurate the simple classification rule using overall similarity is when applied to the reference groups. Accepting the Fordisc classification for unknown remains involves an inference, and inferring that the classification will be correct in a two-way analysis when 50% of the individuals in the reference groups classify correctly is quite different from inferring that the classification is correct when 99% of the individuals in the reference groups classify correctly. In another sense, depending on the inference drawn from a classification, Fordisc is always wrong. Technically, no unknown individual being classified is already a member of a reference group; reference groups are somewhat arbitrary and the same individuals can be placed into different reference groups; the classified individual will not be "correct" in every possible grouping of reference individuals; and a sample from every human population, however defined, is not part of Fordisc.

The nature of human craniometric variation

Because there is geographic patterning in human craniometric variation, samples from every human population are not necessary as long as the anthropologist recognizes that he or she must interpret results from Fordisc and draw inferences before making conclusions. A Laotian male in the FDB is one of the example cases included with Fordisc, and when all forensic groups are selected, he is most similar to the Vietnamese male sample. Was Fordisc wrong? He was not a Vietnamese; he was a Laotian. If the classification from any analysis is taken too literally, and the inferences are very narrow, then the anthropologist may well draw the wrong conclusion. In our experience, law-enforcement personnel take the anthropologist's opinions literally, which can lead to nonidentification of the deceased. In the Laotian case, it is true that he is most similar to the Vietnamese male sample, but the anthropologist's conclusion from the Fordisc results should not be that he was a Vietnamese. Rather, the conclusion from the Fordisc results should be that he most likely came from Southeast Asia.[1]

Analyzing Ancestry in Fordisc 3

DFA is a practical undertaking, and one metaphor for the process could be that of a game, and the goal of this game is to get the highest classification percentage. Of course, as in all games, rules must be followed and there are certain winning strategies. Most importantly, no matter how good the apparent correct classification rate is, if certain rules are broken, there are reasons to doubt such performance will hold when applied to the individual being classified.

Figure 15.7 shows a DFA flowchart for analyzing remains that encompasses the requirements and results that are most important. DFA using Fordisc should be run initially using all possible groups into which an unknown individual may classify. In using all groups one could term it a "naïve" analysis because no assumptions are made as to sex or ancestry. If additional evidence is available that helps narrow down possible groups, it should be used. For example, sex can sometimes be strongly inferred from clothing or articles associated with the remains, or from independent osteological indicators. In such cases, including only male groups, for example, will help facilitate analysis. After the initial run, check the measurements for any possible data mistakes. Also, check the sample sizes of the reference groups: are they at least three times the number of measurements being analyzed? If not, reduce the measurements through deselecting them or through stepwise selection. Once the number of measurements is adjusted, examine the classifications. Are the correct classification rates similar? Look at the extended results and make sure the level of variation in each group is more or less the same through the p value provided. Then remove the most dissimilar group, one at a time, based on pps and tps. A classified individual showing a pp of 0.000 and especially a tp less than 0.01 for a reference group is morphologically very different from that group. Repeat with the remaining groups and keep in mind that groups with small sample sizes may be eliminated, making it possible to use more measurements in further analyses, which is preferable. Continue eliminating groups until the number of reference groups is two to four, based on pps and tps. Classifications into two to four groups are expected to be more accurate than those involving many more groups, but the classifications involving more groups need to come first in order to establish which groups are most similar. When the final classification is run, it will be an acceptable classification if the VCMs are similar enough, the sample sizes are large enough for the number of measurements used, and the tps are above 0.05 for the most similar group. We strongly suggest that you report the pps and tps for the final groups used, because both probabilities are a measure of statistical certainty for the classification of the unknown individual. Fordisc also keeps a record of all analyses run during a session, which can be accessed by clicking the Log tab. The text from the log can be pasted into a forensic report appendix and thereby document all analyses that were run.

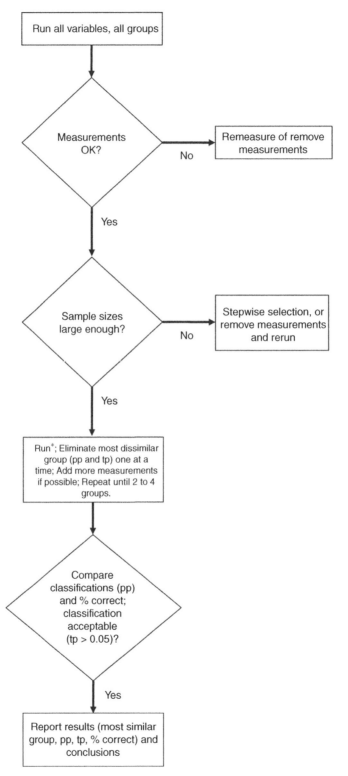

Figure 15.7 DFA flowchart. pp, posterior probabilities; tp, typicality probabilities. *Check for unequal VCMs and outliers.

SUMMARY

Fordisc uses DFA, a multivariate procedure that requires careful evaluations by the practitioner at every stage of analysis, in order to produce the most accurate estimations of sex and ancestry. Before measurement, the remains should be examined for any morphological changes due to pathological conditions or taphonomic factors. The analyst should be very familiar with the measurement definitions and measurements should be carefully taken and recorded to avoid errors. All appropriate groups should be included in the initial analysis. After each analysis, check measurements for errors; compare sample sizes to the number of measurements being used; look for outliers in the reference groups; check that the level of variation in each group is about the same through a statistical test; check that correct classification rates in all groups are about the same, unless intermediate groups are also being classified; pay attention to tps and pps; do not rely on too few measurements; and narrow down the reference groups to the two to four most similar groups. After the final analyses are run, report the pps and tps of the most similar groups, do not interpret classifications too literally, and remember that as long as certain requirements are met it is a very reasonable inference that the classification function with the highest correct classification percentage is more valid than other classification functions for classifying an unknown individual.

NOTE

1   As an illustration of the arbitrary nature of human group classifications, the Vietnamese sample in Fordisc comes from Ba Chuc, a village in Vietnam near the Cambodian border that was part of the "killing fields" massacres by the Khmer Rouge. The border between Vietnam and Cambodia had been fluid for many years, and did not correspond to ethnic differences. Ironically, the Khmer Rouge assassinated males from Ba Chuc even though the village was linguistically and culturally Cambodian (Kenyhercz et al. 2010).

REFERENCES

Afifi, A.A. and Clark, V.A. (1997). *Computer-Aided Multivariate Analysis*, 3rd edn. Chapman and Hall, New York.

Albrecht, G. (1980). Multivariate analysis and the study of form, with special reference to canonical variate analysis. *American Zoologist* 20: 679–693.

Ayers, H.G., Jantz, R.L., and Moore-Jansen, P.H. (1990). Giles and Elliot race discriminant functions revisited: a test using recent forensic cases. In G.W. Gill and S. Rhine (eds), *Skeletal Attribution of Race* (pp. 65–71). Maxwell Museum of Anthropology, Albuquerque, NM.

Bamshad, M.J., Wooding, S., Watkins, W.S., Ostler, C.T., Batzer, M.A., and Jorde, L.B. (2003). Human population genetic structure and inference of group membership. *American Journal of Human Genetics* 72: 578–589.

Efron, B. and Tibshirani, R.J. (1993). *An Introduction to the Bootstrap*. London: Chapman & Hall.

Efron, B. and Tibshirani, R.J. (1997). Improvements on cross-validation: the .632+ bootstrap method. *Journal of the American Statistical Association* 92: 548–560.

Elliott, M. and Collard, M. (2009). FORDISC and the determination of ancestry from cranial measurements. *Biological Letters* 5: 849–852.

Fisher, R.A. (1936). The use of multiple measurements in taxonomic problems. *Annals of Eugenics* 7: 179–188.

Giles, E. and Elliot, O. (1962). Race identification from cranial measurements. *Journal of Forensic Sciences* 7: 147–157.

Giles, E. and Elliot, O. (1963). Sex determination by discriminant function analysis of crania. *American Journal of Physical Anthropology* 21: 53–68.

Howells, W.W. (1973). *Cranial Variation in Man: A Study by Multivariate Analysis of Patterns of Difference Among Recent Human Populations*. Papers of the Peabody Museum, vol. 67. Peabody Museum of Archeology and Ethnology, Harvard University, Cambridge, MA.

Huberty, C.J. (1994). *Applied Discriminant Analysis*. Wiley and Sons, New York.

Huberty, C.J. and Olejnik, S. (2006). *Applied MANOVA and Discriminant Analysis*. Wiley and Sons, New York.

İşcan, M.Y. and Cotton, T.S. (1990) Osteometric assessment of racial affinity from multiple sites in the postcranial skeleton. In G.W. Gill and S. Rhine (eds), *Skeletal Attribution of Race* (pp. 83–90). Maxwell Museum of Anthropology, Albuquerque, NM.

Jantz, R.L. and Ousley, S.D. (1993). *FORDISC 1.0: Computerized Forensic Discriminant Functions*. University of Tennessee, Knoxville, TN.

Jantz, R.L. and Ousley, S.D. (2005). *FORDISC 3: Computerized Forensic Discriminant Functions*. Version 3.1. University of Tennessee, Knoxville, TN.

Jantz, R.L. and Ousley, S.D. (2012). Introduction to Fordisc 3. In M. Tersigni-Tarrant and N. Shirley (eds), *Forensic Anthropology: An Introduction*. Taylor and Francis, London (in press).

Johnson, R.A. and Wichern, D.W. (2007). *Applied Multivariate Statistical Analysis*, 6th edn. Prentice Hall, Upper Saddle River, NJ.

Kachigan, S.K. (1991). *Multivariate Statistical Analysis: A Conceptual Introduction*, 2nd edn. Radius Press, New York.

Kenyhercz, M.W., Pietrusewsky, M., Damann, F., and Ousley, S.D. (2010).

Borders and boundaries: a clash between culture and biology in Southeast Asia. *Poster presented at the 79th annual meeting of the American Association of Physical Anthropologists*, April 14–17, Albuquerque, NM.

Krzanowski, W.J. (2000). *Principles of Multivariate Analysis*, revised edn. Clarendon Press, Oxford.

Lachenbruch, P.A. and Mickey, M.R. (1968). Estimation of error rates in discriminant analysis. *Technometrics* 10: 1–11.

Legendre. P. and Legendre, L. (1998). *Numerical Ecology*, 2nd edn. Elsevier, Amsterdam.

Mahalanobis, P.C. (1936). On the generalised distance in statistics. *Proceedings of the National Institute of Sciences of India* 2: 49–55.

Manly, B.F.J. (2005). *Multivariate Statistical Methods: A Primer*. Chapman and Hall, New York.

Ousley, S.D. and Jantz, R.L. (1996). *FORDISC 2.0: Personal Computer Forensic Discriminant Functions*. University of Tennessee, Knoxville, TN.

Ousley, S.D. and Jantz, R.L. (1997). The Forensic Data Bank: documenting skeletal trends in the United States. In K.J. Reichs (ed.), *Forensic Osteology: Advances in the Identification of Human Remains*, 2nd edn (pp. 297–315). Charles C. Thomas, Springfield, IL.

Ousley, S.D., Jantz, R.L., and Freid, D.L. (2009). Understanding race and human variation: why forensic anthropologists are good at identifying race. *American Journal of Physical Anthropology* 139: 68–76.

Rosenberg, N.A., Pritchard, J.K., Weber, J.L., Cann, H.M., Kidd, K.K., Zhivotovsky, L.A., and Feldman, M.W. (2002). The genetic structure of human populations. *Science* 298: 2381–2385.

Siegel, N. and Ousley, S.D. (2011). The importance of testing and understanding statistical methods in the age of Daubert: can FORDISC really classify individuals correctly only one percent of the time? Paper presented at the *63rd Annual Meeting of the American Academy of Forensic Sciences*, February 21–26, Chicago, IL.

Tabachnick, B.G. and Fidell, L.S. (2001). *Using Multivariate Statistics*, 4th edn. Allyn and Bacon, Boston, MA.

Thieme, F.P., and Schull, W.J. (1957). Sex determination from the skeleton. *Human Biology* 29: 242–273.

Williams, F.L., Belcher, R.L., and Armelagos, G.J. (2005). Forensic misclassification of ancient Nubian crania: implications for assumptions about human variation. *Current Anthropology* 46: 340–346.

Estimating Stature

Stephen D. Ousley

We estimate, rather than determine, stature from human remains, as we estimate other aspects of the biological profile. We can only say that we determine the stature of an individual if we actually go through the process of measuring the individual, but, as it turns out, an individual's stature will change throughout the day and over time. When we analyze a part of the biological profile, utilizing different aspects of the skeleton may result in different estimates, and it is especially quantifiable in stature estimation: using different bones will result in somewhat different estimates. In this chapter I provide an overview of stature estimation and provide recommendations for its use in forensic anthropology.

Stature estimates are generally calculated using linear regression, a statistical method related to correlation, a measure of association between measurements. Linear regression provides the best estimate for one measurement from another measurement when they have a relatively simple relationship, and numerous resources are available that explain it (e.g., Neter et al. 1985). Due to the imperfect correlation between bone lengths and stature, there are often differences between the predicted stature represented by the regression line and actual stature. In some texts these differences are termed "random errors," but it is more constructive to think of them as simply normal variation: there are many other variable contributions to stature besides the length of a single bone. The regression line has a slope and intercept, and minimizes the average differences due to variation. Because of this averaging, a stature estimate for an individual who comes from the same reference group will be accurate; that is,

A Companion to Forensic Anthropology, First Edition. Edited by Dennis C. Dirkmaat.
© 2012 John Wiley & Sons Ltd. Published 2015 by John Wiley & Sons Ltd.

not be biased by providing estimates that are usually too high or too low. When a different reference group is used to estimate stature, the estimates may show bias.

The regression line represents the point estimate of stature for any given bone length, but more importantly for estimation, regression provides a measure of uncertainty. In general, the larger the bone length, the greater the precision, and a single estimate is preferable to averaging the mean of several estimated statures. Some bones show a higher correlation to stature than others; for example, maximum femur length is more highly correlated to stature than is the length of the calcaneus. This should come as no surprise because femur length provides the single greatest contribution to stature. When estimating stature, we naturally want to use the bone measurements that show the smallest differences between predicted stature and actual stature, and that therefore have the smallest uncertainty and provide the greatest precision of prediction. Uncertainty is expressed using a plus/minus figure with a specific percentage prediction interval (PI), similar to a confidence interval. For instance, a 95% PI for stature means that in the long run, the actual stature will be outside the interval 5% of the time. In other words, the prediction will be incorrect one out of 20 times. The PI becomes slightly larger as a bone measurement gets further from the mean bone length due to the uncertainty in the estimate of the regression slope. With a large sample size, the parameters are better estimated and the differences in the PI may be negligible across the distribution (Ousley 1995; Giles and Klepinger 1988). Also, we should not forget that we can calculate the PI in intervals of tenths of an inch, but the minimum and maximum values will often be rounded to the nearest inch in the USA (or to the nearest centimeter in countries using the metric system), which can erase small PI advantages.

Estimating stature only has value when it provides an estimated stature that is better than a naïve stature estimation from a population sample mean and standard deviation. For example, the average stature of American white males between 30 and 50 years old is approximately 5 feet, 10 inches (1.78 m), and the standard deviation is about 3 inches (7.5 cm). Without measuring any bones there is approximately a 95% probability that the stature of a white male decedent is between 5 feet, 4 inches, and 6 feet, 4 inches (1.62 and 1.93 m). Such a large prediction range, reflecting greater uncertainty, is also characteristic of methods for estimating stature from fragmentary long bones (Simmons et al. 1990; Steele 1970). At first glance, the PI for an estimate, even with large samples, may seem very wide, such as 13 cm, but it reflects the variability in stature for given bone lengths. Such large PIs can nonetheless be valuable in excluding or matching possible identifications, such as in the case of a 3 feet, 11 inches (119 cm) murder suspect (Morse 2010).

Anthropologists recognized early on that using femur and tibia measurements improved the accuracy of predicted statures, but only when the two are added together. Linear regression can be used with more than one bone measurement, but in the case of skeletal measurements there is a problem of multicollinearity: Many bone measurements show high correlations to other bone measurements, and as a result, more than one bone cannot often be used in regression. Linear regression involves calculating a unique solution (a slope and an intercept) to the data, and when variables show high correlations there are too many possible solutions. When measurements are summed, the result is one number, so multicollinearity will not be an issue. Using more than one measurement, however, improves estimates, and generally the more measurements that are added together the smaller the prediction interval, including measurements that do not contribute to stature, such as the lengths of arm bones. The reason is that arm bones are correlated to other components of stature such as the length of the vertebral column. One further improvement in PIs is possible through calculating the principal components (which are uncorrelated with each other) of several measurements and regressing stature on the principal components (Lynn and Ousley 2010). These manipulations can result in small reductions in the PIs, although further research is needed.

Generally, unbiased estimates with the smallest PIs are those that involve just one sex and ethnic group. All things being equal, taller people have relatively longer legs than shorter people, and the distal limb segments (lower leg and forearm) are also relatively longer; however, there are also population differences in limb proportions. For example, there are differences between American blacks and whites in relative leg lengths that affect estimates (Tibbets 1981), and recent Balkan males are taller than Trotter and Gleser's (1952) World War II males, but appear to have relatively shorter legs and relatively longer arms (Ross and Konigsberg 2002). If one is uncertain of the ancestry when providing a stature estimate, it may be best to combine suspected ancestral reference groups, which will usually result in a wider PI, but may also be biased. In addition to sex and ethnic group, the age of the reference group samples is an important consideration. Many groups have experienced secular increases in stature and changes in limb proportions (Meadows and Jantz 1995, 1999; Tibbets 1981), and estimates of modern statures using older samples will be biased.

Due to population differences and temporal changes, estimates of modern American remains should use reference samples from modern Americans. Fordisc 3.1 (Jantz and Ousley 2005) estimates statures and PIs from long-bone lengths using the most up-to-date measurements from the Forensic Data Bank (Ousley and Jantz 1998). The most often cited stature-estimation equations come from Trotter and Gleser (1952, 1958), who used males who died in World War II and the Korean War, and females born in the nineteenth century from the Terry Collection. Their equations involving the tibia should be used with caution, however, because Trotter apparently measured the tibia incorrectly by excluding the medial malleolus (Jantz et al. 1995). In my experience, few anthropologists seem to measure the tibia correctly, following the classic definition for condylomalleolar length (which is admittedly difficult; Moore-Jansen et al. 1994), but measure it closely enough because it seems to result in very small differences in the measurement. One way around the tibia measurement problem is to use the highly correlated maximum length of the fibula, if available. If maximum precision is the main goal of stature estimation, then the best estimate should involve using all skeletal elements that contribute to stature, known as the Fully Method (Fully 1956; Fully and Pineau 1960), which was more recently explained in English by Lundy (1988), with clarifications and improvements by Raxter et al. (2006).

You would think that the relationship between a measured bone and a measured stature would show the highest correlations and therefore show the smallest PI. That is generally the case, but in forensic anthropology the measured statures of missing persons are rarely available. What is more commonly available is a stature from a driver's license or other official source. As Willey and Falsetti (1991) have shown, driver's license statures can be grossly inaccurate and are usually biased upwards in males, but are readily available. Ideally, everyone would have their stature consistently measured when they apply for a driver's license, but in most cases the driver's license stature is given by the applicant. These unmeasured statures are known as forensic statures (FSTATs) to contrast them with measured statures (MSTATs), such as in military or other records. When individuals are repeatedly measured, however, MSTATs are highest just after long periods of sleep or recumbency and decrease by at least 1 inch (2.5 cm) as an individual remains upright for long periods (Ousley 1995). A further complication is found in using cadaver lengths, which have been referred to as CSTATs, because they have an imprecise relationship to MSTATs. We can think of MSTATs and CSTATs as biological statures because they are directly measured, and, as mentioned, theoretically, they should show the highest correlations to bone lengths. However, MSTATs are complicated by a decrease in stature after age 40 or so, which, based on longitudinal studies, accelerates (Friedlaender et al. 1977). One advantage of FSTATs over MSTATs is that while MSTATs change over many years, FSTATs rarely change. Likewise, long-bone lengths do not change over time, so the relationship between them is actually more stable than between long-bone lengths and MSTATs. Using a combination of MSTATs, CSTATs, and FSTATs provides even larger samples and may have little effect on the precision of stature estimates (Wilson et al. 2010), although further research regarding bias and other issues is needed.

In summary, stature estimation can be an important part of the biological profile, but there are numerous considerations in estimation, such as measurement technique, statistical methods, reference sample ethnicity, location, and age, and the kind of stature information available for comparison.

REFERENCES

Friedlaender, J.S., Costa, Jr, E.T., Bosse, R., Ellis, E., Rhoads, J.G., and Stoudt, H.W. (1977). Longitudinal physique changes among healthy white veterans at Boston. *Human Biology* 49: 541–558.

Fully, G. (1956). [A new method of determining stature.] *Annales de Medecine Legale et de Criminologie* 36: 266–273 (in French).

Fully, G. and Pineau, H. (1960). [Determination of stature using the skeleton.] *Annales de Medecine Legale et de Criminologie* 40: 145–154 (in French).

Giles, E. and Klepinger, L.L. (1988). Confidence intervals for estimates based on linear regression in forensic anthropology. *Journal of Forensic Sciences* 33: 1218–1222.

Jantz, R.L. and Ousley, S.D. (2005) *FORDISC 3: Computerized Forensic Discriminant Functions.* Version 3.1. University of Tennessee, Knoxville, TN.

Jantz, R.L., Hunt, D.R., and Meadows, L. (1995). The measure and mismeasure of the tibia: implications for stature estimation. *Journal of Forensic Sciences* 40: 758–761.

Lundy, J.J. (1988). A report on the use of fully's anatomical method to estimate stature in military skeletal remains. *Journal of Forensic Sciences* 33: 534–539.

Lynn, K.S. and Ousley, S.D. (2010). Stature estimation: are there any advantages to using principal component analysis? Poster presented at the *62nd Annual Meeting of the American Academy of Forensic Sciences*, February 22–27, Seattle, WA.

Meadows Jantz, L. and Jantz, R.L. (1995). Allometric secular change in the long bones from the 1800s to the present. *Journal of Forensic Sciences* 40: 762–67.

Meadows Jantz, L. and Jantz, R.L. (1999). Secular change in long bone length and proportion in the United States, 1800–1970. *American Journal of Physical Anthropology* 110: 57–67.

Moore-Jansen, P.H., Ousley, S.D., and Jantz, R.L. (1994). *Data Collection Procedures for Forensic Skeletal Material.* Report of Investigations No. 48. Department of

Anthropology, University of Tennessee, Knoxville, TN.

Morse, D. (2010). Short shooting suspect, with a record and a softer side, eludes Montgomery police. *Washington Post* June 5, www.washingtonpost.com/wp-dyn/content/article/2010/06/04/AR2010060404826.html.

Neter, J., Wasserman, W., and Kutner, M.M. (1985). *Applied Linear Statistical Models*, 2nd edn. Richard D. Irwin, Homewood, IL.

Ousley, S.D. (1995). Should we estimate biological or forensic stature? *Journal of Forensic Sciences* 40: 768–773.

Ousley, S.D. and Jantz, R.L. (1998). The Forensic Data Bank: documenting skeletal trends in the United States. In K.J. Reichs (ed.), *Forensic Osteology: Advances in the Identification of Human Remains*, 2nd edn (pp. 441–458). Charles C. Thomas, Springfield, IL.

Raxter, M.H., Auerbach, B.M., and Ruff, C.B. (2006). Revision of the fully technique for estimating statures. *American Journal of Physical Anthropology* 130: 374–384.

Ross, A.H. and Konigsberg, L.W. (2002). New formulae for estimating stature in the Balkans. *Journal of Forensic Sciences* 47: 165–167.

Simmons, T., Jantz, R.L., and Bass, W.M.

(1990). Stature estimation from fragmentary femora: a revision of the Steele method. *Journal of Forensic Sciences* 35: 628–636.

Steele, D.G. (1970). Estimation of stature from fragments of long limb bones. In T.D. Stewart (ed.), *Personal Identification in Mass Disasters* (pp. 85–97). Smithsonian Institution, Washington DC.

Tibbets, G.L. (1981). Estimation of stature from the vertebral column in American blacks. *Journal of Forensic Sciences* 26: 715–723.

Trotter, M. and Gleser, G.G. (1952). Estimation of stature from long bones of American whites and negroes. *American Journal of Physical Anthropology* 10: 463–514.

Trotter, M. and Gleser, G. (1958). A re-evaluation of estimation of stature based on measurements of stature taken during life and of long bones after death. *American Journal of Physical Anthropology* 16: 79–123

Willey, P. and Falsetti, T. (1991). Inaccuracy of height information on driver's licenses. *Journal of Forensic Sciences* 36: 813–819.

Wilson, R., Hermann, N.P., and Meadows Jantz, L. (2010). Evaluation of stature estimation from the database for forensic anthropology. *Journal of Forensic Sciences* 55: 684–689.

CHAPTER 17

Interpreting Traumatic Injury to Bone in Medicolegal Investigations

Steven A. Symes, Ericka N. L'Abbé, Erin N. Chapman, Ivana Wolff, and Dennis C. Dirkmaat

INTRODUCTION

In a recent examination of forensic anthropology, scientists suggested that stagnation of the field and advancements in DNA technology have made the discipline less applicable to medicolegal authorities and courts of law. Without meaningful change, they posit, this situation could place its practitioners at the precipice of extinction (see Chapter 22 in this volume; Dirkmaat et al. 2008).

Thankfully, the intellectual foundation of forensic anthropology has begun to shift and with it a more practical approach to research has arisen. Causative factors include the US judicial system's increased requirement for incident scene reconstruction, public support of human-rights investigations, and the successful use of anthropological methods in mass disasters and victim identification. The recognition, examination, and interpretation of skeletal trauma in modern human remains forms the cornerstone of this paradigm shift and has become an important element in multidisciplinary death investigations (Galloway et al. 1999; Hart 2005; Johnson 1985; Martrille et al. 2007; Pinheiro 2006; Sauer 1984; Smith et al. 2003; Symes et al. 2002; Symes and Smith 1998).

A Companion to Forensic Anthropology, First Edition. Edited by Dennis C. Dirkmaat.
© 2012 John Wiley & Sons Ltd. Published 2015 by John Wiley & Sons Ltd.

Today anthropologists are held to a much higher standard of credibility than in the past (Dirkmaat et al. 2008). The traditional role of the anthropologist has focused almost exclusively on the biological profile and presumptive identification, but this information is not regularly presented at trial. Instead, the medicolegal system requires evidence that contributes to knowledge of the cause and manner of death, or the postmortem events and timing of a body's disposal. To understand the evolution of bone trauma analysis in both medicolegal and anthropology arenas, a brief history of the field is discussed here.

History and emergence of skeletal trauma analysis

In the last 30 years, anthropologists have written specialized skeletal trauma reports for forensic pathologists, investigators, and lawyers. Despite the need and interest in the discipline, its maturation has been delayed. Why?

There are several reasons for reduced proficiency in the field, including: (i) the manner in which the skeletal elements are traditionally received and processed, (ii) the attitude and personal experience of the anthropologist, (iii) the attitude and personal experience of the forensic pathologist, and (iv) a lack of understanding of bone biomechanics.

Historically, the interpretation of fractured or damaged bones has required a researcher to reference a paleopathology book (e.g., Auferheide and Martin-Rodriguez 1998; Ortner 2003; Ortner and Putschar 1981; Wells 1964). Aside from a few references in medical journals that deal with fractures from a clinical standpoint, basic research on the effects of blunt-force trauma, cutting, stabbing, and sawing with sharp objects, and gunshot impact on the underlying skeletal structures was limited. Additionally, many early scientists and physicians conducted analyses and produced conclusions from osseous tissues subjected to numerous taphonomic agents such as skeletal tissue destruction, animal activity, and burial effects.

The documentation and exhumation of modern human remains was also problematic. Skeletal elements were often haphazardly removed from a site and sent to the laboratory. The lab-based researcher, typically a paleopathologist or anthropologist, had no knowledge of the *in situ* scene and was unfamiliar with the context in which the body was recovered, or how the material was handled from the scene to the lab. Hence, early researchers were unable to distinguish traumatic injuries from taphonomic influences.

Another problem is that many anthropologists adhere to and continue to teach the attitude of autopsy avoidance. Historically, anthropologists were limited to dry bone analysis and did not concern themselves with soft-tissue injuries or with the possibility of providing information on the cause and manner of death. This was viewed to be beyond the expertise of a "bone person." Due to serendipitous research collaborations between anthropologists and pathologists in the USA and abroad, the separation of anthropologists (dry bone) and pathologists (fresh bodies) has changed substantially in the past 30 years (e.g., Kranioti and Paine 2011; Martrille et al. 2007; Pinheiro 2006; Sauer 1984; Smith et al. 1990, 2003).

With the development of forensic anthropology as a distinct discipline in the 1970s, several trained forensic anthropologists found employment in medical examiner's offices throughout the USA. One pioneering force behind this interdisciplinary collaboration was Dr J.T. Francisco, the Chief Medical Examiner in Memphis, TN. In the 1980s, he hired two forensic anthropologists from Dr William Bass' research facility at the University of Tennessee in Knoxville: first Hugh Berryman and later Steve Symes (see Chapter 26 in this volume). Berryman and Symes worked as assistants to medical examiners in Memphis. Their work included conducting scene investigations, assisting at autopsy, and documenting and interpreting all forms of skeletal trauma. Their research in skeletal injuries benefited significantly from observing bone-tissue injuries during autopsy and in further evaluating the fractured bone, both macroscopically and microscopically.

With the synergy of flesh and bone, these researchers discovered common ground in their description of sharp, blunt, and ballistic injuries to bone (Marks et al. 1999; Martrille et al. 2007; Pinheiro et al. 2008; Smith et al. 2003; Symes et al. 1996, 2002). In essence, forensic anthropologists realized that interpreting bone fractures outside of the context in which they occurred was not useful and often resulted in misdiagnosis. Coincidentally, forensic pathologists realized that cases involving traumatic injury benefited greatly from examination of both soft and osseous tissue *in situ* and of the bone itself following soft-tissue removal.

The inevitable connection of soft- and osseous-tissue injuries also led to the examination and further refinement of tool-mark analysis in bone, particularly the possibility of distinguishing between hand and power saws as the offending weapon in cases of dismemberment (Symes 1992). Since that time, anthropologists have contributed substantially to tool-mark analysis and its use in forensic investigations (e.g., Bartelink et al. 2001; Chadwick et al. 1999; Clow 2005; Costello and Lawton 1990; de Gruchy and Rogers 2002; Humphrey and Hutchinson 2001; Lewis 2008; Prieto 2007; Saville et al. 2006; Symes et al. 1998). Research has also focused on implementing better microscopic technology and techniques (Alunni-Perret et al. 2005; Bello and Soligo 2008; Symes et al. 2002; Tucker et al. 2001).

Today, medical examiners and forensic anthropologists often work collaboratively. The National Association of Medical Examiners (NAME; http://thename.org/), which was developed in 1966 and is a means of auditing medical examiner offices in the USA, grants facility accreditation only to those organizations that require a board-certified forensic anthropologist be on call to work with badly decomposed and skeletonized remains.

While the training ground for forensic anthropologists shifted from the bone laboratory and occasional archaeological excavations to the morgue, an increased awareness of the utility of skeletal trauma analysis emerged in other areas and may be attributed to the following.

1. Employment of anthropologists in international human-rights activities. Dr Clyde Snow, a pioneer of human-rights investigations in South America, trained anthropologists to excavate mass graves, analyze skeletal remains to identify victims and detect trauma, and interact with families and communities impacted by human-rights violations. Recent armed conflicts in Eastern Europe, Rwanda, and Iraq led to a need for forensic anthropological skills to recover human remains and associated cultural material (often from mass graves), reconstruct past events, develop a biological profile and, most importantly, establish a chain of custody from the time the remains were discovered to the presentation of evidence in court (Campos-Varela and Morcillo-Méndez 2011; Doretti and Snow 2003; Fondebrider 2009; Kimmerle and Baraybar 2008; Snow et al. 1984; Steadman et al. 2005).

2. Incorporation of archaeological methodology into the forensic sciences. These methods focus on recovering details of the context in which a body is discovered. This perspective, along with participation of a forensic anthropologist at autopsy, has reinvigorated the discipline of human skeletal trauma analysis into a vibrant scientific endeavor capable of withstanding Daubert challenges in court.

3. Improved efforts to document physical abuse of children and the elderly from skeletal remains (e.g., Bilo et al. 2010; Kleinman 1998; Pierce et al. 2004; Schwend et al. 2009). Previously, autopsies of suspected child abuse victims included a visual and cursory radiographic examination of the remains. Healed and recently healed injuries were often missed (Brogdon 1998). Removal of tissues with subsequent macroscopic and microscopic analysis of skeletal elements by a forensic anthropologist has been shown to clarify issues regarding the timing of injuries; for example, healed versus unhealed injuries, and the direction of force of the injury (Maat 2008).

4. Employment of anthropologists following the events of September 11, 2001. Forensic anthropologists were called upon to assist in field recoveries, maintain the chain of evidence in these recoveries, and describe and identify fragmentary and commingled remains. Consequently, additional opportunities for forensic anthropologists were created in the form of full-time jobs in medical examiner's offices (e.g., New York City Office of Cheif Medical Examiner) and military organizations (JPAC-CIL), increased consultation on cases, and research opportunities (e.g., National Institute of Justice-sponsored research).

In this chapter, the origins and future developments of skeletal trauma analysis in the field of forensic anthropology are discussed. As background, we explain the basics of bone biomechanics and then move to a more in-depth discussion of specific classifications of skeletal trauma. Where applicable, case studies are used to explain the importance of combining osteology, biomechanics, observable fracture patterns, and context as a more scientific approach to the analysis of traumatic injury.

SKELETAL BIOMECHANICS

Bone is a heterogeneous material composed of collagen fibers (organic matrix) embedded within crystals of calcium hydroxyapatite (inorganic matrix) (Pearson and Lieberman 2004; Shipman et al. 1986; White and Folkens 2000). The organic matrix provides bone with its elasticity, flexibility, and strength in tension. The inorganic matrix of bone provides it with its rigidity, hardness, and strength in compression (Shipman et al. 1986). This combination of organic and inorganic materials offers bone both strength and elasticity (Pearson and Lieberman 2004), which are important in resisting externally applied forces encountered in day-to-day activities and provides some leeway when extreme stress (i.e., trauma) is placed on the human skeleton.

Medical and anthropology experts often rely on summarizing the work of engineers as a means to explain the biomechanics of bone injury (e.g., Galloway 1999a). However, forensic anthropologists differ from biomechanical engineers with respect to how bone trauma is described and explained. This is primarily due to the different contexts in which they observe traumatic fractures. Forensic anthropologists and forensic pathologists focus on specifics of human skeletal anatomy, the function of the human body, and a description of traumatic bone injuries within a living body. In contrast, engineers focus on the physics of fractures and use experiments to evaluate the material properties and failure points of isolated bone sections outside of their anatomical function. These two professions rarely share middle ground. Trauma assessment reports often appear with either an anthropological or an engineering perspective, but rarely both.

In this chapter, the biomechanical explanation of trauma has been written for a forensic anthropologist. A brief description is provided on: (i) intrinsic and extrinsic forces acting on materials, (ii) the configuration of bone and its response to stress, (iii) bone failure, and (iv) fracture morphology in living and nonliving tissue.

Intrinsic and extrinsic forces on material

Extrinsic and intrinsic forces need to be considered as bone fractures are based on the interplay between intrinsic forces of the body and how they react with externally applied forces. Force is defined here as any mechanical disturbance or load. Force acting on an object can deform it, change its state of motion, or both (Özkaya and Nordin 1999). Extrinsic forces act on the body, whereas intrinsic forces hold the body together and react to the basic biomechanical properties of the skeletal system accordingly.

Several types of directional force, namely tension, compression, torsion, and shearing, have been identified and are suggested to be responsible for certain types of bone injury (Gozna 1982). Tensile forces pull bone apart. Compressive forces, push and squeeze bone material together. Here is one area of disagreement between engineers and anthropologists. Anthropologists do not observe pure tension and compression fractures. The reason is that these forces, while present in a controlled biomechanical study, do not necessarily occur in real-life injuries.

In real-life bone fractures, a combination of tension, compression, shearing, and torsion are often involved in a single fracture event. According to Özkaya and Nordin (1999: 127) "The extent of [bone] deformation will be dependent upon many factors including the magnitude, direction, and duration of the applied force, the material properties of the object, the geometry of the object, and the environmental factors such as heat and humidity." Therefore, the concept of describing an injury based on a single contributing force is not useful in trauma analysis. See Table 17.1 for a summary and definition of the above mentioned forces.

The response of bone to load (strain) is dependent upon the velocity and magnitude of that force. Therefore, the speed at which a load impacts bone can be divided into two categories: slow and rapid. A slow load is described here for demonstrative purposes in terms of kilometers per hour and includes, but is not limited to, beatings, motor vehicle accidents, falls from heights, and airplane crashes. In contrast, a rapid load is described in terms of meters per second and is attributed to ballistic injuries, which in this chapter refers to any physical trauma sustained from the discharge of arms, munitions, or explosives. The recording of the location, appearance, and morphological attributes of fractures are necessary and may provide interpretative clues to speed and direction of load.

Factors involved in producing bone fractures are quite complex such that single or even multiple skeletal fractures may only provide basic clues as to the direction and magnitude of load, the position of the body at the time of injury, and the biological

Table 17.1   Summary of tension, compression, and shear forces on bone. Note the capability of bone to withstand the three types of external loading (information adapted from Hildebrand and Goslow 2001; images reproduced from Hildebrand and Goslow 2001).

Force	Definition	Strength of Fresh Compact Bone	Diagram
Compression	Force is applied toward an object. The object becomes shorter in the direction of the applied force; strain is negative.	$1330-2100 \, kg/cm^3$	
Tension	Force directed away from an object. The object becomes longer; strain is positive.	$620-1050 \, kg/cm^3$	
Shear	Force is applied parallel to a surface but in opposite directions; slides one part of a material crosswise to adjacent parts.	As low as $500 \, kg/cm^3$ = force applied parallel to grain. As high as $1176 \, kg/cm^3$ – force applied crosswise to the bone	

composition of the skeletal elements (Gozna 1982; Gurdjian et al. 1950; Reilly and Burnstein 1974). Definitions of terminology important to the discussion and interpretation of skeletal injuries include (Hildebrand and Goslow 2001; Özkaya and Nordin 1999):

- *force* (or *load*): any mechanical disturbance that causes an object to deform, change its state of motion, or both;
- *magnitude*: the area (and possibly weight, and possibly speed) of the force being applied; the property of *relative size* or *extent*;
- *direction*: the line along which the magnitude of the load travels;
- *stress*: force per unit of area;
- *strain*: the relative deformation (change in length, volume, or angle).

Configuration of bone and its response to stress

Bone is described as heterogeneous, anisotropic, viscoelastic, brittle, and weak in tension (Gozna 1982; Reilly and Burnstein 1974). Bone is heterogeneous in the following ways: shape and configuration (e.g., long bones, cranial bones, hand bones); structure within a single bone according to location (e.g., diaphysis versus metaphysis in the ratio of cortical and trabecular bone); and bone cell types and their arrangement in cortical and trabecular bone (Gozna 1982; Pearson and Lieberman 2004).

Bone is described as an anisotropic structure because it responds differently to loads according to the direction of load and the impact location on the material (Gozna 1982). The configuration of the structure itself also affects the response of the bone. Due to the typical construction of the shafts of long bones in which collagen fibers primarily run longitudinally, long bones are able to resist axial loads much better than loads applied transversely to the shaft (Frankel and Nordin 2001; Gozna 1982).

Viscoelastic properties include both the viscosity and elasticity of active bone tissue and its relationship to the speed of a load (Gozna 1982; Harkess et al. 1996; Frankel and Nordin 2001; Özkaya and Nordin 1999). When a slow loading force is applied to bone, it can respond by returning to its original shape after the force is removed (elastic deformation), permanently deforming (plastic deformation), or fracturing. However, under rapid loading, such as ballistic trauma – referred to in this chapter as any form of sustained impact force from the discharge of a firearm or munition – bone resists the force until it shatters. When evaluating the mechanism of a bone injury, material properties of bone may differentiate loads on bone.

Because bone is brittle, it may fracture differently than other surrounding tissues. Because it bends only slightly before breaking, and because it is viscoelastic, bone is important for evaluating shock loads. Brittleness is directly attributed to bone's high mineral content. It will break before other tissues because it is a poor absorber of shock waves and rapid loads.

Bone demonstrates a weakness in tension when subjected to bending loads. The elastic, or collagen (organic), component of bone is responsible for the yielding and ductile nature of the material while the inorganic component comprises approximately 65% of bone, making it nearly twice as strong in compression as it is in tension (Frankel and Nordin 2001; Harkess et al. 1996; Hildebrand and Goslow 2001; Özkaya and Nordin 1999). When stressed, bone is stronger in compression, failing first in tension. This axiom forms the basis for interpreting most bone injuries.

Bone reacts to stress uniquely because of the abovementioned properties. Taken as a whole, these attributes can create a variety of responses to stress from mechanical loads on human bones. For this reason, the specific bone, location of injury on the bone, type of bone affected (e.g., spongy versus cortical), an individual's age, and the presence of prior pathological conditions provide important implications for discussing bone's resistance to injury as well as the mechanism of that injury (Frankel and Nordin 2001; Harkess et al. 1996; Özkaya and Nordin 1999; Reilly and Burnstein 1974).

Bone fractures and speed of a force

Living human bone will fracture when it is subjected to abnormal loads that are beyond the capacity of the bone to resist, bend, and return to a normal shape. Biomechanists routinely use Young's modulus of elasticity (stress/strain curves; Figure 17.1) to explain the response of materials, such as soft steel, glass, and aluminum, as well as wet and dry bone, to different stresses. Elasticity is defined as the capacity of a material to completely return to its original shape after a load has been removed. In the elastic phase, all deformations are recoverable upon the removal of extrinsic forces. In contrast, plasticity is defined as the threshold at which the elastic limit has been reached and at least some permanent deformation occurs (Frankel and Nordin 2001; Hildebrand and Goslow 2001; Özkaya and Nordin 1999). If the stress

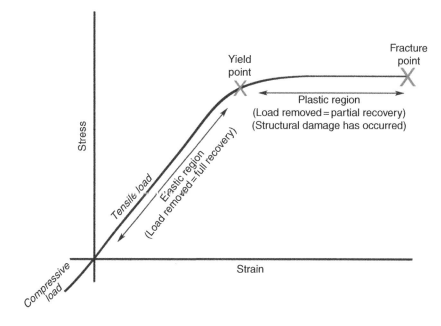

Figure 17.1 A generalized stress/strain curve created for illustration and descriptive purposes (not associated with any particular type of material). The small x represents the yield point and transition from the elastic phase to the plastic phase. The larger X represents the point of complete failure.

is removed and the material returns to its normal shape, it was stressed within its elastic limit (Frankel and Nordin 2001; Gozna 1982; Reilly and Burnstein 1974). However, if the load is removed and the material remains mis-shapen, it falls within the plastic phase. In this phase, the material is compromised microscopically and applying additional load can cause it to completely fail; that is, fracture.

The most important determinant to the morphology recognition of a fracture is speed. As decribed above, we can illustrate this concept by examining the impacting object in terms of speed, either slow (kilometers per hour) or rapid (meters per second). With respect to moving objects, kinetic energy is equal to one half the mass (m) of the object multiplied by the velocity of the object (v) squared ($E_k = \frac{1}{2}mv^2$). If one compares bone fractures from slow and rapid loads, differences in fracture morphology become apparent and can be viewed with the naked eye, or under low magnification. Bone is better able to resist rapid-loaded than slow-loaded forces. The reason is that slow-loaded forces put bone under stress for a longer period of time than a rapid-loaded force. With a slow-loaded force, the bone is physically exhausted as it endures both an elastic and plastic phase prior to failure. In contrast, bone impacted at a high rate of speed resists to a point and then shatters, with little or no plastic deformation demonstrated.

The basic biomechanical principle of rapid versus slow loading provides the foundation through which sharp and blunt injuries (kilometers per hour) are differentiated from ballistic (meters per second) trauma in bone. A slow-loaded force (kilometers per hour) permits bone to respond and compensate for an increase in stress. When the

Table 17.2 Description of different types of bone fracture (Gozna 1982; Tencer 2006).

Fracture	Description
Transverse	Bone fractures across the diaphysis of the bone at 90° angle. A pure tension fracture occurs in experimental situations but not in real-life injuries.
Oblique (transverse)	Bone fractures at approximately a 45° angle, consequence of bending and compression.
Butterfly	Bone fractures as a result of tension, compression, and bending forces. Bone is stronger in compression then tension. Bone breaks into two pieces with one triangular fragment and two segmental fragments.
Spiral	Bone experiences extreme torsion. The fracture encircles the bone shaft and creates an oblique-like break around the bone axis.
Comminuted	Bone fractures into two or more pieces.

stress is removed, the bone may: (i) return to its original shape, (ii) remain deformed (plastic deformation), or (iii) fail. Within these circumstances, bone acts as a viscoelastic material. In contrast, bone is highly resistant to a rapid-loaded force (meters per second). In this instance, bone does not experience bending or adjustment: it merely shatters due to energy absorbance (Gurdjian et al. 1950; Özkaya and Nordin 1999; Reilly and Burnstein 1974). During rapid loading, bone reacts in a way more similar to glass, or other brittle substances, than to viscoelastic material.

The types of fracture that can be produced are also dependent on the direction load. In general, it has been demonstrated that specific types of load will produce characteristic fracture patterns. Several of these fractures have been described and include transverse, oblique, butterfly, spiral, and comminuted fractures (Gozna 1982; Tencer 2006) (see Table 17.2).

Fracture morphology in wet and dry bone

An important component of trauma analysis is the ability to estimate when the traumatic event(s) occurred. A deciding factor in this process is often based on the composition and decomposition of bone tissue. After death, the loss of organic material in bone affects its ability to resist stress. For this reason, fresh or "wet" and dry (decomposed) bone present different fracture morphologies.

If stress exceeds strain, dry bone fractures immediately. Dry bone may resist stress to a higher level, but failure occurs immediately at the end of the elastic phase or the beginning of the plastic phase. Wet bone can pass through an elastic and plastic phase prior to failure (Özkaya and Nordin 1999: 210).

Postmortem fractures are unique and can be associated with either drying or weathering of bone or willful/accidental damage. When skeletal remains are exposed to sun, air, soil wind, and fire, they lose moisture and collagen fibers. As a consequence to this loss of the organic components, bones begin to shrink and warp. During the process of shrinking, numerous cracks in the outer cortical layer of bone occur, producing the classic mosaic appearance (patina), and eventually pieces of the external cortical bone layers will flake away. Longitudinal cracks may also appear in the bone and may be confused with a fresh break to the untrained eye. However, postmortem cracks appear dry, follow the grain of the bone, and, in some instances, there may be a difference in color between the outer and innermost layers of bone.

An important point to remember is that anthropologists and medical examiners differ considerably on the use of perimortem and postmortem terminology (Nawrocki 2009; Symes et al. 2002). A pathologist defines the perimortal period as the process of death, which may be from injury and/or disease to the cessation of somatic life. In contrast, the anthropologist evaluates the response of bone tissue, fresh or dry, to an external mechanical stress (e.g., blunt, ballistic, etc.). When an anthropologist uses the term perimortem, they are implying that an external stress had been placed on wet or greasy (fresh) bone. Due to varying circumstances in decomposition rates, this can extend for an indefinite period of time and clearly overlaps into the clinical postmortem period (Nawrocki 2009; Symes et al. 2002). The use of discipline-specific terminology needs to be clarified in a trauma analysis report, especially when an anthropologist and forensic pathologist may testify on the same evidence.

IDENTIFICATION OF SKELETAL TRAUMA

Skeletal trauma is defined as a modification of bone based on a slow- or rapid-loaded impact with an object. Accurately interpreting the role humans play in the creation of traumatic injuries is key for the forensic anthropologist. This interpretation requires a detailed analysis of fracture morphology and an understanding of the context of the scene in which the body was recovered.

In writing a comprehensive skeletal trauma analysis, the analyst must differentiate the effects of human-induced modifications, such as medical and criminal, from modifications associated with environmental factors attributed to animals, decomposition, and fire. These are defined as taphonomic agents and should be discussed separately from human-induced trauma. The recovery of human remains and evidence in context is described in Part II of this volume.

For the past 20 years, the first author has conducted research on extrinsic traumatic injury to bone and has classified these injuries according to frameworks that are valuable to forensic pathologists during autopsy and which eventually contributed to medical examiners'/coroners' interpretations of cause and manner of death (COD, MOD). Within this framework, three major classes of human-induced skeletal trauma are defined in Table 17.3. The classifications include blunt-force trauma (BFT), sharp-force trauma (SFT), and gunshot trauma (GST). Descriptions of bone modifications associated with past trauma (healed defects) are also discussed as differential patterns of healing can be used to interpret potential cases of abuse or torture. Burning or thermal destruction of soft and osseous tissue is a common method employed by perpetrators in an attempt to conceal perimortem injuries and identification. Therefore, postmortem bone modification related to thermal destruction is an important research avenue in trauma analysis as it permits the anthropologist to accurately interpret a fire-altered death scene and the fatal-fire victim.

Much information can be obtained from the analysis of traumatic injury in bone (e.g., Baraybar et al. 2008; Symes et al. 1998; Brogdon 1998; Galloway et al. 1999;

Table 17.3 Three major classes of skeletal trauma with descriptions of the types of force involved as well as morphological manifestations.

Trauma Types	Sharp	Blunt	High Velocity (Ballistics)
Impacting Action (Load)	1. Stabbing 2. Cutting 3. Sawing	1. Impact with a blunt object (no straight line or incised wounds) 2. Automobile accidents 3. Fall from heights	1. Gunshot wounds 2. Explosives
Load Velocity Expected Fracture Types	Slow (km/h) 1. Impact (with at least one incised, or straight line cut, edge) 2. Radiating 3. Concentric (in) 4. Tension/ Compression (butterfly fracture)	Slow (km/h) 1. Impact 2. Radiating 3. Concentric (in) 4. Tension/ Compression (butterfly fracture)	Rapid (m/s) 1. Plug and spall 2. Radiating 3. Concentric heaving (out)
Signature Observations	1. Straight line incisions 2. Simply blunt force trauma with a sharp object	1. Delamination of bone 2. Plastic deformation (fractures as a bone in layers) 3. Bevel (internal)	1. Fractures as a uniform material 2. Bevel (internal and external) 3. Termination points of pre-existing fractures (due to the speeds at which fractures travel).

Kleinman 1998; Maat 2008; Martrille et al. 2007; Pinheiro 2006; Sauer 1984; Smith et al. 2003). More importantly, in medicolegal cases of skeletal trauma all the bones need to be considered as evidence. For the best and most accurate results, trauma analysis needs to be viewed as a team effort that involves both the forensic pathologist and the forensic anthropologist in the collection of data from the body. This evidence, along with consideration of the context, provides a very effective and court-defensible reconstruction of past events.

BLUNT-FORCE TRAUMA

Blunt-force trauma (BFT) is the most common form of injury observed at autopsy (Spitz and Spitz 2006). BFT results in abrasions, contusions, lacerations, and/or bone fractures. It can represent a single event that led to death such as a significant impact from an object or a fall from height, or it can represent a series of events from which the individual had recovered that is indicated by evidence of healing bone. If the individual recovered from multiple traumatic episodes, as evidenced by healing sites on multiple bones, a pattern of abuse or torture may be indicated.

BFT is difficult to assess because of: (i) the multiplicity of ways BFT characteristics appear on a body, such as falls from heights, vehicular accidents, and external blows, etc.; (ii) the variety of objects that impact and fracture bone; and (iii) issues related to whether the body or a body part was supported or unsupported during impact.

Forensic pathologists focus primarily on the soft-tissue modifications associated with BFT, which often gives the best information as to the event frequency, tool, or mechanism used to produce the injuries (Spitz and Spitz 2006). If the force to the body is significant, the underlying skeletal structures may be impacted and fractured. In certain situations, the circumstances of the incident and the context of the scene can lessen the need for a detailed trauma analysis. However, in cases of unexplained deaths, child and elderly abuse, and situations of torture, the forensic anthropologist with significant training in trauma analysis can make a contribution to the determination of cause and manner of death (e.g., Maat 2008; Pinheiro 2006; Symes et al. 1996).

How does bone break with BFT?

BFT is defined from a biomechanical standpoint as a slow-loaded impact to a focal area of bone. The focal area may vary from broad to small. Although the literature is replete with examples of objects that produce blunt trauma such as baseball bats, gym weights, bricks, and hammers (Byers 2002), the key concept in defining BFT is not the object that was used to impart the stress, but rather the amount of kinetic energy that was transferred to the bone from that object. The more kinetic energy transferred from the impacting object to the bone, the more damage occurs to the area. For example, the baseball player Albert Pujos can swing a bat powerfully and the resulting damage would be more significant than that of a less agile person. However, regardless of who is wielding the bat, the velocity, for teaching purposes, can be visualized in terms of kilometers per hour and will never reach speeds of ballistic trauma, which cannot be visualized and are expressed in meters per second.

Additional characteristics to consider in blunt-force injuries include the size of the contact surface area and the bone's ability to resist and absorb the impact. A large contact surface area will permit dispersion of the kinetic energy through the material, resulting in less damage than would be the case when the surface area of contact is small. For example, the curved surface of the human skull results in a limited contact surface area. Despite the added strength to a curved surface, the kinetic energy transfer to a smaller focal point may results in more damage than would be found on a flat surface.

A bone's ability to disperse kinetic energy and resist fracture is dependent upon a rapid and even distribution of that force (Frankel and Nordin 2001). With respect to humans, bones of younger individuals are generally more resistant to fractures because of the amount of cartilage and organic material in the long bones (epiphyseal plates) and the cranium (membranous tissues between the cranial elements) as compared to the amount of inorganic material. On the other hand, general aging and disease processes in the elderly typically result in thinner cortical bone, alterations in the organic-mineral content of bones, and the fusion of bones (e.g., cranial suture synostosis) that results in a lessened ability to resist fractures from impact injuries. In other words, the bones of younger individuals are more elastic than those of older individuals (Pearson and Lieberman 2004).

General analysis of blunt-force injuries and tool marks

First, the analyst determines whether the damage to the bones indicates BFT or GST. This is based on fracture morphology and possible evidence of bone deformation on the skeletal remains (e.g., Berryman and Symes 1998). In order to make this decision, the researcher needs to perform macroscopic, microscopic, and radiographic examinations that include the following general procedures.

First, cranial or postcranial fragments need to be fitted together and conjoined with nonpermanent glue or tape. Provide an illustration of major and minor fracture morphology as a sketch. Next, macroscopic analysis includes a description of fracture propagation, fracture length (combined lengths of all fractures exhibited), and characteristics of the fractures in cross-section. These patterns may be illustrated on an outline drawing in multiple views. The diagrams will be useful for estimating the minimum number of impacts.

Finally, microscopic analysis involves the use of a low-magnification (power × 3–40) stereoscope. Low magnification is useful in determining impact site(s), areas of tension and compression, as well as sequence of blows. Microscopic analysis may also yield other physical evidence left within the wound, such as fibers, hair, metal, or pieces of the implementing tool. Most importantly, the researcher needs to examine the internal and external areas of fracture propagation.

Once the point of impact is identified, a careful analysis, including a microscopic examination of the area around the impact site, may provide evidence of tool impact and even an impression of the tool. If possible, determine the general class of the object used.

There is a common misconception that scanning electron microscopy (or SEM) is necessary for the proper analysis of skeletal trauma. However, it is the authors' strong opinion that only a stereomicroscope with low magnification is necessary to observe all significant fracture characteristics. In fact, a great deal of information is lost when using high magnification such as with scanning electron microscopy. When the above-mentioned steps are completed, it may be possible to assess:

- the point of impact for each blow;
- the minimum number of blows;
- the sequence of impacts;
- and, occasionally, the general class of tool or implement used.

General analysis of blunt injuries: documenting tool marks in BFT

Determining the tool or instrument used to impact bone in BFT is neither easy nor in many circumstances possible. In some instances, the tool will come into direct contact with the bone and leave a mark on the bone. In other cases the force of impact is sufficient to penetrate the outer cortical bone layer. In these cases it might be possible to determine an approximate angle of impact.

Often anthropologists go "beyond the data" and attribute marks on bones to a specific instrument. Another common practice involves fitting the suspected specific tool into the (actual) bone defect. Not only can this be misleading (it may represent a hammerhead impact, but not necessarily the specific hammer found at the scene), but may introduce additional defects to the bone that may be later misinterpreted (e.g., under magnification) and is definitely not recommended.

A better approach is to suggest that the mark is consistent or not consistent with a particular instrument in question (Berryman and Symes 1998). This can be done by determining the impact site(s) and evaluating whether the impact surface area was broad (diffuse) or focused (focal). A positive cast of both the suspected tool and the tool mark (in the bone) should be manufactured to make direct comparisons without further compromising the bone.

BFT to the cranium

The human cranial vault is a rounded structure comprising three layers: an outer and inner layer of cortical bone with a layer of spongy (diploë) bone sandwiched between (White and Folkens 2000). In addition to being curved, cranial bone thickness varies across the vault. In the frontal and occipital bones, a thicker layer of bone, or buttressing, is present. Despite this complexity, the biomechanics and characteristics of BFT trauma to this region of the body are somewhat predictable.

At the initial point of impact, compressive forces on the external vault force a portion of the cranial bone inward, sometimes crushing the diploë, but often creating tensile forces on the internal table (Özkaya and Nordin 1999). Since bone commonly fails in tension before compression (Figure 17.2), fractures first occur on the endocranium (e.g., Frankel and Nordin 2001; Özkaya and Nordin 1999). With enough energy, fractures radiate from the impact site and create wedge-shaped bone fragments (Kroman et al. 2011). Radiating fractures can be followed back to the impact site (Figure 17.2, bottom). Concentric fractures may develop between radiating fractures, and thus circumscribe the impact site (Figure 17.2, top).

Concentric fractures are attributed to a change in the location of compressive and tensile forces on the bone, due to initial failure. Concentric fractures initially fail externally in tension and compressive forces appear on the inner table just distal to the point of failure (see Figure 17.2, bottom). The cross-section morphology of concentric fractures is different from radiating fractures (Berryman and Symes 1998; Hart 2005). In order to accurately assess fracture characteristics in the skull, the endo- and ectocranial vaults need a meticulous examination.

Fracture propagation often follows the path of least resistance and is diverted away from areas of heavy buttressing, such as the midline frontal squamous, the petrous portion of the temporal bone, and the mastoid process (Berryman et al. 1991; Berryman and Symes 1998). Instead, fractures tend to traverse through adjacent orbits on the side of the cranium.

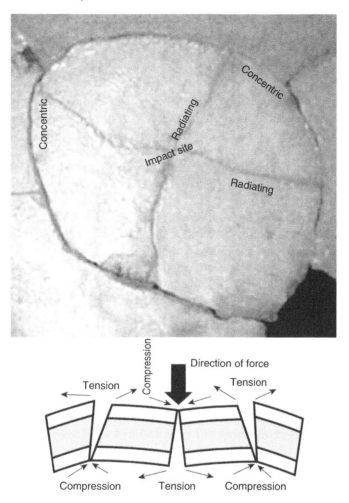

Figure 17.2 BFT cranium. Top: wedge-shaped plates caused by radiating and concentric fractures. Bottom: cross-section illustration of tensile and compressive forces at the point of impact (radiating fractures) and with subsequent concentric fractures.

Blunt-force injuries: determining sequence of impacts

The concept of fractures following the path of least resistance is useful in determining sequence of BFT impacts in cases involving two or more blows. The first impact site commonly results in radiating fractures which traverse in multiple directions and which end when the kinetic energy has dissipated. As mentioned above, concentric fractures may also be produced when enough kinetic energy is transferred to the bone. When a second impact to bone occurs, Puppe's law of sequence, or the phenomenon of intersecting fracture lines, can be applied to determine fracture order (Symes et al. 1996). Following Puppe's law of sequence, radiating fractures from a second impact site continue until they intersect a previous fracture, whereby most of the kinetic energy is dissipated (Berryman and Symes 1998; Smith et al. 1987; Symes et al. 1996). In other words, the fracture from the second impact will terminate in a previously existing fracture. By carefully documenting all of the fractures and looking for a few key attributes, one can determine the sequence of multiple blows.

BFT to long bones

Shafts of human long bones are tubular in shape and consist of relatively thick cortical bone with little or no spongy bone. When force is applied in the transverse plane, more or less perpendicular to the shaft, compressive forces occur on the side of the bone that is impacted. The opposite side of the tubular structure is then subjected to tensile forces (e.g., Frankel and Nordin 2001; Özkaya and Nordin 1999). Again, bone resists compressive forces better than tensile forces and fails first in tension on the opposite side of the impact site.

The transition from tension, through shear, to compression redirects the fracture(s) because compressed bone inhibits a continuous transverse fracture. The fractures are redirected around the area of compression and the cortical bone until they reach the original side of impact, which is recognized as the concave side of the fracture (Berryman et al. 1991). The cross-sectional morphology of these fractures is reminiscent of a butterfly and they are colloquially described as "butterfly fractures." Butterfly fractures are diagnostic of BFT in long bones. Characteristic microscopic features of a butterfly fracture include a bone tear, a breakaway spur or notch, possibly minor fracture lines, and an area of shear between tension and compression. Bone tears can be noted on the tension side and appear mottled and billowy. Break-away spurs are jagged in appearance; one bone fragment exhibits a "dog-eared" notch while the other piece retains a bone extension. This area of bone is where compressive forces reach its maximum force. Unless strong evidence is present to clearly define the point of impact, the key concept in analyzing butterfly fractures and determining direction of impact is to carefully analyze and interpret the areas of compressive and tensile stress within the bone, and the direction in which the bone was bending (concave surface) prior to failure (e.g., Berryman and Symes 1998; Symes et al. 1996).

Special case: fractures of the hyoid

A controversial topic in forensic pathology and anthropology is the interpretation of damage to the hyoid bone. In many skeletal cases, if the hyoid is fractured it is often assumed to have resulted from strangulation. This may not be the case. Fractures of the hyoid bone and thyroid cartilage have been documented in cases of strangulations, hangings, nonlethal throttling (Ubelaker 1992), sports injuries, motor vehicle accidents, and profuse vomiting (Dalati 2005; de la Grandmaison et al. 2006; Gross and Eliashar 2004; Gupta et al. 1995). While it is accepted that a direct force to the anterior aspect of the neck is needed to fracture this bone, the location of these fractures does not differ between the various abovementioned mechanisms of injury. Currently, without soft tissue or other contextual information a fracture of the hyoid bone can only indicate that an injury occurred to the anterior neck.

Summary of BFT

Evidence for BFT impact(s) sometimes includes identification of a point of impact, as well as radiating and concentric fractures. The presence of plastic deformation is a key feature that is used to indicate slow-load BFT. As mentioned above, velocity of impact, as opposed to the shape of the offending tool, is usually the most telling characteristic in BFT analysis. Obviously, determining a specific offending tool would be preferred, but it is rarely possible. When forensic specialists become focused on determining the class of the tool rather than the biomechanics of the fractures, the trauma assessment runs the risk of overanalysis and error.

Special Class of BFT: Evidence for Abuse/Torture

The role of the forensic anthropologist in the medicolegal assessment of child abuse focuses on a description and analysis of acute and healing bone fractures for indications of biomechanic forces and timing of injuries. A forensic anthropologist's knowledge of the processes of bone fracture and bone healing are key components in the assessment of possible child and elderly abuse cases. Considerations here include:

- the number and location of acute traumatic fractures;
- the number and location of healing or healed injuries: often multiple and repetitive;
- the context of the trauma episode (indoor versus outdoor, other individuals present, etc.).

Child abuse

Cases of suspected child abuse are difficult to identify and even more difficult to prove beyond reasonable doubt. Forensic pathologists, coroners, and anthropologists are obligated to ensure there is a strong case for abuse whenever pursuing suspected incidents of abuse. In most cases of child death involving injury, the caregiver often suggests that the injuries resulted from an accident or rough play with older children, for example. If only one injury (fracture) is involved, it is difficult to rule abuse. However, when findings during the postmortem examination include a combination of soft-tissue damage with unhealed and healing lesions on the bones, which cannot be attributed to car accidents or other types of serious injury, or the death event itself, a child abuse case is highly plausible. In these cases, evidence of a healed or healing injury on the bone provides a good record of previous injury and, perhaps, repetitive assaults.

Physiology of bone fractures

Since bone is a living tissue, significant trauma that results in the fracturing of bone will also be associated with damage to soft tissues, including muscles and blood vessels. While this is not macroscopically observable, these processes may be assessed microscopically. Traumatic injury to bone also elicits a nonspecific inflammatory response (Sherwood 2006). After a bone fractures, mast cells release histamine that induces a localized vasodilation of the arterioles. The consequence is an increase in blood flow to the fracture site, which includes more phagocytic leukocytes (white blood cells) and plasma proteins. The capillaries' permeability increases to permit plasma proteins that normally remain isolated in the blood to escape into the inflamed tissues. Ultimately, the accumulation of this leaked plasma raises the interstitial fluid-colloid osmotic pressure (Sherwood 2006) and, together with the increased capillary blood pressure, results in a localized edema. All of these responses will cause the characteristic redness, heat, pain, and swelling of soft tissues around the site of a fracture (Sherwood 2006). Basic knowledge of physiological and histological healing processes is important for estimating a time for bone fractures and bone healing.

At the site of injury, the neutrophils are the first to arrive followed by the slower-moving leukocytes and monocytes, the latter of which mature into macrophages upon reaching the fracture. These phagocytic cells, along with the phagocyte-secreted chemicals, mediate the inflammation response. The tissues around the fracture site are cleared to allow tissue repair to take place. Along with inflammation, clot formation also occurs as a consequence of damage to the blood vessels at the fracture site.

Repairing bone fractures

Macroscopically, the repairing of injured bone can be described in three sequential phases: inflammatory, reparative, and remodeling (Ogden 1991). With respect to duration of the phases, healing rates are faster in growing infants than in children, and faster in children than in adults (Pierce et al. 2004).

Inflammatory phase

During this period, a fibrocartilaginous callus is formed around the injured area. In adults, cellular phagocytosis, osteoclastic activity, and the deposition of bone spicules have been noted 2–10 days after injury (Maat 2008: 246). This cartilaginous structure is the framework upon which woven bone will be laid down. In approximately 6 weeks in adults, a large amount of woven bone is deposited onto this lattice framework, such that an external callus is visible (macroscopically and radiographically) on the skeletal structure (Barbian and Sledzik 2008; Maat 2008). In a child, this callus may be visible as early as 2 weeks after injury (Ogden 1991; Pierce et al. 2004).

Reparative phase

In the reparative healing phase, the circumferential woven bone around the injured area serves to stabilize and aid in repair of the injured site. Within the internal callus, if the defect is not greatly displaced, bridging of bone across the fracture defect will commence. During this phase, the fracture may be described as being in *clinical union*, in which manipulation of the bone would indicate a relatively stable condition. In most adults, within 2–3 months the bones will be united and the callus stabilized in size (Maat 2008).

Remodeling phase

The last, and slowest, of the general healing phases is remodeling. Remodeling occurs when the unorganized bone in the callus is reorganized into normal cortical bone via osteoclastic and osteoblastic activity. In other words, the quickly laid-down woven bone (Phases 1 and 2) is replaced by organized lamellar bone. In adults, a remodeled callus may be observed for years after the initial injury. On the other hand, a healing callus may last only a few months in children due to the rapid healing abilities of immature bone. The success of the resorption and smoothing of the callus process is

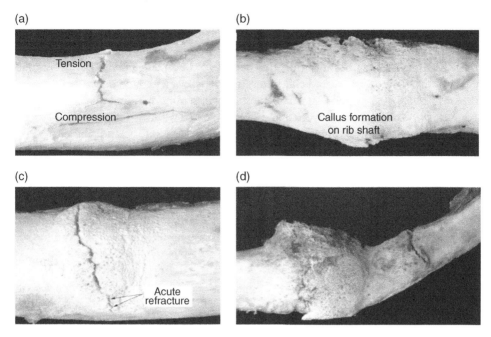

Figure 17.3 Different stages of healing of rib fractures in children. (a) No macroscopic healing. An incomplete butterfly fracture formed with external compression (concave bending) on rib 3. (b) Callus formation on the internal surface. Macroscopically, the rib is within clinical union (rib 7). (c) Callus formation with a refracture (internal surface). The bone was in clinical union but additional injuries have caused destabilization. (d) Healing callus with multiple refractures (internal surface). The callus is far larger than the original bone shaft, which may suggest numerous episodes of refracture and misalignment.

based on the subsequent patterns of stress that are placed on the skeletal element. If the area is immobilized and no unique stress placed on the site, remodeling will proceed to a point whereby the area is stronger than it was prior to the fracture. If the healing process is interrupted by another bout of significant stress, new fractures may result. The three stages of healing, with corresponding photographs of unhealed, partially healed, and refractured child ribs, are presented in Figure 17.3.

Rate of bone healing

Besides age, the location of the fracture, and the type of bone that is fractured (e.g., cranial, long bone, flat bone) has a major influence on bone healing. The rate at which these sequential stages occur is also dependent on sex and disease profile of the person (Bilo et al. 2010; Kleinman 1998; O'Connor and Cohen 1998; Maat 2008; Schwend et al. 2009). The age of the individual is the most important variable for evaluating healing rates in bone, especially during early childhood. During infancy and childhood, there is a high rate of osteogenic activity in both the periosteum and endosteum. With each year of childhood, cellular activity becomes progressively less active until reaching young adulthood. After this point, the rate remains relatively constant from early adult life to old age. This slow decline in cellular activity is directly related to slowing in the rate of healing in adults compared to children.

In general, adults exhibit early phases of healing around the injured area within 4 days to 8 weeks after the injury. Resorption of the initial fracture line, the creation of a soft callus (4–7 days) and the deposition of woven bone within this callus (Bilo et al. 2010; Kleinman 1998) occur within this period of time. An example of differential healing rates between adults and children can be illustrated with comparable fractures at the mid-shaft of the femur. At birth the shaft will reunite within 3 weeks; at 8 years of age this process will take 8 weeks; at 12 years old, 12 weeks is required; and at age 20, the process will be completed within 20 weeks (O'Connor and Cohen 1998; Salter 1980: 190).

Skeletal evidence of child abuse

Both infants and children have lower bone mineral content than adults, a condition that makes their skeletal structure more resistant to fracture, especially comminuted injuries (Pierce et al. 2004). Studies have shown that fractures due to common accident scenarios, such as falls from heights less than 170 cm, are very rare in children under 4 years of age (Kleinman 1998). In this age group, a skeletal fracture is likely to be indicative of much stronger external stressors, which are often associated with abusive activities inflicted upon the child (Brogdon 1998; Kleinman 1998). Continuing in this vein, multiple and repetitive acute fractures in different locations of the body are extremely uncommon and, regardless of the presumed context of the injury, are clear indications of abuse.

Three contextual clues may be used to establish a case for child abuse distinct from accidental injury:

1. the type and location of the fracture(s);
2. the age and development of the child; and
3. the manner in which the injury was sustained (Bilo et al. 2010; Kleinman 1998; Schwend et al. 2009).

Injuries associated with child abuse are most commonly found in the skull, the chest (ribs), and the long-bone diaphyses and growth plates (Bilo et al. 2010; Brogdon 1998; Kleinman 1998). In infants less than 1 year of age, traumatic injury is often focused on the rib cage, metaphyseal joints, and the cranium, whereas cranial and long-bone fractures are more commonly observed in older children (Kleinman 1998). This difference can be attributed to the manner in which the injury is delivered. For instance, an infant is more likely to be shaken and/or squeezed at a young age, whereas pulling, twisting, or punching is more commonly noted in abuse cases of older children. This evidence of trauma is often associated with other issues of neglect such as poor dental health, poor milestone development, and a lack of growth in the child (Bilo et al. 2010; Kleinman 1998; Schwend et al. 2009; Walker et al. 1997). In essence, an abused child is often not provided with an opportunity to thrive in their environment.

One key attribute in child abuse cases is the finding of multiple injuries in different stages of healing. Repetitive injuries can also be seen as disruptions in the healing process on the same bone and in the same area, such as a fracture within a healing callus (Figure 17.3c and d). Even in cases of clinical union, it does not necessarily mean that the bone healed to its original strength (Ogden 1991). With continual insults, an acute and unstable refracture may occur and cause disruption to the original callus in the healing process. In Figure 17.3c, a fully formed callus with a refracture is shown. The rib appeared to have been in clinical union before it was destabilized with another assault to the area.

With continual refracture to the same structure, the healing callus continues to remodel, expands in size beyond that of the original callus, and becomes considerably larger than the normal circumference of the bone shaft (Figure 17.3d). The reactive bone tissue continues to lay down new bone within the injured callus as a means to protect the integrity of the structure. Thus the relative size of the callus provides good evidence of bony response to repeated injury.

The forensic anthropologist also needs to be cognizant of nonabusive skeletal modifications which mimic abuse injuries. These include those resulting from birth trauma, metabolic disorders (vitamin C and D deficiencies), normal periostitis (especially in 6–8-month-old infants), anomalies of collagen formation (*osteogenesis imperfecta*), tumors, and leukemia (Bilo et al. 2010; Kleinman 1998; Schwend et al. 2009; Walker et al. 1997).

Documenting child abuse

Documentation and analysis of child abuse includes macroscopic analysis, radiographic examination, and low-magnification microscopic analysis. Following the autopsy and forensic pathological examination of the victim, a thorough description of the soft-tissue damage, and comprehensive radiographic series, the area(s) exhibiting healed and unhealed injuries are removed from the body and macerated. The following analyses can then be completed:

- note the specific location of the injuries to each bone affected via a skeletal diagram (macroscopic);
- examine the lesions from multiple perspectives via fine-grained radiographs;
- record the general stage of healing for the fractures;
- attempt to provide a biomechanic explanation for the trauma;
- attempt to determine the *relative* timing of the trauma in healing bone, using extreme caution.

Case study 1: child abuse

In the fall of 2009, the first author (SAS) was contacted to assist in the examination of the rib cage of a 19-month-old child who was suspected to have been a victim of child abuse. As noted in the forensic pathological examination, insults to the chest had been repetitive, excessive, and severe. In fact, six traumatic bone injuries in different stages of healing were noted on a single rib. Many bony calluses had been refractured and had become grossly enlarged (Figure 17.3).

The case study is used to demonstrate the manner in which an anthropologist may be able to establish both a pattern and general timing of repetitive injuries by focusing on incompletely healed bone associated with either acute and/or completely healed fractures. It is also used to demonstrate repetitive abusive behavior. In order to observe and document these injuries, the skeletal tissues were culled and the soft tissue removed. The rib cage was examined macroscopically, through radiographs, and a visual assessment was completed microscopically with a Leica MZ16A stereomicroscope. The chest was retained in its anatomical order, with meticulous slow processing that removed the tissues from the bone shafts, but left the bone joints intact. After the chest was examined and photographed in this state, the bones were disarticulated for close examination under magnification and additional radiographs. It is recommended to disarticulate and refrain from cutting a bone in autopsy, especially those bones that may be injured or previously injured and frail.

The chest in young children is susceptible to axial loading and when bending, compression, and tensile forces are placed on the rib cage and butterfly fractures often occur due to BFT. The sternal end of the ribs – which contain a greater proportion of spongy to cortical bone – provide an exception to this rule. In this case, buckling fractures – or initial failure of the bone under compressive stress – often appear (Love and Symes 2004).

Analytical results
A total of 45 injuries were recorded on the ribs of this individual and included 35 partially healed and 10 acute fractures. The various injuries noted on the rib cage were described and classified as follows: acute fractures without healing ($n = 10$), healing injuries without refracture ($n = 24$), and healing injuries with refracture ($n = 11$). No injuries were observed on left ribs 1, 2, and 12 as well as right ribs 1 and 2. The acute fractures, which exhibited no evidence of healing, can be described as tension/compression bending fractures of either the rib shaft/head or an axial load to the costal cartilage junction.

When discussing age of an antemortem injury, it is necessary to realize that calluses form very quickly in children, often cited in the literature as less than 2 weeks (Ogden 1991: 75). The continual abuse endured by this toddler over a period of weeks or months was extreme and may be considered tantamount to torture. However, the legal definition of "torture" is variable and is often tied to other issues, so the anthropologist may be better off avoiding the term unless they are prepared to defend the concept.

Other forms of abuse
In the case of elderly abuse, knowledge of context is imperative as the aged are more likely to suffer from degenerative bone disorders (i.e., osteoporosis) that make their bones more susceptible to fracturing from simple accidents such as falls. With regard to torture, a consideration of the lifestyle and day-to-day job of the individual, the context of the death event, and other factors must be considered. The principle of skeletal analysis for documenting physical abuse is to focus on recording multiple injuries in various stages of healing, documenting the specific locations of the damaged bone, and attempting to provide a biomechanical explanation for these injuries.

SHARP-FORCE TRAUMA

Sharp-force trauma (SFT) from a forensic pathology perspective involves an impact from a narrow-edged implement that results in either cutting of skin and incised damage to the underlying bone. From an osteological point of view, SFT is defined as a narrowly focused, dynamic, slow-loaded, compressive force with a sharp object that produces damage to hard tissue in the form of an incision (broad or narrow) (Symes et al. 2002). Unlike BFT, SFT usually results in a penetrating defect to soft tissue and bone. SFT can involve a variety of weapons and tools. Any tool with an edge bevel can produce incised wounds. Anthropologically, the two most common sharp tools examined are knives and saws.

Incised wounds

The first criterion for the recognition of sharp-force injury is that the offending tool must be able to incise bone, so it must have an edge bevel (Symes et al. 1999, 2002, 2007). Edge bevel is defined as the border of the blade that is an acute angle. Often, this bevel is created in the blade to prevent splintering of the material that is being cut. Within this definition, machetes, box cutters, razor blades, and axes are considered knives, but other blades such as bush hogs, augers, tree chippers, many ornamental swords, letter openers, and boat propellers are not (Symes et al. 2002). The latter instruments have blades with squared edges (90° angle).

SFT patterns: instruments with and without an edge bevel

Instruments lacking an edge bevel can scrape, chisel, shave, scratch, and crush but are not able to incise, cut, or saw. The trauma patterns produced with these square-edged instruments, as discussed above, are more consistent with BFT than SFT. In an analysis of skeletal injuries associated with boat-propeller accidents, Semeraro et al. (2009) noted that the dull, square edges of a propeller caused straight, parallel, narrowly focused defects on the skin and superficial soft tissues. The forensic pathologist commonly classifies these types of skin wounds as sharp trauma. However, the underlying bone exhibits characteristics of BFT, including crushed bone, butterfly fractures, and cortical bone delamination. In this instance, the soft tissues are difficult to interpret grossly and they mask the true characteristics of the skeletal trauma. These results emphasize the need to consider both hard- and soft-tissue damage in the multidisciplinary analysis conducted by the forensic pathologist, coroner, and forensic anthropologist.

Knives

Knives have thin blades that sometimes terminate in a point. Knives commonly exhibit blade bevel in which both sides of the blade are cut at an angle. In this case, the blade tapers from the back of the blade to the edge. All knives have at least one area of edge bevel (sharpened edge) on the blade (Figure 17.4a). Within these guidelines, box cutters, razor blades, or machetes may be classified as knives.

Knives come in a wide variety of shapes, sizes, and utilities. Cheap, household kitchen knives are often used in cases of SFT. These knives are either straight-edged or exhibit a serrated edge (Figure 17.4b). A serrated knife has teeth manufactured into the blade. The serrated edge of the knife is typically thinned (beveled) on one side to enhance a sharp cutting surface. Straight-edged knives can be separated into single-edged or double-edged (Figure 17.4c).

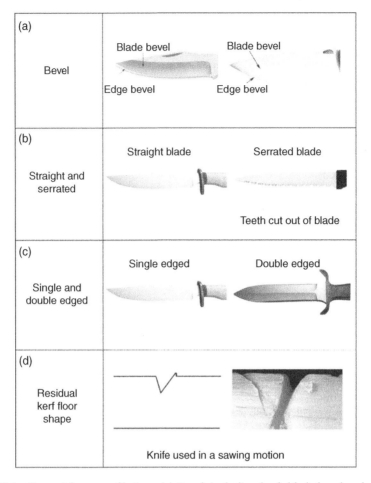

Figure 17.4 General features of knives. (a) Bevel, including both blade bevel and edge bevel. (b) Serrated versus straight-edged blades. (c) Single- and doubled-edged blades. (d) V-shaped kerf in cross-section created by a knife when used in a sawing motion. This differs from the kerf shape created by a saw.

Residual marks in bone left by knives and other tools with an edge bevel

Knife-wound analysis has received some attention in forensic investigation (Chadwick et al. 1999; Costello and Lawton 1990; Reichs 1998; Symes et al. 1999). Even though knife stab wounds are a major cause of violent death (Ambade and Godbole 2006), the widespread use of superficial and misleading descriptors such as "sharp," "single-edged blade," and "hesitation mark" (which erroneously implies behavior) are common and may result in serious misinterpretation by attorneys, judges, and juries. These meaningless descriptors, along with the lack of tested and validated standards, demonstrate the need for advancement in skeletal trauma research (Symes et al. 2002).

Knife blades can be used in a sawing motion. A knife creates a V-shaped kerf floor when viewed in cross-section, regardless of whether there are teeth manufactured into the blade or not (Symes 1992) (Figure 17.4d).

Knife cut wounds

Knife wounds to soft tissue are commonly referred to as knife stab wounds (KSWs). Anthropologists, in particular, often misuse the term KSW, since most wounds they examine are without soft tissue. Many of these wounds to bone are knife-incised (cut) wounds, but cannot necessarily be attributed to stabbing. Using the term knife cut (incised) wound (KCW) instead of KSW is more accurate and is inclusive of many blade actions. A KCW in bone is indicated when a sharp-edged tool superficially incises bone while traversing over the surface of the bone. While a nonstabbing KCW often follows the contour of a bone, a stab may puncture, nick, or gouge a bone as it enters the body. If cut marks follow the contour of a bone, or if a knife is used in a reciprocating motion, these features are likely indicative of dismemberment (Symes et al. 2002).

In case study 2, medical personnel have difficulties in distinguishing a postmortem defect from a knife wound on a human rib. In this situation, a forensic anthropologist must examine the clean dry bone under magnification to reliably distinguish between a lethal stab wound and possible rodent activity; as expected, this is a critical component in establishing a valid cause and manner of death.

Case study 2: pseudotrauma

Following a police chase, two suspects ran into a wooded area at night in the middle of winter; only one suspect came out. Several years later, skeletal remains of the second suspect were discovered, worse for wear and scattered and chewed by carnivores.

When medical personnel evaluated the skeletal remains, they identified a defect on one of the victim's ribs, which they suggested was a knife stab wound and hence the potential cause of death. The investigating officers were concerned with the ruling of homicide, as they felt the victim had probably succumbed to exposure complicated by high drug levels, and not a stab wound. They requested a second opinion of the injury, specifically to determine whether the defect was a knife stab wound and, if it was, what type of blade inflicted the wound.

While the defect to the mid-rib appears to the naked eye to have resulted from an incised wound (Figure 17.5, arrow), magnification of the injury revealed that the bone did not exhibit an incised defect; rather the bone was crushed. Color differences between the undisturbed and disturbed areas of bone are visible, which indicate that the defect occurred both after the body had skeletonized and after the bone discolored from the associated decomposition, soil, and other environmental factors. In addition, multiple score marks (animal tooth marks) on the outer bone surface were identified. Therefore, this defect is consistent with carnivore scavenging rather than a perimortem injury.

Puncture (gouging) wounds

A penetrating wound from the tip (or point) of a sharp instrument that impacts bone will result in compression of the outer table of the bone to produce a gouge impact. Certain bones, like the cranial vault, may capture a rather accurate replica of the knife tip. These negative images can be cast and measured for length, thickness, and depth. The measurements can be applied to any confiscated tools in a class comparison of size and shape of a tool.

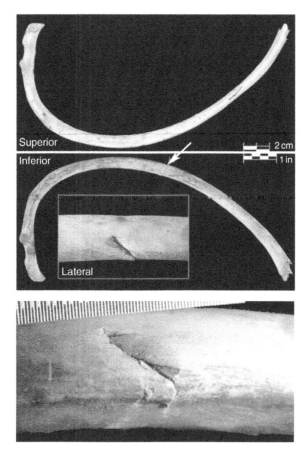

Figure 17.5   Defect noted in a rib (arrow) that was initially assumed to be SFT. Note the rough edges and depressed bone in the magnified image inset as well as the close-up image at the bottom. This defect was caused by animal activity and not by SFT.

Chopping wounds

Many sharp-edged tools are designed to chop material, for example machetes, axes, and hatchets. In some instances, the force of the blow of the sharp object may result in fracturing of the bone. This is directly related to the force of the load where, for instance the heavy weight of an axe in a chopping motion coming over the shoulder will probably generate more force than any bone can bear. However, in order to be classified as SFT, the instrument must have an edge sharp enough to incise the bone. Fractures associated with sharp injuries are the result of blunt-force failure (Symes et al. 2002). Thus, an impact from a sharp object with excessive force begins with an incised wound in bone and may end in a tension/compression fracture. In this case, the bone trauma can be characterized as *BFT with a sharp object* and all rules pertaining to the interpretation of blunt-force injuries may be applicable to this impact. The notion that sharp-force injuries may end in blunt-force fractures was recently described and defined by human-rights specialists in Central and South America. A definition of these injuries by the Guatemalan Forensic Anthropology Foundation is translated below:

Cortocontundente (TCC) are those injuries (fractures) caused by a sharp object using excessive force. They normally appear as complete fractures but this may not always be the case. Other characteristics are: sharp edges, smooth walls and loss of bone tissue. (Fundacion de Antropologia Forense Guatemala 2009)

SFT analysis

Tool-mark identification is potentially attainable and particularly informative in KCW and knife and saw dismemberment and mutilation cases. This often requires knowledge of how tools are manufactured, as well as how tools react to bone (Blumenschine et al. 1996; Burd and Kirk 1942; Burd and Greene 1957; Burd and Gilmore 1968). The ability to diagnose class characteristics of cut marks to bone into categories as simple as knives, axes, or saws is essential to a criminal investigation. What follows is a brief outline of the general analytical steps needed for a SFT analysis.

The first step is to determine whether the implementing tool is a knife, a saw, or another type of tool (e.g., axe, machete). Whether a knife is used to saw, stab, or chop, incisions to bone produce a V-shaped kerf floor. In cross-section residual striations will be visible and these indicate the direction of the blade motion. An axe or hatchet will leave a similar tapered kerf floor, but the kerf walls should indicate a thick tool. A saw will commonly produce a relatively wide square or W-shaped kerf floor in cross section. Residual striations can be observed on the kerf walls. In some cases the incised bone may also be associated with BFT. This is often indicative of a sharp object that was heavy and capable of producing both incised bone and BFT. Once the general class of tool is diagnosed, more detailed class characteristics may be available. With knives, the obvious next question is whether the blade is serrated or not serrated. With saws, the next step is to examine the cut bone for indicators of saw size, set, shape, direction of blade motion, and saw power (Symes 1992; Symes et al. 2002). While these broad indicators do not identify a particular saw, they combine to give the investigators, and eventually juries, a narrowed description of the potential offending tools.

Saw cuts to bone

A saw is simply defined as a blade with teeth. The working edge of the saw blade contains projections (teeth) that act like chisels and produce slicing or shaving actions. Saws are generally designed to cut a wider swath than the blade width. This is accomplished with lateral bending of every other tooth, termed *tooth set*, and is designed to prevent binding of the blade within the cut it is producing (Andahl 1978). Saws come in a great variety of styles and functional applications. It is important to understand a few basic concepts about saws and saw blade action before attempting to interpret saw marks in bone.

How do saws work?

Saws work through a chiseling action that creates a broad incision in the outer cortical bone and then compresses and removes bone from in front of the teeth in the direction the blade is moving (i.e., the stroke direction). Saw strokes are classified into cutting and passive strokes. A cutting stroke is the direction of sawing action that creates the most effective removal of bone. Passive strokes are in the opposite direction of the cutting strokes. Cutting strokes are responsible for a majority of the cutting while passive strokes do not utilize the saw tooth to its fullest potential. Because most Western saws are designed to cut on the push, most people in the USA are familiar with the push cutting stroke, as opposed to saws that cut on the pull stroke (Symes 1992).

The act of sawing is essentially pushing, pulling, or rotating the teeth of a saw blade in such a manner as to either cut (when the teeth are needle-pointed, like a knife) or chisel (when the teeth are designed similar to a flat-bottomed wedge) through material. To understand residual characteristics of saw cuts, it is necessary to examine the action of the saw blade. Saw action includes the slicing or shaving of material by a knife blade or chisel tooth, as well as the actions of the banks of teeth working in unison or in opposition to the blade. The principles of cutting action rely on blade and tooth design, and the manner in which energy is transferred from the tool to the material (Andahl 1978; Symes and Berryman 1989; Symes 1992; Symes et al. 1998, 2007). Since the saw teeth perform the cutting, actions of each tooth and the combinations of these teeth on a blade must be examined. Saw actions contribute to size, set, shape, power, and direction of cut (Symes 1992; Symes et al. 1998, 2002).

Characteristics of saw blades and teeth
Blade and tooth size refers to the minimum width of a saw tooth impression left in the sawed (bone) material or the minimum width of the complete saw cut trough. This saw impression is termed a kerf. Saw kerf widths are dependent upon the blade thickness and the amount of tooth set. Saws are more commonly classified by two standard measurements: (i) points per inch (PPI) or (ii) teeth per inch (TPI) (Andahl 1978; Nagyszalanczy 2003; Rae 2002; Self 2005; Wilson 1994). PPI represents the number of points or the tip of a tooth that are represented in 1 inch of the blade. On the other hand, TPI represents the number of complete teeth on 1 inch of the blade (Figure 17.6a). The number of PPI is generally one value greater than the number of teeth per inch. It is important to note that all metric assessments of saw-tooth point distances in residual bone striations refer to TPI.

Blade and tooth shape refers to the shape of the individual teeth and the saw blade. Shape refers to the contour of the teeth, the angle in which the teeth are filed, and the contour of the blade itself (Figure 17.6b). Handsaws and many mechanically powered saws are classified into two basic types based on tooth shape: rip and crosscut. These tooth shape types are important in that each function in a different manner to effectively cut different types of material. The rip saw is designed to use chisel-shaped teeth to "rip" along the grain of the wood. The crosscut saw uses filed teeth to "cut" across the grain of the wood.

Differences between crosscut and rip teeth are illustrated in Figure 17.6b. Rip saws exhibit flat teeth and are designed to cut in a chiseling fashion, pushing out debris at the end of the stroke (Cunningham and Holtrop 1974: 82; Lanz 1985). The front of rip teeth project from the blade to form a raker angle of 90° (perpendicular to the plane of the teeth), then trail off to the back side of the tooth and form a gullet angle of about 60° with the front of the next tooth (Figure 17.6b). The rip design cuts quickly and roughly through material.

Crosscut teeth are smaller and bite less material with each stroke than a rip saw, as the teeth are rotated back 15°. Typically, crosscut teeth are filed on the cutting edge

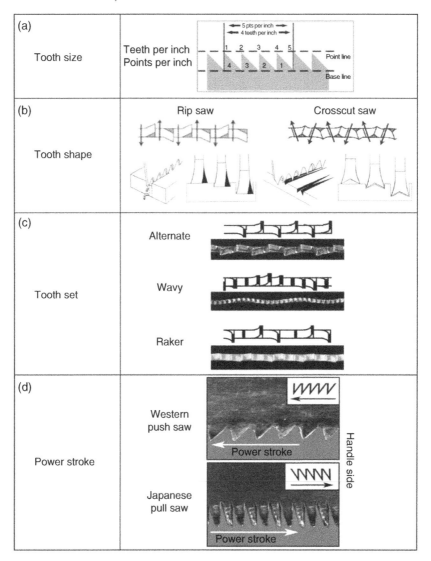

Figure 17.6 Important class characteristics of saws. (a) Tooth size, teeth per inch (TPI) versus points per inch (PPI). (b) Tooth shape: rip and crosscut saws. (c) Tooth set: alternating, wavy, and raker. (d) Push versus pull: Western saws are designed to have their power stroke on the push and Japanese saws are designed to have a power stroke that cuts during the pull motion.

at about a 60–75° angle. The front of each tooth is similar to a knife edge and forms a needle point, rather than a chisel (Jackson and Day 1978: 76; Nagyszalanczy 2003). This filing creates a tooth that terminates in a point, or wedge, and essentially takes on the shape of a sharpened blade that cuts rather than chisels material. Therefore, crosscut teeth are often the same shape as rip teeth, but the front side of the tooth is noticeably sloped back (actually rotated) on the blade, rather than aligned perpendicularly to the blade as seen with rip teeth. Each tooth progresses through wood fibers with a sharp edge, and slices instead of chisels (Figure 17.6b).

The shape of the tooth of a saw is designed to make the cutting action more efficient, or effective relative to the direction of force applied in the cutting action; this is
specific to whether the saw is cutting on the push or pull stroke (Figure 17.6d). The
typical Western hand saw cuts produces more material on the push stroke while the
passive stroke essentially glides over the material with minimal cutting. The major
alternate design to push saws is the Japanese pull saw (Figure 17.6d), where the teeth
are angled back toward the handle. Because forces are exerted on the pull, tension can
be maintained even on thinner blades with a minimal set of hardened teeth that create
a narrower kerf, Japanese pull saws waste less material, and demand less effort for the
same job. Because of the efficient design and mass production, pull saws are very
common today in the USA.

Other examples of saw shape include fixed-radius blades (e.g., circular saws), curved
blades (e.g., pruning saws), and flexible blades (e.g., Gigli saws).

Tooth set is the lateral bending of the teeth of the blade, which reduces blade binding in material and has been an integral part of saw design that has existed for close to
2000 years (Andahl 1978; Disston 1922). In most saw blades, the distal portions of
the teeth deviate laterally from side to side in an alternating pattern. In other saws, the
teeth alternately project laterally. While saw blade tooth set is essential to the effectiveness of most saws, it is also not required, as serrated knives, metacarpal saws, and
many flexible blade saws produce sawing actions without tooth set.

Teeth are generally set according to their size. As a rule, the kerf does not exceed
1.5 times the thickness of the blade (Cunningham and Holtrop 1974: 84; Jackson
and Day 1978: 75–76). If the set is greater, the teeth bend laterally to the extent that
a central section of the material in the kerf floor will be untouched in the midline as
the tooth reaches its greatest flare.

The three most common types of tooth set in saws are: alternating, raker, and wavy.
The most common saw set pattern is the alternating set, which indicates that the teeth
in the blade bend laterally in an alternating pattern from side to side (Figure 17.6c).

Raker set is a specialized tooth on a blade (usually in conjunction with alternating
sets) designed to "rake" material or imperfections from the kerf floor rather than cutting the wall of the material, essentially cleaning up after other teeth while troughing
the floor (Figure 17.6c). Rakers are teeth with no set, designed to rake all debris out
of the kerf. Rakers are generally not placed between every tooth but rather appear
every third, fourth, or fifth tooth. This design alters kerf floor shape and reduces blade
drift that is often noted in alternating set blades without raker teeth. Blade drift is
defined as the blade movement from side to side as each consecutive alternating tooth
enters material, pulling the blade toward or away from the midline with every introduction of a new tooth. Having a raker in the tooth set allows for more lateral bending of the individual teeth. A raker set is designed for softer material, such a soft wood
(e.g., pruning saws) (Salaman 1975: 405) or for small teeth designed to cut hard but
ductile materials (e.g., hacksaws) (Symes 1992).

The wavy-set saw blade teeth pattern is unique, yet cuts material along the same
principle as alternating-set teeth (Figure 17.6c). Wavy-set blades generally have small
teeth that deviate from side to side in groups rather than by individual teeth. Each
wave, composed of many small teeth, functions like a single tooth. This set is typically
found in blades with a high number of TPI. Because it is impractical, inefficient, and
difficult to bend each tooth laterally in an alternating fashion, the manufacturer

essentially bends a group of teeth one direction then laterally bends them in the opposite direction.

Saw power is defined as how a tool is powered; whether it is mechanically powered or hand-powered. Human-powered saw cuts show great variation in cutting action and residual tool marks on bone due to differences in speed and force of the cut, the direction and orientation of the cut, and the skill and strength of the person yielding the saw. Mechanically powered saws have become more common due to mass production of low-quality power saws. However, from a forensic perspective, the use of hand-powered saws is more commonly seen in dismemberment cases (Symes et al. 2002).

Hand-powered saws have variable blade and tooth designs. In general, they have thinner blades than mechanically powered saws. Mechanical saws powered by gas, electricity, or pneumatics all but eliminate human variation from saw strokes while adding speed and cut uniformity. These saws are designed to work either through reciprocating (back and forth) action (e.g., sawzall), or through continuous cutting (e.g., circular saw) and may be supported by a frame or be hand-held. Mechanically powered saws differ in design from hand-powered saws. Increasing the power generally requires that the teeth be short and wide. The width of the blade is increased in comparison to hand-powered saws in order to stabilize the blade due to increased speed and torque. Power saws typically waste more material. The exception is the band saw, which has a thin blade supported by two pulleys that cuts in a continual motion.

Residual marks left in bone by saws

There are numerous residual characteristics and traits that are commonly observed in saw-cut bone. This chapter does not allow for an in-depth discussion of each of these traits, but they can be found elsewhere (Symes et al. 1998).

As saw teeth cut into bone, a groove or *kerf* is formed. Saw marks to bone are classified as sharp trauma because there is always some portion of a saw tooth that is incising bone. The result of these incising activities can be observed on the kerf walls and floors. Kerf floors offer the most information about saw type because evidence of teeth set and teeth per inch (TPI) is best exhibited here. Striae in the kerf walls offer information about the sides of the teeth and, in particular, those teeth set to that particular side. In addition, the shape, depth, and frequency of these striae may present information regarding the shape of the blade, the amount of energy transferred to the material, and the motion in which the blade traveled through the bone.

Direction of cut is the final major classification of saw mark analysis. Establishing the direction of cut in bone is feasible and contributes to the interpretation of how a saw is used. However, "direction" used may be misleading unless it is clearly defined. Direction of cut indicates two separate saw actions, the *direction of blade progress*, and the *direction of blade stroke*. Indicators of the direction of saw progress center on the false start and breakaway spur or notch. False starts are defined as impressions in the bone in which individual saw blade teeth strike and chisel bone material, or where actual kerfs are started but abandoned for another cut. The breakaway spur is defined as the projection of bone at the floor of a terminal cut where the bone fractures, leaving a projection of bone at the base of the kerf floor. The breakaway notch is the resulting void in the bone left by a breakaway spur (Symes et al. 1998). The plane formed between the false-start entrance and the breakaway spur or notch exit usually provides the precise direction of saw blade progress. Direction of blade progress is essentially perpendicular to stroke and tooth striae that illustrate the direction of the cutting stroke.

Saw-mark analysis

Saw-mark analysis basically involves the examination of saw-cut kerfs and characteristics present on kerf walls and floors (Andahl 1978; Bello and Soligo 2008; Symes 1992). Information potentially retrievable (as discussed above) includes width of the kerf cut (thickness of the blade teeth set), configuration of the kerf floor (flat, curved, or W-shaped), striations on the kerf walls, and entrance and exit chipping of bone. These tool marks, if enough bone and bone surface is available, permit reconstruction of: (i) whether a power or hand saw was involved, (ii) direction and orientation of cut episodes (especially with a hand-held saw), (iii) morphology of the teeth (including set configuration and TPI), and (iv) the shape of the teeth. Thus, it is possible to determine at least the general class of saw used in many cases. This reconstruction is more likely possible in cases where cuts occur in a long bone, with thick cortical bone and both sides of the cut present. Cuts to the cervical and lumbar vertebrae, wrists, knees, and ankles, while common in dismemberment and mutilation, usually offer less information due to the preponderance of trabecular bone.

Saw class (not individual) characteristics also serve to narrow the list of potential saws utilized (Saville et al. 2006; Symes 1992). Identifying individualizing characteristics and the narrowing to a specific tool or weapon is often not possible, not to mention risky. However, it may be possible to eliminate a tool, suggest that it may be consistent (or not consistent) with a particular tool, or readily admit that results are insufficient to provide a conservative and defendable attribution to a particular tool.

BALLISTIC TRAUMA

Ballistic trauma is described as any damage to biological tissue from a fast-loaded force, such as a bullet fired from a gun. Since the most common cause of ballistic trauma is associated with gun-related injuries, it is often referred to as gunshot trauma (GST), a term that will be used here. However, it must be noted that ballistic trauma also includes damage to tissues as a result of the explosion itself or the debris expelled by explosives (bombs). The term "gunshot wound" (GSW) is often used by forensic pathologists and anthropologists. GSWs to the cranium are always associated with bone fracture. GSWs to the thorax and abdomen, while fatal, may not injure bone (de la Grandmaison et al. 2001).

The key concept in distinguishing GST from other types of trauma (BFT/SFT) is the velocity of the impacting object prior to striking tissue. Objects moving at rapid (ballistic) speed – meaning meters per second, rather than kilometers per hour – impart tremendous kinetic energy to the biological structure. As discussed earlier, these levels of kinetic energy do not allow the viscoelastic (bending) properties of bone to be expressed before the bone breaks. Instead, the bone acts as a brittle material and fracture is nearly instantaneous (Özkaya and Nordin 1999). These general properties permit the initial assessment of damaged bone. However, final determinations require careful documentation of the extent of the damage, including the number and paths of fractures, bone-deformation evidence, and even the microscopic consideration of bone cross-sections and fracture edges.

Despite the significant amount of research that has been conducted on the biomechanics of gunshot wounding (e.g., Berryman and Symes 1998; Di Maio 1999; de la Grandmaison et al. 2001; Langley 2007; Smith et al. 1987; Spitz and Spitz 2006), GST is often misunderstood and is a frequent topic of controversy. There appears to be an enormous disjoint in communication between ballistics engineers and physicians who specialize in traumatic injuries (Fackler 1988) and medicolegal specialists (i.e., medical examiners, coroners, anthropologists, and criminal lawyers) who routinely encounter fatal GSWs. While Fackler (1988) insists that ballistics engineers and specialized trauma physicians overemphasize velocity, the authors contend that many medicolegal practitioners still underestimate or misunderstand the relevance of bullet velocity on the resulting tissue damage. Anthropologists are no exception to this misunderstanding, despite the fact that they are frequently exposed to fleshed bodies. In general, the relationship between soft- and hard-tissue injuries associated with ballistic trauma is rarely considered during autopsy. Therefore, the interpretation and assessment of high-velocity impacts to osseous tissue remains in its infancy.

As always, the context in which the skeletal remains are discovered is essential. If the anthropologist is not present at autopsy, crucial information may be lost. Therefore, complete documentation and good communication between the pathologist and anthropologist are necessary for a successful and accurate interpretation of the injury or injuries.

Effects of gunshots on biological tissue

While discussions in this section focus on GST, the basic principles may also apply to other traumatic injuries produced when external forces impact bone at ballistic trauma speeds, such as those generated by bomb blasts.

When a bullet strikes and/or enters biological tissue (soft or osseous), two simultaneous actions occur. Upon impact, the bullet crushes the tissue. With enough kinetic energy, the bullet continues into the body and disrupts tissue by pushing it aside until it either exits the body or becomes embedded in the tissue after loosing energy. The linear defect created when a bullet traverses through soft tissue is referred to as a "permanent cavity" (Di Maio 1999; Fackler 1988). Around this structure, a "temporary cavity" may form and is responsible for further crushing, displacement, and stretching of nearby tissues (Di Maio 1999; Fackler 1988).

Following an impact with tissue, especially bone, a bullet's trajectory may be disrupted. It may deviate along the long axis (yaw), or it may tumble, deform, or fragment. In most cases, these alterations result in an increased surface area for the projectile, which contributes to an increase in the transfer of kinetic energy and subsequent tissue damage. If the bullet impacts and traverses through bone, small bone fragments can be displaced into the permanent cavity or expelled through the exit wound (see Smith in Symes et al. 1996). When displaced and put into motion, these fragments may act as secondary missiles and cause additional damage to organs and tissues. The velocities of these secondary missiles are slower than the bullet itself; therefore, the resulting damage may be described as BFT, not GST (Fackler 1988; Smith et al. 1987).

Blunt trauma associated with ballistic wounds

Some fractures associated with GSWs may not be the result of ballistic force. If a bullet loses most of its kinetic energy prior to impact, the consequential damage is more likely to exhibit blunt force rather than ballistic features (Smith et al. 1987). In Figure 17.7, anterior–posterior (right) and lateral (left) radiographs of a suicide victim with a self-inflicted GSW to the mouth are shown. In the lateral radiograph (Figure 17.7, left), the opaque bullet fragments make it possible to track the bullet in the permanent cavity. After entering the palate, the bullet progressed upward and slightly posterior. The final resting point, however, was along the midline of the cranial vault, posterior to the original trajectory path and contrary to the expected route.

According to the basic principles of ballistics, a bullet will travel in a straight line at a high rate of speed until it loses its kinetic energy. Therefore, the displaced final resting place of the bullet fragments and the trajectory through the skull can be directly attributed to a loss of kinetic energy. In other words, the bullet was no longer travelling at a ballistic speed, but at a blunt-force speed.

In this particular case, evidence that the bullet had lost most of its velocity and kinetic energy is indicated by the fact that the bullet had not exited the top of the neurocranium and had redirected from its original trajectory. Even at a reduced speed, one might expect that the impact of the bullet to the internal aspect of the cranial vault would result in bone damage. Upon closer inspection of the frontal bone, injuries are observed on both the internal and external surfaces and the bone is reacting as expected in BFT, where the slow impact bent the bone outward in elastic deformation with an eventual failure in tension externally.

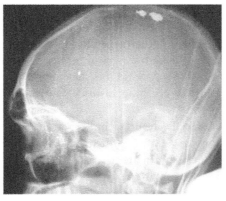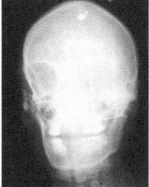

Figure 17.7 Lateral (left) and anterior–posterior (right) radiographs of a self-inflicted GSW. The bullet was found posterior to the track wound, which is indicated by opaque bullet shards. The bullet lost velocity and kinetic energy. This is evident by the fact that the bullet did not exit the neurocranium and was redirected from its original trajectory.

Contributions of an anthropologist to GST analysis

Anthropologists can contribute to medicolegal investigations of GSWs by describing:

- details of the patterns of the bone damage in ballistics injuries;
- direction and orientation of the bullet trajectory; and
- sequence of impacts (Berryman et al. 1995; Huelke and Darling 1964; Huelke et al. 1968).

However, many anthropologists focus on other variables, such as the distance of the shooter from the victim, or the caliber of the bullet. Except in cases where soot is imprinted on the bone from a contact or near-contact GSW, the distance between weapon and victim is not determinable from evidence of bone damage. Determining bullet caliber is not as straightforward as expected.

GST and the cranium

Smith et al. (1987) provided an informative description of fracture patterns resulting from bullets striking human cranial bone. The primary characteristics of GST to the skull include: (i) plug-and-spall bone fragments, and (ii) radiating and (iii) concentric heaving fracture patterns. These patterns often help predict (within a range) the caliber of the bullet, direction of the shot, and the number and sequence of impacts.

Plug-and-spall (entrance and exit defects)

Typically, when a bullet impacts bone a circular entrance wound is created with approximately the same diameter as that of the caliber of the bullet. However, a direct association between bullet caliber size and plug size in bone has been shown to be misleading due to variation in a bullet's velocity, projectile design, and angle of impact, as well as the strength (or resistance) of the target (Berryman et al. 1995; Ross 1996).

Unlike entrance wounds, exit defects tend to be larger and more irregular in shape due to a number of factors, including the potential of bullet deformation or fragmentation and the potential loss of much of the bullet's kinetic energy within the cranium (see Smith in Symes et al. 1996). Biomechanics of bone fracture assist in cranial trauma interpretations. Radiating fractures associated with entrance wounds can travel at a rate of thousands of meters per second, much faster than the speed of the bullet. Therefore, radiating fractures from the entrance defect may cross the area of the skull exit location before the bullet exits (Berryman and Symes 1998). Because of the loss of kinetic energy, exit wounds will exhibit a lower magnitude of damage than entrance wounds (see Smith in Symes et al. 1996).

Upon impact with the cranium, the bullet shears a plug of bone in front of it into the brain (Figure 17.8, bottom). There is spalling around the plug of bone, creating an internally beveled edge (Smith in Symes et al. 1996) (shown in Figure 17.8, upper left). This beveling feature can be used to indicate the direction of the bullet. If the bullet exits the skull, the outer table of bone is spalled-off and produces an external bevel (Figure 17.8, upper right).

In three separate, but related, studies on entrance and exit wounds in dry bone, Quatrehomme and İşcan (1997, 1998, 1999) suggested that the beveling of entrance and exit wounds were inconsistent with the direction of fire and, therefore, could not

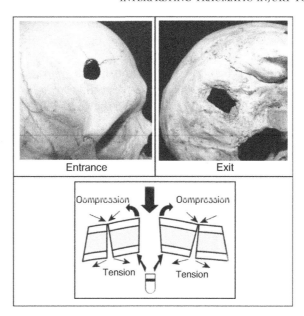

Figure 17.8 Examples of entrance (top left) and exit (top right) wounds exhibiting clearly defined internal and external beveling. The illustration at the bottom shows a cross-section of cranial bone with points of compression and tension following GST.

be relied upon to determine the direction of a bullet. The results are counterintuitive to most research and practice, which has shown that entrance and exit beveling is by far the most useful and accurate indicator of the direction of a bullet in the cranium (e.g., Berryman and Symes 1998; Di Maio 1999; Smith et al. 1987). While the authors agree that the curvature of a skull may somewhat diminish the appearance of beveling, careful analysis of the bullet-impact area and beveling evidence always provides a good indication of the direction the bullet was traveling relative to the bone. The one exception is when the bone is thin, since thin cortical bone is uniform in construction and the bone is probably incapable of bevel, at least in terms of gross examination. It is assumed that this exception is what Quatrehomme and İşcan discovered.

Bullets often strike bone at an angle, producing tangential bullet defects. In the skull, this impact creates a unique fracture pattern, which has been termed a "keyhole" defect (Figure 17.9). As a bullet strikes bone at a steep angle (tangentially), the initial defect is typically oval in shape (e.g., Berryman and Symes 1998; Di Maio 1999; Dixon 1982). As ballistic projectiles enter the skull, fractures are often created, radiating in the direction of greatest resistance. As the first author (SAS) likes to report, a pair of fractures radiate where the bullet goes on edge with the bone, correlating to the point of most resistance to the bullet. (Figure 17.9). Since the angled bullet is progressing immediately under the surface of the bone, a concentric heaving fracture is often formed between the two radiating fractures. This creates an outward levering (spalling) of bone and produces an external bevel at the point of "most resistance." Together, the entrance oval defect and the eventual fan-shaped spalled bone area resemble an old-fashioned keyhole (Berryman and Symes 1998; Dixon 1982; Smith in Symes et al. 1996). The primary morphological attribute of the entrance keyhole defect is that internal beveling is exhibited. It should

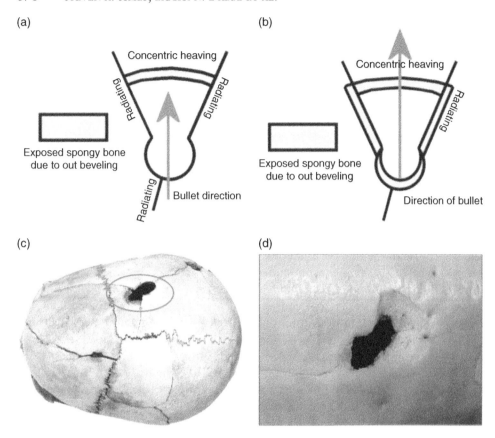

Figure 17.9   Schematic of "keyhole" defect production. (a) External surface. (b) Internal surface. (c) Superior view of skull with a keyhole defect (circle). (d) Close-up view of skull with a keyhole defect.

be noted that exit wounds might also present a morphology that mirrors an entrance keyhole (Berryman and Symes 1998). In this case, the bevel is noted around the entire periphery of the defect.

Fracture morphology: radiating and concentric heaving fractures
Depending on the magnitude of pressure, fractures radiate from the point of impact and extend until the energy is absorbed. In some cases, they may travel from the point of impact to the other side of the skull *before* the bullet exits. In cross-section, radiating fractures may be either perpendicular or stair-stepped (Smith et al. 1987; Figure 17.9) and the perpendicular expanding (tension) fractures fit the biomechanical terminology of "hoop" stresses (Smith in Symes et al. 1996).

If kinetic energy is sufficient, or when the intracranial pressure is not relieved with the radiating fractures, concentric heaving fractures, or tertiary fractures, may appear (Smith et al. 1987). Concentric fractures extend in roughly arched patterns between pre-existing radiating fractures. Since concentric fractures are the result of levering-out of cranial bone, the inner bone table fails first in tension followed by compressive failure on the outer table (Berryman and Gunther 2000; Hart 2005; Symes et al. 1996). These fracture patterns are usually best observed on the inside of the cranium. The internal surface of a nonhomogenous material is usually the most accurate surface to examine.

Concentric fractures in GSWs always exhibit external bevel, regardless of whether the defect was due to the entrance or exit of the bullet. In general, the greater number of radiating and concentric fractures that appear in multiple levels, the more kinetic energy was imparted to the bone upon impact. In the case of multiple GSWs to osseous tissue, Puppe's law of sequence (described above) can be used to determine fracture order (Madea and Staak 1988).

GST and the postcranial skeleton

Entrance wounds in long bones differ in morphology depending on the type of bone (cancellous or compact) that is impacted. In the proximal and distal ends of long bones, vertebrae, and hands and feet, entrance wounds may appear as smooth, round defects because of the thin cortical bone. In contrast, in areas of dense bone, such as the mid-shaft of the femur, comminuted fractures and extreme fragmentation are more likely to be found. Exit wounds in postcranial bones are more destructive, exhibit irregular fracture patterns, and are less defined than entrance wounds. This has been well demonstrated by descriptions of GST to ribs and long bones (Berryman and Gunther 2000; Huelke and Darling 1964; Huelke et al. 1968; Langley 2007).

When long-bone fragments are reconstructed, the fracture patterns are similar to butterfly fractures that are commonly observed in BFT (Figure 17.10). In fact, some anthropologists and pathologists refer to these GSW injuries as "butterfly fractures" (e.g., Huelke et al. 1968). This is an incorrect classification of the fracture because the biomechanics of this type of injury are completely different than fractures caused by BFT. This term should be reserved for the trauma patterns resulting from BFT injuries (described above), which have been standardized previously in the biomechanical, medical, and anthropological literature. As described above, BFT to long bones involves a slow-loaded force that results in bending and compression of the bone, with the bone failing in tension prior to compression. BFT to long bones involves a slow-loaded force that results in bending and compression of the bone, with the bone failing in tension opposite to the area of impact (or compression). In contrast, during a fast-loaded (high-velocity) force, no bending is involved and the bone reacts as a brittle substance, and shatters. The consequence is that radiating

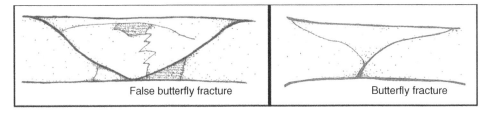

Figure 17.10   A false butterfly produced by GST (left) and a butterfly fracture produced by BFT (right). In GST, the bone immediately shatters and the radiating fractures produce a V shape. In BFT, the bone undergoes plastic deformation and fails in tension on the opposite side of compression. Illustrations by Dr A.M. Kroman.

fractures produce a V shape as failure first occurs at the impact site (Figure 17.10 left bottom). Only in GST injuries in tubular bone can radiating fractures be followed back to the plug and spall of a bullet impact. This is in contrast to blunt-force injuries in which failure first occurs in tension and opposite to the impact site (Figure 17.10, right).

Summary of GST

Forensic pathologists focus on soft-tissue damage to provide general descriptions of GST with respect to type and caliber of bullet (small versus large), and direction of impact in three planes (right to left, front to back, superior to inferior). By cleaning and reconstructing the bones, forensic anthropologists can enhance the pathologist's description by providing much more detail, including more precise determination of direction of the bullet(s) in three planes, general estimation of caliber, and sequence of shots.

TRAUMA ANALYSIS OF FATAL-FIRE VICITIMS

Another area of forensic investigation in which the forensic anthropologist can provide critical input relates to the fatal-fire victim. A common method for attempting to hide evidence of foul play is to burn the victim, either the body by itself, or within a structure. The assumption is that fire will destroy most, if not all, evidence of identification (soft tissue) and perimortem trauma (gunshot, stab wounds, broken bones). However, skeletal evidence often remains after fires (described in Chapter 6) and a biological profile can be constructed. In many cases, DNA can be extracted and a profile obtained. In addition, recent research by forensic anthropologists (e.g., deGruchy and Rogers 2002; Hermann and Bennett 1999; Marciniak 2009; Pope and Smith 2004; Symes et al. 2008; Thompson 2004) has shown that evidence of perimortem trauma is still retrievable after heat and fire alteration. As with all other forms of trauma, the proper analysis and interpretation of human-induced damage to bones begins at the scene with the understanding of the contextual setting.

Another component of the proper scientific interpretation of skeletal trauma in fire-altered human remains is the ability to distinguish between the effects of fire on bones from perimortem damage. This requires knowledge of the sequence of soft- and hard-tissue destruction, body postioning, postmortem fracture morphology, and observable color changes in the osseous material (Symes et al. 2008).

With respect to general recognizable patterns of heat and fire on the human body, the five-stage Crow–Glassman scale (Glassman and Crow 1996) was an early morgue-based practical observation-based study. This research created a means to evaluate the sequence of destruction and has the categories "recognizable," "possibly recognizable," "nonrecognizable," "extensive burn destruction," and "cremation." The first two categories focus on early soft-tissue changes like blistering and early charring sequences, while the last three refer to destruction of soft tissue and the alteration of hard tissue into cremated remains. Symes et al. (2008) cautioned that one or more of these categories might be scored on any one element in the body. However, a consistent sequence of these events on fresh remains (e.g., burning of soft prior to hard issue) can be recorded and used to reconstruct the progression of bone destruction and possibly reveal subtle information concerning traumatic injury that may have occurred to the body prior to the fire (Glassman and Crow 1996; Symes et al. 2008).

When exposed to heat and flame, all muscles of the human body, but especially the large, antagonistic muscles, shrink and contract. Contraction of these muscles causes hyperextension of the neck, abduction of the shoulders, arching of the back, and flexion of the elbow, wrist, hand, fingers, hip, knee, ankle, and toes into what has been termed a pugilistic, or boxer, posture. This posture influences the observable pattern of burning and postmortem fractures. The absence of a typical burn pattern on a body in the pugilistic pose may indicate restriction of the body or perimortem injury (Symes et al. 2008). In addition, the anthropologist has to remember that the pugilistic posture and its associated burn pattern can be observed only during the simultaneous burning of soft and hard tissues. After a major degree of decomposition has occurred or skeletal remains alone are burned, the remains present a different, often less variable, color and fracture pattern.

Color changes in heat-altered bone

During later stages of fire exposure, color change in burned bone can be separated into stages (Symes et al. 2008), from unaltered fresh bone to a heat line, a heat border, charred, and calcined bone (Figure 17.11a). Unaltered fresh bone is material that is insulated or protected from heat or direct flame and does not show the effects of

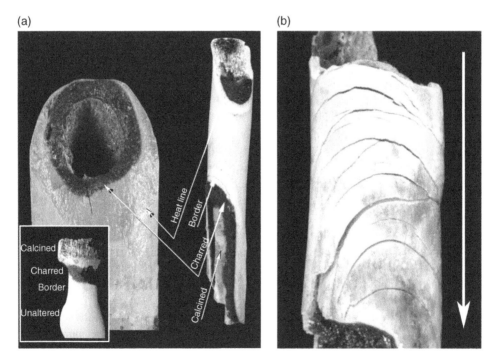

Figure 17.11 (a) Range of burned bone colors (charred to calcined); note the heat line and border. (b) An example of curved transverse fractures. The large arrow on the right indicates the progress of heat and fire.

exposure to thermal destruction. Thin heat lines may occur where the initial evidence of heat damage is visible next to unburned bone. The heat border is the initial obvious chemical change "border" between unburned and burned bone. The border is characterized by brown or white bone of variable width and probably dependent on overall temperature and tissue depth. The border indicates where collagen has been permanently altered or destroyed by heat. Charred bone is black and indicates an advanced stage of burning, likely in direct contact with fire and heat. The demarcation between bone border and charred bone is unmistakable. Each area represents altered bone; the former is protected by tissues but is still influenced by convection heat, the latter is a product of direct confrontation of heat and smoke. Calcined bone ranges from gray to white and suggests a destructive modification of inorganic bone (Mayne Correia 1997; Symes et al. 2008; Thompson 2004; Thompson 2005; Ubelaker 2009).

Fracture patterns in heat-altered bones

Several unique fracture patterns and bone-surface alterations can be observed – namely longitudinal, transverse, patina, and curved transverse – in thermally altered bone. Burn-line defects, external bone-surface patina, as well as splintering and delamination are also noted (Hermann and Bennett 1999; Symes et al. 2008). These postmortem changes result from dehydration and loss of the organic component of bone and leads to a combination of shrinkage, warping, and fracturing (Thompson 2004; Thompson 2005). The length of time the body is exposed to fire, the positioning of the body in the fire, the type of bone affected (e.g., cortical or spongy), and the condition of the bone prior to burning (fresh, decomposed, and dry) also affects the appearance of these fractures.

Longitudinal fractures typically run along the grain (the longitudinal axis) of the bone; step and transverse fractures may extend from them or may be created independently. The analysis of these fractures found within the defined heat borders, in combination with color patterns, may be used to evaluate whether the remains had burned with or without flesh. Patina fractures result from the destruction and warping of the outer cortical layer of bone, which may also lead to splintering and delamination of the inner and outer bone layers (Hermann and Bennett 2004; Mayne Correia 1997; Symes et al. 2008).

Even though these patterns are heavily dependent upon muscle shrinkage, body morphology, and body positioning, their recognition allows the investigator to separate heat-related fractures from those produced by other forces, such as blunt or ballistic trauma. It is important to note that fractures that are generated as a consequence of heat do not enter, or travel, into unaltered bone. In comparison, perimortem fractures can and will traverse beyond the burned area. Therefore, it is possible to assess whether the fracture had occurred before or after the burning. Patterns of color change are also important for evaluating changes in perimortem or postmortem destruction (Hermann and Bennett 2004).

Case study 3: Necklacing in South Africa

In 2009, the partially burned body of a middle-aged African male was discovered on a national highway in the Kimberley District, South Africa. The remains were in an early stage of decomposition and were initially taken to the morgue in Kimberley and then transferred to the Department of Anatomy, University of Pretoria, for maceration and anthropological analysis. According to the South African Police Service (SAPS), the individual had been a victim of "necklacing." Necklacing is a form of execution that involves placing two or more tires (filled with fuel) around a victim's chest and arms and then setting the tires, and the person, on fire (Bornman et al. 1998). The practice began in the mid-1980 and 1990s as a form of vigilante justice against criminals and conspirators of the apartheid regime (Bornman et al. 1998). In 2008, this practice briefly reappeared with the xenophobic attacks on immigrants from Zimbabwe. No further information regarding the victim was provided and the context in which the body had been discovered was not recorded.

The investigating officers requested a biological of the remains and an estimate of the time since death. They also sought to know whether the victim had been assaulted, or perhaps killed, prior to being burned. No information was available on the number of tires or the location of the tires on the body. The sternum, right scapula, hands, sacrum, lower leg, and feet had not been recovered. Due to the paucity of the recovered elements, no autopsy was conducted.

A normal burn pattern was observed on the skeletal elements from the mandible, thorax, upper limbs, back, and left hip. Substantial deviation from this pattern was noted, however, on the cranium (skull) and the right hip joint (os coxa and femur). The abnormal pattern may have been caused by the position of the tires, perimortem injuries, or both.

In the skull, the parietal, frontal, and occipital regions were charred whereas the facial bones were unaltered. Separating these two areas was a distinct heat border and line that contained a predictable pattern of postmortem cracks and fractures. The abnormal burn pattern may have been due to the location of the tire, which *could have* shielded the face from the flame as well as preventing the head from entering into a pugilistic position; other possibilities are related to unobservable damage to the muscular structures of the head and neck.

Perimortem, blunt-force injuries were noted in the upper body and the face (left orbit and maxilla), the mandible, and the left second rib. The minimum number of blows to the cranium was five. Both mandibular condyles exhibited butterfly fractures and another fracture was noted on the left mandibular body. A discontinuity between the burned and unburned parts of the mandibular body suggested that the fracture had occurred prior to the burning event and prior to the handling of the material (Figure 17.12).

An additional radiating fracture, which traversed outside the burned area, was noted on the right os coxa. This evidence, combined with an abnormal burn pattern on the proximal right femur, can be used to suggest that the fracture had most likely occurred prior to the burn event.

Although human remains are often burned as a means to conceal a crime, in this case the burning event was the criminal act. With regard to evaluating the cause of death with burned victims, a forensic anthropologist needs to defer to the forensic pathologist and the autopsy report. However, an anthropologist can use skeletal evidence to describe the sequence of events. In this case, the victim had been assaulted prior to being set on fire. The cause of death is not known; however, the manner of death is homicide.

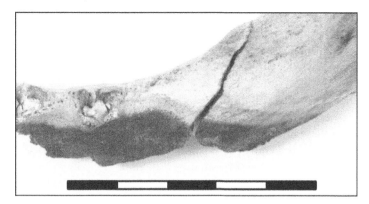

Figure 17.12   A normal burn pattern was observed on the mandible. Note the discontinuity between the burn lines; this indicates that the fracture occurred prior to the burn event. Scale bar: centimeters.

DISCUSSION

While the traditional role of the anthropologist has focused on the assessment of the biological profile and ultimately contributions to identification of the victim, this information is often not a source of contention in the courtroom. However, the courts are looking for contributing factors of cause, manner, and postmortem disposal or mutilation of a body. This requires a higher expectation of anthropologists, and, among other issues, requires going beyond the biological profile, and investigating scene context and especially interpretation of bone trauma (Dirkmaat et al. 2008).

Until recently, anthropological contributions to bone trauma analysis have been limited. A recent flurry of research has led to new awareness of skeletal trauma and a paramount need for forensic anthropologists to be involved. However, there are ramifications to the rapid interest and growth in bone trauma analysis. The main issue revolves around becoming an expert. Any analysis without training, testing, and experience creates the potential for misdiagnosis and inaccurate results. Unlike other areas of physical anthropology, the repercussions of irresponsible actions and unsubstantiated opinions are enormous. Whereas hypothesis formulation and testing is part of academia, the judicial system is the last place to formulate a hypothesis.

The hardest lesson to learn for a forensic anthropologist is to only make comments based on facts, not conjecture. The forensic anthropologist cannot mistake their job responsibilities for those of the medical examiner or coroner. Medical practitioners are paid to make accurate decisions; however, they are not held to the same accuracy as the anthropologist. Anthropologists should only offer opinions when they are strongly supported with factual evidence. The medical examiner/coroner is prepared to amend their report in the event that the death investigation produces more information, but the anthropologist, less constrained by time to make a decision, should rarely amend a report.

In skeletal trauma analysis the mistake of inferring too much from too little is one of the greatest problems in an anthropologist's routine analysis of bone injuries. In the absence of soft tissue or a lack of knowledge regarding the context in which the remains were discovered, the anthropologist needs to be conservative in trauma interpretations (Dirkmaat and Adovasio 1997; Dirkmaat et al. 2008). As in a court of law, the interpretation of either the cause and or manner of death from one or multiple skeletal elements has tremendous implications for the accused not to mention the career of the anthropologist.

Anthropologists must be willing to wear both academic and investigative hats. The anthropologist needs to have a willingness to look beyond the bone and into the context in which these injuries occurred. Thus, without a doubt, the training ground for trauma analysis is the morgue. Without autopsy experience working beside a skilled forensic pathologist, trauma analysis is difficult and erroneous conclusions are likely. With knowledge of soft-tissue injury, hard tissues need to be processed clean and examined in a dry state. Even with these measures, accurate trauma assessments are complicated.

In the last 20 years, anthropologists and pathologists have developed advanced methods and techniques for trauma analysis that are applicable to many aspects of a medicolegal investigation of death. However, these disciplines need to work together to fully realize the potential of the examination, removal, analysis, and retention of injured tissues as evidence in the interpretation of the mechanics of patterned traumatic injury and, ultimately, admissibility in a court of law (Cunha et al. 2008; Dirkmaat et al. 2008; Martrille et al. 2007; Pinheiro 2006; Smith et al. 1990, 2003; Symes et al. 1998). In trauma analysis, multidisciplinary efforts always produce the best results.

ACKNOWLEDGMENTS

ENL would like to acknowledge Natalie Keough of the University of Pretoria for use of the forensic anthropology report on the necklaced victim from Kimberley. Marius Loots photographed the case on necklacing. Natalie Uhl, Katrina Young, Kathleen Waldock, Katelyn Rusk, Sara Getz, Devin Rivers, Sara Fredette, Erin Franks, Michael Kenyhercz, Jolandie Myburgh, and Kathi Staaf offered editorial advice.

REFERENCES

Alunni-Perret, V., Muller-Bolla, M., Laugier, J.P., Lupi-Pégurier, L., Bertrand, M.F., Staccini, P., Bolla, M., and Quatrehomme, G. (2005). Scanning electron microscopy analysis of experimental bone hacking trauma. *Journal of Forensic Sciences* 50(4): 796–801.

Ambade, V.N. and Godbole, H.V. (2006). Comparison of wound patterns in homicide by shapr and blunt force. *Forensic Science International* 156(2–3):166–170.

Andahl, R.O. (1978). The examination of saw marks. *Journal of Forensic Science Society* 18: 31–36.

Auferheide, A. and Martin-Rodriguez, C. (1998). *The Cambridge Encyclopedia of Human Paleopathology*, 2nd edn (pp. 19–47). Cambridge University Press, Cambridge.

Baraybar, J.P., Cardoza, C.R., and Parodi, V. (2008). Torture and extra-judicial execution in the peruvian highlands: forensic investigation in a military base. In E.H. Kimmerle and J.P. Baraybar (eds), *Skeletal Trauma: Identification of Injuries Resulting from Human Rights Abuse and Armed Conflict* (pp. 255–262). CRC Press-Taylor and Francis Group, Oxford.

Barbian, L.T. and Sledzik, P.S. (2008). Healing following cranial trauma. *Journal of Forensic Sciences* 53(2): 262–268.

Bartelink, E.J., Wiersema, J.M., and Demaree, R.S. (2001). Quantitative analysis of sharp-force trauma: an application of scanning electron microscopy in forensic anthropology. *Journal of Forensic Sciences* 46(6): 1288–1293.

Bello, S.M. and Soligo, C. (2008). A new method for the quantitative analysis of cutmark micromorphology. *Journal of Archaeological Science* 35(6): 1542–1552.

Berryman, H.E., Symes, S.A., Smith, O.C., and Moore, S.J. (1991). Bone fracture II: gross examination of fractures. *43rd Annual Meeting of the American Academy of Forensic Sciences*, Anaheim, CA, p. 150.

Berryman, H.E., Smith, O.C., and Symes, S.A. (1995). Diameter of cranial gunshot wounds as a function of bullet caliber. *Journal of Forensic Sciences* 40(5): 751–754.

Berryman, H.E. and Symes, S.A. (1998). Recognizing gunshot and blunt cranial trauma through fracture interpretation. In K.J. Reichs (ed.), *Forensic Osteology: Advances in the Identification of Human Remains*, 2nd edn (pp. 333–352). Charles C. Thomas, Springfield, IL.

Berryman, H.E. and Gunther, W.M. (2000). Keyhole defect production in tubular bone. *Journal of Forensic Sciences* 45(2): 493–487.

Bilo, R.A.C., Robben, S.G.F., and van Rijn, R.R. (2010). *Forensic Aspects of Paediatric Fractures: Differentiating Accidental Trauma from Child Abuse* (pp. 1–24). Springer-Verlag, Berlin.

Blumenschine, R.J., Marean, C.W., and Capaldo, S.D. (1996). Blind tests of interanalyst correspondence and accuracy in the identification of cut marks, percussion marks, and carnivore tooth marks on bone surfaces. *Journal of Archaeological Science* 23: 493–507.

Bornman, E., van Eeden, R., and Wentzel, M. (eds) (1998). *Violence in South Africa: a Variety of Perspectives* (pp. 155–172). Human Sciences Research Council, Pretoria.

Brogdon, B.G. (1998). Child abuse. In B.G. Brogdon (ed.), *Forensic Radiology* (pp. 281–314). CRC Press, New York.

Burd, D.Q. and Kirk, P.L. (1942). Tool marks: factors involved in their comparison and use as evidence. *Journal of Criminal Law and Criminology* 32: 679–686.

Burd, D.Q. and Greene, R.S. (1957). Toolmark examination techniques. *Journal of Forensic Sciences* 2: 297–310.

Burd, D.Q. and Gilmore, A.E. (1968). Individual and class characteristics of tools. *Journal of Forensic Sciences* 13: 390–396.

Byers, S.N. (2002). *Introduction to Forensic Anthropology: A Textbook*, 2nd edn (pp. 339–355). Pearson Education, Boston, MA.

Campos-Varela, I.Y. and Morcillo-Méndez, M.C. (2011). Dismemberment: cause of death in the Colombian armed conflict. *63rd Annual Meeting of the American Academy of Forensic Sciences*, Chicago, IL, 17: 379.

Chadwick, E.K.J., Nicol, A., Lane, J.V., and Gray, T.G.F. (1999). The biomechanics of knife stab attacks. *Forensic Science International* 105: 35–44.

Clow, C.M. (2005). Cartilage stabbing with consecutively manufactured knives: a response to *Ramirez v. State of Florida*. *Association of Firearms and Tool Mark Examiners Journal* 37(2): 86–116.

Costello, P.A. and Lawton, M.E. (1990). Do stab-cuts reflect the weapon that made them? *Journal of Forensic Science Society* 30(2): 89–95.

Cunha, E., Badal, J.V., Líryo, A., Pinheiro, J., and Symes, S.A. (2008). How easily can we derive cause and manner of death on the basis of dry bones? Lessons derived from Coimbra identified skeletal collections. *Proceedings of the 60th Annual American Academy of Forensic Sciences*, 14: 370.

Cunningham, B.M. and Holtrop, W.F. (1974). *Woodshop Tool Maintenance*. Chas. A. Bennett, Peoria, IL.

Dalati, T. (2005). Isolated hyoid bone fracture: review of an unusual entity. *International Journal of Oral Maxillofacial Surgery* 34: 449–452.

de Gruchy, S. and Roger, T.L. (2002). Identifying chop marks on cremated bone: a preliminary study. *Journal of Forensic Sciences* 47(5): 933–936.

de la Grandmaison, G.L., Brion, F., and Durigon, M. (2001). Frequency of bone lesions: an inadequate criterion for gunshot wound diagnosis in skeletal remains. *Journal of Forensic Sciences* 46: 593–595.

de la Grandmaison, G.L., Krimi, S., and Durigon, M. (2006). Frequency of laryngeal and hyoid bone trauma in nonhomicidal cases who died after a fall from a height. *American Journal of Forensic Medicine and Pathology* 27(1): 85–86.

Di Maio, V.J.M. (1999). An introduction to the classification of gunshot wounds. In *Gunshot Wounds: Practical Aspects of Firearms, Ballistics, and Forensic Techniques*, 2nd edn (pp. 65–122). CRC Press, New York.

Dirkmaat, D.C. and Adovasio, J. (1997). The role of archaeology in the recovery and interpretation of human remains from an outdoor forensic setting. In W.D. Haglund and M.H. Sorg (eds), *Forensic Taphonomy: the Postmortem Fate of Human Remains* (pp. 39–57). CRC Press, New York.

Dirkmaat, D.C., Cabo, L.L., Ousley, S.D., and Symes, S.A. (2008). New perspectives in forensic anthropology. *Yearbook of Physical Anthropology* 51: 33–52.

Disston, Henry and Sons, Inc. (1922). *The Saw in History: a Comprehensive Description of the Development of this most Useful of Tools from the Earliest Times to the Present Day*, 6th edn. Keystone Saw, Tool, Steel and File Works, Philadelphia, PA.

Dixon, D.S. (1982). Keyhole lesions in gunshot wounds of the skull and direction of fire. *Journal of Forensic Sciences* 27(3): 555–566.

Doretti, M. and Snow, C.C. (2003). Forensic anthropology and human rights. In D.W. Steadman (ed.), *Hard Evidence: Case Studies in Forensic Anthropology* (pp. 290–310). Pearson Education, Old Tappan, NJ.

Fackler, M.L. (1988). Wound ballistics: a review of common misconceptions. *Journal of the American Medical Association* 259(18): 2730–2736.

Fondebrider, L. (2009). The application of forensic anthropology to the investigation of 67 cases of political violence: perspectives from South America. In S. Blau and D.H. Ubelaker (eds), *Handbook of Forensic Anthropology and Archaeology* (pp. 67–76). Research Handbooks in Archaeology Series. Left Coast Press, Walnut Creek, CA.

Frankel, V.H. and Nordin, M. (2001). Biomechanics of bone. In M. Nordin and V.H. Frankel (eds), *Basic Biomechanics of the Musculoskeletal System*, 3rd edn (pp. 26–59). Lippincott Williams and Wilkins, Philadephia, PA.

Fundacion de Antropologia Forense Guatemala (2009). *Standard Operating Procedures*. d. Cortocontundente (TCC). Fundacion de Antropologia Forense Guatemala-FAFG, Guatemala [in Spanish].

Galloway, A. (1999a). The biomechanics of fracture production. In A. Galloway (ed), *Broken Bones: Anthropological Analysis of Blunt Force Trauma* (pp. 35–62). Charles C. Thomas, Springfield, IL.

Galloway, A., Symes, S.A., Haglund, W., and France, D.L. (1999). The role of forensic anthropology in trauma analysis. In A. Galloway (ed), *Broken Bones: Anthropological Analysis of Blunt Force Trauma* (pp. 5–31). Charles C. Thomas, Springfield, IL.

Glassman, D.M. and Crow, R.M. (1996). Standardization model for describing the extent of burn injury to human remains. *Journal of Forensic Sciences* 41(1): 152–154.

Gozna, E.R. (1982). Biomechanics of long bone injuries. In E.R. Gozna and I.J. Harrington (eds), *Biomechanics of Musculoskeletal Injury* (pp. 1–24). Williams and Wilkins, Baltimore, MD.

Gross, M. and Eliashar, R. (2004). Hyoid bone fracture. *Annals of Otology, Rhinology, Laryngology* 113(4): 338–339.

Gupta, R., Clarke, D.E., and Wyer, P. (1995). Stress fracture of the hyoid bone caused by induced vomiting. *Annals of Emergency Medicine* 26(4): 518–521.

Gurdjian, E.S., Webster, J.E., and Lissner, H.R. (1950). The mechanism of skull fracture. *Radiology* 54(3): 313–338.

Harkess, J.W., Ramsey, W.C., and Harkess, J.W. (1996). Principles of fractures and dislocations. In C.A. Rockwood, D.P.

Green, R.W. Bucholz, and J.D. Heckman (eds), *Rockwood and Green's Fractures in Adults*, 4th edn (pp. 3–120). Lippincott-Raven, Philadelphia, PA.

Hart, G.O. (2005). Fracture pattern interpretation in the skull: differentiating blunt force from ballistics trauma using concentric fractures. *Journal of Forensic Sciences* 50(6): 1276–1281.

Hermann, N.P. and Bennett, J.L. (1999). The differentiation of traumatic and heat-related fractures in burned bone. *Journal of Forensic Sciences* 44(3): 461–469.

Hildebrand, M. and Goslow, G. (2001). *Analysis of Vertebrate Structure*, 5th edn (pp. 380–400). John Wiley and Sons, New York.

Huelke, D.F. and Darling, J.H. (1964). Bone fractures produced by bullets. *Journal of Forensic Sciences* 9(4): 461–469.

Huelke, D.F., Harger, J.H., Buege, L.J., Dingman, H.G., and Harger, D.R. (1968). An experimental study in bioballistics: femoral fractures produced by projectiles. *Journal of Biomechanics* 1: 97–105.

Humphrey, J.H. and Hutchinson, D.L. (2001). Macroscopic characteristics of hacking trauma. *Journal of Forensic Sciences* 46(2): 228–233.

Jackson, A. and Day, D. (1978). *Tools and How to Use Them: an Illustrated Encyclopedia* (pp. 74–111). Alfred A. Knopf, New York.

Johnson, J. (1985). Current developments in bone technology. *Advances in Archaeological Method and Theory* 8: 157–235.

Kimmerle, E.H. and Baraybar, J.P. (eds) (2008). *Skeletal Trauma: Identification of Injuries Resulting from Human Rights Abuse and Armed Conflict*. CRC Press, Boca Raton, FL.

Kleinman, P.K. (1998). Skeletal trauma: general considerations. In P.K. Kleinman (ed.), *Diagnostic Imaging of Child Abuse*, 2nd edn (pp. 8–26). Mosby, St. Louis, MO.

Kranioti, E. and Paine, R.R. (2011). Forensic anthropology in Europe: an assessment of current status and application. *Journal of Anthropological Sciences* 89: 71–92.

Kroman, A., Kress, T., and Porta, D. (2011). Fracture propagation in the human cranium: a re-testing of popular theories. *Clinical Anatomy* 24(3): 309–318.

Langley, N.R. (2007). An anthropological analysis of gunshot wounds to the chest. *Journal of Forensic Sciences* 52(3): 532–7.

Lanz, H. (1985). *Japanese Woodworking Tools*. Sterling Publishing, New York.

Lewis, J.E. (2008). Identifying sword marks on bone: criteria for distinguishing between cut marks made by different classes of bladed weapons. *Journal of Archaeological Science* 35(7): 2001–2008.

Love, J.C. and Symes, S.A. (2004). Understanding rib fracture patterns: incomplete and buckle fractures. *Journal of Forensic Science* 49(6): 1153–1158.

Maat, G.J.R. (2008). Case study 5.3: dating of fractures in human dry bone tissue. The Berisha ase. In E.H. Kimmerle and J.P. Baraybar (eds), *Skeletal Trauma: Identification of Injuries Resulting from Human Rights Abuse and Armed Conflict* (pp. 245–254). CRC Press-Taylor and Francis Group, Oxford.

Madea, B. and Staak, M. (1988). Determination of sequences of gunshot wounds to the skull. *Journal of the Forensic Science Society* 28(5–6): 321–328.

Marciniak, S.M. (2009). A preliminary assessment of the identification of saw marks on burned bone. *Journal of Forensic Sciences* 54(4): 779–785.

Marks, M., Hudson, J.W., and Elkins, S.K. (1999). Craniofacial fractures: collaboration spells success. In A. Galloway (ed), *Broken Bones: Anthropological Analysis of Blunt Force Trauma* (pp. 258–286). Charles C. Thomas, Springfield, IL.

Martrille, L., Cattaneo, C., Symes, S.A., and Baccino, E. (2007). Bones in aid of forensic pathology: trauma isn't only skin deep. *Proceedings of the 60th Annual American Academy of Forensic Sciences*, 14: 335.

Mayne-Correia, P.M. (1997). Fire modification of bone: a review of the literature. In W.D. Haglund and M.H. Sorg (eds), *Forensic Taphonomy: the Postmortem Fate of Human Remains* (pp. 275–293). CRC Press, New York.

Nagyszalanczy, S. (2003). Tools that saw. In *The Homeowner's Ultimate Tool Guide:*

Choosing the Right Tool for Every Home Improvement Job (pp. 78–107). Taunton Press, Newtown, CT.

Nawrocki, S.P. (2009). Forensic taphonomy. In S. Blau and D.H. Ubelaker (eds), *Handbook of Forensic Anthropology and Archaeology*. World Archaeological Congress: Research. Handbooks in Archaeology (pp. 284–295). Research Handbooks in Archaeology Series. Left Coast Press, Walnut Creek, CA.

O'Connor, J.F. and Cohen, J. (1998). Dating fractures. In P.K. Kleinman (ed.), *Diagnostic Imaging of Child Abuse*, 2nd edn (pp. 168–177). Mosby, St. Louis, MO.

Ogden, J.A. (1991). The uniqueness of growing bone. In C.A. Rockwood, K.E. Wilkins, and R.E. King (eds), *Rockwood and Wilkins' Fractures in Children*, 3rd edn (pp. 5–80). Lippincott, Williams and Wilkins, Philadelphia, PA.

Ortner, D.J. (2003). Trauma. In *Identification of Pathological Conditions in Human Skeletal Remains*, 2nd edn (pp. 119–177). Smithsonian Institution Press, Washington DC.

Ortner, D.J. and Putschar, W.G.J. (1981). Trauma. In *Identification of Pathological Conditions in Human Skeletal Remains* (pp. 55–81). Smithsonian Contribution to Anthropology No. 28. Smithsonian Institution Press, Washington DC.

Özkaya, N. and Nordin, M. (1999). *Fundamental of Biomechanics: Equilibrium, Motion and Deformation*, 2nd edn (pp. 17–21; 125–147; 206–210). Springer Science and Business Media, New York.

Pearson, O.M. and Lieberman, D.E. (2004). The aging of Wolff's law: ontogeny and responses to mechanical loading in cortical bone. *American Journal of Physical Anthropology* 39: 63–99.

Pierce, M.C., Bertocci, G.E., Vogeley, E., and Moreland, M.S. (2004). Evaluating long bone fractures in children: a biomechanical approach with illustrative cases. *Child Abuse and Neglect* 28(5): 505–524.

Pinheiro, J. (2006). Introduction to forensic medicine and pathology. In A. Schmitt, E. Cunha, and J. Pinheiro (eds), *Forensic Anthropology and Medicine: Complementary Sciences from Recovery to Cause of Death* (pp. 13–38). Humana Press, Totowa, NJ.

Pinheiro, J., Lyrio, A., Cunha, E., and Symes, S.A. (2008). Cranial bone trauma: misleading injuries. *Proceedings of the 60th Annual Meeting of the American Academy of Forensic Sciences* 14: 363.

Pope, E.J. and Smith, O.C. (2004). Identification of traumatic injury in burned cranial bone: an experimental approach. *Journal of Forensic Sciences* 49(3): 431–440.

Prieto, J. (2007). Stab wounds: the contribution of forensic anthropology: a case study. In M. Brickley and R. Ferllini (eds), *Forensic Anthropology: Case Studies from Europe* (pp. 19–37). Charles C. Thomas, Springfield, IL.

Quatrehomme, G. and İşcan, M.Y. (1997). Beveling in exit gunshot wounds in bone. *Forensic Science International* 89(1–2): 93–101.

Quatrehomme, G. and İşcan, M.Y. (1998). Analysis of beveling in gunshot entrance wounds. *Forensic Science International* 3(1): 45–60.

Quatrehomme, G. and İşcan, M.Y. (1999). Characteristics of gunshot wounds in the skull. *Journal of Forensic Sciences* 44(3): 568–576.

Rae, A. (2002). *Hand Saws. Choosing and Using Hand Tools* (pp. 188–201). Lark Books, New York.

Reichs, K. (1998). Postmortem dismemberment: recovery, analysis and interpretation. In K.J. Reichs (ed.), *Forensic Osteology: Advances in the Identification of Human Remains*, 2nd edn (pp. 353–388). Charles C. Thomas, Springfield, IL.

Reilly, D.T. and Burstein, A.H. (1974). The mechanical properties of cortical bone. *Journal of Bone Joint Surgery* 56A(5): 1001–1021.

Ross, A.H. (1996). Caliber estimation from cranial entrance defect measurements. *Journal of Forensic Sciences* 41: 629–633.

Salaman, R.A. (1975). *Dictionary of Tools used in the Woodworking and Allied Trades, c. 1700–1970*. Scribner, New York.

Salter, R.B. (1980). Birth and pediatric fractures. In R.B. Heppenstall (ed.), *Fracture*

Treatment and Healing (pp. 189–234). W.B. Saunders, Philadelphia, PA.

Sauer, N.J. (1984). Manner of death: skeletal evidence of blunt and sharp instrument wounds. In T.A. Rathbun and J.B. Buikstra (eds), *Human Identification* (pp. 176–184). Charles C. Thomas, Springfield, IL.

Saville, P.A., Hainsworth, S.N., and Rutty, G.N. (2006). Cutting crime: the analysis of the "uniqueness" of saw marks on bone. *International Journal of Legal Medicine* 121: 349–357

Schwend, R.M., Blakemore, L.C., and Lowe, L. (2009). The orthopeadic recognition of child maltreatment. In J.H. Beaty and J.R. Kasser (eds), *Rockwood and Wilkins' Fractures in Children*, 7th edn (pp. 192–219). Lippincott, Williams and Wilkins, Philadelphia, PA.

Self, C. (2005). Blade runner: knowing your saw blades will improve your shop's efficiency. *Wood Shop News* March 3: T29–T32.

Semeraro, D.S., Passalacqua, N.V., Symes, S.A., and Gilson, T.P. (2009). Patterns of blunt force trauma induced by motorboat and ferry propellers as illustrated by three known cases from Rhode Island. Paper presented to the *61st meeting of the American Academy of Forensic Sciences*, Denver, CO, 15: 318–319.

Sherwood, L. (2006). The blood and body defenses. In *Fundamentals of Physiology: a Human Perspective*, 3rd edn (pp. 315–363). Thomson Learning, Belmont, CA.

Shipman, P., Walker, A., and Birchell, J. (1986). *The Human Skeleton*. Harvard University Press, Cambridge, MA.

Smith, O.C., Berryman, H.E., and Lahern, C.H. (1987). Cranial fracture patterns and estimate of direction of low velocity gunshot wounds. *Journal of Forensic Sciences* 32: 1416–1421.

Smith, O.C., Berryman, H.E., and Symes, S.A. (1990). Changing role for the forensic anthropologist. *Proceedings of the 42nd Annual Meeting of the American Academy of Forensic Sciences*, Cincinnati, OH, p. 132.

Smith, O.C., Pope, E.J., and Symes, S.A. (2003). Look until you see: Identification of trauma in skeletal material. In D.W.

Steadman (ed.), *Hard Evidence: Case Studies in Forensic Anthropology* (pp. 138–154). Pearson Education, Old Tappan, NJ.

Snow, C.C., Levine, L., Lukash, L., Tedeschi, L.G., Orrego, C., and Stover, E. (1984). The investigation of the human remains of the "disappeared" in Argentina. *American Journal of Physical Anthropology* 5(4): 297–299.

Spitz, W.U. and Spitz, D.J. (2006). *Spitz and Fischer's Medicolegal Investigation of Death: Guidelines for the Application of Pathology in Crime Investigation*, 4th edn (pp. 175–382). Charles C. Thomas, Springfield, IL.

Steadman, D.W., Wolfe, L.D., and Haglund, W.D. (2005). The scope of anthropological contributions to human rights investigations. *Journal of Forensic Sciences* 50(1): 1–8.

Symes, S.A. (1992). *Morphology of Saw Marks in Human Bone: Identification of Class Characteristics*. PhD dissertation, Department of Anthropology, University of Tennessee, Knoxville, TN.

Symes, S.A. and Berryman, H.E. (1989). Dismemberment and mutilation: general saw type determination from cut surfaces of bone. *Proceedings of 41st Annual Meeting of the American Academy of Forensic Sciences*, Las Vegas, NV, p. 102.

Symes, S.A. and Smith, O.C. (1998). It takes two: combining disciplines of pathology and physical anthropology to get the rest of the story. *Proceedings of the 50th Annual Meeting of the American Academy of Forensic Sciences*, 4: 208.

Symes, S.A., Smith, O.C., Berryman, H.E., Peters, C.E., Rockhold, L.A., Haun, S.J., Francisco, J.T., and Sutton, T.P. (1996). Bones: bullets, burns, bludgeons, blunderers, and why. Workshop conducted at the *48th Annual Meeting of the American Academy of Forensic Sciences*, Nashville, TN.

Symes, S.A., Berryman, H.E., and Smith, O.C. (1998). Saw marks in bone: introduction and examination of residual kerf contour. In K.J. Reichs (ed.), *Forensic Osteology: Advances in the Identification of Human Remains*, 2nd edn (pp. 389–409). Charles C. Thomas, Springfield, IL.

Symes, S.A., Smith, O.C., Gardner, C.D., Francisco, J.T., and Horton, G.A. (1999). Anthropological and pathological analyses of sharp trauma in autopsy. *Proceedings of the 51st American Academy of Forensic Sciences*, 5: 177–178.

Symes, S.A., Williams, J.A., Murray, E.A., Hoffman, J.M., Holland, T.D., Saul, J.M., Saul, F., and Pope, E.E. (2002). Taphonomical context of sharp force trauma in suspected cases of human mutilation and dismemberment. In W.D. Haglund and M.H. Sorg (eds), *Advances in Forensic Taphonomy: Method, Theory and Archaeological Perspectives* (pp. 403–434). CRC Press, New York.

Symes, S.A., Rainwater, C.W., and Myster, S.M.T. (2007). Standardizing saw and knife mark analysis on bone. Paper presented to the *59th Annual Meeting of the American Academy of Forensic Sciences*, San Antonio, TX, 13: 336.

Symes, S.A., Rainwater, C.W., Chapman, E.N., Gipson, D.R., and Piper, A.L. (2008). Patterned thermal destruction of human remains. In C.W. Schmidt and S.A. Symes (eds), *Analysis of Burned Human Remains* (pp. 15–54). Elsevier Press, New York.

Tencer, A.F. (2006). Biomechanics of fixation and fractures. In C.A. Rockwood, D.P. Green, R.W. Bucholz, and J.D. Heckman (eds), *Rockwood and Green's Fractures in Adults*, 6th edn (pp. 3–42). Lippincott-Raven, Philadelphia, PA.

Thompson, T.J. (2004). Recent advances in the study of burned bone and their implications for forensic anthropology. *Forensic Science International* 146S: S203–S205.

Thompson, T.J. (2005). Heat-induced dimensional changes in bone and their consequences for forensic anthropology. *Journal of Forensic Sciences* 50(5): 1008–1015.

Tucker, B.K., Hutchinson, D.L., Gilliland, M.F.G., Charles, T.M., Danial, H.J., and Wolfe, L.D. (2001). Microscopic characteristics of hacking trauma. *Journal of Forensic Sciences* 46(2): 234–240.

Ubelaker, D.H. (1992). Hyoid fracture and strangulation. *Journal of Forensic Sciences* 37(5): 1216–1222.

Ubelaker, D.H. (2009). The forensic evaluation of burned skeletal material: a synthesis. *Forensic Science International* 183(1–3): 1–5.

Walker, P.L., Cook D.C., and Lambert, P.M. (1997). Skeletal evidence for child abuse: a physical anthropological perspective. *Journal of Forensic Sciences* 42(2): 196–207.

Wells, C. (1964). *Bones, Bodies, and Disease*. Thames and Hudson, London.

White, T.D. and Folkens, P.A. (2000). *Human Osteology*, 2nd edn. Academic Press, London.

Wilson, S. (ed.) (1994). *Popular Mechanic's Encyclopaedia of Tools and Techniques*. Hearst Books, New York.

PART IV Developments in Human Skeletal Trauma Analysis

Introduction to Part IV

Dennis C. Dirkmaat

As described in Chapter 1, and later in Chapter 26, the pioneering work and employment of forensic anthropologists in medical examiner's offices, particularly in Memphis, TN, USA, in the early 1990s, lead forensic pathologists to the realization that the diversity of skill that forensic anthropologists bring to the job was of utility to their work. Forensic osteological analysis of varied types of skeletal material at the morgue, in the laboratory, and even during the field recovery of remains, provided a perfect partnership in the medicolegal analysis of human victims in forensic scenarios. The most significant revelation was the role that forensic anthropology could play in the analysis of skeletal trauma. Forensic pathologists conducting the postmortem examination rarely have the time to thoroughly analyze the damage to bone caused by a variety of traumatic situations, including gunshot and blunt-force trauma (BFT). At best, attempts are made to roughly piece together bone fragments, with adhering soft tissue and blood, in cranial trauma in order to determine the number of gunshot wounds and provide a general description of gunshot trajectory direction. Employing a forensic anthropologist to gather and clean the bones, join them correctly, and thoroughly analyze the patterns of bone damage represented a huge leap forward in the interpretation of trauma injuries in medicolegal situations. New research and understanding of skeletal trauma ensued. Steve Symes' dissertation research on saw cuts to bone as a result of dismemberment helped solidify the forensic anthropologist's position as a human skeletal trauma expert. This is also true when considering broken human bones in bioarchaeological and paleoanthropological situations. Historically, paleopathologists were thought to provide the best interpretations and descriptions of skeletal trauma. This is no longer the case. Adding flesh and blood to the analysis of trauma provides 'close to the bone' reconstruction of the traumatic event.

In Chapter 17 Symes et al. provide a comprehensive review of the current state of skeletal trauma research. It is clear from their descriptions that forensic anthropological perspectives and analytical methods provide an invaluable, and previously missing, component to trauma analysis within the medicolegal investigation. These analyses complement forensic pathological analysis and interpretation of soft-tissue damage and provide a more "accurate" picture of the traumatic event.

Symes et al. divide trauma into three major categories: BFT, sharp-force trauma (SFT), and gunshot trauma. They indicate that the focus is not initially on the tool used in the traumatic event; rather, the proper approach is to look at trauma from the perspective of (i) the state of the bone (living or decomposing), (ii) the speed of the impacting object (fast in terms of meters per second, or slow in terms of kilometers per hour), and (iii) type (class) of object impacting the tissue. By examining the bone-fracture biomechanics, in combination with analysis of the alteration and modification of the bones, it is possible to separate out damage that is done prior to death, around the time of death, and long after death.

Another key component to proper trauma analysis is the effective documentation of the remains at the time of discovery through archaeological recovery protocols. This serves to negate the effects of damage to bone due to recovery, transport, and subsequent handling, all factors that have adversely affected the proper analysis of trauma in the past. For example, picking up a skull by the eye orbits or nasal cavity can result in breakage of the delicate facial bones. Pulling bones from the ground in a burial, as opposed to excavating the remains, may result in new damage to the bones. The best example of recovery and transport effects on bones is with respect to fatal-fire victims. Bones altered by fire will be further damaged when removed from the scene, placed in body bags, examined at the autopsy table, and the remains manipulated by other forensic specialists. Careful notation of the condition of the bones in the field through written notes, photographs, and maps will dramatically improve the accuracy of the skeletal trauma analysis, especially whether perimortem trauma is indicated. The primary focus in the medicolegal death investigation is the reconstruction of past events and the interpretation of trauma. With the proper documentation of the context of the remains, the analysis of the bones is enhanced in regard to taphonomic modification. It is also critical to identify whether broken and damaged bones are a result of animal modification. In the interpretation of trauma in a fatal-fire victim, it is important to distinguish the normal effects of heat and fire on bones from trauma that occurs prior to the fire.

Importantly, the renewed consideration of taphonomic, skeletal biological, and biomechanical factors impacting the bone, as well as the analysis of acute (unhealed) trauma versus past trauma (healed and healing), has led to dramatic improvements in the identification, documentation, and interpretation of child abuse.

Symes and coauthors provide guidelines and even analytical pointers of how to recognize and interpret the various classes of skeletal trauma. This is only the tip of the iceberg in terms of the explosion of new research and information coming from human skeletal trauma investigations. One only needs to look to the recent significant improvements in the investigation of child abuse, fatal-fire victims, and human-rights abuse cases to understand the importance of this research.

In Chapter 18, Berryman, Lanfear, and Shirley document the current state of gunshot and ballistic trauma research. They suggest that Daubert court standards for presenting scientific analysis and interpretation requires the field of human skeletal trauma analysis to conduct new research that focuses on providing validation of current methods. Trauma analysis has previously relied on deriving information from the autopsy-room experiences and from laboratory-based biomechanical engineering research. New validation studies are needed, especially under controlled experimental conditions. To that end, they describe exciting new research including finite element analysis (FEA) in which computer modeling of complex shapes is used to determine the details of strain distribution on materials during trauma events. FEA was originally used to study the effects of automatic restraint systems (seatbelts) on human ribs and thoraxes, and recently expanded to include effects of terminal ballistic impact to tubular long bones and crania. They also discuss new ways to more definitively differentiate between gunshot trauma or BFT, and taphonomic modification. Particulate analysis of primer-derived gunshot residue on bone is examined using scanning electron microscopy in combination with energy-dispersive X-ray analysis.

In Chapter 19, Passalacqua and Fenton review the history of human skeletal trauma analysis in the USA and document current trends. They provide a discussion of some of the key issues facing the field, including (i) the most appropriate definition of perimortem versus postmortem time interval when talking about skeletal analysis of trauma, since forensic pathologists define these terms a bit differently; (ii) the role played by forensic taphonomy in trauma analysis; and (iii) the significance of fast versus slow loading in interpreting and describing skeletal trauma.

As discussed in Chapter 17 and 18, trauma analysis has evolved to a stage whereby new research needs to be conducted to validate past results and interpretations. In response to that call, Passalacqua and Fenton describe some of their recent research in BFT. Using fetal pigs as an animal model for infants, they were able to document specific biomechanical details of cranial-fracture propagation in infants under strict laboratory conditions.

CHAPTER 18

The Biomechanics of Gunshot Trauma to Bone: Research Considerations within the Present Judicial Climate

Hugh E. Berryman, Alicja K. Lanfear, and Natalie R. Shirley

Introduction

Trauma interpretation is a dynamic field within forensic anthropology that has undergone changes in theory and practice through innovative research and will continue to change in response to demands placed on scientific rigor by rulings such as *Daubert v. Merrell Dow Pharmaceuticals* (1993). This ruling demands that forensic experts use methods that satisfy a set of criteria in order to be allowed as testimony in a court of law. Namely, the methods must be rigorously tested, subject to peer review and publication, have a known error rate, have standards controlling the technique's operation, and be accepted within the relevant scientific community. Among the areas of forensic anthropology most challenged by these admissibility demands is trauma interpretation.

Traditionally, skeletal trauma interpretation has been based on descriptive categories (i.e., blunt, sharp, gunshot, and thermal) established centuries earlier by coroners. These categories were based upon known events that resulted in gross injury characteristics that readily lent themselves to identification. Although identification of trauma type remains the initial objective of death investigation, courts increasingly expect more

A Companion to Forensic Anthropology, First Edition. Edited by Dennis C. Dirkmaat.
© 2012 John Wiley & Sons Ltd. Published 2015 by John Wiley & Sons Ltd.

precise interpretation (e.g., number of blows, bullet trajectory, amount of force required, bullet caliber). Until recently, such interpretations have derived largely from hands-on experience at autopsy or were surmised from osteological observations within an archaeological context. Bone trauma interpretation has not routinely been derived from controlled experiments that examine the complex interactions of intrinsic and extrinsic factors affecting the expression of bone trauma. This is not to imply that observation-based interpretations are incorrect or that experience is irrelevant in trauma analysis, but rather that conclusions reached in this manner do not satisfy Daubert criteria.

Biomechanical research can produce information about the repeatability of interpretations and the associated error rates needed to satisfy these more rigorous Daubert standards. Biomechanics brings a deeper understanding of the etiology of trauma than is possible through observations alone. However, empirically based trauma analysis requires an established body of literature beyond those found in basic bone biomechanics or trauma surgery journals. Research designed to address specific questions involving forensic analyses needs to be disseminated through the literature for trauma interpretation to address Daubert specifications.

The need for this type of research was not ignored completely by early practitioners. A number of controlled bone trauma experiments on embalmed human cadaveric specimens and on dry bones were conducted as early as the mid-1900s using strain gauges, a "stress coat" technique, and, in some studies, high-speed photography to examine fracture-producing impacts to the cranium, mandible, femur, pelvis, and spine (Evans 1952; Gurdjian and Lissner 1945, 1947; Gurdjian et al. 1947, 1949, 1950a, 1950b; Huelke 1960a, 1960b, 1961a, 1961b, 1961c). These experiments quantified the energy required to produce a skull fracture and suggested a mechanism of linear fracture production from blunt impacts. However, recent research has suggested that Gurdjian's proposed mechanism of skull fracture is flawed (Kroman et al. 2005; Kroman 2007). Gurdjian et al. (1950a, 1950b) suggested that linear skull fractures originate some distance from the impact area and travel back toward the site of impact. Controlled experiments using high-speed video, a steel "drop tower" structure, and a weight with an attached instrumented load cell could not verify this mechanism (Kroman et al. 2005; Kroman 2007). This study clearly showed that the fractures originate at the point of impact and travel away from this site. These recent experiments were conducted at impact speeds consistent with those of low-velocity trauma and on bone surface areas consistent with that affected by blunt and sharp impacts. This research led Kroman (2007: vii) to conclude that "Bone trauma is best viewed as a continuum (rather than discrete independent categories), with the variables of force, acceleration/deceleration, and surface area of impacting interface governing the appearance of the resulting fractures. The application of this new way of thinking will allow anthropologists to better understand bone fracture and injury to the body as a relationship between the engineering inputs and the anatomical outputs." To understand the position gunshot trauma takes on this continuum, the basics of bone biomechanics must be understood.

BASIC BONE BIOMECHANICS

Knowledge of biomechanics and, in particular, the intrinsic properties of bone is needed to determine the boney response to various loading situations. Currey (1970) eloquently describes the mechanical properties of bone, illustrating that despite the complicated nature involved in analyzing bone, much of the basic knowledge is available. He notes that "...the study of bone mechanics is entering the beginning of an extremely interesting second phase: we are now in a position to start to understand bone" (Currey 1970: 230).

The mineral content of bone is the main determinant of differences in mechanical properties (Currey 2002). As bone material properties change during growth, the architecture is modified to provide the most advantageous behavior of the bone (Currey 2002). According to the general meaning of Wolff's law, the bone functional adaptation model, organisms possess the ability to adapt their structure to new living conditions, and bone cells are capable of responding to local mechanical stresses (Ruff et al. 2006). In other words, during modeling and remodeling, bone is organized to resist loads imposed by functional activities, meaning that bone is laid down where it is needed and removed where it is not needed.

Additionally, bone is a viscoelastic material in that it can behave with a ductile or brittle response depending on the velocity, rate, duration, and direction of the force(s) involved. The physical and biomechanical properties of ductile materials are such that they can absorb more energy before failure than brittle materials. The viscoelastic properties of bone derive from the material and structural properties of this specialized connective tissue, which in turn account for the biomechanical differences in various types of bone tissue. In general, cortical bone comprises about 80% of bone mass in healthy individuals, and trabecular bone constitutes the remaining 20%. Approximately 60% of cortical bone weight comes from calcium compounds that provide stiffness and strength, 30% is comprised predominantly of Type I collagen that provides tensile strength, and about 10% comes from water which provides strength against compressive forces and maintains bone health. In addition, cortical bone tissue is comprised of dense populations of Haversian systems, or osteons, arranged around vessels for blood and lymph supply.

In contrast, trabecular bone is comprised of a porous network of trabeculae, or lamellar bone, filled with bone marrow. The porosity of trabecular bone is the primary determinant of its stiffness and strength; however, trabecular architecture can vary among anatomical sites and with age, further influencing its biomechanical properties. An analysis of the mechanical properties of bone can range from the microscopic collagen fibrils to trabecular or cortical bone architecture to the structural properties of an entire bone. Consequently, it is important to understand the contribution of each of these attributes to the overall behavior of bone under various mechanical stresses and to incorporate these principles into trauma interpretation (Currey 2002).

The material properties of a ductile material can be shown using a stress/strain curve. Three significant regions can be seen along the stress/strain curve: elastic deformation, plastic deformation, and failure. In the elastic region, stress (force per unit area) is proportional to strain (amount of deformation under loading), and the material retains its original shape properties upon the release of loading. Once elastic deformation is surpassed, the material enters the region of plastic deformation where the molecular properties of the material are rearranged to the point of permanent shape deformation. Failure is the threshold at which maximum stress causes the material to fracture or crack; at this point the material cannot absorb further energy.

Toughness is defined by the area under the stress/strain curve up to the point of failure and can be thought of as the material's strength. Metal is one of the most ductile of materials, and glass is one of the most brittle, as it undergoes no deformation prior to failure. For example, bone typically behaves as a brittle material when subjected to high-velocity trauma such as gunshot injuries. With high-velocity gunshot wounds the rate of loading is so rapid that the bone fails suddenly before deformation can occur. On the other hand, bone behaves more like a ductile material and deforms prior to failure when subjected to slow-loading trauma such as blunt trauma, or where a bullet has lost most of its energy, or wounding capacity, prior to impacting bone.

Common loading situations for bone are compression, torsion, and bending. Bone is strongest in compression and weakest in tension and shear. Bone is also anisotropic, meaning that it responds differently to loads applied from different directions. The anisotropic properties of cortical bone differ from those of trabecular bone. Cortical bone is stiffer than trabecular bone and is much stronger in compression than in tension and shear. Trabecular bone is also stronger in compression than in tension but has a higher energy-storage capacity prior to failure than cortical bone (Keaveny et al. 1994). In addition, the stiffness of trabecular bone varies greatly with differences in density. These differences, combined with the microscopic differences in cortical and trabecular bone (i.e., osteons versus lamellae), translate to differences in fracture biomechanics and appearance in various anatomical locations.

A thorough trauma analysis must take numerous variables into account, including the material properties of bone and the numerous extrinsic factors that influence fracture and injury production (e.g., velocity, rate, duration, and magnitude of the force). In order to have a better understanding of the complex relationships of these variables, it is imperative that trauma research in forensic anthropology becomes focused on their interaction in a controlled, experimental setting. In addition, accurate interpretation of bone trauma necessitates a multidisciplinary approach that integrates knowledge of bone histology and morphology, injury biomechanics and, with gunshot trauma, terminal ballistics. The interplay of extrinsic factors associated with gunshot trauma is complex. Gunshot trauma interpretation is complicated by higher velocities with greater rates of loading and by varying bullet masses and bullet designs.

GUNSHOT WOUND TRAUMA

The behavior of a bullet upon impact with a target and its interaction with that target has been covered extensively in the literature under a subfield of ballistics called *terminal ballistics* (Berlin et al. 1979; DiMaio 1999; Dodd 2005; Harvey et al. 1962). In essence, a projectile transfers kinetic energy ($E_k = \frac{1}{2}mv^2$, where m is the mass and v is the velocity of the projectile) to the body tissues upon impact. This energy transfer results in the formation of a temporary cavity that undulates for 5–10 ms before settling into its final configuration, termed the permanent cavity or wound tract. The dimensions of the temporary cavity are largest at the point where maximum kinetic energy loss occurs (Marshall and Sanow 2001). Many factors affect the size and shape of the temporary cavity, including bullet design, bullet velocity, yaw, and elasticity and cohesiveness of the affected tissue. As is evident from the equation, the amount of kinetic energy that a projectile possesses increases exponentially as velocity increases; therefore, high-velocity projectiles have the potential to produce greater tissue destruction than low-velocity projectiles. Also, penetrating trauma (i.e., gunshot wounds in which the bullet does not exit) from a high-velocity bullet has the potential to cause massive tissue destruction because all of the projectile's kinetic energy is transferred to the surrounding tissues. However, with perforating trauma (i.e., wounds with an entrance and exit) from a high-velocity bullet, a portion of the kinetic energy is transferred to the body and a portion leaves with the projectile upon exit, producing less tissue destruction. In some instances bullets may fragment, and the secondary projectiles may cause additional injuries to surrounding tissues.

Various weapon classes produce distinctive wound characteristics in both soft tissue and bone (Marshall and Sanow 2001). Given the same firing distance and scenario, small-caliber handguns are less destructive than large-caliber handguns and rifles. Handguns and rifles fire a single projectile and the rifling of these weapons imparts a spin to the projectile, providing gyroscopic stability and greater accuracy. Shotguns, on the other hand, have a smooth bore with no rifling and fire multiple pellets (or, occasionally, a single slug), producing wounds that vary widely in appearance depending on the gauge, choke, and proximity of the weapon to the victim (Guerin 1960). At close ranges, the wadding and the gases produced by a shotgun upon firing can cause additional tissue injury. Close-range shotgun wounds are highly destructive, since the pellets enter the body as a single mass, causing a single entrance defect and massive fragmentation, particularly in the skull. At more distant ranges, the pellets separate and may enter the body individually, thereby creating a number of separate impact defects or pellet wipes (areas of discoloration on the bone produced by pellet strikes). From very distant ranges, the pellets may possess the energy to penetrate soft tissues but fail to perforate the underlying bony tissue.

A thorough analysis should consider the complex set of extrinsic variables that the weapon and ammunition contribute to the final appearance of the trauma. While it is generally possible to distinguish weapon class and, in some instances, to discern small-versus large-caliber weapons, correlating wound dimensions with a specific bullet caliber is not advisable (Berryman et al. 1995; Ross 1996). DiMaio (1999) maintains that, "the size of an entrance in bone cannot be used to determine the caliber of the bullet that perforated the bone though it can be used to eliminate bullet calibers." Depending on the scenario, the resulting bony defect may be larger or smaller than the bullet caliber (Symes et al. 1996). For example, multiple gunshot wounds in a cranium caused by a single weapon may vary widely in size. The initial wound may exhibit more damage than subsequent wounds, in which the intracranial pressure has been reduced by the first gunshot wound, and the integrity of the cranial vault has been compromised. Furthermore, subsequent gunshot wounds can enter and/or exit existing fractures or open sutures. Gunshot wound appearance varies with bullet velocity, distance between the muzzle and victim, body position at the time of impact, angle of entry/exit, and the presence or absence of clothing (DiMaio 1999). Furthermore, bullet shape and surface treatment, strength characteristics, tangential impacts, and intermediate targets influence the size and appearance of gunshot wounds (Berryman et al. 1995). Although consistencies can be observed under controlled conditions, predicting bullet caliber or bullet characteristics in a clinical case is not practical, and represents an area where research is needed to satisfy the demands of Daubert.

CURRENT GUNSHOT RESEARCH AND FUTURE DIRECTIONS

Fractures do not occur randomly, but strictly obey the laws of physics with pattern resulting from the culmination of both intrinsic factors (e.g., bone elasticity, plasticity, thickness, presence of sutures, shape, compromised bone integrity) and extrinsic factors (e.g., caliber or gauge, velocity, mass, bullet characteristics, distance from target). However, at autopsy the identification and interplay of these factors may be impossible to discern, making accurate interpretation implausible. In a laboratory setting, intrinsic and extrinsic factors can be tightly controlled and varied to observe resulting gunshot fracture patterns and entrance- and exit-wound characteristics. Entrance-wound dimension and morphology is a common and understandable focus in gunshot research, specifically to ascertain bullet direction and caliber. An often-overlooked extrinsic factor is the dynamic effects of bullet velocity, especially obvious with gunshot wounds to the cranium. Intuitively, the entrance-wound defect produced in a cranial vault by a .22-caliber handgun should be basically the same as a .223-caliber rifle, as both bullets measure twenty-two one hundredths of an inch in diameter. However, differing velocities of these two ammunition types will produce vastly different temporary cavities with highly disparate fracture patterns to the vault. The muzzle velocity of the .223 caliber ammunition may exceed 915 m/s, while the muzzle velocity of a .22 long rifle bullet is less than 450 m/s. Although the muzzle velocity of a .223 bullet is twice that of a .22 bullet, the magnitude of difference in the transfer of kinetic energy is exponential, as can be seen in the formula ($E_k = \frac{1}{2}mv^2$), thereby resulting in much more destruction to bone.

The affects of differing muzzle velocities on fracture production, as well as other extrinsic factors, can be laboratory tested using fleshed human crania or ballistic gelatin with real or simulated bone. However, new technologies, such as finite element analysis (FEA), offer the opportunity for virtual testing. FEA is a computer method for determining forces, such as strain, in loaded models and is particularly useful in determining strains in bones of complex shapes (Richmond et al. 2005). This method reduces complex geometry into a finite number of elements, each comprised of simpler geometries, allowing strain to be modeled throughout the entire structure in addition to its surface. The specimen must be digitized in some fashion and material properties of the bone inputted. Loads are then added, which results in an output of the distribution of strains produced in the model. Each step makes several assumptions that can be of concern, primarily the knowledge of the material properties of bone. However, in the case of fracture mechanics these assumptions can, within reason, be accepted (Richmond et al. 2005) where the aim is a general understanding of the strain experienced by bone under a particular load.

Panagiotopoulou (2009) and Richmond et al. (2005) present excellent reviews of the application and limitations of FEA. FEA trauma research started in car-accident investigation and crash-testing research and is published predominantly in Highway Traffic Safety Administration papers and the *Stapp Car Crash Journal*. Most of these FEA models were developed for testing automatic restraint systems and, consequently, deal with the ribs and thorax (Li et al. 2010; Roberts et al. 2007; Shen et al. 2008; Stepanskiy and Seliktar 2007; Zhao and Narwani 2005). Zhao and Narwani (2005), in particular, integrated several finite element models, including the thorax, abdomen, shoulder, and head/neck, to predict chest deflections and shapes of the fourth and eighth rib sections. Comparisons with autopsy data showed that the model could estimate bone-fracture and organ-failure locations.

Shen et al. (2008) also developed and validated a finite element model of swine and human thoraxes and abdomens. This model had subject-specific anatomies that were able to predict body response to blunt impact. Li et al. (2010) investigated whether a finite element model comprised of both cortical and trabecular bone in human ribs would provide more accurate simulations. They found that a more simple model using constant cortical thickness did not significantly change reaction forces but increased the failure displacement and time. As the simpler model is less taxing to compute, Li et al. (2010) suggest using this when accuracy in failure displacement and time are not the main concern of the investigation.

Berryman et al. (2003) presented one of the first papers that introduced FEA to the forensic anthropology community in a study that examined fracture formation in tubular bone. However, the use of FEA to investigate terminal ballistic trauma is still in its infancy. Mota et al. (2003) were among the first to use FEA to simulate firearm injury to the human cranium. Their finite element model simulated the impact of a bullet with a human parietal bone. High- (1000 m/s) and low- (600 m/s) velocity impacts were simulated, both of which resulted in comminuted fragments similar to those observed in forensic medicine contexts. However, the Mota et al. (2003) model does not include soft tissues. They argue that soft tissues would play a dissipative role because the soft tissue would absorb some of the impact energy and reduce the wound diameter observed. Despite several simplifying assumptions of their finite element model, Mota et al. (2003) demonstrate that this method provides a unique tool for investigating the mechanics of firearm injuries, including the effects of caliber, trajectory, and speed on the geometry and severity of injury.

More recently, Chen at al. (2010) used a finite element model to simulate ballistic impact to the mandibular angle of a pig. They found no statistically significant difference between the finite element simulation and the experimental study in the transferred energy from the bullet to the mandible or in the surface area of the entrance wound. The mean surface area of the exit wounds was significantly different between the finite element simulation and the experimental study, with the experimental exit wounds being larger than those of the finite element simulation (Chen et al., 2010). Interestingly, the simulation indicated that trabecular bone had less stress concentration and lower speed of stress propagation compared to cortical bone (Chen et al., 2010). FEA will lead to a better understanding of the biomechanics of gunshot trauma to bone.

A bullet can fracture bone without leaving the characteristic circular entrance wound which can lead to misinterpretation, or taphonomic processes can modify bone, impeding accurate assessment. Techniques are needed to differentiate between gunshot trauma and damage produced by blunt trauma or taphonomy. One promising study uses the presence of primer-derived gunshot residue (GSR) on bone as an indicator of gunshot trauma (Berryman et al. 2010). Unlike the soot observed around close-range entrance wounds, primer-derived GSRs are unique in their particle morphology and elemental composition. This makes them distinct from other environmental particulates, including occupational particles such as lead aerosols, automobile exhaust, and condensation fumes that may contain one or more elements of primer-derived GSR (Basu, 1982). These particles are unique to the shooting environment and in the study by Berryman et al. (2010) they were found deep within the wound tract of pork ribs that were shot at distances ranging from 30 cm to 1.5 m. As these particles range in diameter from 0.1 to 10 μm, a scanning electron microscope (SEM) must be used in order to visualize primer-derived GSR. An energy-dispersive X-ray analysis (EDXA) is also necessary to verify the particles' elemental composition. Unfortunately, this research faces several practical limitations. The cost and expertise necessary to run SEM-EDXA limits the number of individuals who can utilize the method. Additionally, further research is necessary to evaluate whether these primer-derived GSR particles remain after decomposition, and at what gun-to-target distance these particles are observable on bone.

CONCLUSIONS

Changes in the judicial climate related to the forensic sciences demand a rigorous research agenda and innovative thinking in all areas of forensic anthropology. In trauma analysis, this has meant a rethinking of traditional, category-based interpretations and a shift toward a more holistic understanding of skeletal trauma. Researchers profit by incorporating ideas and principles from materials engineering, biomechanics, medical sciences, and terminal ballistics into their research designs. The application of research approaches involving FEA, SEM, and EDXA point to a new era that promises to align trauma analysis in general and gunshot trauma in particular with the expectations set forth by today's scientific and legal environment.

REFERENCES

Basu, S. (1982). Formation of gunshot residues. *Journal of Forensic Sciences* 27(1): 72–91.

Berlin, R., Janzon, B., Kokinakis, W., Eiseman, B., Albrecht, M., and Owen-Smith, M. (1979). Various technical parameters influencing wound production: a panel discussion. *Acta Chirurgica Scandinavica* 489 Suppl.: 103–120.

Berryman, H.E., Smith, O.C., and Symes, S.A. (1995). Diameter of cranial gunshot wounds as a function of bullet caliber. *Journal of Forensic Sciences* 40(5): 751–754.

Berryman, J.F., Berryman, H.E., LeMaster, R.A., and Berryman, C.A. (2003). Numerical simulation of fracture propagation in a test of cantilevered tubular bone. *Proceedings of the 54th Annual Meeting of the American Academy of Forensic Sciences*, Chicago, IL, IX: 260.

Berryman, H.E., Kutyla, A.K., and Davis, J.R. (2010). Detection of gunshot primer residue on bone in an experimental setting—an unexpected finding. *Journal of Forensic Sciences* 55(2): 488–491.

Chen, Y., Miao, Y., Xu, C., Zhang, G., Lei, T., and Tan, Y. (2010). Wound ballistics of the pig mandibular angle: A preliminary finite element analysis and experimental study. *Journal of Biomechanics* 43(6): 1131–1137.

Currey, J.D. (1970). The mechanical properties of bone. *Clinical Orthopedic* 73: 210–231.

Currey, J.D. (2002). *Bones: Structure and Mechanics.* Princeton University Press, Princeton, NJ.

Daubert v. Merrell Dow Pharmaceuticals, 509 US 579, 113 S.Ct. 2786, 125 L.Ed. 2d 469 (US June 28, 1993) (No. 92-102).

DiMaio, V.J. (1999). *Gunshot Wounds: Practical Aspects of Firearms, Ballistics, and Forensic Techniques*, 2nd revised edn. Elsevier Science Publishing Company, New York.

Dodd, M. (2005). *Terminal Ballistics: A Text and Atlas of Gunshot Wounds*. CRC Press, Boca Raton, FL.

Evans, F.G. (1952). Stress and strain in the long bones of the lower extremity: instructional course lectures. *American Academy of Orthopaedic Surgeons* 9: 264–271.

Guerin, P.F. (1960). Shotgun wounds. *Journal of Forensic Sciences* 5: 294–318.

Gurdjian, E.S. and Lissner, H.R. (1945). Deformation of the skull in head injury: a study with the "stresscoat" technique. *Surgery Gynecology and Obstetrics* 81: 679–687.

Gurdjian, E.S. and Lissner, H.R. (1947). Deformations of the skull in head injury as studied by the "stresscoat" technique. *American Journal of Surgery* 53: 269–281.

Gurdjian, E.S., Lissner, H.R., and Webster, J.E. (1947). The mechanism of production of linear skull fractures. *Surgery Gynecology and Obstetrics* 85: 195–201.

Gurdjian, E.S., Webster, J.E., and Lissner, H.R. (1949). Studies on skull fracture with particular reference to engineering factors. *American Journal of Surgery* 78: 736–742.

Gurdjian, E.S., Webster, J.E., and Lissner, H.R. (1950a). The mechanism of skull fracture. *Journal of Neurosurgery* 7: 106–114.

Gurdjian, E.S., Webster, J.E., and Lissner, H.R. (1950b). The mechanism of skull fracture. *Radiology* 54: 313–338.

Harvey, E.N., McMillen, J.H., Butler, E.G., and Puckett, W.O. (1962). Mechanism of wounding. In J.B. Coates Jr and J.C. Beyer (eds), *Wound Ballistics*. US Government Printing Office, Washington DC.

Huelke, D.F. (1960a). Experimental studies on the mechanisms of mandibular fractures. *Journal of Dental Research* 39: 694.

Huelke, D.F. (1960b). Deformation studies of the mandible under impact. *Anatomical Record* 136: 214.

Huelke, D.F. (1961a). The production of mandibular fractures: a study in high speed cinematography. *Journal of Dental Research* 40: 743–744.

Huelke, D.F. (1961b). Mechanisms involved in the production of mandibular fractures: a study with the "stresscoat" technique. Part I. symphyseal impacts. *Journal of Dental Research*, 40: 1042–1056.

Huelke, D.F. (1961c). High speed photography of mandibular fractures. *Journal of the Biological Photographic Association* 20: 137–144.

Keaveny, T.M., Wachtel, E.F., Ford, C.M., and Hayes, W.C. (1994). Differences between the tensile and compressive strengths of bovine tibial trabecular bone depend on modulus. *Journal of Biomechanics* 27: 1137–1146.

Kroman, A.M. (2007). *Fracture biomechanics of the human skeleton*. Dissertation, University of Tennessee, Knoxville, TN.

Kroman, A., Kress, T., and Symes, S. (2005). Mandible and cranial base fractures in adults: experimental testing. *Proceedings of the American Academy of Forensic Sciences Annual Meetings*, New Orleans, LA, XI: 289.

Li, Z., Kindig, M.W., Kerrigan, J.R., Untaroiu, C.D., Subit, D., Crandall, J.R., and Kent, R.W. (2010). Rib fractures under anterior-posterior dynamic loads: experimental and finite-element study. *Journal of Biomechanics* 43: 228–234.

Marshall, E.P. and Sanow, E.J. (2001). *Stopping Power: A Practical Analysis of the Latest Handgun Ammunition*. Paladin Press, Boulder, CO.

Mota, A., Klug, W.S., Ortiz, M., and Pandolfi, A. (2003). Finite-element simulation of firearm injury to the human cranium. *Computational Mechanics* 31: 115–121.

Panagiotopoulou, O. (2009). Finite element analysis (FEA): Applying an engineering method to functional morphology in anthropology and human biology. *Annals of Human Biology* 36(5): 609–623.

Richmond, B.G., Wright, B.W., Grosse, I., Dechow, P.C., Ross, C.F., Spencer, M.A., and Strait, D.S. (2005). Finite element analysis in functional morphology. *Anatomical Record* 258(A): 259–274.

Roberts, J.C., Merkle, A.C., Biermann, P.J., Ward, E.E., Carkhuff, B.G., Cain, R.P., and O'Connor, J.V. (2007). Computational and experimental models of the human torso for non-penetrating ballistic impact. *Journal of Biomechanics* 40: 125–136.

Ross, A.H. (1996). Caliber estimation from cranial entrance defect measurements. In *Journal of Forensic Sciences* 41(4): 629–633.

Ruff, C., Holt, B., and Trinkaus, E. (2006). Who's afraid of the big bad Wolff?: "Wolff's Law" and bone functional adaptation. *American Journal of Physical Anthropology* 129: 484–498.

Shen, W., Niu, Y., Mattrey, R.F., Fournier, A., Corbeil, J., Kono, Y., and Stuhmiller, J.H. (2008). Development and validation of subject-specific finite element models for blunt trauma study. *Journal of Biomechanical Engineering* 130: 012022.

Stepanskiy, L. and Seliktar, R. (2007). Predicting fracture of the femoral neck. *Journal of Biomechanics* 40: 1813–1823.

Symes, S.A., Smith, O.C., Berryman, H., Peters, C., Rockhold, L., Huan, S., and Francisco, J.T. (1996). Bones: bullets, burns, bludgeons, blunders, and why. *Proceedings of the 48th Annual Meeting of the American Academy of Forensic Sciences*, Nashville, TN.

Zhao, J.Z. and Narwani, G. (2005). Development of a human body finite element model for restraint system R&D applications. *The 19th International Technical Conference on the Enhanced Safety of Vehicles (ESV)*, Washington DC, paper no. 05-0399.

CHAPTER 19

Developments in Skeletal Trauma: Blunt-Force Trauma

Nicholas V. Passalacqua and
Todd W. Fenton

INTRODUCTION TO BLUNT-FORCE TRAUMA

The goal of this chapter is to briefly discuss the history of blunt-force trauma (BFT) research in forensic anthropology, and then present some of the current developments in the field. The draft guidelines on trauma analysis generated by the Scientific Working Group for Forensic Anthropology (SWGANTH) in 2011 states that "blunt force trauma is produced by low velocity impact from a blunt object (e.g., a beating, motor vehicle accident, concussive wave) or the low velocity impact of a body with a blunt surface (e.g., fall)" (SWGANTH 2011) As it concerns the forensic anthropologist, BFT can be understood as skeletal fractures that occur from a means of (relatively) slow-load application to bone. Slow loading results in more or less typical bony fracture characteristics in terms of the types and directions of forces applied that usually produce gross bony deformation. In contrast, projectile (sometimes referred to as "ballistic") trauma is generated by (relatively) *rapid*-load application (see Chapter 18 in this volume, and Berryman and Symes 1998 for further discussion). The difference between how bone reacts to BFT compared with projectile trauma has to do with the viscoelastic properties of fresh (wet) human bone (Harkness et al. 1984). Biomechanically speaking, when a force acts on a bone, the bone will react in predictable and consecutive stages: *stress*, the force applied to the bone; *strain* (*elastic deformation*), forces passing through the bending bone; *strain* (*plastic deformation*), forces bending the bone with permanent deformation; and *failure*, fracture of the bone (Gozna 1982). These stages occur at both the microscopic and macroscopic

A Companion to Forensic Anthropology, First Edition. Edited by Dennis C. Dirkmaat.
© 2012 John Wiley & Sons Ltd. Published 2015 by John Wiley & Sons Ltd.

levels (see Berryman and Symes 1998; Smith et al. 2003 for further discussion). As noted by Smith et al. (2003), bony responses to blunt-force injuries will depend on a great number of both intrinsic factors (bone morphology, density, buttressing, microstructure, etc.) and extrinsic factors (object shape, weight, loading rate, loading duration, etc.), all of which must be taken into account when conducting a trauma analysis.

The difference in slow- versus rapid-load applications can be understood as how much time the bone has to bend and react (deform) before it fails. In slow-loading trauma, there is more time for the bone to bend, which is why significant deformation is so typical of BFT. Rapid loading, on the other hand, tends to have minimal deformation of skeletal tissue, which is why it is so typical of gunshot and blast injuries (and why reconstructions of these bony injuries fit together so much better).

Concerning the propagation of fractures in bone under blunt-force conditions, Kulin et al. (2008) tested equine bone at quasistatic and dynamic strain rates. In this case, the quasistatic (slower) rates produced much more tortuous fracturing throughout the bony microstructure as the fracture propagated. This basically occurs as a result of how the microstructural features of the bone deflect the propagating fracture away from the most direct crack-path, in essence limiting the overall fracture length and creating a very rough fracture surface as energy is dissipated. Under dynamic (faster) loading conditions, the fractures were much straighter with less peripheral damage and a smoother fracture surface. These results suggest a rate-dependent change in the properties of collagen from ductile to brittle as rates increase, the significance for forensic anthropology being the demonstration of the rate-dependent nature of fracture propagation.

Generally speaking, blunt-force injuries can be examined in terms of tension/compression for directionality of fracture (Figure 19.1). As bone is stronger in compression than tension (Curry 1970), bone usually fails under tension except under some unique circumstances (Love and Symes 2004). These simple principles often allow for fracture directionality to be determined; however, there are many other factors that can affect the direction and shape of bone fractures, and caution should be used whenever interpreting bone trauma (Galloway 1999; Gozna 1982; Smith et al. 2003).

Differences in fracture characteristics are also important for the determination of timing of injury, especially when determining whether the fractures occurred perimortem or postmortem. Sauer's (1998) article on the timing of injuries distinguishes among antemortem, perimortem, and postmortem trauma. Sauer (1998) states that "any injury directly associated with the manner of death is considered a perimortem injury." This view aligns closely with the use of the term by forensic pathologists, where perimortem is defined as "at or around the time of death." Nawrocki (cited as a personal communication in Symes et al. 2008) argues that the *perimortem interval* lasts until the skeletal remains exhibit dry-bone characteristics. Nawrocki's viewpoint is based on the fact that as skeletal remains lose their organic content and viscoelastic properties, they change from a wet (perimortem) to dry (postmortem) state (also see Wieberg and Wescott 2008). This is complicated, however, as different elements in a single body may be able to shift from wet to dry phases at different times depending on differential decomposition or heat alteration (Symes et al. 2008).

The disagreement on the definition of perimortem is an issue that the SWGANTH has grappled with in the trauma analysis draft document (SWGANTH 2011). The

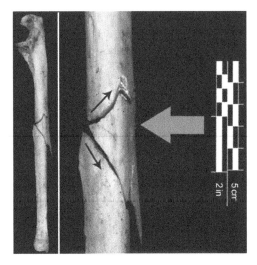

Figure 19.1   Butterfly fracture on a right ulna (often referred to as a "parry fracture" suggesting it is a defensive wound; Ortner 2003: 137). The large arrow indicates direction of impact. Compression forces occur at the impact site, whereas tension forces (small arrows) occur on the opposite side. The bone fails first in tension (at the body of the butterfly), then sheers towards the site of impact (the butterfly wings). See also Berryman and Symes (1998).

first draft document posted on the SWGANTH website in February of 2010 stated "Perimortem trauma refers to an injury occurring around the time of death (i.e., slightly before or slightly after). Within the anthropological realm, perimortem is determined on the basis of evidence of the biomechanical characteristics of fresh bone and *does not take into consideration the death event*" (SWGANTH 2011). Using the term "perimortem" in this way, however, potentially creates a problematic situation where the forensic anthropologist's definition of "perimortem" differs from that of the forensic pathologist's.

This is a serious concern because there may be occasions when a fracture identified as perimortem by the forensic anthropologist is misinterpreted by the forensic pathologist as being related to cause of death when it is not. Forensic anthropologists are not experts in death, but in how bones break. We can determine whether a fracture occurred in fresh bone and exhibits healing, if it occurred in fresh bone with no healing, or if the fracture occurred when the bone was in a dry state. Certainly, most fresh bone fractures with no healing occur perimortem; however, there are unique instances where bones remain fresh long past the death event. Thus, the use of the term "perimortem," which specifically refers to the death event, may misrepresent the type of analysis a forensic anthropologist is capable of performing on skeletal remains. This has prompted some forensic anthropologists to suggest that we abandon the term perimortem as it relates to bone fracture, and instead refer to it as "fresh bone fracture."

The issue of postmortem interval then becomes additionally cloudy as most taphonomic modification to human remains occurs in terms of slow loading. While some taphonomic modification is easy to identify, such as carnivore chewing or rodent gnawing (see Haglund 1997, 1997), other factors may be more difficult, especially with fragile remains. For instance, Ubelaker and Adams (1995) noted that butterfly fractures also occur in dry bone. Further, Wheatley (2008) examined deer (*Odocoileus virginianus*) long bones and found that fresh bone required more energy to produce fractures and that fresh bone absorbed more energy at impact than dry bone. Important to note is that while bone may fail in slightly different ways due to differences in organic content present, the acting forces are still the same. Tension and compression will always be present and bone fractures will always propagate in the same directions. Thus, failure and fracture of the bone occurs in predictable ways because there are only so many ways a bone can fail.

What may be important to note is the fact that any antemortem or perimortem damage to skeletal tissues must first have damaged the overlying soft tissues. Skeletal tissues are by definition much more robust in the nature of traumatic data that they record. Due to this fact, soft tissue may often be more accurate in recording data such as minimum number of blows, but skeletal tissue may better preserve tool marks (or general object shape) (Bonte 1975), or the sequence of blows. Unfortunately, while the sequence of blows may often be determined from skeletal remains, in BFT it is a challenging endeavor. Further, the estimation of first, second, and third impact in BFT is not frequently important and thus acts as a piece of information that, while possible to estimate, is associated with too much uncertainty with regard to its evidentiary value. In addition, while bone can record tool marks, without direct association of the blunt instrument involved or unique class characteristics present, tool marks are often too general to be indicative of blunt-instrument shape or type.

HISTORICAL TRENDS IN TRAUMA ANALYSIS

Trauma analysis did not become a common component in forensic anthropology casework until surprisingly late in the development of the field. This is most easily inferred by looking at the most prominent textbooks of the past – Krogman (1962) and Stewart (1979) – where the job of the forensic anthropologist was described as identifying human bone and developing a biological profile of an individual for purposes of *identification*. When bone trauma or "damage" was found, it was superficially described and measured. For instance, Stewart states: "a forensic anthropologist should simply describe any evidence of bone damage, point out its location in relation to vital centers, explain the possibility of its having been sustained at the time of death or otherwise, and discuss the likely objects that produced the damage" (Stewart 1979: 76). Only with the introductory text by Byers (2002) did the analysis of trauma take an equal footing with age, sex, ancestry, and stature in a textbook on forensic anthropology.

What may be helpful is to consider the evolution of trauma analysis in our discipline using the "Eras of Forensic Anthropology" as roughly defined by Stewart (1979) and Sledzik et al. (2007).

1. *Pioneering era*: pre-World War II. Characterized by anatomists and physical anthropologists consulting on forensic cases without formal instruction, research, or positions.
2. *Formative era*: 1940s–early 1970s. Forensic anthropology as a *subfield* of anthropology begins to emerge with recognition by government agencies and the

medicolegal community (may also be considered as Stewart's "Modern Period"; Stewart 1979: 11).

3. *Professionalization era*: late 1970s–1990. This era begins with the founding of the Physical Anthropology section of the American Academy of Forensic Scientists (AAFS) and the creation of the American Board of Forensic Anthropology (ABFA), but is better characterized by an increase in professional forensic anthropology training, research, and practice.

4. *Standardization era*: 1990–present (2010). This era is characterized by the establishment of forensic anthropology as its own *discipline* and the broadening of the scope of forensic anthropological work (also referred to as Sledzik et al.'s "Fourth Era"; Sledzik et al. 2007).

It was not until the professionalization era that trauma analysis in forensic anthropology casework began to become mainstream. This delay in taking a systematic and scientific examination of bone trauma may be explained by the evolving role of the forensic anthropologist in the medicolegal death investigation. As it is the responsibility of the forensic pathologist to determine cause and/or manner of death, the idea that forensic anthropologists had important knowledge to contribute to understanding cause and manner was slow in coming. The first PhD programs training students in forensic anthropology, established in the late 1970s and early 1980s, played an important role in expanding the scope of forensic anthropology in medicolegal death investigation. Namely, programs at the University of Tennessee, the University of Arizona, the University of Florida, and the University of New Mexico (under the directorships of Dr Bill Bass, Dr Walt Birkby, Dr Bill Maples, and Dr Stan Rhine, respectively) established strong bonds with the local medical examiner offices and worked closely with the forensic pathologists in those offices on a large range of cases, including cases involving skeletal trauma. Over time in those offices it became protocol for the forensic anthropologist to consult on most cases involving skeletal trauma, whether the case was a skeleton or a fresh body. This was certainly the case for Dr Birkby at the University of Arizona, where he averaged 100 cases a year during the 1980s and 1990s, with many of those cases involving the analysis of skeletal trauma.

During this period, the earliest publications on trauma by forensic anthropologists are case-study-based (e.g., Kerley 1976, 1978; Maples 1986; Maples et al. 1989). But beginning in the late 1980s systematic skeletal trauma research began to take shape in forensic anthropology, pioneered by the work of Hugh Berryman, Steve Symes, and O.C. Smith and colleagues (based on a collaboration of forensic pathologists and forensic anthropologists in the medical examiner's office in Memphis, TN). Spurred on by their casework at Memphis, Berryman and Symes published the landmark 1998 book chapter titled "Recognizing gunshot and blunt cranial trauma through fracture interpretation" (Berryman and Symes 1998). Based on their overall contributions that helped to establish the analysis of trauma in forensic anthropology, we see the late 1980s and 1990s as the Berryman–Symes era.

On the heels of the first wave of research on skeletal trauma came the first major volume dealing with blunt-force skeletal trauma, the book *Broken Bones: Anthropological Analysis of Blunt Force Trauma* (1999) edited by Alison Galloway. This publication, often referred to as the "bible" of BFT, is an essential reference on the fundamentals of BFT description and interpretation, including fracture biomechanics, fracture patterns, and the various circumstances of BFT.

Today, with the increase in graduate programs focusing on forensic anthropology and increasing collaborations between forensic anthropologists, forensic pathologists, and biomechanical engineers, the analysis of skeletal trauma by the forensic anthropologist has become an accepted practice in many jurisdictions. An outcome of the early research by Berryman, Symes, and Smith is the current framework that most forensic anthropologists use when identifying the mechanism of perimortem skeletal trauma, classifying trauma into these categories: blunt force, projectile, sharp force, and thermal. This medicolegal classification scheme of skeletal trauma uses biomechanical, pathological, and anthropological principles to classify causes of bony fracture. This is also the framework used in the SWGANTH draft guidelines on trauma analysis (SWGANTH 2011).

The trend of increasing research and education in skeletal trauma is clearly displayed in a review of past issues of *Journal of Forensic Sciences* (1972–2009). There is a large increase in articles involving BFT and projectile trauma research beginning in the early 1990s, with trauma research diversifying as time goes on. Over the years, research on BFT was almost always the leading type of trauma. Additionally, there is an obvious increase in trauma research by students in relation to PhDs. There is also an increase in experimental versus actualistic (case) studies, accompanied by a large increase in nonhuman models for bone trauma research. These same trends are found in surveys of abstracts from proceedings of annual meetings of the American Academy of Forensic Scientists.

RESEARCH IN BFT

The history of BFT research in forensic anthropology is a confusing issue, mainly because the first experimental researchers were not forensic anthropologists, but multidisciplinary teams of engineers led by the neurosurgeon and anatomist E.S. Gurdjian, beginning in the 1940s (Gurdjian 1975; see also Kroman 2003). Gurdjian and colleagues published numerous manuscripts on the subject of blunt-force cranial trauma generated from human cadaver experiments (summarized in Gurdjian 1975). Gurdjian and colleague's theories were largely based on experimental research using the "stress-coat technique," which involved preparation of cranial bones and the application of a brittle lacquer, which would fracture under their loading conditions. Among this productive team's conclusions were that there were no biomechanical differences between dry, fresh, and living bone, and that fractures initiate away from the point of impact in areas of outbending and then propagate back towards the impact site (Gurdjian 1975). These studies were quite groundbreaking in scale, as well as documentation and scientific rigor, and they still set the standard for experimental trauma research today.

Another avenue of bone trauma research involves the application of biomechanics to either known trauma cases (e.g., Fenton et al. 2003), or fracture initiation or propagation experimentation (e.g., Gurdjian 1975; Kroman 2007; Baumer et al. 2010). These studies often have a more theoretical focus and attempt to explain larger questions (i.e., *bone fractures like this because…*; not, *bone will tend to fracture like this in this scenario...*). However, this experimental research is also problematic for various reasons, one of the greatest obstacles being the fact that human skeletal tissue is difficult to obtain without a high price tag and animal models do not easily correlate to human morphology. Further, these highly controlled laboratory experiments generate large amounts of data which often do not recreate observed circumstances of human fracture patterns, the greatest example being the Gurdjian versus Kroman debate (see Kroman 2003 for a full discussion).

As forensic anthropologists began to be employed in medical examiner's offices, forensic blunt trauma cases with known circumstances did not appear to agree with the Gurdjian finding of fracture initiation in areas away from the point of impact (Berryman and Symes 1998). Kroman and colleagues retested the Gurdjian theories using adult human cadaver heads and found fractures initiating at the site of impact (in this case, the center of the parietal), and propagating away from this area (Kroman 2004, 2007).

While Gurdjian found no differences in fracture production between fresh and dry bone, the utility of rehydrating bone and testing it assuming "fresh" conditions is questionable (Rappazzo 2006) attempted to rehydrate nonhuman ribs in a saline solution in order to examine differences in fresh-fractured versus rehydrated-fractured bones. Here the rehydrated bones all fractured in atypical patterns; however, another study by Daegling et al. (2008) applied a similar method with more encouraging results.

While most methods developed to generate the components of the biological profile of an unknown individual involve statistical evaluations of their reliability as required by the Daubert criteria (*Daubert v. Merrell Dow Pharmaceuticals* 1993), research on skeletal trauma is still in its infancy. Although some work has been done attempting to generate statistics in order to back up autopsy observations of trauma (e.g., Hart 2005; Love and Symes 2004; Quatrehomme and İşcan 1999; Tomczack and Buikstra 1999), large sample sizes of known trauma cases are only starting to be compiled (Fenton et al. 2011; Kimmerle et al. 2009). Further, these observational studies are more difficult to analyze, as there are many independent variables that must be indirectly approached by the practitioner unless circumstances of injury are known. This observational research, while exceedingly important, tends to generate qualitative results, which are helpful only to match patterns of trauma, not to explain the how and why of the resulting fracture patterns.

BFT RESEARCH USING NONHUMAN ANIMAL MODELS

Recent experimental blunt-force research using infant porcine skulls (*Sus scrofa*) was carried out by a team of scientists from Michigan State University consisting of forensic anthropologists and biomechanical engineers. This research used infant porcine crania (specifically the parietal) as a nonhuman model for testing both material properties and fracture patterns. The results suggest that infant pig cranial material properties are similar to those of infant humans (Baumer et al. 2009; Coats and Margulies 2006; Margulies and Thibault 2000). These results are limited, however, because, as seen in bony measurements acquired from human infant computed tomography (CT) scans and comparable gross bony measurements of the porcine crania, cranial shape differs significantly as age increases; thus the infant porcine skull may only be comparable up to human infants aged to about 22 days (Baumer et al. 2010).

A primary goal of the Michigan State team was to establish baseline data on fracture initiation and propagation from experimental impacts with a flat surface to the center of the parietal. There were several significant findings using the immature porcine animal model. First, fracture initiation consistently occurred away from the site of impact at the surrounding suture margins and propagated back toward the site of impact (Baumer et al. 2010). Second, a single impact commonly produced two or more separate fractures away from the site of impact. Third, with increased energy, more fractures (in terms of overall fracture length) are produced. Fourth, for a given impact site, fractures occur at common locations across individuals of the same age and interface. Fifth, blunt-force impacts from rigid and compliant interfaces display different amounts of fracturing that change with age. And sixth, analysis of material properties of the developing porcine cranial vault suggests an age correlation of days in the pig to months in the human infant.

It is currently unclear how these data correlate to human infants, as both age and geometry (gross head shape) are key variables in the mechanism and pattern of bony fracture. This is, however, an important baseline in the experimental study of fracture patterns, which will only grow with the acquisition of infant human tissue for experimental research. Other research utilizing an animal model includes Marceau (2007), who examined the cortical density and cross-sectional geometry of human (*Homo sapiens*), pig (*S. scrofa*), and deer (*O. virginianus* and *Odocoileus hemionus*) long bones. Marceau concluded that "both deer and pig skeletons can serve as suitable models in forensic experiments based on their geometric and densitometric similarities to human bone" (Marceau 2007: 63). Interestingly, Marceau then compared taphonomic weathering characteristics of each species and found differences in the patterns of deer and pigs, with deer long bones acting in a way more similar to human bone. This example of using a nonhuman model is quite pertinent as nonhuman models are the norm in experimental trauma and taphonomic research due to the many problems of acquiring human tissue for destructive purposes.

Calce and Rogers (2007) attempted to examine the affects of taphonomy when interpreting blunt-force impacts to juvenile pig heads (juveniles because they were livestock animals and were thus slaughtered before they reached maturity). Ten pig heads were used, five of which were defleshed prior to impacting and exposure. All the specimens were then "impacted" using a hammer on the right and left parietals until the ectocranial surface was "perforated," or assumed perforated in the case of the fleshed specimens. The heads were then exposed for 52 weeks in different microenvironments, but within the same enclosure in southern Ontario, Canada. Unfortunately the taphonomic variables in this study are hardly quantified, and it is unclear how certain variables such as freeze–thaw cycles were differentiated from others like presence or weight of rain or snow (variables that are very interdependent). More problematic is the fact that the authors claim that freeze–thaw cycles and soil erosion have the ability to "completely obliterate" evidence of BFT after an exposure period of 52 weeks (Calce and Rogers 2007: 522, 524). Other research into freeze–thaw cycles and cryoturbation using deer long bones (*O. virginianus*) found that bone will exhibit taphonomic weathering patterns only *after* it has passed from the wet to dry state, thus making it possible to differentiate between perimortem trauma and postmortem damage even after 52 weeks of exposure (Passalacqua and Rainwater 2006). In addition, Herrmann and Bennett (1999) were able to discern BFT after severe burning of porcine remains in a majority of their cases with only partial recoveries, and Symes et al. (2008) were able to distinguish perimortem blunt-force fractures from postmortem thermal damage in forensic cases, demonstrating that bone will retain characteristics of BFT even after the thermal alterations of calcification and fragmentation.

Other research has focused on artificial models of infant and adult human crania with limited success. A full-body infant model of a 1.5-year-old was developed by Coats and Margulies (2008) and demonstrated similar data to previous impact experiments; however, the correlation of impact load to fracture initiation and propagation has yet to be made. Similarly Desantis et al. (2002) and Roth et al. (2007, 2008) have also failed to successfully build a computational model that can be used as a substitute for human crania.

LOOKING TO THE FUTURE

Skeletal trauma analysis is one of the most challenging tasks confronting the forensic anthropologist. Looking back, it becomes apparent that the field of forensic anthropology emerged with the need for bone specialists in the medicolegal field. As forensic anthropology casework progressed, deficiencies in knowledge regarding bone trauma began to be filled first by actualistic case studies, then by experimental research, verifying and refining earlier work. These same themes continue today. As we encounter more trauma cases and increase our knowledge of bone biomechanics, we are constantly looking to refine our ability to make confident statements about the mechanisms of injury and contribute to the cause of death via the skeletal remains. Current research demonstrates that while we know more about BFT to the human skeleton than ever before, we are still confronted by many challenges that require extensive additional research. Finally, it is our charge to further the field of forensic anthropology by developing collections, references, standards, and practices for analyzing skeletal trauma, and that we do so through scientific rigor and validation built upon actualistic data and not via post hoc assumptions or untested models. We have to move beyond our reliance on experience as the only way to inform us in the analysis of trauma, and develop a strategy for the future of trauma analysis in order to advance our abilities in this area. We need new directions in trauma interpretation to include further experimental work in which fracture patterns are better understood and thus utilized for future trauma interpretation. What we are suggesting is a paradigm shift in how we approach skeletal trauma analysis: moving toward a science of trauma that involves the testing of hypotheses and the establishment of baseline parameters on how bones break in numerous scenarios. We are convinced this paradigm shift needs to include scientists from outside of forensic anthropology, such as biomechanical engineers, and computer learning experts. We also need validation where we set up systematic tests to validate our techniques in trauma analysis.

REFERENCES

Baumer, T., Powell, B., Fenton, T., and Haut, R. (2009). Age dependent mechanical properties of the infant porcine skull and a correlation to the human. *Journal of Biomechanical Engineering* 131(11): 111006–1/6.

Baumer, T., Passalacqua, N.V., Powell, B., Newberry, W., Smith, W., Fenton, T., and Haut R. (2010). Age-dependent fracture characteristics of rigid and compliant surface impacts on the infant skull – a porcine model. *Journal of Forensic Sciences* 55(4): 993–997.

Berryman, H.E. and Symes, S.A. (1998). Recognizing gunshot and blunt cranial trauma through fracture interpretation. In K.J. Reichs (ed.), *Forensic Osteology: Advances in the Identification of Human Remains*, 2nd edn (pp. 333–352). Charles C. Thomas, Springfield, IL.

Bonte, W. (1975). Toolmarks in bone and cartilage. *Journal of Forensic Sciences* 20(2): 315–325.

Byers, S.N. (2002). *Introduction to Forensic Anthropology.* Allyn & Bacon, Boston, MA.

Calce, S.E. and Rogers, T.L. (2007). Taphonomic changes to blunt force trauma: A preliminary study. *Journal of Forensic Sciences* 52(3): 519–527.

Coats, B. and Margulies, S.S. (2006). Material properties of human infant skull and suture at high rates. *Journal of Neurotrauma* 23(8): 1222–1232.

Coats, B. and Margulies, S.S. (2008). Potential for head injuries from low-height falls. *Journal of Neurosurgery: Pediatrics* 2: 321–330.

Curry, J.D. (1970). The mechanical properties of bone. *Clinical Orthopedics and Research* 73: 210–231.

Daegling, D.J., Warren, M.W., Hotzman, J.L., and Self, C.J. (2008). Structural analysis of human rib fracture and implications for forensic interpretation. *Journal of Forensic Sciences* 53(6): 1301–1307.

Daubert v. Merrell Dow Pharmaceuticals, 509 US 579, 113 S.Ct. 2786, 125 L.Ed. 2d 469 (US June 28, 1993) (No. 92-102).

DeSantis, K.K., Hulbert, G.M., and Schneider, L.W. (2002) Estimating infant head injury criteria and impact response using crash reconstruction and finite element modeling. *Stapp Car Crash Journal* 46: 165–194.

Fenton, T.W., deJong, J.L., and Haut, R.C. (2003). Punched with a fist: the etiology of a fatal depressed cranial fracture. *Journal of Forensic Sciences* 48(2): 1–5.

Fenton, T.W., Haut, R.C., Jain, A., and deJong, J.L. (2011). Pediatric fracture printing: creating a science of statistical fracture signature analysis. Number 2011-DN-BX-K540. National Institute of Justice, Washington DC.

Galloway, A. (ed.) (1999). *Broken Bones: Anthropological Analysis of Blunt Force Trauma.* . Charles C. Thomas, Springfield, IL.

Gozna, E.E. (1982). *Biomechanics of Musculoskeletal Injury.* Williams & Wilkins, Baltimore, MD.

Gurdjian, E.S. (1975). *Impact Head Injury: Mechanistic, Clinical and Preventive Correlations.* Charles C. Thomas, Springfield, IL.

Haglund, W.D. (1997a). Dog and coyotes: postmortem involvement with human remains. In W.D. Haglund and M.H. Sorg (eds), *Forensic Taphonomy: The Postmortem Fate of Human Remains* (pp. 367–382). CRC Press, Boca Raton, FL.

Haglund, W.D. (1997b). Rodents and human remains. In W.D. Haglund and M.H. Sorg (eds), *Forensic Taphonomy: The Postmortem Fate of Human Remains* (pp. 383–394). CRC Press, Boca Raton, FL.

Harkness, J.W., Ramsey, W.C., and Ahmadi, B. (1984). Principles of fractures and dislocations. In C.A. Rockwood and D.P. Green, *Fractures in Adults*, vol. 1 (pp. 1–18). Philadelphia: JB Lippincott Company.

Hart, G. (2005). Fracture pattern interpretation in the skull: differentiating blunt force from ballistics trauma using concentric fractures. *Journal of Forensic Sciences* 50(6): 1276–1281.

Herrmann, N.P. and Bennett, J.L. (1999). The differentiation of traumatic and heat related fractures in bone. *Journal of Forensic Sciences* 44(3): 461–469.

Kerley, E.R. (1976). Forensic anthropology and crimes involving children. *Journal of Forensic Sciences* 21(2): 333–339.

Kerley, E.R. (1978). Identification of battered-infant skeletons. *Journal of Forensic Sciences* 23(1): 163–168.

Kimmerle, E.H., Okoye, M.I., Obefunwa, J.O., Bennet, T.L., and Mellen, P.F. (2009). Skeletal fracture patterns in documented cases of torture, assault, abuse and accidents. *Proceedings of the American Academy of Forensic Sciences* February, Denver, CO, H109 363

Krogman, W.M. (1962). *The Human Skeleton in Forensic Medicine*. Charles C. Thomas, Springfield, IL.

Kroman, A.M. (2003). *Backward and Forward: the Gurdjian Theory and Blunt Force Impact of the Human Skull*. Thesis, University of Tennessee, Knoxville, TN.

Kroman, A.M. (2004). Experimental study of fracture propagation in the human skull: a re-testing of popular theories. *Proceedings of the 56th Annual Meeting of the American Academy of Forensic Sciences*, February 16–21, Dallas, TX.

Kroman, A.M. (2007). *Fracture Biomechanics of the Human Skeleton*. Dissertation, University of Tennessee, Knoxville, TN.

Kulin, R.M., Jiang, F., and Vecchio, K.S. (2008). Aging and loading rate effects on the mechanical behavior of equine bone. *Journal of Minerals, Metals and Materials* 6: 39–44.

Love, J.C. and Symes, S.A. (2004). Understanding rib fracture patterns: incomplete and buckle fractures. *Journal of Forensic Sciences* 49: 1153–1158.

Maples, W.R. (1986). Trauma analysis by the forensic anthropologist. In K.J. Reichs (ed.), *Forensic Osteology: Advances in the Identification of Human Remains* (pp. 218–228). Charles C. Thomas, Springfield, IL.

Maples, W.R., Gatliff, B.P., Ludeña, H., Benfer, R., and Goza, W. (1989). The death and mortal remains of Francisco Pizarro. *Journal of Forensic Sciences* 34: 1021–1036.

Marceau, C.M. (2007). *Bone Weathering in a Cold Climate: Forensic Applications of a Field Experiment using Animal Models*. MA thesis, University of Alberta.

Margulies, S.S. and Thibault, K.L. (2000). Infant skull and suture properties: measurements and implications for mechanisms of pediatric brain injury. *Journal of Biomechanical Engineering* 122(4): 364–371.

Ortner, D.J. (2003). *Identification of Pathological Conditions of the Human Skeleton*. Academic Press, San Diego, CA.

Passalacqua, N.V. and Rainwater, C.W. (2006). *The affect of freeze and thaw cycles on bone*. Presented at *Annual Meeting, Mountain, Swamp, & Beach (MSB)*, Radford, VA.

Quatrehomme, G. and İşcan, M.Y. (1999). Characteristics of gunshot wounds in the skull. *Journal of Forensic Sciences* 44(3): 568–576.

Rappazzo, K. (2006). Fracture pattern characteristics of deer ribs: fresh tissue versus rehydrated tissue. Paper presented at *Modification of the Human Skeleton*, Mercyhurst College, Erie, PA.

Roth, S., Raul, J.S., Ludes, B., and Willinger, R. (2007). Finite element analysis of impact and shaking inflicted to a child. *International Journal of Legal Medicine* 121(3): 223–228.

Roth, S., Raul, J.S., Ludes, B., Willinger, R. (2008). Influence of the benign enlargement of the subarachnoid space on the bridging veins strain during a shaking event: a finite element study. *International Journal of Legal Medicine* 122(4): 337–340.

Sauer, N. (1998). The timing of injuries and manner of death: Distinguishing among antemortem, perimortem, and postmortem trauma. In K.J. Reichs (ed.), *Forensic Osteology: Advances in the Identification of Human Remains*, 2nd edn (pp. 321–332). Charles C. Thomas, Springfield, IL.

Scientific Working Group for Forensic Anthropology (SWGANTH) (2011). *Trauma Analysis*. Draft, February 16, 2010. http://swganth.startlogic.com/Trauma.pdf.

Sledzik, P.S., Fenton, T.W., Warren, M.W., Byrd, J.E., Crowder, C., Drawdy, S.M., Dirkmaat, D.C., Galloway, A., Finnegan, M., Hartnett, K. et al. (2007). The fourth era of forensic anthropology: examining the future of the discipline. *Presented at the 59th annual meeting of the American*

Academy of Forensic Sciences, San Antonio, TX, p. 350.

Smith, O.E., Pope, E.E., and Symes, S.A. (2003). Look until you see: identification of trauma in skeletal material. In D.W. Steadman (ed.), *Hard Evidence: Case Studies in Forensic Anthropology* (pp. 138–154). Pearson Education, Old Tappan, NJ.

Stewart, T.D. (1979). *Essentials of Forensic Anthropology-Especially as Developed in the United States*. Charles C. Thomas, Springfield, IL.

Symes, S.A., Rainwater, C.W., Chapman, E.N., Gipson, D.R., and Piper, A.L. (2008). Patterned thermal destruction of human remains in a forensic setting. In C. Schmidt and S.A. Symes (eds), *The Analysis of Burned Human Remains* (pp. 15–54). Elsevier, London.

Tomczack, P.D. and Buikstra, J.D. (1999). Analysis of blunt trauma injuries: vertical deceleration versus horizontal deceleration injuries. *Journal of Forensic Sciences* 44(2): 253–262.

Ubelaker, D.H. and Adams, B.J. (1995). Differentiation of perimortem and postmortem trauma using taphonomic indicators. *Journal of Forensic Sciences* 40(3): 509–512.

Wheatley, B.P. (2008). Perimortem and postmortem bone fractures? An experimental study of fracture patterns in deer femora. *Journal of Forensic Sciences* 53(1): 69–72.

Wieberg, D.A.M. and Wescott, D.J. (2008). Estimating the timing of long bone fractures: Correlation between the postmortem interval, bone moisture content, and blunt force trauma fracture characteristics. *Journal of Forensic Sciences* 53(5): 1028–1034.

CHAPTER **20** Advances in the
Anthropological
Analysis of
Cremated Remains

*Traci L. Van Deest, Michael W.
Warren, and Katelyn L.
Bolhofner*

INTRODUCTION

Cremation has been used for millennia as a means of disposal of dead bodies, with the claims for the earliest cremation dating to 20 000–26 000 years ago in Australia (Bowler et al. 2003). The practice was common among Native American groups (Miller 2001) and Ancient Romans (Nock 1932) as well as numerous other groups. In the USA, burial has always surpassed cremation as the prominent means for disposition of the dead. Prior to the middle of the twentieth century, cremations accounted for only approximately 3% of all deaths, while today nearly 35% of deaths are followed by cremation instead of burial (Cremation Association of North America 2008). Similarly, the prevalence of the involvement of forensic anthropologists in modern cremation cases is increasing as well. This chapter will address the forensic analysis of cremated remains, and specifically modern cremation involving commercial retorts and processors. The cremation process results in fragmentary remains morphologically recognizable as human skeletal elements, which then are reduced further to small, often unrecognizable fragments, and ash. This process negates the use of standard osteological identification techniques and greatly increases the difficulty of forensic analysis, including but not limited to conclusions as to the identity of the individual remains.

A Companion to Forensic Anthropology, First Edition. Edited by Dennis C. Dirkmaat.
© 2012 John Wiley & Sons Ltd. Published 2015 by John Wiley & Sons Ltd.

Although cremation has been practiced for centuries, not all previous scholars saw the value of cremated remains from the archaeological record, as was exemplified by one museum curator stating that he would not waste precious curation space on cremains (Gejvall 1970). More recently, the value of cremated remains has been rediscovered. Both in the archaeological record and in more recent and current contexts, cremains have come into the spotlight because information can be derived from the calcined bone fragments, no matter the resulting size. The application of methods in chemistry, radiologic sciences, and molecular advances to cremated remains has allowed new questions to be asked, in addition to moving us closer to answering past lingering questions. Previous research in burned bone and cremated remains was extensively reviewed by Mayne Correia (1997), in the edited volume by Schmidt and Symes (2008), and in a recent literature review by Ubelaker (2009). These reviews have focused more broadly on the analysis of burned remains, while this chapter will specifically address the analysis of commercially cremated remains.

Forensic anthropologists examine modern commercially cremated remains for a variety of reasons. Primary among them is the suspicion of misconduct by the crematorium or funeral home, resulting in doubt as to the identity of the remains (e.g., Brooks et al. 2006). As mentioned above, the use of the processor to reduce the fragment size of the cremated remains excludes the possibility of standard osteological techniques in the forensic analysis of those remains. Thus, in using some of the techniques discussed in this chapter, even though it is not possible to reach a conclusive identification, the analysis can determine whether the remains are consistent with the life history, mortuary events, and accompanying materials for the purported decedent. Information about the condition of the remains at the time of cremation, such as clothing, jewelry, and other personal effects, can be used to support or undermine an opinion that the remains are consistent with the purported decedent. These types of artifact can also support the opinion as to whether or not the remains are consistent with human or nonhuman cremated remains, presumptively linking the remains to a human individual through a series of analyses.

CREMATION PROCESS

The cremation process has been extensively reviewed in numerous publications (e.g., Murad 1998; Schultz et al. 2008), hence it will only be discussed briefly here. The commercial cremation process begins in a crematorium, with a large retort or cremation furnace. The remains of the decedent are placed within a combustible container and positioned feet-first in the retort. Temperatures during the cremation process may range from 760 to 980 °C, with the possibility of reaching 1150 °C. The length of time that the remains are cremated depends on several factors, including the size of the decedent, the brand of equipment used, fuel type, and whether the retort has been preheated by a previous cremation. In general, most cremationists employ a cycle lasting approximately 2 h with an additional cooling time. The remains are then removed, often by sweeping them into a receptacle situated below the retort door. Subsequently, the cremated remains are placed in a cooling tray, where all metal and other non-osseous artifacts are removed. The only device removed prior to cremation is a cardiac pacemaker, which contains an energy source that may explode if heated. The cremated remains are finally placed in a processor, variously known as a cremulator or pulverizer, to reduce the remains. This final reduction process insures that the remains are suitable for inurnment or scattering, often reducing the fragments until they are anatomically unrecognizable. A cremation tag – that looks similar to a pet's collar tag – follows the remains throughout the entire process. A number inscribed on this tag corresponds to paperwork completed by the funeral director, thus insuring that no clerical errors cause the cremated remains to be returned to the wrong family.

Crossing between Archaeological and Modern Cremation Studies

Cremation studies traditionally have focused on archaeological cremations and the cremation ritual. Early emphasis was the analysis of relatively large fragments of burned and calcined bone. Previous work focused on distinguishing between human and nonhuman (Cattaneo et al. 1999; Whyte 2001), perimortem versus heat-induced fractures (Herrmann and Bennett 1999; Mayne 1990), and the condition of the remains prior to cremation (Baby 1954; Binford 1963; Buikstra and Swegle 1989; Webb and Snow 1945). While these factors can be of concern in cremated remains analyses, the remains after processing generally do not allow for their analysis. Thus, contemporary forensic analyses focus primarily on the identification of the remains and issues of erroneous or potentially negligent practices by the funeral home or crematory staff. However, some earlier studies that focused on heat changes in bone and reconstruction of ancient funerary practices can be informative in the analysis of modern, commercially cremated remains.

Baby (1954) divides the stages of the change in heat-altered bone into three categories. The first level (I) is complete incineration and calcination, evident by a white to blue-gray color and deep checking, with no organic material present. The second level (II) is considered smoked or incompletely incinerated, with blackened fragments frequently containing bits of organic material. The third level (III) is the nonincinerated or normal stage, where fragments are not highly affected by heat with some "smoking" or charring around the edges. Due to both the temperature and the duration of cremation processes today, the remains are expected to be reduced almost completely to Baby's Level I classification. This state of incineration is equivalent to Scale 5 of the Crow–Glassman standards (Glassman and Crow 1996). If the remains are not consistent with complete calcination, the process may have been shortened or it may be an indication that other issues are present.

Persistence of Diagnostic Fragments

Although the aforementioned research is insightful into the changes in bony tissue caused by heating, it is of little help when dealing with commercially cremated remains due to the reduction of the fragments into smaller fragments and often powder. Even though the remains are reduced into primarily nondiagnostic fragments, most cremated remains yield at least some remnants of teeth and osseous fragments identifiable by a well-trained osteologist. In a recent set of cremated human remains examined by the authors, cranial fragments approximately 4 mm in size were recognized by the presence of the inner and outer tables with interposed diploë. One cranial fragment

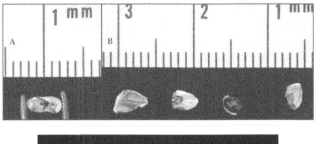

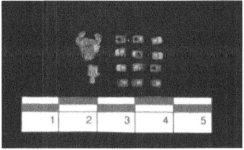

Figure 20.1 Top: tooth root (A) and enamel (B) fragments. Bottom: personal artifact of a zipper found with human cremated remains. Scales are in centimeters.

also exhibited a clear vascular groove on the endocranial surface. In cases where the decedent retained at least some of their natural dentition, it is likely that portions of the teeth may be present and able to be recognized. Fragments of enamel and roots are distinguishable from the remainder of the cremated remains, as seen in Figure 20.1. This illustrates the need for careful and close examination of all of the fragments, as diagnostic fragments may exist.

ARTIFACTS IN CREMATION

The presence of artifacts within the cremated remains can be the most important line of evidence as to the identity of the decedent. Information as to the artifacts which may be present can be ascertained through a questionnaire for the family or funeral home, such as that suggested by Warren (2008). The material present in a set of cremated remains can be divided primarily into osseous and non-osseous materials. Those non-osseous artifacts can be used as evidence for identity, determine whether commingling has occurred, and provide a link between the cremated remains and antemortem medical records. Artifacts can be informative as to two separate phases of the decedent: life history and funerary-related materials. Warren and Schultz (2002) divide cremation artifacts into five separate categories, discussed individually below.

Medical and dental artifacts can provide evidence of procedures performed and events during the course of the life of the decedent. Medical artifacts consist of medical devices, orthopedic implants, and surgical metals such as vascular clips and surgical staples. If there are surgical implements not known to belong to the decedent, then there are the possibilities of commingling and that the remains are not those of the named individual. Although commingling on some level is inevitable, the nature of civil litigation is such that the members of the court will decide an acceptable amount of commingling (Warren 2008). Dental artifacts, such as dental implants, crown, dentures, root canal material, and other dental implements, will survive the cremation process. These materials may have melted or altered shape due to the intense heat, but nonetheless may be recognizable. In cases where dental artifacts are present, it is advantageous to confirm the type and model through the involvement of a forensic odontologist.

Artifacts related to the funerary event are personal and mortuary items. Personal artifacts are those that would accompany the decedent on the outside of the body, such as jewelry and remnants of clothing. Mortuary artifacts are items linked to mortuary practices performed on the remains prior to cremation.

Radiographic imaging can be useful in identifying and locating both osseous and non-osseous material (Warren et al. 1999). When radiographic images are created, it may be possible to locate fragments of dentition, ear ossicles, and diagnostic bone fragments. Radiographic images also can be used to locate cultural artifacts. After separating the remains by size, they are placed within shallow boxes and overlain with a metal grid. The grid serves as a reference from which radiolucent objects can be located easily within the gridded areas.

ADVANCES IN THE STUDY OF CREMATION

Chemical analysis

Understanding chemical reactions and changes within bone as a result of heat is essential to the interpretation of elemental analysis of cremated remains. Research has shown that heat will alter the chemical signature of skeletal remains. Although carbon isotopes are not affected significantly by heat, nitrogen isotopes significantly increase when a bone is exposed to heat (Munro et al. 2008; Schurr et al. 2008; Schurr and Hayes 2008). In one 2006 study, heating was shown to release trace elements from their position in the molecular structure of bone mineral, thus making them more available in the natural environment, which also solidified the role of chemical analysis in determining changes induced by heat throughout the process (Danilchenko et al. 2006b).

Recent chemical analyses of cremated remains has shown great promise in understanding the validity of the remains (i.e., whether they are human remains or other material) and in detecting chemical signatures and elements which may be linked to the medical and life histories of the decedent. Brooks et al. (2006) examined the chemical signature by means of inductively coupled plasma optical emission spectroscopy (or ICP-OES) analysis of human remains, concrete, and mixtures of various ratios of cremated remains and concrete to establish whether remains from the infamous Tri-State Crematory in northern Georgia, USA, were human, other material, or a combination of human and other material. Bergslien et al. (2008) used X-ray fluorescence (XRF) to separate cremated remains, geological apatite, and other materials visually similar to human cremated remains, but found crystal structure more reliable.

The presence of dental restorations can be detected by XRF analysis and used as a line of evidence for the legitimacy of the purported cremated remains. Bush et al. (2007) used XRF to determine whether resin from dental fillings could be detected among cremated remains. Donated cadavers were analyzed for dental work, which was removed if present, and then five different resins were used to place restorations within the available dentition. Following cremation, the resins were examined using the XRF and the brand of resin was determined.

Elemental analysis additionally provides evidence for the life history of an individual. Brooks and colleagues (2006) also established that previous life events of the decedents could be detected through chemical analysis. Two of the known human cremations within the study had previously experienced gunshot wounds, one with a lead bullet and the other with a jacketed bullet, and showed elevated levels of lead and copper and of magnesium, respectively.

Chemicals introduced during medical procedures are visible during elemental analyses of cremated remains. Warren et al. (2002) revealed elevated levels of barium, used in radiographic contrast studies of the gastrointestinal system, in known human cremated remains. Gadolinium, injected intravenously as a contrast medium for magnetic resonance (MRI) or computed tomography (CT) scans can be absorbed by body tissues (White et al. 2006). Schultz and colleagues (2008) showed that gadolinium was present within cremated remains of a decedent known to have undergone diagnostic imaging.

Additionally, it is possible to detect the presence of medically implanted devices through elemental analysis. Warren et al. (2002) showed increased levels of titanium in one of the samples of known human cremated remains. They concluded the increase in titanium resulted from the presence of a surgical ligation clip in the sample material; the composition of the ligation clip was confirmed later by the authors as pure titanium. Denton et al. (1984) examined the uptake of metal ions, particularly titanium and aluminum, from medical implants on the systemic level by examining soft tissue of rabbits. While this research is not directly applicable to the study of human cremated remains and bone, it illustrates the systemic uptake of metal ions from foreign objects in the body, making it available for substitution in all tissues. According to Ribiero et al. (2006), titanium ions substitute into the hydroxyapatite crystal latticework. Since bone constantly rejuvenates itself, new bone tissue growth, maintenance, and bone repair will incorporate the currently available substances into the newly forming tissue, including metal ions, titanium, and other elements ingested through the diet, by external environmental exposure, and as a result of medical procedures.

Changes in microstructure

During cremation, heat alters the microstructure contemporaneously with the changes happening at the macrolevel. With the incineration of the organic phase of bone, the inorganic material remains present, although in an altered form. In one of the first studies on changes in bone crystal structure with heat, Shipman et al. (1984) illustrated an increase in the size of the crystal with increased temperature. Stiner et al. (1995) stated that the recrystallization of the inorganic phase of bone was a natural process of diagenesis happening over a time span of millennia during the fossilization process. Exposure to fire and heat does not create a new and unique outcome on the microstructural level, but rapidly accelerates a natural process (Stiner et al. 1995).

X-ray diffraction (XRD) has been used to document the crystal changes and recrystallization of hydroxyapatite in calcined bone, indicating that even after heat treatment the cremated remains should be distinguishable still as bone as opposed to other materials (Shipman et al. 1984). Holden et al. (1995) related crystal changes at differing temperature levels to the age of the decedent. The authors concluded that it was possible to differentiate the age of the individual based on the appearance of the microstructure after heat exposure, given both a known or reliably estimated duration and an approximate temperature of the exposure (Holden et al. 1995).

Confirmation of the early studies by Shipman et al. (1984) and Holden et al. (1995) was accomplished using various methods. XRD measures used by Piga et al. (2009), Danilchenko et al. (2006a), and Rogers and Daniels (2002) relate changes to both the temperature and duration of the event. Small- and wide-angle scatter analysis also confirms reports of increase in crystal size with heat exposure (Hiller et al. 2003). Hiller et al. (2003) propose that the change in size coincides with a change in structure, resulting in a more perfectly formed lattice as opposed to the poorly organized crystal structure created through biomineralization.

Crystal structure analysis can determine whether purported cremated remains are human remains or other materials. Results confirm the ability for the XRD technology to distinguish between human cremated remains and concrete, and a combination of the two from one another (Bergslein et al. 2008; Bodkin and Mies 2008). When comparing modern calcined bone, concrete, and sheetrock, and other common material purported to be used as filler in cremated remains, the crystal structure of each material was distinguished easily from one another, with the close appearance of biological and geological apatite (Bergslein et al. 2008).

Cremation weights

According to Heymsfield et al. (1997), less than 1% of total body calcium is found in soft tissue, indicating that the measurement of calcium provides for an accurate meas-ure of bone mineral *in vivo*. The products of cremation also will reflect differences in the amount of inorganic material present in life. As noted earlier in this chapter, after cremation, the remaining material reduces to its inorganic components; that is, bone mineral (Bonucci and Graziani 1975; Stiner et al. 1995; Walker et al. 2008). In a study by Walker et al. (2008), remains treated in a muffle furnace at temperatures over 600 °C at durations of 1–3 h did not retain any unaltered or nonpyrolyzed collagen, the primary organic component of bone tissue. Stiner et al. (1995) supported this finding, after observing no organic matrix in completely calcined bone.

Weight as a means of information of cremated remains can be traced to a presenta-tion by Sonek (1992), whose values were reported by Murad (1998), initially proposed as a means to determine whether or not commingling had occurred. By presenting average weights of cremated remains for males and females, he made the case that weights of cremated remains falling outside of the expected range might suggest the possibility of commingling of more than one set of decedents. Case reports have also used weight as an analytical tool (Murray and Rose 1993).

Research in the weight of cremated remains was conducted in Great Britain to determine what proportion of a set of remains was present in archaeological collections of cremated remains (McKinley 1993). While the weights obtained within this study were consistent with those reported by Warren and Maples (1997), a problem arose. Primarily, seven of the 15 individuals were dissecting-room cadavers and no documentation of the completeness of the skeletal remains at the time of cremation is available (McKinley 1993). If skeletal elements were dissected or portions removed, then the weights obtained for those individuals would not reflect that which would be expected for a complete individual from the general population.

Warren and Maples (1997) examined how body dimensions and the weight of cremated remains related within a Florida population. According to this study, stature had the strongest correlation with the weight of cremated remains among the variables of sex, age, stature, cadaver weight, and estimated skeletal mass. Bass and Jantz (2004) addressed the weight of cremated remains to determine whether values were determined regionally or if the values from Florida and southern California were applicable also to Tennessee. The Tennessee sample was significantly heavier than that collected in Florida, leading the authors to conclude that the difference was likely due to secular differences in nutrition and body composition between Tennessee and Florida. Van Deest (2007) examined a large sample in northern California to determine a regional comparison of the weight of human cremated remains and its utility in the estimation of sex and age. Regional differences in the weight of cremated remains persist in the analysis by Van Deest (2007). Work also has been conducted in Thailand to determine which parameters of the biological profile relate to the weight of the remains (Chirachariyevej et al. 2006). Age and weight of cremated remains have an inverse relationship, while body weight and weight of cremated remains have a direct relationship in that one can be predicted when given a known amount of the other (Chirachariyevej et al. 2006). Continued research into the relationship between the weight of cremated remains and aspects of the biological profile will help determine the mechanisms behind the differences in average weights between regions and populations.

Particulate sizes

An area of recent consideration is the size of fragments of cremated remains that should be expected after processing. While it has long been a practice to separate cremated remains by size (Schultz et al. 2008), the proportion of the remains expected in each size category after the standard industry processing time of 30s only recently received attention. A preliminary study by Bolhofner (2008) illustrated that the proportion of the cremated remains in each particulate size category was not changed significantly by extending the processing time from 30 to 60s (Table 20.1). Although this study was limited in its scope due to the small number of cases with cremated remains analyzed, the results are informative in the expected outcome of processing fragments of cremated remains. Additional research is needed to more completely understand the effect that processing the remains has on fragment size and the proportion of a size category within the cremated remains.

While a rotary-blade processor is used by most crematoria, in cases where ball-and-hammer processors or hand processing is used fragment size generally will be larger. In the ball-and-hammer method, the remains are processed until the fragments fit through a screen with holes measuring 4 mm across (Warren and Schultz 2002), whereas the rotary-blade processor continues to break fragments down while the machine is running. The size of the fragments after sorting following Warren et al. (1999) can be seen in Figure 20.2. The size and proportion of the fragments expected within processed cremated remains may be used as

Table 20.1   Break down of particulate size as an average percentage of the total human cremated remains with a processing time of either 30 or 60 s (after Bolhofner 2008).

| | Percentage of total remains | | | |
| | 30 s | | 60 s | |
Fragment size	Male	Female	Male	Female
>4 mm	12.29	7.62	8.30	5.00
>2 mm	15.84	16.08	14.71	13.88
>1 mm	12.01	12.98	11.89	12.96
<1 mm	59.86	63.33	65.09	68.09

Figure 20.2   Particulate size comparison of human cremated remains. Scale is in centimeters.

indications of the legitimacy of the remains in cases involving allegations of misconduct or remains suspected to be nonhuman.

DNA in cremated remains

While there are those who purport the possibility of extracting and testing DNA from cremated remains, scientific research has not supported these claims. An article published in the cremation industry's publication advocates the use of mitochondrial DNA (mtDNA) testing as a means to identify whether cremated remains are human or nonhuman and to establish presumptive identity (Raees 2003). The article claims successful extraction of mtDNA from two commercially cremated bone fragments. Instead of supporting the claim, the article focuses on why mtDNA is better than nuclear DNA when dealing with remains previously subjected to extreme environments, such as cremation. Although this is true due to the vast number of copies of mtDNA in comparison to the pair of nuclear DNA strands in each cell, organic material must be present to extract DNA of any kind.

Two examples in the literature claim to find DNA in cremated and heat-altered remains. In research by Brown et al. (1995), DNA was extracted from their samples. The problems associated with this claim are that the cremated remains examined in this study were ancient and of archaeological origin, and thus have the problem of DNA degradation due to time as well as heat exposure. Although the authors claim their results are not from contamination due to the preventative procedures used, it is difficult to rule out the possibility of DNA left by at least one individual who handled those remains during the original cremation process, excavation, processing, and curation before the DNA analysis was conducted. Ye et al. (2004) claim to have established a method to readily extract DNA from old and burned remains. As with other studies in the literature their claim of extracting genetic material from burned remains does not discuss to what extent the remains were heat-damaged. Thus, if the burned bone was not altered to the point of calcination, the remaining organic matrix may have been sufficient to extract genetic material.

Experimental and case studies indicate DNA cannot be extracted from commercially cremated remains. Cattaneo et al. (1999) researched mitochondrial DNA in cases of burned bone as a means to determine human versus nonhuman origin of the remains. For comparison with precremation buccal swabs, von Wurmb-Schwark and colleagues (2004) collected remains after cremation for DNA analyses on 10 individuals. Of the 10 sample pairs, none produced a match between the two samples taken from the same individual. Although the authors did find DNA material, it was assumed to be from the postcremation processing and handling of the remains, thus being contaminated. In a review of skeletal cases sent for DNA analysis, Nelson and Melton (2007) stated that none of the five cases of cremated remains yielded results. Rees and Cox (2010) examined the survivability of DNA in porcine molars treated at temperatures of up to 600 °C. In experimental heating within the laboratory, no DNA could be extracted using PCR methods. However, DNA could be extracted using PCR after fleshed pig heads burned in open-air environments up to temperatures of 625 °C for 15 min. They equate these results to the protection of the dentition by the soft tissue (Rees and Cox 2010). The temperature and duration are well below that of commercial cremation, indicating the futility of DNA analysis in commercially cremated remains.

Due to the increasing reliance on DNA identification methods and the public perspective that DNA will always provide a definitive answer, the understanding of its limited utility concerning cremated remains will help the public understand why only limited conclusions can be made about the remains. It is necessary to educate the general public about the limitations and uses of DNA and other biomolecular techniques, particularly in relation to heat-altered and cremated remains.

CONCLUSIONS

Advances in technology, the rise of new questions, and the revisiting of older questions allow for the progression of the study of cremated remains and expand our understanding of how skeletal remains and heat interact with one another. The use of chemical analysis can shed light on the life history of the individual and potentially answer questions about the identity of the cremated remains. Radiographic imaging can be used to locate and identify osseous versus non-osseous material while other radiographic sciences methods, such as XRD and XRF, can illuminate the crystal structure and the chemical signature of the cremated remains. Although there are claims that DNA of different forms can be extracted from cremated remains for comparison to family members, scientific research has yet to support this claim in modern, commercially cremated remains. As the understanding of the impact of heat and the commercial cremation process on remains progresses, the application of new and increasingly advanced technologies can continue to improve the foundation for the opinions drawn from commercially cremated remains.

ACKNOWLEDGMENTS

The data collection performed by Katelyn L. Bolhofner for the chapter was done with support provided by two grants from the University of Florida: the University Scholars Program Grant for Undergraduate Research, and the Wentworth Travel Scholarship for Undergraduate Research.

REFERENCES

Baby, R. (1954). *Hopewell Cremation Practices* 1: 1–7. Papers in Archaeology. The Ohio Historical Society, Columbus, OH.

Bass, W. and Jantz, R. (2004). Cremation weights in East Tennessee. *Journal of Forensic Sciences* 49(5): 901–904.

Bergslein, E.T., Bush, M., and Bush, P.J. (2008). Identification of cremains using X-ray diffraction spectroscopy and a comparison to trace element analysis. *Forensic Science International* 175: 218–226.

Binford, L.R. (1963). An analysis of cremations from three Michigan sites. *Wisconsin Archaeologist* 44: 98–110.

Bodkin, T. and Mies, J. (2008). X-Ray diffraction (XRD) analysis of human cremains and concrete. *Proceedings of the Annual Scientific Meeting of the American Academy of Forensic Sciences*, H59: 334.

Bolhofner, K. (2008). *The Particulate Size of Cremated Remains*. Undergraduate honor's thesis, Department of Anthropology, University of Florida.

Bonucci, E. and Graziani, G. (1975). Comparative thermogravimetric, X-ray diffraction, and electron microscope investigations of burnt bones from recent, ancient, and prehistoric age. *Atti Memorie Academia Nazionale die Lincei Scienze, Fisiche, Matematiche Naturali, Ser. 8, Sec. 2A (Roma)* 59: 517–534.

Bowler, J.M., Johnston, H., Olley, J.M., Prescottk, J.R., Roberts, R.G., Shawcross, W., and Spooner, N.A. (2003). New ages for human occupation and climatic change at Lake Mungo, Australia. *Nature* 42: 837–840.

Brooks, T.R., Bodkin, T.E., Potts, G.E., and Smullen, S.A. (2006). Elemental analysis of human cremains using ICP-OES to classify legitimate and contaminated cremains. *Journal of Forensic Science* 51(5): 967–973.

Brown, K., O'Donoghue, K., and Brown, T.A. (1995). DNA in cremated bones from an early Bronze Age cemetery cairn. *International Journal of Osteoarchaeology* 5(2): 181–187.

Buikstra, J.E. and Swegle, M. (1989). Bone modification due to burning: experimental evidence. In R. Bonnichsen and M. Sorg (eds), *Bone Modification* (pp. 247–258). Center for the Study of the First Americans, Orono, ME.

Bush, M.A., Miller, R.G., Prutsman-Pfeiffer, J., and Bush, P.J. (2007). Identification through X-ray fluorescence analysis of dental restorative resin materials: a comprehensive study of noncremated, cremated, and processed cremated individuals. *Journal of Forensic Sciences* 52(1): 157–165.

Cattaneo, C., DiMartino, S., Scali, S., Craig, O.E., Grandi, M., and Sokol, R.J. (1999). Determining the human origin of fragments of burnt bone: a comparative study of histological, immunological and DNA techniques. *Forensic Science International* 102: 181–191.

Chirachariyevej, T., Amnueypol, C., Sanggarnjanavanich, S., and Tiensuwan, M. (2006). The relationship between bone and ash weight to age, body weight, and body length of thai adults after cremation. *Journal of the Medical Association of Thailand* 89(11): 1940–1945.

Cremation Association of North America, Market Research and Statistics, SmithBuckin Corporation (2008). Final 2006 Statistics and Projections to the Year 2025 & 2007 Preliminary Data. *The Cremationist of North America* 44(4): 12–23.

Danilchenko, S.N., Koropov, A.V., Protsenko, Y.I., Sulkio-Cleff, B., and Sukhodub, L.F. (2006a) Thermal behavior of biogenic apatite crystals in bone: an X-ray diffraction study. *Crystal Research Technology* 41(3): 268–275.

Danilchenko, S.N., Kulik, A.N., Pavlenko, P.A., Kalinichenko, T.G., Bugai, A.N., Chemeris, I.I., and Sukhodub, L.F. (2006b). Thermally activated diffusion of magnesium from bioapatite crystals. *Journal of Applied Spectroscopy* 73(3): 437–443.

Denton, J., Freemont, A.J., and Ball, J. (1984). Detection and distribution of aluminum in bone. *Journal of Clinical Pathology* 37: 136–142.

Gejvall, N.-G. (1970). Cremations. In D. Brothwell and E. Higgs (eds), *Science in Archaeology: A Survey of Progress and Research* (pp. 468–479). Praeger Publishers, New York.

Glassman, D.M. and Crow, R.M. (1996). Standardization model for describing the extent of burn injury to human remains. *Journal of Forensic Sciences* 44(1): 152–154.

Herrmann, N.P. and Bennett, J.L. (1999). The differentiation of traumatic and heat related fractures in burned bone. *Journal of Forensic Sciences* 44(3): 461–469.

Heymsfield, S.B., Wang, Z., Baumgartner, R.N., and Ross, R. (1997). Human body composition: advances in models and methods. *Annual Review of Nutrition* 17: 527–558.

Hiller, J.C., Thompson, T.J.U., Evan, M.P., Chamberlain, A.T., and Wess, T.J. (2003). Bone mineral change during experimental heating: an X-ray scattering investigation. *Biomaterials* 24: 5091–5097.

Holden, J.J., Phakey, C.C., and Clement, J.G. (1995). Scanning electron microscope observations in heat treated human bone. *Forensic Science International* 74: 29–45.

Mayne Correia, P.M. (1997). Fire modification of bone: a review of the literature. In W.D. Haglund and M.H. Sorg (eds), *Forensic Taphonomy: The Postmortem Fate of Human Remains* (pp. 275–293). CRC Press, Boca Raton, FL.

Mayne, P.M. (1990). *The Identification of Precremation Trauma in Cremated Bone.* MA Thesis, University of Alberta, Department of Anthropology.

McKinley, J. (1993). Bone fragment size and weights from modern British cremations and the implications for the interpretations of archaeological cremations. *International Journal of Osteoarchaeology* 3(4): 283–287.

Miller, J. (2001). Ashes ethereal: cremation in the Americas. *American Indian Culture and Research Journal* 25(1): 121–137.

Munro, L.E., Longstaffe, F.J., and White, C.D. (2008). Effects of heating on the carbon and oxygen-isotope compositions of structural carbonate in bioapatite from modern deer bone. *Palaeogeography, Palaeoclimatology, Palaeoecology* 266: 142–150.

Murad, T. (1998). The growing popularity of cremation versus inhumation: some forensic implications. In K. Reichs (ed.), *Forensic Osteology: Advances in the Identification of Human Remains* (pp. 86–105). Charles C. Thomas, Springfield, IL.

Murray, K.A. and Rose, J.C. (1993). The analysis of cremains: a case study involving the inappropriate disposal of mortuary remains. *Journal of Forensic Sciences* 38(1): 98–103.

Nelson, K. and Melton, T. (2007). Forensic mitochondrial DNA analysis of 116 casework skeletal samples. *Journal of Forensic Sciences* 52(3): 557–561.

Nock, A.D. (1932). Cremation and burial in the Roman Empire. *The Harvard Theological Review* 25(4): 321–359.

Piga, G., Thompson, T.J.U., Malgosa, A., and Enzo, S. (2009). The potential of

X-ray diffraction in the analysis of burned remains from forensic contexts. *Journal of Forensic Sciences* 54(3): 534–539.

Raees, S. (2003). New DNA method to identify cremation remains. *The Cremationist of North America* 39(1): 4–25.

Rees, K.A. and Cox, M.J. (2010). Comparative analysis of the effects of heat on the PCR-amplification of various sized DNA fragments extracted from *Sus scrofa* molars. *Journal of Forensic Sciences* 55(2): 410–417.

Ribeiro, C.C., Gibson, I., and Barbosa, M.A. (2006). The uptake of titanium ions by hydroxyapatite particles: structural changes and possible mechanisms. *Biomaterials* 27: 1749–1761.

Rogers, K.D. and Daniels, P. (2002). An X-ray diffraction study of the effects of heat treatment on bone mineral microstructure. *Biomaterials* 23: 2577–2585.

Schmidt, C.W. and Symes, S.A. (eds) (2008). *The Analysis of Burned Human Remains.* Academic Press, London.

Schultz, J., Warren, M.W., and Krigbaum, J. (2008). Analysis of human cremains: gross and chemical methods. In C. Schmidt and S. Symes (eds), *The Analysis of Burned Human Remains* (pp. 75–94). Academic Press, London.

Schurr, M.R. and Hayes, R.G. (2008). Stable carbon- and nitrogen-isotope ratios and electron spin resonance (ESR) *g*-values of charred bone: changes with heat and a critical evaluation of the utility of *g*-values for reconstructing thermal history and original isotope ratios. *Journal of Archaeological Science* 35: 2017–2031.

Schurr, M.R., Hayes, R.G., and Cook, D.C. (2008). Thermally induced changes in the stable carbon and nitrogen ratios of charred bones. In C. Schmidt and S. Symes (eds), *The Analysis of Burned Human Remains* (pp. 95–108). Academic Press, London.

Shipman, P., Foster, G., and Schoeninger, M. (1984). Burnt bones and teeth: an experimental study of color, morphology, crystal structure, and shrinkage. *Journal of Archaeological Science* 11: 307–325.

Sonek, A. (1992). The weight(s) of cremains. *Proceedings of the 44th Annual Meeting of the American Academy of Forensic Science,* February 17–22, New Orleans, LA, pp. 169–170

Stiner, M., Kuhn, S., Wiener, S., and Bar-Yosef, O. (1995). Differential burning, recrystalization, and fragmentation of archaeological bone. *Journal of Archaeological Science* 22: 223–237.

Ubelaker, D.H. (2009). The forensic evaluation of burned skeletal remains: a synthesis. *Forensic Science International* 183: 1–5.

Van Deest, T.L. (2007). *Sifting through the Ashes: Age and Sex Estimation based on Cremains Weight.* MA thesis, Department of Anthropology, California State University, Chico.

von Wurmb-Schwark, N., Simeoni, E., Ringleb, A., and Oehmichen, M. (2004). Genetic investigation of modern burned corpses. *International Congress Series* 1261: 50–52.

Walker, P.L., Miller, K.W.P., and Richman, R. (2008). Time, temperature, and oxygen availability: an experimental study of the effects of the environmental conditions on the color and organic content of cremated bone. In C. Schmidt and S. Symes (eds), *The Analysis of Burned Human Remains* (pp. 129–135). Academic Press, London.

Warren, M.W. (2008). Detection of commingling in cremated human remains. In B. Adams and J. Byrd (eds), *Recovery Analysis and Identification of Commingled Human Remains* (pp. 185–198). Humana Press, Totowa, NJ.

Warren, M.W. and Maples, W.M. (1997). The anthropometry of contemporary commercial cremation. *Journal of Forensic Sciences* 42: 417–423.

Warren, M.W. and Schultz, J. (2002). Postcremation taphonomy and artifact preservation. *Journal of Forensic Sciences* 47(3): 656–659.

Warren, M.W., Falsetti, A.B., Hamilton, W.F., and Levine, L. (1999). Evidence of arteriosclerosis in cremated remains. *American Journal of Forensic Medicine and Pathology* 20(3): 277–280.

Warren, M.W., Falsetti, A.B., Kravchenko, I.I., Dunnam, F.E., Van Rinsfelt, H.A., and Maples, W.M. (2002) Elemental analysis of bone: proton-induced X-ray emissions testing in forensic cases. *Forensic Science International* 125: 37–41.

Webb, W.S. and Snow, C.E. (1945). *The Adena People.* University of Kentucky

Reports in Anthropology and Archaeology, vol. VI. Department of Anthropology and Archaeology, University of Kentucky, Lexington, KY.

White, G.W., Gibby, W.A., and Tweedle, M.F. (2006). Comparison of Gd(DTPA-BMA) (Omniscan) versus Gd(HP-DO3A) (Prohance) relative to gadolinium retention in human bone tissue by inductively coupled mass spectroscopy. *Investigative Radiology* 41(3): 272–278.

Whyte, T.R. (2001). Distinguishing remains of human cremations from burned animal bones. *Journal of Field Archaeology* 28(3/4): 437–448.

Ye, J., Ji, A., Parra, E.J., Zheng, X., Jiang, C., Zhao, X., Hu, L., and Tu, Z. (2004). A simple and efficient method for extracting dna from old and burned bone. *Journal of Forensic Sciences* 49(4): 754–759.

PART V   Advances in Human Identification

Introduction to Part V

Dennis C. Dirkmaat

Personal identification of human skeletal and dental remains found in outdoor settings typically involves a series of investigative phases. The first phase is the production of skeletal biological profiles to narrow age and stature estimates, and to produce more definitive estimates of sex and ancestry. These estimates are reported to law enforcement who will use the profiles to narrow down the missing persons list. Reduced lists lead to searches for specific antemortem records and primarily dental records, but also a few medical and radiographic records. These antemortem records are used by forensic odontologists and other relevant forensic specialists to conduct antemortem to postmortem comparisons and eventually provide an assessment of identity (positive, negative, or undetermined). In cases in which law enforcement has an idea of who the victim might be, usually based on contextual clues such as location of body in house or car, lengthy anthropological analysis of the bones can often be bypassed because the forensic odontologist can provide an assessment of identification rather quickly from the examination of the dentition alone. Considering this investigative scenario, forensic cases in which forensic anthropologists become involved to produce a bioprofile were often limited to body dumps along roads or in other "out-of-context" scenes. On occasion, if no teeth or dental records are available for comparison, the forensic anthropologist may be called upon to compare medical records or radiographs to the skeletal evidence relative to a prosthetic device, a unique skeletal feature (e.g., frontal sinuses), or healed previous trauma. Still, the role of the forensic anthropologist in the process of victim identification is often very limited. And then, to add misery to neglect, along came DNA.

A Companion to Forensic Anthropology, First Edition. Edited by Dennis C. Dirkmaat.
© 2012 Blackwell Publishing Ltd. Published 2012 by Blackwell Publishing Ltd.

It was argued in Dirkmaat et al. (2008) and again in Chapter 1 of this volume, that unless definitions, practices, roles, and duties of forensic anthropology were redefined, the field faced possible extinction. DNA analysis could provide *positive* identifications and do it fast, and increasingly faster. This fact is driven home by Cabo's discussion in Chapter 22 in which he provides a succinct history of research in DNA analysis as related to the production of positive identifications in medicolegal investigations. Historically, the cost in both money and time for sending DNA samples away to be analyzed was generally prohibitive for most law-enforcement and medicolegal organizations. Typically, DNA analysis was a break-the-budget, last resort after other standard identification methods were exhausted and did not yield a positive identification. That will not be the case in the future. As presented in Chapter 22, enhanced speed and reduced cost of DNA analysis, in conjunction with the availability of huge comparison databases, will allow DNA to be collected and analyzed automatically for all identification cases and likely will serve as the *first* line of victim-identification attempts. The need to conduct fingerprint and dental comparisons will be greatly diminished. And relevant to forensic anthropology, the need to produce a biological profile through skeletal analysis as a means to reduce the missing person list would no longer be justified. However, this rather bleak assessment of the future of forensic anthropology must be tempered by the fact that the field has been able to reinvent itself through new definitions, goals, and activities (see Chapter 1), providing what would appear to be a relatively bright light at the end of the tunnel.

During the investigation of "typical" forensic cases involving remains from outdoor contexts, DNA may be used to confirm other positive identification assessments or as the sole tool for identification efforts. In these situations, it is usual for the forensic anthropologist to be involved in the preparation of samples to be submitted to DNA laboratories. Therefore, it is useful to possess basic knowledge of the Dos and Don'ts of handling, sampling, and submitting DNA samples. Boyer in Chapter 23 provides a very good primer with respect to these issues. He addresses the effects of taphonomic agents such as weathering and burning on DNA, optimal samples, and best methods for collecting and prepping samples.

One important area of investigation in which forensic anthropologists may become involved relative to personal identification issues is an area in which DNA will not help: the analysis of human cremated remains, garnered from commercial crematoria. Since DNA is long destroyed by the heating process, the analysis of the human remains focuses on the search for retained skeletal features that might provide some clues to biological profile. However, with methods in place to "process" the burned bones after cremation – essentially to crush the material to "unrecognzable" condition – very little skeletal feature information remains. Fortunately, as discussed in Chapter 20 by Van Deest, Warren, and Bolhofner, there are a number of analytical methods that have recently become available to analyze these cremated remains. Since the heating and burning process alters nearly all of the materials to similar color schemes (primarily, blacks, grays, and whites), new chemical analyses have been developed to sort bone from other material in the burned assemblage. In addition, analytical techniques have been developed to study cremated remains including weight, radiographic images, bone microstructure, retained medical and dental artifacts, and even bullet residue from previous gunshot trauma through elemental analysis. These methods and techniques have resulted in the ability to obtain a wealth of information from highly altered human remains.

In Chapter 21, Sauer, Michael, and Fenton discuss the latest techniques in skull–photo superimposition and forensic image comparison that have been used to address false identifications based on photos and identify the recently deceased. This is especially useful in areas of the world with poor antemortem records.

REFERENCE

Dirkmaat, D.C., Cabo, L.L., Ousley, S.D., and Symes, S.A. (2008). New perspectives in forensic anthropology. *Yearbook Physical Anthropology* 51: 33–52.

CHAPTER 21

Human Identification Using Skull–Photo Superimposition and Forensic Image Comparison

Norman J. Sauer, Amy R. Michael, and Todd W. Fenton

INTRODUCTION

The human face is an enigma for people involved in human identification. While each face is unique (certainly less so for identical twins), numerous studies comment on the notorious inaccuracy of eyewitness facial identification (Wells and Olson 2003). Even close relatives and acquaintances occasionally misidentify victims whose faces are relatively undistorted.

Identification by anthropologists using the human face tends to fall into two categories: skull–photo superimposition and forensic image comparison. The former is normally applied to the identification of the skull of an unknown decedent while the latter is more typically used to identify a living person, possibly a suspect, from some kind of photo, often a surveillance image. While the two procedures share some of the same equipment and analytical methods, the bases upon which judgments are made are usually quite different. In this chapter we will deal first with the skull–photo superimposition and then address forensic image comparison.

SKULL–PHOTO SUPERIMPOSITION

That there is a strong relationship between the morphology of the facial skull and the soft features of the human face is well known. This relationship forms the foundation

A Companion to Forensic Anthropology, First Edition. Edited by Dennis C. Dirkmaat.

of two forensic techniques. Dating back at least to the early twentieth century, forensic artists have used the shape of the skull to reconstruct or approximate a face (Rogers 2005). Artists may form three-dimensional faces using clay on actual skulls or casts or they may create two-dimensional faces by drawing or with the use of a computer. Because these procedures are normally probative and not involved in an actual identification they are beyond the scope of this chapter and will not be discussed further. However, the close relationship between a skull and a face may be used for identification purposes when a semitransparent image of a face is systematically superimposed on a skull.

Forensic anthropologists have used image comparison and skull–photo superimposition techniques to provide identifications of decedents through exclusion. Fenton et al. (2008) introduced the use of skull–photo superimposition in the cases of individuals migrating to the American Southwest by way of unpatrolled borders. In these instances, antemortem medical records are generally unavailable or nonexistent, requiring the forensic anthropologist to rely upon other means of identification. Like Dorion (1983), Austin-Smith and Maples (1994), and Glassman (2001) commented on before them, Fenton et al. (2008) believe that the use of skull–photo superimposition is best suited for the exclusion of individuals based on facial anatomy and skull features. That is, skull–photo superimposition by itself should not be considered a positive identification technique, but rather a powerful analytical tool capable of proving exclusion or providing corroborative evidence that supports a positive identification. These circumstantial identifications or exclusions can greatly aid law enforcement in directing their investigations.

The reliance on skull–photo superimposition for human identification varies significantly around the world. Writing from India, for example, Jayaprakash and colleagues state: "Skull–photograph superimposition continues to be the most prevalent method employed for identifying a skull recovered in a criminal case as that belonging to a putative victim whose face photograph is available" (Jayaprakash et al. 2001: 121).

The difference between the heavy reliance on skull–photo superimposition in some parts of the world and its much rarer usage for identification in the USA likely has to do with the availability of missing persons' antemortem data, particularly X-rays. In the USA many citizens have multiple dental radiographs, often every year during routine dental exams. Further, medical X-rays of the head and body are probably more available in the USA than many other countries of the world. The affect of this disparity is that the skull–photo superimposition techniques are often more refined outside of the USA.

During the late nineteenth and early twentieth centuries, skull–photo superimposition techniques were employed by anatomists to identify the skulls of historically significant individuals (Rogers 2005). One of the earliest recorded uses of a skull–photo superimposition method in a judicial case is that of the Ruxton case of 1935 (Glaister and Brash 1937). Dr Buck Ruxton murdered his wife and housekeeper and removed physically identifying traits such as fingerprints and ears. He dismembered both bodies and disposed of them in various locales. Dr John Glaister made the first known attempt to identify the victims through the use of skull–photo superimposition. It was determined that the antemortem photo was a match to Mrs Ruxton's remains and Dr Ruxton was convicted of the crime. This first case paved the way for the use of the technique in other challenging forensic cases. Skull–photo superimposition was employed intermittently in the USA until the 1980s when there was a burgeoning of the field (Brown 1983; Bastiaan et al. 1986; Brocklebank et al. 1989; Delfino et al. 1986; Lan and Cai 1993; Yoshino et al. 1997). Krogman and İşcan (1986) reviewed a number of cases involving the comparison of antemortem photographs to postmortem anatomy throughout the twentieth century, noting in particular that photographs are often not aligned in a transverse or sagittal plane. Arrangement of the skull in concordance with the available photographs has long been cited as the primary concern in skull–photo superimposition cases.

Validation studies in the USA have not yet demonstrated that a match made during photo–photo or skull–photo analysis is representative of one and only one person (Austin-Smith and Maples 1994). Nonetheless, skull–photo superimposition is accepted as a method of positive identification in India and, increasingly, in China (Dong-Sheng et al. 1989; Ubelaker et al. 1992). Chinese researchers have employed large sample sizes to demonstrate the exclusivity of facial features among Han Chinese populations (Dong-Sheng et al. 1989).

Methods

There are several categories of method involved in skull–photo superimposition. They may involve a paired camera system and a mixer, some device to position and adjust the magnification of the skull and photograph, and a systematic technique designed to arrive at a statement of identification or exclusion. In the early days the method often involved superimposing a negative of an unknown skull over a facial photograph of a known decedent using a combination of viewing boxes and projectors (Aulsebrook et al. 1995). Later, researchers employed video equipment, including cameras, a mixing device, and a monitoring screen (Aulsebrook et al. 1995; Glassman 2001; Iten 1987). In the 1980s researchers increasingly began to utilize computer programs to quantify the relationship between a two-dimensional photograph and a three-dimensional skull (Aulsebrook et al. 1995; Ubelaker et al. 1992). Additionally, computer software has become available to manipulate images for comparison. Limitations persist in these methods; namely the difficulties in positioning the skull for comparison and issues related to different camera types and lens systems.

Skull mounting and manipulation systems vary greatly in construction and elaborateness. At one end of the spectrum is a simple ring or pad that allows a skull to be positioned and reoriented without much "wobble." At the other end, an elaborate portable Chinese system uses a skull-holding mechanism adapted from a halo device originally developed to stabilize the head in facial trauma cases, an orientation or "pan-and-tilt" device, and a camera mount, all attached to aluminum rails (Brocklebank and Holmgren 1988).

In our laboratory at Michigan State University we use a dual-camera system with a monitor, recording device, and a video mixer. The skull is mounted either on a ring or a movable clamping device inserted into the foramen magnum. The cameras are affixed to sliding mounts (heavy-duty copy stands) and each is equipped with a zoom lens. One camera is positioned above the skull and the other above a photograph. Adjacent and connected to these cameras are a video mixer, a video-cassette recorder, and a monitor. Of course, all of this instrumentation may be digital (Figure 21.1).

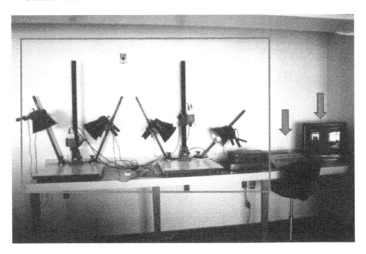

Figure 21.1   Superimposition and image analysis set-up at Michigan State University Forensic Anthropology Laboratory. Shown here are the monitor and mixer (marked by arrows) and the dual-camera system (outlined in the text).

Orienting and sizing the skull with respect photographs is normally done simultaneously by the trial-and-error method in the following sequence (Fenton et al. 2008). Several landmarks are identified and marked on the skull and photograph:

- left and right ectocanthion (face) and Whitnall's tubercles (skull) (Stewart 1983);
- subnasal point (skull and face);
- gnathion (skull) and the inferior most point on the mandible in the midsagittal plane (face);
- rods (Q-tips) are affixed to the external auditory meati.

With these markers in place, it is possible to carefully size and orient the skull with respect to the fixed image of the face. The inability to align these features after several attempts at resizing and orienting the skull may signal that the skull and image represent different people. One has to be cautious, however, since small variations may be due to distortion caused by different camera lenses. Since the distance between camera and object as well as focal length play a significant part in shaping a three-dimensional object, the trial and error should include changing the distance between the camera and the object (skull) and accommodating with the zoom lens and focus.

At the Forensic Anthropology Laboratory at Michigan State University we employ the dynamic orientation method, which allows for on-screen, live manipulation of the skull orientation. Once the skull and photo are satisfactorily matched, then individual features are examined in sequence. We normally make use of the guidelines presented by Austin-Smith and Maples (1994). The comparison of images may also be facilitated by the placement of tissue-depth markers on specific points on the skull (Ubelaker et al. 1992; Fenton et al. 2008). There are a variety of sources for tissue-depth marker data (Rhine and Campbell 1980; Rhine and Moore 1982; Taylor 2001).

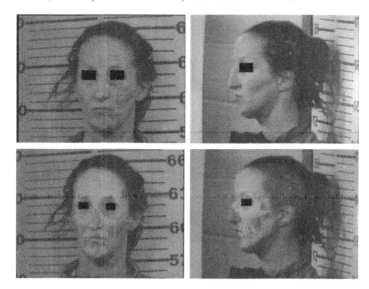

Figure 21.2 Frontal and lateral views of the suspected decedent (top). These antemortem photographs were provided by law enforcement. Frontal and lateral views of suspected decedent with the unknown skull superimposed over the antemortem images (bottom).

Case Study

Human remains were discovered in a shallow grave in the US state of Michigan in 1996. The circumstances of the scene and skeletal trauma indicated that manner of death was homicide. Of course, establishing the identity of the victim was a paramount issue. We (NJS and TWF), using standard forensic anthropological methods, generated the following biological profile:

Sex:	Female
Age:	33–49 years
Ancestry:	European
Height:	165–175 cm

A missing person from the same area of the state had the following listed as her personal data:

Sex:	Female
Age:	35 years
Ancestry (race):	White
Height:	170 cm

Because of the corresponding personal characteristics and circumstances of the case an attempt was made to positively identify the above missing person using mitochondrial DNA and then skull–photo superimposition. Dental or medical X-rays were never recovered for the missing person. The mitochondrial DNA comparison produced a haplotype in common with about 1 in 5000 people. To avoid any questions about positive identification, the prosecutor and medical examiner requested an anthropological skull–photo superimposition analysis using frontal and lateral police booking photos (Figure 21.2).

The analysis was carried out in the Michigan State University Forensic Anthropology Laboratory using the instruments described above: a dual-camera set-up, a mixer, and a monitor. Following our normal procedure for a frontal photograph comparison, four markers were placed on the skull: one each at the right and left Whitnall's tubercles (Whitnall's tubercle is a bony eminence on the lateral margin of the eye orbit that aligns with the lateral attachment of the upper and lower eyelids; ectocanthion); one at the subnasal point (which corresponds to the subnasal point on the face); and one at the gnathion (the lowermost point in the chin in the midline). Next, Q-tips are placed in each external auditory meatus. The positioning of these markers and the Q-tips allow for the skull to be oriented (inclination, declination, and rotation) with respect to the face (Figure 21.2, bottom left). A lateral view of the skull was similarly positioned to compare to a left side booking photo (Figure 21.2, bottom right).

The following morphological features were observed according to the guidelines published by Austin-Smith and Maples (1994).

Frontal view

1. The length of the skull from bregma to menton fits within the face. Bregma is usually covered with hair. *(Our case is consistent.)*
2. The width of the cranium fills the forehead area of the face. *(Our case is consistent.)*
3. The temporal line can sometimes be distinguished on the photograph. If so, the line of the skull corresponds to the line seen on the face. *(Temporal line is not visible.)*
4. The eyebrow generally follows the upper edge of the orbit over the medial two-thirds. At the lateral superior one-third of the orbit the eyebrow continues horizontally as the orbital rim begins to curve inferiorly. *(Our case is consistent.)*
5. The orbits completely encase the eye including the medial and lateral folds. The point of attachment of the medial and lateral palpebral ligaments can usually be found on the skull. These areas align with the folds of the eye. *(Our case is consistent.)*
6. The lacrimal groove can sometimes be distinguished on the photograph. If so, the groove observable on the bone aligns with the groove seen on the face. *(Lacrimal groove is not visible.)*
7. The breadth of the nasal bridge on the cranium and surrounding soft tissue is similar. In the skull the bridge extends from one orbital opening to the other. In the face, the bridge spreads between the medial palpebral ligament attachments. *(Our case is consistent.)*
8. The external auditory meatus opening lies medial to the tragus of the ear. The best way to judge this area is to place a projecting marker in the ear canal. On superimposition, the marker will appear to exit the ear behind the tragus. *(Our case is consistent.)*
9. The width and length of the nasal aperture falls inside the borders of the nose. *(Our case is consistent.)*
10. The anterior nasal spine lies superior to the inferior border of the medial crus of the nose. With advanced age the crus of the nose begins to sag and the anterior nasal spine is located further superiorly. *(Our case is consistent.)*
11. The oblique line of the mandible (between the buccinators and the masseter muscles) is sometimes visible in the face. The line of the mandible corresponds to the line of the face. *(The oblique line of the mandible is not visible.)*

12. The curve of the mandible is similar to that of the facial jaw. At no point does the bone appear to project from the flesh. Rounded, pointed, or notched chins will be evident in the mandible. *(Our case is consistent.)*

Lateral view

1. The vault of the skull and the head height must be similar. *(Our case is consistent.)*
2. The glabellar outline of both the bone and the soft tissue must have a similar slope although the line of the face does not always follow the line of the skull exactly. There may be slight differences in soft tissue thicknesses that do not relate to nuances in the contour of the bone. *(Our case is consistent.)*
3. The lateral angle of the eye lies within the bony lateral wall of the orbit. *(Our case is consistent.)*
4. The glabella, nasal bridge, and nasal bone area are perhaps the most distinctive. The prominence of glabella and the depth of the nasal bridge are closely approximated by the soft tissue covering this area. The nasal bones fall within the structure of the nose and the imaginary continued line, composed of the lateral nasal cartilages in life, will conform to the shape of the nose except in cases of noticeable deformity. *(Our case is consistent.)*
5. The outline of the frontal process of the zygomatic bones can normally be seen in the flesh of the face. The skeletal process can be aligned with the process seen in the face. *(Our case is consistent.)*
6. The outline of the zygomatic arch can be seen and aligned in those individuals with minimal soft tissue thickness. *(Our case is consistent.)*
7. The anterior nasal spine lies posterior to the base of the nose near the most posterior portion of the lateral septal cartilage. *(Our case is consistent.)*
8. The porion aligns just posterior to the tragus, slightly inferior to the crux of the helix. *(Our case is consistent.)*
9. The prosthion lies posterior to the anterior edge of the upper lip. *(Our case is consistent, although prosthion has receded due to long-term edentula of the central incisors.)*
10. The pogonion lies posterior to the indentation observable in the chin where the obicularis oris muscle crosses the mentalis muscle. *(Our case is consistent.)*
11. The mental protuberance of the mandible lies posterior to the point of the chin. The shape of the bone (pointed or rounded) corresponds to the shape of the chin. *(Our case is consistent.)*
12. The occipital curve lies within the outline of the back of the head. This area is usually covered with hair and the exact location may be difficult to judge. *(Our case is consistent.)*

IMAGE COMPARISON

There are many similarities between skull–photo superimposition identification and image comparison. Both methods compare features of the face; both methods rely to some extent on underlying skeletal structure; and both methods typically employ a morphological as well as a superimposition component. There are significant differences, however, between these two methods as well. In skull–photo comparisons the unknown is the skull. The quality of the images of the unknown is limited only by the available imaging instruments and the skill and experience of the observer. Images supplied by those who know a missing person are often quite clear. In image–image assessments, the evaluator is often handed a VHS surveillance tape or a still picture of an unknown offender that has been captured from such a tape. The quality of surveillance images is often poor, which is why experts commonly are often invited to assist with the identification. If a high quality face-on image is of a loved one or well-known person, such as Barack Obama, a visual assessment would likely be as strong as any formal laboratory analysis, whether or not it is considered a positive identification. One of the initial challenges for the forensic investigator then, is to obtain the best possible image of the unknown. This may simply involve informing a law-enforcement agency that an original or high-quality image is required for the analysis. In some cases, however, an agency may be uncooperative and the original image may require legal intervention. The general public or law-enforcement community may not understand that valuable information is lost in each generation of video-tape copies. In fact, on more than one occasion, our lab has been presented with a Xerox copy of a still image taken from a poor-quality surveillance tape. In such cases a reliable superimposition is normally not possible. With the recent increased use of high-quality digital surveillance cameras, obtaining clear images of offenders less of a problem than in the past.

As is the case with skull–photo comparisons, a great deal of attention must be paid to orientating known and unknown images. There is an exception to this rule, however. Over the years, our lab has been asked to assist with the identification of an image of a person who is deceased or who is not available for live recording. In such cases superimposition is often not possible and identifications and exclusions are carried out solely by morphological comparison.

In those cases where there is a living suspect, then a dynamic orientation can be carried out either in or away from the laboratory. The process of face-to-photo orientation is similar to skull–photo, except that in skull–photo matches it is the skull that is manipulated to match the known photograph. In image analysis, it is the known person or (more often) suspect whose photograph is manipulated to match the unknown photograph. Therefore a living person is the subject of the orientation and sizing process. There are several methods published to facilitate the orientation of images of a live subject to a still photograph. Yoshino et al. (2000), for example, present a "3-D physiognomic range finder" (p. 225) to assist with face orientation and identification. Their results are compelling, but the specialized equipment they employ is generally unavailable to most forensic and anthropology labs.

In our laboratory we use essentially the same equipment for skull–photo and image identification. We use the same mixer but a different camera and monitor for the process. We normally work closely with a local production company, which supplies a high-resolution digital camera, portable studio lighting, a portable monitor, and a skilled camera operator. All of the equipment can be packed in two cases and carried on commercial aircraft (Figure 21.3).

Once comparable images are obtained, the process of identification or exclusion begins. In our lab, we normally use two procedures when comparing known and with unknown images, morphological comparisons, and superimpositions. Morphological comparison is a step-by-step evaluation of established individual landmarks. There are several lists of morphological characteristics available, including an exhaustive one

Figure 21.3   Camera and monitor system used during the dynamic orientation technique with analysts and suspect (seated).

presented by İşcan and Hellmer (1993). İşcan and Hellmer's list includes 55 facial features with a number of options to describe varying descriptions for each one. A sample from the list follows:

Facial forms
   Elliptical
   Round
   Oval
   Pentagonal
   Square
   Trapezoid
   Wedge-shaped
   Double concave
   Asymmetrical

Forehead height
   Low
   Medium
   High

Baldness
   Absent
   Slight
   Advanced
   Complete

Eyebrow thickness
   Slight
   Small
   Average
   Large

Eyebrow shape
   Straight
   Wavy
   Arched

Eyebrow density
   Sparse
   Thick
   Bushy

Nasion depression
   Trace
   Slight
   Average
   Deep
   Very deep

Bony profile (nose)
   Straight
   Concave
   Wavy
   Convex

Bridge height
   Small
   Medium
   High

In addition to these features there are descriptors for the shape of the lips and mouth, cheeks, chin, ear, and several others. There are also opportunities to record unusual or unique features like moles, birthmarks, scars, and asymmetry.

Other sources present details of specific areas that may be useful for individualization. For example, in his book, *System of Ear Identification*, Iannarelli (1964) details 13 different features of the external ear. van der Lugt's 2001 volume, *Earprint Identification*, lists 14 ear features, many the same as Iannarelli: crus of the helix; helis; lobe; upper and lower crus of the anti-helix; anti-helix; triangular fossa; scaphoid fossa; tragus; anti-tragus; inter-tragus notch; anterior knob; anterior notch; auricular tubercle; and posterior auricular furrow. Together, these sources provide a list of specific items that can be checked between the known and unknown images. It is important to note that in no case will all of the features of the face or ear be usable. The applicability of specific features will depend on the orientation and quality of the unknown image. In most cases, we have control over the quality and orientation of the images.

Several years ago our lab was asked to identify or exclude a hand of a probable offender in a computer image that involved child pornography. We were successful in excluding four suspects but were unable to exclude a fifth. The fifth suspect shared a number of detailed morphological similarities with the hand captured on the video. That fifth suspect was later convicted and sentenced for illegally producing and distributing pornography.

Some individuals display unusual features (e.g., visible moles, nose deformities, or misshapen ears) that may assist with identification or exclusion. If it turns out that an unknown offender has a deep nasal depression or an attached ear lobe and the suspect has a trace of a nasal depression or a clearly detached ear lobe, then the likely conclusion is that they are different people and the suspect would be excluded. Some individuals display unusual features (like visible moles, nose deformities, or misshapen ears) that may assist with identification or exclusion. On the other hand if no differences are detected that cannot be explained by factors such as like poor lighting or a bad camera angle and the suspect cannot be excluded on the basis of the morphological comparison, a superimposition may be warranted.

The goal of superimposition in image comparisons of the living is similar to skull–photo superimposition; that is, to evaluate the alignment and position of features relative to one another. In other words, like inconsistent attributes, an inability to superimpose a known image on an unknown image so that the features themselves superimpose exactly on both images may be grounds for exclusion. The biggest challenges with either skull–photo or image superimposition are orientation and cameral distortion. Camera distortion can usually be controlled by changing the distance from the focal plane to the subject and zooming in or out to match image sizes.

Similar to the method we use for skull–photo superimposition, dynamic orientation to the known and the unknown images is used. An important difference is that with skull–photo comparisons we have the skull in our lab and are able to manipulate it on a copy stand with a fixed camera. With image comparisons, the known is a living person who must cooperate and sit in front of a camera while appropriate images are captured normally by adjusting the camera. Often the known is an incarcerated suspect or defendant and our team (with equipment) must travel to a correctional facility. Once we are in the same room with the suspect, we must ask him or her to cooperate and sit while we closely record images for later comparison. The suspect must be told that the process is as much for exclusion as it is for identification. Images are recorded digitally on a movie camera while the analyst studies the live superimposition of the suspect onto each of the images captured from the surveillance tape (or other source). By capturing hundreds of pictures from a distance and angle that approximate the same distance and angle of the unknown image, the likelihood of obtaining good decent usable images is quite high.

Once the known and unknown images are selected, they can be superimposed in a manner similar to the skull–photo comparison. By using vertical and horizontal wipes and dissolves, the analyst can compare features for similarities and differences in shape as well as position with respect to one another. Many of the same features can be used in both skull–photo and image comparisons; however, for the latter the comparisons are direct. There is no need to project from skull to photograph. Again, the analyst attempts to exclude the suspect, in this case by evaluating similarities in shape and position. If there is not a perfect match between features that cannot be explained by slight variations in orientation or lighting, then exclusion is likely.

DISCUSSION

Several regions of the skull have been used with varying degrees of success in skull–photo superimposition. Features of the mid-face and dentition (De Angelis et al. 2007; Klonaris and Furue 1980; McKenna et al. 1984) have been employed in super-imposition analyses of unknown individuals.

In regions of the world with poor health infrastructure or lack of access to medical care records after death, forensic anthropologists need to be aware of other useful methods in identifying the deceased. De Angelis et al. (2007) argued that a protocol for the analysis of dentition in skull–photo superimposition should be employed. It is assumed here that family members or acquaintances of the decedent can produce antemortem photographs with visible teeth that can be compared to dental casts. The authors provide a scoring procedure that may be used for quantifying the fit of the relationship between the cast and the photo. Further studies featuring larger sample sizes, genetically homogenous populations or individuals, and inter- and intra-observer error rates should be considered in the future. Additionally, dental pathologies and their effect on the dentition over time should be addressed.

The principal concern in using superimposition in a court of law is that of a false identification. Austin (1999) found that there was a slight incidence of misidentification when comparing skulls to photographs, even when both profile views and full-face views were involved. These error rates lead Austin (1999) to recommend that the anthropologist secure multiple photographs of the subject from various angles before resolving a superimposition case. Yoshino et al. (1995) also found that positive identification of an unknown decedent was possible when two or more photographs of the face taken from various angles were superimposed on the skull. These authors advised that frontal, as well as oblique and or lateral, photographs be employed in forensic cases. In a more recent publication Stephan (2009) confirmed that a minimum of two differently oriented antemortem photographs be made be available to the practitioner and, ideally, the dentition should be visible in one or more of the photographs. In a comparison of three skulls and 100 police photographs, Austin-Smith and Maples (1994) discovered that when one photograph was compared to a skull an error rate of approximately 9% occurred; however, false positives were drastically reduced drastically when two or more photographs taken from different viewpoints were used. Sekharan (1993: 105) suggested that flexion or extension of the head, lateral flexion of the head, and rotation of the head are critical in the assessment of a photograph. The researcher must evaluate the positioning of the head in the available photographs in order to correctly arrange the skull for superimposition analysis. Aulsebrook et al. (1995) stated that common features of the skull and soft tissues of the face must be quantified in order to make more sophisticated objective comparisons. Furthermore, the authors argued that anthropologists must be able to convince the court of true anatomical matches between the skull of the decedent and the photograph(s); this is achieved by a statistical analysis of shape and fit (Aulsebrook et al. 1995).

Austin (1999) quantified rates of misidentification using superimposition methods in a study of 25 skull–photo cases from the C.A. Pound Laboratory. At an accuracy rate of 99%, Austin (1999) asserted that this method may be used with confidence in the court room. Other researchers have used mathematical models to quantify the rate of association between a skull and a photograph (Palhares et al. 1995).

CONCLUSION

Skull–photo superimposition and image comparison are two important tools for human identification. In many missing persons cases or criminal investigations the only available link between an unknown and a known person is a photograph. In the case of skull–photo superimposition, facial images, often obtained from family members or through previous police records, are typically "placed over" carefully aligned and sized skull images and the position of specific points is evaluated. The evaluation is facilitated by systematically removing the photograph and exposing the underlying skull through a series of horizontal and vertical wipes or dissolves. Identity is assessed by evaluating the correspondence between specific landmarks on the face and skull. It is up to the investigator to decide whether the points match sufficiently to establish identification at some level of probability or whether nonmatches are sufficient for exclusion.

Image analysis actually has some advantages over skull–photo superimposition, because soft tissue structures are compared directly. In other words, there is no need to extrapolate from the underlying skull to the image of the face. While superimposition plays a role in image identification, specifically in establishing the relative position and size of features, it is subordinate to a morphological comparison. A morphological comparison is essentially a detailed list of traits that characterize, in both images, facial features like the nose, mouth, eye, and eyebrows. Again, a match or exclusion depends on the judgment of the investigator. We, and most investigators in the USA, consider neither skull–photo superimposition nor image comparison sufficient evidence, by themselves, for position identification. In other countries, perhaps those with fewer opportunities for more established identification methods like antemortem/ postmortem X-ray comparisons or less access to DNA facilities, positive identification routinely relies on skull–photo and image comparisons. Nevertheless, both methods are powerful tools for supporting other associated evidence for positive identification and for exclusion.

Current and future research will likely deal with a number of issues relating to skull–photo superimposition and image comparison, including the role of measurement and statistics and with scientific validation. For image comparisons, analysts will consider the uniqueness of the expression of specific features. It would be useful to know, for example, how likely or unlikely it is that a person will express a widow's peak on the hairline or an auricular tubercle on the ear. Finally, scientists must agree on what will be the best practices and procedures for skull–photo superimposition and image comparison and the appropriate manner by which to report results.

REFERENCES

Aulsebrook, W.A., İşcan, M.Y., Slabbert, J.H., and Becker, P. (1995). Superimposition and reconstruction in forensic facial identification: a survey. *Forensic Science International* 75: 101–120.

Austin, D. (1999). Video superimposition at the C.A. Pound Laboratory 1987 to 1992. *Journal of Forensic Sciences* 44(4): 695–699.

Austin-Smith, D. and Maples, W.R. (1994). The reliability of skull/photograph superimposition in individual identification. *Journal of Forensic Sciences* 39: 446–455.

Bastiaan, R.J., Dalitz, G.D., and Woodward, C. (1986). Video superimposition of skulls and photographic portraits – a new aid to identification. *Journal of Forensic Sciences* 31: 1373–1379.

Brocklebank, L.M. and Holmgren, C.J. (1988). Development of equipment for the standardization of skull photographs in personal identifications by photographs in personal identifications by photographic superimposition. *Journal of Forensic Sciences* 34(5): 1214–1221.

Brocklebank, L.M. and Holmgren, C.J. (1989). Development of equipment for the standardization of skull photographs in personal identifications by photographic superimposition. *Journal of Forensic Sciences* 34: 1214–1221.

Brown, K.A. (1983). Developments in craniofacial superimposition for identification. *Journal of Forensic Odontostomatology* 1: 57–64.

De Angelis, D., Cattaneo, C., and Grandi, M. (2007). Dental superimposition: a pilot study for standardizing the method. *International Journal of Legal Medicine* 121: 501–506.

Delfino, P.V., Colonna, M., Vacca, E., Potente, F., and Introna, Jr, F. (1986). Computer-aided skull/face superimposition.

American Journal of Forensic Medicine and Pathology 7: 201–212.

Dong-Sheng, C., Yu-Wen, L., Cheng, T., Run-Ji, G., Yong-Chuan, M., Jian-Hai, F., Wei-Dong, W., and Jiang, Z. (1989). A study on the standard for forensic anthropologic identification of skull-image superimposition. *Journal of Forensic Sciences* 34(6): 1343–1356.

Dorion, R.B.J. (1983). Photographic superimposition. *Journal of Forensic Sciences* 28(3): 724–734.

Fenton, T.W., Heard, A.N., and Sauer, N.J. (2008). Skull-photo superimposition and border deaths: identification through exclusion and failure to exclude. *Journal of Forensic Sciences* 53(1): 34–40.

Glaister, J. and Brash, J.C. (1937). *Medico-legal Aspects of the Ruxton Case*. William Wood and Company, Baltimore, MD.

Glassman, D.M. (2001). Methods of superimposition. In K.T. Taylor (ed.), *Forensic Art and Illustration* (pp. 477–498). CRC Press, Boca Raton, FL.

Iannarelli, A.V. (1964). *System of Ear Identification*. The Foundation Press, Brooklyn, NY.

İşcan, M.Y. and Helmer, H.P. (1993). *Forensic Analysis of the Skull*. Wiley-Liss, New York.

Iten, P.X. (1987). Identification of skulls by video superimposition. *Journal of Forensic Sciences* 32(1): 173–188.

Jayaprakash, P.T., Srinivasan, G.J., and Amravaneswaran, M.G. (2001). Cranio-facial morphanalysis: a new method for enhancing reliability while identifying skulls by photo superimposition. *Forensic Science International* 117: 121–143.

Klonaris, N.S. and Furue, T. (1980). Photographic superimposition in dental identification. Is a picture worth a thousand words? *Journal of Forensic Sciences* 25(4): 859–865.

Krogman, W.M. and İşcan, M.Y. (1986). *The Human Skeleton in Forensic Medicine*, 2nd edn. Charles C. Thomas, Springfield, IL.

Lan, Y. and Cai, D. (1993). Technical advances in skull-to-photo superimposition. In M.Y. İşcan and R.P. Helmer (eds), *Forensic Analysis of the Skull* (pp. 119–129). Wiley-Liss, New York.

McKenna, J.J.I., Jablonski, N.G., and Fearnhead, R.W. (1984). A method of matching skulls with photographic portraits using landmarks and measurements of the dentition. *Journal of Forensic Sciences* 29(3): 787–797.

Palhares, F.A.B., Silveira, M.A.M., Olivier, S.L., Tozzi, C.L., Tommaselli, A.M.G., and Hasegawa, J.K. (1995). A model for evaluating the correspondence in the craniofacial identification process. In B. Jacob and W. Bonte (eds), *Advances in Forensic Sciences: Proceedings of the 13th Meeting of the International Association of Forensic Sciences*, August 22–28, 1993, Dusseldorf. Koster, Berlin, 7: 335–340.

Rhine, S. and Campbell, H.R. (1980). Thickness of facial tissues in American Blacks. *Journal of Forensic Sciences* 25: 847–858.

Rhine, S. and Moore, C.E. (1982). *Facial Reproduction: Tables of Facial Tissue Thickness of American Caucasoids in Forensic Anthropology*. Maxwell Museum Technical Series, No. 1. University of New Mexico, Albuquerque, NM.

Rogers, N.L. (2005). The first use of a composite image in forensic facial superimpostion: the case of John Paul Jones, 1907. *Journal of Forensic Identification* 55(3): 312–326.

Sekharan, P.C. (1993). Positioning the skull for superimposition. In M.Y. İşcan and R.P. Helmer (eds), *Forensic Analysis of the Skull* (pp. 105–118). Wiley-Liss, New York.

Stewart, T.D. (1983). The points of attachment of the palpebral ligaments: their use in facial reconstructions on the skull. *Journal of Forensic Sciences* 28: 858–863.

Stephan, C.N. (2009). Craniofacial identification: techniques of facial approximation and craniofacial superimposition. In S. Blau and D.H. Ubelaker (eds), *Handbook of Forensic Anthropology and Archaeology* (pp. 304–321). Left Coast Press, Walnut Creek, CA.

Taylor, K.T. (2001). *Forensic Art and Illustration*. CRC Press, Boca Raton, FL.

Ubelaker, D.H., Bubniak, E., and O'Donnell, G. (1992). Computer-assisted photographic superimposition. *Journal of Forensic Sciences* 37(3): 750–762.

van der Lugt, C. (2001). *Earprint Identification*. Elsevier, Amsterdam.

Wells, G.L. and Olson, E.A. (2003). Eyewitness testimony. *Annual Review of Psychology* 54: 277–295.

Yoshino, M., Imaizumi, K., Miyasaka, S., and Seta, S. (1995). Evaluation of anatomical consistency in cranio-facial superimposition images. *Forensic Science International* 74: 125–134.

Yoshino, M., Matsuda, H., Kubota, S., Imaizumi, K., Miyasaka, S., and Seta, S. (1997). Computer-assisted skull identification system using video superimposition. *Forensic Science International* 90: 231–244.

Yoshino, M., Matsuda, H., Kubota, S., Imaizumi, K., and Miyasaka, S. (2000). Computer-assisted facial image identification system using a 3-D physiognomic range finder. *Forensic Science International* 3: 225–237.

CHAPTER 22 DNA Analysis and the Classic Goal of Forensic Anthropology

Luis L. Cabo

No, I am not scared, and neither should you be. (Muhammad Saeed al-Sahhaf, "Bagdad Bob," Iraqi Information Minister)

INTRODUCTION

In a famous scene from *Raiders of the Lost Ark*, a formidable Arab swordsman confronts Indiana Jones at an Egyptian market. Staring at Jones with an insolent sneer on his face, the warrior proceeds with an impressive display of swordsmanship, twirling his heavy scimitar with the ease and dexterity of a majorette flying her baton. Visibly unconcerned, Indy casually pulls his gun from its holster and shoots the swordsman dead. The lesson is clear: it does not matter how good you are with your saber, the gun is simply a better option. During this chapter I will discuss why, if the goal is solely victim identification, it is not forensic anthropology, but molecular genetics that is holding the right weapon, and how this issue is bound to affect the scope and future development of forensic anthropology.

As discussed briefly in Chapter 1 in this volume, at present the average forensic anthropologist justifiably sees DNA analysis more as an ally than as a rival, and is typically both unaware of and unconcerned with where the rapid advances in genetics are leading DNA analysis. When it comes to the assessment of biological profile (the key component of the anthropological contribution to victim identification), the forensic anthropologist is used to an evolutionary landscape characterized by gentle

A Companion to Forensic Anthropology, First Edition. Edited by Dennis C. Dirkmaat.
© 2012 John Wiley & Sons Ltd. Published 2015 by John Wiley & Sons Ltd.

slopes and gradual changes. Improvements in assessment methods for age, sex, stature, or ancestry typically consist of small modifications and fine-tuning of existing central estimates or confidence intervals. Most often these changes are simply based on the examination of new and better reference samples or on the introduction of more powerful but not necessarily all that novel statistical methods to analyze the same old traits and techniques. In the more risqué cases the innovation may come from looking at different anatomical areas, but even in these relatively rare examples drastic revolutions have been very sparse and, in general terms, we are still looking mostly at the same classic skeletal markers as we did half a century ago.

From a professional point of view, rather than from drastic improvements of our basic techniques, the growth of forensic anthropology in the last decades can be better explained in terms of an increased awareness and recognition of the utility of forensic anthropology by other forensic and law-enforcement professionals. This has translated into a steady but consistent growth in professional opportunities and, consequently, in the ranks of practitioners who could make a living mostly from their forensic anthropology skills.

From this tradition and mindset, forensic anthropologists may be tempted to assume that the evolution of DNA analysis progresses in similar ways. After all, following the process from the consumer end, the main improvements in the genetic field witnessed by practicing forensic anthropologists have been with respect to an increase in the number of DNA laboratories to which they can submit their samples, decreased fees and turnaround times, and a greater willingness by investigative agencies to risk some more money by giving forensic DNA analysis a try; in other words, an increased awareness in the forensic community, more laboratories and personnel and, with this, an increased investment and demand, reduced costs, and more refined techniques. Just as with forensic anthropology.

However, there is much more than business volume to the changes wrought by DNA. The observed improvements in costs, precision, analytical possibilities, and turn-around times of DNA analysis are not merely the result of increased economies of scale, derived from decreasing marginal costs as DNA tests escalated toward mass production and demand (the principle allowing sports apparel companies to manufacture progressively cheaper sneakers as they increase their production). Although DNA analysis has certainly become a multibillion dollar industry (another factor to keep in mind), the qualitative improvements of genetic techniques during the past few decades have been at least as dramatic, if not more so, as their quantitative growth and generalization, having developed at a pace that we could not foresee even in our wildest dreams.

Keeping with our arms race analogy, while we were sharpening our biological profile scimitar and trying to figure out new ways of using it to cut our daily bread, the development of DNA analysis was probably equivalent to the evolution of firearms all the way from the harquebus to machine guns, only this time in a matter of just a few decades, rather than four centuries. And it is not likely to slow down in the near future.

In this chapter we will briefly examine the challenges successfully overcome by molecular genetics in the past half century, and those that it must still confront before forensic DNA analysis truly becomes a routine technique, faster and cheaper (and, of course, more precise in terms of attached probabilities) than anthropological analysis. The chapter is also intended to serve as a very basic primer on basic DNA analysis for forensic anthropologists, stressing those aspects that I believe to be of particular importance to understand the imbrication of DNA analysis and forensic anthropology, as well as the future of their relationship. With this in mind, I will adopt an historic perspective to sequentially review the main steps of DNA sequencing and analysis, discussing (i) how the importance of forensic anthropology techniques came largely from their ability to reduce some of the original practical limitations in these steps and (ii) how the conquest of most of these limitations by genetic techniques may diminish the importance of forensic anthropology in its classic goal of victim identification.

REPLICATION

The first synthetic nucleic acids (RNA) were obtained by Severo Ochoa in the late 1950s (Kornberg 2001). Shortly thereafter, working with Ochoa at New York University, Arthur Kornberg identified and isolated the first DNA polymerase, the enzyme responsible for building the new DNA strands during replication (Kornberg 2001). Both scientists received the 1959 Nobel Prize in Physiology or Medicine, for their discovery of the mechanisms in the biological synthesis of ribonucleic acid and deoxyribonucleic acid. This triggered a frantic and fruitful race to unveil the secrets of genetic materials during the following decades. The ability to synthesize primers and replicate DNA and RNA sequences was a key component in identifying and analyzing genetic sequences of interest. However, DNA replication *in vitro* was a slow and delicate process. Even in the simplest prokaryotes, natural DNA replication involves a large and intricate enzymatic complex, particularly to separate the twin DNA strands and keep them in place while the DNA polymerase runs through the replication fork. This enzymatic process proved to be virtually impossible to duplicate artificially and had to be left mostly to natural organisms. When some degree of accuracy was required, a particular sequence was tagged with a chemical marker (a heavier isotope, fluorescent molecule, or radioactive element) to distinguish it from the genetic material of the host organism, inserting it in the genome of a virus, plasmid, or bacterium, and letting these organisms replicate naturally. Long incubation times were required to obtain just very small amounts of cloned DNA.

AMPLIFICATION

This situation would change drastically with the development of the *polymerase chain reaction* (PCR) method by Kary Mullis, in the mid-1980s. With a background in chemistry rather than biology, Mullis' basic idea was using heat to completely substitute the enzymes involved in the initiation phase of natural DNA replication.

As mentioned above, the key function of all these enzymes is separating the complementary strands of DNA and giving the DNA polymerase access to them. The two complementary strands are kept together by hydrogen bonds, and thus strand separation requires energy to break these bonds. During natural replication this energy is obtained from breaking ATP molecules enzymically. Simply applying heat can also break the hydrogen bonds, in a process called *denaturation*. However, the temperature required for denaturation is also too high for the natural DNA polymerase to bond to the DNA template and start the elongation process.

Mullis' solution was instead to utilize the DNA polymerase of *Thermus aquaticus*, a bacterium living in hot springs and thus resistant to high temperatures (*thermophilic*). The new enzyme (known as *Taq* DNA polymerase) was thermostable and could still operate at the high temperatures necessary to keep the DNA strands apart, after initial separation at even higher temperatures. The resulting method (PCR) allows the researcher to produce large quantities of DNA replicas in the lab (*DNA amplification*), with minimum ingredients, which are basically the DNA template to be replicated, Taq DNA polymerase, a primer sequence to start the elongation, and the free nucleotides (deoxyribonucleotide triphosphates or dNTPs), in an aqueous solution. The first application of the method was published in 1985 (Saiki et al. 1985), and the first comprehensive description of the protocol 3 years later (Saiki et al. 1988). Mullis received the Nobel Prize in Chemistry in 1993 for this contribution.

From a forensic point of view, one of the immediate impacts of the ability to copy (amplify) DNA molecules almost *ad infinitum* through PCR is the possibility of detecting and analyzing even extremely small amounts of DNA evidence that might have gone unnoticed or would have been impractical in the past. But, more importantly, modern PCR techniques allow for a virtually endless number of analyses and comparisons of this evidence to be performed, a property that will be extremely important in the following steps of DNA analysis.

SEQUENCING

Once the problem of amplification was solved, the next limiting factor of DNA analysis became *sequencing* and *comparison*. The former relies on the different molecular weights and electric charges of the DNA nucleotides. As a consequence, different DNA sequences will also have different molecular weights and slightly different charges (although, generally speaking, their polarity will always be negative, due to the negative charge of the phosphate backbone). A family of techniques known generically as *electrophoresis* takes advantage of these properties by placing DNA in an electric field. The different DNA molecules will then migrate through a supporting medium to the positive electrode (anode), at a rate depending on their size and charge (i.e., their DNA sequence). In the case of large molecules, like most DNA sequences, this rate will depend mostly on their molecular weight, primarily determined by their length, while the influence of the exact nucleotide composition (and thus our ability to discriminate between sequences of similar lengths) increases as molecular mass decreases.

Separation between different sequences is promoted by placing the sample on a medium able to interact with the DNA molecule through its electric charge and pore diameter, further slowing the molecule's migration at a rate also proportional to molecular mass and, to a much lesser extent, charge (i.e., the migration rate will depend on the nucleotide sequence of the DNA fragment, especially in terms of how long is this sequence). The classic electrophoretic media are gels of polyacrylamide (more precise with small molecules) or agarose (requiring larger DNA fragments, but also cheaper and, unlike polyacrylamide, nontoxic). The result in these media is a ladder of discrete bands, each composed of a collection of DNA molecules of approximately the same length. At this level, if the original DNA was unaltered and contained full DNA molecules, the technique would theoretically allow for separation of chromosomes without further treatment. For example, a diploid female would display a single band for the sexual chromosomes (corresponding to the two X chromosomes), while a male would have an additional band for chromosome Y. However, full chromosomes are extremely large molecules, that can be separated and identified even by centrifugation, much like in the famous Meselson–Stahl experiment (Meselson and Stahl 1958), but do not work well in electrophoresis. As mentioned above, they would be grossly sorted according to their overall length, their massive sizes leaving small room for the detection of the relatively small differences derived from their exact sequences. In order to be able to sequence our DNA and precisely compare sequences, we will need to break it into smaller fragments. To compare different samples, the fragments must also be homologous (corresponding to the same loci in all samples), so we also need to break our molecules by cutting (cleaving) them at very precise locations. A group of enzymes called *restriction nucleases* are employed in this task.

Restriction nucleases were first utilized to produce specific fragments of a DNA molecule in 1971, as Kathleen Danna and Daniel Nathans, of Johns Hopkins University in Baltimore, MD, USA, utilized *endonuclease R* in combination with electrophoresis techniques to analyze a simian virus (Danna and Nathans 1971). The enzyme had been discovered by Hamilton Smith and Kent Wilcox (Smith and Wilcox 1970; Roberts 2005), also of Johns Hopkins University. Nathans and Wilcox would receive the 1978 Nobel Prize in Physiology or Medicine for these achievements, showing how authorship order must not be overstated. Danna and Smith were Nathans' and Wilcox's graduate students, the common practice in most natural science fields being that the person directing the research will appear as the last author.

Restriction enzymes serve to illustrate both the importance of being able to generate large amounts of amplified DNA, and the difficulty of early sequencing techniques. For example, obtaining the exact base sequence of a DNA molecule requires cutting a particular DNA molecule into all possible fragments, using successive combinations of restriction enzymes (a step that is generally known as *subcloning*). As restriction enzymes cleave the molecule at known base sequences, we always know the first and last nucleotide (adenine, thymine, guanine, or cytosine) of the fragment generated by a particular restriction nuclease. Thus, if we treat different samples each with a different restriction enzyme, and order the resulting fragments according to their length (producing what are known as *nested arrays*), we can obtain the exact DNA sequence. The three-nucleotide fragment starts with A and ends with T, the two-nucleotide one starts with A and ends with G, the sequence is AGT.

Rather than sequentially treating each sample with an enzyme at a time, the sample can be treated with a complex restriction enzyme "cocktail," adding fluorescent or radioactive markers that attach themselves to a specific base (A, T, G, C) at the 3′ or 5′ end of each sequence in the nested array. In this way, we will be able to identify the last base in each sequence of the nested array, and ordering these bases from the largest to the smallest fragment in the array will give us the precise sequence of the initial DNA fragment. In the initial protocols marker detection was usually attained through photographic or radiographic techniques. The analyst would photograph or radio-graph the array "ladder" on its gel support, process the film and look in the negatives for the bright spots from the radioactive marker. Fluorescence techniques later allowed one to better distinguish between shades of color for different bases, and thus to tag more than one base at a time in a single sample. As commented below, this opened the gate for progressive automation, but in its initial stages the process was still slow and artisanal, depending largely on the skill and visual abilities of the analyst.

Of course, DNA fragments of interest are typically much, much longer than three nucleotides. You can probably guess how the number of restriction enzymes, fragment combinations, and DNA copies increase with the number of nucleotides in the sequence and, thus, how our ability to produce and process the nested arrays in a timely manner weighs on the time and monetary costs of the analysis.

As new loci and sequences were identified, and the ability to synthesize them in the lab increased, the sequencing task was simplified by the ability to compare the sample with libraries of these known sequences. The cleaved sample can be exposed to marked synthetic fragments of known sequence. If the correct complementary sequence matching the one in our sample is present in the solution, both strands will pair themselves. In this way, we will be able to identify the exact sequences present in our sample simply by identifying the chemical tags attached to the hybrid fragments, composed of a marked and an unmarked strand. However, once again this last step was a slow and hand-crafted one, treating each sample manually and individually.

COMPARISON AND MATCHING

If the mental picture that you drew as you read the preceding paragraphs was one of a painstakingly time-consuming and tedious process, you were right on the money. And the worst part is that, until very recently, sample comparison (*matching*) also had to be done "by hand." A myriad of useful and increasingly sophisticated sequencing techniques were developed during the 1990s and 2000s, aimed at taking full advantage of the new possibilities opened by PCR (Hartwell et al. 2006, part III, pp. 301–465 provides a very useful and meaningful review of all main techniques and their application; for a briefer but still excellent introduction to basic electrophoresis and sequencing techniques see Alberts et al. 2004, pp. 323–364). However, at the end of the process the analyst still had to visually compare side by side the obtained sequence or electrophoresis output with those of all potential matches, basically one by one.

PRESORTING: MEET THE BAYONET

With all these factors in mind, it is easy to understand the importance of the role played by forensic anthropology at this junction. DNA tests were first slow and expensive and each new required comparison added a considerable burden to the process. The extra work was not just derived from having to sequence the problem sample, but also from obtaining the comparative sample to which the former would be matched.

In the most common scenario, obtaining the comparative sample requires obtaining individual samples from several close relatives of each potential victim. In a common investigation, this means first creating a list of potential victims, for example people reported missing in the area in recent years. Then, once the potential victim list is in place, we must identify a second list of close relatives of each of the victims, appropriate for DNA comparison and matching. Finally, it is necessary to contact these families, asking them to contribute DNA samples, either from tissues collected from some of the victim's belongings (e.g., hair from a hairbrush) or from themselves. Each additional potential victim included in the initial list represents one more family contacted, asked to contribute samples for a specific comparison (after informing them that you suspect that their loved one is likely dead), and ultimately disappointed with the results. In the best-case scenario, only one of those families will get closure after going through the process. Hence, making a large number of comparisons posed an important logistical problem and, consequently, any technique that led to some presorting of the samples, and reduced the number of comparisons required, represented a very useful tool. It is much more efficient to compare our sample with just males within a particular age and height range, and maybe even with some particular type of antemortem trauma, rather than with all missing persons in the area or even the whole target population. Sometimes you do not need to choose between the knife and the gun, you can have both.

Consequently, forensic anthropological analyses found an important role in reducing the list of potential victims, so that other identification professionals, like forensic odontologists or, now, DNA analysts could focus on obtaining and analyzing just a small number of antemortem or family records. This task became more relevant when long lists of potential victims were involved, such as in mass disasters, human-rights investigations, or "John Doe" cases in which the victim could not be satisfactorily matched to a particular individual on a short list of recent missing persons in the area. In regular forensic cases, involving one or just a few victims, pre-sorting based on the biological profile also reduced the number of comparative samples to be obtained and sequenced. The difference between obtaining and analyzing samples from the close relatives of a single potential victim, instead of from 20 or 30 potential victims, resulted in an extreme reduction of the investigation and analysis efforts, costs, and times.

The problem posed by the high costs and long processing times of forensic DNA analyses was further aggravated (and thus the role of forensic anthropology enhanced) as prosecutors and law enforcement became more aware of the availability and evidentiary value of DNA analysis, and therefore started submitting more and more samples to be sequenced and matched. Forensic labs with DNA capabilities were soon overwhelmed with requests, and a significant backlog of DNA samples started to build rapidly. The problem became so pressing that, following other previous efforts, the US House of Representatives finally passed the DNA Analysis Backlog Elimination Act of 2000 (H.R. 4640, 106th Cong., Dec. 7 2000, 2nd S., H.R. Rep. 106–900, Part 1, 2000). The purpose of this piece of legislation, as specified in its opening statement, was "To make grants to States for carrying out DNA analyses for use in the Combined DNA Index System of the Federal Bureau of Investigation, to provide for the collection and analysis of DNA samples from certain violent and sexual offenders for use in such system, and for other purposes" (ibid). The Act resulted in the development of a number of federal initiatives and funding programs, which in 2005 would crystallize in the Forensic DNA Backlog Reduction Program of the National Institute of Justice (US Department of Justice). This program provides funds both to analyze existing samples and to improve the analytic capabilities of laboratories (a comprehensive description of the program can be accessed at www.dna.gov/funding/backlog-reduction).

Who Is Afraid of the Big, Bad Wolf?

With these antecedents, far from a threat, forensic DNA analysis appears to represent an optimal niche for forensic anthropology to reinforce and fulfill its classic goal of aiding in victim identification. Actually, within this landscape of slow and costly analyses, sample submission frenzy, and laboratory backlogs, the introduction and popularization of DNA analysis may have been one of the leading forces behind the resurgence and growth of forensic anthropology as a recognized professional activity during the last couple of decades. The inclusion and extraordinary contribution of forensic anthropologists in high-profile mass-identification efforts such as the various human-rights investigation teams across the world, or the US Disaster Mortuary Operational Response Teams (DMORT; Saul and Saul 1999; Sledzik 1996), seem to exemplify how fruitful can be the marriage of DNA analysis and forensic anthropology. DMORT was constructed in the mid-1990s to serve as rapidly deployable multidisciplinary human-identification teams, involving the whole spectrum of forensic identification professionals in cases of mass fatalities overwhelming local resources. Since its implementation, DMORT teams have proven effective in a wide variety of mass disaster scenarios, from plane crashes to mass suicides and large-scale floods (see for example Sledzik and Hunt 1997; Ubelaker et al. 1995). Given that most of the biological remains at these sites typically consist of commingled, fragmented, and often burned or badly decomposed tissues, it is only natural that the contribution of participant forensic anthropologists soon became vital in most of these scenarios (Sledzik and Rodriguez 2002).

Thus, where is all the huffing and puffing? If anything, the presumptive big, bad DNA wolf seems to be giving us a hand with the roofing and mortgage refinancing of our hay house. However, it is precisely through these high-profile disaster response teams, in which the marriage of DNA analysis and forensic anthropology seem to be particularly fruitful for victim identification, that we get the first indication that there is actually some cause for concern. The recent emplacement of DNA-collection teams within DMORT and, especially, the steep increase in the biological items subject to DNA analysis (rapidly nearing 100% in many scenarios), suggest that the role of all forensic specialists in these mass disaster teams may change dramatically within the next few years (see Chapter 7 in this volume). If all body fragments are going to be analyzed anyway, it is obvious that presorting and sample-selection tasks are losing some importance.

When it comes to everyday forensic cases, we would not be concerned if this trend toward an increase in the number of samples submitted for forensic DNA analysis was just derived from policy changes, or a greater willingness by institutions to invest more money and time in victim identification in these high-profile cases. Unfortunately, this is not the case: in the following sections I will discuss how it is also derived from a rapid and constant decrease in costs and turnaround times for DNA analysis. We can process more DNA samples because it is much cheaper and faster now.

Before going ahead with this point, it is important to remember again that, as explained above, the current relevance of forensic anthropology for victim identification is largely derived from its ability to shorten and simplifying some steps of DNA analysis by providing a biological profile from the bones, thus reducing the number of DNA samples to be sequenced and the number of comparison samples required. Savings in time and cost are appreciable ... today. However, DNA by itself contains all the information necessary for establishing *positive* victim identification, plus information related to other very important components of the biological profile, namely sex and ancestry. Completely solving issues of lengthy DNA sequencing times and substantial cost, by making mass sequencing cheaper and faster than anthropological assessments of the biological profile and preassembling genetic profile databases already containing all potential matches, may render forensic anthropological work completely irrelevant in the identification process. In the following sections I will discuss how both changes are already on their way.

AUTOMATING DNA PROCESSING

Long gone are the heroic efforts of the pioneers described above. DNA sequencing has become fully automated and sequence comparison is extremely close to being so. The old agarose and polyacrylamide gels have been replaced almost completely by capillary electrophoresis. This procedure sorts the DNA fragments along the electric gradient based on their ability to migrate through the capillary diameter depending on their size. Basically, it means running each sample through a single pore, rather than the complex pore microstructure of the classic gels. This translates into the ability to simultaneously run dozens of samples through tightly packaged parallel capillaries in the modern automatic sequencers. The fluorescent tags in the last nucleotide of each sequence in the nested array can be read by an optic sensor connected to a computer, which translates and stores them, sorted according to the molecular weight of each fragment of the array. This means that the machine can produce the exact sequence of our DNA sample just as we ran multiple electrophoreses. Moreover, the libraries of identified sequence templates described above now can also be incorporated into increasingly cheaper computer chips, which can be read directly by computerized systems and produce not only exact sequences, but match statistics. DNA specialists still play a role in the interpretation of the results, by checking for ambiguities and missing bases in the output, but this role is becoming increasingly diminished and simplified, as powerful computer algorithms for sequence coding, disambiguation, and comparison are developed.

What are the results of these advances? We can try to quantify the global progress by taking a look at the times, resources, and costs associated with the sequencing of the first few complete human genomes. The Human Genome Project (HGP), aimed at identifying and mapping all loci in the human genome, was the first necessary step to be able to sequence complete individual human genomes. The HGP took 13 years (from 1990 to 2003), an investment of US$3 billion, and the efforts of more than 39 major institutions from all around the world (International Human Genome Sequencing Consortium 2004). As a matter of fact, the HGP was initially scheduled for completion in 15 years, but technological advances permitted that timeline to be cut down by 2 years. Building on the results of HGP, sequencing of the full genome of J. Craig Venter, published just 4 years after completion of the former, required only 5 months, with an investment of $100 million, distributed among four laboratories in the USA, Canada, and Spain (Levy et al. 2007). Less than 1 year later the full genome of James Watson was sequenced in only 4 months by a single US laboratory for $1.5 million (Wheeler et al. 2008). The first Asian full genome was also sequenced later in 2008 in less than 2 months and with a cost of less than half a million dollars (Wang et al. 2008). Nowadays it is rare for a month to pass without announcements of new individual genomes having been sequenced.

In other words, in the 18 years from 1990 to 2008, the time required to sequence a full human genome was reduced by 7800%, and the cost by 600 000%. More realistically, given the mammoth dimensions of the pioneering HGP, just since 2008 the costs have been reduced by one third and the processing times by one half. With the times and costs reduced to a level in which complete sequencing can be tackled by single laboratories, with budgets and timelines well within the range of conventional research funding sources, 2008 also witnessed the birth of initiatives like the 1000 Genomes Project, an international collaboration aimed at producing an extensive catalogue of human genetic variation, by sequencing about 2000 unidentified individuals from 20 populations around the world, thus moving the goal to the population level (1000 Genomes Project Consortium 2010). A 2010 survey by *Nature* revealed that, as compared to the handful of complete human genomes that had been sequenced by 2008, the genomes of approximately 30 000 people are expected to be known by the end of 2011 (Nature News 2010).

Of course these high-end advances do not impact forensic applications right away, but they offer a clear illustration of the speed at which DNA techniques are evolving, and the rate of improvement that we can expect from forensic DNA analysis in the near future. In addition, they bring us realistically close to achieving other goals, such as being able to routinely infer characteristics such as skin, hair, or eye color, as well as ancestry probabilities from genetic materials. In a 2000 report from the National Commission on the Future of DNA Evidence (National Institute of Justice 2000: 31, 35, and 61), these advanced capabilities were indeed targeted as an explicit goal, and optimistically predicted to be available as soon as 2010. This is to say, DNA analysis may not just be able to identify the victim in a very cheap and quick way, but also to produce very soon some of the key components of the biological profile itself, without any help from forensic anthropological techniques. Is the DNA bride considering becoming independent and living her own life? If that were the case, she has all of the positive attributes to be the one retaining and running the family business, while the forensic anthropology groom is left out in the unemployment line.

Powering and Simplifying Profile Matching

The good news (for the forensic anthropologist) is that there are several factors currently preventing the abovementioned development from happening in the very immediate future. The first and probably most important one is the presence of the significant case backlog described earlier, which is mostly related to the collection and analysis of comparative samples. In order to match a DNA sample from a crime scene with a particular individual, a library of DNA profiles and identities structured in a database is still needed. In the USA, that database library is the National DNA Index System (NDIS), and the database and communications system supporting the NDIS is the Combined DNA Index System (CODIS), of the Federal Bureau of Investigation

(FBI). According to the FBI website (www.biometriccoe.gov/Modalities/CODIS.htm), as of October 2010, 170 US laboratories and more than other 40 forensic laboratories in 25 countries used CODIS to support their forensic DNA libraries. The NDIS/CODIS system (and thus most forensic DNA analyses) currently focuses on information from 13 short tandem repeat (STR) loci. These are short and highly variable sequences. The former of these characteristics serves to speed sequencing and matching, while their high variability translates to very low probabilities that two individuals, taken at random, would share the same sequence for these loci. These probabilities are in the range of one in several quadrillions (10^{15}), when all 13 loci are available. The sex of the individual can also be currently inferred from the *Amelogenin* locus, which has different lengths in the X and Y chromosomes. There is an almost overwhelming wealth of resources describing the NDIS/CODIS system and its applications, including the FBI web pages for the resource (which can be currently accessed from www.biometriccoe.gov/Modalities/DNA.htm), but I have found Dale et al. (2006) to be a particularly informative starting point.

At present, according to the FBI CODIS brochure (www.fbi.gov/about-us/lab/codis/codis_brochure) the bulk of the DNA profiles contained in the US CODIS system corresponds to convicted offenders and arrestees (as of July 2010, close to 9 million profiles) and crime-scene samples (*forensic profiles*, totaling more than 300 000 CODIS entries at the time of writing). The profiles corresponding to unidentified human remains, missing persons, and their biological relatives represents a very small, and actually not specified fraction.

There are several reasons to explain this minute representation of profiles that are relevant for victim identification. The first and most evident one is that, fortunately, the number of offenders of one type or another dwarfs that of murder victims. In 2007, 153 public crime laboratories in the USA reported receiving around 1 million requests to process offender samples, compared with around 140 000 forensic samples obtained from crime scenes (Hurst and Lothridge 2010). As of 2003, all US states required DNA from sex offenders and murderers, but 45 of them also required DNA from burglary convictions, 36 from some drug convictions, and 31 from all felony convictions, with a trend toward increasing the number of offenses requiring DNA collection in most states (Zedlewski and Murphy 2006; Lovrich et al. 2004). In a country with more than 7 million individuals under some form of correctional supervision (Glaze and Bonczar 2009) there is still quite some catching up to do. However, as discussed below, in spite of the increase in requests, the offender backlog is actually decreasing.

The second reason for the small number of victim profiles is that murder cases and suspected violent deaths actually represent a rather small proportion of the forensic cases for which DNA analysis is requested. As reported by the Division of Criminal Justice Services of the State of New York (http://criminaljustice.state.ny.us/forensic/typesofcrimesfirst1000hits.htm), of the first 1000 matches obtained in the New York State DNA Databank, only 7% were related to murder cases, including attempted murders, as compared to the 72% for rape cases and 18% for the combined categories of burglary, robbery, and assault. DNA analysis is no longer used just for high-profile cases (i.e., suspected homicides), and there is a strong thrust to include DNA in the investigation of a growing number of "minor" crimes (Zedlewski and Murphy 2006).

Still, the number and case types of DNA samples submitted is still increasing at a faster pace than the improvements in the analytical capabilities of the laboratories, thus adding to the previously discussed case backlog. In the same 2007 poll described above (Hurst and Lothridge 2010), the 153 laboratories surveyed were able to process around 124 000 of the 140 000 forensic samples received during that year. However, the survey also revealed rapid improvements in processing times, with most laboratories completing DNA casework requests within 90 days. As an indication of the kind of improvement that can be expected in a very short term, 12 of these laboratories already reported average turnaround times of 30 days or less (Hurst and Lothridge 2010). Even more importantly, the study also showed how the Forensic DNA Backlog Reduction Program was obtaining very quick and promising results: laboratories were able to process 1.2 million offender profiles that year, while receiving only around 1 million new requests, thus resulting in a backlog reduction from 841 847 to 657 165 pending offender profiles.

Conversely, as the improvements in funding and laboratory capabilities are proving extremely fruitful, attention is increasingly shifting toward increasing the databases of unidentified human remains and relatives of missing persons. When describing the main objectives of NDIS/CODIS for the immediate future, the FBI web page states that "a considerable focus during this time will be to enhance kinship analysis software for use in the identification of missing persons. This next generation of CODIS will utilize STR and [mitochondrial DNA] information as well as meta data (such as sex, date of last sighting, age, etc.) to help in the identification of missing persons" (www.fbi.gov/about-us/lab/codis/codis_future). How long will it take until family DNA samples are routinely taken and stored in CODIS in all cases involving missing persons? This question is extremely relevant for the future focus of forensic anthropology, as this development will allow for rapid comparison of the victim's DNA with the profiles of all reported missing persons in the country, thus making reduction of the potential victim list through the generation of a skeletal biological profile a lesser, if not completely irrelevant, task. The inclusion of mitochondrial DNA in the records is equally relevant, as this genetic material is better suited for the analysis of poorly preserved remains.

DISCUSSION: WHERE WE WERE AND WHERE WE ARE

At this point you will probably have realized why the advances in DNA analysis have been explained in such detail. In order to acquire an adequate perspective of the present and future of the field of forensic anthropology, it is particularly important for forensic anthropologists to understand the extent and nature of DNA advances and how they will come to impact their profession. What we have learned from the previous paragraphs is that the utility of the classic anthropological profile generated for aiding positive identification, especially through DNA analysis, was largely related to some key limitations of the latter. In particular, presorting the list of potential victims, to whose profiles the sample DNA would be compared *à la carte*, resulted in a exponential reduction in the number of sample comparisons.

However, these limitations were basically technical ones, rather than related to the potential and information contained in DNA itself and, thus, far from permanent.

The extension and popularization of forensic DNA analysis is not merely the result of the increase in the number of laboratories capable of performing it, following the development of PCR. On the contrary, we have also learned that its fundamental initial technical limitations have already been overcome. Following our initial analogy with the development of firearms, we could say that the basic weapon is already being manufactured and distributed, and we have entered the phase in which the race is just between manufacturers to develop cheaper and faster production processes, models, and accessories. The industrial comparison is not a *boutade*. Massive parallel sequencing methods allow for the sequencing of millions of bases per hour, making it potentially possible to sequence the complete genome of an individual in just a few days (Rogers and Venter 2005). These advances in technology and bioinformatics will certainly make their way progressively into conventional forensic laboratories. Some of the figures provided above also call attention to another aspect to be taken into account: DNA analysis has become a multibillion dollar industry, the resources of which, in terms of both personnel and budgets, dwarf those of forensic anthropological research.

Accompanying these technological and scientific advances, many governments and law-enforcement agencies were prescient enough to start developing forensic DNA programs and databases early enough to take full advantage of the new developments. At present it appears that comprehensive victim DNA datasets will be available in a matter of years, rather than decades. When these data are available, anthropological presorting efforts may be completely unnecessary in most cases, just as they are nowadays when, for example, fingerprints and fingerprint records are simultaneously available.

Abandon All Hope?

The fundamental question to close this chapter is whether forensic anthropology will still be relevant as an identification technique when the new DNA resources are fully implemented. I believe so, although I suspect that its weight within the forensic sciences will be very significantly diminished. In cases involving multiple victims, such as mass fatality incidents or human-rights investigations, solving commingling issues through forensic anthropological techniques and eventually reducing the number of forensic samples to be processed will still be extremely useful, as sample extraction and specimen preparation will probably still be the most limiting factor in terms of time and costs. In these and regular forensic cases, a biological profile will provide a useful safeguard for identifying cases of false identities (including misplaced paternities, as might occur with family records that include unfaithful spouses), misplaced samples, or missing records. Quick initial profile assessments will also serve for rapid identification or to guide the first steps of the investigation, until the DNA analyses arrive, especially in cases where there is already a strong suspicion of the identity of the victim (e.g., when dental records are readily available from the local odontologist).

Still, all of these scenarios reflect an essential change: the assessment of the biological profile through forensic anthropological techniques will no longer be an essential technique that must necessarily precede DNA analysis, but an accessory technique that may be helpful in some cases. If this subtle difference does not appear that important to you, just stop again to think about how often a skeletal biological profile is requested in fully fleshed victims or when fingerprints can be recorded. Only DNA is even more precise and versatile than fingerprints, and can be recovered in many more cases than the latter, especially after the new emphasis on collecting and storing more mitochondrial DNA data in mainstream databases.

Does this mean that forensic anthropology is bound to disappear or become a secondary or even completely irrelevant discipline? Should we refund the money that you spent on this book? Absolutely not (especially, the latter!). But I believe that the developments in DNA analysis will shift the scope of the discipline from mere victim identification to other fortunately already blossoming subjects, such as trauma analysis, forensic archaeology (scene recovery), and forensic taphonomy. Ultimately, moving away from the limitations imposed by its initial definition as only a laboratory field provides a perfect opportunity to highlight very promising new lines of research within the field. Forensic anthropology techniques can actually reveal aspects of the biological profile not contained in the genetic information; specifically, the age, stature, and physical constitution of the victim. Refining our estimates and techniques for these parameters requires placing a higher stress on subjects like human physiology, allometric scaling, and biomechanics, both in our research efforts and in the academic training of future practitioners. In other words, it means paying more attention to process analysis than to classification techniques or, if you prefer, to basic biology and physiology rather than to social science approaches.

REFERENCES

Alberts, B., Bray, D., Hopkin, K., Johnson, A., Lewis, J., Raff, M., Roberts, K., and Walter, P. (2004). *Essential Cell Biology*, 2nd edn. Garland Science, Taylor & Francis Group, New York.

Dale, W.M., Greenspan, O., and Orokos, D. (2006). *DNA Forensics: Expanding Uses and Information Sharing*. NCJ 217992. SEARCH, The National Consortium for Justice Information and Statistics, Sacramento. http://bjs.ojp.usdoj.gov/content/pub/pdf/dnaf.pdf.

Danna, K. and Nathans, D. (1971). Specific cleavage of simian virus 40 DNA by restriction endonuclease of *Hemophilus influenzae*. *Proceedings of the National Academy of Sciences USA* 68(12): 2913–2917.

Glaze, L.E. and Bonczar, T.P. (2009). *Probation and Parole in the United States, 2008*. NCJ 228230. Bureau of Justice Statistics. http://bjs.ojp.usdoj.gov/index.cfm?ty=pbdetail&iid=1764.

Hartwell, L.H., Hood, L., Goldberg, M.L., Reynolds, A.E., Silver, L.M., and Veres, R.C. (2006). *Genetics. From Genes to Genomes*. McGraw Hill, New York.

Hurst, L. and Lothridge, K. (2010). *2007 DNA Evidence and Offender Analysis Measurement: DNA Backlogs, Capacity and Funding*. Final report for NIJ grant 2006-MU-BX-K002. NCJ 230328. www.ncjrs.gov/pdffiles/nij/grants/230328.pdf.

International Human Genome Sequencing Consortium (2004). Finishing the euchromatic sequence of the human genome. *Nature* 431: 931–945.

Kornberg, A. (2001). Remembering our teachers. *Journal of Biological Chemistry* 276(1): 3–11.

Levy, L., Sutton, G., Ng, P.C., Feuk, L., Halpern, A.L., Walenz, B.P., Axelrod, N., Huang, J., Kirkness, E.F. et al. (2007). The diploid genome sequence of an individual human. *PLoS Biology* 5(10): e254.

Lovrich, N.P., Pratt, T.C., Gaffney, M.J., Johnson, C.L., Asplen, C.H., Hurst, L.H., and Schellberg, T.M. (2004). *National Forensic DNA Study Report*. Final report for NIJ grant 2002-LT-BX-K003. NCJ 203970. www.ncjrs.gov/pdffiles1/nij/grants/203970.pdf.

Meselson, M. and Stahl, F.W. (1958). The replication of DNA in *Escherichia coli*. *Proceedings of the National Academy of Sciences USA* 44: 671–682.

National Institute of Justice (2000). *The Future of Forensic DNA Testing: Predictions of the Research and Development Working Group*. NCJ 183697. National Commission on the Future of DNA Evidence. www.ncjrs.gov/pdffiles1/nij/183697.pdf.

Nature News (2010). Human genome: genomes by the thousand. *Nature* 467: 1026–1027.

Roberts, R.J. (2005). How restriction enzymes became the workhorses of molecular biology. *Proceedings of the National Academy of Sciences USA* 102(17): 5905–5908.

Rogers, Y.H. and Venter, J.C. (2005). Genomics: massively parallel sequencing. *Nature* 437: 326–327.

Saiki, R.K., Scharf, S., Faloona, F., Mullis, K.B., Horn, G.T., Erlich, H.A., and Arnheim, N. (1985). Enzymatic amplification of β-globin genomic sequences and restriction site analysis for diagnosis of sickle cell anemia. *Science* 230: 1350–1354.

Saiki, R.K., Gelfand, D.H., Stoffel, S., Scharf, S.J., Higuchi, R., Horn, G.T., Mullis, K.B., and Erlich, H.A. (1988). Primer-directed enzymatic amplification of DNA with a thermostable DNA polymerase. *Science* 239: 487–491.

Saul, F.P. and Saul, J.M. (1999). The evolving role of the forensic anthropologist: as seen in the identification of the victims of the Comair 7232 (Michigan) and the KAL (Guam) air crashes [abstract]. *Proceedings of the American Academy of Forensic Sciences* 5: 222.

Sledzik, P.S. (1996). Federal resources in mass disaster response. *Cultural Resource Management* 19(10): 19–20.

Sledzik, P.S. and Hunt, D.R. (1997). Disaster and relief efforts at the Hardin Cemetery. In D.A. Poirier and N.B. Bellantoni (eds), *In Remembrance: Archaeology and Death* (pp. 185–198). Bergin and Garvey, Westport, CT.

Sledzik, P.S. and Rodriguez, W.C. (2002). Damnum fatale: the taphonomic fate of human remains in mass disasters. In W.D. Haglund and M.H. Sorg (eds), *Advances in Forensic Taphonomy: Methods, Theories and Archaeological Perspectives* (pp. 321–330). CRC Press, Boca Raton, FL.

Smith, H.O. and Wilcox, K.W. (1970). A restriction enzyme from *Hemophilus influenzae*. I. Purification and general properties. *Journal of Molecular Biology* 51: 379.

1000 Genomes Project Consortium (2010). A map of human genome variation from population-scale sequencing. *Nature* 467: 1061–1073.

Ubelaker, D.H., Owsley, D.W., Houck, M.M., Craig, E., Grant, W., Woltanski, T., Fram, R., Sandness, K., and Peerwani, N. (1995). The role of forensic anthropology in the recovery and analysis of Branch Davidian compound victims: recovery procedures and characteristics of the victims. *Journal of Forensic Science* 32: 335–340.

Wang, J., Wang, W., Li, R., Li, Y., Tian, G., Goodman, L., Fan, W., Zhang, J., Li, J. et al. (2008). The diploid genome sequence of an Asian individual. *Nature* 456: 60–65.

Wheeler, D.A., Srinivasan, M., Egholm, M., Shen, Y., Chen, L., McGuire, A., He, W., Chen, Y.J., Makhijani, V. et al. (2008). The complete genome of an individual by massively parallel DNA sequencing. *Nature* 452: 872–876.

Zedlewski, E. and Murphy, M.B. (2006). DNA analysis for "minor" crimes: a major benefit for law enforcement. *NIJ Journal* 253: 2–5.

23 DNA Identification and Forensic Anthropology: Developments in DNA Collection, Analysis, and Technology

David Boyer

INTRODUCTION

DNA technology has advanced rapidly over the past quarter century and has risen to prevalence in scientific circles. Mapping the human genome and cloning animals are among these notable accomplishments. In the field of forensic sciences, DNA technology has evolved during this same time as a credible and accepted method for solving the issue of human identification. Public awareness of DNA science also has increased significantly as the technology has progressed. Since its first application to crime solving in a 1983 rape/murder investigation in England (Wambaugh 1989), DNA has become a recurring topic in the news and primetime television programming. Movies, television series, and reality television shows frequently have depicted DNA analytical processes; often embellished, commonly oversimplified, and on some occasions even inaccurate.

DNA test results have a variety of applications. Comparing DNA profiles of questionable origins to those of known origins has been useful in solving paternity issues and crimes, and in mass fatality identification. DNA laboratories have used a wide variety of biological materials for analysis to unravel questions of identity. Preferred biological materials include blood, soft tissue, teeth, and bone (National Institute of Justice 2005). The human skeleton has proven to be a highly reliable

A Companion to Forensic Anthropology, First Edition. Edited by Dennis C. Dirkmaat.
© 2012 John Wiley & Sons Ltd. Published 2015 by John Wiley & Sons Ltd.

source for DNA material and in some cases the only source for viable DNA specimens. This chapter will discuss the use of skeletal material for DNA analysis and human identification.

DNA in the Human Skeleton

The characteristics of bone make it ideal as a protective vessel for the fragile biological composition of DNA molecules. While DNA is present in all bone cells, it is best protected in the dense structure of compact bone found in the tubular units of osteons (Mader 2008). Environmental conditions such as heat, humidity, and ultraviolet light destroy DNA rapidly in body fluids and soft tissue, rendering them useless for DNA analysis. However, DNA protected within the structure of bone can withstand adverse environmental conditions and survive for a considerably longer length of time.

Consider these examples. In 1864, the CSS Hunley, an American Civil War submarine, was the first to sink an enemy ship in combat. Subsequent to the attack the CSS Hunley also sank and its crew perished. The wreckage and trapped crew remained submerged in the Charleston Harbor for 136 years before recovery in 2001. In 2004, DNA analyst Jackie Raskins-Burns, from the Armed Forces DNA Identification Laboratory, reported the successful match of a mitochondrial DNA (mtDNA) profile from the remains of the CSS Hunley to a known family reference (Correia 2004). In addition, scientists have reported successful DNA sequence results from Neanderthal skeletal remains dating back thousands of years (Ovchinnikov and Goodwin 2001). In summary, DNA can withstand extreme conditions over lengthy time periods. Therefore, any skeletonized remains have potential for successful DNA typing and must be handled appropriately.

There are two separate and distinct forms of DNA in the human body, nuclear DNA (nDNA) and mtDNA. Both DNA forms are consistent throughout the body and do not change with age. However, they each possess distinctly different characteristics and each presents particular advantages depending on the circumstances. They differ in inheritance, number of copies, amount of information, degradation (Houch and Siegel 2006), and the time and cost involved in their analysis.

nDNA, also referred to as chromosomal DNA, is found in all nucleated cells in the human body. It is inherited from each biological parent, half from the mother and half from the father. The structure of nDNA is a double-helix ladder formed by complementary base pairs aligned along two backbones of long sugar and phosphate molecules. These base pairs align in short tandem repeats, are unique to each individual (except identical twins), and are the foundation for DNA analysis. There are 3 billion base pairs of nDNA in a nondividing nucleated human cell, which can be highly discriminatory. The probability of two individuals having identical DNA is unlikely to the degree that matching an unknown nDNA sample to a known reference can produce a positive match (identification) of scientific certainty, to the exclusion of all others.

mtDNA differs from nDNA in that it is inherited through maternal lines only. Therefore, maternal lines across several generations can be explored for obtaining a suitable family reference. Family references will be discussed in more detail later in this chapter. mtDNA is a small, circular genome found within the mitochondria, the energy-producing cellular organelle residing in the cytoplasm. Because each cell can contain hundreds of mitochondria, there can be up to several thousand mtDNA molecules in each cell. A molecule of mtDNA comprises about 16 659 base pairs (Butler 2005). Only the noncoding region of the mtDNA structure is used for identification. Due to the limited amount of information available in an mtDNA sequence, the positive match from a successful mtDNA comparison does not constitute a positive identification. It is, however, a presumptive conclusion and it coupled with other circumstantial information can lead to a finding of scientific certainty to the exclusion of all other matches.

Nuclear DNA does not survive typically as long as mtDNA in a postmortem state due to its low copy number. mtDNA can survive for many years, as illustrated by the examples above. The amount of recoverable DNA decreases with time after death and the probability of successfully extracting DNA diminishes over time as well. Also, it should also be noted that the level of effort required by scientists to isolate the mtDNA molecule can be very timely and costly.

Determining Appropriate DNA Samples

The term "appropriate sample" requires some explanation. The particular circumstance of an identification issue dictates the definition of appropriateness. The spectrum of possibilities can run from a single set of unidentified remains in pristine postmortem condition to an unearthed mass grave containing dozens of skeletonized remains that have been buried for many years. Completeness (intactness) of the remains or commingling of remains creates special concerns. Environmental influences, postmortem trauma, and other altering factors such as animal activity can severely limit the options for sample collection as well. These are among the many factors that influence decisions on what constitutes an appropriate biological sample from a given set, or sets, of human remains. A definition of "appropriate" will be determined given the circumstances of each specific situation.

Determining the appropriate sample is first dependent upon the condition of the remains. As time passes beyond the time of death, the organic constituents of the remains degrade gradually and sampling options decrease accordingly. Decomposition of the human body begins shortly after the onset of death and cessation of bodily functions. Environmental conditions such as air temperature, humidity, and exposure to water, fire, or other destructive conditions may accelerate the decomposition process. As decomposition progresses the DNA in cells is destroyed. Caution must be exercised to select biological material that is sure to produce successful laboratory results. When questions arise as to the viability of a particular item under consideration for sampling, it is a better option to disregard that potential sample and progress to another option with a greater likelihood of success. Often there is only one chance of obtaining a DNA sample from a set of human remains, and therefore decisions about selection criteria must be based on sound judgment.

A key determining factor in DNA sampling is the completeness of the remains. When a complete set of human remains is present for examination, the collector cycles through a decision progression based on the level of ease in the collection process. However, frequently unidentified human remains are anatomically incomplete, which restricts sampling possibilities. The challenges of DNA sample collection increase when human remains are highly fragmented, disassociated, or commingled. Fragmented remains, frequently the case in aircraft crashes or surface scatters, bring into consideration the added concern of potential re-association. Identification becomes even more challenging in cases where circumstances suggest commingling of body fragments. The 2002 Tri-State Crematory case in Walker County, GA, involved a 6.5 ha crime scene with 339 commingled remains (Adams and Byrd 2008). More than 400 bone samples were collected for identification and re-association in that particular case.

Another key element to the decision process is the number of bodies requiring identification. An incident with a small number of identification cases does not require special consideration. However, an identification project with a large number of sample requirements needs significant consideration. The ability of the supporting DNA laboratory to manage a surge of samples is aided by the consistency of the samples being provided for analysis. For example, if every sample collected is a section of long bone from a specific skeletal element, such as a tibia, the servicing DNA laboratory can repeat the processing steps over and over without changing the processing method to accommodate multiple specimen forms.

DNA laboratories define "appropriate DNA sample" as the best sample available. Simply stated, it is the biological specimen that is the least difficult to obtain, the least challenging for the laboratory to process, and the most likely to produce results. The aspect of deciding the least challenging sample to process relates to monetary cost as well. Less time and effort for sampling reduces labor costs and has a direct impact on overall expenses.

Ease of sample collection is determined by the amount of time and effort it takes to obtain a single sample. Samples frequently are collected in less than ideal circumstances. These efforts sometimes occur in open, remote geographic locations, at disaster sites where infrastructure has been destroyed, or in makeshift mortuaries. Blood and tissue samples are easier to obtain than bone samples but are not always available or reliable. When decomposition has advanced sufficiently to make tissue and blood samples suspect, bone specimens become the preferred choice. Selecting which bone to sample is based upon ease of access and time required to extract DNA from the sample. Experience from a variety of mass-casualty incidents has demonstrated that when dealing with complete sets of human remains it is preferable to obtain the DNA sample from the medial surface of the tibia. The tibia bone can be exposed rapidly with little effort using a scalpel and cut with a Stryker autopsy saw quickly and safely.

Processing of bone samples in a laboratory require more effort than blood or tissue but are frequently the preferred option because, as mentioned above, they retain DNA significantly longer. The best bone samples are obtained from robust, dense structures such as the cortical bone of long bones, cranial bones, and ribs. Vertebrae and other irregularly shaped bones are less desirable as DNA sources due to their more fragile bone structure. It should be noted that not all DNA laboratories are capable of processing bone samples and the laboratory conducting the DNA analysis should be consulted prior to any collection effort. Also, the more laboratory work involved in extracting each DNA sample, the fewer samples that can be processed, and the longer and more expensive the overall process.

The biological sample most likely to produce successful DNA results is also that which is least degraded. Whole blood and tissue from a body that has been dead for more than a few days without cold storage is not likely to yield sufficient quantities of DNA. Until recently, skeletal remains that were more than a few years old were not

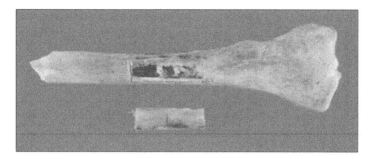

Figure 23.1   Window of bone cut from the diaphysis of a long bone.

likely to contain any useable nDNA either and conducting mtDNA testing was preferred. However, recent developments with mini-short tandem repeats (mini-STRs) have extended the limits for nDNA testing.

BONE SAMPLE COLLECTION

The preferred sample of bone for most DNA laboratory testing is a window of bone from the diaphyses of long bone such as a femur, tibia, or humerus (Figure 23.1); or cranial vault, ilium, or phalanges. As previously stated, vertebrae and other irregularly shaped bones are less desirable as DNA sources because their structure is less dense and more fragile. The standard bone sample for laboratory analysis submission is typically $1-2\,cm \times 4-6\,cm \times 0.5-1\,cm$, weighing $15-25\,g$ (DoD DNA Registry 2008). Not all laboratories have identical standards and even the same laboratory is likely to update their protocols and change the required amount of bone per bone sample with time. Therefore, the laboratory conducting the testing should always be consulted prior to the specimen-collection effort.

DNA sampling of human remains is a destructive process and care has to be taken to ensure the sample being collected does not destroy or alter the characteristics of the remains critical for identification by scientific means. Anthropological features used for classification of stature, age, sex, and race should be safeguarded and left undisturbed. Improper bone sampling that alters these critical features may invalidate the findings of other forensic scientists.

After the selection process has been decided a written protocol is established to document the process. A protocol establishes standard procedures for agencies to follow each time a requirement is necessary or a one-time procedure to support a single event with multiple identification challenges, such as with mass fatality incidents. The protocol is determined prior to commencing the actual sample collection. Circumstances influencing protocol decisions include but are not limited to degree of degradation of the remains, number of bodies, availability of medical equipment and facilities, funding and time constraints, and safety issues. Regardless of the influential circumstances, once a protocol is adopted the sample-collection process remains consistent throughout the project.

Whenever possible, DNA sample collection is accomplished using a team approach. A single individual working alone in an isolated environment collecting DNA samples has greater potential for making errors. A team approach provides a measure of quality assurance to help ensure mistakes are not made (National Institute of Justice 2005). Regardless of the number of personnel used in the collection effort, anyone handling or examining human remains is required to follow accepted safety precautions when contacting human remains. Remains are handled as though they are infectious biological material in any case to safeguard against those situations where the remains are actually infectious. Collectors are required to wear appropriate personal protection equipment (Pan American Health Organization 2004).

The mechanics of collecting bone specimens as DNA samples usually requires the aid of a Stryker autopsy saw and a bone-cracking instrument to successfully harvest a specimen from its host. Once the bone sample is extracted its surfaces are cleaned of residual soft tissue and foreign debris prior to being stored. This normally can be accomplished by wiping the bone surfaces with a paper towel or surgical sponge ($10\,cm \times 10\,cm$). Extra effort in the collection phase reduces handling time for DNA technicians and analysts in the laboratory. When possible, the collected bone sample should be refrigerated or frozen to prevent the growth of bacteria.

It is more cumbersome to collect a bone sample from an intact set of human remains than from barren skeletal remains. Collectors must first expose the selected harvest site and free it of masking soft tissue. In the case of a femur it can be a formidable task to expose the bone from beneath the mass of surrounding dense muscle.

Any disposable equipment and supplies used in the process of collecting biological samples are treated as biohazardous material and disposed of accordingly. Used disposable scalpels are discarded into sharps containers. Disposable scalpels are not cleaned and reused due to the risk of injury. The disposable personal protection equipment worn by personnel collecting the samples is disposed of in a like fashion.

In any collection effort involving biological samples contamination is a concern. When collecting DNA samples there are several steps employed to guard against the possibility of sample cross-contamination. Disposable bench-top sheets are placed on a hard specimen-cutting surface and changed between sample collections. Disposable scalpels are used and discarded after each sample is collected or scalpel blades are changed between sample collections from different human remains. When a Stryker autopsy saw is used, the blade is cleaned between samples with a 10% bleach solution. Gloves are changed after handling each item of biological material or wiped clean with a 10% bleach solution before moving on to the next item.

Specimen collection requires strict attention to detail when documenting the collection process and throughout the packaging and labeling processes. Any biological sample collected for DNA testing is segregated and safeguarded from coming in contact with other samples. This is often accomplished by packaging in a primary and secondary container. A sterile, 50 ml conical tube with screw cap is used frequently for packaging DNA specimens. The hard, clear plastic tube provides a protective shell without obscuring the sample from view and provides a smooth surface for writing specimen nomenclature or attaching an evidence identification label. The primary container is placed subsequently in a polyethylene bag, which is sealed by either tape or heat. Samples are never placed in formaldehyde or other preservatives that can damage the DNA molecules. As mentioned above, temporarily freezing a bone sample is an adequate safeguard to protect the integrity of the DNA until it can be tested in the laboratory.

A chain of custody is established to properly maintain the integrity of DNA samples. The custody chain creates an audit trail of the location where the sample was obtained, the individual responsible for its collection, and the time and date the sample was collected. The custody document also provides a description of the item sampled and identifies its origin. Most agencies have their own pre-printed evidence form for tracking the chain of custody. Multiple samples are documented on a single custody form and numbered consecutively.

SUITABLE DNA COMPARISON REFERENCES

Simply developing a DNA profile from an unidentified source is not sufficient to make an identification. The DNA "fingerprint" generated from laboratory testing must be compared to a known reference and matched for positive identification or excluded as a possibility, thereby eliminating the source of the known reference.

Known references may be obtained from several sources. DNA databases exist for this purpose. In 1991, the US Armed Forces created a DNA repository to store known reference samples for all military personnel and select civilian employees working for the US Department of Defense. The samples are dried bloodstain cards stored in individual vacuum-sealed pouches; the inventory has surpassed the 5 000 000 mark. Federal law (US Code, Title 42) established DNA reference sample collections of all felons convicted of prescribed offenses. The resulting database is the Combined DNA Index System (CODIS) and every state and federal prison is mandated to submit DNA profiles of their felony convicts to the Federal Bureau of Investigation (FBI). Maintaining a searchable DNA database allows for blind searches which may produce cold hits from unidentified DNA evidence of crime scenes. However, use of DNA samples from suspects in criminal investigations requires either voluntary submission of a DNA sample or court-ordered submission.

Personal belongings also can serve as a known source for DNA testing. An example can be drawn from a victim of an airplane crash. Once a person has been confirmed as a passenger on board a crashed airplane, surviving family members may be approached to produce personal effects that contain DNA from the missing individual. Items such as combs, toothbrushes, and clothing with sweat bands are representative of suitable sources. Similarly, many people undergo medical examinations or surgical procedures that can serve as a known DNA source. A paraffin block, biopsy slide, or stored pap smear are all excellent sources for known references.

The other most common type of comparison reference originates from family members who voluntarily surrender a biological specimen for DNA typing. Blood, saliva, or buccal swabs are the most common types of family reference. Family references are typically required in unusual circumstances such as mass fatality incidents resulting in multiple identification issues of recovered remains. These fatalities can be caused by a natural disaster (i.e., Hurricane Katrina) or a other event such as the multiple terrorist attacks in the USA on September 11, 2001. In either case, it is imperative to identify appropriate family members for reference collection.

These family reference DNA profiles are fed then into a searchable database and software programs run preliminary comparisons to select the most likely matches. Trained DNA analysts manually must perform the final comparison to develop matches between unknown and known samples. The conclusions of an analyst must be validated by other analysts before results are reported.

DNA Laboratory Analysis

Laboratories conducting DNA tests and analyses have various protocols for processing samples but all processes follow the same basic steps: extraction, quantification, amplification, and analysis. These basic steps apply to both nDNA and mtDNA processes. In order to satisfy credentialing requirements, forensic DNA laboratories follow operating guidelines set forth by the American Society of Crime Laboratory Directors (ASCLAD)/Laboratory Accreditation Board.

Before any attempt to isolate DNA molecules in bone, a bone sample (unlike blood, tissue, or other biological fluid samples) requires an extra step at the beginning of the process. All exterior surfaces are ground to remove potential contamination. The remaining bone material is then crushed into a coarse powder. Only then can the chemical extraction process that frees DNA from the host bone proceed. Once extracted the DNA is quantified and measured to determine the amount of DNA present. Next, amplification of the isolated DNA molecule is accomplished through a polymerase chain reaction (PCR) conducted in a thermocycler, which replicates the DNA for future analyses. Repetitive cycles of heating and cooling DNA manipulates the DNA molecules through a process of denaturing, annealing, and extension. Scientists identify STRs at specific locations on the DNA strands. Examination of STRs at as many as 16 locations on a DNA strand creates a unique profile for the tested sample.

Analysis of the DNA profile is the final step in the DNA identification process. It is the common digital representation of DNA sequences (the DNA profile) that allows automated searching and a comparison against other stored profiles (Colosimo et al. 2009). DNA profile results compared to known samples (direct or indirect) yield findings that are matching, exclusionary, or inconclusive.

Summary

DNA analysis has become invaluable in answering identity issues of mass fatalities, crimes against humanity, and other forensic matters. Skeletal remains are a primary source of DNA material and frequently the best, if not only, viable source of DNA material from unidentified remains.

Whether a laboratory is attempting to type for nuclear or mitochondrial DNA the general collection guidelines remain constant. Collection protocols are established prior to commencing the sample collection, appropriate sample criteria are determined in advance, and a logical progression for choosing the best available sample is followed. Bone sample selection processes are aided by establishing standardized harvesting criteria to provide sample consistency and to ease laboratory processing. Precautions are exercised to prevent sample cross-contamination and care is taken to establish and maintain a strict chain of evidence of custody.

DNA identification efforts cannot succeed without appropriate references for comparison. Without skeletal remains as a source for DNA, the scope of DNA identification would be limited severely. Regardless of the biological sample being tested, laboratory processes must be conducted in accordance with the accepted scientific standards prescribed by accrediting agencies. Results must be validated and capable of replication by independent sources.

REFERENCES

Adams, B. and Byrd, J. (2008), *Recovery, Analysis, and Identification of Commingled Human Remains*. Humana Press, Totowa, NJ.

Butler, J. (2005), *Forensic DNA Typing*. Academic Press, New York.

Colosimo, M., Graef, R., Lampert, S., and Peterson, M. (2009). *State of the Art Biometrics Excellence Roadmap, Technology Assessment: Volume 3, DNA*. The Mitre Corporation, McLean, VA. www.biometric coe.gov/_doc/SABER-Vol-3-pdf.

Correia, K. (2004). *DNA Match Positively Identifies Crewmember as Joseph Ridgaway of Talbot County, Maryland*. Friends of the Hunley, North Charleston, SC. www.hunley.org/main_index.asp?CONTENT=press&ID=126.

DoD DNA Registry (2008). *Guidelines for the Collection of Specimens Requiring DNA Analysis*. Armed Forces Institute of Pathology, Rockville, MD.

Houch, M. and Siegel, J. (2006). *Fundamentals of Forensic Science*. Academic Press, London.

Mader, S. (2008). *Human Biology*. McGraw Hill, .

National Institute of Justice (2005). *Special Report, Mass Fatality Incidents: A Guide for Human Forensic Identification*. US Department of Justice, Washington DC.

Ovchinnikov, I. and Goodwin, W. (2001). *The Isolation and Identification of Neanderthal Mitochondrial DNA, Profiles in DNA*. http://bioplein.nl/attachments/ File/Humane_3poren/Profilesin DNA-402-09.pdf.

Pan American Health Organization (2004). *Management of Dead Bodies in Disaster Situations, Disaster Manuals and Guidelines Series, No 5*. www.paho.org/English/DD/PED/DeadBodiesBook.pdf.

US Code, Title 42, Chapter 136, Subchapter IX, Part A, 14135a, *Collection and Use of DNA Identification Information from Certain Federal Offenders*.

Wambaugh, J. (1989). *The Blooding: True Story of the Narborough Village Murders*, Bantam Press, London.

CHAPTER 24

Current Research in Forensic Taphonomy

Marcella H. Sorg, William D. Haglund, and Jamie A. Wren

INTRODUCTION

Overview

Taphonomy is the science of postmortem processes, and forensic taphonomy brings that science to bear on solving forensic problems. In fact, taphonomy science potentially underlies the understanding and forensic investigation of all human deaths, focusing on the condition of the remains in relationship to their postmortem environment.

The term taphonomy was originally coined by paleontologist Ivan Efremov (Efremov 1940) to refer to the study of any "death assemblage" as it makes the transition from being a living organism to becoming a fossil. Although he did not envision applications in forensic work, Efremov explicitly included both contemporary and ancient animal remains. The goal was to separate postmortem changes seen in the remains from those characteristics that were present when it was living. Taphonomic methods have been used by paleontologists to study plants and animals in both terrestrial and aquatic environments. Paleoanthropologists and archaeologists (Lyman 1994) utilize taphonomic methods and theory to interpret postmortem processes in the archaeological record, including research on weathering, scavenger modification, and water transport.

Applications of taphonomy to forensic sciences began in the 1980s when Sorg (1986) and Haglund (Haglund 1991; Haglund et al. 1988, 1989) applied research on scavenger modification of remains[1] and decomposition to forensic casework. In 1993, a symposium at the American Academy of Forensic Sciences brought together an interdisciplinary range of research on postmortem processes, culminating in two edited volumes (Haglund and Sorg 1997; Haglund and Sorg 2002). As illustrated in

A Companion to Forensic Anthropology, First Edition. Edited by Dennis C. Dirkmaat.
© 2012 John Wiley & Sons Ltd. Published 2015 by John Wiley & Sons Ltd.

those publications, forensic taphonomy applies that body of interdisciplinary method and theory to the early postmortem period (usually under 50 years, with most under 5 years) and focuses particularly on human remains. As we emphasize in our second volume, forensic taphonomy can provide actualistic research which is meaningful to archaeologists and paleoanthropologists, particularly in understanding the role decomposition and scavenging may play in site formation and distribution of skeletal elements, and in identifying the effects of trauma (Haglund and Sorg 2002a).

During the last decade, forensic taphonomy has become a central feature of research and analysis not only by forensic anthropologists but also by scientists in related forensic disciplines, such as entomology, soil science, and botany. As efforts have expanded in human-rights investigations and mass disaster fatalities, data from mass graves and other large-scene forensic recoveries have provided an expanding research base.

With the development of forensic taphonomy has come a paradigm shift that places more emphasis on studying the remains within the context of their discovery; that is, at the forensic scene. Because of this focus on context, forensic archaeology, and the involvement of anthropologists in the recovery of remains, has taken on an enhanced role (Dirkmaat et al. 2008). Taphonomy emphasizes the critical importance of collecting environmental and provenience data as part of the scene investigation, utilizing exacting archaeological methods. Increasingly, more jurisdictions are including forensic anthropologists as essential participants in forensic scene processing and recovery of remains, including expanded roles in evidence collection and death investigation. They often direct such recovery operations. Nevertheless, forensic taphonomy is an interdisciplinary field, involving many forensic disciplines, particularly anthropology, pathology, entomology, botany, and geology (including soil scientists).

This chapter focuses on selected features of the growing body of research in forensic taphonomy during the past decade. We first present a forensic problem focus, explaining how forensic taphonomy research is associated with essential forensic taphonomy tasks (Sorg and Haglund 2002) (Table 24.1). After an overview of the forensic taphonomy perspective, we will briefly discuss the advantages and disadvantages of research using an experimental approach, compared to actualistic research using findings from forensic cases. Following that, we will go into more depth in selected topics where the research has been strongest, and the findings most useful in forensic casework, including basic taphonomic processes and contexts, agents of taphonomic modification, and application of forensic taphonomy research to critical forensic problems. The discussion is restricted to basic taphonomic topics, excluding research on unusual microenvironments, or extreme temperatures, which also produce taphonomic changes.

Applied research

Forensic taphonomy is an "applied research" field. That is, research and analysis in forensic taphonomy is directed toward solving forensic problems that are basic to death investigation. Generally, these problems include deciding who the decedent is, where they died, when they died, how they died, and what caused their death. With each of these aspects, the focus of the research and analysis is on the decedent or decedents in the forensic case, rather than other types of evidence or other parts of the investigation. Forensic taphonomy research and analysis ultimately supports the work of the medical examiner or coroner who must discover these details in order to record

Table 24.1 Problem-focused areas of forensic taphonomy research.

Forensic task	Taphonomic research topics
Reconstruct events surrounding and following the death, including sequences, location, and effects of human modification of the decedent(s) in the perimortem and postmortem periods	Techniques to maximize recovery of remains and evidence Techniques to reconstruct sequences and characteristics of events, reflected in the spatial association of remains to each other and to artifacts Techniques to locate places remains were deposited and underwent decomposition; i.e., cadaver dogs, mass spectroscopy, chemical soil analysis Multidisciplinary techniques that can localize plants, animals, soils, and chemical signatures
Identify individuals and/or groups	Techniques to maximize recovery of remains and to focus recovery on those remains that will help in identification of individual persons, age or sex-specific groups, or ethnic groups Techniques to reconstruct scatter patterns of scavengers to maximize recovery of remains Remote-sensing techniques to maximize location of buried and scattered remains
Estimate the postmortem interval	Modeling the process and timing of the natural processes of decomposition, consumption, dispersal, and assimilation in terrestrial and aquatic contexts using human and nonhuman analogues, as well as biological "clocks" (e.g., necrophagous insects and marine invertebrates) Understanding the role of key biotic and abiotic modifiers of decomposing remains, including heat, moisture, plants, animals, and geological activity
Identify trauma and other effects of human modification of decedents in the perimortem and postmortem periods	Techniques for reconstructing human taphonomic modifications of remains (e.g., analysis of trauma, dismembering, commingling, burial, reburial, and transport, as well as modifications associated with recovery and examination) Differentiating trauma from other postmortem processes, such as scavenger modification

them on a death certificate. It also supports the work of the criminal justice system, which may investigate a death when criminal acts are suspected.

Taphonomic studies in forensic work are most often directed toward determining where and when someone died, but they can also be helpful in identifying the decedent or developing evidence regarding the cause of death. Taphonomy also includes the identification of human behavior that may modify a corpse, such as embalming or dismemberment (Lyman 2010). An understanding of how bodies decompose or how they are scattered by scavengers may also contribute to the location of human remains during a search process. Table 24.1 provides an overview of the topics of taphonomic research and analysis connected to basic forensic tasks.

As experts from these fields and others lend their knowledge to understanding postmortem processes, the evidence base for forensic taphonomy grows. The field has developed from more historic or anecdotal approaches and become more rigorous. Recently the Scientific Working Group in Forensic Anthropology (SWGANTH; www.swganth.org) has selected taphonomy as a key area of skills and knowledge, demanding the development of evidence-based consensus guidelines for the field. This chapter reviews basic concepts in forensic taphonomy, including recent advances in the field during the past decade.

The forensic taphonomy perspective

By studying the remains in context, forensic taphonomy adopts an ecological perspective, which focuses on human remains as the "centerpiece of a newly emerging microenvironment" (Sorg et al. 1997). Taphonomy integrates information about ecological processes of decomposition, consumption, dispersal, and assimilation involving plants, animals, and microorganisms that become associated with the decomposing body. All of these changes create a new microenvironment immediately surrounding the body that continues to change through time.

Ultimately, as a result of these natural processes, once-living organisms are reduced to their constituent molecules after death, and ultimately even to the chemical elements from which they were originally composed. The term for this is decomposition, viewed in a broad sense, undoing the "composition" organisms had in life. This includes natural recycling processes of chemical breakdown (autolysis), as well as processing by other living organisms in the surrounding ecological community (putrefaction by microorganisms; consumption and assimilation by animal scavengers including insects; and chemical or physical breakdown caused by plants). In addition, nonbiological processes such as erosion and desiccation can break down tissues, and heat and moisture can speed or influence chemical reactions. It is important to note that these processes occur on land as well as in water environments.

Due to the wide ranging effects of biological, chemical, and physical agents, taphonomy is interdisciplinary, potentially requiring a broad range of expertise spanning the natural sciences. For example, entomologists contribute knowledge about the identification of insects associated with the body and the timing of their developmental states, which allows them to estimate how much time has passed since the death. Botanists indentify plants associated with the remains, which may reveal, for example, that a body was moved from one location to another.

In all forensic applications, the taphonomic approach views the human remains within the context of discovery. This means that collecting details about the location and environmental characteristics of the site where the remains were found are often just as important as information about the condition of the remains themselves. For example, bodies decompose more slowly in cooler temperatures; so knowing the temperature of the area where the body was found would contribute to understanding its taphonomy. Careful and comprehensive taphonomic documentation of the scene should include a wide range of ecological information about the microenvironment. Particularly important are those factors that either concentrate or limit the body's access to heat, moisture, and scavenging.

Experimental and actualistic research designs

Starting with the earliest applications of taphonomy in paleontology and paleoanthropology, researchers have designed experiments to test hypotheses (experimental approach) as well as studying actual processes in nature (actualistic approach). For example, in some of the earliest efforts to study the effects of moving water on bone, paleoanthropologists (Boaz and Behrensmeyer 1976) constructed experiments with artificial waterways ("flumes") into which they placed different types of bone element in order to study the influence of bone shape on waterborne trajectories. In contrast, Lyman (1989) conducted actualistic research when he studied the damage to wild animals killed by the 1980 Mount St. Helens volcano eruption in Washington state in the USA (Lyman 1989). Understanding how natural processes affect bone is helpful in interpreting the condition of skeletal remains in forensic cases.

Both experimental and actualistic approaches have been applied to forensic problems (Haglund and Sorg 1997, 2002). Much of the experimental taphonomic research in forensics has focused on soft-tissue decomposition. For example, experimental research designs have been used to test methods of detecting decomposition chemicals (Vass 2008), and to study environmental variation in the insects attracted to decomposing organisms (Hobischak et al. 2006). The actualistic approach in forensic taphonomy research, in contrast, usually involves studies of real forensic cases, either in the form of an individual case study or a study of a series of cases (Haglund et al. 1989).

Both approaches have advantages and disadvantages. The actualistic approach provides real-world examples of the variation seen in postmortem change. Forensic cases are unique historical events in which researchers can search for repeating patterns. But, because the forensic investigator arrives after those processes have already occurred, some of the circumstances may be unknown. Nevertheless, this method contributes knowledge about the wide range of variation in both the condition of the remains and the contexts in which forensic cases are found. In contrast, the experimental approach in forensic taphonomy has the advantage of allowing scientists to focus on known factors affecting postmortem change, control some of the causal variables, and observe and measure the outcomes. The disadvantage of the experimental approach is that the controlled situation in a laboratory setting may not reveal the full range of variation that one might see in the real world.

Taphonomic Processes in Context

Overview

In every death a variety of taphonomic agents act in combination to modify a body during the postmortem period. Some agents are simply the environmental factors present at the time and place the body is deposited, such as the ambient temperature and moisture levels. These can be termed "independent variables," characteristics of the environment that significantly affect the progress of decomposition or preservation. Insect or mammalian scavenging would also be considered independent variables. The resulting changes in the condition of the remains are "dependent variables"; for example, bloating of the body or the length of time it takes to become skeletonized. Despite the fact that the basic process of decomposition is fairly well understood, each forensic case has unique characteristics. There can be variation in the combination or strength of any of the independent variables, which can in turn change the progress, timing, or outcome of any given case.

The postmortem context includes independent variables introduced by the general climate, ecology, and geology of the area in which the remains were found. This may include a broad area where scattered remains are found, or it can be a room in a building. The context can be terrestrial or aquatic, or a combination of the two. The ideal forensic taphonomy investigation requires a thorough documentation of the microenvironmental factors that can influence the condition of the body. For example, in an outdoor case, this includes recent weather (temperature, precipitation, and moisture), relevant vegetation and access to shade or sun, scavenger access, and physical characteristics of the site, particularly those features directly associated with the remains, or potentially associated with decomposition processes. Important local geological features might include, for example, the slope and potential for erosion, the soil types and porosity, and the location of the water table and waterways. The most important taphonomic factors are likely temperature and scavenging.

Temperature is usually the most important driver of decomposition, since higher temperatures (generally those above freezing point but below boiling point) accelerate chemical reactions associated with cell breakdown as well as bacterial growth. Assessment of outdoor environmental factors that may affect heat can include the presence of shade from structures, trees, or other vegetation, and the presence of hills or valleys that shield the remains from sunlight. Indoor factors might include access or barriers to heaters or sunny windows.

Scavengers of all types can alter the pace and patterning of decomposition. Insect scavengers, particularly certain types of fly and beetle, reproduce quickly and greatly speed the defleshing process, so long as the temperature stays above 4°C. If the body is exposed, flies arrive almost immediately and lay hundreds of eggs. The larval form, termed the first instar, emerges from the eggs and begins to consume the decaying flesh. The larvae grow and develop into the second and third instar forms, as they continue to consume the soft tissue. When they are fully grown they migrate a short distance away from the body to form a pupal case within which they develop into the adult form. Beetles generally arrive later than flies. Because the timing of fly and beetle metamorphosis is known for many species, the presence and form of these insects can be not only an agent of taphonomic change (soft-tissue consumption) but in the early postmortem period they can serve as a taphonomic indicator of how much time has passed since the body was exposed. This is often helpful in estimating the time of death.

Many species of mammal, bird, crustacean, and fish are common scavengers of recently dead or decaying animals in North American environments. Some groups, such as canids, felids, raptors, and ursids are hunters, and scavenge decaying flesh only when other food sources are unavailable. Others, such as the catharthids, some murids, procyonids, and corvids, consume it regularly.[2] Other animal predators, such as the mustelids, may consume relatively fresh animal carrion, but are less likely to scavenge decaying flesh. More than one scavenger species may be involved at once or in sequence; the species combination within a local ecosystem is referred to as the "scavenger guild."[3]

Competing with the scavengers are the microbes: microscopic bacteria and fungi, usually termed "decomposers." Their involvement with a body in the postmortem period changes its chemistry, even creating toxins, which reduces appeal and safety for scavengers.[4] Microbes recycle waste products from plants and animals, breaking down large molecules, and recycling the nutrients. They and play a critical role in nature's carbon, nitrogen, and phosphorus cycles. This process, called "putrefaction," is an essential part of all decomposition. In the living human body, as with other animals, the gastrointestinal system is full of bacteria essential to the digestion and breakdown of plant and animal foods to free up nutrients needed for survival. When an animal dies, these bacteria, increasingly dominated by anaerobic forms, continue to proliferate and break down the body's proteins, fats, and carbohydrates into simpler chemicals: organic acids and gases.

In addition to the chemical changes caused by bacteria, some changes are simply a byproduct of the loss of oxygen from respiration and the resulting changes in pH. Cell walls break down and release enzymes, which stimulate the breakdown of more body tissues. This process is called "autolysis." Autolysis and putrefaction occur simultaneously.

New decomposition chemistry research

Over the last 10 years, research on decomposition chemistry has included experimental approaches both in laboratories and in outdoor settings. Gas chromatography mass spectroscopy (Vass 2008; Vass et al. 2002, 2004; Dekeirsschieter 2009) has been used to identify the types of chemical byproduct that characterize a decomposing body, mainly those chemicals that become volatile (volatile organic compounds or VOCs), and may be part of decomposition odor. Organic compounds identified by portable mass spectroscopic equipment in the air and in the soil, can signal the presence of a clandestine grave, for example. Degradation of organic compounds in the body may also provide clues to the length of time a body has been decomposing (Vass 2010; Vass et al. 2002).

One of the most critical outcomes of decomposition in forensic work has to do with how decomposition affects the DNA needed to do identifications, one aspect of molecular taphonomy. Recent experimental laboratory studies have examined the impact of tissue degradation on DNA integrity. One such study demonstrates the damage done to DNA if the remains are boiled for maceration (Iwamura et al. 2005). It was found that exposed bone can retain bone cells (osteocytes) with DNA even after it has begun to demineralize, pointing to the necessity of taking samples before cleaning.

Although bone color is often used as an indicator of taphonomic sequences, it is extremely variable and subject to rapid changes; for example, ue to sun bleaching. In an experimental study of 40 pig humeri with 20 controls, researchers examined color associated with burial and surface exposures (Huculak and Rogers 2009). They found that several factors contribute to bone color, including sun, hemolysis (blood-cell breakdown), fungi, soil, and decomposition, although the latter only produces slight staining. Bones buried for 4 weeks and then exposed on the surface for 4 weeks and those buried 4 weeks after exposure on the surface for 4 weeks were similarly colored when compared in cross-section using Munsell color charts. The presence of fungus on the surface is an indicator of surface exposure. No soil analysis was reported for this study.

Environmental and microenvironmental factors

Environmental factors such as season, weather, regional vegetation, general topography, and biogeographic region can all influence the pattern and pace of decomposition. Although this overall environmental context is very important, local microenvironmental features in the body's immediate surroundings can cause an individual set of remains to deviate substantially from the expected patterns. For example, local site characteristics that emphasize shade, heat, weather exposure, or chemistry (for example, bogs that are anaerobic and acidic) can dramatically affect the decomposition process, and should be accounted for, particularly in estimations of postmortem interval.

Forensic entomologists routinely measure temperatures at a forensic scene in order to calibrate local temperatures with the temperatures at the nearest appropriate weather station in order to estimate postmortem interval. They identify the maximum metamorphic stage achieved by insect species associated with body, and how many accumulated degree-days (ADD[5]) it takes for that species to reach that stage under normal conditions. They can then utilize the adjusted weather station temperatures going back in time to estimate the same number of ADDs, and pinpoint the likely day the insects first colonized the body. Recent research cautions against simply using the nearest weather station (Dabbs 2010). It is important to try to match characteristics that may influence temperature (such as altitude or nearby vegetation patterns) with those at the scene; the most comparable weather station location may not be the nearest.

A number of recent experimental studies using pig cadavers delve into variation in regional, seasonal, and microenvironmental characteristics and their various effects on decomposition rates. In an Australian study 20 newborn pigs were placed exposed in a damp forest setting with a temperate four-season climate, five in each season (Archer 2004). Decomposition speed, as indicated by loss of body mass and later decay stage, was generally greater in seasons with higher temperatures and also when there was more rainfall; the rate of decomposition differed from year to year due to overall weather fluctuations. Higher moisture burial contexts can speed loss of mass and decomposition, unless there is high clay content (Carter et al. 2010; Jaggers and Rogers 2009), or adipocere formation (Duraes 2010). Acid soils can also speed decomposition, probably due to microbial activity (Haslam and Tibbett 2009). In a northern England experiment, pigs were buried in three upland habitats (pasture, moorland, deciduous woodland) and recovered 6, 12, and 24 months later (Wilson et al. 2007). The study revealed both seasonal and microenvironment differences, even within the same habitat and within the same pig. The authors attribute part of the variation to differences in soil type and structure within each grave.

A number of studies have focused on sun/shade differences in both decomposition rates and insect succession. In an experimental study in Saskatchewan, Canada, 18 pig cadavers were placed on the surface, in either sun or shade, in a prairie ecozone, separated into three seasons (Sharanowski et al. 2008). Besides the expected seasonal differences with greater decomposition in warmer times of the year, sun-exposed cadavers also had more associated insect species. In terms of impact on the rate of decomposition, only in the spring did pigs show a significant sun/shade difference in the rate of decomposition, with the pigs in sun decomposing faster. Although one might conclude that decomposition in the sun results in faster insect development due to the warmer temperature in the sun, one West Virginia study has shown that the faster insect development persists, even when controlling for temperature (expressed as accumulated degree-hours, or ADH) (Joy et al. 2006).

Most taphonomic research in the literature concerns soft tissue and bone, but one recent study focuses on teeth, demonstrating that it is possible to distinguish between cracking that occurs antemortem or perimortem and heat-induced postmortem cracking (Hughes and White 2009). This experimental laboratory study used 36 pig teeth and exposed one third of them to a dry heat (41–49°C) for 9 h/day over 14 days. Heat-induced desiccation cracks begin within the internal dentine, propagating through to the enamel, whereas *in vivo* cracks begin at the external enamel and move inward, often stopping at the dentine/enamel junction. A high-powered microscope was used to view the crack patterns.

Terrestrial contexts

Soil science research in forensic taphonomy has expanded in recent years (Tibbett and Carter 2008; Vass et al. 1992). Major areas of research include: detection of decomposition byproducts, estimation of postmortem interval, and the utility of these techniques in locating and assessing buried remains. Research studies have utilized experimental research designs involving both rat and pig cadavers to study variation in decomposition associated with different soil types, time frames, and moisture levels. Most soil research in forensic taphonomy has focused on buried remains rather than remains on the surface.

When a cadaver is left undisturbed in one area, the fluids discharged during the decomposition process will stain, and modify the localized soil, forming what is known as gravesoil. Carter et al. define gravesoil as "...any soil associated with cadaver decomposition, regardless of species or surface or burial deposition" (Carter et al. 2007). As gravesoil forms, the soil chemistry changes, significantly increasing pH, nitrogen, and phosphorus (Benninger et al. 2008). Other compounds commonly elevated within gravesoil are ammonia, calcium, carbon, magnesium, and potassium, among numerous others (Tibbett and Carter 2008). Out of all identified compounds, nitrogen (tested in the lab as nihydrin-reactive nitrogen) is believed to be the best indicator of decomposition both in burials and on the surface (Carter et al. 2008; Van Belle et al. 2009). Detection of these compounds can be useful in locating areas in which a body may have decomposed.

With buried bodies, the condition of the body at the time of burial as well as the nature of the burial microenvironment will impact the ultimate condition of the remains at the time of exhumation (Mant 1950; Mant 1987). For example, open trauma may speed decomposition, whereas wrappings and clothing are likely to retard it. Nevertheless, a recent experimental study with pig muscle demonstrated that freezing prior to burial does not increase the decomposition rate (Stokes et al. 2009). Condition of the body at the time of exhumation also depends upon the time between death and burial, and the length of time buried.

The depth of burial is an important factor in decomposition (Weitzel 2005). Shallow graves are more likely to be affected by fluctuations of temperature above ground, including the amount and duration of exposure of the grave to the sun, the type of soil surface exposed, its susceptibility to heat absorption, and the season of burial. If the soil over the body ("overburden") is loosely compacted and contains rocks or pebbles that heat up, it will be more likely to fluctuate in temperature. The deeper the burial, the more constant the temperature of the grave becomes. Shallow graves are also more vulnerable to invasion by insects and other scavengers. Bodies within shallow graves may even generate their own heat when decomposition is advanced or the remains are infested by quantities of maggots.

The relationship between the grave microenvironment and the decomposing body is mutual, and will persist. That is, the grave environment affects the body's decomposition, and the decomposition in turn changes the grave environment, including the microbial load, pH, moisture, and changes in oxygenation (Wilson et al. 2007). In one recent study, gravesoils and nongrave control soils were compared after burial of three pigs at depths of 0–15 and 15–30 cm for 430 days (Hopkins et al. 2000). Gravesoils contained more total carbon, microbial biomass carbon, and total nitrogen, and showed increased rates of respiration and nitrogen mineralization compared to the control soils. The gravesoils were also more alkaline.

Graves with more than one individual in contact with each other are termed "mass graves," and are more taphonomically complicated. Close contact of the bodies promotes preservation that does not occur in individual graves (Mant 1950, 1987). Bodies near the core tend to be better preserved than those at the periphery. Historically, mass graves have occurred in a variety of circumstances that include customary burial practices, mass fatalities from natural disasters or accidents, or to hide crimes by perpetrators.

Attention to large-scale mass-grave exhumations in the recent past was triggered by war in the Former Yugoslavia and Rwanda. Of these contemporary mass graves, the most intensely investigated were those under the auspices of the International Criminal Tribunal for the former Yugoslavia (ICTY), involving as many as 8100 missing Bosnian Muslim men and boys from the "safe haven" of Srebrenica in early July 1995. Most of these men and boys had been killed and put in mass graves to conceal the crimes and to confound investigators.

Initial exhumations in 1996 revealed some undisturbed, "primary graves" containing relatively complete remains in anatomical position. Some primary graves had subsequently been "robbed" in an attempt to remove remains for reburial elsewhere in "secondary graves," in order to shield them from discovery. Attempts to empty graves typically involved blindly digging for remains using large earth-moving equipment. While some remains were removed, others were left behind. The result was that remains of single individuals might be distributed over multiple sites, and partial remains of many individuals became commingled, especially in the secondary graves.

In the investigation of mass graves associated with war crimes and crimes against humanity, the large-scale, postmortem burial, movement, and reburial of bodies are themselves taphonomic processes. In most forensic cases the goal is to strip away postmortem effects in order for the medical examiner/coroner to identify the cause and manner of death. With criminal mass graves, the taphonomic process itself must be identified and described to provide evidence of crimes against humanity. Resolution of commingling becomes a tool to identify and describe the taphonomic process of mass burial and reburial.

Sorting out the commingling is important, not only to identify individuals and discover how they died, but also to document the mass killing and subsequent attempts to hide the contents of the primary mass graves by mass reburial. With such large numbers of decedents it becomes even more important to use efficient and effective methods of commingling resolution. Although methods that measure and compare shape and size of skeletal elements, "osteometric sorting," can be a powerful tool (Ubelaker 2002), authors in one article point out disadvantages (Byrd and Adams

2003). For example, osteometric sorting is of less value when applied to individuals of a similar general size.

As it turns out, the majority of commingling resolutions for Srebrenica mass graves have been made by DNA testing administered by the International Commission for Missing Persons (ICMP). In August 2011 the ICMP announced that it had identified 6616 of those reported missing from Srebrenica (ICMP 2011). DNA has been the primary tool for commingling resolution, and hence for reconstruction of the taphonomic history. DNA is playing a critical role worldwide in helping to resolve mass-grave identifications, even in historic settings (Rios et al. 2010; Zupanic et al. 2010).

Aquatic contexts

Understanding aquatic taphonomic patterns is an essential component of forensic taphonomy. The majority of the earth's surface is covered with water, and many cases are found either in the water or at the water's edge. Not only do water environments have different characteristics according to their type (e.g., lake or river or ocean), they tend to have shifting, three-dimensional dynamics that differ in chemistry, currents, and temperature according to region, season, depth, bottom type, and topography (Sorg et al. 1997). As a result, there is great variation in the key taphonomic factors such as temperature and scavenging, as well as movement and dispersal of bodies and body parts.

Both aquatic and terrestrial forensic taphonomy studies are concerned with decomposition, disarticulation, dispersal, and scavenging. Just as on land, it is critical to understand the microenvironmental context in which a body is found. Since bodies in water are more likely to have moved, it is also important to reconstruct the potential sequences of the body's location. Floating bodies are actually subject to taphonomic processes associated not only with water (e.g., currents and water-based scavengers), but also with land (e.g., bird and insect scavengers, sun exposure).

Water environments with little or no current tend to foster the formation of adipocere during decomposition, which can foster preservation of soft-tissue morphology. Adipocere, sometimes referred to as grave wax or corpse wax , can form in both terrestrial and aquatic microenvironments, but its development is dependent upon the presence of moisture and body fat (O'Brien and Kuehner 2007; Ubelaker and Zarenko 2011). Other factors that promote adipocere include anaerobic conditions, warm temperatures, alkaline pH, and anaerobic bacteria (Forbes 2005; Forbes et al. 2004). As in other decomposition processes temperature acts as a major variable.

Adipocere can preserve forensic evidence, as well as soft-tissue pathology or trauma, but it makes estimation of the postmortem interval very difficult. (Ubelaker and Zarenko 2011). Although adipocere can begin to form within hours of death (Yan et al. 2001), it usually takes weeks. It can persist for hundreds of years, but frequently breaks down over time (Frund and Schoenen 2009), and can even occur simultaneously with desiccation (Schotsmans et al. 2011). Caution is needed for studies using pig analogues to study adipocere; experimental research shows that its formation in pigs takes slightly longer than for humans and that it has slightly different chemistry (Notter et al. 2009).

Although adipocere formation occasionally preserves the human soft-tissue anatomy, most water environment cases decompose and skeletonize (Haglund and Sorg 2002b; Sorg et al. 1997). Research in the US Pacific Northwest with pigs tethered at 7.6 and 15.2 m in depth revealed decomposition stages similar to previous marine taphonomy studies from the Washington state coastal waters (Anderson and Hobischak 2004b; Haglund and Sorg 2002b). Anderson and Hobischak (2004) include in their article a comprehensive table comparing terrestrial, marine, and freshwater decomposition rates in that region.

Rates of decomposition may be similar within a single region, but they differ substantially across regions, depending on the ecologies of those regions (Dumser and Türkay 2008), including bottom type, body covering, and scavenger access. Dumser and Türkay document a Namibian body recovered from the East Atlantic in a uniform as a complete skeleton after a 3-month postmortem interval from a depth of 580 m; it was associated with scavenging lyssianasids (also called amphipods, shrimp-like crustaceans). Another body in uniform was recovered in the Mediterranean Sea near Sicily after 1 month; most of the body was intact, with skin slippage and some adipocere, and no associated scavengers.

Research on pig cadavers decomposing in an intertidal region of Hawaii, at the terrestrial and marine interface, reveals a complex set of decomposition states, including insect infestation and tidal action: fresh, buoyant/floating, deterioration/ disintegration, buoyant remains, and scattered skeletal (Davis and Goff 2000).

Due to the putrefaction process and the resulting gas formation in the gut, many bodies float, and may be carried by currents to jurisdictions remote from where they entered the water (Ebbesmeyer and Haglund 2002). Complicating efforts to predict their trajectories is the fact that currents vary due to weather, season, depth, and shoreline terrain and vegetation. Nevertheless, patterns can be found. Ebbesmeyer and Haglund studied a mass fatality ferry collision and several forensic cases and found that computer simulations were very helpful in predicting where bodies might float in the Puget Sound area of Washington state, USA. A Portuguese study of seven victims of a bus accident (Blanco Pampín and Lupez-Abajo Rodrìguez 2001) documented recovery on the Spanish coast. Even though victims of the bus accident were pushed by the currents as much as 380 km in only 60 h, all were eventually recovered from a single area of the coast of Spain, arriving at different times.

Insects

Insects play three main roles in forensic taphonomy (Anderson and Cervenka 2002; Goff 2009; Haskell et al. 1997). First, necrophagous (carrion-eating) insects can be important taphonomic agents, consuming soft tissue and skeletonizing a body. Secondly, they can be used as biological clocks to estimate postmortem interval, or more precisely postexposure interval, particularly in the first few weeks. Insects associated with the body may also provide an indicator of movement of the body from one location to another. For these reasons, insect collection is a critical feature of taphonomic methodology. Taphonomic scene investigation includes collecting samples of insects associated with the body at all metamorphic stages: eggs, larvae, pupating larvae, and flies.

Despite the widespread distribution of many taphonomically important insect species, however, there is evidence that they may behave differently depending on the local environment or season (Hobischak et al. 2006; Sharanowski et al. 2008). Experimental studies, mostly using pigs, are now being published for many areas that document regionally specific data, for example South China (Wang et al. 2008), West Virginia in the USA (Joy et al. 2006), Alberta (Hobischak et al. 2006) and Saskatchewan in Canada (Sharanowski et al. 2008), Germany (Schroeder et al. 2003), and the Hawaiian intertidal zone (Davis and Goff 2000).

Carrion is a critical food source for insects, but also for many other animal groups, including many species of mammal, bird, crustacean, and fish. Recent experimental studies in England point to the dominance of insects in decomposition in some terrestrial environments (Bachmann and Simmons 2010; Simmons et al. 2010a, 2010b). Nevertheless, in colder climates insects are only active during about half of the year (Sorg and Wren 2011). Terrestrial cases in northern New England in the USA tend to feature either insect infestation or mammalian scavenging, but not both. Exceptions include mammals (such as bear and skunk) and birds (such as turkey vultures) that eat insect larvae, and may be attracted to insect-infested carrion.

Mammalian and avian scavenging

Scavenging includes modification by large and small animals as well as birds. Mammalian and avian scavengers compete with insects and microbial decomposers for access to food from animal carcasses. In forensic cases, scavengers may not only alter bodies but also may change scenes or obliterate evidence. Applying knowledge of typical scavenger "signatures" sequences may help in judging relative postmortem interval (Haglund 1997a), as well as locating missing skeletal elements that have been scattered (Haglund 1997b). In addition, an understanding of scavenger patterns can be helpful in discriminating between trauma from the death event and that caused by postmortem scavenger modification (Calce and Rogers 2007; Moraitis and Spiliopoulou 2010).

Recent forensic science research on terrestrial scavenging has included several actualistic studies focused on identifying modification patterns, often reporting details about one or more forensic cases. One recent study reports a case of indoor scavenging by two large pet dogs in a heated home after the owner died, likely about a month earlier, during winter (Steadman and Worne 2007). Very few parts of the skeleton remained (calvaria, two long bone shafts, and some bone shaft splinters). There was no evidence of perimortem trauma, blood, decomposition, or insect infestation. Carson and colleagues present a case-series comparison of seven black-bear-scavenged cases from New Mexico (Carson et al. 2000), which are compared to earlier published studies of canid- and polar-bear-scavenged cases (Haglund 1997a; Merbs 1997). They conclude that bears are more likely to modify the bones of the axillary skeleton, but they caution that such patterns may not be strictly diagnostic. In another forensic study of zoo animals, carnivore canines and carnassial teeth were measured and documented to identify the probable species responsible for a bite mark on a human victim (Murmann et al. 2006). A related study from the ecology literature includes a calculation of bite force across species (Christiansen and Wroe 2007).

A number of experimental forensic studies of scavenging have been done using pig cadavers as surrogates for human bodies and focusing on scatter patterns. For example, Kjorlien and colleagues found patterns of scatter in a study of 24 clothed and unclothed pig cadavers in Alberta; 12 were placed in woodland settings and 12 in open grassland (Kjorlien et al. 2009). They report a high level of variation with some basic patterns. Unclothed and grassland pigs were scavenged earlier. Scattering trajectories followed game trails and tended to go away from human activity. In an Australian study of scavenging, O'Brien and colleagues placed pig cadavers with controls in four environmental contexts and observed scavenging using 24-h trail cameras (O'Brien et al. 2010). They detail involvement by avian and mammalian carnivores, documenting variation among sites, seasons, and hour of the day. The scavenger guild in Australia has only small and medium-sized carnivores. Avian scavengers were more common during the day and mammals at night. In addition, the intensity of scavenging was higher for the bloat stage and later stages, compared to the fresh stage. In a northern New England woodland study a wide-ranging scavenger guild was responsible for complete skeletonization of pigs deposited in late fall (Sorg and Wren 2011). Two of three pigs were largely consumed by small and medium-sized scavengers while under 60 cm of snow; access to the bodies was afforded by snow tunnels built by raccoons, which were also used by other species (bobcat, raven, and turkey vulture). Low temperatures precluded insect infestation. One pig, untouched for 5 months over winter, was consumed by turkey vultures in April when temperatures rose above 4°C, in combination with heavy insect infestation.

TAPHONOMIC APPLICATIONS TO FORENSIC PROBLEMS

Search strategies

Most improvements in forensic search strategies in the past decade have to do with experimental research focused on detecting chemicals associated with decomposition. These chemicals might be associated with a remains buried in a clandestine grave, with remains now scattered by scavengers, or with a specific area where a body may have been previously located, such as an empty grave, or the now-empty trunk of a car (Bull et al. 2009; Oesterhelweg et al. 2008). Mass spectroscopy is used for such detection, including volatile organic compounds, usually fatty acids (Vass 2008; Vass et al. 1992, 2002, 2004), as well as solid fatty acids (such as those found in adipocere) (Fiedler et al. 2004), and chemicals (such as ninhydrin-reactive nitrogen) found in soil or adherent to soil or other substrates (Carter et al. 2008; Lovestead and Bruno 2010; Van Belle et al. 2009).

Cadaver dogs can also detect those same volatile organic chemicals, which make up an airborne "scent cone" propelled by wind or the act of sniffing. They can be trained to distinguish between human and nonhuman decomposition odors and to localize those odors on the landscape (Lasseter et al. 2003; Rebmann et al. 2002; Sorg et al. 1998). In one experiment cadaver dogs were able to detect which carpet squares had been exposed to human remains for 2 min 3–10 days previously (Oesterhelweg et al. 2008).

In order to use mass spectroscopy for detection, it has been necessary to document which chemicals are known to result from decomposition (Vass 2008). As research has developed in this area, an appreciation has also developed for the variation in decomposition chemicals depending on the environment and on the postmortem interval, including methods to discriminate human and nonhuman patterns (Hoffman et al. 2009). Experiments in the lab and in the field have included donated human bodies as well as pigs.

Some research has focused on developing better ground-penetrating radar capacity to locate burials (Schultz 2008). In the latter Florida study of 12 pig graves, the deeper the grave, the longer the duration of the postmortem interval in which the ground-penetrating radar could detect it.

Other studies have looked at patterns of deposition (by the perpetrators) and scatter (by scavengers). Manhein and colleagues' review of a series of Louisiana cases using geographic information systems (GIS) showed that bodies were frequently dumped along rural roads, about 1.5–6 km from the road (Manhein et al. 2006). In Kjorlien's study of clothed and unclothed pig cadaver scatter in wooded and grassland areas near Edmonton, Alberta in Canada (Kjorlien et al. 2009), some patterns were found. Scatter tended to go along game trails and away from human presence, but there was much variation. Scavengers were more likely to select pigs that were unclothed and/ or were located in open, rather than wooded areas. Particular scavenger species were not identified.

Estimating the postmortem interval

Plants, and fungi and algae directly associated with human remains can function as "biological clocks," indicating by their growth patterns and life cycles the elapsed time since the body was accessible. Courtin and Fairgrieve report a case in which the body was positioned over a tree branch that could be assessed in terms of tree rings to indicate the number of years since the body was positioned that way (Courtin and Fairgrieve 2004). Hitosugi et al. (2006) suggest that the colonization time for *Penicillium* and *Aspergillus* (3–7 days after attaching to the skin) can be used as a time marker in forensic cases. Similarly Haefner et al. (2004) suggests that algae growth might be quantified in order to estimate postmortem submersion interval in some cases.

Preliminary studies with rat cadavers suggest soil chemistry can also be used to estimate postmortem interval, particularly for intervals beyond 30 days (Benninger et al. 2008). The levels of different substances peak at different postmortem intervals when compared under constant laboratory conditions. When refined and replicated, these techniques may be very helpful in extending postmortem interval precision beyond the usual range for insect-based estimates. Nevertheless, these researchers report that soil type and moisture content can influence decomposition rates, with slowed rates at highest and lowest moisture levels (Carter et al. 2010).

Several studies assess systematic changes in chemical status of bone during the postmortem period, including the effects of weathering (Janjua and Rogers 2008), and the loss of calcium citrate (Schwarcz et al. 2010). One study (Wieberg and Wescott 2008) attempts to pin down moisture-loss characteristics associated with the perimortem period. Bone increasingly loses its moisture in the postmortem period, which ultimately causes it to appear different, particularly in terms of fracture angles and surface morphology when it breaks. Anthropologists use this characteristic to indicate whether a bone break occurred either in the perimortem or postmortem period. Thus, when applied to traumatic damage, the term "perimortem" refers to bone freshness, and only indirectly indicates a defined time period. Wieberg and Wescott's (2008) experimental study of pig bones assessed moisture loss and fracture appearance over a 141-day period. They found that bone did indeed look and fracture significantly differently as it lost moisture, but also found that the bones retained some "fresh" characteristics, even after that much time. This study points to the need to look at other chemical changes besides moisture, and possibly to extend the time period under scrutiny.

Two studies have attempted to correlate the condition of the remains in terms of soft tissue lost to decomposition to ADDs (Megyesi et al. 2005; Vass 2010). In 1992, Vass and colleagues monitored seven bodies at a decomposition study facility in eastern Tennessee over four seasons as they decomposed and became skeletal, testing for volatile fatty acids (VFAs) released into the soil. He converted the postmortem interval to ADDs based on the maximum and minimum daily temperature at the site, tracking the association of ADDs with decomposition changes. The bodies became skeletonized by 1285 ± 110 ADDs. This ADD total needed for skeletonization has been used as a baseline for more recent studies. In another study 68 forensic cases with known postmortem intervals were assessed and given a "Total Body Score" (TBS), which was associated with their postmortem interval and calibrated against the Vass et al. (1992) total of 1285 ADDs, producing a linear regression. The TBS is performed for each region of the body, assessing its decomposition phase, and assigns an ordinal score (higher score for later decomposition phase). The resulting regression formula produces an estimate of ADDs that can be converted to postmortem interval using local temperature data. It is important to note that these bodies were not significantly scavenged, but lost their soft tissue through decomposition.

The study of Vass (2010) took a somewhat similar approach, scoring the body with a single estimate of percentage decomposed, calibrated to scales suggested for warmer and cooler temperature regimes. Bodies at the eastern Tennessee research facility were used to develop and test the formulas, providing estimates for both buried and surface remains. In order to apply these formulas, it is necessary to know the humidity reading. The formula provides a value for ADDs, which must be converted to postmortem interval using local site temperatures (or weather station data, adjusted for local site microenvironmental temperature differences from the weather station) for the suspected postmortem interval. These formulas are for bodies in which flesh is removed by insects and decomposition processes, but not mammalian and avian scavenging.

Trauma detection and identification

Soft-tissue damage can be mistaken for trauma, for example confusing surgical wounds with perimortem sharp-force trauma (Byard et al. 2006). Two recent studies report cases in which ants produced postmortem linear abrasions in skin, as well as holes and substantial hemorrhage (Campobasso et al. 2009; Ventura et al. 2010).

Care must also be taken to avoid confusing postmortem bone damage and scavenging for trauma (MacAulay et al. 2009; Moraitis and Spiliopoulou 2010). The effects of postmortem freezing, erosion, and precipitation can potentially disguise blunt-force trauma (Calce and Rogers 2007). Fractures due to blunt-force trauma (produced by gunshot or homicidal blows) can usually be differentiated from scavenger modification by the presence of additional "signature" marks made by dentition or claws. These include pits and punctures, U-shaped grooves, scalloping, and scat associated with the remains.

There is some controversy regarding the potential for trauma presence to speed the pace of decomposition, as shown in previous studies (Mann et al. 1990). Recent experimental research on penetrative trauma in pig cadavers suggests, however, that traumatic wounds may not affect insect infestation locations or decomposition speed (Cross and Simmons 2010).

CONCLUSIONS

During the last decade research on forensic taphonomy topics has flourished, particularly in understanding the chemistry of decomposition and the impact of regional variation and local site microenvironments, including both terrestrial and aquatic contexts. Recent research is supporting a move away from taphonomic universals and toward developing techniques to control for contextual variability in temperature, humidity, exposure, and scavenging, among other factors. The field will continue to benefit from a combination of experimental and actualistic research designs and deliberate interdisciplinary collaboration.

NOTES

1   Throughout this chapter we frequently use the term "remains" rather than other terms (e.g., body, corpse, carcass). The taphonomic process often includes disarticulation and scattering, whereby bone or dental elements (or fragments) are separated from one another. "Remains" refers to dead bodies or carcasses or any portion thereof.

2   Members of the canid family include coyotes, wolves, domestic dogs, and foxes. Members of the ursid family include all bear species. Members of the felid family include bobcat, lynx, and domestic cat. The most common procyonid is the raccoon. Members of the catharthid family include vultures and condors. The mustelid family includes weasels, mink, fisher, and marten. Members of the murid family include rats and mice. The crustacean scavengers include amphipods, isopods, crabs, shrimp, and lobsters.

3   Many other groups contain species that scavenge full or part time; this discussion has mentioned the ones more typically associated with forensic cases.

4   Some scavengers, such as vultures, are better able to withstand these toxins due to a highly acidic digestive system.

5   Accumulated degree-days (ADDs) are generally calculated by averaging the maximum and minimum temperatures for each day and summing those averages for all the days in the time range. This converts accumulated heat for the site into one number.

REFERENCES

Anderson, G.S. and Cervenka, V.J. (2002). Insects associated with the body: their use and analysis. In W.D. Haglund and M.H. Sorg (eds), *Advances in Forensic Taphonomy: Method, Theory, and Archaeological Perspectives* (pp. 173–200). CRC Press, Boca Raton, FL.

Anderson, G.S. and Hobischak, N.R. (2004). Decomposition of carrion in the marine environment in British Columbia, Canada. *International Journal of Legal Medicine* 118(4): 206–209.

Archer, M.S. (2004). Rainfall and temperature effects on the decomposition rate of exposed neonatal remains. *Science and Justice* 44(1): 35–41.

Bachmann, J., and Simmons, T. (2010). The influence of preburial insect access on the decomposition rate. *Journal of Forensic Sciences* 55(4): 893–900.

Benninger, L.A., Carter, D.O., and Forbes, S.L. (2008). The biochemical alteration of soil beneath a decomposing carcass. *Forensic Science International* 180(2–3): 70–75.

Blanco Pampín, J. and Lupez-Abajo Rodrìguez, B.A. (2001). Surprising drifting of bodies along the coast of Portugal and Spain. *Legal Medicine* 3(3): 177–182.

Boaz, N. and Behrensmeyer, A.K. (1976). Hominid taphonomy: transport of human skeletal parts in an artificial fluviatile environment. *American Journal of Physical Anthropology* 45(1): 36–50.

Bull, I.D., Berstan, R., Vass, A., and Evershed, R. (2009). Identification of disinterred grave by molecular and stable isotope analysis. *Science and Justice* 49(8): 142–149.

Byard, R.W., Gehl, A., Anders, S., and Tsokos, M. (2006). Putrefaction and wound dehiscence: a potentially confusing postmortem phenomenon. *American Journal of Forensic Medicine and Pathology* 27(1): 61–63.

Byrd, J.E. and Adams, B.J. (2003). Osteometric sorting of commingled human remains. *Journal of Forensic Sciences* 48(4): 717–724.

Calce, S.E. and Rogers, T.L. (2007). Taphonomic changes to blunt force trauma: a preliminary study. *Journal of Forensic Sciences* 52(3): 519–527.

Campobasso, C.P., Marchettie, D., Introna, F., and Colonna, M.F. (2009). Postmortem Artifacts Made by Ants and the Effect of Ant Activity on Decompositional Rates. *American Journal of Forensic Medicine and Pathology* 30(1): 84–87.

Carson, E.A., V.H. Stefan, J.F. Powell. (2000). Skeletal Manifestations of Bear Scavenging. *Journal of Forensic Sciences* 45(3): 515–526.

Carter, D.O., Yellowlees, D., and Tibbett, M. (2007). Cadaver decomposition in terrestrial ecosystems. *Naturwissenschaften* 94(1): 12–24.

Carter, D.O., Yellowlees, D., and Tibbett, M. (2008). Using ninhydrin to detect gravesoil. *Journal of Forensic Sciences* 53(2): 397–400.

Carter, D.O., Yellowlees, D., and Tibbett, M. (2010). Moisture can be the dominant environmental parameter governing cadaver decomposition in soil. *Forensic Science International* 200: 60–66.

Christiansen, P. and Wroe, S. (2007). Bite forces and evolutionary adaptations to feeding ecology in carnivores. *Ecology* 88(2): 347–358.

Courtin, G.M. and Fairgrieve, S.L. (2004). Estimation of postmortem interval (PMI) as revealed through the analysis of annual growth in woody tissue. *Journal of Forensic Sciences* 49(4): 781–783.

Cross, P. and Simmons, T. (2010). The influence of penetrative trauma on the rate of decomposition. *Journal of Forensic Sciences* 55(2): 295–301.

Dabbs, G.R. (2010). Caution! All data are not created equal: the hazards of using National Weather Service data for calculating accumulated degree days. *Forensic Science International* 202(1–3): 49–52.

Davis, J.B. and Goff, M.L. (2000). Decomposition patterns in terrestrial and intertidal habitats on Oahu Island and Coconut Island, Hawaii. *Journal of Forensic Sciences* 45(4): 836–842.

Dekeirsschieter, J. (2009). Cadaveric volatile organic compounds released by decaying pig carcasses in different biotopes. *Forensic Science International* (189): 46–53.

Dirkmaat, D.C., Cabo, L.L., Ousley, S.D., and Symes, S.A. (2008). New perspectives in forensic anthropology. *Yearbook of Physical Anthropology* 51: 33–52.

Dumser, T.K. and Türkay, M. (2008). Postmortem changes of human bodies on the bathyal sea floor—two cases of aircraft accidents above the open sea. *Journal of Forensic Sciences* 53(5): 1049–1052.

Duraes, N. (2010). Comparison of adipocere formation in four soil types of the Porto (Portugal) District. *Forensic Science International* 195: 168–173.

Ebbesmeyer, C.C. and Haglund, W.D. (2002). Floating remains on Pacific Northwest waters. In W.D. Haglund and M.H. Sorg (eds), *Advances in Forensic Taphonomy: Method, Theory, and Archaeological Perspectives* (pp. 219–240). CRC Press, Boca Raton, FL.

Efremov, I. (1940). Taphonomy: a new branch of paleontology. *Pan-American Geologist* 74: 81–93.

Fiedler, S., Schneckenberger, K., and Graw, M. (2004). Characterization of soils containing adipocere. *Archives of Environmental Contamination and Toxicology* 47(4): 561–568.

Forbes, S.L., Stuart, B.H., Dadour, I.R., and Dent, B.B. (2004). A preliminary investigation of the stages of adipocere formation. *Journal of Forensic Sciences* 49(3): 566–574.

Forbes, S.L., B.H. Stuart, B.B. Dent. (2005). The effect of burial environment on adipocere formation. *Forensic Science International* 154: 24–35.

Frund, H.C. and Schoenen, D. (2009). Quantification of adipocere degradation with and without access to oxygen and to the living soil. *Forensic Science International* 188: 18–23.

Goff, M.L. (2009). Early postmortem changes and stages of decomposition in exposed cadavers. *Experimental and Applied Acarology* 49: 21–37.

Haefner, J.N., Wallace, J.R., and Merritt, R.W. (2004). Pig decomposition in lotic aquatic systems: the potential use of algal growth in establishing a postmortem submersion interval (PMSI). *Journal of Forensic Sciences* 49(2): 330–336.

Haglund, W.D. (1991). *Application of Taphonomic Models to Forensic Investigations*. Dissertation, University of Washington, Seattle, WA.

Haglund, W.D. (1997a). Dogs and coyotes: postmortem involvement with human remains. In W.D. Haglund and M.H. Sorg (eds), *Forensic Taphonomy: The Postmortem Fate of Human Remains* (pp. 367–381). CRC Press, Boca Raton, FL.

Haglund, W.D. (1997b). Scattered skeletal human remains: Search strategy considerations for locating missing teeth. In W.D. Haglund and M.H. Sorg (eds), *Forensic Taphonomy: The Postmortem Fate of Human Remains* (pp. 383–394). CRC Press, Boca Raton, FL.

Haglund, W.D. and Sorg, M.H. (eds) (1997). *Forensic Taphonomy: The Postmortem Fate of Human Remains*. CRC Press, Boca Raton, FL.

Haglund, W.D. and Sorg, M.H. (eds) (2002a). *Advances in Forensic Taphonomy: Methods, Theory, and Archaeological Perspectives*. CRC Press, Boca Raton, FL.

Haglund, W.D. and Sorg, M.H. (2002b). Human remains in water environments. In W.D. Haglund and M.H. Sorg (eds), *Advances in Forensic Taphonomy: Method, Theory, and Archaeological Perspectives* (pp. 201–218). CRC Press, Boca Raton, FL.

Haglund, W.D., Reay, D.T., and Swindler, D.R. (1988). Tooth mark artifacts and survival of bones in animal scavenged human skeletons. *Journal of Forensic Sciences* 33(4): 985–997.

Haglund, W.D., Reay, D.T., and Swindler, D.R. (1989). Canid scavenging/disarticulation sequence of humans in the pacific northwest. *Journal of Forensic Sciences* 34(3): 587–606.

Haskell, N.H., Hall, D.W., Cervenka, V.J., and Clark, M.A. (1997). On the body: insects' life stage presence, their postmortem artifacts. In W.D. Haglund and M.H. Sorg (eds), *Advances in Forensic Taphonomy: Method, Theory, and Archaeological Perspectives* (pp. 415–448). CRC Press, Boca Raton, FL.

Haslam, T.C.F. and Tibbett, M. (2009). Soils of contrasting pH affect the decomposition of buried mammalian skeletal muscle tissue. *Journal of Forensic Sciences* 54(4): 900–904.

Hitosugi, M., Ishii, K., Yaguchi, T., Chigusa, Y., Kurosu, A., Kido, M., Nagai, T., and Tokudome, S. (2006). Fungi can be a useful forensic tool. *Legal Medicine (Tokyo)* 8(4): 240–242.

Hobischak, N.R., VanLaerhoven, S.L., and Anderson, G.S. (2006). Successional patterns of diversity in insect fauna on carrion in sun and shade in the Boreal Forest Region of Canada, near Edmonton, Alberta. *Canadian Entomologist* 138(3): 376–383.

Hoffman, E.M., Curran, A.M., Dulgerian, N., Stockham, R.A., and Eckenrode, B.A. (2009). Characterization of the volatile organic compounds present in the headspace of decomposing human remains. *Forensic Science International* 186(1–3): 6–13.

Hopkins, D.W., Wiltshire P.E.J., Turner, B.B. (2000). Microbial characteristics of soils from graves: an investigation at the interface of soil mircobiology and forensic science. *Applied Soil Ecology* 14: 283–288.

Huculak, M.A. and Rogers, T.L. (2009). Reconstructing the sequence of events surrounding body disposition based on color staining of bone. *Journal of Forensic Sciences* 54(5): 979–984.

Hughes, C.E. and White, C.A. (2009). Crack propagation in teeth: a comparison of perimortem and postmortem behavior of dental materials and cracks. *Journal of Forensic Sciences* 54(2): 263–266.

International Commission for Missing Persons (ICMP) (2011). *US Ambassador to Bosnia and Herzegovina Patrick Moon visits ICMP*. Press release, August 2011. www.ic-mp.org/press-releases.

Iwamura, E.S., Oliveira, C.R., Soares-Vieira, J.A., Nascimento, S.A., and Munoz, D.R. (2005). A qualitative study of compact bone microstructure and nuclear short tandem repeat obtained from femur of human remains found on the ground and exhumed 3 years after death. *American Journal of Forensic Medicine and Pathology* 26(1): 33–44.

Jaggers, K.A. and Rogers, T. (2009). The effects of soil environment on postmortem interval: a macroscopic analysis. *Journal of Forensic Sciences* 54(6): 1217–1224.

Janjua, M.A. and Rogers, T.L. (2008). Bone weathering patters of metatarsal v. femur and the postmortem interval in Southern Ontario. *Forensic Science International* 178: 16–23.

Joy, J.E., Liette, N.L., and Harrah, H.L. (2006). Carrion fly (Diptera: Calliphoridae) larval colonization of sunlit and shaded pig carcasses in West Virginia, USA. *Forensic Science International* 164(2–3): 183–192.

Kjorlien, Y.P., Beattie, O.B., and Peterson, A.E. (2009). Scavenging activity can produce predictable patterns in surface skeletal remains scattering: observations and commments from two experiments. *Forensic Science International* 188: 103–107.

Lasseter, K.J., Farley, R., and Hensel, L. (2003). Cadaver dog and handler team capabilities in the recovery of buried human remains in the Southeastern United States. *Journal of Forensic Sciences* 48(3): 1–5.

Lovestead, T.M. and Bruno, T.J. (2010). Detecting gravesoil with headspace analysis with adsorption on short porous layer open tubular (PLOT) columns. *Forensic Science International* 204(1–3): 156–161.

Lyman, R.L. (1989). Taphonomy of cervids killed by the 18 May 1980 volcanic eruption of Mount St. Helens, Washington, U.S.A. In R. Bonnichsen and M.H. Sorg (eds), *Bone Modification* (pp. 149–167). Center for the Study of Early Man, University of Maine, Orono, ME.

Lyman, R.L. (1994). *Vertebrate Taphonomy*. Cambridge University Press, Cambridge.

Lyman, R.L. (2010). What taphonomy is, what it isn't, and why taphonomists should care about the difference. *Journal of Taphonomy* 8: 1–16.

MacAulay, L.E., Barr, D.G., and Strongman, D.B. (2009). Effects of decomposition on gunshot wound characteristics: under cold temperatures with no insect activity. *Journal of Forensic Sciences* 54(2): 448–451.

Manhein, M.H., Listi, G.A., and Leitner, M. (2006). The application of geographic information systems and spatial analysis to assess dumped and subsequently scattered human remains. *Journal of Forensic Sciences* 51(3): 469–474.

Mann, R.W., Bass, W.M., and Meadows, L. (1990). Time since death and decomposition of the human body variables and observations in case and experimental field studies. *Journal of Forensic Sciences* 35(1): 103–111.

Mant, A.K. (1950). *A Study of Exhumation Data*. University of London, London.

Mant, A.K. (1987). Knowledge acquired from post-war exhumations. In A. Boddington, AN. Garland, and R.C. Janaway (eds), *Death Decay and Reconstruction: Approaches to Archeology and Forensic Science* (pp. 10–21). Manchester University Press, London.

Megyesi, M.S., Nawrocki, S.P., and Haskell, N.H. (2005). Using accumulated degree-days to estimate the postmortem interval from decomposed human remains. *Journal of Forensic Sciences* 50(3): 618–626.

Merbs, C. (1997). Eskimo skeleton taphonomy with identification of possible polar bear victims. In W.D. Haglund and M.H. Sorg (eds), *Forensic Taphonomy: The Postmortem Fate of Human Remains* (pp. 249–262). CRC Press, Boca Raton, FL.

Moraitis, K. and Spiliopoulou, C. (2010). Forensic implications of carnivore scavenging on human remains recovered from outdoor locations in Greece. *Journal of Forensic and Legal Medicine* 17(6): 298–303.

Murmann, D.C., Brumit, P.C., Schrader, B.A., and Senn, D.R. (2006). A comparison of animal jaws and bite mark patterns. *Journal of Forensic Sciences* 51(4): 846–860.

Notter, S.J., Stuart, B.H., Rowe, R., and Langlois, N. (2009). The initial changes of fat deposits during the decomposition of human and pig remains. *Journal of Forensic Sciences* 54(1): 195–201.

O'Brien, T.G. and Kuehner, A.C. (2007). Waxing grave about adipocere: soft tissue change in an aquatic context. *Journal of Forensic Sciences* 52(2): 294–301.

O'Brien, R.C., Forbes, S.L., Meyer, J., and Dadour, I. (2010). Forensically significant scavenging guilds in the southwest of Western Australia. *Forensic Science International* 198(1–3): 85–91.

Oesterhelweg, L., Krober, S., Rottmann, K., Willhoft, J., Braun, C., Thies, N., Puschel, K., Silkenath, J., and Gehl, A. (2008). Cadaver dogs - a study on dectection of contaminated carpet squares. *Forensic Science International* 174: 35–39.

Rebmann, A.J., David, E., and Sorg, M.H. (2002). *Cadaver Dog Handbook: Training and Tactics for the Forensic Recovery of Human Remains*. CRC Press, Boca Raton, FL.

Rios, L., Ovejero, J.I., and Prieto, J.P. (2010). Identification process in mass graves from the Spanish Civil War, I. *Forensic Science International* 199(1–3): e27–e36.

Schotsmans, E.M.J., Van de Voorde, W., De Winne, J., and Wilson, A.S. (2011). The impact of shallow burial on differential decomposition to the body: a temperate case study. *Forensic Science International* 206(1): e43–e48.

Schroeder, H., Klotzbach, H., and Puschel, K. (2003). Insects' colonization of human corpses in warm and cold season. *Legal Medicine (Tokyo)* 5: S372–S374.

Schultz, J.J. (2008). Sequential monitoring of burials containing small pig cadavers using ground penetrating radar. *Journal of Forensic Sciences* 53(2): 279–287.

Schwarcz, H.P., Agur, K., and Jantz, L.M. (2010). A new method for determination of postmortem interval: citrate content of bone. *Journal of Forensic Sciences* 55(6): 1516–1522.

Sharanowski, B.J., Walker, E.G., and Anderson, G.S. (2008). Insect succession and decomposition patterns on shaded and sunlit carrion in Saskatchewan in three different seasons. *Forensic Science International* 179(2–3): 219–240.

Simmons, T., Adlam, R.E., and Moffatt, C. (2010a). Debugging decomposition data—comparative taphonomic studies and the influence of insects and carcass size on decomposition rate. *Journal of Forensic Sciences* 55(1): 8–13.

Simmons, T., Cross, P.A., Adlam, R.E., and Moffatt, C. (2010b). The influence of insects on decomposition rate in buried and surface remains. *Journal of Forensic Sciences* 55(4): 889–892.

Sorg, M.H. (1986). Scavenger modification of human remains. *Proceedings of the Annual Meeting of the American Academy of Forensic Sciences*.

Sorg, M.H. and Haglund, W.D. (2002). Advancing forensic taphonomy: purpose, theory, and practice. In W.D. Haglund and M.H. Sorg (eds), *Advances in Forensic Taphonony: Method, Theory and Archaeological Perspectives* (pp. 3–29). CRC Press, Boca Raton, FL.

Sorg, M.H. and Wren, J.A. (2011). Taphonomic impacts of scavenging in northern New England. *National Institute of Justice 2011 Conference*, Arlington, VA.

Sorg, M.H., Dearborn, J.H., Monahan, E.I., Ryan, H.F., Sweeney, K.G., and David, E. (1997). Forensic taphonomy in marine contexts. In W.D. Haglund and M.H. Sorg (eds), *Forensic Taphonomy: The Postmortem Fate of Human Remains* (pp. 587–604). CRC Press, Boca Raton, FL.

Sorg, M.H., David, E., and Rebmann, A.J. (1998). Cadaver dogs, taphonomy, and postmortem interval in the Northeast. In K.J. Reichs (ed.), *Forensic Osteology*, 2nd edn (pp. 120–144). Charles C. Thomas, Springfield, IL.

Steadman, D.W. and Worne, H. (2007). Canine scavenging of human remains in an indoor setting. *Forensic Science International* 173(1): 78–82.

Stokes, K.L., Forbes, S.S., and Tibbett, M. (2009). Freezing skeletal muscle tissue does not affect its decomposition in soil: evidence from temporal changes in tissue mass, mircobial activity and soil chemistry based on excised samples. *Forensic Science International* 183: 6–13.

Tibbett, M. and Carter, D.O. (eds) (2008). *Soil Analysis in Forensic Taphonomy: Chemical and Biological Effects of Buried Human Remains.* CRC Press, Boca Raton, FL.

Ubelaker, D.H. (2002). Approaches to the study of commingling in human skeletal biology. In W.D. Haglund and M.H. Sorg (eds), *Advances in Forensic Taphonomy: Method, Theory, and Archaeological Perspectives* (pp. 331–351). CRC Press, Boca Raton, FL.

Ubelaker, D.H. and Zarenko, K.M. (2011). Adipocere: what is known after two centuries of research. *Forensic Science International* 208: 167–172.

Van Belle, L.E., Carter, D.O., and Forbes, S.L. (2009). Measurement of ninhydrin reactive nirtogen influx into gravesoil during above-ground and belowground carcass decomposition. *Forensic Science International* 193: 37–41.

Vass, A.A. (2008). Odor analysis of decomposing human remains. *Journal of Forensic Sciences* 53(2): 384–391.

Vass, A.A. (2010). The elusive universal post-mortem interval formula. *Forensic Science International* 204(1–3): 34–40.

Vass, A.A., Bass, W.M., Wolt, J.D., Foss, J.E., and Ammons, J.T. (1992). Time since death determinations of human cadavers using soil solution. *Journal of Forensic Sciences* 37(5): 1236–1253.

Vass, A.A., Barshick, S.A., Sega, G., Caton, J., Skeen, J.T., Love, J.C., and Synstelien, J.A. (2002). Decomposition chemistry of human remains: a new methodology for determining the postmortem interval. *Journal of Forensic Sciences* 47(3): 542–553.

Vass, A.A., Smith, R.R., Thompson, C.V., Burnett, M.N., Wolf, D.A., Synstelien, J.A., Dulgerian, N., and Eckenrode, B.A. (2004). Decompositional odor analysis database. *Journal of Forensic Sciences* 49(4): 760–769.

Ventura, F., Gallo, M., and Stefano, F.D. (2010). Postmortem skin damage due to ants. *American Journal of Forensic Medicine and Pathology* 31(2): 120–121.

Wang, J., Li Z, Chen, Y., Chen, Q., and Yin, X. (2008). The succession and development of insects on pig carcasses and their significances in estimating PMI in south China. *Forensic Science International* 179(1): 11–18.

Weitzel, M.A. (2005). A report of decomposition rates of a special burial type in Edmonton, Alberta from an experimental field study. *Journal of Forensic Sciences* 50(3): 641–647.

Wieberg, D.A. and Wescott, D.J. (2008). Estimating the timing of long bone fractures: correlation between the postmortem interval, bone moisture content, and blunt force trauma fracture characteristics. *Journal of Forensic Sciences* 53(5): 1028–1034.

Wilson, A.S., Janaway, R.C., Holland, A.D., Dobson, H.I., Baran, E., Pollard, A.M., and Tobin, D.J. (2007). Modelling the buried human body environment in upland climes using three contrasting field sites. *Forensic Science International* 169: 6–18.

Yan, F., McNally, R., Kontanis, E.J., and Sadik, O.A. (2001). Preliminary quantitative investigation of postmortem adipocere formation. *Journal of Forensic Sciences* 46(3): 609–615.

Zupanic, P.I., Gornjak, P.B., and Balazic, J. (2010). Molecular genetic identification of skeletal remains from the Second World War Konfin I mass grave in Slovenia. *International Journal of Legal Medicine* 124(4): 307–317.

PART VI Forensic Taphonomy

Introduction to Part VI

Dennis C. Dirkmaat and Nicholas V. Passalacqua

Forensic taphonomy was identified by Dirkmaat et al. (2008) as a key development in the field of forensic anthropology in the last 15 years. In this section of the book, the history of the field and its significant impact on forensic anthropology is explored further. Here some of the basic concepts of the field are outlined. Beary and Lyman in Chapter 25 provide an excellent, in-depth discussion of the development and unique role of taphonomy in forensic contexts, as well as general principles of forensic taphonomy and recent developments in forensic taphonomic research and methods.

Paleontologists were the first to analyze skeletal assemblages carefully and attempt to address issues of missing bone elements, congregations of skeletal deposits, surface damaged bones, and other issues. These original efforts later led to attempts to go beyond the bones and reconstruct populations as a whole, individual behaviors, habitat occupation, and ecological distribution, among other issues. Early pioneers in the field embraced the title *taphonomy* (from the Greek *taphos*, meaning burial), as originally presented by Ivan Efremov, a Soviet paleontologist, and also a science fiction author, in the early 1940s (Efremov 1940; Chudinov and Sokolov 1987). Defined as the study of what becomes of an organism after burial, it focused on the conversion of bone to fossil, dealt with preservation issues, and represented a great leap forward in geological and paleontological research. These general concepts spilled over into human paleontology (paleoanthropology) in the 1970s with the work of Brain, Behrensmeyer, and others, dealing primarily with early African hominin sites. Paleontology mostly was used to argue for or against the existence of an osteodontokeratic culture of hominin tool assemblages (Shipman and Phillips-Conroy

A Companion to Forensic Anthropology, First Edition. Edited by Dennis C. Dirkmaat.

1977) and for examining bone weathering to "give specific information concerning surface exposure of bone prior to burial and the time periods over which bones accumulated" (Behrensmeyer 1978: 161).

At the onset, forensic anthropology was a discipline almost solely concerned with providing accurate reconstructions of biological profiles of individuals represented by skeletal remains found in forensic settings. The purpose was to provide identifications of the deceased. If requested to address issues of bone surface modification (exfoliation, gnawing, burning, staining, root-etching, and sun-bleaching), broken and fractured bones, missing elements, scattered elements, and a wide range of other topics, the forensic anthropologist generally looked to research conducted in other disciplines including fossil hominid taphonomy, paleopathology, and bioarchaeology. The key component linking the subdisciplines was that context was carefully noted in all cases through archaeological recovery methods. Given that limited or no notation of the original context (aside from a few pictures provided by law enforcement) was usually associated with the bones in forensic cases, determinations of time since death and specific reasons for scattered, missing, and damaged bones was left to "educated guesses," based on "years-of-experience" justifications.

In the 1980s a small core of forensic anthropologists, most of whom were trained in archaeological sciences, recognized that this gap in producing a scientific product was similar to the state of the discipline of paleontology in the 1940s, and paleoanthropology in the 1970s; that is, trying to address questions regarding the whys, whens, and hows of the patterns of skeletal remnants of individuals without contextual grounding. The forensic anthropologists saw very analogous situations and attempted to address similar questions by turning to a taphonomic perspective and roughly constructed a subdiscipline that was simply termed *forensic taphonomy*.

The succinct definition of forensic taphonomy provided by Haglund and Sorg (1997) serves well: "the study of postmortem processes which affect (1) the preservation, observation, or recovery of dead organisms, (2) the reconstruction of their biology or ecology, or (3) the reconstruction of the circumstances of their death" (1997: 13). Forensic taphonomy can then be understood as a backward process attempting to use other previous actualistic research and case studies to inform us about the causes of modification noted on the remains. However, forensic taphonomy embraces the central principles of taphonomy with minor modifications, that account for important observations in a forensic setting. Paleontological taphonomy focuses on what happens to the remains after burial and most of the discussion is related to how bones fare in the process of being buried, and, ultimately, conversion to mineral matrices. Forensic taphonomy focuses on what happens to the body immediately after death until recovery. This includes the study of the patterns and rates of soft-tissue decomposition including factors played by insect activity, bacteria, etc., the effects of animals on the remains in terms of scattering, damaging, and removing elements, the effects of plants on the remains, and even the effects of soil conditions, plant cover, shade factors, and temperature regimes. Forensic cases involving the recently deceased provide the best laboratory to study these changes and not paleontology.

In addition, it should be noted that recovery refers not only to the excavation of the remains but also to postrecovery transport, autopsy examination, and any other forensic examination up until the remains arrive in the forensic anthropology laboratory. Thus any modifications of the remains during the recovery and transport to the laboratory should be considered as taphonomic events, as should any modifications occurring during laboratory analyses, such as processing, reconstructing, and labeling of the remains. Forensic taphonomy thus incorporates *all* modifications to the remains after death.

Today we can recognize two primary foci of taphonomic reconstructions as they relate to human decedents associated with suspicious deaths: (i) providing a scientifically based determination of time since death and (ii) determining whether there is postdeath human altering of the remains at the scene.

Time-since-death estimates require a very strong multidisciplinary effort. Forensic entomology and the study of the insect-fauna associated with the decomposing remains often provide the best determination of a postmortem interval, in terms of precision and accuracy, when dealing with advanced stages of decomposition. Forensic botany comes into play when dealing with tree shade issues, understory conditions, plant recovery, root growth, decay rates of plants found buried in association with the remains, and soil composition and pH levels, as they all influence bone quality via diagenesis and temperature regimes. The concept of degree-days (more thoroughly discussed in Megyesi et al. 2005) is useful in providing more accurate estimates of a postmortem interval as they impact the life cycles of insects attracted to the mini ecosystem of a decomposing body as well as general soft-tissue decomposition. Geographic information systems (GIS) are also beginning to be used in forensic recovery in a multitude of ways, not just for providing maps of the scene, but also for generating models for body deposition (Manhein et al. 2006). The potential utility of GIS in forensic applications is just beginning, with the ability to incorporate detailed information on scene setting into the taphonomic analysis. GIS applications can provide data on tree cover, roads, water sources, topographic information, and a wide range of additional useful information, thus increasing the ability to understand taphonomic factors which may be affecting the remains. Finally, it is useful to include information on the fauna in the area, especially relative to carnivores, scavengers, rodents, and any taphonomic agents whose activities might have an impact on the body.

Issues related to postdepositional movement and the condition of the remains are included in taphonomic analyses in an effort to determine (i) why skeletal remains are not in their original anatomical position, and who or what is responsible, and (ii) whether modifications of the remains are due to trauma or taphonomic damage, and when this modification to the remains occurred (i.e., during the perimortem or postmortem interval). Here the ultimate goal is, through a process of elimination, to determine the role of humans or other agents in moving/removing portions of the remains and otherwise modifying the condition of the remains at the time of original deposition. In a sense, forensic taphonomists are asking the question: What occurred between the original deposition and recovery to create the current condition and arrangement of the remains?

After decomposition begins, remains may be displaced from their original anatomical position through a number of disruptive factors including movement by creatures (large scale) or movement by other natural agents (small scale) such as soil bioturbation, soil cryoturbation, plant root envelopment, worm activity, and diagenesis. Nonhuman modification of the remains results from carnivore, rodent, vulture and other scavenger interaction that may leave diagnostic marks on the bones (Haglund 1997a, 1997b). When confronted with isolated skeletal elements or partial sets of remains, determining possible causes for differential recovery of remains begins with macroscopic and microscopic (though only via limited magnification) examination of bone surfaces. Carnivore chewing, rodent gnawing, or evidence of dismemberment through saw or knife cuts, provide the most diagnostic features.

Scattered human remains are also common and interpretations are best approached following a proper archaeological recovery through the examination of landscape topography and environmental conditions (e.g., fluvial activity). The best way to document the context of remains at a scene during recovery is through archaeological methodology, description, and analysis; generating a *detailed map* with supplementary photographs and notes works well (Chapter 2, this volume). Contextual evidence is lost if archaeological recovery is not followed, which results in incomplete collection of the information necessary for an accurate taphonomic reconstruction and skeletal analysis.

The final goal of forensic taphonomic analysis is to identify all taphonomic agents potentially impinging upon the human remains in order to focus on those related to human intervention, which establishes forensic significance. Only after postmortem factors are explained can the perimortem events be reconstructed.

REFERENCES

Behrensmeyer, A.K. (1978). Taphonomic and ecologic information from bone weathering. *Paleobiology* 4: 150–162.

Chudinov, P.K. and Sokolov, B.S. (1987). *Ivan Antonovich Efremov, 1907-1972.* Nauka: Moscow.

Dirkmaat, D.C., Cabo, L.L., Ousley, S.D., and Symes, S.A. (2008). New perspectives in forensic anthropology. *Yearbook of Physical Anthropology* 51: 33–52.

Efremov, I.A. (1940). Taphonomy: a new branch of paleontology. *Pan-American Geologist* 74: 81–93.

Haglund, W.D. (1997a). Dog and coyotes: postmortem involvement with human remains. In W.D. Haglund and M.H. Sorg (eds), *Forensic Taphonomy: The Postmortem Fate of Human Remains* (pp. 367–382). CRC Press, Boca Raton, FL.

Haglund, W.D. (1997b). Rodents and human remains. In W.D. Haglund and M.H. Sorg (eds), *Forensic Taphonomy: The Postmortem Fate of Human Remains* (pp. 383–394). CRC Press, Boca Raton, FL.

Haglund, W.D. and Sorg, M.H. (1997). Method and theory of forensic taphonomic research. In W.D. Haglund and M.H. Sorg (eds), *Forensic Taphonomy: The Postmortem Fate of Human Remains* (pp. 13–26). CRC Press, Boca Raton, FL.

Manhein, M.H., Listi, G.A., and Leitner, M. (2006). The application of geographic information systems and spatial analysis to assess dumped and subsequently scattered human remains. *Journal of Forensic Sciences* 51(3): 469–474.

Megyesi, M.S., Nawrocki, S.P., and Haskell, N.H. (2005). Using accumulated degree-days to estimate the postmortem interval from decomposed human remains. *Journal of Forensic Sciences* 50(3): 1–9.

Shipman, P. and Phillips-Conroy, J. (1977). Hominid tool making versus carnivore scavenging. *American Journal of Physical Anthropology* 46: 77–86.

CHAPTER **25**

The Use of Taphonomy in Forensic Anthropology: Past Trends and Future Prospects

Mark O. Beary
and R. Lee Lyman

The more bones are exposed to air, the more quickly they disintegrate. The quantity of precipitation, the number of days below freezing, covering with clay, burial in sand or loam – all these factors play an important role in forensic medicine. (Weigelt 1927, 1989)

With many cases, taphonomic assessment represents the most important contributions made by anthropologists. This is especially true in the interpretation of skeletal evidence of foul play. (Ubelaker 1997)

Forensic taphonomy has dramatically changed the entire playing field of forensic anthropology. (Dirkmaat et al. 2008)

INTRODUCTION

As the quotes above illustrate, Weigelt (1927) foresaw the importance of forensic taphonomy early on; 70 years later, Ubelaker (1997) reaffirmed the now recognized importance of forensic taphonomy. Most recently, Dirkmaat et al. (2008) proclaimed the revolutionizing effect that taphonomy has had on the field of forensic anthropology as a whole. Clearly, any comprehensive text covering forensic anthropology must now contain one or more chapters on taphonomy.

A Companion to Forensic Anthropology, First Edition. Edited by Dennis C. Dirkmaat.
© 2012 John Wiley & Sons Ltd. Published 2015 by John Wiley & Sons Ltd.

In the history of formalized scientific pursuits, taphonomy is a relative newcomer. It originated in paleontology and eventually was adopted by paleoanthropology and archaeology. The application of taphonomy in forensic settings came later as forensic anthropologists, who were called on to assist with the identification of skeletonized human remains, saw the benefits of investigating postmortem alterations of deceased individuals. Given that taphonomy has become an integral part of forensic investigation for many forensic anthropologists, it is relevant to pause and assess where we have been, where we are, and where we are heading. Furthermore, although the utilization of taphonomic investigation has become commonplace for some forensic anthropologists, there remains a need to identify its importance to others within the field and, simultaneously, to law-enforcement investigators and researchers in other fields.

We have five goals in this chapter. First, we will provide a brief history of taphonomic research outside of forensic work so that those with a background in fields other than paleontology or anthropology may understand how the field developed prior to its emergence in the service of jurisprudence. Second, we will discuss some of the differences between paleozoological taphonomy and forensic taphonomy. Third, we will review the introduction and use of taphonomy in criminal investigations via an overview of taphonomy-related publications in the *Journal of Forensic Sciences*. Fourth, we will discuss general principles in forensic taphonomy and highlight recent advances in the field. Finally, we will provide our thoughts on the future of forensic taphonomic research and how it relates to other areas of taphonomic study.

History of Taphonomy Research

Here we provide only the highlights of the history of taphonomy research. Readers interested in the development of taphonomy should consult other historical overviews of taphonomic research for details (e.g., Dodson 1980; Olson 1980; Cadée 1990; Lyman 1994). Many historians suggest that Efremov's 1940 article in the *Pan-American Geologist*, in which he introduced the word *taphonomy* as "the science of the laws of embedding," the chief aim of which was "the study of the transition (in all its details) of animal remains from the biosphere in to the lithosphere," constitutes the birth announcement of the field. Such views, however, neglect earlier taphonomic work. Efremov was attempting to integrate and synthesize concepts that had been previously considered, and he acknowledged that "neither the problems nor methods are new" (Efremov 1940: 92). He was concerned with the ways that information related to the biological and geological aspects of embedding (a process ultimately leading to fossilization) was being utilized. While his immediate purpose was to establish a new subfield of paleontology that would unite a variety of pre-existing research areas, a secondary goal was to position paleontology in such a way that it could glean from the fossil record more information about past ecologies and faunal communities, thus ultimately providing more insightful interpretations of evolutionary histories. Efremov approached taphonomy as a means to better understand biasing processes – how and why the paleobiological record differs from a biological or living faunal community – and thereby to provide more information about past organisms and their paleoecologies and phylogenies (Efremov 1958).

Prior to Efremov's (1940) formulation of taphonomy, Weigelt (1927; see also the 1989 English translation by J. Schaefer; Weigelt 1989), and Richter (1928) produced relevant works. We find these studies to be particularly important because the type of investigations they undertook are being carried out by forensic taphonomists today. What is sometimes referred to as "neotaphonomy" or "actuopaleontology" involves observation (hence "actually seen") of modern transitions (hence "neo") of animal carcasses from the biosphere to the lithosphere in order to have a uniformitarianist base from which to interpret prehistoric animal remains. At the same time that Weigelt and Richter were conducting their investigations, other researchers were undertaking similar types of study. Efremov's aim was to unite these endeavors under one subdiscipline that he called taphonomy. Not only did he provide arguments for the formulation of this subfield, but he also provided a guide to its implementation. This implementation was divided into two foci: (i) the study of the geological processes that enable the transition of animals from the biosphere to the lithosphere and (ii) the study of transitions of contemporary living assemblages of animals (biocoenoses) to death assemblages (thanatocoenoses) to taphonomic assemblages (taphocoenoses) so as to provide actualistic information that sheds light on the paleoecological and evolutionary meaning of fossil assemblages, where an assemblage is a group of animal remains usually stratigraphically associated and representing multiple individual organisms and multiple taxa. To say that Efremov was responsible for *introducing* the whole of what we today call taphonomy would not be accurate, even by his own account; however, the fact that the field of taphonomy has grown to the extent that it has signifies that he deserves credit for *defining* and *delineating* the whole of the field.

Despite the early insights of Weigelt, Richter, Efremov, and others, taphonomic research did not immediately become commonplace in paleontology. In 1961, mammalogist and paleontologist George Gaylord Simpson noted that "there is great need for better understanding of the factors that act between the living fauna and the preservation of part of it in fossil state, as well as factors involved in the formation of fossil deposits in general" (Simpson 1961: 1683). Things were changing even as Simpson wrote, however, and within a decade several key studies in paleontology jump-started the field. These included Michael Voorhies' (1969) monograph on an early Pliocene site in Nebraska that had produced more than 20 000 bones and teeth from over 500 individual mammals, including extinct forms of horse and antelope. About the same time, David Lawrence (1968, 1971) detailed how taphonomic processes formed a biased paleozoological record (one that was not an accurate reflection of past faunal communities) but that with sufficient knowledge and work, the "taphonomic overprint" that biased the fossil record could be stripped away and the prehistoric faunal community revealed. This *taphonomy as biasing* perspective dominated paleontology until it was recognized that taphonomic processes could not only remove potential information but might also add information obtainable in no other way (e.g., Wilson 1988). Primary examples of the latter concern predation marks such as bore holes in mollusk shells and gnawing marks from carnivores on prey bones (Kowalewski and Kelley 2002). Today, taphonomic analysis is part of many paleobiological investigations.

Taphonomic research in an archaeological context is easily construed as having started in two contexts. First, during the early twentieth century in the Old World, medical doctor Raymond Dart suggested that the animal remains he found associated with the first recognized remains of Australopithecines (Plio-Pleistocene hominids) were the result of the latter using the former as tools (Dart 1949, 1957). Dart's proposed "osteodontokeratic culture" was controversial, and by the late 1960s C.K. "Bob" Brain was doing neotaphonomic research and demonstrating that the proposed culture was an unlikely explanation for the particular skeletal parts and fragment types found (e.g., Brain 1969). Subsequent neotaphonomic research by others has not changed the outcome of this controversy (Pickering et al. 2007). The New World was the second context for the beginnings of taphonomic research. There, paleozoologist Theodore White suggested that the different frequencies of skeletal parts of prey animals in archaeological sites reflected the fact that some portions of carcasses had been preferentially transported by prehistoric human hunters from the kill site to the consumption or habitation site whereas other portions had been left at the kill site (see Lyman 1985 for review). This explanation was also evaluated with some intensity in the 1970s, and the evaluation carried over into investigations in the Old World. In both contexts, some researchers argued that different frequencies of skeletal parts are the result of differential transport by human predators, whereas other researchers have argued the frequencies are the result of differential, density mediated destruction of some parts. The most recent steps towards a final resolution were taken in the series of articles in the second volume of the young *Journal of Taphonomy*, published in 2004 in Spain.

There were also parallels in both arenas. In particular, by about 1980, paleontologists (e.g., Schäfer 1972) and archaeologists (e.g., Behrensmeyer 1984; Shipman 1981) with interests in the paleoecological implications of faunal remains had begun intensive taphonomic analyses of both modern carcasses and collections of prehistoric remains in an effort to gather the most information possible of an unbiased nature from collections of faunal remains. A major modern research endeavor in this regard concerns "fidelity studies"; these involve assessment of "the quantitative faithfulness of the [fossil] record of morphs, age classes, species richness, species abundance, trophic structure, etc., to the original biological [community]" (Behrensmeyer et al. 2000: 120). These studies are undertaken by comparing living faunas to collections of remains of relatively recent age (either a few months, years, decades, or centuries old). Such studies underscore the general uniformitarian nature of taphonomy.

Today, taphonomy has its own textbooks for paleontologists (Martin 1999; Rogers et al. 2007) and for zooarchaeologists (Lyman 1994). Unfortunately, such division of effort reifies disciplinary autonomy, which in turn has negative influences on both. For example, some taphonomic variables are discovered or analytical techniques designed in one field and then (re)discovered a decade or so later in the other. Due to virtually nonexistent interdisciplinary communication, the second discovery not only involves the metaphorical reinvention of the wheel, but must also be accompanied by the same steep learning curve that attended the first discovery. It is our fervent hope that chapters such as this one will encourage more interdisciplinary communication between forensic scientists and paleozoological taphonomists so that we are not constantly faced with reinvented, yet initially square, wheels. By this, however, we do not intend to imply that paleontological taphonomy and forensic taphonomy are identical, or even that they have similar research questions. To underscore this point, we turn next to differences between the two.

DIFFERENCES BETWEEN PALEOZOOLOGICAL TAPHONOMY AND FORENSIC TAPHONOMY

We suspect many would say that any research field labeled "forensic" is simply that field applied in a legal setting. This is, by and large, a true statement, but it fails to capture how a forensic field differs from its mother discipline. For example, it is commonly held that a good forensic anthropologist must first be a good physical anthropologist. Although this is true, a good forensic anthropologist should also be a good archaeologist if she/he intends to be involved in field recoveries. Also, since forensic anthropologists routinely encounter trauma scenarios that nonforensic anthropologists do not (e.g., gunshot wounds, dismemberment with power tools), they ought to have some training in analyzing these scenarios. Furthermore, as taphonomy becomes more common within forensic anthropology analyses, forensic anthropologists ought to be trained in taphonomy as well as physical anthropology. This, then, raises the question of how forensic taphonomy differs from archaeological and paleontological taphonomy. Clearly, paleontological taphonomists will not encounter situations that forensic taphonomists do, such as patterns of bone fragmentation created by a mechanical wood chipper (Williams 2007), but are there more fundamental differences? For simplicity's sake, in the following we use the term "paleozoology" and derivatives to signify the study of zooarchaeological remains (animal remains recovered from archaeological sites and with associated artifacts) and/or paleontological remains (animal remains recovered from paleontological sites and without associated artifacts).

Temporal differences: the issue of the microenvironment

Lyman (1994) indicates that taphonomic histories begin when an animal dies, but acknowledges that paleozoologists generally have little need to consider the immediate postmortem changes to a deceased organism due in part to the long temporal span separating the investigator from the specimen under study. However, in forensic taphonomy, pinpointing the actual time of death is often a matter of utmost importance. Therefore, one of the most obvious differences between paleozoological and forensic taphonomy is the length of time between an organism's death and initiation of the study of that organism's remains. In forensic taphonomy, the lifespan of the investigator often overlaps the lifespan of the decedent. Thus, the forensic investigator is generally granted access to information that typically is unavailable to a paleozoologist: direct environmental data. The absence of environmental, particularly microenvironmental, data precludes the paleozoologist from giving much consideration to immediate postmortem changes.

A clear example of this can be seen in Lyman and Fox's (1997) review of Behrensmeyer's (1978) work on weathering stages. Behrensmeyer's actualistic research on bone weathering led to the formulation of the commonly employed six-stage description of subaerial weathering. Since Behrensmeyer had knowledge of the time of death of the animals in her study, the weathering stages were simply a function of taphonomic time measured in years since death:

$$WS = f(YD) \tag{25.1}$$

where WS is the weathering stage and YD is the years since death. Lyman and Fox (1997) indicated that Behrensmeyer's stages could not readily be applied to prehistoric bone assemblages because in those situations the determination of weathering stages involves a much more complex function:

$$WS = f(YD, SE, TX, ME, ED, AH) \tag{25.2}$$

where SE is skeletal element, TX is taxon, ME is depositional microenvironment, ED is exposure duration, and AH is accumulation history. Although determining SE and TX are relatively easy, ME, ED, and AH are generally unknowable in prehistoric contexts. Knowledge of ME, ED, and AH is, however, accessible in modern forensic scenarios. Knowledge of TX and SE is important because weathering rates vary across taxa and skeletal elements (Lyman and Fox 1997). However, in forensic settings we are generally only concerned with human remains, so variation in TX is not an issue; additionally, AH is typically not an issue at most forensic scenes. Instead of TX, we might add ontogenetic variation, but this variable has not yet been studied in either paleozoological taphonomy or forensic anthropology. Both ME and ED are critical to the formula. Therefore, the bone-weathering equation in a forensic setting would be:

$$WS = f(YD, SE, ME, ED) \tag{25.3}$$

Differences between the second and third equations illustrate that forensic investigators are not as constrained as paleozoologists when making inferences about postmortem changes to skeletal remains. Paleozoologists rely on actualistic studies which they then must use to draw analogies of the past according to uniformitarian principles to explain properties of prehistoric faunal assemblages. Forensic investigators also conduct actualistic research; although they also rely on the uniformity of nature to construct their inferences, this is much less of a slippery interpretive slope because they generally operate within the same time period as the decedent. Thus, not only will they often have environmental data that assists with the interpretation of the remains, but the shorter time span between the death of the organism and the recovery of its remains means that there has been much less opportunity for taphonomic processes to alter the remains and obscure or destroy traces of mortality agents and the like.

Temporal differences: antemortem, perimortem, postmortem

By definition, taphonomy concerns everything that happens to a carcass after an organism dies. Pathologies of soft and skeletal tissues are not taphonomic features, but forensic and paleozoological taphonomists are both interested in modifications to bodily tissues that are perimortem, especially if those modifications reveal the agent or means of mortality. Further, both forensic and paleozoological taphonomists are interested in postmortem traumas related to the crime or to how an animal carcass was rendered into usable or consumable parts, respectively. Paleozoological taphonomists use a plethora of terms to denote stages or states between living organisms and the pile of bones and teeth on a laboratory table with which they are concerned. We have arrayed some of these terms against those most familiar to forensic taphonomists in Figure 25.1. Note in the figure the various terms under paleozoological taphonomy; these terms signify what paleozoological taphonomists take to be critically important stages and states in a taphonomic history. The terms reflect the fundamental notion that taphonomic processes take place over time; more time generally means more processes have influenced the carcass, and more modifications of more kinds have affected animal remains. Thus, because many paleozoological taphonomists, like

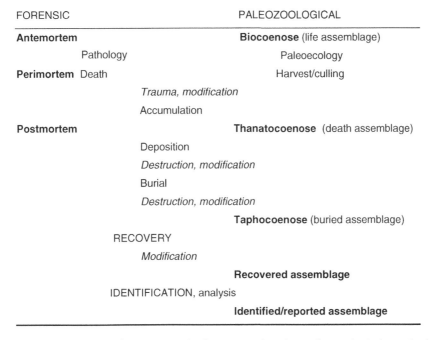

FORENSIC	PALEOZOOLOGICAL
Antemortem	**Biocoenose** (life assemblage)
Pathology	Paleoecology
Perimortem Death	Harvest/culling
Trauma, modification	
Accumulation	
Postmortem	**Thanatocoenose** (death assemblage)
Deposition	
Destruction, modification	
Burial	
Destruction, modification	
	Taphocoenose (buried assemblage)
RECOVERY	
Modification	
	Recovered assemblage
IDENTIFICATION, analysis	
	Identified/reported assemblage

Figure 25.1   Forensic taphonomy terminology arrayed against paleozoological terminology. Time passes from top to bottom.

Efremov before them, are ultimately interested in what the live animal (represented by bones and teeth) reveals about past ecology and evolution, the analytical target is often the living faunal community, or biocoenose, because it reflects paleoecology. Further, paleozoologists are well aware that they typically sample the remains of a community, and that only some subsample of what is recovered will be identifiable to skeletal element and to taxon represented. Thus, the temporal span of the existence of an animal bone of interest to a paleozoologist encompasses the time from its creation to the time it (at least a description of it) ends up in the literature.

Forensic taphonomists take as their focal analytical target a much narrower range of that time span; they are interested in what criminally generated perimortem trauma and immediate postmortem taphonomy can tell them about the crime. Forensic anthropologists are generally only concerned with antemortem features to the extent that they might provide a positive identification or reveal pre-existing patterns of abuse. This is not to deny the utility of using pathologies to assist with victim identification, or to deny the importance of distinguishing noncriminal postmortem modifications from those related to criminal activities aimed at the victim. The difference between forensic taphonomy and paleozoological taphonomy with respect to the schematic time range in Figure 25.1 originates in the different questions asked of a sample of bones. Paleozoologists tend to ask a great many more questions of widely disparate types than those aimed more narrowly by forensic taphonomists at the perimortem and immediately postmortem eras. Thus, paleozoologists record and recognize many kinds of modification to bones and teeth that are seldom considered by forensic taphonomists. This is not to imply that one field or set of questions is somehow better or more thorough than the other; rather, it is to say that, although there is considerable overlap between the two fields, there are also considerable differences resulting from dissimilarity in the questions asked.

Epistemological differences: the issue of "knowing" taphonomy

We have indicated that forensic taphonomy and paleozoological taphonomy both depend on actualistic research to provide interpretive bases for understanding why the remains under study are where they are, and in the condition in which they are found. This means that one must establish cause–effect relationships between processes or mechanisms and the traces they create in an organism's tissues. In the best of all possible worlds, every kind of effect would have but a single cause. Were that the case, one could speak of taphonomic *signatures* that simply by their presence implicate a particular and singular taphonomic process or agent. The procedure for doing taphonomic research is, then, different than that of the archetypical experimental sciences such as chemistry and physics. Rather, taphonomy is a kind of historic science given the particularistic contingencies and context of each individual organism. This does not mean empirical generalizations cannot be derived, nor that these empirical generalizations cannot become rather like explanatory theories. Taphonomic research is, then, very much like other historical sciences such as geology, archaeology, and evolutionary paleobiology. There is no possibility of observing a particular cause and its attendant effect in these research endeavors; the cause took place in some remote prior time, and all that remains is its effect or result. Thus, in historical sciences the search is on for a "smoking gun," a trace or result that provides an unambiguous indication of a particular cause or family of causes (Cleland 2001).

Sometimes the smoking gun or signature trace is not tightly diagnostic of a single cause and instead suggests several different possible causes. Such ambiguity is subsumed under the concept of equifinality: Ludwig von Bertalanffy's notion that in open systems (those which exchange matter or energy with their environment) different processes can produce similar results (Lyman 2004). Much modern actualistic research in paleozoological taphonomy seeks to clear up such ambiguities, and while some success is attained every year, the research often better points out the limits of what we can know about the past. It is now appreciated that information that is readily observable and knowable to an ecologist or zoologist working with live animals simply cannot be accessed by studying the paleozoological record that is made up of (usually) dead animals that are to varying degrees anatomically incomplete. However, this makes paleozoology and taphonomy no less scientific than any other science. We cannot, for example, claim that mammalogy is unscientific because it cannot tell us the lithology of a fossilized bone. Every research field has boundaries in terms of the questions it asks and the data it can muster to answer those questions. Taphonomy is no different. This is one area, therefore, where we think forensic taphonomists might benefit from greater knowledge of paleozoological taphonomy. In both fields of inquiry, there are some questions that lie beyond our knowledge or analytical skills and cannot be answered. Having said this, forensic taphonomists must (as some do) also see taphonomy as a scientific pursuit if they are going to define the boundaries of what knowledge can or cannot be gained through the use of taphonomy.

Perspective differences: the issue of bias

As noted above, when paleontologists first began to think about taphonomy, the general notion was that it concerned biases and how a collection of prehistoric bones and teeth is different than the faunal community from which it derived. This is not surprising because it was precisely those differences – those biases – that were the concern of Efremov, Weigelt, Richter, and other early taphonomists. But as more was learned about the processes of transition from the biosphere to the lithosphere, it became clear that those processes were not always biasing. Issues of bias relate to the research question asked of a collection. Ungulate teeth tend to preserve rather well, whereas some ungulate long bones do not preserve well at all. If, on the one hand, the research question concerns the demography of the herd of ungulates from which a fossil collection derives, chances are good that an accurate estimate of that demography can be made on the basis of the teeth. On the other hand, that same collection may not provide a good estimate of skeletal part abundances if the ungulate carcasses underwent carnivore attrition prior to recovery. The collection will be unbiased with respect to demography but biased with respect to skeletal part frequencies as carcasses will, to varying degrees, be anatomically incomplete.

Forensic taphonomists seem to be following the same path as paleozoological taphonomists, but in the opposite direction. We say this because the former are increasingly recognizing that taphonomic processes that are active later during the postmortem era can distort or destroy the traces created by processes that were active early during that time. Fire is the exemplary late process. Burning destroys soft tissue and can distort and eventually destroy skeletal tissue. In the process of burning, other pre-existing taphonomic evidence, such as tool marks, that could have had a positive impact on the resolution of a forensic case might be lost; thus, from the forensic anthropologist's perspective with regard to the trauma, the fire represents a negative biasing taphonomic process. Similarly, degradation of DNA evidence due to environmental processes or the cleaning of remains in the laboratory might be seen as a bias when the DNA is necessary for making a positive identification. Examples of these types of research are discussed below.

Taphonomy in Forensics

History of anthropology in the *Journal of Forensic Sciences*

Ellis Kerley is credited with being one of the central figures in the formation of the Physical Anthropology section of the American Academy of Forensic Sciences (Snow 1982; Ubelaker 2001). At the 1971 meeting of the Academy in Phoenix, Kerley approached the executive committee of the Academy with a proposal to create this new section. As the story goes (Snow 1982; Ubelaker and Hunt 1995), after getting the committee to acquiesce, Kerley located Clyde Snow and William Bass, and the trio convened to Kerley's hotel room where, with the help of Kerley's scotch and Snow's Federal Aviation Administration "emergency-only" credit card, they proceeded to contact colleagues across the country who would serve as the section's first members. Kerley reported back to the executive committee the following day that he had signed up the required number of potential members and the committee approved the new section, which held its first session the following year (Snow 1982).

As a means of tracking the introduction of taphonomy into forensic anthropology, we reviewed the *Journal of Forensic Sciences* (*JFS*), the official publication of the American Academy of Forensic Sciences. Since the Physical Anthropology section of the Academy convened for the first time in 1972, we selected that year as our starting point. Incidentally, 1972 was also the year that William Bass and his graduate students at the University of Tennessee began collecting data on human decay rates (Bass 1997). We determined the number of forensic anthropology contributions to *JFS* as well as those which qualify as taphonomically oriented contributions within forensic anthropology for the years 1972 through 2008. To qualify as a forensic anthropology contribution, the contribution must have had *either* (i) an anthropologist as one of its authors or (ii) the terms "physical anthropology" and/or "forensic anthropology" as keywords. To qualify as an anthropology-taphonomy contribution, the submission must have had *both*: (i) an anthropologist as an author and (ii) a primary topic that is an issue of taphonomic relevance. Excluded from this category were submissions that would be considered taphonomic but were not directly anthropological in nature, such as articles related to entomology and pathology. Although anthropology, pathology, taphonomy, and entomology are all interconnected within forensic investigations, our main purpose was to track the development of forensic anthropology and, within forensic anthropology, the development of forensic taphonomy. Qualifying submissions included articles, case reports, technical notes, and book reviews; excluded were letters to the editor and commentary pieces.

History of taphonomy in *JFS*

The first article published in the *JFS* that approximates a taphonomically oriented contribution by an anthropologist was the 1976 Morse, Crusoe, and Smith article on forensic archaeology (Morse et al. 1976). Although the primary goal of the article was to illuminate the role of the archaeologist in the recovery of human remains, the article focused on burials and included a section on dating the time of burial based on the condition of the remains, an issue that remains central in forensic taphonomy today. The remainder of that decade saw a 1977 article by Kirkham et al. on postmortem pink teeth and a 1978 article by Zimmerman that dealt with mummified remains.

After 1978, the journal did not see another taphonomically oriented article from an anthropologist until 1983, when Bass published a case report on two Japanese trophy skulls and Rodriguez and Bass published an entomological article (Bass 1983; Rodriguez and Bass 1983). Since that year, the journal has seen at least one taphonomically oriented anthropology article every year, with the exception of 2000 (Figure 25.2). Anthropologists have made an increasing number of contributions to *JFS* since the inception of the Physical Anthropology section in 1972. As demonstrated by Figure 25.2, taphonomic articles began to appear consistently about a decade after the formation of the section. These data indicate that, although the trend for forensic anthropology articles in general has been to increase in frequency, those related to taphonomic issues have remained fairly stable.

Although the theme of several articles in the late 1970s and early 1980s was taphonomic, the first use of the word "taphonomy" in the text of a *JFS* article was in Haglund et al. (1988), an article about animal scavenging. The authors introduce taphonomy and list numerous references from paleontology and archaeology. At that

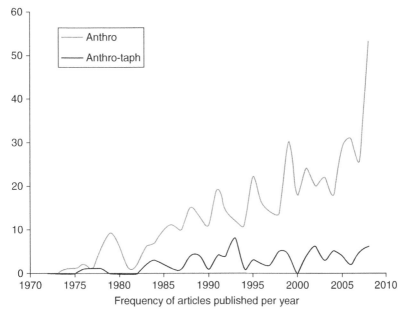

Figure 25.2 Frequency of articles published per year in the *Journal of Forensic Sciences*. Larger spikes in the forensic anthropology line indicate years in which symposia related to a special topic or honoring a particular person were held. Anthro, anthropology; Anthro-Taph, anthropology-taphonomy.

time, William Haglund was working as the Chief Investigator for the King County Medical Examiner's Office in Seattle, Washington. The consulting physical anthropologist for forensic cases at the time was Daris Swindler of the University of Washington. Exposure to cases involving skeletonized human remains and the techniques that forensic anthropologists utilize, along with the encouragement of Swindler, led Haglund to conduct graduate work at the University of Washington. Haglund's involvement in the Green River Murder investigations (Haglund et al. 1987, 1990) provided experience in numerous cases of decomposed human remains and consultation with another forensic anthropologist, Clyde Snow. Haglund's experience with animal-scavenged skeletal remains, in combination with his educational pursuits in anthropology, provided him the prerequisites for introducing taphonomy to the *JFS*. These early works on animal scavenging were influenced by Marcella Sorg's (1986) contributions to forensic taphonomy. This mutual and early interest in forensic applications of taphonomy by Haglund and Sorg would result in two coedited volumes (Haglund and Sorg 1997, 2002) on forensic taphonomy that together currently represent the most comprehensive coverage of the subject.

GENERAL PRINCIPLES AND TRENDS IN FORENSIC TAPHONOMY

Forensic taphonomy has many commonalities with the taphonomic research carried out by paleozoologists. However, the questions that are asked are often quite different. Whereas taphonomic research of prehistoric remains might be used to answer

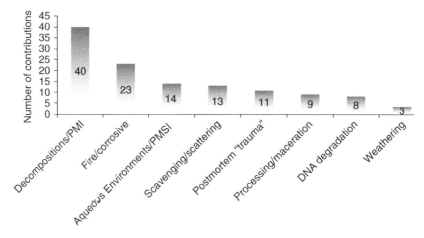

Figure 25.3   Taphonomic contributions from the *Proceedings of the American Academy of Forensic Sciences*, 2002–2009.

questions of hominid subsistence patterns or to assist in reconstructing paleoecological conditions (Lyman 1994), taphonomic research in a forensic setting has different goals. We perceive five basic goals in forensic taphonomy:

1. determining whether recovered remains are of forensic significance or not;
2. determining/estimating the time since death or postmortem interval (PMI);
3. determining how the remains have come to rest at their location of discovery;
4. determining what actions may have been taken to conceal the victim's identity or the entire crime;
5. determining what taphonomic factors have been operating on the remains and thus understanding how these factors affect (positively or negatively) the investigator's ability to glean information about the victim or the crime.

In classifying contributions to *JFS*, we use eight categories that cover the breadth of these goals and provide a brief overview of research that has taken place in each. Much of what we cover below comes from our review of the *JFS*. Additionally, to identify the most recent trends within forensic taphonomy, we tracked contributions listed in the *Proceedings of the American Academy of Forensic Sciences* (*PAAFS*) from 2002 to 2009 (Figure 25.3). These contributions indicate the direction of taphonomic investigations that may not have yet made it to publication. This time frame also represents works that have been contributed since the publication of the most recent edited volume on forensic taphonomy (Haglund and Sorg 2002). Studies on decomposition/PMI are by far the most numerous, indicating the importance of this topic within the field (Figure 25.3). Weathering studies are the least represented, suggesting that researchers currently see little utility in them.

Decomposition/PMI

Decomposition studies in forensic contexts are frequently related to determinations of the PMI and therefore are considered in a single category here. This category includes the most taphonomy contributions to *JFS*, which we believe is a direct result of the desire to determine the PMI in most cases seen by forensic anthropologists. Historically, most research dealing with determination of the PMI has focused on gross observations of soft tissues. Since decomposition of human remains often involves the feeding activities of insects, forensic anthropology has developed a close relationship with forensic entomology (e.g., Rodriguez and Bass 1983; Schoenly et al. 1991; Komar and Beattie 1998; Megyesi et al. 2005). Of course, decay rates will vary by region and climate; therefore, reliance on gross observations has led to numerous reports of decomposition in specific environments. For example, Rodriguez and Bass (1983, 1985), in two separate articles, examined decay rates of four and six human cadavers, respectively, at the Anthropology Research Facility (ARF) located in Knoxville, TN, USA; Mann et al. (1990), also working at the Anthropology Research Facility, reviewed general observations from approximately 150 decompositions; Galloway et al. (1989) reviewed 189 cases of human decomposition in arid southern Arizona; Komar (1998) reviewed 20 cases of human decomposition and Weitzel (2005) investigated decomposition rates of pigs (*Sus scrofa*), both in the vicinity of Edmonton, Alberta, Canada; Shean et al. (1993) examined differential decomposition of shaded and exposed pig carcasses near Olympia, Washington; and Spennemann and Franke (1995) reported on decomposition of bodies on a small coral atoll of the Marshall Islands.

Gross observations of human decomposition in specific regions, along with experiments set up to test differences between microenvironments, are a necessary step in gaining information on the variables that affect decomposition. These studies provide insight to variables that influence the rate of decomposition. Such variables include temperature, humidity/aridity, rainfall, soil pH, trauma, insects, burial and depth, carnivore and rodent activity, size and weight of the body, surface placement, clothing, and embalming (Mann et al. 1990); they can be categorized as either extrinsic or intrinsic to the decedent. More recent work examining extrinsic variables includes: the effects of subaerial temperature on surface finds (Megyesi 2002), subsurface temperature on buried remains (Weitzel 2006), the effects of lime on buried remains (Thew and Nawrocki 2002), decomposition in mass graves (White 2007), the contribution of nonenteric microbes to decomposition (Carter et al. 2008), the effect of light on decomposition in indoor settings (Franicevic 2008), and the effects of body covering such as plastic and blankets (Dautartas 2009) or clothing (Miller 2002). Recent work focusing on the intrinsic aspects of decomposition include the effects of human body mass (Stuart 2003), the effects of body fat (Chibba 2008) and carcass weight (Brand 2008) on pigs (*S. scrofa*), and the influence of penetrative trauma (Cross 2008).

Although identifying the factors that influence the rate of decomposition is useful in adjusting time-since-death estimates, developing methods of estimating the PMI is the necessary first step. Bass (1997) provides an overview of his experience and lessons learned through his, and his students', years of research at the University of Tennessee Anthropology Research Facility. In doing so, he provides an overview of gross changes that one might expect to see from the first day up to the first decade after death. The gross characteristics of decomposition that Bass outlines are most likely the characteristics that most forensic anthropologists are using today to provide estimates of PMI. This is because there are few alternatives available; however, some work has sought to provide options. Early work by Castellano and colleagues (1984) attempted to determine PMI via analyses of total lipids, triglycerides, cholesterol, free fatty acids, total proteins, zinc, iron, manganese, and phosphorous. Vass et al. (1992) analyzed volatile fatty acids and various anions and cations extracted from soil solutions gathered from beneath decomposing corpses to determine PMI. The stated accuracy of this method for remains older than 1 year is ±2 weeks. A later study by Vass et al. (2002) also focused on decomposition chemistry. In this analysis, decompositional byproducts of several organs were studied to determine their utility in estimating the PMI and the results suggested that accurate determinations can be made when temperature data are available. In a different approach, Perry et al. (1988) tracked the autodegradation of DNA in human ribs to determine its relationship with the PMI. While their results were promising, they have, so far as we can tell, yet to be tested on actual case material. Other approaches include more recent investigations of radioisotopes and trace elements (Rea and Back 2002; Ubelaker et al. 2006; Howard and Meyer 2007), as well as factors related to postmortem tooth loss (McKeown and Bennett 1995; Dabbs 2009).

DNA degradation

Although DNA degradation could technically be included under decomposition studies, the purpose of many DNA studies is to determine how environmental factors affect cellular integrity and, therefore, under what circumstances DNA can or cannot be collected for determining the identification of the deceased; for that reason it is given its own category. In an early study, Duffy et al. (1991) reported on rates of putrefaction of dental pulp from humans and pigs in British Columbia, Canada. Although they focused on sex chromatin, they noted that their findings could have implications for DNA analysis. Fisher et al. (1993) examined the need to decalcify bone prior to DNA extraction. They found that decalcification was unnecessary and succeeded at extracting and amplifying mitochondrial DNA from the 125-year-old bones of a US Civil War soldier. Rankin et al. (1996) investigated the ability to derive DNA from compact bone exposed to a variety of environments for restriction fragment length polymorphism (or RFLP) analysis. More recently, contributions to the study of DNA preservation include the works of Damann et al. (2002) and Kontanis (2003) on predicting DNA recovery based on soft tissue and skeletal preservation; Latham's (2003) work on detecting the presence of polymerase chain reaction (PCR) inhibitors in skeletal DNA so that they may be removed through purification; Curtis and colleagues' (2003) investigation into using restriction enzymes to reduce the inhibitory properties of soil microorganism DNA when attempting PCR on buried human bone; and, finally, the works of Mundorff and Bartelink (2007), Fredericks and Simmons (2008), and Jans et al. (2008) on recovering DNA from bones subjected to mass fatality incidents, burning, and diagenesis, respectively.

Scavenging/scattering

Scavenging of human remains by animals often results in the scattering of the remains and therefore these two topics are typically linked. Note that entomological species are also scavengers; however, their effects are traditionally subsumed within the decomposition processes. As previously indicated, the word taphonomy first entered *JFS* in Haglund and colleagues' (1988) work on the distribution and patterning of animal scavenging features (e.g., gnaw marks) on skeletal remains. The following year, Haglund et al. (1989) and Willey and Snyder (1989) discussed the impacts that canid scavenging and disarticulation have on PMI estimates. Pickering (2001) drew attention to carnivore voiding (defecation and regurgitation) as a potential source of forensic information. Haglund's (1992) early work provided case-study examples of rodent scavenging and Klippel and Synstelien's (2007) later work focused on brown rats and gray squirrels as scavengers and their taphonomic imprints as possible estimators of time since death.

Other contributions have included investigations of scavenging by rats (Leher 2004; Kiley et al. 2006), raccoons (Synstelien et al. 2005), vultures (Reeves 2008), and dogs (Pinheiro et al. 2009). Reports of scavenging in aqueous environments include mollusks (Scott 2009) and sharks (Allaire and Manhein 2009; Rathburn and Rathburn 1984); in the latter, it is not entirely clear whether the animals attacked the decedent while they were still alive or truly scavenged their remains after death. Barnacles (Johnson et al. 2009) have also been observed inhabiting the occlusal surfaces of teeth; although these animals are not scavenging human remains, they do attach themselves to the remains for the purpose of feeding.

Although scattering is often associated with the scavenging activities of animals, scavengers are not required for skeletal remains to become scattered. Manhein et al. (2003) and Dabbs (2008) investigated the role of landscape formations in the scattering of remains. Gravity is a simple yet significant and ubiquitous taphonomic mechanism.

Aqueous environments/fluvial transport/PMSI

Research in this category can be assigned to one of three types: (i) research on decomposition in aquatic environments and determinations of the postmortem submersion interval (PMSI); (ii) research on the formation of adipocere; and (iii) research related to the movement of human bodies and remains in aquatic systems.

As with terrestrial sites, aquatic decomposition is affected by location and numerous environmental variables. Because decay processes in aquatic environments differ significantly from those on land, the term PMSI has become common. Although decomposition in aquatic environments is often said to be delayed relative to that of terrestrial environments, as Shepherd et al. (2009) indicate, the proverbial "two weeks in the water equals one week on land" leaves much to be desired. To remedy this, several researchers have reported on specific cases or reviewed pre-existing cases to help isolate critical variables. Haglund (1993) reported on disarticulation sequence patterns of 11 human remains recovered from sounds, rivers, and lakes in the US Pacific Northwest, and from New York City. Reports from the southern USA that provide baseline data there include Shepherd et al.'s (2009) work in Florida and the efforts of Hurst and Manhein (2002) and Shamblin and Manhein (2003) in Louisiana.

In the UK, Heaton and Simmons (2007) reviewed 135 cases of decomposition in the River Clyde; Brewer (2005) reviewed 103 cases from the River Thames; and Lagden and Simmons (2008) reviewed 49 cases from the River Mersey and compared the findings to the previous two studies. In the USA, Perry (2005) reviewed 69 cases of decomposition in the Detroit River. All of these investigators provided information on variables related to decomposition and estimation of the PMSI in fluvial environments.

Only a few reports have been made of remains from deep-sea settings. Dumser and Türkay (2008) reported on remains recovered from deep water (540–580 m) off the coasts of Sicily and Namibia, and Di Vella et al. (2003) discussed the case of 52 bodies recovered from a sunken ship (at 800 m) in the Adriatic Sea.

The formation of adipocere, or "grave wax," receives attention because, similar to mummification, it is the result of a transformational process whereby putrefaction processes are precluded in the presence of certain environmental factors. One of the prerequisites of adipocere formation is an aquatic environment; however, it should be noted that adipocere can form in locations other than open bodies of water. Cotton et al. (1987) provided an early overview of adipocere formation in relation to a case of two bodies pulled from a harbor after 5 years of submersion. Yan et al. (2001) reported on experimental work with pigs to determine whether the formation of adipocere could be useful in determining the PMI. Their quantification of the degradation of oleic acid during adipocere formation should provide avenues for future investigation. O'Brien and Kuehner (2007) recently provided a thorough overview of the chemistry of adipocere formation and reported on the results of some experimental work conducted with human cadavers. Their findings reinforce the so-called "Goldilocks phenomenon," whereby the presence of certain bacteria and water within a specific temperature range creates conditions that are "just right" for adipocere formation.

Although some features of aquatic transport of human remains may be applicable across all cases, such as which skeletal elements will travel the farthest following disarticulation, other aspects of transport may be specific to certain locations simply because of specific features of particular bodies of water. Reports on aquatic transport include the movement of remains in the lower Mississippi River (Bassett and Manheim 2002), the tidal River Thames (Brewer 2005), the Magdalena River (Guatame-Garcia et al. 2008), and Puget Sound (Ebbesmeyer and Haglund 1994). Herrmann et al. (2004) studied the effects of transport in smaller creeks where the flow of water is more seasonal and associated with high-velocity waters due to heavy rains.

Burning and corrosion

Research focused on taphonomic alterations arising from episodes of burning can be categorized into four distinct areas; these include (i) descriptions of changes to bone, both macroscopic and microscopic, brought on by exposure to fire; (ii) the effects of fire as a biasing agent; (iii) studies of cremated human remains; and (iv) studies related to chemical burning. We include the effects of corrosive substances here because this type of alteration also results in destruction of the material and is often considered "chemical burning." We briefly review contributions to each of these areas.

Reports on thermally induced changes to bone ultimately add to our knowledge of how variables related to the bones (e.g., element, density, amount of soft tissue coverage, etc.) react with regard to variables related to the fire (i.e., temperature, duration, etc.) so that anthropologists are more informed about the fire-related cases they investigate. Christensen (2002) provided an explanation for cases often referred to as "spontaneous human combustion" and determined that similarities in these cases included low bone density and high body fat. Huxley and Kósa (1999) examined the extent of shrinkage of human fetal diaphyses from fresh bone to both carbonized and calcined states as these changes can severely affect age estimates based on bone measurements. Thompson (2005) also investigated heat-induced dimensional changes, discussing not only shrinkage but also potential expansion in bone, and described the impact of these changes on anthropological techniques that frequently utilize measured dimensions of unmodified bones.

More recently, investigations into the possible biasing aspect of fire, both at the macroscopic and microscopic level, have begun to appear in the literature. Herrmann and Bennett (1999) tested for differences between traumatic fractures and those produced during burning. They found that although sharp trauma was readily identifiable postincineration, signatures of blunt-force trauma were more difficult to identify after burning. De Grunchy and Rogers (2002) carried out a similar experiment when they examined the effects of incineration on bones inflicted with chopping trauma from a cleaver. Again, they found that these chops marks could be identified postincineration. Whereas the pervious authors burned nonhuman animal bones in their studies, Pope and Smith (2004) burned 40 cadaver heads with various blunt, sharp, and gunshot traumas. Their observations indicated that evidence of each type of trauma to the cranium is capable of being detected postincineration.

Bradtmiller and Buikstra (1984) investigated the effects of burning on the microstructure of bone to determine if microscopic methods of age determination could be applied to cases involving burning. Their preliminary findings indicated that, contrary to prior predictions, osteon size increased and that, ultimately, bone that had burned at temperatures up to 600 °C could still provide accurate information for aging methods. However, Nelson (1992) found that osteon size decreased with burning and suggested caution when applying aging techniques to burned bones until additional research has been conducted. Finally, Quatrehomme et al. (1998) utilized scanning electron microscopy to examine specimens of bones burned at various temperatures. Their results indicated that, although there were notable alterations to the bone as temperature increased, correlating those changes to specific temperature ranges was difficult.

A handful of studies have focused on analyzing cremains. Murray and Rose (1993) reviewed a case in which two sets of cremains, purported to have originated from the same individual, were examined. Warren and Maples (1997) collected baseline data (e.g., weight, stature, etc.) on individuals prior to cremation and then developed correlations between those variables and the final weight of the cremains. Their results also indicated that males and females could be distinguished based on cremains weight, although there was considerable overlap between them. Warren and Schultz (2002) followed up with a discussion of methods and tools employed at crematoriums to pulverize the osseous remains after firing is complete. Whereas hand processing and ball-and-hammer mill processing both resulted in some diagnostic fragments being recovered, newer rotary-blade processors were consistently capable of pulverizing cremains to an extent that no diagnostic fragments remained.

Very few studies have been carried out on the effects of corrosive agents on human remains, most likely reflecting the low number of cases where this type of taphonomic agent is involved. Ubelaker and Sperber (1988) reviewed a case in which a caustic substance [most likely lye (sodium hydroxide)] had been poured over the face of a murder victim in an attempt to conceal the individual's identity. More recent research (e.g., Dupras et al. 2002; Fulginiti et al. 2009) has focused on experimenting with chemicals commonly available in household cleaning agents to determine which are most effective at destroying soft and hard tissues; in all instances, hydrochloric acid was found to produce the most deterioration.

Postmortem "trauma"

The category of postmortem "trauma" is somewhat ambiguous. Trauma is generally taken to refer to injuries inflicted upon living tissues. However, forensic anthropologists regularly speak of postmortem trauma versus ante- and perimortem types. The situation is further complicated since the perimortem interval is also ambiguous because bones of a deceased individual will continue to respond to traumatic impacts in the same manner as living bone for a considerable amount of time after death. From a taphonomic perspective, this creates difficulty in classification because taphonomic histories commence once an individual ceases to live. Here we have chosen to exclude the trauma most often associated with the perimortem period (blunt, sharp, and gunshot trauma) from "trauma" that occurs during the postmortem period (e.g., dismemberment). The latter is clearly taphonomic, whereas the former may or may not be taphonomic.

Determining whether trauma is perimortem or postmortem can have significant ramifications for forensic cases, and therefore this issue has generated some discussion. Ubelaker and Adams (1995) reviewed a case in which "butterfly" fractures, a type of fracture often associated with perimortem trauma, were demonstrated to have occurred during the postmortem period. Wheatley (2008) experimentally tested for various characteristics typically associated with both perimortem and postmortem trauma and found that, in many instances, characteristics typically reported to be associated with either perimortem or postmortem samples could be found in both. Although some differences between the samples (e.g., smoother fracture surfaces and more fracture lines in "wet" than "dry" bones) were statistically significant, most interpretations of individual bones remain problematic. Wieberg and Wescott (2008) investigated several variables (e.g., bone moisture content, fracture angle, and surface morphology) to determine whether correlations to the PMI could be drawn. Their findings illustrated that nonliving bone will continue to produce perimortem-like fractures long after death and that the duration of this perimortem-like period is also likely dependent on microenvironmental condition. Although they were able to regularly distinguish between bones broken near death and those broken 5 months after death, bones broken at various points between those intervals were ambiguous with regard to the duration of the PMI.

Aside from the question of the timing of trauma, other researchers have detailed the characteristics associated with varying types of postmortem trauma or the implements used to inflict such trauma. Tucker et al. (2001) indicated that the class of chopping weapon (machetes, axes, and cleavers) could be determined based on microscopic features of the cut surface. Kahana et al. (2005) reviewed 14 cases of dismemberment, highlighting information on the intent of the perpetrator, common anatomical locations of severing cuts, and the various tools used to inflict such trauma. Symes et al. (2006) discussed tool mark recognition and interpretation and provided descriptions of five features (saw cut direction, saw power, saw design, saw tooth size, and saw tooth set) to assist forensic anthropologists in classing tools used in dismemberments.

Finally, Calce and Rogers (2007) examined the effects of multiple taphonomic variables on pre-existing blunt force trauma. They concluded that some processes (e.g., freeze–thaw cycles) have the potential to disguise perimortem trauma.

Weathering

Whereas decomposition refers to the degeneration of soft tissues, weathering refers to any process that alters the chemical or physical nature of the hard tissues (bones and teeth), whether they are located in a subaerial or subsurface context. Very little research has been carried out with regard to weathering in forensic contexts; there are perhaps two reasons for the dearth of these investigations. First, weathering studies take time and, as Bass (1997) noted, one reason why information on longer PMIs is absent is because graduate students simply lack the time necessary to complete these research projects. Perhaps as graduate students with forensic taphonomy interests move into academic positions, results of more long-term studies will become available. Second, there is a conception that weathering studies may not provide resolution on a scale useful for forensic investigation. As Schoenly et al. (1991) suggest, "a major difficulty faced by both forensic pathologists and anthropologists is that, as the PMI increases, the accuracy of its determination necessarily decreases" (p. 1395). However, without the necessary experimentation, this notion remains speculative. Simply put, the degree to which weathering can be used as an estimator of PMI prior to degradation remains largely unknown.

Two published works relate directly to weathering. Huxley (1998) investigated postmortem shrinkage, through desiccation, of fetal bones. She found that shrinkage is present, especially in the earlier lunar age groups when there is less inorganic structure to the bone (Huxley 1998). This study has a clear application in methods for calculating gestational age from diaphyseal lengths and should be utilized by anyone attempting to do so. Tersigni (2007) examined the effects of freezing on bone and questioned whether freezing would leave histological artifacts that might indicate its occurrence and, if freezing had occurred, what its effects might be on age assessments using histological indicators. She found that histological artifacts of freezing (cracks propagating from Haversian systems) are discernable when using scanning electron microscopy analysis; however, the effects of freezing do not appear to affect histomorphological analyses relevant for aging.

Other recent contributions to weathering investigations include Hughes and White (2008) and Beary and Cabo (2008). Hughes and White (2008) examined both the macroscopic and microscopic effects of dehydration on the propagation of postmortem cracks in teeth, information that can be used to discern between perimortem and postmortem alterations. Beary and Cabo (2008) utilized both actualistic and controlled laboratory experiments to develop a regression formula for accurately estimating the exposure duration of sun-bleached bones. Their research involved correlating spectrophotometric measurements of bone surface color with environmental variables such as incident radiation and humidity.

Recovery, transport, and processing

This section includes all measures taken to recover human remains from the crime scene, transport them to the morgue or laboratory, and process them during autopsy and/or analysis. We have come across only two published studies that relate to this topic. In the first, Rennick et al. (2005) conducted an experiment to determine what the effects of three skeletal preparation techniques (boiling bone in water, in bleach, and in sodium carbonate) would be on subsequent DNA processing. They found that bleach cleaning resulted in statistically significant reduction in DNA yields compared to the both water and sodium carbonate (detergent) cleaning. In the second published report, Adlam and Simmons (2007) sought to determine the effects of researchers as taphonomic agents and the accuracy of the models built from decomposition studies. Of interest was how repeated physical disturbances of the remains under study ultimately affected the results of such studies. While no overall differences were notable between the repeatedly disturbed samples and the control samples, they did observe some differences in temperature and weight loss between the treatment and control groups. Thus, researchers ought to be cognizant of their own impacts when carrying out taphonomic research.

Despite the lack of published material on this topic, proceedings from recent meetings indicate research in this type of taphonomic research is growing. For example, Latham et al. (2003) and Steadman et al. (2003) investigated the effects of heated maceration on the ability to amplify DNA from processed remains. Edson et al. (2006) sought to determine how various long-term storage conditions affect the degradation of mitochondrial DNA of skeletal remains. Baker et al. (2008) found that using instant glue (cyanoacrylate) to help preserve teeth did not ultimately affect subsequent DNA analyses. Kemp et al. (2009) assessed the effects of papain (a protease derived from unripened papaya fruit) and ethylenediamine tetra-acetic acid (EDTA; a common chelating agent used in commercially available papain) on the processing of remains; whereas EDTA was found to be destructive to bone, papain alone was found to be an effective defleshing agent. Finally, Pope (2009) investigated a variety of taphonomic agents that are regularly involved in collecting remains from fire scenes, ranging from efforts to suppress the fire to search and recovery of the remains to the effects of transporting these fragile specimens.

THE FUTURE OF FORENSIC TAPHONOMY

Hazarding a guess as to the future position of forensic taphonomy is an exercise in dead reckoning. Although we might chart the course based on where we have been, no one knows for sure what unexpected gale might blow in to push us off into uncharted territory. Thus, we offer the following based on the trends we are currently witnessing, in addition to areas that we believe need additional exploration.

While all of the categories of forensic taphonomic research that we outlined above are important and will continue to generate research, we believe certain areas hold significant potential to stimulate a good deal of research. First, decomposition studies are likely to remain paramount as questions related to the PMI take on more importance in forensic anthropology. Although observational studies and case reports offer useful regional information, future studies that control for the multiple variables of decomposition will shed additional light on these processes. Second, forensic anthropology should continue to focus on the biasing aspects of taphonomy. Although forensic anthropology should be applauded for illustrating the positive aspects of taphonomic research, researching how certain taphonomic factors (e.g., fire, laboratory processing) might obscure the information of primary interest (e.g., positive identification, trauma analysis, PMI) is a necessity and will only strengthen the field. Third, there is a continued need for additional research related to trauma analysis. Determining whether a traumatic fracture is perimortem or postmortem is a question that continues to elude both paleozoological and forensic taphonomists. While there is no doubt that there will be boundaries as to what information can be gleaned from these traumata, these boundaries need to be delineated. Finally, weathering is relatively unexplored in the forensic literature. Although there appears to be a perception that weathering data will be of little use, or that it will take too much time to learn what weathered bones will mean forensically, this perception is untested. Weathering need not imply decades of exposure; bones will begin to "weather" as soon as they are exposed. Thus, short-term weathering patterns may, in fact, produce meaningful results; the work simply needs to be done.

CONCLUSION

Paleozoological taphonomy originated because the fossil record was a reflection, at best imperfect and at worst analytically misleading, of a past faunal community (biocoenose). The fossil record had a distinct mode of occurrence relative to a biocoenose. Since paleobiologists wanted to do (paleo)biology (behavioral or evolutionary) and (paleo) ecology, they had to determine how to make the fossil record amenable to such analyses. That requirement resulted in the emergence and development of taphonomy early in the twentieth century. Paleozoological taphonomy continues to grow as we discover how to learn what a collection of bones and teeth means in terms of paleoecology and evolution, and also in terms of learning what we cannot know or ever hope to know, given the unique mode of occurrence of the fossil record. Forensic taphonomy represents a significant expansion in forensic science from a focus on victim identification to deciphering criminal acts perpetrated against a victim, criminal efforts to obscure and destroy evidence, and distinguishing postmortem modifications that have nothing to do with a crime from those that will prove central to criminal prosecution.

The expansions of paleozoological taphonomy and of forensic taphonomy are, in hindsight, what we might expect as we learn more about what happens to bodies – human and nonhuman – once death has arrived. The expansions are logical outgrowths of all such empirical kinds of inquiry. To date, they have been largely independent of one another. Yet this is, we think, precisely where the two kinds of taphonomic inquiry can benefit one another. Paleozoological taphonomists need to pay attention to forensic taphonomists if for no other reason than to constantly remind themselves that the pile of bones on the lab table was once part of an animate organism whose behaviors likely influenced the taphonomy of its remains. Similarly, forensic taphonomists need to pay attention to developments in paleozoology given its relatively broad focus on everything that happens to a carcass between death and publication. We believe the cross-pollination of the two fields will be of major benefit to each.

REFERENCES

Adlam, R.E. and Simmons, T. (2007). The effects of repeated physical disturbance on soft tissue decomposition – are taphonomic studies an accurate reflection of decomposition? *Journal of Forensic Sciences* 52: 1007–1014.

Allaire, M.T. and Manhein, M.H. (2009). Shark-inflicted trauma on human skeletal remains [abstract]. Proceedings of the 61st Annual Meeting of the American Academy of Forensic Sciences, February 16–21, Denver, p. 317, abstract H34.

Baker, L.E., Meadows, Jantz, L., Baktash, Y.M., and Pearce, J.R. (2008). Preservation of skeletal collections: the viability of DNA analysis after the application of chemical preservative [abstract]. *Proceedings of the 60th Annual Meeting of the American Academy of Forensic Sciences*, February 18–23, Washington DC, p. 337, abstract H64.

Bass, W.M. (1983). The occurrence of Japanese trophy skulls in the United States. *Journal of Forensic Sciences* 28: 800–803.

Bass, W.M. (1997). Outdoor decomposition rates in Tennessee. In W.D. Haglund and M.H. Sorg (eds), *Forensic Taphonomy: the Postmortem Fate of Human Remains* (pp. 181–186). CRC Press, Boca Raton, FL.

Bassett, H.E. and Manheim, M.H. (2002). Fluvial transport of human remains in the lower Mississippi River. *Journal of Forensic Sciences* 47: 1–6.

Beary, M.O. and Cabo, L.L. (2008). Estimation of bone exposure duration through the use of spectrophotometric analysis of surface bleaching and its applications in forensic taphonomy [abstract]. *Proceedings of the 60th Annual Meeting of the American Academy of Forensic Sciences*, February 18–23, Washington DC, p. 315, abstract H24.

Behrensmeyer, A.K. (1978). Taphonomic and ecologic information from bone weathering. *Paleobiology* 4: 150–162.

Behrensmeyer, A.K. (1984). Taphonomy and the fossil record. *American Scientist* 72: 558–566.

Behrensmeyer, A.K., Kidwell, S.M., and Gastaldo, R.A. (2000). Taphonomy and paleobiology. In D.H. Erwin and S.W. Wing (eds), *Deep Time: Paleobiology's Perspective* (pp. 103–147). Paleontological Society, Lawrence, KA.

Bradtmiller, B. and Buikstra, J.E. (1984). Effects of burning on human bone microstructure: a preliminary study. *Journal of Forensic Sciences* 29: 535–540.

Brain, C.K. (1969). The contribution of Namib Desert Hottentots to an understanding of australopithecine bone accumulations. *Scientific Papers of the Namib Desert Research Station* 39: 13–22.

Brand, H.J. (2008). The effects of carcass weight on the decomposition of pigs (*Sus scrofa*) [abstract]. *Proceedings of the 60th Annual Meeting of the American Academy of Forensic Sciences*, February 18–23, Washington DC, p. 324, abstract H41.

Brewer, V.L. (2005). Observed taphonomic changes and drift trajectory of bodies recovered from the tidal Thames, London England: a 15-year retrospective study [abstract]. *Proceedings of the 57th Annual Meeting of the American Academy of Forensic Sciences*, February 21–26, New Orleans, LA, p. 286, abstract H7.

Cadée, G.C. (1990). The history of taphonomy. In S.K. Donovan (ed.), *The Process of Fossilization* (pp. 3–21). Columbia University Press, New York.

Calce, S.E. and Rogers, T.L. (2007). Taphonomic changes to blunt force trauma: a preliminary study. *Journal of Forensic Sciences* 52: 519–527.

Carter, D.O., Yellowlees, D., and Tibbett, M. (2008). The reliability of cadaver decomposition: can non-enteric microbes rapidly contribute to cadaver breakdown in soil? [abstract]. *Proceedings of the 60th Annual Meeting of the American Academy of Forensic Sciences*, February 18–23, Washington DC, p. 316, abstract H26.

Castellano, M.A., Villanueva, E.C., and von Frenckel, R. (1984). Estimating the date of bone remains: a multivariate study. *Journal of Forensic Sciences* 29: 527–534.

Chibba, K.N. (2008). The influence of body fat on the rate of decomposition in traumatized pigs [abstract]. *Proceedings of the 60th Annual Meeting of the American Academy of Forensic Sciences*, February 18–23, Washington DC, p. 321, abstract H37.

Christensen, A.M. (2002). Experiments in the combustibility of the human body. *Journal of Forensic Sciences* 47: 466–470.

Cleland, C.E. (2001). Historical science, experimental science, and the scientific method. *Geology* 29: 987–990.

Cotton, G.E., Aufderheide, A.C., Goldschmidt, V.G. (1987). Preservation of human tissue immersed for five years in fresh water of known temperature. *Journal of Forensic Sciences* 32: 1125–1130.

Cross, P.A. (2008). The influence of penetrative trauma on the rate of decomposition [abstract]. *Proceedings of the 60th Annual*

Meeting of the American Academy of Forensic Sciences, February 18–23, Washington DC, p. 316, abstract H27.

Curtis, J., Turk, C.M., and Ritke, M.K. (2003). Using restriction enzymes to reduce inhibitory properties of bacterial DNA on PCR amplification of human DNA sequences [abstract]. *Proceedings of the 55th Annual Meeting of the American Academy of Forensic Sciences*, February 17–22, Chicago, IL, p. 254, abstract H29.

Dabbs, G.R. (2008). Predicting the location of scattered human remains: when will heads roll and where will they go? [abstract]. *Proceedings of the 60th Annual Meeting of the American Academy of Forensic Sciences*, February 18–23, Washington DC, p. 325, abstract H43.

Dabbs, G.R. (2009). Decomposition of Sharpey's fibers in estimating postmortem interval [abstract]. *Proceedings of the 61st Annual Meeting of the American Academy of Forensic Sciences*, February 16–21, Denver, CO, p. 299, abstract H3.

Damann, F.E., Leney, M., and Bunch, A.W. (2002). Predicting mitochondrial DNA (mtDNA) recovery by skeletal preservation [abstract]. *Proceedings of the 54th Annual Meeting of the American Academy of Forensic Sciences*, February 11–16, Atlanta, GA, p. 218, abstract H9.

Dart, R.A. (1949). The predatory implemental technique of the australopithecines. *Am J of Phys Anthropol* 7: 1–16.

Dart, R.A. (1957). The osteodontokeratic culture of *Australopithecus Prometheus*. Transvaal Museum Memoir No. 10. Pretoria.

Dautartas, A.M. (2009). The effects of coverings on the rate of human decomposition [abstract]. *Proceedings of the 61st Annual Meeting of the American Academy of Forensic Sciences*, February 16–21, Denver, CO, p. 300, abstract H5.

De Gruchy, S. and Rogers, T.L. (2002). Identifying chop marks on cremated bone: a preliminary study. *Journal of Forensic Sciences* 47: 1–4.

Dirkmaat, D.C., Cabo, L.L., Ousley, S.D., Symes, S.A. (2008). New perspectives in forensic anthropology. *Yearbook of Physical Anthropology* 51: 33–52.

Di Vella, G., Campobasso, C.P., and Introna, F. (2003). Peculiar marine taphonomy findings: preservation of human remains as a result of submersion in sequestered environments [abstract]. *Proceedings of the 55th Annual Meeting of the American Academy of Forensic Sciences*, February 17–22, Chicago, IL, p. 255, abstract H31.

Dodson, P. (1980). Vertebrate burials. *Paleobiology* 6: 6–8.

Duffy, J.B., Skinner, M.F., and Waterfield, J.D. (1991). Rates of putrefaction of dental pulp in the Northwest Coast environment. *Journal of Forensic Sciences* 36: 1492–1502.

Dumser, T.K. and Türkay, M. (2008). Postmortem changes of human bodies on the Bathyal Sea Floor – two cases of aircraft accidents above the open sea. *Journal of Forensic Sciences* 53: 1049–1052.

Dupras, T.L., Lang, J.E., Reay, H.L., Schultz, J.J., Falsetti, A.B., and Palma, N.A. (2002). Masking identity: the effects of corrosive household agents on soft tissue, bone, and dentition [abstract]. *Proceedings of the 54th Annual Meeting of the American Academy of Forensic Sciences*, February 11–16, Atlanta, GA, p. 218, abstract H10.

Ebbesmeyer, C.C. and Haglund, W.D. (1994). Drift trajectories of a floating human body simulated in a hydraulic model of Puget Sound. *Journal of Forensic Sciences* 39: 231–240.

Edson, S.M., Barritt, S.M., Leney, M.D., and Smith, B.C. (2006). MtDNA from degraded human skeletal remains: is quality affected by storage conditions? [abstract]. *Proceedings of the 58th Annual Meeting of the American Academy of Forensic Sciences*, February 20–25, Seattle, WA, p. 322, abstract H84.

Efremov, I.A. (1940). Taphonomy: new branch of paleontology. *Pan-American Geologist* 74: 81–93.

Efremov, I.A. (1958). Some considerations on biological bases of paleozoology. *Vertebrat Palasiatic* 2: 83–98.

Fisher, D.L., Holland, M.M., Mitchell, L., Sledzik, P.S., Wilcox, A.W., Wadhams, M., and Weedn, V.W. (1993). Extraction, evaluation, and amplification of DNA from decalcified and undecalcified United States Civil War bone. *Journal of Forensic Sciences* 38: 60–68.

Franicevic, B. (2008). Basement bodies: the effect of light on decomposition in indoor settings [abstract]. *Proceedings of the 60th Annual Meeting of the American Academy of Forensic Sciences*, 2008 February 18–23, Washington DC, p. 318, abstract H31.

Fredericks, J.D. and Simmons, T. (2008). DNA quantification of burned skeletal tissue [abstract]. *Proceedings of the 60th Annual Meeting of the American Academy of Forensic Sciences*, February 18–23, Washington DC, p. 323, abstract H39

Fulginiti, L.C., Di Modica, F., Hartnett, K., and Karluk, D. (2009). Sealed for your protection II: the effects of corrosive substances on human bone and tissue [abstract]. *Proceedings of the 61st Annual Meeting of the American Academy of Forensic Sciences*, February 16–21, Denver, CO, p. 352, abstract H90.

Galloway, A., Birkby, W.H., Jones, A.M., Henry, T.E., and Parks, B.O. (1989). Decay rates of human remains in an arid environment. *Journal of Forensic Sciences* 34: 607–616.

Guatame-Garcia, A.C., Camacho, L.A., and Simmons, T. (2008). Computer simulation for drift trajectories of objects in the Magdalena River, Columbia [abstract]. *Proceedings of the 60th Annual Meeting of the American Academy of Forensic Sciences*, February 18–23, Washington DC, p. 319, abstract H33.

Haglund, W.D. (1992). Contribution of rodents to postmortem artifacts of bone and soft tissue. *Journal of Forensic Sciences* 37: 1459–1465.

Haglund, W.D. (1993). Disappearance of soft tissue and the disarticulation of human remains from aqueous environments. *Journal of Forensic Sciences* 38: 806–815.

Haglund, W.D. and Sorg, M.H. (eds) (1997). *Forensic Taphonomy: the Postmortem Fate of Human Remains*. CRC Press, Boca Raton, FL.

Haglund, W.D. and Sorg, M.H. (eds) (2002). *Advances in Forensic Taphonomy: Method, Theory, and Archaeological Perspectives*. CRC Press, Boca Raton, FL.

Haglund, W.D., Reay, D.T., and Snow, C.C. (1987). Identification of serial homicide victims in the "Green River murder" investigation. *Journal of Forensic Sciences* 32: 1666–1675.

Haglund, W.D., Reay, D.T., and Swindler, D.R. (1988). Tooth mark artifacts and survival of bones in animal scavenged human skeletons. *Journal of Forensic Sciences* 33: 985–997.

Haglund, W.D., Reay, D.T., and Swindler, D.R. (1989). Canid scavenging/disarticulation sequence of human remains in the Pacific Northwest. *Journal of Forensic Sciences* 34: 587–606.

Haglund, W.D., Reichert, D.G., and Reay, D.T. (1990). Recovery of decomposed and skeletal human remains in the "Green River murder" investigation: Implications for medical examiner/coroner and police. *American Journal of Forensic Medicine and Pathology* 11: 35–43.

Heaton. V,G. and Simmons, T. (2007). The decomposition of human remains recovered from the River Clyde, Scotland: a comparative study of UK fluvial systems [abstract]. *Proceedings of the 59th Annual Meeting of the American Academy of Forensic Sciences*, February 19–24, San Antonio, TX, p. 367, abstract H76.

Herrmann, N.P. and Bennett, J.L. (1999). The differentiation of traumatic and heat-related fractures in burned bone. *Journal of Forensic Sciences* 44: 461–469.

Herrmann, N.P., Bassett, B., and Jantz, L.M. (2004). High velocity fluvial transport: a case study from Tennessee [abstract]. *Proceedings of the 56th Annual Meeting of the American Academy of Forensic Sciences*, February 16–21, Dallas, TX, p. 282, abstract H24.

Howard, S.J. and Meyer, J. (2007). Estimating time since death from human skeletal remains by radioisotope and trace element analysis [abstract]. *Proceedings of the 59th Annual Meeting of the American Academy of Forensic Sciences*, February 19–24, San Antonio, TX, p. 375, abstract H87.

Hughes, C.E. and White, C.A. (2008). Taphonomy and dentition: understanding postmortem crack propagation in teeth [abstract]. *Proceedings of the 60th Annual Meeting of the American Academy of Forensic Sciences*, February 18–23, Washington DC, p. 307, abstract H11.

Hurst, S.L. and Manhein, M.H. (2002). Aquatic decomposition rates in south central Louisiana [abstract]. *Proceedings of the 54th Annual Meeting of the American Academy of Forensic Sciences*, February 11–16, Atlanta, GA, p. 236, abstract H43.

Huxley, A.K. (1998). Analysis of shrinkage in human fetal diaphyseal lengths from fresh to dry bone using Petersohn and Köhler's data. *Journal of Forensic Sciences* 43: 423–426.

Huxley, A.K. and Kósa, F. (1999). Calculation of percent shrinkage in human fetal diaphyseal lengths from fresh bone to carbonized and calcined bone using Petersohn and Köhler's data. *Journal of Forensic Sciences* 44: 577–583.

Jans, M.M.E., Tyrrell, A.J., Loreille, O., and Kars, H. (2008). Early diagenesis of bone and DNA preservation [abstract]. *Proceedings of the 60th Annual Meeting of the American Academy of Forensic Sciences*, February 18–23, Washington DC, p. 323, abstract H40.

Johnson, A., Bytheway, J.A., and Pustilnik, S.M. (2009). Skeletal remains in a fluvial environment: microscopic evidence of glycoproteinous adhesive of *Balanus improvises* on the occlusal surface of mandibular teeth [abstract]. *Proceedings of the 61st Annual Meeting of the American Academy of Forensic Sciences*, February 16–21, Denver, CO, p. 355, abstract H95.

Kahana, T., Aleman, I., Botella, M.C., and Hiss, J. (2005). Dismembered bodies – who, how, and when [abstract]. *Proceedings of the 57th Annual Meeting of the American Academy of Forensic Sciences*, February 21–26, New Orleans, LA, p. 332, abstract H84.

Kemp, B.J., Siegel, M.I., Judd, M.A., Mooney, M.P., and Cabo-Pérez, L.L. (2009). The effects of papain and EDTA on bone in the processing of forensic remains [abstract]. *Proceedings of the 61st Annual Meeting of the American Academy of Forensic Sciences*, February 16–21, Denver, CO, p. 303, abstract H10.

Kiley, S.A., Parr, N.M., and Nawrocki, S.P. (2006). Extensive rat modification of a human skeleton from central Indiana [abstract]. *Proceedings of the 58th Annual Meeting of the American Academy of Forensic Sciences*, February 20–25, Seattle, WA, p. 306, abstract H56.

Kirkham, W.R., Andrews, E.E., Snow, C.C., Grape, P.M., and Snyder, L. (1977). Postmortem pink teeth. *Journal of Forensic Sciences* 22: 119–131.

Klippel, W.E. and Synstelien, J.A. (2007). Rodents as taphonomic agents: bone gnawing by brown rats and gray squirrels. *Journal of Forensic Sciences* 52: 765–773.

Komar, D.A. (1998). Decay rates in a cold climate region: a review of cases involving advanced decomposition from the medical examiner's office in Edmonton, Alberta. *Journal of Forensic Sciences* 43: 57–61.

Komar, D.A. and Beattie, O. (1998). Postmortem insect activity may mimic perimortem sexual assault clothing patterns. *Journal of Forensic Sciences* 43: 792–796.

Kontanis, E.J. (2003). Nuclear DNA preservation in soft and osseous tissues [abstract]. *Proceedings of the 55th Annual Meeting of the American Academy of Forensic Sciences*, February 17–22, Chicago, IL, p. 264, abstract H46.

Kowalewski, M. and Kelley, P.H. (eds) (2002). *The Fossil Record of Predation*. Paleontological Society Special Paper No. 8. Yale Printing Service, New Haven, CT.

Lagden, A.C. and Simmons, T. (2008). Decomposition scoring as a method for estimating the postmortem submersion interval of human remains recovered from United Kingdom rivers – a comparative study [abstract]. *Proceedings of the 60th Annual Meeting of the American Academy of Forensic Sciences*, February 18–23, Washington DC, p. 326, abstract H46.

Latham, K.E. (2003). Using amplification of bacteriophage lambda DNA to detect PCR inhibitors in skeletal DNA [abstract]. *Proceedings of the 55th Annual Meeting of the American Academy of Forensic Sciences*, February 17–22, Chicago, IL, p. 264, abstract H45.

Latham, K.E., Zambrano, C.J., and Ritke, M. (2003). The effect of heat associated with maceration on the DNA preservation in skeletal remains [abstract]. *Proceedings of the 55th Annual Meeting of the American Academy of Forensic Sciences*, February 17–22, Chicago, IL, p. 253, abstract H28.

Lawrence, D.R. (1968). Taphonomy and information losses in fossil communities. *Geological Society of America Bulletin* 79: 1315–1330.

Lawrence, D.R. (1971). The nature and structure of paleoecology. *Journal of Paleontology* 45: 593–607.

Leher, T.L. (2004). Home is where the bones are: rat nesting behavior as a tool in forensic investigations [abstract]. *Proceedings of the 56th Annual Meeting of the American Academy of Forensic Sciences*, February 16–21, Dallas, TX, p. 283, abstract H27.

Lyman, R.L. (1985). Bone frequencies: Differential transport, *in situ* destruction, and the MGUI. *Journal of Archaeological Science* 12: 221–236.

Lyman, R.L. (1994). *Vertebrate Taphonomy*. Cambridge University Press, New York.

Lyman, R.L. (2004). The concept of equifinality in taphonomy. *Journal of Taphonomy* 2: 15–26.

Lyman, R.L. and Fox, G.L. (1997). A critical review of bone weathering as an indication of bone assemblage formation. In W.D. Haglund and M.H. Sorg (eds), *Forensic Taphonomy: the Postmortem Fate of Human Remains* (pp. 223–247). CRC Press, Boca Raton, FL.

Manhein, M.H., Listi, G., and Leitner, M. (2003). The landscape's role in dumped and scattered remains [abstract]. *Proceedings of the 55th Annual Meeting of the American Academy of Forensic Sciences*, February 17–22, Chicago, IL, p. 256, abstract H32.

Mann, R.W., Bass, M.A., and Meadows, L. (1990). Time since death and decomposition of the human body: variables and observations in case and experimental field studies. *Journal of Forensic Sciences* 35: 103–111.

Martin, R.E. (1999). *Taphonomy: a Process Approach*. Cambridge University Press, Cambridge.

McKeown, A.H. and Bennett, J.L. (1995). A preliminary investigation of postmortem tooth loss. *Journal of Forensic Sciences* 40: 755–757.

Megyesi, M.S. (2002). The effects of temperature on the decomposition rate of human remains [abstract]. *Proceedings of the 54th Annual Meeting of the American Academy of Forensic Sciences*, February 11–16, Atlanta, GA, p. 216, abstract H7.

Megyesi, M.D., Nawrocki, S.P., and Haskell, N.H. (2005). Using accumulated degree-days to estimate the postmortem interval from decomposed human remains. *Journal of Forensic Sciences* 50: 1–9.

Miller, R.A. (2002). The role of clothing in estimating time since death [abstract]. *Proceedings of the 54th Annual Meeting of the American Academy of Forensic Sciences*, February 11–16, Atlanta, GA, p. 237, abstract H44.

Morse, D., Crusoe, D., and Smith, H.G. (1976). Forensic archaeology. *Journal of Forensic Sciences* 21: 323–332.

Mundorff, A.Z. and Bartelink, E.J. (2007). DNA preservation of skeletal elements from the World Trade Center disaster: some recommendation for mass disaster management [abstract]. *Proceedings of the 59th Annual Meeting of the American Academy of Forensic Sciences*, February 19–24, San Antonio, TX, p. 385, abstract H105.

Murray, K.A. and Rose, J.C. (1993). The analysis of cremains: a case study involving the inappropriate disposal of mortuary remains. *Journal of Forensic Sciences* 38: 98–103.

Nelson, R. (1992). A microscopic comparison of fresh and burned bone. *Journal of Forensic Sciences* 37: 1055–1060.

O'Brien, T.G. and Kuehner, A.C. (2007). Waxing grave about adipocere: soft tissue change in an aquatic context. *Journal of Forensic Sciences* 52: 294–301.

Olson, E.C. (1980). Taphonomy: its history and role in community evolution. In A.K. Behrensmeyer and A.P. Hill (eds), *Fossils in the Making* (pp. 5–19). University of Chicago Press, Chicago, IL.

Perry, P.A. (2005). Human decomposition in the Detroit River [abstract]. *Proceedings of the 57th Annual Meeting of the American Academy of Forensic Sciences*, February 21–26, New Orleans, LA, p. 285, abstract H6.

Perry, W.L., Bass, W.M., Riggsby, W.S., and Sirotkin, K. (1988). The autodegradation of deoxyribonucleic acid (DNA) in human rib bone and it relationship to the time interval since death. *Journal of Forensic Sciences* 33: 144–153.

Pickering, T.R. (2001). Carnivore voiding: a taphonomic process with the potential for the deposition of forensic evidence. *Journal of Forensic Sciences* 46: 406–411.

Pickering, T., Schick, K., and Toth, N. (eds) (2007). *Breathing Life into Fossils: Taphonomic Studies in Honor of C. K. (Bob) Brain.* Stone Age Institute Press, Gosport, IN.

Pinheiro, J., Cunha, E., Pissarra, H., and Real, F.C. (2009). Eaten or attacked by his own dogs? From the crime scene to a multidisciplinary approach [abstract]. *Proceedings of the 61st Annual Meeting of the American Academy of Forensic Sciences*, February 16–21, Denver, CO, p. 322, abstract H41.

Pope, E.J. (2009). From scene to seen: Post-fire taphonomic changes between the *in situ* context and the medicolegal examination of burned bodies [abstract]. *Proceedings of the 61st Annual Meeting of the American Academy of Forensic Sciences*, February 16–21, Denver, CO, p. 366, abstract H114.

Pope, E.J. and Smith, O.C. (2004). Identification of traumatic injury in burned cranial bone: an experimental approach. *Journal of Forensic Sciences* 49: 1–10.

Quatrehomme, G., Bolla, M., Muller, M., Rocca, J., Grévin, G., Bailet, P., and Ollier, A. (1998). Experimental single controlled study of burned bones: contribution of scanning electron microscopy. *Journal of Forensic Sciences* 43: 417–422.

Rankin, D.R., Narveson, S.D., Birkby, W.H., and Lai, J. (1996). Restriction fragment length polymorphism (RFLP) analysis on DNA from human compact bone. *Journal of Forensic Sciences* 41: 40–46.

Rathbun, T.A. and Rathburn, B.C. (1984). Human remains recovered from a shark's stomach in South Carolina. *Journal of Forensic Sciences* 29: 269–276.

Rea, C.N. and Back, H.O. (2002). Determining postmortem interval: a preliminary examination of postmortem thorium, actinium, and radium isotopes in bone [abstract]. *Proceedings of the 54th Annual Meeting of the American Academy of Forensic Sciences*, 2002 February 11–16, Atlanta, GA, p. 238, abstract H47.

Reeves, N.M. (2008). Taphonomic effects of vulture scavenging [abstract]. *Proceedings of the 60th Annual Meeting of the American Academy of Forensic Sciences*, February 18–23, Washington DC, p. 319, abstract H32.

Rennick, S.L., Fenton, T.W., Foran, D.R. (2005). The effects of skeletal preparation techniques on DNA from human and non-human bone. *Journal of Forensic Sciences* 50: 1–4.

Richter, R. (1928). Aktuopalaontologie und Palaobiologie, eine Abgrenzung. *Senckenbergiana* 10: 285–292.

Rodriguez, W.C. and Bass, W.M. (1983). Insect activity and its relationship to decay rates of human cadavers in east Tennessee. *Journal of Forensic Sciences* 28: 423–432.

Rodriguez, W.C. and Bass, W.M. (1985). Decomposition of buried bodies and methods that may aid in their location. *Journal of Forensic Sciences* 30: 836–852.

Rogers, R.R., Eberth, D.A., and Fiorillo, A.R. (eds) (2007). *Bonebeds: Genesis, Analysis, and Paleobiological Significance.* University of Chicago Press, Chicago, IL.

Schäfer, W. (1972). *Ecology and Palaeoecology of Marine Environments.* University of Chicago Press, Chicago, IL.

Schoenly, K., Griest, M.D., and Rhine, S. (1991). An experimental field protocol for investigating the postmortem interval using multidisciplinary indicators. *Journal of Forensic Sciences* 36: 1395–1415.

Scott, A.L. (2009). Taphonomic degradation to bone through scavenging by marine mollusks of the class Polyplacophora [abstract]. *Proceedings of the 61st Annual Meeting of the American Academy of Forensic Sciences*, February 16–21, Denver, CO, p. 355, abstract H94.

Shamblin, C. and Manhein, M. (2003). Two miles and nine years from home: the taphonomy of aqueous environments [abstract]. *Proceedings of the 55th Annual Meeting of the American Academy of Forensic Sciences*, February 17–22, Chicago, IL, p. 251, abstract H24.

Shean, B.S., Messinger, L., and Papworth, M. (1993). Observations of differential decomposition on sun exposed v. shaded pig carrion in coastal Washington State. *Journal of Forensic Sciences* 38: 938–949.

Shepherd, K.L., Walsh-Haney, H.A., Rao, V.J., Wardak, K.S., Bulic, P., and Roberts, C. (2009). Decomposition variables: a comparison of skeletal remains recovered after long-term submersion in Florida aquatic environments [abstract]. *Proceedings of the 61st Annual Meeting of the American Academy of Forensic Sciences,* February 16–21, Denver, CO, p. 354, abstract H93.

Shipman, P. (1981). *Life History of a Fossil: an Introduction to Taphonomy and Paleoecology.* Harvard University Press, Cambridge, MA.

Simpson, G.G. (1961). Some problems of vertebrate paleontology. *Science* 133: 1679–1689.

Snow, C.C. (1982). Forensic anthropology. *Annual Review of Anthropology* 11: 97–131.

Sorg, M.H. (1986). Scavenger modifications of human skeletal remains in forensic anthropology. *Paper presented at the 38th annual meeting of the American Academy of Forensic Sciences,* February 10–15, New Orleans, LA.

Spennemann, D.H.R. and Franke, B. (1995). Decomposition of buried human bodies and associated death scene materials on coral atolls in the tropical Pacific. *Journal of Forensic Sciences* 40: 356–367.

Steadman, D.W., Wilson, J., Sheridan, K.E., and Tammariello, S. (2003). Impact of heat and chemical maceration on DNA recovery and cut mark analysis [abstract]. *Proceedings of the 55th Annual Meeting of the American Academy of Forensic Sciences,* February 17–22, Chicago, IL, p. 250, abstract H23.

Stuart, J. (2003). The effects of human body mass on the rate of decomposition [abstract]. *Proceedings of the 55th Annual Meeting of the American Academy of Forensic Sciences,* February 17–22, Chicago, IL, p. 257, abstract H35.

Symes, S.A., Kroman, A.M., Myster, S.M.T., Rainwater, C.W., and Matia, J.J. (2006). Anthropological saw mark analysis of bone: what is the potential of dismemberment interpretation? [abstract]. *Proceedings of the 58th Annual Meeting of the American Academy of Forensic Sciences,* February 20–25, Seattle, WA, p. 301, abstract H48.

Synstelien, J.A., Klippel, W.E., and Hamilton, M.D. (2005). Raccoon (*Procyon lotor*) foraging as a taphonomic agent of soft tissue modification and scene alteration [abstract]. *Proceedings of the 57th Annual Meeting of the American Academy of Forensic Sciences,* February 21–26, New Orleans, LA, p. 333, abstract H86.

Tersigni, M.A. (2007). Frozen human bone: microscopic investigation. *Journal of Forensic Sciences* 52: 16–20.

Thew, H.A. and Nawrocki, S.P. (2002). The effects of lime on the decomposition rate of buried remains [abstract]. *Proceedings of the 54th Annual Meeting of the American Academy of Forensic Sciences,* February 11–16, Atlanta, GA, p. 237, abstract H45.

Thompson, T.J.U. (2005). Heat induced dimensional changes in bone and their consequences for forensic anthropology. *Journal of Forensic Sciences* 50: 1–8.

Tucker, B.K., Hutchinson, D.L., Gilliland, M.F.G., Charles, T.M., Daniel, H.J., and Wolfe, L.D. (2001). Microscopic characteristics of hacking trauma. *Journal of Forensic Sciences* 46: 234–240.

Ubelaker, D.H. (1997). Taphonomic applications in forensic anthropology. In W.D. Haglund and M.H. Sorg (eds), *Forensic Taphonomy: the Postmortem Fate of Human Remains* (pp. 77–90) CRC Press, Boca Raton, FL.

Ubelaker, D.H. (2001). Contributions of Ellis R. Kerley to forensic anthropology. *Journal of Forensic Sciences* 46: 773–776.

Ubelaker, D.H. and Sperber, N.D. (1988). Alterations in human bones and teeth as a result of restricted sun exposure and contact with corrosive agents. *Journal of Forensic Sciences* 33: 540–548.

Ubelaker, D.H. and Adams, B.J. (1995). Differentiation of perimortem and postmortem trauma using taphonomic indicators. *Journal of Forensic Sciences* 40: 509–512.

Ubelaker, D.H. and Hunt, D.R. (1995). The influence of William M. Bass III on the development of American forensic anthropology. *Journal of Forensic Sciences* 40: 729–734.

Ubelaker, D.H., Buchholz, B.A., and Stewart, J. (2006). Evaluation of date of death through analysis of artificial radiocarbon in distinct human skeletal and dental tissues [abstract]. *Proceedings of the 58th Annual Meeting of the American Academy of Forensic Sciences*, February 20–25, Seattle, WA, p. 316, abstract H74.

Vass, A.A., Bass, W.M., Wolt, J.D., Foss, J.E., and Ammons, J.T. (1992). Time since death determinations of human cadavers using soil solution. *Journal of Forensic Sciences* 37: 1236–1253.

Vass, A.A., Barshick, S., Sega, G., Caton, J., Skeen, J.T., Love, J.C., and Synstelien, J.A. (2002). Decomposition chemistry of human remains: a new methodology for determining the postmortem interval. *J Foresnic Sci* 47: 542–553.

Voorhies, M.R. (1969). *Taphonomy and Population Dynamics of an Early Pliocene Vertebrate Fauna, Knox County, Nebraska*. Contributions to Geology, Special Paper No. 1. University of Wyoming, Laramie, WY.

Warren, M.W. and Maples, W.R. (1997). The anthropometry of contemporary commercial cremation. *Journal of Forensic Sciences* 42: 417–423.

Warren, M.W. and Schultz, J.J. (2002). Post-cremation taphonomy and artifact preservation. *Journal of Forensic Sciences* 47: 656–659.

Weigelt, J. (1927). *Rezente wibeltierleichen und ihre paläobiologische bedeutng*. Max Weg Verlag, Leipzig.

Weigelt, J. (1989). *Recent Vertebrate Carcasses and their Paleobiological Implications* (English translation of Weigelt 1927, by J. Schaefer). University of Chicago Press, Chicago, IL.

Weitzel, M.A. (2005). A report of decomposition rates of a special burial type in Edmonton, Alberta from and experimental field study. *Journal of Forensic Sciences* 50: 1–7.

Weitzel, M.A. (2006). Temperature variability in the burial environment [abstract]. *Proceedings of the 58th Annual Meeting of the American Academy of Forensic Sciences*, February 20–25, Seattle, WA, WA, p. 299, abstract H45.

Wheatley, B.P. (2008). Perimortem or postmortem bone fractures? An experimental study of fracture patterns in deer femora. *Journal of Forensic Sciences* 53: 69–72.

White, R.E. (2007). Decomposition in a mass grave and the implications for past mortem interval estimates [abstract]. *Proceedings of the 59th Annual Meeting of the American Academy of Forensic Sciences*, February 19–24, San Antonio, TX, p. 367, abstract H75.

Wieberg, D.A.M. and Wescott, D.J. (2008). Estimating the timing of long bone fractures: correlation between the postmortem interval, bone moisture content, and blunt force trauma fracture characteristics. *Journal of Forensic Sciences* 53: 1028–1034.

Willey, P. and Snyder, L.M. (1989). Canid modification of human remains: implications for time-since-death estimation. *Journal of Forensic Sciences* 34: 894–901.

Williams, J.A. (2007). Bone fragmentation created by a mechanical wood chipper [abstract]. *Proceedings of the 59th Annual Meeting of the American Academy of Forensic Sciences*, February 10–15, New Orleans, LA, p. 326, abstract H10.

Wilson, M.V.H. (1988). Taphonomic processes: information loss and information gain. *Geoscience Canada* 15: 131–148.

Yan, F., McNally, R., Kontanis, E.J., and Sadik, O.A. (2001). Preliminary quantitative investigation of postmortem adipocere formation. *Journal of Forensic Sciences* 46: 609–614.

Zimmerman, M.R. (1978). The mummified heart: a problem in medicolegal diagnosis. *Journal of Forensic Sciences* 23: 750–753.

CHAPTER 26

Forensic Anthropologists in Medical Examiner's and Coroner's Offices: A History

*Hugh E. Berryman
and Alicja K. Lanfear*

The second Luetgert murder trial in 1894 and the expert testimony of George A. Dorsey of the Columbia Field Museum in Chicago launched physical anthropology as a valuable judicial tool (Stewart 1978). Throughout the early and middle twentieth century, physical anthropologists employed in academic settings were increasingly called upon by law-enforcement agencies to examine skeletal remains to ascertain identity (Stewart 1972). Wilton Krogman's 1939 publication in the *FBI Law Enforcement Bulletin* entitled "A guide to the identification of human skeletal material" stimulated interest in the medicolegal potential held by physical anthropology, and served as the earliest material evidence of a new specialty taking root. Krogman's expanded version of this bulletin appeared in 1962 under the title of *The Human Skeleton in Forensic Medicine*, and provided a formal scholarly toolkit for early practitioners (Krogman 1939, 1962).

In 1972, the American Academy of Forensic Science introduced Physical Anthropology as a new section and during the 1970s forensic anthropology courses began to appear on many campuses and graduates began to look beyond academia for employment. Austin and Fulginiti (2008) noted 19 forensic anthropologists in full-

A Companion to Forensic Anthropology, First Edition. Edited by Dennis C. Dirkmaat.
© 2012 John Wiley & Sons Ltd. Published 2015 by John Wiley & Sons Ltd.

time positions with medical examiner's offices in the USA and an additional nine who were employed with shared duties. But it was in the late 1970s when physical anthropologists were first being considered for employment in medical examiner's and coroner's offices, although their employers often struggled for justification considering the infrequent and sporadic nature of traditional anthropology casework. The information contained here was first presented in 2009 as a paper at the 61st Annual Meeting of the American Academy of Forensic Sciences in Denver, CO, USA (Berryman 2009). Due to the varied titles, and nature of work undertaken by forensic anthropologists in medical examiner's and coroner's offices, it was necessary to interview these anthropologists directly. Most of the information provided in this chapter was updated through e-mail contact with those anthropologists presented herein, which is indicated by the term "personal communication" and the date of contact. This chapter presents the positions in temporal order from earliest to most recent and identifies the corresponding anthropologists, their degree upon employment, and the duties they accepted.

In September 1980, Hugh Berryman, PhD ABD ("all but dissertation"), was employed as Director of the Shelby County Medical Examiner's Morgue and the University of Tennessee Hospital Morgue in Memphis, TN. A position for an anatomist or physical anthropologist with a PhD was advertized and the interviews for the position were conducted by a hospital pathologist and a forensic pathologist. The forensic pathologist was Dr J.T. Francisco, Chief Medical Examiner for Shelby County and the State of Tennessee. Dr Francisco had gained notoriety as the pathologist who had carried out the autopsies of Dr Martin Luther King and Elvis Presley. When Berryman was hired for this position it marked the first full-time employment of a forensic anthropologist in a medical examiner's office in the USA. Since forensic anthropologists were commonly only associated with skeletal analysis, early employers were faced with the challenge of identifying a job title encompassing duties that would keep the anthropologist busy between bone cases. In Memphis, the position of Morgue Director was upgraded by hiring an individual with a degree with the purpose of bringing an academic presence to the morgue and to the medical examiner's office. With administrative duties providing justification, this position insured the presence of a forensic anthropologist for the sporadic skeletal casework that would arise from time to time. As with most of the anthropologists in these early positions, the range of skills they demonstrated was soon adapted for additional duties and responsibilities. Berryman was hired with a PhD ABD in September 1980 and graduated in Spring 1981 with his PhD.

In 1981, the New Jersey State Medical Examiner's Office hired Donna Fontana, MS, as the first *state* forensic anthropologist in the nation, but since there was no New Jersey civil service title of forensic anthropologist, her first title was "Forensic Microscopist." Ms Fontana provided forensic anthropology services for 21 counties, and eventually took on added responsibilities "...such as histology, radiology, fingerprinting, facial reconstruction, crime scene excavation, lecturing, and completing National Crime Information Center forms for entry by law enforcement." With time her role "expanded to include all aspects and processing of unidentified remains for identification" (D. Fontana, personal communication, 2011).

Ms Fontana's current title is "Research Scientist I" within the New Jersey State Police (NJSP), where she is Director of the first Forensic Anthropology Laboratory in

New Jersey, a laboratory that she designed. This laboratory includes the human-remains repository for the long-term curation of unidentified remains and "…provides forensic anthropological assistance to the NJSP Office of Forensic Sciences FBI-partnered Regional Mitochondrial DNA Laboratory." Ms Fontana "initiated the creation of a central dental repository that houses the dental records and radiographs of all unidentified and missing persons in New Jersey. At the present time, a Dental Initiative, utilizing the expertise of 17 volunteer forensic odontologists, digitizes and codes dental records for entry into national databases such as National Crime Information Center and NamUs" (D. Fontana, personal communication, 2011).

David W. Jones, MD, retired from the Kentucky Medical Examiner's Office in 2002 after 38 years as the Executive Director. In 1982, Dr Jones hired Dr David Wolf as the Commonwealth of Kentucky's first full-time forensic anthropologist. In an e-mail correspondence (personal communication, 2011) Dr Jones describes Dr Wolf and the circumstances of his employment as follows.

My interest in employing a forensic anthropologist for our office came about in 1977. On May 28, 1977 the Beverly Hills Nightclub in Southgate Kentucky caught on fire. The fire became one of the most deadly fires in the history of the USA resulting in numerous injuries and fatalities. I was requested by Dr Fred Stine, the Campbell County Coroner, to come and assist him with the body recovery and establishing a temporary morgue. We were having some trouble identifying a couple of the bodies that were badly burned. Dr Robert Reichert the Coroner in Kenton County said that if we had a forensic anthropologist they would be very helpful in the identification of some of our unidentified bodies. I asked Dr Reichert if he knew where we could locate a forensic anthropologist. He thought there was one in Knoxville, Tennessee but he did not know his name. We eventually identified all 167 bodies from the Beverly Hills Nightclub Fire and I returned to the Kentucky Medical Examiner's Office in Frankfort.

I was determined that the next time the Kentucky Medical Examiner's Office had to identify skeletal or charred human remains we would have a forensic anthropologist to assist us. I was told that if I would contact Dr Clyde Snow he may know of a forensic anthropologist that might be interested in working in Kentucky. When I contacted Dr Snow he told me there was a forensic anthropologist already working in Kentucky. He was not working as a forensic anthropologist but was employed at the University of Kentucky working on a longevity project between the USA and Russian governments.

I called Dr Wolf at the University of Kentucky Department of Anthropology and ask if he might be interested in providing forensic anthropology services the Kentucky Medical Examiner's Office. He said he was very interested and he would like to meet with me to discuss this possibility. In September 1977 the State of Kentucky had a contract with Dr David J. Wolf to provide forensic anthropology services on a case-by-case basis.

Dr Wolf's duties were to perform forensic anthropology services for the Kentucky Medical Examiner's Office in accordance with the provisions in the Kentucky coroner and medical examiner statutes. While serving in this capacity he would conduct medicolegal death investigations on skeletalized, burned, decomposed, and/or mutilated bodies in order to ascertain the identity and cause and manner of death. At the conclusion of any medicolegal death investigation conducted he would submit a copy of his report to the coroner having jurisdiction and an additional copy to the Kentucky Medical Examiner's Office.

In July 1982 Dr Wolf became a full-time employee of the Kentucky Medical Examiner's Office. His duties remained the same as they were in his state contract. The only additional duties were to provide training in forensic anthropology to the Kentucky coroners and Kentucky law enforcement agencies.

Dr Wolf worked for the Kentucky Medical Examiner's Office from 1982 till his untimely death from colon cancer in 1992. Life was never dull when David was around. He loved life and he loved to work a challenging death investigation. The coroners of Kentucky were very fond of David. They thought so much of him that after his death the Kentucky Coroners Association inaugurated an award in his name that is considered the most prestigious award presented by the Kentucky Coroners Association. This award is presented annually to an individual that the Association feels has contributed to improving the medicolegal death investigation system in Kentucky.

In 1983, Berryman reorganized the Shelby County Morgue, eliminating two autopsy technician positions, and hired Craig Lahren, a Master's-level physical anthropologist, as Assistant Director. This position served as an interface with the physicians and technicians, enhanced the anthropology influence at autopsy, and facilitated research in the morgue. It also represented the first employment of more that one anthropologist in a medical examiner's office.

In 1984, William Rodriguez, III, PhD, completed his pioneering work at the Anthropology Research Facility, University of Tennessee, Knoxville, and was employed as Deputy Chief Coroner and Forensic Anthropologist for the Caddo Parish Coroner's Office in Louisiana. Between the Caddo Parish office and their private office in Bossier City, this system provided autopsy services for two-thirds of the state. A high murder rate coupled with the subtropical climate produced a large number of decomposed and skeletal remains in Louisiana providing the impetus for a forensic anthropologist. Of all early anthropologists working with medical examiner's and coroner's offices, Dr Rodriguez's duties were perhaps the most eclectic. In addition to forensic anthropology casework, he oversaw investigations including death scenes and rape examinations, and as Deputy Chief Coroner he was also responsible for psychiatric commitments. One of Dr Rodriguez's most memorable psychiatric commitments was a lady known as the "Monkey Woman." This woman would on occasion wonder around the courthouse totally nude with her pet monkey on her shoulder. In 1986, Dr Rodriguez left Louisiana for Syracuse, New York, to become Forensic Anthropologist and Chief of Operations for the Onondaga County Medical Examiner's Office. The medical examiner's office provided services for a large number of counties and it was their workload involving decomposed and skeletal cases that created a need for this position (W. Rodriguez, personal communication, 2011).

Also in 1986, Craig Lahren left Memphis for Chattanooga, TN, to become Coordinator of Forensic Services for the Hamilton County Medical Examiner's Office. Hamilton County's medical examiner, Dr Frank King, had worked as a forensic fellow in the Memphis morgue and was familiar with Mr Lahren's administrative abilities, his experience at autopsy, and his skills as a forensic anthropologist. Dr King hired Mr Lahren in 1986 shortly after becoming medical examiner for Hamilton County.

In Memphis, the position for Assistant Morgue Director vacated by Craig Lahren was immediately filled by Robert Mann, MA, who, after 8 months, left for a position at the Smithsonian Institution in Washington DC. In 1986 Steve Symes, MA, had been hired as Morgue Director for the Metropolitan Nashville/Davidson County Medical Examiner. On his first day of work he discovered that his desk was literally located next to the autopsy table. Dr Berryman replaced Bob Mann, at the Regional Forensic Center, with Steve Symes as Assistant Morgue Director in 1987. In 1990, the Shelby County Medical Examiner's Morgue became the Regional Forensic Center and Berryman and Symes worked there together until 2000 when Berryman left Memphis for Nashville to do private consulting and Dr Symes – who had completed his doctorate on saw marks in May 1992 – became Forensic Anthropologist for the Regional Forensic Center, Memphis, TN.

In 1979, William Haglund was hired as a medical investigator at the King County Medical Examiner's Office in Seattle, WA. At the time of his hiring, he had a degree from the California College of Mortuary Science, Los Angeles, CA, and a BS in Biology from the University of California, Irvine. As a medical investigator, Haglund responded to all deaths in his jurisdiction including sudden undetermined or suspicious deaths, homicides, suicides, accidental deaths, and deaths not under the care of a physician. His responsibilities included scene responses – except for homicides – collection of evidence, family interviews, death notifications, and report writing. In 1983, he became Chief Medical Investigator and was responsible for the supervision and training of medicolegal death investigators and autopsy support staff, along with budgetary and facilities maintenance responsibilities. It was the 1982 investigations of the Green River Killings that kindled his interest in forensic anthropology (Haglund et al. 1990). After obtaining his Master's degree in physical anthropology from the University of Washington in 1988, forensic anthropology duties were added to his responsibilities. In 1990 he received his PhD and in late 1995 Dr Haglund resigned from the King County Medical Examiner's Office to take a position as Senior Forensic Advisor for the United Nations (W. Haglund, personal communication, 2009).

In 1989, Dr William Rodriguez left Syracuse to accept a position as Chief Forensic Anthropologist and Deputy Chief Medical Examiner for Special Investigations for the Armed Forces Institute of Pathology, Washington DC. The position was necessitated by the need for a forensic anthropologist that would be able to travel at any time to handle high-profile military and federal cases, as well as mass fatalities. Unlike other forensic anthropology positions in medical examiner's offices, this one required all professional staff to be weapon-qualified annually and they were often issued a weapon for many forensic missions abroad (W. Rodriguez, personal communication, 2011).

The hiatus in new forensic anthropology positions between 1990 and 1993 was broken in 1994 when Dr Emily Craig filled the position of Kentucky State Forensic Anthropologist, left open by the untimely death of Dr David Wolf in 1992. Dr Craig's academic and experiential background included a MS degree in 1976 from the Medical College of Georgia, and from 1976 to 1991 she worked as a professional medical illustrator and photographer at the Hughston Sports Medicine Foundation in Columbus, GA. Dr Craig received her PhD in anthropology from the University of Tennessee, Knoxville, where she enjoyed the tutelage of Dr William M. Bass. Immediately upon graduating in 1994, she was hired by Dr George Nichols, Kentucky's Chief Medical Examiner, who valued her academic training in forensic anthropology, her familiarity with human anatomy and medicine, her talent as a medical illustrator (applied to accurately illustrating evidence), and her expertise in facial sculpture. Her duties included providing traditional and expected forensic anthropology services for the entire state, mass fatality planning and response, musculoskeletal trauma analysis in fresh cases, and maintenance of the state's database of unidentified dead. In December 2010 Dr Craig retired and fiscal restraints prevent the position from being filled at this time (E. Craig, personal communication, 2011). Dr Craig notes that "…after Wolf died it was budget constraints that kept that job 'closed' for 2 years. During that time some major cases made authorities realize the value of a forensic anthropologist. Hopefully this job will … again open up when the economy improves. It is [was] one of the best jobs a forensic anthropologist could wish for" (E. Craig, personal communication, 2011).

In 1994 shortly after graduating from the University of Tennessee, Knoxville, Gwen Haugen, MA, became a full-time forensic investigator/anthropologist for the St. Louis County Medical Examiner's Office. Her skills as a forensic anthropologist represented an advantage when she applied for the forensic investigator vacancy. Additionally, she provided forensic anthropology consultation for the surrounding Franklin, Jefferson, and St. Charles County Medical Examiner's Offices. In 1997, she left after accepting a position as forensic anthropologist with the Joint Prisoners of War, Missing in Action Accounting Command Central Identification Laboratory (JPAC-CIL) in Hawaii and then returned to the St. Louis County Medical Examiner's Office in 2007. Her duties require her to conduct field recovery and forensic analysis of recovered osseous remains. In addition, she routinely provides skeletal trauma consultations during autopsy at the request of the forensic pathologists. Typically, excavation and/or surface collection of remains and supporting evidence is made in conjunction with law-enforcement personnel. In addition to providing basic forensic anthropology analysis, Ms Haugen establishes and maintains chain of custody for all recovered/received evidence and compiles detailed reports of recovery and analysis for distribution to involved agencies. Ms Haugen also inputs all unknown cases into the NamUs database and manages the collection and submission of all DNA evidence for identification. While Ms Haugen was at JPAC, from 1997 to 2007, two forensic anthropologists filled this position: Gina Overshiner, MA (1997–2001) and Matt Venemeyer, MA (2001–2005) (G. Haugen, personal communication, 2011).

In June 1996, the Tarrant County Medical Examiner's Office, Fort Worth, TX, advertized for a primary forensic anthropologist and part-time trace analyst. Dr Dana Austin was hired in the Criminalist I position where she advanced to her current position as Senior Forensic Anthropologist. Earlier two other anthropologists had been employed as trace analysts – Max Houk in 1993 and Teresa Woltanski in 1995 – who occasionally worked forensic anthropology cases. Originally, the anthropology caseload was not thought to warrant a full-time anthropology position, so Dr Austin was initially asked to train in trace analysis. By 2003 the anthropology caseload was well established and she was formally released from trace analysis duties. In 2006, the anthropology line was officially moved to the newly created Human Identification Laboratory, which is headed by a forensic odontologist and includes two fingerprint specialists (D. Austin, personal communication, 2011).

In 1985 forensic anthropology services were established at the University of Massachusetts by Dr Ann Marie Mires to provide consultation services to law enforcement and medical examiners. In 1996, Dr Mires created the position of Forensic Anthropologist at the Office of Chief Medical Examiner (OCME) in Boston and was employed to cover the entire state. She was instrumental in developing the Identification Unit and her duties were those typically associated with a forensic anthropologist, including scene recovery, skeletal profiling, coordinating identification procedures, trauma interpretation, and courtroom testimony. Additionally, Dr Mires served as a "…liaison to law enforcement, the Massachussetts Historical Commission, and Native American Communities." During her tenure at the OCME, Dr Mires "… worked to develop standard operating procedures in identification and emergency response while coordinating OCME participation with local, state, and national agencies." Her position was reduced to a half-time appointment with benefits in 2005 and continued with the state until 2009 (A.M. Mires, personal communication, 2011).

Craig Lahren, MA, left Chattanooga in the summer of 1997 and was hired as Administrator and Forensic Anthropologist for the North Dakota State Forensic Examiner's Office in Bismarck, ND. Mr Lahren functioned in an administrative role as well as providing forensic anthropology services in much the same way he had done in Chattanooga (C. Lahren, personal communication, 2008). Craig Lahren passed away on April 16, 2009.

In July 1997, Tom Bodkin, MA, replaced Craig in Chattanooga as Investigative Assistant to the Medical Examiner. His activities included forensic anthropology cases, autopsy prosection, radiography, photography, external body examination, death investigation, scenes, discussion of cases with families (including nonanthropology cases), research, and public lectures, and he also functioned as Occupational Safety and Health Administration (OSHA) compliance officer. Mr Bodkin left the position in December 2007 (T. Bodkin, personal communication, 2011).

In 1997 Laura Fulginiti, PhD, was hired as Forensic Anthropologist for the Maricopa County Medical Examiner's Office in Phoenix, AZ, where she had been working as a contract employee for 5 years. One year later her salary was needed to augment a forensic pathologist position, and she was placed back under contract as a consultant. In October 2010, Dr Fulginiti and forensic odontologist Dr John Piakis were hired as half-time salaried contract employees (19.5 hours per week) (L. Fulginiti, personal communication, 2011).

Also in 1997, Fran Wheatley, MA, was employed as a death investigator and forensic anthropologist for the Metropolitan Nashville/Davidson County Medical Examiner. Ms Wheatley had just graduated the previous year from the University of Tennessee, Knoxville, with Dr William M. Bass as her mentor. Her duties required her to investigate death scenes, as did other death investigators, but she would also work all of the basic bone scenes for several counties in the Middle Tennessee area. Dr Hugh Berryman served as her backup and provided the board certification needed for National Association of Medical Examinations (NAME) accreditation. In 2006, Ms Wheatley moved to Memphis to become the Administrator of the Shelby County Medical Examiner's Office (F. Wheatley, personal communication, 2011). Dr Berryman provided forensic anthropology consultation for the Nashville office in Ms Wheatley's absence. In 2010, Ms Wheatley returned to the Metropolitan Nashville/Davidson County Medical Examiner's Office as Chief Administrator, but has limited duties involving forensic anthropology. Dr Berryman continues to work the forensic anthropology cases for this office on a consultancy basis.

In 1999 Amy Mundorff completed her Master's degree and became the first full-time forensic anthropologist for the OCME, New York City. During her Master's degree at California State University, Chico, she spent the summer as an unpaid intern at the Medical Examiner's Office in New York City. When she graduated she wrote the Chief informing him that it was time he hired a forensic anthropologist and strongly recommended herself for the position. He hired Ms Mundorff, but had her cross train in serology in the Department of Forensic Biology out of concerns that the anthropology caseload would not justify the position. However, the bone cases were sufficient, preventing her from working a single serology case. Initially, she worked on skeletal cases from the five boroughs (Manhattan, Brooklyn, Queens, the Bronx, and Staten Island) and during down times she worked on backlogged cases. Ms Mundorff slowly began to educate the medical examiners on the fact that forensic anthropologists have valuable skills in the autopsy room. She became involved in grand rounds and eventually began to remove trauma specimens at autopsy and provide trauma interpretation. Ms Mundorff provided testimony in New York, New Jersey, and in federal courts. She won the favor of law enforcement by identifying nonhuman cases while at the scene, thus eliminating unnecessary work, and eventually began to respond to all found cases. She further demonstrated her utility by providing lectures to medicolegal investigators, fellows, law enforcement, district attorneys, and the public. Her position as supervising anthropologist was integral following the World Trade Center (WTC) disaster of September 11, 2001, where she assisted in establishing standards and procedures for the WTC Human Identification Project. She also established procedures for and participated in the victim identification process following the 2001 crash of American Airlines flight 587 and the 2003 Staten Island Ferry crash. Mundorff resigned in 2004 to continue her education (A. Mundorff, personal communication, 2011). She secured the New York position with initiative, innovation, and by continually expanding her duties.

Jennifer C. Love, PhD, worked at the Regional Forensic Center, Memphis, from 2001 to 2006 under several job titles. Initially her tenure overlapped with that of Dr Steve Symes, and from September 2001 to July 2003 she worked as autopsy assistant and assistant forensic anthropologist performing autopsies and bone trauma casework under his supervision. From July 2003 to approximately August 2004 she was involved in skeletal and bone trauma analysis, scene response, report writing, testimony, and maintenance of the evidence collection under the title of Forensic Anthropologist. From August 2004 to July 2006 Dr Love, as Morgue Supervisor and Forensic Anthropologist, oversaw training, scheduling, standard operating procedure development, and quality assurance for the morgue staff of 13, served as the safety officer, provided scene response and skeletal and bone trauma analysis, issued reports, testified, and maintained the evidence collection. From July 2006 to August 2006, she worked under her final title of Forensic Anthropologist and was responsible for providing skeletal and bone trauma analysis, scene response, report writing, testimony, and maintenance of the evidence collection (J. Love, personal communication, 2011).

In 2002 the Georgia Bureau of Investigation's Medical Examiner's Office employed Rick Snow, MA, as Forensic Anthropologist after more than 300 bodies were discovered at Tri-State Crematorium in Noble, North Georgia. The position was initially funded through a 2-year grant to analyze the skeletal collection housed there; however, his duties further consisted of assisting law enforcement in recovery and analysis of current cases. At the termination of the 2-year grant-funded position his employment continued on a full-time basis. Mr Snow's responsibilities at the Georgia Bureau of Investigation were to help law enforcement in the location, recovery, and analysis of human remains and trauma analyses at autopsy. While with the Georgia Bureau of Investigation he "...worked over 300 forensic cases for more than 80 law-enforcement agencies, over 100 of which were homicides ... [and] conducted more than 60 recoveries of human remains including the excavation of eight wells....

[He also] took a long-neglected skeletal collection that was located in a storage room, organized it, analyzed the remains, and entered them into National Crime Information Center and NamUs" (R. Snow, personal communication, 2011). Mr Snow completed his PhD while at the Bureau, but state fiscal problems resulted in his being laid off in January 2009 (Kris Sperry, Chief Medical Examiner for the State of Georgia, Georgia Bureau of Investigation Division of Forensic Science, personal communication, 2009; R. Snow, personal communication, 2011).

Also in 2002 Gina Hart, MA, was employed as Forensic Investigator and Forensic Anthropologist with the Regional Medical Examiner's Office in Newark, NJ, a position likely motivated by the September 11, 2001 attacks. This office was under the indirect jurisdiction of the state medical examiner, was responsible for a four-county area, and was one of the two regional offices in the state. Her day-to-day duties included forensic investigations, scenes, maceration, and skeletal analysis. Ms Hart is also involved in the state medical examiner's mass disaster planning (G. Hart, personal communication, 2011).

In 2004, the King County Medical Examiner's Office in Seattle, WA, placed Katherine Taylor, PhD, in a newly established full-time forensic anthropology position. Motivation for the position was influenced by the arrest of Gary Ridgeway, the Green River Killer, and his agreement to take law enforcement to the dumpsites for 48 of his victims. Dr Taylor had worked with the King County Medical Examiner's Office since August 1996 as an investigator with a specialty in anthropology and then in 2004 as a full-time anthropologist. Dr Taylor noted that the "arrest of Gary Ridgeway did spark interest in having a full-time anthropologist but really our chief medical examiner had been lobbying for a position for over 5 years. He recognized that having an anthropologist on staff in a different position (investigator) meant having to pull the anthropologist away from regular duties when anthro cases came up and that was not fair to the anthropologist or to the other investigators. It was also necessary to have someone in charge of the … [cold cases]. In addition, King County has a long history of prominent skeletal remains cases, first with Ted Bundy and then with Gary Ridgeway … so the powers that be are more cognizant of the need for … [a forensic anthropologist]. The timing of the position creation (2004) does suggest that Ridgeway's confession in 2003 was a prominent player but really it was only a small part of the puzzle. The activity around … [Gary Ridgeway] made the need more obvious to the King County Council, but the medical examiner management had known, and had been fighting for the position, long before" (K. Taylor, personal communication, 2011). It is noteworthy that Dr Richard Harruff, King County's Chief Medical Examiner, had previously worked with Berryman, Lahren, Mann, and Symes in Memphis.

Dr Taylor's duties include scene processing, skeletal analysis, trauma analysis, specific identification, skeletal inventory and curation, cold-case identification involving coordination with state and federal agencies, providing training to law enforcement, crime laboratory personnel, King County Medical Examiner's Office fellows, and other community partners, and mass fatality response planning. In 2011, due to budget constraints, the position was cut to three-quarters time (K. Taylor, personal communication, 2011).

In 1997 Bradley Adams accepted a position at the Central Identification Laboratory in Hawaii and worked at this facility until 2004 when he filled the Director vacancy left in the New York City Medical Examiner's Office by Amy Mundorff. Over the course of time, Dr Adams hired seven forensic anthropologists to assist in various aspects of the New York Medical Examiner's Office operation (B. Adams, personal communication, 2011). Jeannette Fridie, MA (hired 2004, switched full-time to the Anthropology Unit in 2005), is primarily involved with forensic anthropology case analysis. Christian Crowder, PhD (hired 2006), is the Assistant Director of the Forensic Anthropology Unit. He is also the WTC Site Coordinator and is responsible for managing OCME's monitoring operations at the WTC site. Benjamin Figura, PhD (hired 2006), is responsible for oversight of the unclaimed and unidentified WTC remains. He also assumed the role of Director of the Identification Department in 2009. Scott Warnasch, MA (hired 2006), works primarily with WTC duties, especially the archaeological monitoring of construction activities at the WTC site for evidence of human remains. Jennifer Godbold, BA (hired 2006), works primarily with the unidentified and unclaimed WTC remains. She also manages information related to New York City's unidentified decedents. Christopher Rainwater, MS (hired 2007), is primarily involved with forensic anthropology case analysis. He also assumed the role of Director of Photography in 2009. Kristen Hartnett, PhD (hired 2007), is primarily involved in forensic anthropology case analysis. She also assists with the archaeological monitoring of construction work at the WTC site for evidence of human remains. Dr Adams states, "The roles and responsibilities of the Anthropology staff are very diverse and span into other operational areas. Regarding strictly anthropological responsibilities, the Forensic Anthropology Unit serves numerous functions, which are shared between all members of the department: (i) consultations for the medical examiners in the five borough offices; (ii) scene assistance with Medicolegal Investigators in the five boroughs, especially with outdoor scenes or when skeletal remains are discovered; (iii) management and oversight of the WTC remains temporarily housed at the OCME; (iv) daily monitoring of construction around the WTC site for potential human remains and for the oversight of excavation and/or sifting of material potentially containing WTC human remains; (v) critical members of OCME's disaster preparedness and special operations capabilities; (vi) data collection and information management/dissemination regarding New York City's unidentified decedents; (vii) training OCME staff, law enforcement, and death investigators on scene recovery techniques and anthropological analyses" (B. Adams, personal communications, 2011). Without question, the large number of forensic anthropologists at New York City's OCME is due primarily to the work associated with the 2001 terrorist attacks of the WTC and the agency's subsequent recognition of the important role that forensic anthropologists can play in medical examiner operations.

In 2000, Bruce Anderson, PhD, began work as a medicolegal investigator for the Pima County Office of the Medical Examiner (PCOME) in Tucson, AZ, where for 30 years Dr Walt Birkby had functioned as a contract forensic anthropologist. In 2003, Dr Anderson's duties as medicolegal investigator were expanded to include forensic anthropology consultations. According to Dr Anderson, the expanded duties resulted from the drastic increase in border-crossing deaths in southern Arizona between 2000 and 2005. Motivated by the large number of cases, a full-time forensic anthropologist position was funded in 2006 and Dr Anderson was hired. His duties included forensic anthropology consultation for the Forensic Science Center and numerous other counties in Arizona. Dr Anderson maintains a missing persons file and routinely makes comparisons between it and identifications he has made. Dr Anderson has "…performed no fewer than 100 forensic anthropology consults each year since 2006, and … over 150 consults in 2010…. The PCOME now has over 500 unidentified cases dating back 10 years … [and has] currently identified approximately 1500 of the 2000 border-related deaths…. The PCOME now has more than 800 unresolved missing persons reports for missing migrants." Dr Anderson states that because there are "…many cases for a sole practitioner to perform on an annual basis, I am pleased to report that in 2011 I will receive assistance from three pre-doctoral and one post-doctoral positions in forensic anthropology" (B. Anderson, personal communication, 2011).

The Harris County Institute of Forensic Sciences (HCIFS) Forensic Anthropology Division (FAD) in Houston, TX – created in September 2006 – is staffed by three full-time PhD anthropologists and one postdoctoral fellow. The FAD Director of Forensic Anthropology is Dr Jennifer C. Love, Forensic Anthropologist I/Mass Fatality Coordinator is Dr Jason M. Wiersema, and Agency Coordinator/Forensic Anthropologist is Dr Sharon M. Derrick. The FAD is tasked with providing basic forensic anthropology services (developing biological profiles, trauma analysis, and establishing identifications through radiographic and other techniques), processing scenes, and processing unidentified and unclaimed remains. FAD anthropologists provide expert testimony, conduct independent research, and provide training. Decedent descriptions are provided to local law enforcement and media, and data are entered into the National Crime Information Center, the National Missing and Unidentified Persons System, and, when appropriate, the National Center for Missing and Exploited Children. They are involved with "…coordinating the submission of a DNA sample for analysis and entry into Combined DNA Index System; and coordinating dental examinations by board-certified forensic odontologists. The division is also responsible for following-up on any leads generated through case dissemination" (J. Love, personal communication, 2011).

Dr Sharon M. Derrick was hired in February 2006 as Agency Coordinator/Anthropologist for the Harris County Institute of Forensic Sciences. She is a physical anthropologist with experience in contract skeletal recovery and analysis and public health surveillance methods. Her employment came after meeting with the chief medical examiner in a professional setting and suggesting that she would be available for skeletal recovery and analyses should a position open with his office. After being hired as Agency Coordinator, she was able to merge anthropology into her scope of work with original duties consisting of analyses of skeletal cases and report writing, and skeletal scene recovery, as well as epidemiological research and support, and acting as liaison with the organ- and tissue-procurement agencies. After on-the-job training and experience in forensics, Dr Derrick's role changed to include more anthropology: she is responsible for one-third of the forensic anthropological casework (S. Derrick, personal communication, 2011).

As Director of FAD, Dr Love oversees daily operations, the development, implementation, and review of standard operating procedures, and the development of the training program. She has successfully created a forensic anthropology fellowship position and developed a summer internship program through grant funding (J. Love, personal communication, 2011).

In September 2006, Dr Wiersema was initially hired as Forensic Anthropology Investigator at the HCIFS FAD. His role included the completion of forensic anthropological casework and medicolegal death scene response. His job title and role changed in 2007 to Forensic Anthropologist I/Disaster Preparedness Coordinator. In this capacity he is responsible for one-third of the forensic anthropological casework and for the HCIFS disaster and mass fatality preparedness efforts (J. Wiersema, personal communication, 2011).

In 2007, as previously discussed, Gwen Haugen returned to the St. Louis County Medical Examiner's Office, while in 2008 Murray Marks, PhD, left an Associate Professor position in the Department of Anthropology, University of Tennessee, Knoxville, and accepted a dual appointment in the Departments of Pathology and Oral Surgery (General Dentistry) at the University of Tennessee, Graduate School of Medicine in Knoxville, Tennessee. The Oral Surgery position involves directing dental histology and pathology research projects and teaching forensic dentistry identification methods to residents. Pathology is his home department where he introduces residents to the interpretation of bone trauma at autopsy. His appointment in the Pathology Department provides an office and two laboratories at the Regional Forensic Center housed at the hospital. He had been providing forensic anthropology services since 2000 and now has three to five cases per week. Dr Mileusnic-Polchan, Chief Medical Examiner for Knox County, created the anthropologist position in the Department of Pathology. According to Dr Marks, "a critical factor in establishing the position was National Association of Medical Examiners (NAME) accreditation, which requires a board-certified forensic anthropologist" (M. Marks, personal communication, 2011).

In 2009, Debra Prince-Zinni, PhD, was hired in a half-time position listed as Environmental Analyst IV by the OCME for Massachusetts. The position title is misleading: Dr Prince-Zinni is the state forensic anthropologist. The main office is headquartered in Boston, MA, but has several regional offices. Dr Prince-Zinni's duties include recovery, analysis, identification, and occasionally trauma assessment (D. Prince-Zinni, personal communication, 2011).

A number of interesting observations can be made from this relatively brief history that begins in 1980 when the Shelby County Medical Examiner's Office in Memphis, TN, hired Hugh Berryman in the first full-time forensic anthropology position. Since then there has been a continual increase in forensic anthropologists hired in such positions until the recent economic downturn. In the 1980s, six Master's-level people and five with PhDs were hired for a total of 11 positions (see Figure 26.2). In the 1990s, five Master's-level and five PhD-level positions were filled for a total of 10. Since 2000, six Master's-level people and 13 with PhDs were hired, as well as one Bachelor's degree person, for a total of 20 positions. Since 1980, a total of 41 full-time positions in forensic anthropology have been filled.

Of the 20 medical examiner/coroner's offices that hired anthropologists, at present 11 have eliminated these positions; however, most offices continue to use anthropologists on a contractual or "as-needed" basis. Four of the 20 offices show a discontinuity in service ranging from 1 to 9 years. The number of offices with full-time anthropologists working at any one time can be seen in Figure 26.1. In 1980 there was one office with one forensic anthropologist, in 1985 there were a total of four offices (with five anthropologists), in 1990 six offices (with seven anthropologists), in

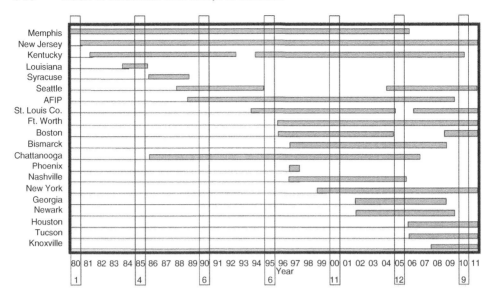

Figure 26.1 Medical examiner's and coroner's offices in the USA employing a forensic anthropologist since 1980 (horizontal bars), and total number of forensic anthropology positions in all offices in the USA (vertical bars). AFIP, Armed Forces Institute of Pathology.

1995 six offices (seven anthropologists), in 2000 11 offices (11 anthropologists), in 2005 12 offices (13 anthropologists), and in 2010 there were nine offices (with 19 anthropologists). Although complex, the stimulus for the development of many of these positions can be associated with specific incidents. Kentucky's state forensic anthropology position – originally held by David Wolf and succeeded by Emily Craig – may be traced to the 1977 Beverly Hills Supper Club disaster in Southgate, Kentucky. The Green River Killer may have been a factor in securing the position in Seattle. The tragedies associated with the waves of illegal aliens who come across the southern border of the USA is likely responsible for the position held by Bruce Anderson in Tucson, AZ. The increase in offices hiring forensic anthropologists correlates with mass fatality incidents such the 1993 first attack on the WTC and in 1995 with the Oklahoma City Bombing (Figure 26.2). The most dramatic increases in the number of offices hiring anthropologists correspond to more recent events such as September 11, 2001 when the WTC was hit for the second time, and, to a lesser extent, the Tri-State Crematorium in Noble, GA, in 2002 (Figure 26.2). In addition, until 2008 the economy, as reflected by the DOW Jones annual averages, showed a steady increase, suggesting that a strong economy is also needed to support the creation of new positions for forensic anthropologists. With the dramatic economic decline in 2008 (Figure 26.2, gray line), there was a marked decrease from 12 offices in 2005 to nine offices in 2011. However, during these past 6 years the number of anthropologists employed in these offices has increased from 13 in 2005 to 18 today. The two offices primarily responsible for these higher numbers are the HCIFS FAD in Houston with three anthropologists and New York City Medical Examiner's Office with eight.

In conclusion, over the past three decades there has been an overall increase in the number of forensic anthropologists hired full-time in medical examiner's and coroner's offices. In addition, of those offices that created full-time positions, 45% have

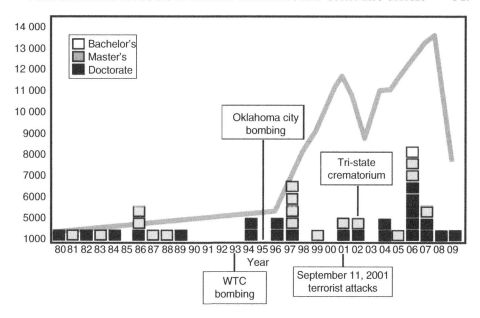

Figure 26.2 Number of forensic anthropology positions (by degree level) in medical examiner's and coroner's offices in the USA by year. Significant national events are noted, as is the DOW Jones average during this time period (gray line).

retained them. There has always been a mix of people with Master's and PhD degrees employed, but very recently a Bachelor's-level position has been filled by the New York office (Figure 26.2). Factors that have influenced the creation of full-time medical examiner/coroner positions for anthropologists include a good economy, mass disasters or high-profile complex cases, and the realization that forensic anthropologists can contribute far beyond skeletal analysis. Scene recovery, taphonomic analysis, and trauma analysis are but a few of the tools the anthropologist can offer a forensic pathologist today. But perhaps the most overlooked and under-appreciated factor responsible for creating full-time employment involves personal initiative, hard work, creativity, and determination.

ACKNOWLEDGMENTS

We thank the forensic anthropologists and pathologists whose personal communication through e-mails and phone conversations made the writing of this chapter possible.

REFERENCES

Austin, A. and Fulginiti, L. (2008). The forensic anthropology laboratory in a medical examiner setting. In M.W. Warren, H.A. Walsh-Haney, and L.E. Freas (eds), *The Forensic Anthropology Laboratory* (pp. 23–46). CRC Press, Boca Raton, FL.

Berryman, H.E. (2009). Full time employment of forensic anthropologists in medical examiner's/coroner's offices in the

United States—a history [abstract]. *Proceedings of the 61st Annual Meeting of the American Academy of Forensic Sciences,* Denver, CO.

Haglund, W.D., Reichert, M.A., Reay, D.G., and Donald, T. (1990). Recovery of decomposed and skeletal human remains in the "Green River Murder" investigation. *American Journal of Forensic Medical Pathology* 11: 35–43.

Krogman, W.M. (1939). A guide to the identification of human skeletal material. *FBI Law Enforcement Bulletin* 8(8): 3–31.

Krogman, W.M. (1962). *The Human Skeleton in Forensic Medicine.* Charles C. Thomas, Springfield, IL.

Stewart, T.D. (1972). *Essentials of Forensic Anthropology: Especially As Developed in the United States.* Charles C. Thomas, Springfield, IL.

Stewart, T.D. (1978). George A. Dorsey's role in the Luetgert case: a significant episode in the history of forensic anthropology. *Journal of Forensic Science* 23(4): 786–791.

PART VII

Forensic Anthropology Beyond Academia

Introduction
to Part VII

Dennis C. Dirkmaat

Most of the forensic anthropology casework completed in the United States is done by academics who serve as consultants to law enforcement or coroner/medical examiner offices. Given past (and some current) definitions of the field in which forensic anthropologists are called upon to primarily, or exclusively, provide clues to the identity of the deceased through laboratory production of skeletal biological profiles, the number of cases in any given jurisdiction in which the skills of the forensic anthropologist are called upon are few and far between. A full-time position for a forensic anthropologist cannot often be justified. However, in a best-case scenario reconsideration of the role of forensic anthropology in medicolegal investigations, forensic anthropologists are not sitting in their laboratory waiting for the dry bones of individuals who cannot be identified through any other identification means (teeth, tissue, or clay) to come up with a better biological profile. They are performing a multitude of tasks: directing well-organized and thorough searches for unlocated remains; providing an important role in the evaluation of forensic significance in real time; leading documentation and recovery efforts at all outdoor crime-scene scenarios from surface scatters to large-scale disaster scenes; conducting laboratory analyses of skeletal remains to provide interpretations of a biological profile and human skeletal trauma; and providing forensic taphonomic analyses to estimate a postmortem interval (PMI) and ultimately "accurate" reconstructions of past events.

The skill sets possessed by most forensic anthropologists are quite expansive (Table VII.1) and encompass not only skeletal analytical and archaeological recovery skills, but photography, radiography, scientific writing, and editing skills, to name a few.

A Companion to Forensic Anthropology, First Edition. Edited by Dennis C. Dirkmaat.
© 2012 Blackwell Publishing Ltd. Published 2012 by Blackwell Publishing Ltd.

Table VII.1   Current skill sets of forensic anthropologists.

Skill sets	Fields involved
Visual identification of complete human remains and, importantly, fragmentary human remains	Osteology, anatomy, physical anthropology
Differentiate complete and fragmentary animal bones from human bones (simple exclusion by elimination not always acceptable in medicolegal contexts)	Zooarchaeology, paleontology, vertebrate zoology
Understand and be experienced with human skeletal variation at many different levels (variation due to chronological age, sex, ancestry, epigenetic, and stochastic/random factors)	Physical anthropology, vertebrate zoology, biometry
Ability to conduct effective and efficient searches for unlocated scenes	Archaeology, surveying (optimal search theory)
Ability to thoroughly document context at outdoor crime scenes through written and photographic means, and by three-dimensional Cartesian coordinate mapping	Geography (archaeology, field ecology, sedimentology)
Ability to archaeologically recover a wide variety of physical evidence (including human remains) from a wide variety of outdoor scenes from surface scatters to buried bodies	Archaeology, paleontology (taphonomy)
Ability to interpret bone-surface modification (the result of a wide variety of taphonomic agents) including animal chewing and swallowing, surface erosion, sun exposure, burial, and burning	Taphonomy, paleontology, oestology, biomechanics
Ability to sample and analyze variables and materials relevant to all environmental factors and conditions that may effect such modification	Field ecology, edaphology , histology, biochemistry, genetics
Ability to separate intentional from natural modification	Taphonomy, paleontology, osteology, biomechanics
Ability to reconstruct human modification, including the class and individual characteristics of the tools employed to produce such modification	Taphonomy, paleontology, osteology, biomechanics, criminalistics

These skills are beneficial in the forensic anthropology laboratory and at the outdoor crime scene. Additionally, forensic anthropologists can work side by side with the forensic pathologist in the preliminary analysis of: (i) human and nonhuman recovered bones, (ii) significantly decomposed remains, (iii) radiographs of tissue, and (iv) skeletal trauma. They are especially valuable in dealing with fragmentary skeletal remains, and taphonomically modified bones, including those that have been altered by heat and fire. All in all, the forensic anthropologist provides a useful sidekick to the forensic pathologist. Given all of these unique skill sets, a full-time position at most middle-sized to large medical examiner's offices is certainly justified.

However, these benefits were not realized until the early 1980s when William Bass promoted the hiring of some of his graduates in the Nashville, TN, Medical Examiner's office and then at the Memphis, TN, Medical Examiner's office. As described in Chapter 26 of this volume, forensic anthropologists prove to be a valuable asset for the medical examiner's office in many different venues, but especially with respect to the analysis of trauma to bone.

Although great gains were made by forensic anthropology work in Memphis by Hugh Berryman and Steve Symes, it was little duplicated throughout the country. However, following the destruction of the World Trade Center buildings in New York City in September 2001, it was realized that forensic anthropology was particularly valuable with respect to dealing with highly fragmented remains. New full-time positions were created in the New York City Medical Examiner's office where today seven full-time forensic anthropologists are employed. The diversity of skills that they employ in their day-to-day jobs is described in Chapter 27 in more detail. The work ranges from field and laboratory work on forensic cases, assisting law enforcement and forensic pathologists, to photographic consultations, histology, expert testimony, and skeletal biology research. In addition, forensic anthropologists are providing leadership roles in the continued search and excavations for human remains associated with the destruction of the World Trade Center buildings. They are not sitting around waiting for dry bones to appear so that their full-time positions can be justified.

Berryman and Lanfear (Chapter 26 in this volume) provide a rather comprehensive history of positions created for trained forensic anthropologists outside academia. Starting with Berryman's position in Memphis in 1980, forensic anthropologists have slowly but surely been able to find gainful employment in medical examiner's offices. Most of the time it is the result of a forensic pathologist realizing the limitations of pathology to address skeletal issues, or the potential benefit of having a forensic anthropologist to deal with fragmented, commingled, and burned bones. Significant upturns in posted positions resulted from other catastrophic events such as the Oklahoma City bombing and the Tri-State Crematorium in northern Georgia. Over 40 full-time positions have been created around the country since 1980, with over 20 in coroner/medical examiner's offices.

The history of forensic anthropology contains many references to the work conducted by physical anthropologists for the military services during wartime. During non-wartime years, this role was no longer needed and the physical anthropologists returned to their day jobs in academia and as curators in museums. However, in the 1990s there was a renewed effort to recover all of the unaccounted war dead in foreign countries. The military expanded their central laboratory in Hawaii primarily to include the identification of often highly fragmented and degraded skeletal tissue recovered from plane crashes, battlefield burials, and even secondary burial features throughout Southeast Asia, but also to include other US military conflicts around the world. Initially, forensic anthropologists were used exclusively in the laboratory to work with the skeletal remains, but it was quickly realized that recoveries conducted by military personnel unfamiliar with archaeological principles and practices was not the answer. Forensic anthropologists trained in forensic archaeology practices were absolutely critical to maximizing field recovery of the war dead. Today the Joint Prisoners of War, Missing in Action Accounting Command-Central Identification Laboratory (JPAC-CIL) is the largest employer of forensic anthropologists is the world and conducts recovery missions with a mixture of anthropologists and military personnel. As described in Chapter 28, forensic anthropologists have assumed leadership roles in the organization as "Recovery Leaders," and even as upper-level management. They explain, via a number of recent cases, what skills are required to provide a role as Recovery Leader and offer a useful template for forensic anthropology students aspiring to work at JPAC-CIL.

CHAPTER **27**

Forensic Anthropology at the New York City Office of Chief Medical Examiner

*Christopher W. Rainwater,
Christian Crowder, Kristen
M. Hartnett, Jeannette S.
Fridie, Benjamin J. Figura,
Jennifer Godbold,
Scott C. Warnasch,
and Bradley J. Adams*

INTRODUCTION

The Office of Chief Medical Examiner of New York City (OCME-NYC) has developed the largest dedicated staff of forensic anthropologists outside of the US Department of Defense owing to New York City's high volume of cases and the complex investigations associated with the September 11, 2001 World Trade Center (WTC) disaster. The Forensic Anthropology Unit (FAU) currently consists of eight full-time anthropologists. The FAU is available to assist the medical examiners with gross, histologic, and radiographic analysis of bone and cartilage. The FAU is also available on call 24 h a day to assist medicolegal investigators (death investigators) with scene search and recovery. While the FAU covers all five boroughs in New York City (Brooklyn, the Bronx, Manhattan, Queens, and Staten Island) with a population of over 8 million people, consultations for medical examiner's offices and various law-enforcement agencies in other jurisdictions are common.

A Companion to Forensic Anthropology, First Edition. Edited by Dennis C. Dirkmaat.
© 2012 John Wiley & Sons Ltd. Published 2015 by John Wiley & Sons Ltd.

T. Dale Stewart classically defined forensic anthropology as "that branch of physical anthropology which, for forensic purposes, deals with the identification of more or less skeletonized remains known to be, or suspected of being, human" (Stewart 1979: ix). By this definition, the forensic anthropologist is a specialist in the reconstruction of the biological profile from skeletal material. Given that mostly skeletonized material represents a minority of forensic cases, Stewart went so far as to suggest, "the word 'branch' in the above definition can be replaced by the word 'sideline' for it is still rare for a physical anthropologist to have fulltime employment in the forensic field" (Stewart 1979: x). It was not long after this publication, however, that anthropologists began securing employment in medical examiner's/coroner's offices (see Chapter 26 in this volume). Furthermore, the understanding that the forensic anthropologist only works with dry skeletal material, or in isolation with remains that cannot benefit from an analysis performed by a forensic pathologist, is no longer a valid concept. As Dirkmaat and colleagues noted, there has indeed been a paradigm shift in the conceptual framework by which the forensic anthropologist operates with specific reference to forensic taphonomy, forensic archaeology, and trauma analysis (Dirkmaat et al. 2008). This shift has opened up the possibility for increased anthropological work in the medical examiner's setting. Additionally, the presence of anthropologists in the medical examiner's office has, in turn, resulted in advancing the classic goal of identification in specialized ways, namely disaster-victim identification. The continually growing role of the forensic anthropologist is prevalent in the work performed by the FAU at the OCME-NYC.

This chapter is intended as an overview of forensic anthropological roles and operations in a medical examiner's office with particular attention to anthropology at the OCME-NYC. As noted by Austin and Fulginiti (2008), anthropologists frequently perform roles in addition to forensic anthropology and this is no different in New York City. Although members of the FAU do not act as death investigators,[1] a common role for forensic anthropologists in many other jurisdictions, the roles and responsibilities of the FAU are diverse and reach into other operational areas. The content of this chapter might be considered as guidelines for potential roles that may be presented to an agency when promoting an anthropology position in a medical examiner's/coroner's office or be beneficial to someone preparing for a career as a professional forensic anthropologist.

History of Forensic Anthropology at the NYC-OCME

Prior to 1999, the OCME-NYC did not have a full-time forensic anthropologist. Various anthropologists performed forensic anthropology on a consultant basis: Leslie Eisenberg in the 1980s and Peggy Caldwell in the 1990s. In 1999, Amy Mundorff became the first full-time forensic anthropologist for the OCME-NYC. In 2004, she left the position to pursue her doctorate degree and Dr Bradley Adams was hired as the Director of Forensic Anthropology. With the forensic anthropology casework growing, Dr Adams began developing the FAU, hiring Jeannette Fridie in 2005 as the second full-time anthropologist. Due to the increasing demand on the unit for forensic anthropological analysis and the continued need for managing WTC victim-identification operations, four more anthropologists were added to the unit.

Dr Christian Crowder, Dr Benjamin Figura, Jennifer Godbold, and Scott Warnasch were hired in 2006. During this period a renewed search and recovery operation began at the WTC site, resulting in the need for further staff to manage the multiple operations established to recover and sift thousands of cubic meters of material, while maintaining the daily forensic anthropology caseload (Sledzik et al. 2009). To alleviate this need, Dr Kristen Hartnett and Christopher Rainwater were hired in 2007.

Over the past 10 years the FAU has grown from initially working out of a small office with one laboratory space in the morgue to currently having three lab spaces, including a hard-tissue laboratory developed by Dr Crowder, and an administrative office. In addition to the expansion of facilities for the FAU, various members of the FAU have been assigned additional roles – Dr Benjamin Figura is the Director of Identification and Christopher Rainwater is the Director of Photography – while maintaining their traditional role as forensic anthropologists. All the members of the FAU are also members of the Medical Examiner's Special Operations Response Team (MESORT) and are certified hazardous materials technicians. Overall, the FAU employs four anthropologists with PhD degrees (two of whom are currently board-certified), three with Master's degrees, and one with a Bachelor's degree.

ROLES OF THE FAU IN NEW YORK CITY

Since 2005, the FAU has seen a general increase in the number of cases each year (Figure 27.1). In 2005, there were a total of 72 cases, including 52 cases involving human skeletal remains and 20 cases that were determined to be nonhuman in origin. Of the 52 human skeletal cases, nine were classified as anatomical or not forensically significant. By 2010, the number of anthropology cases had nearly doubled, totaling 143, 116 of which involved human skeletal remains and 27 cases that were determined to be nonhuman. Of the 116 human skeletal cases, five were classified as not forensically significant. Along with an increasing caseload, the types of analysis have also expanded owing to the increase in staff members, the expertise that each anthropologist brings to the unit, the acquisition of specialized equipment and technology, and the growing scope of forensic anthropology. The number of crime scene visits has

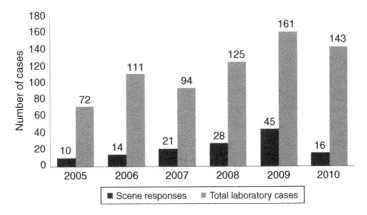

Figure 27.1 Number of cases worked on by the FAU from 2005 to 2010.

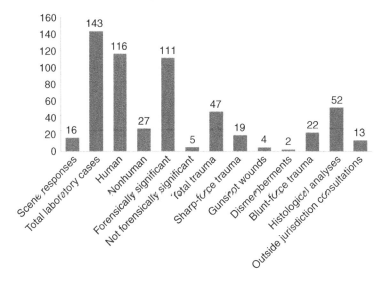

Figure 27.2   Types of case submitted to the FAU in 2010.

fluctuated over the years and is case-dependent; however, on average, the FAU investigates potential crime scenes around two times each month. The FAU is also available for consultation outside of the jurisdiction of New York City. These cases generally require a memorandum of understanding through the legal department. Currently, the OCME-NYC can provide DNA and anthropology services to any jurisdiction in New York State at no cost to the requesting agency through a National Institute of Justice Missing Person's grant. In addition to providing standard DNA and anthropological analyses, NamUs profiles can be entered or updated for jurisdictions that do not have the capability to do so. Figure 27.2 illustrates the types of case received in 2010, and is fairly representative of an average year of anthropological casework in NYC. A more detailed description of the types of cases that are received in the FAU is presented below.

Consultations for medical examiners

Generally, the FAU performs analyses and writes reports at the request of the medical examiner. Cases that may warrant anthropological consultation include badly decomposed or skeletonized remains; remains of questioned human or nonhuman origin or forensic significance; unidentified individuals; skeletal trauma including blunt force, sharp force, gunshot, dismemberment, and thermal alteration; and skeletal pathologies, anomalies, or healed trauma. At the OCME-NYC there are several ways a medical examiner can request an anthropological consultation. The first is a direct request for the anthropologist to be present during an autopsy. Second, a medical examiner removes the tissue during autopsy and submits it to the FAU for analysis. The final method involves reviews of photographs, radiographs, or other documents from a case. Even though the FAU's home office and laboratory space is located in Manhattan in New York City, members of the FAU routinely travel to assist the medical examiners in the outer boroughs with skeletal analysis, interpretation, and decedent identification. Below are descriptions of the typical analyses performed.

Nonhuman cases and nonforensically significant human remains

When skeletal remains are encountered, the FAU is notified. In most cases, a member of the FAU visits the scene in order to determine whether the remains are human and of medicolegal significance. In some circumstances, photographs taken by police or death investigators may be reviewed to alleviate the need for a scene visit. If remains believed to be nonhuman are received at any of the morgue facilities without an anthropological scene visit or review of photographs taken at the time of discovery, an anthropological assessment of the remains will still be completed. The same is true for human cases where the remains appear to be ceremonial, from a cemetery, or from a teaching collection. In the event that the remains are determined to be nonhuman, non-osseous, or not of medicolegal significance, the FAU will generate a short report stating the results. Nonhuman remains are found throughout the city and are often discovered during construction activities around Lower Manhattan associated with the rebuilding efforts of the WTC. A quick determination to confirm that remains are nonhuman in these occurrences is of particular importance considering the ongoing recovery efforts at the WTC site (see the section on WTC operations, below). An assessment of the archaeological context is also important with these finds to verify that the excavated area does not involve potential WTC debris. Regardless of the situation in which nonhuman or non-osseous remains are found, a rapid response and determination by a forensic anthropologist will save time and resources for those impacted by the find (e.g., construction crews, law enforcement, home owners).

Additionally, the FAU often receives cases that involve human remains that are determined to not be forensically significant, such as historical remains and remains that were formerly associated with anatomical teaching collections. The FAU receives on average six cases per year (or 5% of the total annual number of cases) that are human but not forensically significant. Often, these remains are recovered in homes of people in the medical profession or industry, artists, and religious practitioners. New York City has a relatively large number of practitioners of the Afro-Caribbean syncretic religions Santeria and Palo Mayombe (Gill et al. 2009). Ritualistic offerings may include animal remains, beads, shells, coins, metal items, and human remains, among other items (Figure 27.3).

Unidentified individuals

On average, over 20 decedents are received at the OCME-NYC each year that remain unidentified for an extended period of time. Prior to the FAU's involvement in reviewing unidentified cases, the number of long-term unidentified cases averaged 50 individuals per year. In order to provide the most accurate information regarding these unidentified individuals, anthropological input regarding the biological profile (age, sex, ancestry, and stature) is warranted. For badly decomposed or skeletonized bodies, this information may not be attainable without an anthropological consultation. Even for well-preserved bodies, it may be appropriate to consult the forensic anthropologist to assist with an estimation of age at death (as opposed to an estimate based solely on the individual's overall appearance). The general rule at the OCME-NYC is that an unidentified body should not be released for burial without an attempt to accurately determine age, sex, ancestry, and stature. In addition, the FAU ensures that all relevant case information pertaining to the unidentified individual – such as

Figure 27.3   Items recovered from a Palo Mayombe ceremonial site, including human and nonhuman remains.

fingerprints, DNA, dental charting, and full-body X-rays – is accurately and completely collected prior to release for burial. The forensic anthropologist coordinates consultations with forensic odontologists and works closely with the New York Police Department's (NYPD's) Morgue Unit (a subdivision of the NYPD's Missing Persons Division) to ensure that fingerprints are searched through all available databases (including the Department of Homeland Security).

Age, sex, ancestry, and stature are determined through standard anthropological procedures that are chosen based on the state of preservation of the decedent. Specimens for estimating age at death, however, can be easily harvested at autopsy regardless of the stage of decomposition. Before an unidentified decedent is released for burial, a member of the FAU removes the medial clavicles, sternal ends of the fourth ribs, and pubic symphyses, as well as both sixth ribs and the right femur midshafts for histological age estimation. If no soft tissue or blood is available for DNA analysis, the FAU works with the OCME-NYC's Forensic Biology department to determine the most appropriate skeletal element for sampling. The anthropologist would need to thoroughly document the element prior to sampling and may recognize taphonomic processes that could hinder a sufficient DNA yield.

The FAU is also responsible for electronically compiling data on NYC's long-term unidentified decedents, both for current cases as well as archival reviews of past cases that have not yet been identified. Archived case files are reviewed and any information that may be helpful in identifying the decedent – biological profile, descriptions of scars, tattoos, or piercings, evidence of past surgeries, and any clothing or jewelry recovered with the decedent – is recorded. These data are entered into national missing/unidentified databases such as the National Missing and Unidentified Persons System (NamUs) and the National Crime Information Center (NCIC).

NamUs is an online database created by the US National Institute of Justice (NIJ) and contains data on both unidentified decedents and missing persons from across the country. The anthropologist enters all the information collected for each unidentified decedent, as well as recording whether fingerprints, dental X-rays, or DNA are available. Forensic odontologists enter any dental charting recorded from the decedent. This information is then automatically compared to the missing persons section of NamUs, flagging any cases that potentially match a decedent. Additionally, NamUs is available for use by the public, who can search through the database looking for potential matches to an individual who has gone missing. To date the FAU has entered over 1050 cases into NamUs going back to 1990. As a result of these efforts, numerous cold cases have been identified within the past 2 years.

Following a positive identification of remains evaluated by the FAU analysts will submit anonymized case information to the Forensic Anthropology Data Bank (FDB). The FDB was started in 1986 by Dr Richard Jantz at the University of Tennessee at Knoxville, TN, USA, and compiles forensic case information, notably demographic and metric data, from practitioners throughout the world. By protocol at the OCME-NYC, every skull is digitized using a Microscribe with the 3Skull software created by Steven D. Ousley and the data are submitted to the FDB when a decedent becomes identified. In the past few years, the FAU averages approximately 10 submissions to the FDB per year.

In addition to the analysis of unidentified individuals, the FAU is responsible for providing a review of every disinterment performed by the OCME-NYC. A disinterment is warranted when a previously unknown individual is identified and the family requests to bury them elsewhere, or when additional information for investigations is needed from the buried remains. When a decedent is exhumed from a cemetery or City Burial, the remains are transported to the OCME-NYC in the coffin. Upon arrival to the OCME-NYC, the FAU performs a disinterment review, which includes a review of the case records, an inspection of the integrity of the coffin, an examination of the remains present and photographic documentation of the condition of the coffin and remains. Once the review is complete, a member of the FAU compares the case records to the disinterred body to ensure that they are consistent.

Skeletal trauma analysis

The FAU is integral in assisting the medical examiners with skeletal trauma interpretations. Trauma analysis occurs at both the informal level (consultations at autopsy not requiring a report) and at the formal level when specimens are transferred to the custody of the FAU and a forensic anthropology report is generated. Trauma analysis and interpretation constitutes the majority of the annual casework performed by the unit and consists of the following types: blunt force, sharp force, gunshot, and burned (thermal alteration). The anthropologist also assists with classifying whether the trauma is antemortem (healed), perimortem (occurring around the time of the death event), or due to postmortem breakage. If an analysis of trauma is requested, the FAU may assist the medical examiner with harvesting the relevant specimens to evaluate the entire impacted area before any portions are removed. This would be preferred to ensure that the peripheral damage around the impact location is evaluated and the context of the injury is observed. In cases with evidence of sharp-force trauma, the anthropologist should be present during the autopsy and removal of the defect(s) to prevent or observe any autopsy artifacts that may be caused by a scalpel blade or oscillating saw.

Sharp-force and dismemberment trauma constitute the majority of the trauma consultations (approximately 40%) submitted to the FAU. Typically, the analysis of sharp-force trauma involves an assessment of class characteristics (not individualizing) left by an impacting tool, usually a knife or saw, although it may also be helpful to determine the direction and orientation of impacts, when questionable. Similarly, the FAU's analysis of blunt-force and gunshot trauma in bone can assist in the analysis of direction, sequence, and number of impacts. Blunt-force trauma cases are variable and include cases involving impacts with blunt objects, motor vehicle versus pedestrian accidents, suspected strangulation cases (hyoid, thyroid, cricoid fractures), possible elder- (geriatric) and child-abuse cases, and falls from a height. In abuse cases (elder and child abuse) it is of particular importance to document the number, pattern, and timing of injuries to establish a potential history of skeletal trauma. In addition to analysis, the FAU preserves skeletal trauma evidence using high-quality casting procedures (see section on Developing casts and impressions of evidence, below). The preservation of this evidence allows for future analyses and is of such quality that it could be potentially used as an exemplar in court testimony.

In instances of burned remains, the FAU will most likely be called to the scene to assure the maximum recovery of skeletal elements and to document the condition of remains in their primary context. Burned remains are often friable making them susceptible to damage during transport. It is advisable for the forensic anthropologists to be present at the scene, particularly if the body is extensively burned or calcined. Early and frequent assistance from the anthropologist is critical in these types of cases because estimation of age, sex, ancestry, and stature may still be performed and the recognition of perimortem skeletal trauma is possible even when thermal alteration is extensive.

Pathological conditions and anomalies

The medical examiner may request a formal analysis of pathological or anomalous conditions in the skeleton, or may prefer an informal consultation during autopsy. Typically, radiographs are also examined and, if agreed upon by the medical examiner, a specimen may be harvested at autopsy for further analysis. In some cases the anthropologist is asked to assess skeletal heath or fragility to assist in differentiating nonaccidental injuries from possible pathological fractures. This typically occurs in cases involving trauma to children and the elderly to assist with ruling out pathological conditions such as *osteogenesis imperfecta*, brittle-bone disease, or senile osteoporosis. Analysis of idiosyncratic features on the skeleton, including pathologies, anomalies, healed trauma, surgical implants, and prosthetic devices, may also be used to help in identification. For this analysis, a general examination of the skeleton, including radiographs, is conducted and a more in-depth analysis of the affected area is performed. Comparisons of antemortem records (if available) may also be warranted.

Taphonomy and time since death

Interpretations of taphonomic agents and estimating time since death are also commonly requested as part of the casework. Forensic taphonomic interpretations are important in the contextual framework of interpreting skeletal trauma to ensure natural taphonomic influences are not misidentified as potential trauma. New York City is environmentally diverse with its rivers and bays, natural forests and parks, and crowded urban centers. Accordingly, the possible taphonomic influences are also diverse. Time-since-death estimates vary throughout these environments and are reported with necessary caution.

Hard-tissue histological analysis

The study of hard-tissue histology (bone and tooth microstructure) in humans provides information regarding their biology, such as ontogeny, adaptation, movement, age-related changes, and health. Typical histological analyses requested include the determination of human versus nonhuman, estimation of age at death, and evaluation of bone pathological conditions. Histological samples are typically embedded into resin and prepared for evaluation using standard protocols. Depending on the condition of the sample, histological preparation may take several days. The OCME-NYC is one of only a few medical examiner offices in which the forensic anthropologist can offer these services.

Determining human from nonhuman bone is achieved at the histological level through the identification of plexiform bone and other microstructures (e.g., osteon area, osteon circularity). This analysis can be performed on very small bone fragments. The estimation of adult age at death is performed at the histological level using cortical bone wedges or cross-sections from specific skeletal areas. Preferred methods are those that utilize cross-sections from the middle third of the sixth rib and the femoral midshaft (Crowder and Pfeiffer 2010). These samples should remain undecalcified (calcified) and will not typically receive a stain. In some circumstances, cortical bone histology can provide insight into trauma or pathological conditions that affect bone remodeling. The pathologist will first remove the samples that they need to submit for standard histological analysis, where the samples will be decalcified and stained in accordance to the agency's guidelines. The anthropologist and pathologist will often review the histological slides together in these circumstances.

Developing casts and impressions of evidence

The FAU and the OCME-NYC as a whole have a conservative tissue-retention policy. Accordingly, a casting laboratory using high-quality materials was developed. Casts are invaluable from an evidentiary perspective when the actual material is not retained. The FAU uses a high-performance silicone-based rubber to create exceptional-quality negative molds of skeletal trauma, pathologies, and age specimens. The molds can be used to make as many positive casts as needed and will last indefinitely. Positive casts are made using a low-viscosity liquid plastic, producing an accurate and well-detailed positive resin cast. In sharp-force trauma analyses, impressions of tool marks are taken using a polyvinyl siloxane material that can retain toomarks from both bone and cartilage. The preservation of this evidence allows for future analyses and is of such high quality that it can be submitted in court, used as teaching specimens, or retained as exemplars to compare against other cases with similar types of evidence. With casting materials mentioned above, whole bones can be replicated or just small sections of bone with evidence of sharp-force trauma, dismemberment trauma, gunshot-wound trauma, blunt-force trauma and evidence of healing, abnormal bony growths and pathologies, or portions of bone showing various age markers. Most casts are generally accomplished using simple box molds, two-part molds, and paint-layering techniques.

SCENE RESPONSE

The FAU assists the OCME's medicolegal investigators with various types of scene recovery and is available for 24-h scene response in all five boroughs. In most cases, when a medicolegal investigator is notified that a single bone, disarticulated remains (e.g., an outdoor setting with dispersal of elements), or buried bodies are discovered the FAU is contacted to discuss the need for anthropological assistance. These cases may require an immediate scene visit to search for additional remains. If the medico legal investigator responds to the scene and believes that the remains are nonhuman, they may be advised to send digital images to the FAU for a final determination so the scene may be released. Typical scenes requiring anthropological assistance include situations where decomposed, skeletonized, fragmentary, burned, or buried remains are discovered. When the FAU responds to these scenes, the anthropologist and the medicolegal investigators must work in concert with the NYPD, who often conduct their own scene investigations if it is determined to be a crime scene.

At the OCME-NYC, the members of the FAU have extensive training in archaeo-logical techniques and methods. A few members of the FAU have worked in cultural resource management firms (CRM firms), while others have worked as crew chiefs on various archaeology projects or recovery team leaders for the Joint Prisoners of War, Missing in Action Accounting Command's Central Identification Laboratory (JPAC-CIL). Archaeological skills are essential for the aforementioned scenarios to ensure that remains and associated evidence are accurately and thoroughly recovered and documented. The use of improper techniques or untrained individuals risks the loss or damage of human remains and associated evidence. In addition to providing stand-ard hand-drawn maps, the FAU is trained in various mapping technologies including a standard total station, a global positioning system (GPS) total station, and a Leica ScanStation 2, which are deployed depending on the complexity of the scene.

THE FORENSIC ANTHROPOLOGY REPORT

Every case submitted to the FAU for analysis will have a report generated. The FAU generates approximately 140 case reports annually. The final report will be distributed to the Records Department and the responsible medical examiner. The FAU main-tains all related case notes and photographs used in the preparation of the Forensic Anthropology report. The forensic case report is submitted in a timely manner; the FAU strives for a maximum report turnaround of 30 days. This turnaround is consid-erably faster for nonhuman cases and for cases only requiring an estimate of age at death. Often preliminary information may be provided in some circumstances to assist law enforcement with beginning a missing person's search. When this occurs, the anthropologist must be cautious in releasing this preliminary information considering that incorrect information regarding the biological profile may hinder the identifica-tion process. Members of the FAU often assist in the positive identification process for unidentified individuals. While this will usually occur in consultation with the medical examiner assigned to the case, members of the FAU may generate an Identification Report as part of the overall Forensic Anthropology case report or as a separate document.

At the OCME-NYC, the FAU is large enough to institute a multilevel peer-review system. This process involves an initial peer review in which the peer reviewer evaluates the report and examines the remains to ensure the quality and accuracy of the analysis. Once the initial peer review is complete and accepted, the Director or Assistant Director of the FAU performs an executive peer review. Finally, the case report is placed through an administrative peer review to ensure the documents are in order, the report format is correct, and that a final editorial review is performed. The disposition of the remains will also be reviewed and verified.

Expert Testimony

Most members of the FAU have been qualified as expert witnesses and have experience testifying in court. Testimony generally occurs in one of the five boroughs, but testimony has been rendered outside of New York City and in federal cases. Nearly all testimonies are rendered about skeletal trauma cases. Annually, the FAU only testifies on a handful of cases, but this does not adequately represent the involvement in the legal process. Assistant District Attorneys and defense lawyers often request case reviews over the phone or in person. The forensic pathologist may also review an anthropology report prior to testimony if the anthropologist has not been called to testify.

Disaster Preparedness and Special Operations

The FAU is a critical component of OCME's disaster response team. All of the staff are certified hazardous materials technicians and play important roles in both morgue and field response. The OCME-NYC has been instrumental in the design and testing of the Unified Victim Identification System (UVIS), the agency's computer application for managing disaster operations. The program is currently being upgraded and converted into a day-to-day case-management system. The FAU fills several roles in the disaster response team which can be divided into three operational areas: field, morgue, and the family assistance center. In the event of a mass fatality incident, the expertise of many forensic specialists are brought together to assist with disaster-victim identification.

Field operations

Archaeological skills are essential to ensure thorough recovery and documentation of victims from a mass fatality incident. Anthropologists are typically involved in the mapping portion of the joint recovery effort (Figure 27.4). Participation in disaster preparedness has allowed the FAU to stay current with more advanced scene-mapping equipment using both GPS and three-dimensional technologies. The FAU personnel

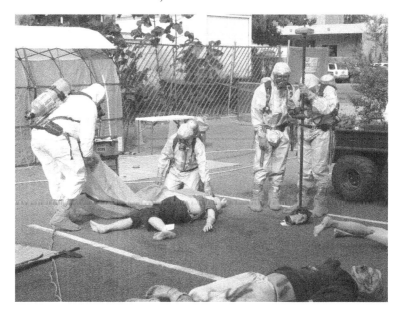

Figure 27.4 The FAU maps human remains and physical evidence as part of the joint recovery effort.

stay acquainted with the mapping technology through monthly training sessions. Additionally, there are semi-annual large-scale, multiagency drills that allow members of the FAU to practice in a realistic context.

Members of the FAU are also involved with managing operations during a mass fatality incident which requires training and an understanding of the Incident Command System (or ICS) used by all levels of government: federal, state, tribal, and local. The Incident Command System is a standardized, on-scene, all-hazards incident-management approach that is applicable across disciplines. The OCME-NYC has been instrumental in raising national awareness regarding the recovery and processing of decedents from an mass fatality incident, as well as developing an operational structure that focuses on fatality management.

Morgue operations

Forensic anthropologists can be utilized in several operational areas of the disaster morgue. Especially for fragmentary scenarios, the skills of the anthropologist are essential for disaster morgue operations at the triage station (e.g., recognition of commingled remains and nonhuman remains) and also for collaboration with medical examiners during autopsy. The purpose of the triage station is to evaluate the remains prior to the complete examination in the disaster morgue. In triage, the remains will be logged, assigned case numbers, reviewed for commingling, and checked for potential hazards. Pertinent information will be entered into the computer system and recorded on data sheets.

In addition to triage, FAU staff members operate the anthropology station in the morgue and have also been employed to monitor the Exit Review of remains processed through the disaster morgue. The purpose of the anthropology station in the disaster morgue is to act in a consulting capacity for the medical examiners. When requested, the FAU will identify the elements present and provide a biological profile of the remains. In addition, the condition of the remains and any trauma present will be described. If appropriate, fragmentary remains will be refitted or reassociated. The Exit Review is the final station in the disaster morgue which serves as a quality-assurance check on all remains before they exit the morgue. Members of the FAU ensure that all necessary analyses have been completed and have been properly documented.

Family assistance center

Family assistance centers (or FACs) are set up following a mass fatality incident and act as a single meeting place for families of victims. Family members can receive assistance and information about the incident prior to its release to the media. Equally important, the family assistance center is also where family members are interviewed to collect antemortem data about the missing person. The disaster-victim identification process involves the reconciliation of antemortem with postmortem data. This process takes place at the victim information center (VIC), which is located in a separate part of the family assistance center. At the OCME-NYC, a member of the FAU serves as the Disaster Victim Identification Coordinator (DVI Coordinator) and manages the disaster-victim identification process. The DVI Coordinator ensures necessary antemortem records are collected based on information provided in the family interview and that all appropriate steps are taken to identify victims. The DVI Coordinator also coordinates the review process for all matches reported by individual identification modalities and heads a multidisciplinary and multiagency Identification Review Board which gives a final recommendation on problematic identifications.

WTC OPERATIONS

In 2006, human remains were unexpectedly discovered around the WTC site and the OCME was designated as the lead agency in the renewed search and recovery efforts (Sledzik et al. 2009). The FAU played a lead role in these efforts and is still actively involved in ongoing WTC operations. The various roles of the anthropologists involved with the WTC operations are described below.

Site monitoring, excavation, and sifting

The FAU has led the city's renewed search and recovery effort since additional remains were discovered in 2006 on the roof of a building adjacent to the WTC site. This discovery ultimately resulted in the search and assessment of several buildings surrounding the WTC site, including a wet-screening operation in two buildings directly south of the south tower (Deutsche Bank and 130 Cedar Street). In October 2006, 187 potential human remains were discovered in a buried utility manhole on the west side of the WTC complex. Due to these finds in the manhole and on the building roof, the FAU were tasked with developing an excavation and assessment strategy to verify the presence or absence of additional WTC remains on nearby building rooftops as well as buried WTC deposits in the surrounding area that may contain potential victim remains. The operational goals were to identify the boundaries of the WTC debris deposits, excavate, and document all material that may contain WTC victims' remains and personal effects in a continued effort to identify victims. An archaeological approach was determined to be the most suitable to achieve these goals. The FAU managed all forensic aspects of the field operation including the determination of an excavation strategy, identification and assessment of soil strata, determination of the maximum horizontal and vertical excavation boundaries, documentation of all potential human remains encountered, feature excavations, establishing chain of custody for material removed, and providing daily and monthly situation (progress) reports.

The field operation was logistically complex and required the assistance of several other city agencies in coordinating the excavation, transportation, tracking, and sifting of the material. Once material was removed from the site the FAU personnel were responsible for the forensic oversight of any sifting operations and the forensic analysis of any potential human remains recovered. The OCME-NYC sifting facility was constructed in an office and warehouse space beneath the Brooklyn Bridge at 11 Water Street. Two elevated sifting decks were designed with input by the FAU and fabricated specifically for this operation. The two decks contained 22 sifting stations. All material was wet-screened through two screens and all discarded material was retained and stored. These efforts have also allowed the FAU to employ dozens of anthropologists and archaeologists on a part-time capacity through several phases of excavation and sifting.

The initial excavation and sifting phase of the operation was completed by the end of 2007; however, due to the nature of the construction activity around the site, some areas were not accessible at the time of the excavation. Considering the localized distribution of WTC material across the site, it was determined that the FAU would continue to maintain a daily presence on site and monitor all potentially sensitive areas. In addition to continued site monitoring, anthropology personnel attend weekly construction coordination meetings and receive daily construction work schedules. Over the subsequent years of monitoring, excavations in several areas around the site resulted in the accumulation of a significant amount of additional material. In 2010, a second sifting operation was conducted to search the additional material recovered. The monitoring of construction activity is projected to continue until 2014.

The excavation, to date, has resulted in the recovery of 1841 potential human remains, 12 new identifications, and the linking of over 700 remains to previously identified victims. The excavation data demonstrate that 83% of the established excavation areas contained victim remains. Although the majority of the remains were related to specific features, evidence may have been lost if a consistent archaeological excavation strategy was not followed. The work performed by the FAU at the WTC site demonstrates the importance of using trained anthropologists and archaeological methods for disaster-scene recoveries.

WTC victim identification

The FAU is responsible for the retention and release of all WTC remains housed at the OCME-NYC. This process consists of the following: remains documentation and intake, maintenance of an accurate inventory, reporting identification statistics and victim lists, storage and packaging of approximately 13 000 remains, identification review, and release of remains to funeral homes. Anthropology personnel are also the point of contact for WTC victims' families. In addition to these duties, Anthropology personnel sample remains for DNA analysis and link remains to known victims based on DNA results.

Any newly discovered WTC remains are immediately documented, analyzed, and entered into the WTC database so that identification work can begin. Remains are assigned unique WTC case numbers that accompany the remains and associated paperwork from this stage forward through all analyses, both internally and externally. Each of the remains is photographed and documented, noting which skeletal elements are represented and their relative degree of completeness. Once the documentation is completed, an assessment of the identification potential of the remains is made to determine which additional analyses should be completed. All remains are submitted for DNA analysis, which is coordinated by the anthropology staff. When a new potential identification is made, there are a series of reviews that the anthropology staff must complete before the match is made official. First, a review of the remains against the profile of the victim is made to ensure there are no inconsistencies. The second step is to perform an anthropological review of the remains to ensure there is no duplication. The remains' descriptions are reviewed and any potential inconsistencies are noted (e.g., two left legs). The physical remains packages may be opened and examined by the anthropologist to resolve any potential issues. Once the matches pass all reviews, the remains' records are linked to the victim.

Memorial Park, located adjacent to the OCME-NYC Manhattan facility, is a temporary resting place for all unclaimed and unidentified WTC remains. The space was dedicated in 2002 by Mayor Michael Bloomberg to give families a place to bring flowers and photos until the permanent memorial is completed at the WTC site. The anthropology staff is responsible for the storage and maintenance of all WTC remains housed in Memorial Park until families decide to claim them or until they are transferred to the permanent memorial.

WTC family interactions

Interactions with the WTC families can be a daily occurrence at the OCME-NYC. The NYPD Missing Persons Division handles notifications for first-time identifications and the next of kin are then instructed to call the OCME-NYC. After this initial notification, the anthropology staff is responsible for all subsequent contact with families. Because of the nature of the WTC disaster, multiple remains can be identified to a single victim, and these identifications may occur multiple times over a period of years. To avoid unwanted notifications, families have been given two options for notification: (i) do not notify or (ii) always notify when an additional match is made. If the family chooses to be notified every time, they provide direction to the OCME as to how they would like notification to occur, whether directly or through a third-party such as a funeral director or member of the clergy. The WTC staff answers general questions families have about identified remains, the identification process, updating contact information, and releasing previously submitted personal effects and dental records, and prepares and provides certified copies of victims' files when appropriate. Furthermore, any corrections or amendments to death certificates, including name changes for the official victims list, are made by the WTC anthropology staff.

WTC reporting

The anthropology staff issue various reports on a periodic basis. The current state of the WTC identification process is reported monthly and includes statistics on the total number of recovered remains, identified/unidentified remains, death certificates issued, new identifications, the breakdown of identification by modality, and other information when requested. Situation reports regarding the ongoing excavation and monitoring status at the WTC are generated daily. In addition, a lengthy and comprehensive archaeological report was created to document the renewed search and recovery efforts that began in 2006. Reporting and accountability are fundamental aspects of a large operation that requires long-term planning and financing from multiple government agencies. It is important that operational awareness is maintained and provided (at various levels) to the involved agencies, families of victims, government officials, and – to some degree – the public.

TRAINING AND PROFESSIONAL DEVELOPMENT

At the OCME-NYC, the FAU is involved with both internal and external training for various forensic practitioners and students. The FAU provides training for the in-house rotating medical residents and students and gives one lecture per month covering the scope of forensic anthropology as part of the OCME monthly lecture series. The FAU also works closely with the OCME-NYC forensic pathology fellows each year, taking the opportunity to show young forensic pathologists how anthropologists can assist them with aspects of their casework. Graduating fellows and pathologists who are slated to take the Forensic Pathology Board Examination are offered reviews of the skeleton and training in anthropological techniques. The training is interactive and typically takes place in the anthropology laboratory; however, the FAU has also traveled to the outer boroughs to offer lectures to medical examiners and investigators.

In addition to training the OCME-NYC staff, the FAU personnel take part in other training courses, lectures, and hands-on workshops to groups such as the NYPD Homicide Detectives, Crime Scene Unit, and FDNY Fire Marshalls. The OCME-NYC also hosts several death investigator courses funded by the NIJ for basic- and advanced-level participants throughout the year, which are open to investigators throughout the country. Members of the FAU provide anthropology lectures for the basic death investigators courses and perform a day-long hands-on laboratory and scene recovery workshop for the advanced death investigators course. Several members of the FAU are often asked to give lectures outside of the agency. The FAU has taught courses in physical anthropology, forensic anthropology, archaeology and recovery, and human osteology at several of the city universities and private universities in New York City. In addition, members of the FAU regularly travel to lecture and participate in training courses throughout the country and abroad.

The FAU developed a Visiting Scientist Program that allows advanced graduate students and professionals to work with the unit at the OCME-NYC for 1 month. The program is designed to provide training and professional development through access to hands-on experience. This program immerses participants in various aspects of day-to-day operations within a medical examiner's office, with specific focus on the role of forensic anthropology. Through this program the FAU selects as many as nine visiting scientists each year from the USA and abroad.

In addition to the Visiting Scientist Program, the FAU mentors several interns per year. Most student interns are selected from the Human Skeletal Biology Master's program in the Anthropology Department at New York University (of which many members of the FAU are affiliated faculty). The goal of the internship is to introduce the students to the daily operations of the FAU in New York City. The interns generally assist with maintaining the labs, processing specimens, and other supervised laboratory duties as they arise. The interns are given the opportunity to visit scenes and participate in some of forensic anthropology training exercises. The interns are encouraged to read articles and keep current in the field, to think of potential research topics, and to utilize all of the resources available within the department during their tenure at the OCME-NYC.

Research in Forensic Anthropology

Most medical examiner's offices with an in-house forensic anthropology position employ one anthropologist and that person has little time to pursue research in the field. This is an unfortunate dilemma as necessary research identified by a forensic anthropologist working in a medical examiner's office will likely have immediate relevance to other practicing forensic anthropologists. The FAU has received several grants to perform research in various areas of forensic anthropology such as sharp-force trauma and hard-tissue histology with the goal of improving anthropological analyses at the OCME-NYC and assist in the development of the field. Portions of salaries are written into research grant proposals, which provide the agency with an incentive to support research activities and allow the FAU staff time to engage in research. The members of the FAU have presented the results of these research projects at various venues, notably the American Academy of Forensic Sciences and the American Association of Physical Anthropologists annual meetings. Additionally, several members of the FAU have served in advisory capacities in student research conducted at the OCME-NYC, and serve as members of Master's thesis committees.

Conclusion

It is apparent from the current work performed by the FAU at the OCME-NYC, as well as by the increased employment opportunities over the past decade for forensic anthropologists in medical examiner's/coroner's offices, that forensic anthropology has grown beyond the scope of its classic definition. As the result of this growth the desired education, qualifications, and experience of those pursuing a career as a forensic anthropologist is also changing. At the OCME-NYC, the FAU consists of diverse group of individuals with broad training in anthropology and archaeology, while also contributing specialized knowledge in their area of expertise (e.g., age estimation, bone histology, historical archaeology, and trauma analysis and interpretation). In addition to the traditional role of skeletal analysis, the FAU exemplifies the additional roles in which the anthropologist may be utilized. Furthermore, the FAU's involvement in mass fatality management, disaster-scene operations, and case tracking for the unidentified demonstrate the changing role of the anthropologist in the medicolegal setting. Owing to the broad four-field approach that many anthropologists receive and the opportunities to manage laboratories, excavation crews, or osteological collections, our holistic educational backgrounds allow us the opportunities to develop the leadership skills necessary to fill these roles.

NOTE

1   The death investigators or medicolegal investigators at the OCME-NYC must be registered nurses or physician assistants.

REFERENCES

Austin, D. and Fulginiti, L. (2008). The forensic anthropology laboratory in a medical examiner setting. In M.W. Warren, H.A. Walsh-Haney, and L.E. Freas, (eds), *The Forensic Anthropology Laboratory* (pp. 23–46). CRC Press, Boca Raton, FL.

Crowder, C. and Pfeiffer, S. (2010). The application of cortical bone histomorphometry to estimate age at death. In K.E. Latham and M. Finnegan (eds), *Age Estimation of the Human Skeleton* (pp. 193–215). Charles C. Thomas, Springfield, IL.

Dirkmaat, D.C., Cabo, L.L., Ousley, S.D., and Symes, S.A. (2008). New perspectives in forensic anthropology. *Yearbook of Physical Anthropology* 51: 33–52.

Gill, J.R., Rainwater, C.W., and Adams, B.J. (2009). Santeria and Palo Mayombe: skulls, mercury, and artifacts. *Journal of Forensic Sciences* 54(6): 1458–1462.

Sledzik, P.S., Dirkmaat, D.C., Mann, R.W, Holland, T.D, Mundorff, A.Z., Adams, B.J., Crowder, C.M., and DePaolo, F. (2009). Disaster victim recovery and identification: forensic anthropology in the aftermath of September 11. In D.W. Steadman (ed.), *Hard Evidence: Case Studies in Forensic Anthropology*, 2nd edn (pp. 289–302). Prentice Hall, Upper Saddle River, NJ.

Stewart, T.D. (1979). *Essentials of Forensic Anthropology*. Charles C. Thomas, Springfield, IL.

CHAPTER **28**

The Many Hats of a Recovery Leader: Perspectives on Planning and Executing Worldwide Forensic Investigations and Recoveries at the JPAC Central Identification Laboratory

Paul D. Emanovsky
and William R. Belcher

INTRODUCTION

Forensic anthropologists and archaeologists, in general, will wear many "hats" to deal with the myriad of tasks outside of typical field and laboratory duties and responsibilities. For example, resource management archaeologists commonly play the roles of payroll officer, administrator, contract negotiator, as well as many other jobs, in addition to digging in the dirt. These additional duties become a critical component of the job

A Companion to Forensic Anthropology, First Edition. Edited by Dennis C. Dirkmaat.
© 2012 John Wiley & Sons Ltd. Published 2015 by John Wiley & Sons Ltd.

description of the forensic anthropologists and forensic archaeologists who work for the Joint Prisoners of War, Missing in Action Accounting Command Central Identification Laboratory (JPAC-CIL) investigating cases of missing US military personnel. The JPAC investigations and excavations take place globally, and often in countries that have a high degree of distrust and animosity towards the USA, in general, and the US military specifically. The purpose of this chapter is to provide some insight into the administrative, personal, and logistical hurdles that forensic anthropologists and archaeologists will have to overcome during these international projects. Professionals interested in undertaking large-scale archaeological recoveries of human remains and evidence in a forensic context (whether at JPAC or at other humanitarian and human-rights organizations and projects) may not be able to visualize what awaits them during the transition from an academic environment into a heavily applied environment. To this end, we will present an overview of the political and scientific context of the vast majority of recovery operations conducted by the JPAC-CIL. We also provide "vignettes from the field" in the hopes that students and other professionals can tailor their academic and professional training to gain the types of experiences that are necessary to thrive in an applied, and occasionally, unusual environment.

Overview of JPAC-CIL and Forensic Anthropology

Currently, the JPAC-CIL is the largest employer of full-time forensic anthropologists and archaeologists in the world. The JPAC-CIL is the US Department of Defense's primary tactical and operational arm, involved in the search for, recovery, and identification of US military personnel missing from past US conflicts since World War II, including the Korean War, the Cold War, and the Vietnam War. Occasionally, the JPAC-CIL will recover and identify service members from previous conflicts, such as World War I and the US Civil War. The JPAC-CIL also provides local and international humanitarian assistance during mass disasters and missing person cases with local and federal agencies in the state of Hawaii and throughout the world. Examples include assistance in the identification of victims of the USS *Cole* bombing in Yemen in 2000, assistance in the identification of victims of the September 11, 2001 disaster at the Pentagon and World Trade Center, as well as local assistance to the Honolulu Police Department and the Department of the Medical Examiner, City and County of Honolulu.

JPAC headquarters and the main laboratory of the CIL are located on Joint Base Pearl Harbor-Hickam, near Honolulu, Hawaii, on the island of Oahu. The JPAC organization employs over 400 military and civilian personnel, including members from every branch of the US military service from the Army, Navy, Marine Corps, and Air Force. The CIL currently employs over 30 forensic anthropologists and archaeologists. This number is expected to expand significantly over the next 5 years as mission and identification efforts increase. Additionally, the CIL employs one civilian forensic odontologist (dentist) and two military dentists to assist in the identification and forensic review process. Table 28.1 provides an overview of the JPAC's organizational structure and some of the key functions performed by each section.

Table 28.1   Organization of JPAC and key functions.

JPAC section	Key functions
Command	Mission oversight, maintaining relationships with external governmental and nongovernmental agencies as well as keeping family members of the missing informed about JPAC's efforts
Central Identification Laboratory (CIL)	Search and recovery, identification, investigations, and research
J1-Manpower and Personnel	Administrative support, human resources, payroll, and travel coordination
J2-Intelligence Directorate	Historical research, investigation and analysis, data integrity and geographic information systems, records archive, and special security
J3-Directorate of Operations	Coordination, integration, and synchronization of all operations, logistics, and planning
J4-Logistics and Supply	Developing, managing, and overseeing procurement operations, maintaining logistical warehouses, and property accountability
J5-Policy and Negotiations	Facilitates access, strategy, and support of foreign governments and US agencies
J6-Information Technology	Information technology, system administration, system integration, communications and network security
Public Affairs Office (PAO)	Provides consistent, clear messages about JPAC missions, operations, and activities through the use of media engagements, in-house-produced marketing, internal information materials, and face-to-face contact with JPAC experts
Forensic Science Academy (FSA)	A forensic anthropology academic/applied program, taught under the auspices of the Department of Defense by FSA and CIL staff
Detachments (DET)	Detachments 1–3 are permanent support detachments which facilitate operations conducted in or traveling through their respective areas of operations, i.e., Thailand (DET 1), Vietnam (DET 2), and Laos (DET 3). Detachment 4 is located in Hawaii (Joint Base Pearl Harbor-Hickam) and consists of the recovery teams; typically JPAC has 18 standing recovery teams
Medical	Provides health-care support to JPAC before, during, and after missions to maintain force health protection, and preserve mission readiness and operational flexibility
Forensic Imaging Center (FIC)	Provides the CIL with high-quality photographic documentation of worldwide excavation operations as well as providing the PAO with relevant and newsworthy imagery of JPAC personnel accomplishing the mission

A Short History of the CIL

The US Government views the identification and return of US military personnel to their families as an important responsibility to the service members and their families. After each major conflict since the end of the Civil War, the US Government (through the Department of War/Department of Defense) has established temporary Central Identification Units or Central Identification Facilities. These facilities were primarily operated by the US Army via the American Graves Registration Command and Mortuary Affairs operated out of the US Army's Quartermaster Corps. Obviously, these early identification efforts relied on forensic techniques that can best be described as rudimentary when compared to today's technical and methodological advances (Eckert 1983; Wood and Stanley 1989).

During World War II, the American Graves Registration Service (AGRS) was tasked to collect, identify, and inter battlefield casualties (Steere 1951). Many of these techniques are still used in the modern battlefield contexts of the global war on terrorism, with the use of central collection points and temporary cemeteries. At the end of World War II hostilities, primarily in Europe, the AGRS began an extensive program of search and recovery for many servicemen in territories formerly held by enemy forces.

In the aftermath of World War II, the War Department created the American Graves Registration Command. Under this command, military recovery teams search globally to locate, recover, and repatriate service personnel remains to central identification points throughout Europe. The human remains were processed and identified primarily through historical documents and forensic methods. During this time period, the War Department depended heavily on professional physical anthropologists for the scientific identification of remains.

In 1946, the War Department hired Dr Harry L. Shapiro of the American Museum of Natural History in New York, and established the first central identification point in Strasborg, France (Eckert 1983; Wood and Stanley 1989). Shapiro refined the collection and records process for World War II casualties in the European theater. This process was repeated at additional central identification points in Belgium and Italy. For the identification of war dead in the Pacific Theater, a much larger geographic area, a Pacific Zone CIL was established in Hawaii at the US Army's Schofield Barracks. Dr Charles Snow served as the Laboratory Anthropologist from 1947 to 1948 (Snow 1948) and departed when his leave of absence from the University of Kentucky was over. Dr Mildred D. Trotter of Washington University (St. Louis, MO) replaced Snow as the Director of the Pacific Zone CIL. Early during her tenure, Trotter gained permission from military officials to collect data from the remains received by the CIL; Trotter felt that this activity was essential to advance the science of human identification (Conroy et al. 1992). With the help of Dr T. Dale Stewart of the Smithsonian Institution (Washington DC), Trotter collected the data and conducted the analyses that led to milestone studies on stature and age estimation (Trotter and Gleser 1952). At this time, no physical anthropologists or archaeologists were employed during the recovery operations, which were seen exclusively as a military function by the AGRS.

During the Korean War (1950–1953), anthropological methods were used in the identification of battlefield casualties at the CIL in Kokura, Japan, beginning in 1951. However, professional anthropologists were not employed until after the 1953 armistice with the hiring of T. Dale Stewart by the US Army Quartermaster Corps (McKern and Stewart 1957). Stewart worked with several other physical anthropologists (Thomas McKern, Ellis R. Kerley, Charles P. Warren, and Tadao Furue) to study and identify the recovered war dead, including over 4000 sets of remains repatriated during a bilateral operation and exchange of war dead between the Democratic People's Republic of Korea (DPRK; also known as North Korea) and the United

Nations (this was termed Operation Glory). Again, it should be noted that no physical anthropologists or archaeologists were utilized on recovery operations.

The US Army Quartermaster Corps maintained the responsibility for the recovery and identification of US service members during the Vietnam War (1966–1973). Rapid recovery of soldiers killed in action (also known as "KIAs") minimized the number of unknown or missing service members seen in previous conflicts. The establishment of the CIL for this conflict, first in Thansenout Airfield in Saigon, then Bangkok (known as the US Army CIL-Thai), and, eventually, on the island of Oahu in Honolulu (the US Army CILHI). Currently the CIL is on Hickam Air Force Base, Hawaii. During this period, physical anthropologists employed by the military basically served as technicians. In fact, the advances made in physical and forensic anthropology during the latter decades of the twentieth century in terms of skeletal biology and identification methods were little used by the US military at this time.

During the mid-1980s, several family members began to question and challenge the scientific veracity of the recoveries and positive identifications provided by the military for servicemen from the battlefields of Southeast Asia. The US Government responded by inviting several outside consultants to examine the operations, including Ellis R. Kerley, William Maples, and Lowell Levine (a forensic odontologist). These consultants found that the CIL was isolated academically and geographically as well as being understaffed, which significantly hampered the laboratory operations (Hoshower 1999). In addition, the lack of outside scientific peer review, the lack of modern scientific equipment, and inadequate training were also identified as significant detrimental issues. The Department of Defense and the US Army Quartermaster responded to these needs by appointing Kerley to serve as the Scientific Director between 1987 and 1991.

Initially, anthropologists were hired strictly for laboratory analysis; however, it quickly was realized that more systematic and scientific field recoveries were critical to the overall success of the identification process. This led to the incorporation of anthropologists on recovery teams. Initially, the anthropologist would serve in an advisory role on field strategies and techniques as well as field assessment of human versus nonhuman bone fragments and teeth. In 1994, Thomas D. Holland became the Scientific Director of the US Army CILHI and the role of the team anthropologist began to transform to a more focused role, including changes in responsibility from advisor to *Recovery Leader*. The Recovery Leader then became the primary expert on the site excavations, field strategies, and leadership on the site area. These responsibilities required that Recovery Leaders possess more than a passing knowledge of archaeology and explicitly required formal archaeological training.

In 2003, the US Army CIL was combined with the Joint Task Force-Full Accounting to become a joint command termed the Joint Prisoners of War, Missing in Action Accounting Command (JPAC). The CIL remained as an independent section directly supervised by the Scientific Director who reported directly to the JPAC Commander. Today, in addition to the CIL's primary mission of recovering and identifying the remains of missing US service personnel, another primary goal is to establish the highest possible level of scientific expertise, competence, and integrity, while maintaining a level of ethical standing that is beyond reproach. The CIL strives to provide leadership within the field of forensic anthropology and forensic archaeology. These disparate goals are maintained through the CIL's accreditation with the American

Society of Crime Laboratory Directors – Laboratory Accreditation Board (ASCLD-LAB) (Holland et al. 2008). In 2003, the CIL became the first Skeletal Identification Laboratory accredited under ASCLD-LAB; in 2008, the CIL advanced that accreditation by becoming the first Skeletal Identification Laboratory to be accredited under the ASCLD-LAB International program. Further detailed discussion of the day-to-day operations within the laboratory portion of CIL can be found in Holland et al. (2008). The focus of the remainder of this chapter is on the field recoveries conducted by JPAC.

Individuals hired for positions within JPAC-CIL as anthropologists and archaeologists must possess the necessary archaeological and interpersonal skills that will allow them to be successful in the field. The following vignettes offer examples of some of the situations in which Recovery Leaders find themselves and the types of knowledge and leadership skills that are necessary when working on a global scale in many different cultural and political environments and settings.

FIELD RECOVERIES CONDUCTED BY CIL

The primary purpose of the field operations organized by JPAC-CIL is to conduct the most comprehensive recovery of the human remains identified at each site investigation as possible. The task is particularly difficult due to a number of factors, including, but not limited to the following.

1. Time elapsed since the loss incident: the vast majority of military causality losses are from World War II (now almost 70 years ago); others from the Vietnam War are now more than 45 years ago. This time gap can cause many issues, including simply trying to identify local informants or find appropriate eye witnesses to the incident.
2. Type of incident: most of the incidents investigated are aircraft crashes, though some incidents involving ground troop burial are also investigated. Most of the aircraft crashes investigated result in significant fragmentation of human remains. The CIL makes a distinction between slow-moving, propeller-driven aircraft (including rotary aircraft, such as helicopters) and high-speed jet-propelled aircraft. During World War II, most aircraft were propeller-driven, but during later conflicts a mix of different types of aircraft were used, although the majority of the post-World War II conflicts involved jet-propelled aircraft. These sites contrast significantly with battlefield casualty sites which are burial features near the death scene and in many cases contain relatively well-preserved and intact bones.
3. Location of incident: in many cases, military plane crashes left to be investigated are found in remote areas, including jungles. At these recovery sites, the sediment and soils are not conducive to bone preservation. The jungle sediments in Southeast Asia, for example, are highly acidic. In other situations, bone is rather well preserved. For example, the karstic soils of Papua New Guinea occur over a base of limestone and coral, which leads to better bone-preservation conditions.
4. Site disturbance: many recovery sites have been scavenged for metal in the past, or have been modified through farming or other activities, making locating crash sites in these cases particularly difficult. Impact craters may have been filled-in with debris, while cultivation plowing also obscures potential recovery sites. Scavenging

for memorabilia and aircraft parts is particularly common for World War II aircraft in Europe and throughout the former Pacific Theater, including New Guinea.

5. Political situation of host countries: during the course of a war or conflict, territories will change and fluctuate numerous times; attempts will be made to investigate sites and recover remains when areas are controlled by friendly forces. However, in some cases, aircraft crash sites or isolated burials remain in territories controlled by nations that are hostile to, or at least suspicious of, the US government. This requires the investigation and recovery teams to enter territories that may be hostile to US citizens; this is especially the case with the DPRK, where the USA is part of a ceasefire agreement.

OVERVIEW OF RECOVERY OPERATIONS

Field recoveries for missing soldiers are conducted in a series of steps or phases, not unlike archaeological work that is part of cultural resource management operations, though some modifications of archaeological techniques and methodologies are required. Specific evidence handling and management procedures employed by JPAC archaeologists ensure that material and biological evidence is collected following proper protection and custody. The primary stages involved in field recovery include: (i) research and analysis, (ii) investigation, and (iii) recovery. These steps eventually lead to the recovery and accession of material and biological evidence into the CIL for analysis and identification.

Research and analysis

The J2 section of JPAC conducts an historical review of pertinent historical records, particularly case files [which include Aircraft Accident Cards (Navy), Missing Aircrew Reports (MACRs; World War II, US Army Air Forces), IDPFs (Individual Deceased Personnel Files), Field Search Case files (Korean War), and Reference Number files (REFNO; individual files related to casualties in Southeast Asia)], personnel and combat unit memoirs, military unit histories, and other conflict-related secondary and tertiary historical sources. Lead sheets are created from these and possible cases are selected on their likelihood of success and put forth for field investigation.

Investigation

In an ideal world, a case is selected based on high probability of locating human remains. Attempts are made to locate eye witnesses; however, due to the length of time that has passed since these incidents occurred, most of these individuals are either deceased or cannot recall many details of the incident. Most of the interviews are conducted with informants who know of a crash site or have found evidence and even human remains during cultivation, for example. The investigation team will also attempt to conduct appropriate field observations to specifically correlate a spot on the ground to a specific loss incident. This can include the exclusion of other nearby incidents, finding a data plate or serial number, such as those on a .50 caliber machine gun, that specifically correlates to a type of aircraft or a specific aircraft, life support (aircrew-related gear that may indicate that the aircrew were in an aircraft at the time of the impact), and, of course, identification media, such as an identification tag. Typically, these initial investigation teams comprise military and civilian specialists such as explosive/ordnance disposal technicians, intelligence analysts and/or military historians, medics, and linguists. In the near future, the anthropologists and archaeologists will become more involved in the site survey and evidence handling aspects of the investigations. Nevertheless, the time spent on a site is very limited.

If a site has been investigated, but a Recovery Leader needs more specific information on the recovery site (such as the distribution of different types of aircraft parts), then a Phase 2 Team (P2T) may be called to gather more information. A P2T is a small, functionally specific team lead by a Recovery Leader with extensive archaeological experience. Other members include a communication specialist, explosive/ordnance disposal (EOD) technician, forensic photographer, and a military team leader. This team typically spends several days on a potential site to conduct both surface and subsurface testing of a recovery site. Thus, more specific information can be gathered by these types of survey and attempts will be made to narrow down a site location which will maximize subsequent recovery efforts and make the operation more efficient. For example, the P2T may, through subsurface testing of a particular area on a site, identify the cockpit or other crew station of an aircraft, allowing the recovery team to focus on those specific areas at a later date.

Recovery operations

Recovery operations include the excavation and recovery of both material and biological evidence from a field setting. This takes place after the investigative process has concluded and all the data are reviewed by the command. In consultation with the historical section, and the CIL, the J3 Operations section creates a decision matrix that objectively assigns a score to a site which relates to the probability of locating remains. From this score, recovery sites are chosen.

Based on the information collected during the investigative effort, the Recovery Leader begins to make important planning decisions concerning the team management and size in conjunction with the JPAC J3 Operations and Planning Section. Is the site of sufficient size to accommodate the normal complement of team members, usually about 10–12 individuals, or is a so-called "plussed-up" or double team needed? These types of decision are related to factors such as site size, limited access by the host government, and imminent danger of site destruction through development. The planning phase also requires estimates of the amount of work that can be completed during the mission time frame, including the number of square meters that can be excavated and whether or not the site will need follow-up missions or deployments to complete excavations. The Recovery Leader also devises an efficient excavation strategy including excavation method (hand versus mechanical), screening type (does the site area support a wet-screening operation?), and determining whether any other specialized equipment or training is needed. This latter concern is especially important for remote or mountainous base camps. Experience in camping and other outdoor activities can supplement any academic experience that an archaeologist or anthropologist may have.

A typical field recovery team consists of about 10–12 members including a Recovery Leader (anthropologist or archaeologist), Team Leader (usually a military logistics officer), Assistant Team Leader (a senior noncommisioned officer who acts as the site foreman), an interpreter or linguist, a forensic photographer, an explosive ordnance disposal technician, a medic, a communications technician, a Life Support Analyst (a military or civilian technician who can identify parts of aircraft or materials that would be associated with the an aircrew's body), and two to three additional military service members who conduct general duties at the recovery scene, including excavation, sediment screening, construction, and other logistical support for the excavation of the site.

Once anthropologists or archaeologists arrive on a recovery scene, they begin to conduct the necessary business of documenting the site prior to excavation with the initial site survey. After initial documentation, a recovery operation begins with an on-site pedestrian survey of the scene that is conducted by the Recovery Leader and key team members. The EOD technician conducts a surface survey with a metal detector, which allows the Recovery Leader to understand the subsurface distribution of metal fragments (particularly important in understanding the distribution of aircraft wreckage) as well as indicating possible unexploded ordnance. Previously collected information from the investigative process is reconciled with the current site condition and distribution of evidence. Additionally, the Recovery Leader may choose to re-interview possible eye witnesses and informants.

As the Recovery Leader begins to understand the structure and distribution of a site, logistical requirements and recovery operation planning will be finalized. Logistical requirements include the location of various aspects of site "furniture." These include a screening station consisting of 10 to 20 screens that are used to process the large amounts of sediment that must be examined for evidence, the location of local worker and US personnel break areas, equipment storage, and latrines. All of these must be located in an area that is close enough to maintain an efficient operation, but not within the site boundaries to effect the excavation. If a base camp is required for the operation, it needs to be located within the shortest distance possible from the recovery scene to ensure that an inordinate amount of time is not spent traveling to and from the recovery scene as well as not impacting the site area. However, sometimes logistical concerns, such as a flat area with sufficient area for a base camp or the location of a helicopter landing zone, will take precedence over travel time and a base camp might be located within a 60- to 90-min hike from a recovery scene.

The first step of the excavation involves the clearing of vegetation in order to increase the archaeological visibility of evidence and identify scene topography. Occasionally, mechanical excavation may be required to identify buried archaeological features, such as formal and expedient burials and crash craters. Standard archaeological procedures for recording provenience (three-dimensional coordinate information for the location of evidence) require the use of a standard archaeological excavation grid. Primarily, the excavation units are marked off with stakes and twine/string that conform to a standard 4 m × 4 m square. This configuration allows the Recovery Leader to subdivide the excavation grid into smaller subunits (a 2 m × 2 m unit or even a 1 m × 1 m unit) if finer and more detailed excavation is required. The establishment of an excavation grid often will require the adaptation of more traditional archaeological methods. Many aircraft impact sites are located in extreme topography and establishing a traditional excavation grid can be challenging and requires flexibility by a Recovery Leader: "And if it is too dangerous to measure in the grid using tapes, why are you excavating there anyway?" (Roskams 2001: 98). Obviously, in order to recover human remains from these sites, a grid may have to be constructed in extreme environments!

The excavation units are then dug to incident sterile sediments, which are defined as sediments that contain no evidence related to the incident, or, in this case, no evidence of disturbance from burial or metal intrusion from an aircraft impact. Sediments removed from the excavation unit are transported to screening stations where they are processed through 6 mm hardware cloth or wire mesh. Sediments are moved to these processing areas using bucket lines (buckets passed in a line of local workers from the recovery scene to the screening station), zip lines (buckets passed using a rope-and-pulley system), primitive chutes, and wheel barrows.

Organizing the Field Recovery

Fieldwork at the JPAC, as with any other large-scale forensic archaeological project, involves dealing with common issues such as balancing limited resources while efficiently and effectively capturing the data and preserving the evidence that will ultimately lead to case resolution. Recovery Leaders must contend with finite resources (time, labor, and money), logistical impediments, and sociopolitical realities such as host-nation restrictions and cultural differences. Thus, Recovery Leaders need to be flexible and creative while working within the boundaries of a detailed, standard operating procedure. The key for successful recovery work is to develop strategies that maximize evidence retrieval, while fully documenting the context and spatial relations of the evidence, as well as maintaining chain of custody from the field to the laboratory. In addition to being the forensic and archaeological subject-matter experts, the Recovery Leaders are responsible for carrying out nonscientific roles, including as primary scientific interpreters to various local, national, and international government officials. Other roles include briefing high-ranking military and government officials, serving as an "ambassador" or representative of the USA in other nations (often characterized as *not* acting like the so-called Ugly American), media representative, and safety officer. All of these responsibilities require that the Recovery Leader must be well organized, manage time well, and be a creative problem solver.

Since every recovery site requires its own evaluation and unique excavation strategy and since JPAC is typically operating in several countries simultaneously around the world, it is somewhat difficult to describe a "typical mission" at JPAC. While many forensic field recoveries in the Southeast Asian and World War II contexts involve military aircraft crashes, each scene is unique in terms of spatial distribution of evidence. Topography, the remoteness of the site, and the preservation potential of a given region often result in a wide variety of recovery strategies developed by individual Recovery Leaders to process the site. While initial site formation processes may have been straightforward (e.g., a fast-moving jet hitting the side of a mountain), the sites continue to be modified and obscured by various taphonomic processes, such as secondary burial by humans, gnawing by carnivores, scavengers, and rodents, and fluvial transport of materials, among others.

Figure 28.1   Overview of the A-4F recovery scene showing lower (triangle) and upper project areas (arrow).

Although the standard site-recovery standard operating procedures are followed, considering the variety of sites excavated by the CIL there are numerous unique obstacles faced by the recovery teams due to post-crash time interval, modification of the scenes, and environmental and topographic conditions, as discussed above. The following case studies present some of these unique obstacles faced by the field operations and Recovery Leader in terms of both excavation and interpretation of the archaeological context. Hopefully these discussions will give the reader a better understanding of what is expected of a JPAC Recovery Leader, and their roles and duties, in this unique nonacademic setting.

Case study 1: an A-4F Vietnam-era aircraft crash

This case involves the crash of an A-4F aircraft over northern Vietnam in 1972. The lone airman in the plane was conducting a nighttime armed-reconnaissance mission. Search of the area of the crash had been conducted on numerous occasions and eventually lead to an aircraft wreckage site presumed to be that of an A-4F aircraft. The area is in karstic topography and includes steep limestone talus slopes bounded by vertical limestone cliffs (Figure 28.1). Plane wreckage was observed both near the summit as well as on the lower talus slopes at the base of the cliff.

Due to the difficulty of the terrain and the steepness of the slopes around the wreckage, the site was initially not recommended for excavation. Eventually, an investigative team accompanied by professional mountaineers and climbers resurveyed the incident location. The team found no life support or pilot-related equipment on either the upper or lower sections of the crash site. Presence of such evidence usually allows the Recovery Leader to plan where to begin excavations. The mountaineers recommended that based on safety and logistical considerations, excavations should begin in the lower section of the site at the base of the cliff. Following consideration of what evidence was being retrieved during this phase of the recovery, the Recovery Leader could reevaluate whether to continue excavation of the crash site at the top of the karst cliff.

Figure 28.2   Photograph of the A-4F recovery area showing upper portion prior to clearing (top) and after clearing (bottom).

Full-scale recovery operations began at the lower locus. While this operation commenced, a small team located a suitable approach to the upper area. The team ascended to the top of the cliff and conducted a pedestrian survey of this locus. Significant quantities of life-support materials were located here. A series of boulders and drops occurred between the upper and lower loci, which would have prevented any of the upper material evidence from moving downslope and being deposited downhill on the lower main recovery area. As expected, excavation of the lower locus produced nothing of significance. Obviously, the upper locus had a much higher potential for yielding significant evidence that included aircraft remains, personal effects and, importantly, human biological remains. Due to budgetary and safety concerns, permission was needed from the governments of both the USA and Vietnam to proceed with full-scale excavation of the upper locus. Permissions were finally granted, and while a portion of the team continued to process the lower locus, other recovery team members fixed lines and ladders to vertical sections of the cliff face, built bridges,

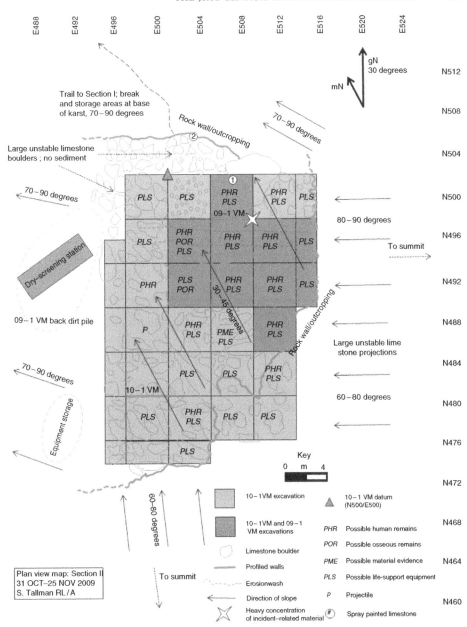

Figure 28.3   Plan-view map at close of excavations of case study 1.

and made the approach to the upper locus safer and more easily traversed, especially in the event of a medical emergency that would require the evacuation of an injured individual.

The upper locus was considerably more defined spatially than the lower locus; however, the upper locus required considerably more safety-related and logistical hurdles, especially in terms of movement of equipment and workers (Figure 28.2). Within the limited space, the screening station needed to be placed in such a location as to ensure that processed sediment would not be screened over incident-related materials and sediments, or that the screened sediments would not fall on personnel in the lower locus!

Significant amounts of life support-related equipment and biological evidence were located in the upper locus (Figure 28.3). However, due to time constraints, the excavation of the entire upper crash site could not be completed, and the site activity was suspended. Although some human remains were recovered, JPAC policy requires a return to a site when feasible until all reasonable and prudent measures have been exhausted and the probability of subsequent recovery is minimal. A subsequent recovery team returned to the upper locus the following year.

Case study 2: a World War II-era P-51D aircraft crash

Case study 2 involves a World War II-era P-51D (Mustang) aircraft that crashed into a cultivated field in Germany. This case illustrates the parallels between skills required and techniques used for processing small-scale (e.g., an isolated burial) and large-scale (e.g., mass graves, crash craters) features. This situation is common with military aircraft crash sites investigated by the CIL; often a body or bodies have been removed from an impact crater and buried in a shallow grave. The impact crater represents a large archaeological feature, while the burial represents a small-scale archaeological feature; both of these require different approaches and techniques to process due to the difference in scale and origin.

Historical and records-based research by the JPAC J2 section identified a potential aircraft crash site location in a German farm field. An investigative team was sent to survey the farm field, interview the landowner, and identify and interview potential eye witnesses and informants in a final assessment of recovery potential. The preliminary investigation of the site included metal detector and pedestrian surveys, which yielded a number of metal concentrations; however, no distinct crash crater or large pieces of wreckage were found in this actively cultivated field. The metal concentrations seemed to provide enough evidence to warrant a full-scale recovery operation in a relatively circumscribed area. Based on the historical information, witness and informant testimony, as well as this physical evidence, a recovery team was deployed to the scene.

Scene processing consisted of two primary phases: (i) discovery phase and an (ii) excavation phase. The discovery phase began with another pedestrian and metal detector survey of the cultivated field, which allowed the team to relocate the metal concentrations previously identified by the investigative team (see Figure 28.4). This strategy allowed the recovery team to establish the boundaries and perimeter of the aircraft debris field. Aircraft wreckage was recorded in a roughly rectangular area, approximately 75 m × 35 m in size, adjacent to an unpaved road running along the edge of the field. After this initial search and identification of a potential excavation area, a backhoe with a toothless bucket (to eliminate gouging of the earth) was used to systematically remove parallel rows of the upper topsoil layer. Each row excavated by the backhoe was inspected systematically for the presence of a possible aircraft impact feature.

At other similar crash sites, a large portion of the aircraft will penetrate the ground upon impact and create an "impact crater" with an associated blast berm. The larger, heavier portions of the aircraft that impact first (the front fuselage or engine for a World War II-type fighter or pursuit aircraft) will be forced below the floor of the impact crater. Subsequent fire and explosions will burn and scatter aircraft components and biological evidence. Over time, the crash crater is filled in through natural processes, as well as through human intervention. Many impact craters were filled with village or town debris or trash to level a field used for farming. The craters also provided a place to deposit debris from houses destroyed during the conflict. *In situ* aluminum aircraft skin and frames embedded into the ground, particularly if they are thermally altered, typically will oxidize and become friable and chalky white/blue in appearance. If inexperienced, a Recovery Leader might overlook this modified metal or mistake it for other materials. In fact, when these substances adhere to other items, they are often mistaken for bone. Ferrous components of the plane oxidize and produce red-brown stains in the soil.

In this case, the P-51D Mustang, a small single-seat pursuit/fighter aircraft, crashed into the field, creating an impact crater with larger components of the plane exposed and present on the ground surface at the time of the incident. Local inhabitants salvaged the more accessible wreckage. Human remains also were likely encountered at the time; however, no witness recalled the removal or burial of human remains to a secondary burial in a local cemetery or churchyard. Over the next 65 years, the crater had been filled during the normal course of farming activities. Fortunately, normal farming activities generally only disturb the plow zone, thus incident-related materials and the crash crater would remain largely undisturbed below the plow zone.

During the backhoe operations, the impact feature was first identified as an area of darker sediment interlaced with small fragments of decomposing and oxidized metals (Figure 28.4, right). Once this potential feature was identified, the plow zone was cleared in the surrounding area. The team then excavated an area approximately 10 m × 10 m in size, and roughly centered in the middle of the crash crater feature.

Figure 28.4 Discovery phase and completed excavation of case study 2. Top left: trenching operations using heavy machinery to strip off the plow-zone level. Right: the first appearance of the crash crater feature in the floor of a test trench. Bottom left: completed excavation.

During excavation, two small exploratory trenches – one approximately 2.0 m × 0.5 m in dimension and 40 cm in depth, and the other approximately 3.0 m × 0.4 m and 50 cm in depth – and running through the impact feature, were hand-excavated to assess the depth and stratigraphic profile of the feature. Initially, the impact site appeared as four separate features due to the uneven subsurface topography; however, as excavation progressed, these individual features coalesced into one primary aircraft impact feature.

The impact feature was then fully excavated by removing only the modified soil from within the feature as defined by the stratigraphic interface between the feature fill and the original, pre-impact matrix. This allowed the Recovery Leader to preserve the original shape of the impact crater. However, while the feature fill provided the focus of the excavation, some areas of the feature were excavated beyond its borders (i.e., beyond the stratigraphic interface) to allow for a more complete stratigraphic assessment and ensure that the area beyond the feature was devoid of material evidence. All sediments from the feature were screened through 6 mm wire mesh. Concurrently, additional test trenches were mechanically excavated with backhoes in areas beyond the crash feature to ensure that no additional features related to the crash incident were located in the vicinity.

The impact feature was mapped via a baseline running parallel to both the adjacent two-track road and the longitudinal orientation of the crash feature (roughly northeast to southwest). Thus, the mapping baseline was oriented relative to the topography of the site area and orientation of the archaeological feature, and not to true or magnetic north. The northeast stake of the baseline was designated the site datum (the primary reference point) and all measurements were obtained from this line. A 1 m × 1 m grid system was established off of this baseline to facilitate the mapping of the crater outline and to systematically record the interface of the impact feature below surface. A transit-and-tape system was used to measure (triangulate) the position of the site datum to semipermanent points on the landscape (in this case, two electrical poles), in the immediate vicinity of the site. While not the best situation for archaeological recovery, these semipermanent points would allow a subsequent team, if necessary, to relocate the site and reestablish the excavation grid system. The transit was then used to map the overall boundaries of the test trenches and the other aspects of the site in reference to the site datum point.

Within the impact crater feature, significant amounts of small aircraft debris were encountered. Some of this debris yielded serialized part numbers that correlated to a P-51D aircraft. Browning M2 .50 caliber aircraft machine guns were found along the perimeter of the impact crater, a position that would be consistent with the wing-mounted aircraft machine guns of a P-51D. A small quantity of osseous material was recovered during the excavation from the deepest (i.e., "punched in") areas of the impact crater, about a meter below the surface. These remains were highly fragmented and degraded after nearly 70 years in the ground; however, the materials still yielded mitochondrial DNA sequences that could be compared to family blood samples.

This case study illustrates the various strategies used to discover a highly modified and transformed site as well as the employment of a site-specific excavation strategy. In addition, there was another component of this job that deserves special mention. In many areas of the world where JPAC Recovery Leaders operate there are cultural resource management or historical preservation laws that require a robust archaeological scope of work to be submitted and approved by various state and, in some cases, federal agencies. In addition, environmental concerns also must be addressed through a permitting process. All of these permits must be finalized and approved prior to the excavation. Finally, the removal of any material evidence (artifacts) and biological materials must be approved by archaeological or historical preservation agencies and, possibly, medicolegal officials. These activities require the JPAC Recovery Leader to be proficient in the historical preservation laws and permitting requirements of a variety of nations throughout the world.

A related topic is that of "wreck hunting" or "plane spotting" activities. Well-meaning historical and World War II enthusiasts locate crash sites and turn over the information to various local and US government officials. Eventually this information will make it back to JPAC's J2 section for investigation into whether it correlates to a known US loss incident. However, many enthusiasts are more interested in the collection of World War II aircraft parts for personal collections as well as for selling in a very lucrative artifact market, and have had access to these easily seen crash sites for over 65 years! Although many of these activities are illegal and unethical, they remain quite common.

The JPAC-CIL teams may also be asked to conduct clean-up activities at disturbed sites to determine whether any biological materials are present. The JPAC teams, particularly in Europe and North America, will receive numerous requests by local enthusiasts to participate in their excavations of these scenes. In these cases the Recovery Leader and others in team leadership must act as public relations managers, site security, safety officers, and ambassadors. When working in a small community, the Recovery Leader attempts to satisfy the local inhabitants' curiosity about the process, promote best practices, discourage looters, and build relations. The goal is to not insult the locals, particularly since they may have reported the site to JPAC originally and their enthusiasm may lead to additional discoveries in the area. Often these local enthusiasts and amateur historians have specialized knowledge of aircraft parts or military history. However, the team leadership needs to understand that, in many cases, there are multiple groups of local enthusiasts who are competing for JPAC's recognition. Problems and difficulties may ensue if the team leadership focuses on one or another of these groups and individuals and it is here that the ambassador role of the Recovery Leader is important.

Case study 3: a naturally disturbed Korean War burial

This case involves a recovery of the remains of several US servicemen in the DPRK. As background, the Korean War was fought during 1951 through 1953 in an attempt to halt the territorial expansion of Socialism throughout the entire Korean Peninsula (Mossman 1990). Over 8100 US servicemen are currently considered missing from this conflict with approximately 5000 in the DPRK, a regime which is openly hostile to the USA and its allies. Between 1996 and 2003, the US Army CILHI (the precursor of the JPAC-CIL) conducted an active program of joint investigation and excavation in the DPRK under the guidance and supervision of the Korean People's Army. In 2003, the USA suspended operations in the DPRK due to safety and communications concerns. Negotiations are ongoing to once again allow the USA access to the DPRK.

Most of the American casualties missing from the Korean War and investigated by JPAC-CIL were ground forces killed on the battlefield. Many of these individuals were hastily buried in fighting positions (so-called foxholes) by both friendly and hostile forces. Other situations encountered include burials in former prisoner-of-war (POW) camps or POW holding areas.

In addition to attempting to work in a relatively hostile political environment, the search for missing soldiers is complicated by the not uncommon practice of secondary reburial in which servicemen graves are excavated and the bones reassembled in a new grave by the Koreans in preparation for American investigations (Holland et al. 1997). The political rationale behind these "faux graves" is unclear or unknown, although the Korean People's Army officials often speak of the recovery of remains as an indicator of the political success of a given recovery mission and to the extraordinary level of cooperation provided by the Korean government to the American government in these affairs.

In many instances, when led to a potential burial location, the Recovery Leader must carefully consider a number of factors, including the location and context of the burial features, and the orientation and positioning of the body, in order to interpret the legitimacy of the grave itself. For example, when excavations reveal a set of remains haphazardly arranged, and not in anatomical position, disturbance and reburial are indicated. This is further corroborated by the location of the grave, the attributes of the grave feature itself, and other factors. Faux burials can range from a burial feature that has been disturbed by the Korean People's Army prior to excavation by a US team (indications of these activities can include a transposed tibia, fresh grass fragments, or even a newspaper fragment from the previous year) to completely manufactured graves in which remains are removed from the original grave and reburied in a new grave elsewhere. This latter type of grave include evidence of attempts to place skeletal elements in an approximate articulated pattern, although the legs and arms are often switched and the vertebral elements piled in the center of the torso area. The key to proper interpretation of these features resides in training and experience in archaeology, geology, and taphonomy to separate natural from artificial configurations. The following case study is offered to provide an example of why a comprehensive understanding of forensic taphonomy, including fluvial transport issues and depositional processes, are key in the interpretation of these scenes.

Investigative research and witness statements led a US Army CILHI investigative team to consider a location, approximately 25 m² in size, on the eastern side of a small valley, below a ridge, as a likely spot for a burial site of several US service members. This area was near a known POW holding area that was used after the Battle of Ch'ongh'chon in late 1950 (Mossman 1990: 61–83). The investigation team identified a small depression (approximately 2.70 m in diameter) adjacent to bedrock outcrops as a likely location for a burial pit. A small test pit was excavated into this depression. Indeed, human remains were encountered along with various material evidence, including most notably a US-manufactured pocketknife in poor condition, and was labeled as Feature 1. At this point, the investigation team ceased operation and the recovery team was brought in to conduct a more rigorous excavation.

Topographically, the site is located in the path of a small, seasonal stream channel, which coursed downward adjacent to this bedrock-bound depression from above (and from the area where the POW holding area was located). Standard archaeological techniques were used to process the possible burial feature and recover additional human remains (Figure 28.5). Feature 1 appeared pit-like in profile and extended 85 cm below the ground surface in its center and feathered upward to only 10–15 cm below the surface along its lateral perimeters. The feature was bounded by coarse granitic bedrock to the north, west, and east.

The recovered human remains consisted of nonarticulated and randomly scattered long bones and six concentrations of cranial elements. The mapping procedure included piece-plotting (and individual numbering) of the human remains and significant pieces of material evidence onto grid paper. During excavation, two arbitrary stratigraphic levels, upper and lower (Figure 28.6) were designated, although they did not represent individual depositional units. Cranial elements appeared to be concentrated along the perimeter of the feature while the long bones were scattered in the center. There was a conspicuous absence of vertebrae, ribs, tarsals, metatarsals, carpals, and metacarpals. Almost all of the long bones lacked the proximal and distal ends and appeared to have been eroded away. The infusion and penetration of roots and rootlets into the cortical bone tissue, as well as root-etching of the bone surfaces,

Figure 28.5 Before (top) and after (bottom) excavation of case study 3. Inset in the bottom panel is another view of Feature 1.

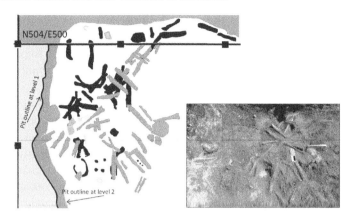

Figure 28.6 Modified sketch map and photo illustrating the commingled human remains and material evidence found within Feature 1 (case study 3). The pattern of deposition and preservation of materials is consistent with fluvial depositional processes.

suggested that the remains had been in this location undisturbed for a substantial period of time (Lyman 1994: 375–377).

After exposure, mapping, and removal of the human remains, excavation continued an additional 20 cm until bedrock was encountered. The deposits within the feature consisted of coarse, granitic sediment that included large decomposed granite pieces infused with crystals of feldspar and quartz. Removal of the remainder of the feature matrix revealed a natural bowl-like depression formed in the granitic bedrock, which explained the surface depression (see Figure 28.6). The "burial" feature lies on a level break in slope, again within the path of the seasonal stream.

This type of bone accumulation has been described in the paleontologic and zoo-archaeological literature as an "active accumulation" which R. Lee Lyman (1994: 162) defines as "...processes which, via transport or movement of skeletal parts (whether or not as complete carcasses/skeletons) significant distances from the location of animal death, result in relatively dense concentrations of bones and teeth in a spatially limited area." Active accumulations are most often associated with physical agencies, such as hydraulic or fluvial activities (Gifford 1981; Hanson 1980; Micozzi 1991; Voorhies 1969; Wood and Johnson 1978).

In addition, the composition of the deposits in the area of the human remains and feature corresponds, in general, to Voorhies' (1969: 69) observations on fluvial transport, redeposition of remains, and, in particular, removal of skeletal elements. The absence of Group I elements (ribs, vertebrae, sacrum, and sternum) seems to represent a winnowed deposit, as described by Behrensmeyer (1975: 471). The human remains assemblage matches many of the indicators of fluvial transport cited in Nawrocki et al. (1997), particularly partial or complete destruction of the facial skeleton, perforation of the thin plates of bone, and abrasion or breakage of exposed edges or processes, among others. The apparent random orientation of the long bones suggests that the fluvial process was not linear, but instead somewhat circular, like a fluvial eddy (a circular swirling water offshoot off the main stream) in a bedrock depression (Voorhies 1969; Shipman 1981; Nawrocki et al. 1997). As discussed above, the angular sediment of the decomposed granite that comprised the matrix of Feature 1 may have been responsible for the extreme abrasion and lack of long-bone epiphyseal regions (Binford and Bertram 1977; Shipman and Rose 1988; Nawrocki et al. 1997).

Given the configuration of the bones within the feature, the eroded nature of the bone ends, and the geological context, the most parsimonious explanation for the concentration of bones is that they were washed out of the original grave feature located somewhere upslope near the POW camp. Through fluvial transport they ended up in an eddy created by the depression in the bedrock. The stream energy was diminished when it entered this depression and the long bones and cranial elements sank to the bottom of the feature (Nawrocki et al. 1997). The remainder of the smaller bones likely continued downstream. Apparently there was still sufficient energy to create a swirling-type effect which resulted in the bones abrading against each other and the granitic sediment and bedrock (see Voorhies 1969; Shipman 1981; Nawrocki et al. 1997), thus explaining the eroded long-bone ends.

Only limited investigation was permitted at the scene and the search for the original burial pit associated with the POW camp was unsuccessful in revealing the location of the mass grave. However, mitochondrial DNA testing of the remains recovered was successful in identifying a number of individuals from this secondary accumulation of human remains (Christensen et al. 2007). Through an understanding of forensic taphonomic processes it was possible to identify likely depositional processes and dismiss others which might have been misinterpreted as a faux burial created by the Korean People's Army (i.e., Holland et al. 1997).

Discussion

The three case studies discussed above serve to illustrate some of the very different operational models and situations that a Recovery Leader can expect to operate under in various cultural and political contexts. In some areas of the world, JPAC recovery teams are left to conduct operations with relative autonomy and little interaction with the host nation's governmental officials. In other investigations, prior governmental approval is necessary, historical preservation issues need to be resolved, land access permits must be obtained, and embassies notified. All of this activity takes place during the pre-excavation and planning phases of the operation. Some countries, such as the DPRK, work only within a framework of military-to-military operations, while others require the team to interface with their civilian counterparts. In still other countries, the freedom of movement, even outside of hotels, is severely restricted and curfews may be enforced.

Preparation for employment with JPAC-CIL: physical, mental, academic

What then are some of the important education skills, work experience, and interpersonal skills necessary to become a JPAC Recovery Leader? Since recoveries are conducted throughout the world and under very different conditions, adaptability, self-reliance, self-confidence, and critical thinking are key. Add to that mix advanced skills in both osteology and archaeology, and the ingredients for a successful application to JPAC-CIL are beginning to emerge. Some of the specific skill sets or background include the following.

1. Forensic osteological skills: while the focus of this chapter has been the Recovery Leader roles and responsibilities, most Recovery Leaders are also proficient in osteological analysis of the human skeleton (Recovery Leaders with a more traditional background in archaeology tend to focus on the analysis of artifacts, or material evidence). Human osteological skills come into play when determining forensic significance of the osteology remains and whether they represent human or animal, and even assessment of biological profile. For example, in the DPRK, Recovery Leaders will frequently be led to burial sites that yield native Korean remains and not missing US service members. Skills in the analysis of forensic taphonomy bone modification are also critical.

2. Archaeological skills: an individual trained in archaeology brings a unique set of skills to forensic anthropological investigation of missing military service personnel, and relate to looking at sites within a geomorphic and cultural landscape, conducting search and survey operations and strategies, and obviously field excavation and site and context interpretations. More traditional site-oriented tasks include excavation grid construction, orienteering skills (topographic map reading, the use of a variety of styles of compasses, the use of a global positioning system receiver), and the use of survey equipment (transits, electronic distance-measuring devices, total stations), as well as an understanding of geology, taphonomy, and stratigraphy. These latter skills allow the Recovery Leader to more properly interpret the site modifications since the time of the incident. In addition, Recovery Leaders require a certain amount of flexibility in the construction of recovery strategies, particularly in terms of being able to adapt excavation methodologies to the specific sites and their topographic/landscape settings. For an archaeologist wishing to become a Recovery Leader, the authors would recommend gaining supervisory experience and developing core leadership and organizational skills in addition to basic archaeological tasks. One may also want to gain experience on a large variety of archaeological surveys as well as different types of archaeological site.

3. Outdoor skills: because the JPAC Recovery Leaders often live and work in austere environments, a background in basic outdoor activity skills is a necessity. Again, knowing how to use a compass and read a topographic map is a useful outdoor as well as archaeological skill to possess. Knowledge of hiking, backpacking, and camping are essential as is training in first aid and cardiopulmonary resuscitation (CPR). Additionally, skills of rock climbing and rope/knot work also are beneficial.

4. Interpersonal skills: the Recovery Leader needs to develop good interpersonal skills. In dealings with military personnel, foreign dignitaries, and foreign nationals from the entire spectrum of socioeconomic status, as well as local and international media, the Recovery Leader needs to display respect and a temperate personality. Public speaking skills and even negotiation skills may be necessary.

5. Writing skills: because a large part of the Recovery Leader's job is spent writing reports, a potential Recovery Leader must have adequate skills, both in note-taking as well as the preparation of final excavation reports. These skills can be gained by the student writing essays and term papers in school, as well as attending writing center workshops at universities or colleges. Another important venue to demonstrate these skills is through academic writing, like a dissertation or Master's thesis as well as publications in peer-reviewed journals or other publications.

According to Sledzik et al. (2007), forensic anthropology has entered the "Fourth Era" where forensic anthropology is recognized as a unique discipline.[1] As such, training in forensic anthropology must include physical anthropology and closely related fields such as human biology and archaeology. Dirkmaat et al. (2008) have suggested that a "paradigm" shift has occurred in the field, as forensic archaeological methods and protocols for acquiring a contextual setting for the skeletons analyzed are interdigitated with forensic osteological methods and techniques. However, many academic programs are stuck in "third gear" in terms of implementing opportunities to acquire rigorous archaeological skills into forensic anthropology-oriented programs; skills that are heavily in demand in applied contexts. As the nature of forensic anthropology transitions from a purely laboratory-based realm into a more fully functional integration of laboratory and field activities (see Chapter 1 in this volume), many academic departments have been slow to integrate requirements for a truly holistic melding of the laboratory-based and field-based expertise that a modern forensic anthropologist needs in order to be successful and contribute in a meaningful way to the changing field.

Many programs do allow for a flexible approach to curriculum building and often put the onus on the student to seek out the proper training, academic work, and field experience. While seemingly ideal, this approach can actually limit one's experience given the significant time demands and economic realities that obtaining practical archaeological experience requires. If one's only experience comes from archaeological course work and a single 6-week field school (as typically required for non-archaeology-focused students) then the depth and breadth of one's knowledge is usually insufficient for employment by the CIL. Additionally, a typical field school curriculum can only present a partial picture of the types of practical skills that are necessary to successfully work for a cultural resource management company, a government agency, or in the international forensic community.

Anthropologists come to JPAC-CIL from a wide variety of life paths. However, some of the most successful are those who have worked on large-scale archaeological projects as part of the field crew and, ultimately, as crew chiefs or principal investigators. These types of position allow the student/employee to gain the appropriate organizational and leadership skills needed to manage large-scale recovery projects. Skills learned for the practical side of archaeology are far different from those used to manage a group of military service members and indigenous field crew. One of the authors (WRB) grew up in a military family and worked on large-scale excavation projects in Pakistan, and so these skills sets were already developed. For others, exposure to leadership roles and responsibilities can come from a variety of sources; professionals and students need to seek out these opportunities and exploit them. For instance, in addition to employment on small- and large-scale cultural resource management projects, there are opportunities such as working on medicolegal recoveries, historic cemetery burial excavations, and international human-rights work. Often, local law-enforcement agencies request assistance when dealing with a clandestine burials or surface scatters of human remains and evidence. Due to the increased visibility of forensic anthropology in the mainstream, many of these agencies may seek expertise from anthropologists and archaeologists at local universities. Become a team member, volunteer for search-and-rescue operations (e.g., skills from archaeology as simple as reading a map and compass and walking transects are applicable to a large number of scenarios). Seek out colleagues with ongoing research-based excavations and volunteer your time as well as gaining employment with local or regional cultural resource management projects. Attendance at regional and national anthropological and archaeological meetings (especially those for the American Association of Physical Anthropology, the Society for American Archaeology, and the American Academy of Forensic Sciences) is particularly useful for networking with other professionals and students.

CONCLUDING REMARKS

The primary goal of JPAC-CIL recovery projects is to recover human remains of missing servicemen that subsequently can be identified through the scientific practices of forensic anthropology. The success of the identification process is dependent on the scientific integrity and success of the excavation and recovery operations. However, to be a successful Recovery Leader, a forensic anthropologist/archaeologist must be able to move freely between scientific, diplomatic and interpersonal responsibilities.

Based on detailed historical research, recovery crews are sent out into foreign locales to corroborate basic information such as the precise location and composition of the debris field and verification of a specific aircraft from debris. This information ultimately will lead to the recovery of the biological remains of missing US service members. These locales often have limited access, over treacherous terrain that may require a 5-hour hike along leech-filled, shallow streams and through triple-canopy jungle. Once at the locale, excavation activities usually encompass standard archaeo-logical techniques, including pedestrian surveys (often with metal detectors), clearing vegetation in the vicinity of the site, setting up excavation grid systems, excavating with various hand and mechanical tools, screening of excavated matrix, and interpreting stratigraphy and archaeological features. Of course, the Recovery Leaders must be comfortable with these skills and adapt these standard techniques to unusual circum-stances, such as extremely steep slopes! The primary means of understanding the difficulties is to realize that many of the sites are unique and pose problems whose solutions cannot be found in an archaeological textbook.

Other skills learned from recreational activities, such as hiking, rock climbing, and photography, can be converted into skills that are required by Recovery Leaders. Outdoor skills such as map reading, orienteering, and camping are also essential. The authors recommend gaining these skills by joining a local hiking or trail-maintenance club, a rock climbing gym, or a climbing club, as well as practicing camping and backpacking skills.

The authors hope that this overview of field operations at JPAC-CIL as well as the case studies convey some of the exciting and challenging aspects of forensic anthropology. A strong background in archaeological recovery techniques and interpretation is needed to adapt standard excavation techniques. Like fieldwork anywhere in the world, the work is often miserable and grimy, but worthwhile in recovering the remains of missing US service members. We should reemphasize here that within all of the vignettes presented the responsibility of maintaining strict timetables, directing workforces, and ensuring scientific integrity of the process is incredibly demanding.

NOTE

1 T. Dale Stewart (1979) and Thompson (1982) both recognized three eras of growth of forensic anthropology as a scientific discipline. The first era occurred before World War II where there was no formal instruction and little work within the medicolegal communities; the second era occurred from the 1940s to the 1970s where forensic anthropology had more interest by the military (and other government agencies) and medicolegal investigative services; and, the third era was characterized by field professionalism with the creation of the Physical Anthropology section of the American Academy of Forensic Sciences and the American Board of Forensic Anthropology.

REFERENCES

Behrensmeyer, A.K. (1975). Taphonomy and paleoecology in the Hominid fossil record. *Yearbook of Physical Anthropology* 19: 63–72.

Binford, L.R. and Bertram, J.B. (1977). Bone frequencies – and attritional processes. In L.R. Binford (ed.), *For Theory Building in Archaeology* (pp. 77–153). Academic Press, New York.

Christensen, A.F., Belcher, W.R., and Bettinger, S. (2007). Analysis of commingled remains using anthropology, archaeology and DNA: a case study from North Korea. Paper presented at the *59th Annual Meeting of the American Academy of Forensic Sciences*, San Antonio, TX.

Conroy, G., Phillips Connroy, J., Peterson, R., Sussman, R., and Molnar, S. (1992). Obituary: Mildred Trotter, Ph.D. (Feb. 2, 1899–Aug. 23, 1991). *American Journal of Physical Anthropology* 87: 373–374.

Dirkmaat, D.C., Cabo, L.L., Ousley, S.D., and Symes, S.A. (2008). New perspectives in forensic anthropology. *Yearbook of Physical Anthropology* 51: 33–52.

Eckert, W.G. (1983). History of the U.S. Army Graves Registration Service (1917–1950s). *American Journal of Forensic Medicine and Pathology* 4: 231–243.

Gifford, D.P. (1981). Taphonomy and paleoecology: a critical review of archaeology's sister disciplines. In M.B. Schiffer (ed.), *Advances in Archaeological Method and Theory*, vol. 4 (pp. 365–438). Academic Press, New York.

Hanson, C.B. (1980). Fluvial taphonomic processes: models and experiments. In A.K. Behrensmeyer and A.P. Hill (eds), *Fossils in the Making* (pp. 156–181). University of Chicago Press, Chicago, IL.

Holland, T.D., Anderson, B.E., and Mann, R.W. (1997). Human variables in the postmortem alteration of human bone: examples from U.S. war casualties. In W.W. Haglund and M.M. Sorg (eds), *Forensic Taphonomy: the Postmortem Fate of Human Remains* (pp. 263–274). CRC Press, Boca Raton, FL.

Holland, T.D., Byrd, J.E., and Sava,V. (2008). Joint POW/MIA accounting Command's Central Identification Laboratory. In M.W. Warren, H.A. Walsh-Haney, and L.E. Freas (eds), *The Forensic Anthropology Laboratory* (pp. 47–63) CRC Press, New York.

Hoshower, L.M. (1999). Dr. William R. Maples and the role of the consultants at the U.S. Army Central Identification Laboratory, Hawaii. *Journal of Forensic Sciences* 44: 568–576.

Lyman, R.L. (1994). *Vertebrate Taphonomy*. Cambridge University Press, Cambridge.

McKern, T.W. and Stewart, T.D. (1957). *Skeletal Age Changes in Young American Males*. Quartermaster Research and Development Center, Natick, MA.

Micozzi, M.S. (1991). *Postmortem Change in Human and Animal Remains*. Charles C. Thomas, Springfield, IL.

Mossman, B.C. (1990). *Ebb and Flow: November 1950 to July 1953*. Center of Military History, United States Army, Washington DC.

Nawrocki, S.S., Pless, J.E., Hawley, D.A., and Wagner, S.A. (1997). Fluvial transport of human crania. In W.W. Haglund and M.M. Sorg (eds), *Forensic Taphonomy: the Postmortem Fate of Human Remains* (pp. 529–552). CRC Press, Boca Raton, FL.

Roskams, S. (2001). *Excavation*. Cambridge Manuals in Archaeology. Cambridge University, Oxford.

Shipman, P. (1981). *Life History of a Fossil: an Introduction to Taphonomy and Paleoecology*. Harvard University Press, Cambridge, MA.

Shipman, P. and Rose, J. (1988). Bone tools: an experimental approach. In S.L. Olsen (ed.), *Scanning Electron Microscopy in Archaeology* (pp. 303–335). British Archaeological Reports, International Series, vol. 452. Oxford University Press, Oxford.

Sledzik, P., Fenton, T.W., Warren, M.W., Byrd, J.E., Crowder, C., Drawdy, S.M., Dirkmaat, D.C., Galloway, A. et al. (2007). The fourth era of forensic anthropology: examining the future of the disipline. *Proceedings of the 59th annual meeting of the American Academy of Forensic Science*, San Antonio, TX.

Snow, C.E. (1948). The identification of the unknown war dead. *American Journal of Physical Anthropology* 6: 323–328.

Steere, E. (1951). *The Graves Registration Service in World War II*, Quartermaster Corps Historical Studies 21. Historical Section, Office of the Quartermaster General. United States Government Printing Office, Washington DC.

Stewart, T.D. (1979). *Essentials of Forensic Anthropology, Especially as Developed in the United States*. Charles C. Thomas Publisher, Springfield, IL.

Thompson, D.D. (1982). Forensic anthropology. In F. Spencer (ed.), *A History of American Physical Anthropology: 1930–1980* (pp. 357–369). Academic Press, New York.

Trotter, M. and Gleser, G.G. (1952). Estimation of stature from long bones of American whites and negroes. *American Journal of Physical Anthropology* 10: 463–514.

Voorhies, M. (1969). *Taphonomy and Population Dynamics of an Early Pliocene Vertebrate Fauna, Knox County, Nebraska*. University of Wyoming Contributions to Geology Special Paper No. 1. Laramie, WY.

Wood, W.R. and Johnson, D.L. (1978). A survey of disturbance processes in archaeological site formation. In M.B. Schiffer (ed.), *Advances in Archaeological Method and Theory*, vol. 1 (pp. 315–381). Academic Press, New York.

Wood, W.R. and Stanley, L.A. (1989). Recovery and identification of World War II dead: American Graves Registration activities in Europe. *Journal of Forensic Sciences* 34: 1365–1373.

CHAPTER 29

European Perspectives and the Role of the Forensic Archaeologist in the UK

Nicholas Márquez-Grant,
Stephen Litherland, and
Julie Roberts

INTRODUCTION

The past two decades have seen an increasing number of publications relating to forensic archaeology, primarily from North America and the UK. These publications, written by practitioners, have had a profound impact on the development of forensic archaeology, providing an insight into the archaeological methods employed in medicolegal or forensic contexts, especially relating to the search, location, and recovery of buried human remains (Morse et al. 1983; Killam 1990; Hunter and Cox 2005; Dupras et al. 2006; Blau and Ubelaker 2009). Recent attention has also increasingly been given to the investigation of mass graves (Cox et al. 2008; Haglund et al. 2001; see also Blau and Ubelaker 2009). Some of the published literature considers archaeological excavation and/or the recovery of human remains under the umbrella of forensic anthropology (Dirkmaat et al. 2008; Scott and Connor 2001; Owsley 2001). In Europe, while forensic anthropology as practiced in different nations has been a topic of publication (Schmitt et al. 2006; Cattaneo 2007; Kranioti and Paine 2011), it is in the UK where most of the publications on forensic archaeology have appeared (Hunter et al. 1994; Hunter and Cox 2005; Hunter 2009), especially in the context of

A Companion to Forensic Anthropology, First Edition. Edited by Dennis C. Dirkmaat.
© 2012 John Wiley & Sons Ltd. Published 2015 by John Wiley & Sons Ltd.

individual suspicious or unexplained deaths; that is to say, in the investigation into individual missing persons rather than the search and excavation of mass graves. Nevertheless, the recognition of forensic archaeology is increasingly rising in other European nations (e.g., Borrini 2007).

Forensic archaeology has been defined as "the application of simple archaeological recovery techniques in death scene investigations involving a buried body or skeletal remains" (Morse et al. 1983: 1). As will be exemplified below, the essence is far broader than the simple *recovery* of human remains. It is, according to Hunter and Cox (2005), the application of archaeological theory (including the principles and its methods) to the search, recovery, and excavation of human remains in death-scene investigations or any other buried material (e.g., weapons, money). Put simply, forensic archaeology can be defined "as the application of archaeological principles and techniques within a medicolegal and/or humanitarian context involving buried evidence" (Blau and Ubelaker 2009: 22).

The objective of this paper is not to present the techniques employed by forensic archaeologists, a topic on which extensive publications already exist (see previous chapters in this volume; see also Killam 1990; Hunter et al. 1996; Hunter and Cox 2005; Dupras et al. 2006; Connor 2007; Cox et al. 2008; Blau and Ubelaker 2009; Holland and Connell 2009; Cheetham and Hanson 2009), nor does it attempt to provide a detailed background of the evolution and development of the discipline (see Hunter et al. 1994). Instead, it focuses on reemphasizing the important role of the forensic archaeologist in a crime scene investigation or for any other legal matter. The chapter reflects the separation but nevertheless the interaction between forensic archaeology and forensic anthropology as it is viewed in the UK, and highlights the skills and expertise that an experienced forensic archaeologist should have. Above all, it aims to change a possible current perception (stemming from discussions at some European conferences) that there is no need for an archaeologist in the field of forensic science since there is only a need to "know how to dig." This chapter will indicate the benefits that experienced archaeologists can provide in the criminal investigation and why they should be used. To illustrate this point of view, the sections below will list the types of request, role, and responsibility experienced by the authors during the course of their police casework in the UK.

This chapter begins by providing a general insight into forensic archaeology in Europe, followed by a focus on the role of forensic archaeology in the UK since this is the geographical region familiar to the authors and where developments made in recent years have helped to promote the discipline. Although some reference will be made to the excavation of mass graves, particularly in Europe (see Haglund et al. 2001; Cox et al. 2008), we will focus primarily on police casework.

FORENSIC ARCHAEOLOGY IN EUROPE

Some material on forensic anthropology in Europe has been published (Kranioti and Paine, 2011; see also Cunha and Cattaneo 2006; Blau and Ubelaker 2009), but the concept of the "forensic archaeologist" in Europe is less well known. When it is used in most European countries it is on occasions clearly associated with the excavation of mass graves, whether Spanish Civil War graves, World War I and II graves, or mass graves from the former Yugoslavia for example. However, not as clear is the role or notion of involving a forensic archaeologist in more recent police crime scene work. Rather than employ a forensic archaeologist, there is mention of the forensic anthropologist, forensic pathologist, or police staff attending the scene and applying archaeological techniques. Some of the differences between countries may relate to different educational backgrounds and also to the legal system in which only actual police staff can be involved in the investigation. The request for an archaeologist to attend a crime scene might also depend on the budget of the individual police force. In other regions, criminals and crime may have a different behavior pattern in relation to urban planning. For example, bodies buried in back gardens may be common to northwestern countries like the UK but uncommon in southern countries such as Spain. Crime rates are also different; in some European countries, homicide rates are extremely low compared to others. In 2007, whereas England and Wales had a total of 784 homicides, and other major European countries such as Germany, France, and Italy surpassed the 500-per-year mark; other countries, such as Cyprus, Austria, Slovenia, Iceland, and Norway, had fewer than 50.[1]

There are a number of questions that are of interest regarding forensic archaeology in Europe. These include whether or not the concept of "forensic archaeology" exists in a particular European country and, if it does, what type of work is done by the forensic archaeologists. Other questions include those relating to the number of forensic archaeologists, where they are based, their background and qualifications, whether there are any forensic archaeology bodies or associations that govern accreditation or quality control, and whether or not there are any national guidelines or policies. A questionnaire[2] was sent out to professionals in Europe and the information below derives from their answers and comments.[3]

In Belgium (see Quintelier et al. 2011), when human remains are unexpectedly found, the Ministry of Justice can request the assistance of a forensic anthropologist or pathologist. If the remains are from a soldier of World War I or II, the situation is a different one. When there is doubt about the context surrounding the remains, the police will request the assistance of an archaeologist, as in cases where there is a need to locate and recover buried remains. Belgium's federal police Disaster Victim Identification (DVI) team (www.polfed-fedpol.be/org/org_cg_dsu_dvi_en.php) carry out excavations performed by a forensic archaeologist where there are clandestine graves. This has been used particularly in the search for missing persons since the Dutroux case in 1996. Since 2007, the DVI team has had an inspector with a qualification in archaeology who has ensured that archaeological principles and techniques are applied to crime scene investigation. Archaeologists from the DVI team may be required to attend court as expert witnesses or to recover bodies using archaeological techniques that they have learned through training or experience. Sometimes training is received abroad. General archaeologists may also be called when human skeletal remains are found and the Ministry of Justice need to establish (alongside perhaps an anthropologist and pathologist) the date of the remains.

In Croatia, although researchers and experts are aware of the potential of forensic archaeology, and despite a number of attempts to generate sufficient interest not only for police casework but also for the discovery of further mass graves, the concept is at present virtually nonexistent.

In Cyprus, the concept is very recent and mainly falls under the umbrella of anthropology, where a combination of human osteology and archaeological techniques are applied to the search, location, and recovery mainly of missing persons from previous war conflicts. This is run by the Committee of Missing Persons (CMP) by Greek-Cypriot and Turkish-Cypriot sides.

In Denmark, only forensic anthropologists (who have knowledge of archaeological techniques) assist in the recovery of remains. The police technical unit may be called to a context with scattered and/or buried remains and will often call the forensic pathologist, who might in turn call the forensic anthropologist. Even more rarely, if there is a confession by a perpetrator or offender about a buried body, the police may ask local archaeologists (e.g., based at museums) to help with the search for recent inhumations. However, with the size of Denmark and around 30–40 homicides a year, these cases are extremely rare and not on a yearly basis. Therefore, it is very difficult to justify the full-time employment of a dedicated forensic archaeologist.

In Finland, the Finnish Association of Forensic Archaeology and Anthropology (FAFAA; www.helsinki.fi/~wperttol/fafaa/index_en.html), established in 2006, aims to promote specialist services, cooperation with other fields, and research in the fields of archaeology and anthropology. This concept of forensic archaeology, however, is in its infancy. It began with the excavation of a site thought to contain the remains of executed Finnish soldiers from World War II, only to prove through archaeology that it was a nineteenth-century mass grave. In police work, at present there is only one member of staff who serves as both forensic archaeologist and anthropologist, with most of the work in the anthropological realm. Other archaeologists with forensic experience are freelance, based at universities or at the National Board of Antiquities. Although there are no national guidelines, training, or expert witnesses in archaeology, buried bodies from criminal cases are rare, perhaps one case a year if any. At present, there is one police officer with archaeological training who has been employed recently to retrieve bodies from clandestine graves. Efforts have also been made by the FAFAA to encourage the government to employ archaeologists where possible.

A recent paper on France (Baccino 2009) provides perspectives on forensic anthropology in the country, but there is no mention of forensic archaeology or the role of forensic anthropologists at crime scenes, although the National Crime Lab of the Gendarmerie has a permanent staff of at least three anthropologists and one osteoarchaeologist, presumably called out to a scene where appropriate when human remains are present. It is the forensic anthropologists applying archaeological techniques who are essential for the reconstruction of the circumstances of burial and decomposition, where much work in France has been done in "field anthropology" (e.g., see Duday and Guillon 2006).

The information obtained about Germany sees the involvement of archaeologists working on cases such as buried dismembered or complete bodies in places like urban parks and allotments. At least one archaeologist was registered for a number of years in the BKA expert database (BKA or *Bundeskriminalamt* is the German equivalent of the US Federal Bureau of Investigation, or FBI). Nevertheless, the concept of forensic archaeology is virtually unknown in Germany, with no university courses. Any use of the term may be erroneously used for archaeological specimens (e.g., Bronze Age murder, etc.). There is an Austrian-German working group (www.akforensik.at), but it operates with slightly more emphasis in anthropology. Forensic archaeologists, when employed, may be required to go to court, or reports may be taken as evidence. There is no training or any national guidelines and when no archaeologists are requested it is usually the police and forensic pathologists who will undertake the search, excavation, and recovery of human remains.

In Greece, although there are no forensic archaeologists, forensic anthropologists who are involved in police work will apply the principles of archaeology. On occasion, the forensic anthropologist will visit the scene for search-and-recovery purposes and, although it is not mandatory to call an anthropologist for the recovery of skeletonized remains, it will depend on the decision of the law-enforcement authorities. It is this forensic anthropologist who provides training (seminars) on search-and-recovery techniques to law-enforcement agencies.

Hungary has no recognized forensic archaeologists undertaking police work. There are forensic anthropologists based at the Institute of Forensic Sciences (www.iszki.hu) and at other departments of forensic medicine (e.g., at universities) who will be involved with war victims buried in mass graves and with victims in police casework. There are three universities that teach archaeology and it may be that graduates are requested to assist in exhumations of recently deceased individuals.

The concept of forensic archaeology in Ireland exists in the form of a forensic anthropologist (with a background and training in archaeology) working with the state pathologist (and employed on a freelance basis by the Department of Justice) when skeletal remains are found in modern criminal investigations. The forensic anthropologist may from time to time read her statements in Coroner's Court. There is also at least one crime officer working at police headquarters who has a degree in forensic archaeology. Since 2007, forensic archaeology has existed in conjunction with the Independent Commission for the Location of Victims' Remains. Independent specialists were asked to advise the Commission. One of the recommendations put forward was to seek the assistance of a forensic archaeologist. As a result, there is at least one full-time employed archaeologist; other archaeologists from the University of Bradford in England have also assisted when required. There are no schemes for assessment, but in general terms there is the Institute of Archaeologists of Ireland, which is a professional body. There are also policies for archaeologists and legislation within which they are required to work (O'Sullivan et al. 2002). There is still a need for police to recognize the needs for the skills of forensic archaeologists and future training will see that the importance of archaeology is highlighted.

In Italy, the term "forensic archaeology" still needs further recognition. In 2001, archaeologists proved to judges and police officers the benefits of using forensic archaeological methods in tasks relating to excavation, recording of stratigraphy and topography, and correct sampling in a case of accidental discovery of human remains (Cattaneo 2009). The case in 2001 provided a stimulus for forensic archaeology in Italy. A publication by Borrini (2007), the only one of its kind in Italy, is solely on forensic archaeology and should help to promote the discipline. At present, there is only one criminal expert witness in forensic archaeology registered at the State Prosecutor Office and Civil Court in Italy. He has applied archaeological methods to the recovery of bodies from actual crime scenes as well as to the recovery of remains of World War II soldiers buried in mass graves. No archaeologists work for the police; there are freelance archaeologists, sometimes attached to universities, who may be called by the State Prosecutor Office or by a Medical Examiner to help in their investigation. Thus, the forensic archaeologist acts as a consultant or auxiliary to the Judicial Police and may also be requested to go to court. If no archaeologists or anthropologists are called, forensic pathologists, the mortuary police, or other police officers (police and *carabinieri*) are left to recover human remains from forensic scenes.

In Kosovo (Schermer et al. 2011), a new law passed and approved in July 2009 (*Law on the Department of Forensic Medicine*, law no. 03/L-137) establishes a new Department of Forensic Medicine within the Ministry of Justice. The Forensic Medical-Anthropology and Archaeology Section is one of the sections established under the Division of Forensic Medicine (Article 4). Among the definitions provided in Article 2, "forensic archaeology" is presented to mean the application of archaeological methods to assist in locating, exhuming, and recording mortal remains to assist in forensic investigation.

In Lithuania, the concept is just emerging due to recent exhumations of mass graves, excavations of war cemeteries (e.g., Jankauskas 2009), and, very seldom, in police work. In 2010 there were only two cases of exhumations where professionally trained archaeologists were requested. There are three persons at least considered competent in such skills in the country, and they are archaeologists or anthropologists who apply archaeological techniques. Otherwise, police perform the task themselves with their own resources and skills.

In The Netherlands, the Netherlands Forensic Institute (NFI; www.forensicinstitute. nl) employs a forensic archaeologist to assist in the recovery of human remains for police casework. Currently, there are two full-time recognized practitioners at this Institute (also attached to a bioarchaeological university department). The minimum qualifications are a Masters in Dutch Archaeology and 6 years of experience in field archaeology. In addition, both archaeologists have NFI-based training in forensic science, physical anthropology and forensic archaeology for approximately 2.5 years, which requires a formal exam, and, after 4 years of practice, an assessment to provide an opinion about the competency of the scientists. The NFI is accredited according to ISO/IEC-17025 norm and the NFI has quality or standard operating procedures for forensic archaeology. The Dutch archaeologists teach awareness courses at the Dutch Police Academy. Some of the work has involved the search (using tools including geographic information systems and total stations) and excavation of clandestine graves (persons or items), exhumation of bodies from formal/legitimate cemeteries for DNA sampling or postmortem examination, the recovery of skeletonized, fragmented, or severely burned human remains, as well as the dating of skeletal remains.

In Norway, the concept of "forensic archaeology" does not exist at present, and there are no forensic archaeologists in the Norwegian police. There are, however, some crime technician police officers that have a degree in archaeology. Apart from these individuals, crime technicians in the police are not trained in archaeological techniques, although they are responsible for recovering bodies from crime scenes.

In Poland, many unknown graves from World War II and the decade that followed continue to be found. Institutions competent for dealing with such discoveries are the main Commission for the Prosecution of Crimes against the Polish Nation (www.ipn. gov.pl/portal/en/5/43/Chief_Commission_for_the_Prosecution_of_Crimes_ against_the_Polish_Nation.html) and its local branches. In Poland there is an increasing interest in the application of archaeological methodology and field research methods in forensic science (sometimes the term "forensic archaeology" is used), so archaeologists are often employed to conduct excavations at this kind of sites. In 2010, a team of Polish archaeologists attended the plane crash scene near Smolensk (Russia) to excavate and recover as much evidence as possible relating to victim identification and the cause of the accident. Further to this increased interest in forensic archaeology,

2010 saw the first interdisciplinary postgraduate studies in forensic archaeology at the University of Wrocław. This is a two-semester module for graduate students of archaeology, history, law, and public administration, and particularly for employees of the Ministry of Internal Affairs. Nevertheless, the relevance and usefulness of archaeological methods in criminal cases is still not widely recognized in Poland, and there are always budgetary problems. In most cases, police officers and forensic pathologists are the only specialists who work at the crime scene, although in certain forensic medicine departments there are some anthropologists employed who have basic knowledge and experience in archaeological techniques. Archaeologists only go to court, however, if it relates to offences against monuments or artifacts of archaeological value.

In Portugal it is the forensic anthropologist who works with police forces, namely in cases of missing persons and homicides, apart from international mass graves from conflict contexts. Forensic anthropologists are consultant professionals to the National Institute of Legal Medicine (*Instituto Nacional de Medicina Legal* or INML) and, since 2008, there has been a full-time technician undertaking forensic anthropology cases. Forensic anthropologists are occasionally trained in archaeological techniques and help the Scientific Police (*Polícia Científica*) with the search and excavation of buried bodies. Forensic anthropologists do not go to court; they inform the police or give their opinion to forensic pathologists who will attend court. Training is provided in forensic anthropology at the Police Academy (*Escola Superior da Polícia*), or police officers may attend the INML.

In Serbia, the excavation of mass graves in a suburb of Belgrade (created during the spring of 1999 at the height of the NATO bombings of Serbia) stimulated forensic professionals to assemble a broad expert team including a small group of skilled archaeologists. In 2001 and 2002 two mass graves were excavated using the stratigraphic method by a team of archaeologists from the National Museum in Belgrade and the University of Belgrade, led by forensic archaeologists from the International Commission on Missing Persons (ICMP). Physical anthropologists from the Laboratory for Anthropology (University of Belgrade) provided significant input in combination with archaeologists at a site with complex burial stratigraphy, commingled remains, taphonomic activity, and observations *in situ* of premortem trauma and any postmortem damage due to the construction of the graves and potential attempts to further destroy the bodies (Tuller and Djurić 2006). Although this experience was positive, leading to the complete retrieval of all evidence, engagement of forensic archaeologists remains isolated in forensic practice in Serbia. Archaeologists are thus not recognized as professionals who are valuable in forensic cases where search and excavation of buried bodies is required. Therefore, they are not requested to attend courts as expert witnesses. Forensic archaeology is not part of the educational system in the country except through insignificant instruction offered in the PhD course in bioanthropology at the School of Medicine, University of Belgrade.

In the Slovak Republic there are no forensic archaeologists, so it is the anthropologist (seldom), forensic pathologist, or medicolegal expert who may help police with findings of human skeletal remains. Sometimes only police undertake the recovery, occasionally resulting in incomplete skeletal retrieval and/or no retention of soil or taphonomic contextual information. Thus, forensic anthropologists try to provide short theoretical lectures (not practical classes) to police officers concerning the location and excavation of human skeletal remains and an introduction to forensic taphonomy.

In Slovenia forensic archaeology as an institutional branch of archaeology or forensic science does not exist. Due to the specific problems of mass graves dating back to the time immediately following the end of World War II (there are around 600 already-discovered graves), archaeological methods were applied in the excavation of these graves (see Jamnik 2008; Leben-Seljak 2007). Now, with the use of forensic and archaeological methods, excavations of graves are carried out on two levels. Firstly, in the 1990s, the Slovenian Police and the prosecution began to investigate hidden World War II mass graves, where archaeological techniques were employed within a forensic context conducted by the Institute of Forensic Medicine of the Medical Faculty in Ljubljana. The head of police investigating the post-World War II killings has promoted the use of suitable archaeological work at these sites. Forensic pathologists provide medical anthropological expertise on the skeletons for court. Secondly, in the same decade (1990s), the Government of the Republic of Slovenia established a commission to deal with issues of postwar grave sites. The Ministry employs an archaeologist, an anthropologist, or a forensic pathologist to conduct the archaeological excavation of mass graves when required.

In Spain "forensic archaeology" is used on many occasions by anthropologists in the context of the excavation of mass graves dating to the Spanish Civil War. A recent paper on Spain (Prieto 2009) does not mention the use of forensic archaeology in police casework. Exhumations are conducted by the forensic pathologist (*médico forense*), police officers, or forensic anthropologists. In Spain, where guidelines in relation to archaeological methodology are mentioned, they tend to be included in books and chapters on forensic anthropology (see Acinas Robledo and Sánchez-Sánchez 2010; Puchalt Fortea 2000), with the understanding that forensic anthropologists are those to provide assistance for the search, location, and recovery of human remains.

It would be incorrect to indicate which countries have and do not have forensic archaeologists. What is clear is that practicing forensic archaeologists involved in recent police casework are present in Belgium, Germany, Ireland, Italy, and The Netherlands. The role or concept of forensic archaeology is emerging in other countries, especially in the context of mass graves and missing persons, as in Cyprus, Finland, Kosovo, Lithuania, Poland, Serbia, Slovenia, and Spain.

In addition, regarding the absence of forensic archaeologists, the use of archaeological techniques is recognized in crime scene investigation in most countries where commonly it is the responsibility of the forensic anthropologist to apply archaeological techniques to the recovery of human remains alongside other police staff. In a talk by the authors at an international physical anthropology conference in 2009, where we presented a paper on forensic archaeology, we were asked why there is a need for forensic archaeologists when "archaeological excavation techniques can be employed by anyone." It seems that in the majority of countries the forensic pathologist or anthropologist is the person to advise upon or undertake the search, location, and recovery of remains. The absence of forensic archaeologists may be due to a number of factors, including a lack of education and training, the judicial system in a given country where the use of external or unofficial experts may not be allowed, or barriers simply relating to budget and the type of work required, crime rates, and criminal behavior that render such personnel unnecessary or fiscally unfeasible. One perception is that the anthropologist must deal with the search and recovery of human skeletal remains using subdisciplines such as forensic archaeology (Cattaneo 2007: 185) so that the anthropologist has an awareness of archaeological techniques (Cattaneo 2007: 186). Nevertheless, even if no archaeologists are thought to be required, the importance of the use of archaeological techniques in the recovery of human remains is at least recognized in most countries, and this is a good start. We aim to change part of this perception because it is not enough to know the archaeological techniques needed to properly recover a body. It is imperative that personnel also know how to apply the techniques, how to record the efforts, and how to accurately interpret the results in the best way possible. In addition to this, there are many other tasks and skills and scenarios where forensic archaeologists will still be required.

A paper by Hunter and colleagues from the UK, Belgium, and The Netherlands (Hunter et al. 2001) attempted to provide some standards and protocols for anthropology as well as archaeology. The authors identified a set of minimum standards of competence for a wide variety of forensic archaeology scenes and a list of basic responsibilities of the forensic archaeologist.

It is perhaps part of this chapter to provide a wider awareness to readers, including police officers, about the need for such techniques and their proper application. As professional forensic anthropologists and archaeologists employed by an independent forensic service provider, we are often asked about our roles and the benefits of using forensic archaeology by police officers and crime scene investigators with whom we work. We are in a good position to demonstrate the benefits directly through our attendance at crime scenes and also by providing formal training to police forces.

The Role of the Forensic Archaeologist in the UK

Background to forensic archaeology in the UK

A number of authors (Hunter et al. 1996; Hunter and Cox 2005; Cox 2009) have provided information on the history, development, educational background, accreditation, and challenges of forensic archaeologists in the UK in recent years (see also Menez 2005); therefore, a brief summary here will suffice. Although archaeological techniques had been employed in the early decades of the twentieth century by Sir Sydney Smith,[4] a pioneer in forensic medicine, and later by A.K. Mant, who carried out a considerable number of exhumations after World War II (see Mant 1987), it was the case surrounding the body of Stephen Jennings in West Yorkshire that set a precedent for the use of archaeological evidence in a British court (Hunter et al. 1994; Cox 2009: 31, 38). The remains of the 3-year-old boy who disappeared in 1962 were discovered partially exposed under rubble in 1988. Archaeologist John Hunter participated in the investigation of the context in which the remains were found and was able to interpret through stratigraphy the sequence of events surrounding the burial (or concealment) of the body.

Since then, archaeology applied to forensic investigation can certainly be regarded as a healthy and growing discipline that is now relatively well accepted among police forces in the UK (Hunter 2009: 364). Awareness talks and training for police forces is certainly helping to emphasize the importance of forensic archaeology, whether at training centres such as the National Policing Improvement Agency (NPIA), courses for crime scene investigators, or talks at police stations and other police conferences. The main role of forensic archaeology is by and large the search, location, and recovery of buried human remains that are part of homicide or major crime investigations (see Hunter et al. 1994). According to Hunter (1994: 758; Hunter et al. 2001: 173) about 9% of homicides (or one in 50 murders) in the UK involve burial. Crime statistics taken for 2008–2009 indicate 651 incidents of homicide in England and Wales (Coleman and Osborne 2010) in a population of almost 55 million (www.statistics. gov.uk). Considering the 9% figure, this would result in approximately 58–59 burials for that year.

In the UK, forensic archaeologists have a degree in archaeology or history and certainly fieldwork experience, commonly after years spent working in commercial archaeology. Undergraduate modules and postgraduate degrees in forensic archaeology are taught at several British universities, including the University of Bradford, University of Bournemouth, Cranfield University, and University College London among others, at times jointly taken with physical or forensic anthropology, with mass-grave investigations and human rights also addressed. In fact, most physical anthropologists in the UK tend to come from an archaeological background, and it is in many archaeology departments that Master's courses in osteoarchaeology, human osteology, funerary archaeology, and forensic anthropology are taught. The advantage of this archaeological background is that it allows the anthropologist to appreciate and interpret the burial environment from which the remains are recovered and to apply archaeological methods to the recording and recovery of remains, especially where they are scattered over a wide area, although at such scenes there may not be the same requirement for interpretation of complex stratigraphy. Conversely, it could be argued that while most forensic archaeologists do not necessarily have an anthropological background in human osteology, there might actually be cases where anthropological input is not considered essential at the scene. Instances of this have occurred when the archaeologist has been the right person to attend the scene for responsibilities such as search and location, for working out excavation strategies, and for carrying out recording and interpreting complex burial environments. In cases of complete fleshed bodies, anthropological input is also desirable, although on occasions it has only been specifically an archaeological job. For archaeological jobs, the anthropologist is not the appropriate person unless he/she has a very strong archaeological background with many years of experience on different types of sites and is able to accurately record and interpret complex stratigraphic sequences if present.

Despite the relatively high number of individuals qualifying with a degree in the discipline, there are approximately around 20 forensic archaeologists in the UK who undertake casework for police forces. These practitioners are based in independent forensic providers such as Cellmark Forensic Services, archaeological units (e.g., Archaeology SouthEast), museums (e.g., Museum of London), universities (e.g., Birmingham, Bournemouth), or nonprofit organizations such as Inforce Foundation. In addition, a small number of police forces such as Greater Manchester Police have their own in-house archaeologist. Other forces have crime scene investigators (CSIs) or scenes of crime officers (SOCOs) who have archaeological degrees and who have had some digging experience. In addition, there are individuals from search teams (police search advisors) or detectives who have an amateur interest in archaeology and promote the use of specialist forensic archaeologists in a particular investigation.

Many of the practicing forensic archaeologists in the UK were recognized by the Council for the Registration of Forensic Practitioners, but this council dissolved in 2009. Currently, most forensic archaeological accreditation, policies, guidelines, and codes of conduct fall under the umbrella of the Institute of Field Archaeologists, within the Forensic Archaeology Special Interest Group, which has developed standard operation procedures (www.archaeologists.net/sites/default/files/node-files/Forensic2010.pdf). Note also that discussions are being held with the Forensic Regulator as part of an overall forensic review in the UK.

The policing system in the UK

Unlike other European countries, there is no national police force in the UK. Rather, there are just over 40 regional forces that operate independently in a number of ways (budget, protocols, etc.). There are also no national governmental laboratories for forensic analysis with the exception of the Forensic Science Service, which is in the process of closing. It was during the mid-1990s when forensic laboratories became privatized that police forces started to outsource most of their work to private and independent forensic providers. A number of the larger police forces such as the Metropolitan Police in London or Greater Manchester Police have some in-house facilities for their own forensic analysis (e.g., fingerprint experts) but overall most of the work is sent to private companies. The reason behind this is partly because it would be too costly, especially for the smaller forces, to employ experts from all disciplines on a full-time basis, since casework may not require certain specialists and equipment regularly throughout the year. Therefore, large private companies such as Cellmark Forensic Services or LGC Forensics may have scientists who are able to undertake DNA analysis, bloodstain pattern analysis, toxicology, drugs examination, e-crime, and other forensic services including archaeology and anthropology. Each company at the same time might have its own external consultants in other relatively smaller disciplines where there may be less demand, such as palynology, soil science or geomorphology, and entomology.

The request for a forensic archaeologist may come from a number of individuals within a police force such as, for example, the senior CSI or SOCO,[5] the scientific support manager, forensic coordinators, the senior investigating officer (SIO), or the officer in charge. The types of request, as will be seen, are varied, although usually within the realm of searching, locating, and recovering buried bodies or items. Sometimes the police may go directly to the forensic science provider and enquire about archaeological services; or it may be a direct call to the archaeologist if the police officer or SOCO has worked with him or her previously and the working relationship was a good one. Teaching, for example at the NPIA, and awareness courses by specialists to different police forces are good ways to communicate to police staff the services of an archaeologist. Alternatively, police forces may contact regional NPIA advisors who can consult a database of registered expert advisors from a wide range of professions, including archaeology.

Since forensic archaeology is integrated into search, location, and recovery strategies, it is in the UK an integral part of the Forensic Search Advisory Group (or FSAG; Hunter and Cox 1996). Moreover, since some police officers, CSIs/SOCOs (mostly civilian police staff), and members of search teams (e.g., police search advisors) have a degree or an amateur interest in archaeology, they are very supportive of the presence of a forensic archaeologist at a scene when they think it appropriate. It is certainly important for scientific support managers to have an awareness of what forensic archaeologists have to offer, as they are in control of the budgets for each force and ultimately have the final say on whether specialists may attend the scene. In Scotland, it is the Procurator Fiscal who has this power. Where it is a noncriminal investigation (e.g., DVI), or where there is a cold case review and exhumation in England, Northern Ireland, and Wales, the coroner makes the decision regarding which specialists to involve.

At a crime scene, forensic archaeologists tend to work with a team of police staff, but mainly liaise with the CSIs and detectives who may decide to exhibit any items, take photographs, etc. The archaeologist will provide advice to the senior SOCO/CSI, and the SIO and any decisions made by the archaeologist regarding, for example, approaches to investigating a scene archaeologically will be done in consultation with the crime scene manager, who will be a senior CSI/SOCO, and the SIO. Further information on the policing system in the UK and the involvement of the archaeologist in a scene of crime can be found in previous literature (e.g., Hunter 2009; Hunter and Cox 2005).

Although forensic archaeology is widely known in the UK and is generally viewed as separate from anthropology, there are still some occasions where the police force will undertake the recovery of remains without a specialist archaeologist or anthropologist. It is training and awareness courses made to police officers, forensic coordinators, crime scene managers, and CSIs that will promote the use of forensic archaeology. These courses are essential for the recognition of the discipline.

THE ROLE OF THE FORENSIC ARCHAEOLOGIST

Some case studies in forensic archaeology have been presented in the academic literature (e.g., Hunter and Cox 2005) and provide an insight into a variety of scenarios in which archaeologists may be involved. Cheetham and Hanson (2009) have summarized the benefits of having archaeological input and involvement in a crime scene in relation to commenting on the dating of the grave, recovering evidence for the identification of the deceased, and identifying ethnic and/or cultural and religious groups, assessing whether the body had been transported after death, reconstructing the crime scene, and linking crime scenes, among other benefits. They carefully point out that these tasks require highly specialized skills.

In this section we present the role of the forensic archaeologist in the UK and at the same time promote the use of forensic archaeologists. Anonymous case studies in which we have been involved are examples taken from our experience of working with a variety of police forces in the UK and abroad. Although it is widely understood that the main role of the forensic archaeologist is the search, detection or location, and recovery of human remains, he or she might also be concerned with ecological evidence types such as soils, pollen, and plants (Killam 1990). The range of requests provides a broader definition to this. The different scenarios where we as archaeologists have been involved are summarized below.

- Search for the body of a named individual who has been missing for some time presumed dead, in which there is very little intelligence information about the whereabouts of the body. On occasion, there is no information regarding the disappearance of an individual. This is probably one of the most common requests made to archaeologists by police forces and, sadly, on many occasions the results are negative in that the body is not found. This search can be undertaken with a variety of methods described in Hunter and Cox (2005; see also an example of methods in Ruffell et al. 2009), and usually the entirety of the suspected area of concealment is explored either manually or via means of a machine (e.g., a mini-digger). The authors have been involved in a number of such cases, most commonly searching in the back gardens of houses. The first example relates to the disappearance of an individual almost 10 years ago. After a series of police investigations, a warrant was obtained to check the house where the individual was living at the time of his disappearance. The archaeologist was requested to work at the rear garden and patio, where the georadar (ground-penetrating radar) and cadaver dogs had identified some anomalies. The archaeologist also undertook a visual inspection of the area, looking for anomalies in the topography, areas of soil disturbance, areas with spoil, broken roots, and so on. The archaeologist targeted these anomalies by means of excavation. These anomalies were negative for human remains[6] and the entire garden was stripped away by machine (a mini-digger). The archaeologist supervised the depth of the excavation to be attained by the machine (down to the natural clay in this case), recording and exploring any features that became apparent. The negative result for human remains after stripping the entire garden led months later to a similar investigation in one of the allotments owned by the wife of the presumed deceased person. Another case, also related to the disappearance of a boy about 15 years ago, involved the excavation of an area within a building. The area was stripped to the eighteenth–nineteenth-century level according to the lead forensic archaeologist, who was experienced in the archaeology of buildings.[7] One case with positive results came from the concern raised by a neighborhood regarding the absence of a neighbor whom they had not seen for several months. After initial investigations it was thought that the house should be searched. Under a layer of concrete in a back garden lay the human remains, and a suspect was arrested.

- To investigate an area where there is (very) good intelligence information regarding allegations that a body has been buried in a particular place. In one such case, we were asked by the Police Search Advisory Team to attend as archaeologists (one senior archaeologist and one anthropologist with an archaeological background) to identify whether there were human remains in the back garden and patio at the rear of a house after some intelligence suggested that there likely was a body there. A team of geophysicists from the police identified an anomaly under an area of concrete. After the removal of concrete by other members of the team, the area was trowelled back by the archaeologists and also by other search team members under the supervision of the archaeologists. An oval feature was identified and this was recorded before excavating one half of the anomaly (half-sectioning the feature). Within this oval feature lay a clothed, skeletonized human body wrapped in fabric. After the removal of the remains, the grave was completely excavated and soil samples were taken from the base of the grave. The entire back garden was also hand-excavated (the area was too small for a mini-digger) with the aim of investigating whether there was any other evidence relating to the identification of the deceased, a murder weapon, or any other information that might be thought relevant. In another scenario, where two of the authors assisted the police, the perpetrator confessed to the murder and indicated where the body was buried. Analysis of the topography and garden saw an area feature with defined edges, a "sunken" fill, spoil around and against the fence and walls of the garden, broken branches in the access routes, and new plant growth in the fill (which aided determination of postmortem interval). The area was trowelled, the grave edges defined and recorded, and then half the feature was excavated. From the position of the body and the dimensions and construction of the grave (tool marks, type of soil, and soil compactness, etc.) it was possible to make some inference about whether the grave was dug and the body deposited by more than one person. These were archaeological investigative skills that were also used to good effect in the interpretation of a deep grave in woodlands in the north of England; indeed, one of the authors was questioned about this extensively in court. Moreover, tool marks were found in the grave cut, which also provided further information in the investigation.

- Search for a potential clandestine burial in a large area (e.g., forest, field, etc.). On occasion, the search for a missing person has to cover large areas and the police search teams will have their own techniques. The archaeologist can be a useful addition to the team, providing additional input regarding any changes in vegetation and soil colour and texture, and targeting any anomalies by excavation of a potential grave-like feature. A number of cases in woodlands have come to us along with some previous pollen and soil evidence (from the suspect's car) that has allowed the search area to be narrowed down. Alongside topography, vegetation, and soil changes, this allows the archaeologist to prioritize the areas to be targeted. In one case, an area that was targeted through previous soil and pollen work revealed a location where the earth had 'sunk' and there was evidence of loose soil and vegetation. On excavation, it turned out to be a grave over 1.5 m deep containing the body of the missing person. After recording and opening one side of the grave to allow easier removal of the body, the archaeologist continued to excavate the fill of the grave and revealed tool marks that would have been used to dig the grave.

- Accidental discovery of human remains, whether buried or not. On occasion, partially exposed bodies have been accidentally discovered, and the archaeologist called to excavate the grave and recover the remains. In addition to this, he or she will be required to record the features at the scene, the human remains (sometimes with the assistance of an anthropologist), and different contexts, while gathering evidence to aid in the reconstruction of events and to help establish an estimated time since death (postmortem interval). Discoveries have occurred on many occasions of partially buried human remains that have been exposed during soil movement, climatic conditions, and so forth. In one particular example the perpetrator, who had been under surveillance, attempted to move the victim's body from the field in which he had buried the victim months previously. He failed to remove the body by pulling on a limb and fled, exposing part of it. Following his arrest, two archaeologists were called in to document, excavate, lift the remains, and recover evidence in an excavation that lasted 32 h without a break (the suspect could be held in custody for 36 h). The body was articulated and over 95% fleshed; it was therefore not necessary to call an anthropologist to identify the remains as human, although the anthropologist did attend the mortuary during the postmortem examination by the forensic pathologist. Another example of buried remains was a recent case where construction workers had found human remains associated with a house and called the police. The police contacted the county archaeologist who indicated that forensic archaeologists and anthropologists should examine the context in which the bones had been found as well as the bones themselves. The forensic archaeologist was able to establish that the remains dated well beyond the time frame of 100 years by looking at the building structure, grave cut, and so on. The regional archaeologist preferred the site to be examined by forensic archaeologists to avoid the risk of inadvertently destroying some evidence in case the investigation proved that the remains were modern. Excluding scatter scenes where the input of an anthropologist is essential, especially if he or she has archaeological training as to recording techniques, there are cases where remains may be found not buried underground but inside water tanks or semiburied under a rubbish tip, for example. One of the authors was called by a police force when a new landlord had found in his attic what he believed to be the skeleton of a baby inside a disused water tank that had been filled with debris. The SOCOs required that an archaeologist attend the scene to excavate the water tank and provide further interpretation of what had happened. This archaeologist, also an anthropologist, confirmed that the remains were human. The water tank was excavated in spits by the archaeologist and the depth of each spit recorded. The material from each of the spits of approximately 5 cm was then sieved and a record of what lay at the top and at the bottom of the water tank was made, followed by a comment on dating and an interpretation of the context in which the remains had been found. Another example was the discovery of a body among a rubbish tip in an urban area. The archaeologist called to attend the scene attempted to understand the "stratigraphy" regarding the body in relation to the surrounding rubbish (evidence for dating, etc.). The authors have also been asked as archaeologists to assist in a number of cases where skeletonized remains have been deliberately deposited within and between the walls of buildings, and they have established postmortem intervals by determining the date the building was constructed, subsequent building phases, and the date of any associated artifacts.

- Potential human burials. Another scenario is where someone is suspicious that there is a grave (e.g., a feature that looks like a grave; there is sunken or disturbed ground; or a witness states that the next-door neighbor buried something in the dark) and this has to be investigated. Although human remains are a possibility, several of these turn out to be a pet burial. One example included a marked grave (with sticks, flowers, and children's drawings and other artifacts) in an area of woodland to which we were called to investigate. After excavation it turned out to be a very recent burial of a pet.

Listed above are probably the most common requests that police forces make of forensic archaeologists. However, there are also a number of other requests that we have worked on in recent years.

- *Exhumation.* Police officers may need to exhume a body in either a cold case review or a current case and they may prefer this to be undertaken by professional archaeologists rather than CSIs/SOCOS. The archaeologist can advise on the safe and practical ways of achieving efficient recovery, particularly in a graveyard where there may be closely spaced burials or where the human remains have been disturbed by a collapsed coffin or root action, among other factors. They can also advise on soil conditions and potential requirement for such things as a mechanical pump to drain out water or the potential for remains to have "shifted" because of geological factors. The archaeologist with some knowledge of taphonomy will also be able to comment on the likely state of preservation of remains and therefore advise on requirements in the mortuary. Such cases have happened where the exhumation (followed by a postmortem examination) was undertaken by both an archaeologist and an anthropologist with archaeological experience. One example was a cold case review where the victim had died 19 years previously and had been kept in a freezer at the mortuary for 3 years, and so had been buried for 16 years; since parts of its body were missing (head and both hands), identification was unsuccessful at the time of his death. The body was exhumed to see if further analysis could be undertaken, including stable-isotope analysis. Another case is exemplified by the need to exhume a neonate that had been buried by her mother as part of a larger case involving other neonates found in the mother's house.

- *Excavating other contexts.* Excavation may not only be focused on clandestine graves under patios, in gardens, or in a field. There have been requests where excavation and archaeological interpretation has been needed during the investigation of possible child abuse and potentially the recovery of human remains in the cellars of buildings. The archaeologist can help in the interpretation of the different layers and dating of the deposition of human remains and other evidence, in addition to recording the areas with its layers or different contexts so that the material can also be sieved separately. Another case was the previously mentioned baby in a water tank found in an attic where, at the request of the SOCO, the archaeologist was asked to dig the fill of the disused water tank and interpret the events surrounding the deposition and possible dating of the remains.

- *Mass-grave excavation.* Work for UK armed forces, government, and nongovernment organizations such as the Foreign and Commonwealth Office and the United Nations has also included the excavation of mass graves abroad (if UK citizens are involved) where search, recovery, and repatriation of bodies is required for humanitarian reasons and criminal investigations.

- *Scatter scenes.* There are many instances where remains are found deposited on the surface (woodland, forest, ditch, etc.), and these tend to be scattered partly due to animal activity. Although this type of scene would suit an anthropologist with knowledge and experience of archaeological recording, archaeologists may also attend the scene to undertake recording (or this may be done by a SOCO), assisting the anthropologist who can identify the bones and describe their orientation and position on the plan and indicate which bones are still missing. It may be that answering the question of whether the body was buried in a shallow grave or observations about the vegetation cover over the body can help assess postmortem interval, in which case the archaeologist may take a botanical sample that is then submitted to the relevant specialist for analysis.

- *Dating.* Archaeologists will be able to advise on dating techniques for human remains and other evidence and may also be able to provide some chronology and dating in relation to the stratigraphic sequence, the interpretation of different phases of building, identifying different types of concrete, and so on. This can provide a *terminus ante-quem* or *terminus post-quem*[8] for the deposition of a body. Archaeological knowledge and skills will also be able to identify when to stop digging if there is a horizon of natural soil or a horizon that can be dated, for example, to the nineteenth century. For example, in the search for the potential body of a person who has been missing for 15 years, by looking at stratigraphy in an urban setting one can provide information on the dating of that layer and in which layer to stop (e.g., a layer of 1930s concrete, a layer of nineteenth-century rubble, etc.). This is again one of the most important reasons for employing archaeologists. Similar concerns apply to many cases of infant remains found within certain house structures, and a good building archaeologist can provide information about the relative dating of these. In the example of a water tank mentioned earlier, the context in which the remains of a baby were lying was from the 1940s as ascertained from analysis of wrappings, newspapers, cigarette packets, etc. The chemical dating of the bones (lead isotopes) provided a date going back to the 1930s–1940s.

- *Fire scenes.* Archaeologists may be called to attend fire scenes, especially in buildings, to ascertain the sequence of events and the distribution of human remains. This is especially worthwhile in cases of multistory buildings.[9] On the one hand, archaeological techniques can help in relation to the search, for example by laying out grids or zones to aid systematic recovery of body parts and fragments of bone. Additionally, understanding the position of objects or remains within their context can help reconstruct the sequence of events that led to a particular disaster and provide an understanding of the chronological sequence of the deposition of different layers and the remains and artifacts within each layer.

- *Scene evaluation and reconnaissance.* Prior to a search or to the exhumation of a body, the archaeologist might take soil samples from the area to be able to recognize any logistical problems (e.g., clay, water drainage) and to provide an idea of what to expect in relation to preservation. Similar tasks include a desk-based assessment prior to attending the scene, where the archaeologists will obtain previous maps, aerial photographs, or satellite images that may help with understanding the history of a particular site and how a particular area has changed over time, potential access routes, and so on. In addition, in a search job, the archaeologist will use his or her skills to visit a scene, comment on any photographs, and advise on different techniques, methods, tools required, number of people, and logistics regarding spoil.

- *Sampling and other ecological evidence.* Archaeologists have been accustomed to taking soil samples in excavations of archaeological interest. Likewise, they are able to apply these skills to obtain soil and pollen samples from crime scenes, although a specialist entomologist or botanist is preferred. Archaeologists should be able to provide advice on which specialist to call to the scene regarding pollen, soil, botany, and other evidence types, and, in collaboration with the specialists, will be able to locate, record, and take samples to be submitted for analysis. These evidence types may link a suspect or a vehicle to a scene, or may provide some information regarding the postmortem interval.

- *Other buried evidence.* Civil cases involving buried evidence (e.g., locating former fence lines and stream courses in boundary disputes) and other material such as money or weapons are also to be searched for, excavated, and recovered by archaeologists and a team of other experts.

- *Training and research.* Archaeologists are also asked to teach/train police staff and to undertake research that has forensic applications. Training provides an awareness of the type of work we do and how the CSI/SOCO can support the archaeologist and the type of skills required. In addition, if ever a specialist is not available to do the work, but the task still needs to be carried out, then the CSI/SOCO would have some skills and awareness to be able to do so, although little – if any – experience. In such a case, the lack of experience, especially in identifying different layers and interpreting the stratigraphic sequence of events, would profoundly hinder the evidence in court. Research may involve experimental archaeology, where, for example, a clandestine grave from casework is replicated in the same area and same soil conditions to understand the time and manpower required to dig a specific grave.

To sum up the points above, the archaeologist is there to advise on the search and, where human remains have been located, to recover the remains, sometimes alongside the anthropologist (it all depends whether the remains are fully fleshed or whether they are semiskeletonized or decomposed remains). The archaeologist will also maximize the recovery of other evidence, especially other ecology-type evidence. Appropriate and correct recording should be undertaken, since archaeology is a destructive process and the records may be used in court by the prosecution and defence. Overall, it is the recording and understanding of the context in which the remains are found that will provide an accurate interpretation of the events surrounding burial and deposition. The understanding of the environment can certainly be useful in interpreting events leading to death, although other experts in pathology, toxicology, fibers, bloodstain pattern analysis, entomology, and other associated evidence are equally if not more important. Dating is also another crucial aspect to which the archaeologist can contribute. These skills and expertise, however, are not widely appreciated, according to Hunter and colleagues (2001: 173). The next section attempts to provide further insight into the skills necessary for conducting a thorough investigation and also to provide further explanation on why a professional forensic archaeologist is essential to undertake the tasks and requests described above.

WHY AN ARCHAEOLOGIST?

There is a difference between an archaeologist and a "nonarchaeologist using archaeological techniques." However, this difference does not necessarily seem clear or important, according to discussions with colleagues in other countries. Questions have arisen among scientists, police staff, and even anthropologists, who fail to appreciate the need or the importance of employing a forensic archaeologist at the scene of a crime, if in fact the term "forensic archaeology" is even acknowledged. Questions have been asked, such as: Can we justify budget spending for an archaeologist? Is an anthropologist not sufficient to attend a scene? Are not the scientific support unit/CSIs capable of conducting the investigation themselves? Anyone can learn to dig and use a trowel; therefore, is there any need for a proper archaeologist? There have also been opinions in which it has been stated that an archaeologist without anthropological skills has no future: we would like to argue that this is not necessarily true.

The archaeologist is present to assist in the search, recording, and excavation of a body or an artifact in the most effective way possible in accordance with the UK Police and Criminal Evidence Act of 1984. From the different requests above, it should be apparent that the work of a forensic archaeologist extends beyond just digging a body buried in soil, and that an experienced forensic archaeologist is often essential. By employing an archaeologist, the body can be located and recovered faster and the evidence better preserved (Hunter 2009).

There are a whole range of skills, abilities, and benefits that the archaeologist can bring to the forensic investigation; some of these are listed below. These archaeological skills and methods are directly transferable to crime scene investigation.

There are schools of thought, frequently discussed in conferences, that hold that archaeological methods are easy to learn and can be learned by anyone, at least in terms of the excavation and recovery of bodies. The question is, then, why is it necessary to spend money on an archaeologist? The authors provide training to police forces and understand that this gives police staff an awareness of the processes involved, particularly if they assist an archaeologist or if an archaeologist in their area cannot attend the scene. However, the difference lies, we believe, not only in the quality of the work, but also in the interpretation of the data, the context, and the stratigraphy, as well as the recording, which is best done by an experienced archaeologist. Owsley (2001) adds that the excavation and recovery must be conducted in such a manner that it preserves evidence for later presentation in court. Moreover, when excavating a grave, it is very easy for the inexperienced to overcut or undercut the grave cut, which will result in a loss of evidence (e.g., tool marks) and soil contamination.

Although scatter scenes need human identification skills and may be much better suited to an anthropologist with experience in archaeological recording, a complete body, especially fleshed, does not need a qualified anthropologist to identify it as human. Moreover, the search and excavation may not necessarily be related to human remains, but rather to hidden weapons, drugs, explosives, or other items, cases in which an anthropologist is clearly of little use.

The archaeologist has been equipped through training and experience with a particular and unique set of skills and experience. He or she also brings a more accurate interpretation of stratigraphy and of the burial or deposition context compared to any other person who is not qualified or experienced. When stripping the topsoil of a garden or other search area, the archaeologist will monitor the excavation (like in an archaeological "watching brief") for any disturbance or features that may translate into a clandestine grave. It is a unique set of skills that allows the archaeologist to dictate at what level the stripping should stop and provides the archaeologist with the ability to investigate any features that may be identifiable to the experienced eye. The archaeologist also assigns context numbers, and attempts to reconstruct the sequence of events. Such skills cannot be underestimated: "Stratigraphic excavation is the cornerstone of the recovery of evidential data in archaeological and forensic archaeological excavation. It is the source of the evidence used to establish, sequence, and demonstrate past events and to independently date these events with a precision that (at present) surpasses all other methods" (Cheetham and Hanson 2009: 146). Since archaeology is a destructive process, any missed statigraphic unit or context can potentially compromise the evidence and interpretations. A lack of stratigraphic recording and understanding will benefit the defence lawyers by allowing them to undermine the archaeologist's support of the prosecution's case and raise an issue that the destroyed evidence could have provided as "'proof' of innocence for the accused" (Cheetham and Hanson 2009: 146). Moreover, since the defence can also appoint an archaeologist to question the prosecution, the employment of an experienced, qualified archaeologist (not an anthropologist) is important.

Moreover, the archaeologist has the skills to see "'hidden' evidence" (Cheetham and Hanson 2009: 141). The grave cut may also indicate premeditation and the type of tools used. Excavation of a grave using archaeological techniques by an experienced forensic archaeologist can elucidate the circumstances surrounding the burial. For example, the archaeologist will have observed how the grave was backfilled, how the body was positioned, whether the cut of the grave gives any indication as to its construction and body position, any subsequent disturbance after the body was buried, and whether tool marks are present. Owsley (2001: 38) points out that one of the values of an archaeologist is the "ability acquired over years of field experience to discern delicately portrayed features in soils of many colors and textures." In a recent case the authors excavated a body buried up to 1.10 m in a back garden in a house in the south of England. The grave cut itself explained the position of the body, the angle of the body, interpretation as to how many individuals would have required to dig the hole and with what tools, and the different phases of the grave: opened first, then enlarged, and then with a step added within the cut to excavate deeper still. The soil was retained and sieved for further evidence. It is the defense lawyers who may want to see how much soil was removed from a grave. It is bad practice if they request the soil to be sieved and this was not been retained. It would be too much of a risk to use another member of staff who is not an experienced archaeologist.

There are, however, a number of situations and prerequisites which the archaeologist should be comfortable with when undertaking forensic casework. These include situations where there are fleshed, decomposed, and often badly smelling remains, the requirement to produce witness statements and act as an expert witness, and the ability to work within budgetary and time constraints. They must also be aware of the validity and usefulness of the evidence, procedures at the crime scene (including exhibiting procedures and the importance of continuity and maintaining the integrity of the evidence), and working within a medicolegal framework (Hunter et al. 1994: 760; Cheetham and Hanson 2009: 141).

As Scott and Connor (2001: 102) note, the archaeologist will have had a breadth and experience of sites and this will help deal with the diverse typology of human body disposal. They also point out that a broad knowledge of material culture is advantageous (e.g., for dating and for different phases in building archaeology). We could not agree more with Owsley (2001), who very clearly highlights the importance of the archaeologist due to training and experience with soils and environmental contexts and their role in a multidisciplinary forensic team. It is a message that cannot be stressed enough. If appropriate, the archaeologist may make a preliminary visit for forward planning, especially in terms of logistics and strategy. Owsley (2001) also adds that the archaeologist is familiar with remote sensing techniques, is able to discern subtle differences in colour soil and texture, will produce plans and statements, and discuss the contexts in which the remains were found. Archaeologists can also interpret aerial photographs where available. In one recent example of this, a body had been found just off a motorway. The county (regional) council and other institutions (National Monuments Records) provided aerial photographs of the area and it was possible to observe how the landscape had changed during the 20-year period that was of interest in this particular case.

Finally, Dupras et al. (2006) indicate a number of skills that the forensic archaeologist should possess, including survey techniques, search methods, site formation and analysis, recording, excavation techniques, use of heavy equipment, artifact collection, sampling, basic knowledge of human and nonhuman anatomy, ability to collect or preserve skeletal remains, and, more importantly, how to apply the skills associated with archaeology to a forensic context.

To summarize, the skills of the archaeologist are essential to carrying out in the most professional way the search, location, recording, recovering, dating, interpretation, and recovery of environmental evidence.

The interaction between archaeology and anthropology

For the reasons given above, our point of view sees archaeology and anthropology as two different disciplines, though clearly interacting and overlapping at different levels in a number of cases. Although in North America both disciplines are highly specialized, it is common for forensic archaeology to be included within the discipline of forensic anthropology (Dupras et al. 2006: 1); a similar situation exists in certain European nations as well. In Britain, many recently graduated anthropologists are trained in archaeological departments. Having an archaeological background can be considered an advantage for scene attendance. In most cases, the anthropologist in the UK will have a background in archaeology and may be able to apply archaeological techniques to the recovery of scattered human remains. On other occasions, both the anthropologist and archaeologist will work in conjunction at the crime scene, where the most experienced archaeologist undertakes the recording and directs the excavation up to the point where human remains (as confirmed by the anthropologist) are discovered and they both work in conjunction to provide the best recovery. Once the body is lifted, the anthropologist will attend the morgue while the archaeologist continues excavating to define the bottom of the grave and any other areas that need to be searched. At times the archaeologist may be at the scene without an anthropologist since, in the case of complete bodies, the archaeologist can excavate the grave and lift the body with police staff assistance. The difference between the two disciplines in the UK still remains unclear, although the Council for the Registration of Forensic Practitioners (dissolved in 2009) tried to define these, reflecting that they are different in competence (Cheetham and Hanson 2009: 142). Developments since 2009 have seen forensic archaeologists and forensic anthropologists forming their own independent societies in the UK: the Forensic Anthropology Group under the umbrella of the Institute of Field Archaeologists and the British Association of Forensic Anthropology under the auspices of the Royal Anthropological Institute. It is a topic that has been dealt with in other works (Hunter 2009). Here, we would like to emphasize the difference and importance of both specialties and the teamwork between archaeologists and anthropologists.

There is a boundary of competence when an archaeologist and an anthropologist work on the same case; both give advice within their speciality and carry out their specific tasks and responsibilities. This strategy is one used by the authors and has worked extremely well in complex cases where there was a need for at least two specialists (an archaeologist and an anthropologist with an archaeological background) at the scene. In addition, a combination of archaeological and anthropological methods can be used to give some idea about the postmortem interval. The anthropologist may also be able to note any damage that occurred to the body through excavation or recovery, or he or she will be able to comment on any abnormalities (pathological conditions) that can be observed *in situ*. The anthropologist can also comment on which bones are missing (e.g., in a case undertaken outside the UK, one of the identifying features of the victim was an amputated finger, and therefore it was vital that total recovery of the hand bones was undertaken; only an anthropologist could ensure this in relation to the number of bones missing, whether some were non-human, etc.). Since most anthropologists in Britain have come from an osteoarchaeological background that provides certain advantages relating to the identification of small fragments and degraded remains (Cox 2009: 31), this should provide a good combination when working with the more experienced archaeologist. In another recent case in the UK, one of us was in charge of directing the search for a body in a back garden with the assistance of an anthropologist (also qualified in archaeology but with less experience than the former). If any bones were found the anthropologist would deal with the bones to see if they were human or not. In the meantime, the archaeologist continued searching and investigating a number of anomalies. Once they uncovered a positive anomaly and recorded it, the anthropologist took over further recording, the lifting, and the transportation of the body.

Contrary to Scott and Connor's (2001: 102) statement that "The archaeologist is also not useful without a strong background in osteology," we want to emphasize that the archaeologist *is* also useful without a strong background in osteology, depending on the crime scene and nature of the investigation. The same could be said about forensic anthropologists in that they are not useful if they have no archaeological background, a statement that would not be true, bearing in mind that some requests only require mortuary attendance for victim identification or fragment identification. Some examples have been mentioned dealing with the excavation of bodies that are fleshed or semiskeletonized. Similarly, the search for other buried items and desk-based assessments do not require bone expertise. Thus, we cannot say that an anthropologist is not useful without a strong archaeological background since many anthropologists will only work in mortuaries.

In fact, to make this distinction clearer between the anthropologist and archaeologist, the tasks and duties can be developed further. There follows a summary of how an archaeologist might contribute at a "typical" search and excavation scene.

The archaeologist will make, where time allows in a prearranged excavation, every effort to understand the area to be searched. This understanding can be acquired from a range of resources, including a consideration of the parameters of the area in question, the underlying geology/soil distribution, past history of the area (revealed through historic maps and aerial photographs), and evidence for potential clandestine graves (e.g., soil disturbance, obvious alterations to flora). At the scene, the search will be the result of a strategy prepared between the archaeologist, other search professionals, and the crime scene manager. A range of different search techniques are generally available to investigators, commonly including police helicopter imagery, thermal imagery, geophysics equipment, cadaver scent dogs, fingertip search teams, and archaeological landform assessments. Full documentation (e.g., scene notes, context recording forms, sketches, detailed plan and section drawings, and photographs) of the archaeological search and investigation is imperative.

Prior to beginning excavation, during a briefing with police officers and CSIs, some of the questions of interest to the archaeologist may include the following: How much time has passed since the investigators believe the body was deposited? How big is the area that will be examined archaeologically? What are the strategies for retaining soil and dumping spoil? What strategy will be adopted with regard to exhibit seizure?

Excavations of graves or potential graves are first carried out by "half-sectioning" the feature (excavating only half of the feature first). This has two advantages. It allows investigators to see what the feature contains without having to dig the entire feature or area. If human remains are detected, a section drawing can then be undertaken to record the fill and interpret the means by which the grave was backfilled. In this case, the section will be drawn and photographed. After the other half is excavated, prior to removal of the body, some further recording is undertaken, including the completion of a body or skeletal recording form. After the body is removed, the grave is further defined (to the bottom of the grave), looking for any potential tool marks and other artifactual or ecological evidence. Context sheets are also completed as necessary to record the context number, the type of context (e.g., structure, layer, deposit, cut), its dimensions, characteristics, and so on. Usually drawings, the designation of context numbers, and the filling-in of recording forms should only be undertaken by a forensic archaeologist. In the case of mass-grave excavation, for example (see Haglund et al. 2001), archaeologists would have the task of documenting the site, making a topographic map, taking photographs, and writing reports.

Reporting and court appearance

It is not the intention here to provide an insight into the legal matters since this has been dealt by Dilley (2005). However, in a recent homicide case at an English Crown Court, one of us was asked about the relevant qualifications and experience, the involvement in the case, the tasks undertaken as requested by the SIO, the techniques employed, and a summary of the findings. This might be challenged by the defence in an attempt to undermine the credibility of the expert witness in the eyes of the jury.

A forensic archaeologist may be required to act as an expert witness in court, and this has occurred in our experience in a limited number of cases. All field notes, sketches, drawings, plans and section drawings, and recording forms will be court-admissible and made available to the defence. If the case is a homicide, it is normally heard in Crown Court and the archaeologist may be expected to give evidence as an expert witness. This evidence may help the court to understand the circumstances surrounding the burial of the body, the accompanying implications in relation to where the victim was killed, and whether or not the body had been moved.

In court, the forensic archaeologist may be requested to set the scene in relation to the search, the way the body was deposited and any implications regarding whether the grave was premeditated, whether it was deep or shallow, what manpower/time investment would have been required to dig the grave, how it had been constructed, and how many offenders might have been involved in transporting the body to the deposition site and/or burying it. The archaeologist may also be asked to comment on any evidence that they have seized.

CONCLUSION

This chapter has not focused on mass graves, which are usually investigated by government and nongovernmental organizations such as the International Criminal Tribunals, the United Nations, and organizations such as Physicians for Human Rights (PHR), where the recovery of remains may be focused on both evidential recovery for future indictments and identification of the deceased. In this chapter we have focused on police casework, drawing mainly from our experience in the UK.

We hope that the chapter has provided a reasonable insight into the tasks of an archaeologist and the figure and concept of the forensic archaeologist, rather than focusing on the archaeological techniques to be used. The importance of a forensic archaeologist, even without any anthropological knowledge, has been highlighted here. It may be that the number of bodies buried in back gardens, under patios, or in forests requires more archaeologists than other countries due to the crime rate, crime behavior, and other cultural aspects (e.g., architecture, urban planning).

In the majority of European countries, forensic archaeology seems to refer simply to archaeological techniques that fall under anthropology or situations in which it is anthropologists (not archaeologists), forensic pathologists, or CSIs that use these techniques (but not skills) to uncover bodies. Anthropologists and archaeologists in Europe tend to have different backgrounds from those in the UK, whereas in the UK many forensic anthropologists derive from an archaeological background, which has its own advantages.

Perhaps it may be said that it is not only important to "know how to dig" and know the techniques, but also that "knowing how to dig well" is necessary and difficult, and the skills involved in recording and interpretation come with years of experience in archaeology. There are roles for both archaeologists and anthropologists, which are sometimes different and often complement one another. It is fair to say that the disciplines are separate but nevertheless entwined, and that only an appreciation of the strengths of both fields can help archaeology to show its potential at maximizing the recovery of evidence at a crime scene.

It is hoped that the figure of the forensic archaeologist has been promoted, and that it is perhaps a possibility that accreditation becomes standardized as a requirement throughout Europe, especially for the search, location, excavation, recovery, and documentation of buried bodies and other evidence types.

ACKNOWLEDGMENTS

We would like to thank Professor John Hunter (UK) for his valuable comments and his encouragement in a final draft of this chapter. Likewise, we would like to thank Dennis Dirkmaat and Luis Cabo for their support, trust, and motivation throughout the writing of this chapter.

NOTES

1   Eurostat data: http://epp.eurostat.ec.europa.eu/statistics_explained/index.php?title=File: Crimes_recorded_by_the_police_-_Homicide,_2002-2008.PNG&filetimesta mp=20110120150450.

2   Does the concept of "forensic archaeology" exist in your country? If yes, in what way (e.g., mass graves, police casework)? Are there any forensic archaeologists working in police work in your country? If yes, what are their roles/tasks or type of work they undertake? Is there a forensic archaeology body or association? How many recognized forensic archaeologists (practitioners) are there in your country? Where are forensic archaeologists based in your country (e.g., university, freelance, within national agencies, within police forces)? What are their qualifications, background, and experience? Is there any scheme for quality assessment or accreditation? Do forensic archaeologists in your country go to court to act as expert witnesses? Are there any facilities for training? Do archaeologists train police? Are there any national guidelines or policies for forensic archaeology? If there are no forensic archaeologists in your country, who helps with the search and excavation of buried bodies? Are they trained in archaeological techniques? Any other comments (e.g., useful links, bibliography, etc.)?

3   We are indebted and would like to thank the following professionals for their contribution in this survey and whose information is given in this section: Kim Quintelier (Belgium), Flemish Heritage Institute (VIOE) and Royal Belgian Institute of Natural Sciences (RBINS), Brussels, Belgium; Mario Šlaus (Croatia), Department of Archaeology, Croatian Academy of Sciences and Arts, Zagreb, Croatia; Xenia-Paula Kyriakou (Cyprus), Committee of Missing Persons on Cyprus Forensic Team (CMP), Nicosia, Cyprus and School of History, Classics and Archaeology, University of Edinburgh, UK; Wesa Perttola (Finland), Department of Archaeology, University of Helsinki, Finland; Eric Baccino (France), Research Laboratory of Legal Medicine and Clinical Toxicology, Faculty of Medicine, University of Montpellier, France; Niels Lynnerup (Denmark), Department of Forensic Medicine, Faculty of Health Sciences, University of Copenhagen, Denmark; Roland Wessling (Germany), Department of Applied Science, Security & Resilience, Cranfield University and Inforce Foundation, UK; Constantine Eliopoulos and Konstantinos Moraitis (Greece), Research Centre in Evolutionary Anthropology and Palaeoecology, School of Natural Sciences and Psychology, Liverpool John Moores University, UK (CE) and Department of Forensic Medicine and Toxicology, School of Medicine, University of Athens, Greece (KM); Niamh McCullagh (Ireland), forensic archaeologist for the Independent Commission for the Location of Victims Remains (ICLVR), Ireland; Matteo Borrini (Italy), School of Archeological Heritage, University of Florence and Papal Theological University 'San Bonaventura,' Italy; Rimantas Jankauskas (Lithuania), Vilnius University and State Forensic Medicine Service, Lithuania; Mike Groen (The Netherlands), Netherlands Forensic Institute, The Hague, The Netherlands; Kenneth Didriksen (Norway), National Authority for Investigation and Prosecution of Economic and Environmental Crime in Norway, Norway; Wiesław Lorkiewicz (Poland), Department of Anthropology, Faculty of Biology and Environmental Protection, University of Łódź, Poland; Eugénia Cunha and Carina Marques (Portugal), Department of Life Sciences, University of Coimbra (EC and CM) and Forensic Sciences Research Centre, Portugal (EC); Marija Djuric (Serbia), Laboratory for Anthropology, School of Medicine, University of Belgrade, Serbia; Soňa Masnicová (Slovak Republic), Department of Criminalistics and Forensic Sciences, Academy of Police Force in Bratislava, Slovak Republic; Pavel Jamnik (Slovenia), Senior Criminal Police Superintendent, General Police Directorate, Criminal Police Directorate, Ljubljana, Slovenia; María del Mar Robledo Acinas and José Antonio Sánchez Sánchez (Spain), School of Legal Medicine, Universidad Complutense de Madrid, Spain.

4   See www.nzedge.com/heroes/smith.html for a biography of Sir Sydney Smith.

5   CSIs or SOCOs are mostly civilian staff. Some may have been police officers but increasingly the majority are civilian staff.

6   In this case the archaeologist had over 10 years of anthropological experience due to previous work as an osteoarchaeologist, and was able to identify any bone as human or not or suggest further analysis or a second opinion. Therefore, some of the options are that the archaeologist can either work alongside an anthropologist, he or she may have a very strong background in osteology, or he or she is an anthropologist with many years experience in general archaeology.

7   Once again, here the senior archaeologist was assisted by an anthropologist with many years experience in osteoarchaeology. In this case, the anthropologist could undertake archaeological tasks such as digging and measuring, while the senior archaeologist was undertaking the recording of stratigraphy, section drawings, plans, and the interpretation of the contexts. At the same time, whatever bones the senior archaeologist would find would be passed to the anthropologist to confirm whether or not they were human.

8   *Terminus ante-quem* is the latest date for something. For example, if something is beneath a 1980s floor and it has not been cut, then anything under it is pre-1980s. *Terminus postquem* is the earliest that something can be. Anything lying above a 1980s floor will have been placed there after the floor (therefore, dating to the 1980s or later).

9   For example, see Waterhouse (2009).

REFERENCES

Acinas Robledo, M.M. and Sánchez-Sánchez, J.A. (2010). [Forensic anthropology. Criminal investigation on human remains]. In M.J. Anadón Baselga and M.M. Robledo Acinas (eds), [*Handbook of Criminalistic and Forensic Sciences. Forensic Techniques Applied to Criminal Investigation*] (pp. 17–52). Editorial Tébar S.L, Madrid [in Spanish].

Baccino, E. (2009). Forensic anthropology: perspectives from France. In S. Blau and D. Ubelaker (eds), *Handbook of Forensic Anthropology and Archaeology* (pp. 49–55). Left Coast Press, Walnut Creek, CA.

Blau, S. and Ubelaker, D. (eds) (2009). *Handbook of Forensic Anthropology and Archaeology*. Left Coast Press, Walnut Creek, CA.

Borrini, M. (2007). [*Forensic Archaeology. Method and Technique for the Recovery of Human Remains in Scientific Investigations*]. Lo Scarabeo Editrice, Torino [in Italian].

Cattaneo, C. (2007). Forensic anthropology: developments of a classical discipline in the new millennium. *Forensic Science International* 165: 185–193.

Catteneo, C. (2009). Forensic anthropology and archaeology: perspectives from Italy. In S. Blau and D. Ubelaker (eds), *Handbook of Forensic Anthropology and Archaeology* (pp. 42–48). Left Coast Press, Walnut Creek, CA.

Cheetham, P.N. and Hanson, I. (2009). Excavation and recovery in forensic archaeological investigations. In S. Blau and D. Ubelaker (eds), *Handbook of Forensic Anthropology and Archaeology* (pp. 141–149). Left Coast Press, Walnut Creek, CA.

Coleman, K. and Osborne, S. (2010). Homicide. In K. Smith and J. Flatley (eds), *Homicide, Firearm Offences and Intimate Violence 2008/09. Supplementary Volume 2 to Crime in England and Wales 2008/09*, 3rd edn (pp. 9–36). Home Office Statistical Bulletin 01/10. National Statistics, Home Office, London.

Connor, M.A. (2007). *Forensic Methods: Excavation for the Archaeologist and Investigator*. Altamira Press, Lanham, MD.

Cox, M. (2009). Forensic anthropology and archaeology: past and present – a United Kingdom perspective. In S. Blau and D. Ubelaker (eds), *Handbook of Forensic Anthropology and Archaeology* (pp. 29–41). Left Coast Press, Walnut Creek, CA.

Cox, M., Flavel, A., Hanson, I., Laver, J., and Wessling, R. (eds) (2008). *The Scientific Investigation of Mass Graves. Towards Protocols and Standard Operating Procedures*. Cambridge University Press, Cambridge.

Cunha, E. and Cattaneo, C. (2006). Forensic anthropology and forensic pathology: the state of the art. In A. Schmitt, E. Cunha, and J. Pinheiro (eds), *Forensic Anthropology and Medicine: Complementary Sciences from Recovery to Cause of Death* (pp. 39–53). Humana Press, Totowa, NJ.

Dilley, R. (2005). Legal matters. In J. Hunter and M. Cox (eds), *Forensic Archaeology: Advances in Theory and Practice* (pp. 177–203). Routledge, London.

Dirkmaat, D.C., Cabo, L.C., Ousley, S.D., and Symes, S.A. (2008). New perspectives in forensic anthropology. *Yearbook of Physical Anthropology* 51: 33–52.

Duday, H. and Guillon, M. (2006). Understanding the circumstances of decomposition when the body is skeletonized. In A. Schmitt, E. Cunha, and J. Pinheiro (eds), *Forensic Anthropology and Medicine: Complementary Sciences from Recovery to Cause of Death* (pp. 117–157). Humana Press, Totowa, NJ.

Dupras, T.L. Schultz, J.J., Wheeler, S.M., and Williams, L.J. (2006). *Forensic Recovery of Human Remains: Archaeological Approaches*. CRC Press, Boca Raton, FL.

Haglund, W.D., Connor, M., and Scott, D.D. (2001). The archaeology of contemporary mass graves. *Historical Archaeology* 35: 57–69.

Holland, T.D. and Connell, S.V. (2009). The search for and detection of human remains. In S. Blau and D. Ubelaker (eds), *Handbook of Forensic Anthropology and Archaeology* (pp. 129–140). Left Coast Press, Walnut Creek, CA.

Hunter, J. (2009). Domestic homicide investigations in the United Kingdom. In S. Blau and D. Ubelaker (eds), *Handbook of Forensic Anthropology and Archaeology* (pp. 363–373). Left Coast Press, Walnut Creek, CA.

Hunter, J. and Cox, M. (2005). *Forensic Archaeology: Advances in Theory and Practice*. Routledge, London.

Hunter, J.R., Heron, C., Janaway, R.C., Martin, A.L., Pollard, A.M., and Roberts, C.A. (1994). Forensic archaeology in Britain. *Antiquity* 68: 758–769.

Hunter, J., Roberts, C., and Martin, A. (1996). *Studies in Crime: an Introduction to Forensic Archaeology*. B.T. Batsford, London.

Hunter, J.R., Brickley, M.B., Bourgeois, J., Bouts, W., Bourguignon, L., Hubrecht, F., De Winne, J., Van Haaster, H., Hakbijl, T., De Jong, H. et al. (2001). Forensic archaeology, forensic anthropology and human rights in Europe. *Science & Justice: Journal of the Forensic Science Society* 41: 173–178.

Jamnik, P. (2008). [Concealed mass graves in Slovenia: the degree of archeological work in the excavation and identification of mass graves]. *Prispevki za novel zgodovino* 48: 173–186 [in Slovenian].

Jankauskas, R. (2009). Forensic anthropology and mortuary archaeology in Lithuania. *Anthropologischer Anzeiger* 67: 391–405.

Killam, E.W. (1990). *The Detection of Human Remains*. Charles C. Thomas, Springfield, IL.

Kranioti, E.F. and Paine, R.R. (2011). Forensic anthropology in Europe: an assessment of current status and application. *Journal of Anthropological Sciences* 89: 71–92.

Leben-Seljak, P. (2007). Anthropological analyses of post-war mass graves in Slovenia. In G. Starc (ed.), *Life in the Time of Conflicts: Book of Abstract*. Slovene Anthropological Society, Ljubljana.

Mant, A.K. (1987). Knowledge acquired from post-war exhumations. In A. Boddington, A.A. Garland, and R.C. Janaway (eds), *Death, Decay and Reconstruction* (pp. 65–78). Manchester University Press, Manchester.

Menez, L.L. (2005). The place of a forensic archaeologist at a crime scene involving a buried body. *Forensic Science International* 152: 311–315.

Morse, D., Duncan, J., and Stoutamire, J. (eds) (1983). *Handbook of Forensic Archaeology and Anthropology*. Rose Printing Co., Tallahassee, FL.

O'Sullivan, J., Hallissey, M., and Roberts, J. (2002). *Human Remains in Irish*

Archaeology: Legal, Scientific, Planning and Ethical Implications. An Chomhairle Oidreachta/The Heritage Council, Kilkenny.

Owsley, D.W. (2001). Why the forensic anthropologist needs the archaeologist. *Historical Archaeology* 35: 35–38.

Prieto, J.L. (2009). A history of forensic anthropology in Spain. In S. Blau and D. Ubelaker (eds), *Handbook of Forensic Anthropology and Archaeology* (pp. 56–66). Left Coast Press, Walnut Creek, CA.

Puchalt Fortea, F.J. (2000). [Discoveries and excavations]. In. J.D. Villalaín Blanco and F.J. Puchalt Fortea (eds), [*Police and Forensic Anthropological Identification*] (pp. 31–38). Tirant Lo Blanch, Valencia [in Spanish].

Quintelier, K., Malevez, A., Orban, R., Toussaint, M., Vandenbruaene, M., and Yernaux, G. (2011). Belgium/België/Belgique/Belgien. In N. Márquez-Grant and L. Fibiger (eds), *The Routledge Handbook of Archaeological Human Remains and Legislation: An International Guide to Laws and Practice in the Excavation and Treatment of Archaeological Human Remains* (pp. 47–60). Routledge, London.

Ruffell, A., Donnelly, C., Carver, N., Murphy, E., Murray, E., and McCambridge, J. (2009). Suspect burial excavation procedure: a cautionary tale. *Forensic Science International* 183: e11–e16.

Schermer, S.J., Shukriu, E., and Deskaj, S. (2011). Kosovo/Kosova. In N. Márquez-Grant and L. Fibiger (eds), *The Routledge Handbook of Archaeological Human Remains and Legislation: An International Guide to Laws and Practice in the Excavation and Treatment of Archaeological Human Remains* (pp. 235–246). Routledge, London.

Schmitt, A., Cunha, E., and Pinheiro, J. (2006). *Forensic Anthropology and Medicine: Complementary Sciences from Recovery to Cause of Death.* Humana Press, Totowa, NJ.

Scott, D.D. and Connor, M. (2001). The role and future of archaeology in forensic science. *Historical Archaeology* 35: 101–104.

Tuller, H. and Djuric, M. (2006). Keeping the pieces together: comparison of mass grave excavation methodology. *Forensic Science International* 156: 192–200.

Waterhouse, K. (2009). *The Use of Archaeological and Anthropological Methods in Fatal Fire Scene Investigations.* Defence Research and Development Canada Centre for Security Science (DRDC CSS TN 2009–04). http://www.css.drdc-rddc.gc.ca/cprc/tr/tn-2009-04.pdf.

PART VIII Forensic Anthropology Outside North America

Introduction
to Part VIII

Dennis C. Dirkmaat

The primary focus of this companion book is on forensic anthropology as it is configured and conducted in the United States. As discussed in earlier chapters, it was formally defined 40 years ago and that definition often involved, almost exclusively, the laboratory analysis of human skeletal remains. Within the last 20 years the relevance of archaeological methods to the documentation and recovery of outdoor crime scenes has been realized and only now are references to recovery of context creeping into overarching definitions of the field.

The recent reevaluation of forensic anthropology in the US discussed in Chapter 1 and Dirkmaat et al. (2008) implies that the field includes both forensic osteology and forensic archaeology configured to address forensic taphonomic issues not answerable by any other forensic science discipline during the medicolegal death investigation. The forensic anthropologist as practioner, then, is primarily an expert in human skeletal biology, fragmentary human osteology, zooarchaeology, and forensic taphonomic modifications to bones. In addition, the well-trained forensic anthropologist should be well versed in archaeological methods, principles, and practices in order to conduct forensic archaeological recoveries. As discussed in Chapter 2, this is an important skill since documentation of context is generally not completed by law enforcement. If the forensic anthropologist is not entirely comfortable or experienced with more complex outdoor scenes, they may enlist the help of an archaeologist to assist in the recovery. However, they both work at the scene at the same time to maximize the understanding of the context of the scene and contribute jointly to the reconstruction of the past history of the remains and the scene. Importantly, while the recovery is being completed,

A Companion to Forensic Anthropology, First Edition. Edited by Dennis C. Dirkmaat.

reasonable hypotheses of the course of past events may be discussed and tested on-scene, especially with respect to ruling out a scenario or two.

Skills in the identification of fragmented human bones and animal bones are also particularly useful in the determination of forensic significance when bones turn up during searches in rural settings, during house remodeling and many other situations. Often these issues can be resolved by means of a picture (with scale) e-mailed to the forensic anthropologist. Also critical in outdoor scene recoveries is the ability to recognize taphonomically altered bones. Burned and heat-altered human remains from fatal-fire situations are particularly difficult to interpret without significant experience with these materials.

However, the US forensic anthropology model is not a universal one. The melding of skills into one forensic anthropologist model promoted in the US is somewhat at odds with the United Kingdom model, which instead maintains separate and distinct disciplines, variably called upon depending on the scenario. Grant and coauthors in Chapter 29 provide a rather exhaustive, well-constructed argument for this separation. Regardless of the philosophical differences, the key concept to be taken from the chapter is that forensic archaeology principles and practices must be a critical component of all investigations of outdoor crime scenes, whether conducting searches in deep woods, or excavating clandestine graves.

Grant et al. also provide an excellent survey and review of the role of anthropology (with a strong focus on forensic archaeology) within medicolegal systems in many European countries. It is clear that forensic anthropology, at least in the sense of the laboratory study of human remains from medicolegal situations, is slowly becoming entrenched in forensic practice, while the perceived need for forensic archaeology lags a bit, perhaps requiring – as it did in England – a critical or high-profile case or two, and more publications of this type.

L'Abbe and Steyn provide an overview of forensic anthropological work in South Africa in Chapter 30. South Africa is one of the most advanced countries on the African continent, but forensic anthropology is still in a relative state of infancy. The model currently in place is one in which the scenes are processed and remains recovered only by law enforcement, who then transport the body or bones to the central forensic anthropology laboratory. In this sense, it is very reminiscent of the US model in the early configuration of the field. As expected, very little or no evidence related to the context of the remains is gathered by law enforcement. Given the high rate of violent crime, the relatively transient nature of the population, and the common lack of evidence related to potential identity of the victim, it is clear that the need for forensic anthropological investigations will not diminish in the near future.

The authors also include an overview of the extensive research potential of the country's museum collections. For example, the Raymond A. Dart Collection at Witwatersrand University in Johannesburg is larger than either the Hamann–Todd Collection at the Cleveland Museum of Natural History or the Terry Collection at the Smithsonian Institution. In addition, a growing, very useful collection of modern forensic case remains can be found in the Forensic Skeletal Collection at the University of Pretoria.

In Chapter 31, Fondebrider discusses the work conducted by the Argentianian forensic anthropology team (the EAAF), the first team in the world assembled specifically to deal with human-rights work, first in Argentina and now throughout the world. With the help of Clyde Snow, teams of non-government-associated, non-law-enforcement-affiliated individuals (actually, most were college students) were trained in forensic osteology and forensic archaeology, primarily in the exhumation of remains from individual or mass graves. Given that the government-sponsored atrocities were conducted a short time before relative democracy was reinstated, families of victims and other interested parties were reticent to allow anyone previously or potentially affiliated with the government to conduct forensic investigations involving their loved ones. Early efforts to merely remove victims from the graves in order to return the victims to families and provide "closure" led to significant problems, including remains being left behind, commingled remains, and misidentifications. The human-rights team provided an impartial group who could be trusted to conduct scientifically valid and unbiased work in the location, excavation, and analysis of the remains of the "disappeared."

Interestingly, in addition to the work at the scene, the EAAF team was also required to conduct the individual case background work, usually completed by lawyers and law enforcement. This perspective therefore led to (i) acquisition of skills in familial research and investigation comparable to those possessed by cultural anthropologists and (ii) strong relationships with the families. Their work is also rather dissimilar to work conducted in the US in that the EAAF members conduct background research to determine what may have happened to a particular individual and where they may be located now. When a clandestine grave or even a mass grave is found, there typically exists a working hypothesis of who will be found before the digging commences.

Clyde Snow provided initial training and has had a continued advisory role since inception, and so the mode of forensic anthropology used by the EAAF is based on the American model and stresses that the forensic investigators must be well versed in forensic anthropology, the detailed and thorough analysis of the skeletal remains recovered, and forensic archaeological excavation skills.

REFERENCE

Dirkmaat, D.C., Cabo, L.L., Ousley, S.D., and Symes, S.A. (2008). New perspectives in forensic anthropology. *Yearbook of Physical Anthropology* 51: 33–52.

CHAPTER **30**

The Establishment and Advancement of Forensic Anthropology in South Africa

Ericka N. L'Abbé and Maryna Steyn

INTRODUCTION

As a nation of approximately 49 million people and 11 official languages, South Africa represents a country of diverse origins, cultures, and belief systems. The majority of people are African (black; 39.1 million, or 79.3%), with 4.5 million whites (9.1%), 4.4 million coloreds (9.0%), and 1.3 million Indians/Asians (2.6%) (Statistics South Africa 2009). Approximately 10 million people reside in the Gauteng Province, which contains two major cities, Pretoria and Johannesburg. In addition, 1 million people legally migrate into South Africa from other African countries each year, while another half a million emmigrate annually to Australia, New Zealand, Europe, Canada, and North America (Statistics South Africa 2009). The continual flow of people, both legally and illegally, across the borders poses a problem for the positive identification of the growing number of unknown persons in the country, many of whom are young to middle-aged males of African origin. A further impediment to their identification is that most of these people never received medical or dental care during their lives.

Law-enforcement personnel and pathologists frequently encounter situations in which the use of mitochondrial or nuclear DNA to obtain a positive identification is initially of little value because there are no available family members with whom to compare this information. In such instances, the expertise of a forensic anthropologist is

A Companion to Forensic Anthropology, First Edition. Edited by Dennis C. Dirkmaat.
© 2012 John Wiley & Sons Ltd. Published 2015 by John Wiley & Sons Ltd.

often requested from either an investigating officer of the South African Police Services or a forensic pathologist. The purpose of this chapter is to highlight the establishment of forensic anthropology as a discipline in South Africa, to provide information on recent international collaborative research projects within the field, and to elucidate on the maintenance of modern skeletal collections in the country for research and teaching.

THE NEED FOR FORENSIC ANTHROPOLOGY IN SOUTH AFRICA

In a study on worldwide violence, homicide, and war, Reza and colleagues (2001) demonstrated that the region of sub-Saharan Africa, which includes Zimbabwe, Namibia, Botswana, South Africa, Zambia, Swaziland, Lesotho, and Mozambique, had the highest percentage of unnatural deaths in the world (8 202 000) due to violence (6%); homicide (2.5%), war (3.3%), and suicide (0.2%) were the greatest contributors (Reza et al. 2001). Only Latin America and the Caribbean are comparable in magnitude (3 009 000), with a homicide rate of 3.4%, suicide rate of 0.7%, and war deaths at 0.7%. In both areas, violence is primarily concentrated among males between 15 and 24 years of age (Norman et al. 2007; Reza et al. 2001). With regard to homicide, these two countries surpass regions such as the USA, India, China, Asia, and the Middle East, all of which have rates of less than 1.2% (Reza et al. 2001).

In 2000, 59 935 unnatural deaths were recorded in South Africa, which is inappreciably different from the 2006 figure of 53 000 (Statistics South Africa 2009; Norman et al. 2007). Both of these values are substantially greater than the unnatural deaths recorded for the entire USA, which was 49 639 in 2003 (Centers for Disease Control and Preventation 2006). In South Africa, leading causes of unnatural death are homicide (46%), motor vehicle accidents (26%), suicide (9%), and fire (7%) (Norman et al. 2007). Deaths associated with homicide contribute to 113 deaths per 100 000 in males, 21 per 100 000 in females, and 65 per 100 000 overall each year. For males between 15 and 29 years of age, interpersonal violence has been shown to be responsible for 184 per 100 000 deaths (Norman et al. 2007). In poverty-stricken townships such as Khayelithsha and Nyanga in the Western Cape, homicide rates are four times higher (451 per 100 000) than the national average for males. Due to the large number of unnatural deaths associated with homicide, South Africa has been perceived as one of the most violent countries in the world (Norman et al. 2007). Reasons for interpersonal violence can be attributed to numerous factors, namely poor education, political strife, socioeconomic disparities, poverty, urbanization, excessive drug and alcohol abuse, unemployment, and corruption, as well as a general disrespect for the law and human dignity (Norman et al. 2007). Violence places a considerable burden on the finances and time of all forensic specialists in the country, yet it also creates a unique opportunity to expand and mature the discipline while providing a much-needed community service.

Since 1993, the Department of Anatomy at the University of Pretoria has received 419 forensic-related cases, the majority of which have been of young to middle-aged black males and females (Figure 30.1; Van Rooyen 2006). The primary reason for the referral of these cases to the Department was to obtain biological information on the deceased. Less common referrals pertain to evaluation of patterns of traumatic injury to bone, such as those associated with child abuse, and evaluation of taphonomic

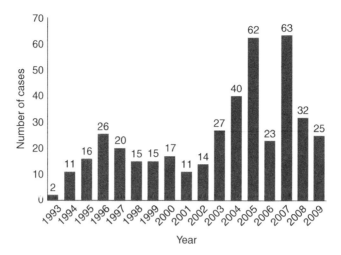

Figure 30.1   Histogram of forensic case work received in the Department of Anatomy at the University of Pretoria between 1993 and 2009. Four anthropologists conducted these analyses: M. Steyn, E.N. L'Abbé, S.R. Loth (deceased 2002), and N. Keough.

influence to the skeleton, such as postmortem damage caused by caged lions. Unusual cases include small-scale commingling (L'Abbé 2005), medicine murder in which human body parts were harvested and used for medicinal purposes (Steyn 2005), and at least three confirmed serial murders that resulted in the cumulative death of approximately 60 females.

Growth and development in the country, through large- and small-scale construction, has also lead to an increase in accidental exposure of skeletal remains of either historic or archaeological origin. The context in which the police retrieved these remains is often not recorded and/or not known. Since the dead are not frequently embalmed in funerary parlors in South Africa, general taphonomic indicators of historic burials are quite broad, and may include the presence of coffin wood and/or nails, potsherds, and miscellaneous artifacts, such as coins or clothing, that can be associated to a particular cultural period. A skeletal feature of African groups that dates to the prehistoric and historic period in the country is dental mutilation. While the presence of certain forms of dental mutilation can be used to exclude a case from having forensic significance, the absence of this feature does not have the opposite implication. Furthermore, several modern groups in the Western Cape are known to remove their teeth as a form of initiation into a gang (Morris 1989). An example of dental mutilation commonly practiced in both males and females during the early to late Iron Age periods in southern Africa can be seen in Figure 30.2, in which the upper central and lateral incisors had been filed into a V shape (Steyn 2003; L'Abbé et al. 2008). From this feature alone it was possible to exclude the remains from having forensic significance; furthermore, a radiocarbon date was obtained on this skull and estimated this person to have lived sometime between AD 1200 and 1700.

While the inadvertent discovery of archaeological or historic remains is not a uniquely South African issue (Rogers 2005), it does highlight the need for better communication among academics and law enforcement so as to mediate the removal

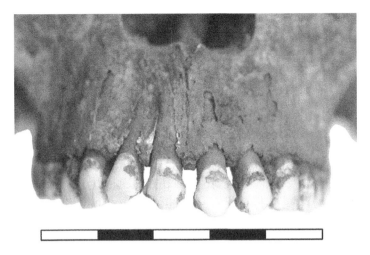

Figure 30.2   V-shaped dental mutilation of the upper central and lateral incisors.
Scale bar, centimeters.

of these skeletons and to prevent the loss of cultural heritage. As a means to address these problems in the northern part of the country, the Forensic Anthropology Research Centre was established in the Department of Anatomy at the University of Pretoria in 2008. Under the guidance of Professor Alan Morris, a similar group has also been formed at the University of Cape Town in the Western Cape.

After the abolishment of apartheid in 1994, a truth and reconciliation council was set up to investigate various humanitarian crimes that occurred during the apartheid era. For the purpose of the excavation and reburial of these skeletal remains, the National Prosecuting Authority chose to elect an impartial forensic anthropology team from Argentina, known as the *Equipo Argentino de Antropología Forense* (Argentine Forensic Anthropology Team), or EAAF. South African universities have investigated other historic cases such as the repatriation of the Pondo rebels and Nontetha Bungu (Nienaber and Steyn 2002).

The status of forensic anthropologists and their possible contributions to analysis of decomposed and skeletonized remains has become more recognized in recent years, as demonstrated by a considerable rise in the number of cases seen by anthropologists in the Department of Anatomy at the University of Pretoria; it is hoped that this trend will continue in the future (see Figure 30.1). However, several problems remain, including poorly excavated clandestine graves and the manner of investigation of crime scenes, as well as the use of inappropriate or outdated anthropological methodology by forensic pathologists and other nonspecialists on skeletal remains.

Past Research

Due to the diversity of its people and the availability of large skeletal collections, a considerable amount of research has been done with regard to population-specific skeletal research. Thus a considerable volume of data is available for age and sex determination, assessment of ancestry, and the calculation of stature for South African populations.

Published results on sex determination include morphology of the skull (e.g., De Villiers 1968), mandible (Loth and Henneberg 1996, 2000), and pelvis (Patriquin et al. 2003). Several researchers have also employed more sophisticated techniques such as geometric morphometrics in shape assessments of cranial and mandibular features (e.g., Steyn et al. 2004; Oettlé et al. 2005; Pretorius et al. 2006; Franklin et al. 2007). A large number of papers have been published on metric sex determination, and include data on the pelvis (Patriquin et al. 2005), humerus (Steyn and İşcan 1999), femur (e.g., Steyn and İşcan 1997; Asala 2001), skull (Kieser and Groeneveld 1986; Steyn and İşcan 1998; Franklin et al. 2005; Dayal et al. 2008), mandible (Franklin et al. 2008), and forearm (Barrier and L'Abbé 2008). Data are also available for smaller and less dimorphic bones such as the patellae (Dayal and Bidmos 2005) and tarsals (Bidmos and Asala 2003, 2004; Bidmos and Dayal 2004).

Less research has been published on the estimation of age at death and ancestry than on sex determination. Population-specific data are available for sternal ends of the fourth ribs (Oettlé and Steyn 2000) and dental maturation sequences (Phillips 2008). Studies have also been conducted on the efficacy of cranial suture closure for these estimations (Dayal 2009) and the accuracy of using histomorphometrics to estimate age at death from the anterior midshaft of the femur (Keough et al. 2009). Five papers have been published with regard to the determination of ancestry using the morphological (De Villiers 1986b) and metric characteristics of the skull (İşcan and Steyn 1999; Franklin et al. 2007) and pelvis (Patriquin et al. 2002). A paper was also published with regard to ancestry estimation from juvenile remains (Steyn and Henneberg 1997).

Population-specific formulas are available for stature estimation of both black (Lundy and Feldesman 1987) and white (Dayal et al. 2008) South Africans. Some publications also deal with the circumference of the skull, short bones, and tarsals (Bidmos and Asala 2005; Bidmos 2006; Ryan and Bidmos 2007) as well as fragmentary bones (Chibba and Bidmos 2007; Bidmos 2009). General information pertaining to the average stature of South Africans can be found in Steyn and Smith (2007).

Publications dealing with personal identification and identification of living individuals are much less common. Some case reports deal with skull–photograph superimposition (Thomas et al. 1986; Steyn et al. 2000), and only one paper has been published that deals with photograph identification (Roelofse et al. 2008).

CURRENT RESEARCH

Various universities continue research into developing population-specific standards for estimating demographic characteristics and stature from skeletal remains. These include age and sex determination from dentition, refinement of stature estimates, investigations into the repeatability of measurements, and the use of hand bones for skeletal identification. A group of researchers at the University of Cape Town are also examining age determination from living individuals (e.g., epiphyseal closure and ossification of hand bones). New developments at the University of Pretoria include the establishment of an outdoor experimental farm to evaluate decomposition in the northern region of the country as well as to record additional taphonomic changes to bone, such as the comparison of burn patterns on decomposed and skeletonized remains. Studies into the techniques of facial reconstruction include the development of tissue thickness values for South African black females and children, the recognizability and accuracy of facial reconstructions, and the patterns of facial growth in black children. An evaluation of the accuracy of skull–photograph superimpositions is also in progress.

The heterogeneous nature of the population is certainly unusual when compared to more traditionally homogeneous groups such as those in Europe. Within the last 500 years various groups of people, including Dutch, French, Malaysian, and Indian, have migrated both willingly and forcedly to South Africa (Schoeman 2007). Over time, this amalgamation of genes and cultures produced the social and biological identity of the country, yet little attention has been given to the anthropological methods used to define these groups or to justify the use of these techniques in forensic case work. While the exact reasons for this disparity in ancestry research are not known, political issues surrounding race in the country may play an important role.

The most commonly defined social groups in South Africa include black, white, colored, Indian, and Chinese. When compiling a skeletal profile, law enforcement often requests that the anthropologist adheres to the best of his/her abilities to this terminology. Yet the traditional approach to the estimation of ancestry does not permit an evaluation of population groups, as requested by law enforcement, but rather forces the anthropologist to use a series of morphological features to choose between three hypothetically distinct groups, namely Mongoloid, Caucasoid, and Negroid (Ousley et al. 2009). In addition to being faulty in its accuracy and precision, these morphological traits were not meant to be applied to populations outside of the southwestern USA (Rhine 1990; Hefner 2009).

Research into the frequency distribution of 13 nonmetric traits, namely nasal bone structure, nasal breadth, nasal overgrowth, anterior nasal spine, inferior nasal margin, interorbital breadth, zygomatic projection, malar tubercle, alveolar prognathism, mandibular torus, palatine torus, incisor shovelling, and Carabelli's cusps, on the crania of 520 white, black, and colored South Africans has recently been investigated (L'Abbé et al. 2011). With the use of ordinal regression analysis, L'Abbé et al. (2011) demonstrated that eight of these traits, namely nasal bone structure, nasal breadth, nasal overgrowth, anterior nasal spine, inferior nasal margin, interorbital breadth, zygomatic projection, and alveolar prognathism, had a statistically significant relationship with ancestry. Of these, nasal bone structure, nasal breadth, and interorbital breadth showed a significant interaction between sex and ancestry, whereas the morphological appearance of the anterior nasal spine and alveolar prognathism had a statistically significant relationship with age at death.

Of the variables that were associated with ancestry, the results showed that there was a considerable amount of variation within a group for the appearance of these traits, and that the pattern of variation found in black and white South Africans was similar to that found in their North American counterparts (Hefner 2009). For example, 78% ($n=143/183$) of black South Africans and 75% ($n=71/95$) of black North Americans had quonset or oval-shaped nasal bone structures, while only 22% ($n=40/183$) and 25% ($n=45/75$) of these groups exhibited variants of this trait that have been associated with European groups (Hefner 2009; L'Abbé et al. 2011). Further research into nonmetric traits analysis and its use in the estimation of ancestry in South Africa is warranted.

As mentioned above, few papers have been published on the estimation of ancestry from the skeleton. In all of these studies, blacks and whites from the Pretoria Bone Collection and/or the Raymond A. Dart Collection were used. However, the major problem with discriminant function analysis is that it is only applicable, and hence accurate, when the unknown person can be associated with the reference sample that was used to create the mathematical formulas. Experience has shown that the unknown skeletal remains from various parts of South Africa do not easily lend themselves to classification into either a black or white group for any of the abovementioned discriminant function formulas. Other than representation, as mentioned above, additional possibilities for these inconsistencies are that only variations for black and white South Africans have been recorded, and that statistical information as to the posterior probabilities and typicalities associated with these discriminant function formulas, which could aid in establishing the probability affiliation of an unknown person into one or more groups, are absent (Jantz and Ousley 2005).

The large amount of diversity within and between groups in the country mandates a new approach to the evaluation and interpretation of ancestry from skeletal remains. A solution is to incorporate our data into the multivariate statistics program Fordisc (Jantz and Ousley 2005). With Dr Stephen D. Ousley, a collaborative craniometric project between Mercyhurst College in Pennsylvania, USA, and the University of Pretoria has been initiated that involves the collection and analysis of craniometric data from modern skeletal collections and identified forensic cases. Craniometry and multivariate statistical analyses are considered the best methods to explore human variation within and between groups as well as to predict regional group classifications (Ousley et al. 2009). Preliminary results have shown that black South Africans can be differentiated from black Americans with regard to both sex and ancestry in that black South Africans were shown to have relatively shorter and wider noses, longer, lower, and narrower crania, and wider interorbital breadths. They also exhibit less sexual dimorphism than their North American counterparts (Ousley and L'Abbé 2010). This study clearly highlights the need for population-specific methodologies and for the use of diverse references samples in forensic casework.

MODERN SKELETAL COLLECTIONS

Since skeletal collections are the foundation for anthropological research into human variation and are needed for testing the reliability and precision of multiple techniques used in forensic practice, the advancement of the discipline in the country necessitates the acquisition and accessioning of human skeletal material. In South Africa, the National Health Act 2003 (Act No. 61 of 2003; South African Parliament 2003) stipulates that the remains of any unclaimed person from a public hospital may be sent to a medical institution to be used for the purposes of medical teaching and research. If a person wishes to donate his/her body to a medical school, then the donation is sent either directly to the anatomy department or as a whole-body donation at the local tissue bank. If the person was over the age of 65 years and/or had died of a particular disease (such as cancer) and the tissues were not harvested, then the remains are often sent to an anatomy department for medical research and training (L'Abbé et al. 2005).

The anatomy departments at the Universities of Pretoria and the Witwatersrand contain the largest and most modern skeletal collections in the country, known as the Pretoria Bone Collection and the Raymond A. Dart Collection, respectively. Whereas these samples have originated in different places and at different times during the twentieth century, the composition of the skeletal material is quite similar, such that they can be considered to represent the same resource pool of unclaimed and donated cadaver material. All unclaimed bodies were acquired from local hospitals in the areas near these schools, and the sample is often over-represented with middle-aged males of African descent. At the other extreme is the donated material, primarily of European origin and often over the age of 60 years (L'Abbé et al. 2005; Dayal et al. 2009). The history of these collections as well as their contents have implications for the interpretation of anthropological research and for the application of that research to forensic case work.

In 1926, Raymond Dart, then head of the Department of Anatomy at the University of Witwatersrand, began collecting skeletal material for the purpose of teaching osteology and gross anatomy (Dayal et al. 2009). This later became the Raymond A. Dart Collection, or the Dart Collection. The initial material was sourced from the dissection room, but additional specimens from other countries as well as from archaeological research sites were added over the years. The collection is larger than the popular Hamann–Todd and Terry research collections in the USA (Hunt and Albanese 2005). Interestingly enough, Raymond Dart formulated the idea to develop this now-famous collection after a year-long visit to the Washington University School of Medicine in St. Louis, Missouri, USA, from 1920 to 1921 (Stewart 1958).

The Dart collection comprises 2500 modern human skeletons of cadaver origin, and contains South African blacks (72%), whites (18%), and a smattering (11%) of other groups; it is also heavily skewed towards males (71%, or 1840) (Dayal et al. 2009). The large proportion of black males has been attributed to the socioeconomic conditions of South Africa, which led many black South Africans to travel from rural areas to find work in the larger cities. Once in the city, these individuals had limited communication with their family members and thus their death often went unnoticed, or the family did not wish to pay for the funeral costs (L'Abbé et al. 2005).

Of more recent origin is the Pretoria Bone Collection, which was established in 1987 in the Department of Anatomy at the University of Pretoria. It is completely cadaver-based. In 2005, similar to the Dart Collection, it contained a large proportion of males (78%, $n = 548/704$ for crania), especially black males (77%, $n = 423/548$). Many of these black males were migrant laborers, and few relatives have returned to claim the remains: between 1996 and 2001, only 4.3% of families reclaimed cadavers or skeletal remains (L'Abbé et al. 2005). To date, this collection houses 1135 complete crania, 816 postcrania, and 399 incomplete remains. The unique aspect of the collection is that it continues to grow at a rate of 40 to 50 skeletons per year, and in the future is to become the largest modern skeletal collection in the country.

Socioeconomics, disease pandemics, and other historical circumstances are to affect the composition of these collections, simply because they are comprised of a portion of natural deaths in the country in a particular year. In the last 20 years life expectancy at birth and adulthood has drastically declined in South Africa, which can be primarily attributed to HIV/AIDS. From 1900 to 1990, life expectancy at birth increased for black South Africans from approximately 37 years to 63 years (L'Abbé et al. 2008). Post-1990, a sharp rise in infant and young adult mortality rates was observed. Life expectancy at birth is conservatively estimated at 51 years, with other sources predicting it to be as low as 49 or 42 years (World Health Organization 2006; Norman et al. 2007; Statistics South Africa 2009).

The dramatic change in life expectancy has had an affect on the demographic structure of the Pretoria Bone Collection, in that a larger number of younger individuals are being accessioned into the collection. Since the inception of the collection in 1987 to 2002, only 65 complete skeletons, who had been less than 50 years of age at the time of their death, were available for research. From 2003 to 2009, 111 young cadavers were added to this sample, with the greatest increases in the 30–39- and 40–49-year age categories. The increase in skeletal material older than 50 years has remained relatively constant – in 2003, 200 skeletons were available – whereas 229 have been added in the past 6 years. While these figures represent significant growth in the collection, they can also be used to indicate a response to the change demographic structure of the country. Furthermore, with an increase in younger skeletons, the collection may become more valuable in evaluating age-at-death studies as well as assessing the issue of secular trends.

In order to be considered representative of the population at large, a collection needs to change with the population, and it needs to reflect problems associated with historical and economical circumstances as well as widespread disease. As the Pretoria Bone Collection expands throughout the twenty-first century, the skeletal composition will permanently reflect the problems surrounding the AIDS pandemic in the country, and without a doubt is to have an effect on future anthropological research and analysis of material considered to be of forensic origin.

Research collections are not the only reason to collect and accession skeletal material. Professional training of postgraduate students requires that they obtain practical experience with poorly preserved or fragmented skeletal elements, with patterns in taphonomic changes to bone, with traumatic injury on the skeleton, and in pathological conditions. This knowledge cannot be obtained from casts or models alone; the student needs exposure to the skeletal material that they are expected to work with in the future. Since 2003, and under the guidelines of the National Health Act (Act 61, 2003; South African Parliament 2003), unidentified skeletal remains of approximately 290 individuals have been accessioned into a Forensic Skeletal Collection at the Department of Anatomy, University of Pretoria, for the purpose of medical education. Various interesting cases can be found in this collection and include, but are not limited to, ballistic trauma, postmortem burning, carnivore activity, blunt-force trauma, and small-scale commingling of remains. Within the framework of the National Health Act (Act 61, 2003), a comparative study into soft- and bone-tissue injuries of the hyoid bone in cases of blunt-force trauma to the neck and body are underway in collaboration with forensic pathologists from the School of Pathology, Division of Forensic Medicine and Pathology, at the University of the Witwatersrand and the Forensic Pathology Services in Johannesburg.

On the archaeological front, numerous collections of African skeletons exist in places around Europe, most of which were collected during the colonial years. These remains are to be the focus of a research project done in collaboration with various South African institutions and the University of Leiden in the Netherlands. While not directly related to forensic inquiries, valuable information with regard to sexual dimorphism and secular trends can be gleaned from these samples, and can be used to interpret the variation found within modern South African populations.

CONCLUSION

With a high rate of violent crime, large, modern, and expanding skeletal collections, and strong collaborative research ties with leading experts throughout Europe and North America, South Africa has the potential to become a leader in the study of human variation and methodology in forensic anthropological techniques applicable to the country specifically, and sub-Saharan Africa in general. Some constraints include the general lack of funding for research, the overwhelming rates of crime and nonrecognition of skills, and the relative isolation of the country from the general academic community.

With rapid globalization, our understanding of within- and between-group variations is to change. Because of this the growth, maintenance, and use of modern skeletal collections, forensic and otherwise, are of the utmost importance for future educators and researchers. These collections can also be used to address continuous problems with the effects of secular trends on the human skeleton.

To keep abreast of international standards an assessment of the applicability of the methodological approaches used to create biological profiles is crucial. Research on the identification of living individuals, be it estimation of age at death or photo identification, should also receive more attention due to problems associated with life expectancy and unidentified remains. This also applies to specific techniques for personal identification, since standard methodology in comparing antemortem and postmortem skeletal and dental radiographs cannot always be used, due to an absence of medical treatment for many victims. DNA analysis is only successful when family members or relatives are actively searching for the missing person.

ACKNOWLEDGMENTS

We would like to thank members of the Forensic Anthropology Research Center, namely Jolandie Myburgh, Coen Nienaber, Yvette Scholtz, Natalie Keough, and Marius Loots. The utmost appreciation is extended to Professor George J.R. Maat of the University of Leiden in the Netherlands, Dr Stephen P. Nawrocki of the University of Indianapolis, and Dr Stephen D. Ousley and Dr Steven A. Symes of the Department of Applied Forensic Sciences, Mercyhurst Archaeological Institute, Mercyhurst College, PA, USA, for their continual collaboration and assistance on research projects and in the education of postgraduate students in South Africa.

REFERENCES

Asala, S.A. (2001). Sex determination from the head of the femur of South African whites and blacks. *Forensic Science International* 117: 15–22.

Barrier, I.L.O. and L'Abbé, E. (2008). Sex determination from the bones of the forearm in a modern South African sample. *Forensic Science International* 179: 85.e1–85.e7.

Bidmos, M.A. (2006). Adult stature reconstruction from the calcaneus of South Africans of European descent. *Journal of Clinical and Forensic Medicine* 13: 247–252.

Bidmos, M.A. (2009). Fragmentary femora: evaluation of the accuracy of the direct and indirect methods in stature reconstruction. *Forensic Science International* 192(1–3): 131.e1–131.e5.

Bidmos, M.A. and Asala, S.A. (2003). Discriminant function sexing of the calcaneus of the South African whites. *Journal of Forensic Sciences* 48: 1213–1218.

Bidmos, M.A. and Dayal, M.R. (2004). Further evidence to show population specificity of discriminant function equations for sex determination using the talus of South African blacks. *Journal of Forensic Sciences* 49: 1165–1170.

Bidmos, M.A. and Asala, S.A. (2004). Sexual dimorphism of the calcaneus of South African blacks. *Journal of Forensic Sciences* 49(3): 1–5.

Bidmos, M.A. and Asala, S.A. (2005). Calcaneal measurement in estimation of stature of South African blacks. *American Journal of Physical Anthropology* 126(3): 335–342.

Centers of Disease Control and Prevention: Morbidity and Mortality Weekly Report (2006). Homicides and Suicides, National Violent Death Reporting System, United States, 2003–2004. *Journal of the American Medical Association* 296: 506–510.

Chibba, K. and Bidmos, M.A. (2007). Using tibia fragments from South Africans of European descent to estimate maximum tibia length and stature. *Forensic Science International* 169: 145–151.

Dayal, M.R. (2009). *Polymorphism of Cranial Suture Obliteration in Adult Crania.*

PhD thesis, University of Adelaide, Adelaide.

Dayal, M.R. and Bidmos, M.A. (2005). Discriminating sex in South African blacks using patella dimensions. *Journal of Forensic Sciences* 50: 1294–1297.

Dayal, M.R., Spocter, M.A., and Bidmos, M.A. (2008). An assessment of sex using the skull of black South Africans by discriminant function analysis. *HOMO* 59: 209–221.

Dayal, M.R., Kegley, A.D.T., Strkalj, G., Bidmos, M.A., and Kuykendall, K.L. (2009). The history and composition of the Raymond A. Dart Collection of Human Skeletons at the University of the Witwatersrand, Johannesburg, South Africa. *American Journal of Physical Anthropology* 140(2): 324–335.

De Villiers, H. (1968a). Sexual dimorphism of the skull of the South African Bantu-speaking Negro. *South African Journal of Science* 64: 118–124.

De Villiers, H. (1968b). *The Skull of the South African Negro.* Witwatersrand University Press, Johannesburg.

Franklin, D., Freedman, L., and Milne, N. (2005). Three-dimensional technology for linear morphological studies: a re-examination of cranial variation in four Southern African indigenous populations. *HOMO* 56: 17–34.

Franklin, D., Freedman, L., Milne, N., and Oxnard, C.E. (2007). Geometric morphometric study of population variation in indigenous Southern African crania. *American Journal of Human Biology* 19: 20–33.

Franklin, D., O'Higgins, P., Oxnard, C.E., and Dadour, I. (2008). Discriminant function sexing of the mandible of indigenous South Africans. *Forensic Science International* 179: 84.e1–84.e5.

Hefner, J.T. (2009). Cranial nonmetric variation and estimating ancestry. *Journal of Forensic Sciences* 54(5): 985–995.

Hunt, D.R. and Albanese, J. (2005). History and demographic composition of the Robert J. Terry Anatomical Collection. *American Journal of Physical Anthropology* 127: 406–417.

İşcan, M.Y. and Steyn, M. (1999). Craniometric determination of population affinity in South Africans. *International Journal of Legal Medicine* 112: 91–97.

Jantz, R.L. and Ousley, S.D. (2005). *FORDISC 3: Computerized Forensic Discriminant Functions.* Version 3.0. University of Tennessee, Knoxville, TN.

Keough, N., L'Abbé, E.N., and Steyn, M. (2009). The evaluation of age-related histomorphometric variables in a cadaver sample of lower socioeconomic status: implications for estimating age at death. *Forensic Science International* 191 (1–3): 114.e1–114.e6.

Kieser, J.Y. and Groeneveld, H.T. (1986). Multivariate sexing of the human viscerocranium. *Journal of Forensic Odontostomatology* 4: 41–46.

L'Abbé, E.N. (2005). A case of commingled remains from rural South Africa. *Forensic Science International* 151(2–3): 201–206.

L'Abbé, E.N., Loots, M., and Meiring, J.H. (2005). The Pretoria Bone Collection: a modern South African skeletal sample. *HOMO* 56: 197–205.

L'Abbé, E.N., Coetzee, F.P., and Loots, M. (2008). A description of Iron Age skeletons from Pilanesberg. *South African Archaeological Bulletin* 63(187): 28–36.

L'Abbé, E.N., Van Rooyen, C., Nawrocki, S.P., and Becker, P.J. (2011). An evaluation of non-metric cranial traits used to estimate ancestry in a South African sample. *Forensic Science International* 209: 195e1–195e7.

Loth, S.R. and Henneberg, M. (1996). Mandibular ramus flexure: a new morphologic indicator of sexual dimorphism in the human skeleton. *American Journal of Physical Anthropology* 99: 473–485.

Loth, S.R. and Henneberg, M. (2000). Gonial eversion: facial architecture, not sex. *HOMO* 51: 81–89.

Lundy, J.K. and Feldesman, M.R. (1987). Revised equations for estimating living stature from the long bones of the South African Negro. *South African Journal of Science* 83: 54–55.

Morris, A.A. (1989). Dental mutilation in historic and prehistoric South Africa. *Quarterly Bulletin of the South African Library* 43: 132–134.

Nienaber, W.C. and Steyn, M. (2002). Archaeology in the service of the community repatriation of the remains of Nontetha Bungu. *South African Archaeological Bulletin* 57(176): 80–84.

Norman, R., Matzopoulos, R., Groene, P., and Bradshaw, D. (2007). The high burden of injuries in South Africa. *Bulletin of the World Health Organization* 85(9): 695–702.

Oettlé, A.C. and Steyn, M. (2000). Age estimation from sternal ends of ribs by phase analysis in South African blacks. *Journal of Forensic Sciences* 45(5): 1071–1079.

Oettlé, A.C., Pretorius, E., and Steyn, M. (2005). Geometric morphometric analysis of mandibular ramus flexure. *American Journal of Physical Anthropology* 128: 623–629.

Ousley, S. and L'Abbé, E.N. (2010). Craniometric variation in South African and American blacks. *Proceedings of the 62nd American Academy of Forensic Sciences;* February 22–27, Seattle, WA.

Ousley, S., Jantz, R., and Fried, D. (2009). Understanding race and human variation: why forensic anthropologists are good at identifying race. *American Journal of Physical Anthropology* 128: 623–629.

Patriquin, M.M., Steyn, M., and Loth, S.R. (2002). Metric assessment of race from the pelvis in South Africans. *Forensic Science International* 127: 104–113.

Patriquin, M.L., Loth, S.R., and Steyn, M. (2003). Sexually dimorphic pelvic morphology in South African whites and blacks. *HOMO* 53: 255–262.

Patriquin, M.L., Steyn, M., and Loth, S.R. (2005). Metric analysis of sex differences in South African black and white pelves. *Forensic Science International* 147: 119–127.

Phillips, V.M. (2008). *Dental Maturation of South African Children and the Relation to Chronological Age.* PhD thesis, University of the Western Cape, Belville.

Pretorius, E., Steyn, M., and Scholtz, Y. (2006). Investigation into the usability of geometric morphometric analysis in assessment of sexual dimorphism. *American Journal of Physical Anthropology* 129(1): 64–70.

Reza, A., Mercy, J.A., and Krug, E. (2001). Epidemiology of violent deaths in the world. *Injury Prevention* 7: 104–111.

Rhine, S. (1990). Non-metric skull racing. In G.W. Gill and S. Rhine (eds), *The Skeletal Attributions of Race* (pp. 9–19). The Maxwell Museum in Anthropology, Albuquerque, NM.

Roelofse, M.M., Steyn, M., and Becker, P.J. (2008). Photo identification: facial metrical and morphological features in South African males. *Forensic Science International* 177: 168–175.

Rogers, T.T. (2005). Recognition of cemetery remains in a forensic context. *Journal of Forensic Sciences* 50(1): 5–11.

Ryan, I. and Bidmos, M.A. (2007). Skeletal height reconstruction from measurements of the skull in indigenous South Africans. *Forensic Science International* 167: 16–21.

Schoeman, K. (2007). *Early Slavery at the Cape of Good Hope, 1652–1778*. Protea Boekhuis, South Africa.

South African Parliament (2003). *National Health Act*, Section 64(1). Act No. 61 of 2003.

Statistics South Africa (2009). *Mid-year Population Estimates. P0302.* Pretoria, South Africa.

Stewart, T.D. (1958). The rate of development of vertebral osteoarthritis in American whites and its significance in skeletal age identification. *The Leech* 28: 141–151.

Steyn, M. (2003). A comparison between pre- and post-colonial health in the northern parts of South Africa, a preliminary study. *World Archaeology* 35: 276–288.

Steyn, M. (2005). Multi-murders from Africa: a case report. *Forensic Science International* 151(2–3): 168–175.

Steyn, M. and Henneberg, M. (1997). Cranial growth in the prehistoric sample from K2 at Mapungubwe (South Africa) is population specific. *HOMO* 48(1): 62–71.

Steyn, M. and İşcan, M.Y. (1997). Sex determination from the femur and tibia in South African whites. *Forensic Science International* 90: 111–119.

Steyn, M. and İşcan, M.Y. (1998). Sexual dimorphism in the crania of the mandibles of South African whites. *Forensic Science International* 98: 9–16.

Steyn, M. and İşcan, M.Y. (1999). Osteometric variation in the humerus: sexual dimorphism in South Africans. *Forensic Science International* 106: 77–85.

Steyn, M. and Smith, J.R. (2007). Interpretation of ante-mortem stature estimates in South Africans. *Forensic Science International* 171(2): 97–102.

Steyn, M., Peens, F., Briers, T., and Meiring, J.H. (2000). Case report: two murder victims identified by means of skull-photo superimposition. *South African Journal of Science* 96(3): 138–140.

Steyn, M., Pretorius, E., and Hutten, L. (2004). Geometric morphometric analysis of the greater sciatic notch in South Africans. *HOMO* 54: 197–206.

Thomas, C.J., Nortje, C.J., and van Ieperen, L. (1986). A case of skull identification by means of photographic superimposition. *Journal of Forensic Odontostomatology* 4: 61–66.

Van Rooyen, C. (2006). *A Description of Forensic Cases Received in the Department of Anatomy from 1993–2005.* Unpublished undergraduate thesis, University of Pretoria, Pretoria.

World Health Organization (2006). *Mortality Country Fact Sheet: South Africa.* www.who.int/entity/healthinfo/statistics/bodgbddeathdalyestimates.xls.

CHAPTER **31**

The Application of Forensic Anthropology to the Investigation of Cases of Political Violence

Luis Fondebrider

INTRODUCTION

The application of forensic anthropology to investigations into cases of political violence (such as human-rights violations), and its development, are closely linked with phenomena of violence that have occurred in several parts of the world since the 1970s. The need to conduct investigations following the disappearance and murder of thousands of people for political, ethnic, and religious reasons became a key element in this process. The search for and exhumation of the remains of victims, usually buried in clandestine graves, their identification, and the establishment of the cause of death are important components in the transition to democracy (Bernardi and Fondebrider 2007; Blau and Skinner 2005; Blau et al. 2010; Doretti and Snow 2003; Fondebrider 2008; Simmons and Haglund 2005; Stover and Shigekane 2002; Tidball-Binz 2006).

Although today forensic anthropology is used in many countries to investigate cases of political violence, two applications have significantly influenced the discipline. In Latin America the year 1984 marked the time when investigations into human-rights violations committed in the region, mostly by military dictatorships (excepting in Colombia and Peru), began (Snow et al. 1984). This approach, described below,

A Companion to Forensic Anthropology, First Edition. Edited by Dennis C. Dirkmaat.
© 2012 John Wiley & Sons Ltd. Published 2015 by John Wiley & Sons Ltd.

contributed to broadening the application of forensic anthropology beyond its classical definition, particularly as it is known in English-speaking countries.

The second application, which arose several years later and is better documented in English language publications, involves the extensive investigations that began in the former Yugoslavia in 1996. Since that time, the International Criminal Tribunal for the Former Yugoslavia (ICTY) has resorted to the services of organizations such as Physicians for Human Rights (PHR) to exhume remains from mass graves scattered mostly across Bosnia. Years later, the International Commission on Missing Persons (ICMP) was engaged to follow through on these tasks, mainly in the field of victim identification (Skinner et al. 2003; Skinner and Sterenberg 2005; Sterenberg 2008).

Most forensic anthropologists currently working in any field of this discipline – whether in ordinary criminal cases or mass disasters – have been involved, at any given time, in the experiences mentioned above, gaining valuable and diverse expertise. For some, these experiences were the springboard for a lifetime commitment to this application. For others, the experiences helped them in their return to more traditional fields of application, such as ordinary criminal cases.

What cannot be denied is that, regardless of the context, this application of forensic anthropology has been developed the most over recent years, producing extensive literature, particularly concerning the archaeological recovery of remains and osteo-logical analysis of bone trauma. In addition, this has enabled many forensic anthro-pologists lacking experience to receive hands-on training. For example, before traveling to Bosnia or Guatemala, most professionals had never seen or excavated a contemporary mass grave containing 40 or 50 corpses. Nor had they had the chance to observe traumas caused by tools such as machetes.

The Political and Legal Framework for Investigations into Political Violence

Some of the 40 countries hard hit by political, ethnic, or religious violence where forensic anthropology has been included in the human-rights investigations are now run democratically, or are on the road to democracy, and are conducting a concurrent review of their past. Well known are the cases of Argentina, Chile, El Salvador, Guatemala, and Peru in Latin America; South Africa, Rwanda, and Sierra Leone in Africa; the former Yugoslavia in Eastern Europe; and the Philippines and East Timor in Asia.

Even though each country came up with a solution of its own, the creation of truth commissions as well as the intervention of local and international courts has been the preferred choice to investigate the recent past of violence. As years have passed, some initial experiences have been adjusted to the local contexts, taking into account the geographical, economic, religious, and ethnic characteristics of each society. Thus, in some, albeit few, countries, emphasis has been placed on identifying the remains found rather than on laying any responsibility for the events on anyone; however, the general trend has been to identify the remains and establish the victims' cause of death as well as to contribute the resulting scientific reports to the relevant criminal proceedings.

The focus of this chapter is on the initial development of the application of forensic anthropology to cases of political violence in Latin America by examining the origins, evolution, and prospects of the *Equipo Argentino de Antropología Forense* (or EAAF, the Argentine Forensic Anthropology Team).

ARGENTINA AND ITS POLITICAL CONTEXT IN 1984

Democracy returned to Argentina on December 10, 1983 after a military dictatorship (1976–1983) came to an end which was marked by severe human-rights violations. The most patent consequence of the dictatorship was the disappearance of 10 000 to 30 000 people and the kidnapping of about 500 babies born to mothers held in captivity.

During the transition to democracy, there was a strong demand from thousands of victims' relatives, as well as from other sectors of society and the new government, to know what had happened to "the disappeared," as they are known, and to condemn the members of the armed and security forces responsible for the crimes.

In 1984 and 1985, the task undertaken by an investigative commission known as the *Comisión Nacional sobre la Desaparición de Personas* (or CONADEP, the National Commission for the Disappearance of People) and by the judicial power was to contribute to confirming the original claims made by thousands of victims' relatives, human-rights organizations, and survivors from clandestine detention centers. In many cases, the work broadened the original claims, thus gaining a better and more concrete insight into the systematic plan adopted by the military regime to cause the disappearance of people.

WHERE ARE THE DISAPPEARED?

During the investigation process, learning what had happened to the disappeared, knowing whether they were dead or alive, finding out where they were or where they had been buried, became vital questions that needed to be answered. Unfortunately, the evidence pieced together proved that thousands of people who had not recovered their freedom were dead and that two main killing methods had been employed: (i) throwing the living victims from airplanes into the Atlantic ocean, or a river, or (ii) executing them extrajudicially (sometimes staging fake confrontations). Their bodies were subject to one of three fates: if the remains of victims thrown into the ocean turned up along the Argentine or Uruguayan coasts (this happened in about 60 cases only), they were buried in nearby cemeteries. Those individuals whose bodies were never recovered (the number of which is still unknown) were more than likely lost due to the action of sea animals and the tides. The bodies of the victims shot dead were buried in the cemetery closest to the sites where they were found or (in only two documented cases to date) in areas under police or military control.

Therefore, at the beginning of the democratic transition, the challenge was to exhume unidentified, skeletonized bodies buried in individual or mass graves in cemeteries, try to identify them, and determine the cause of death. However, it became increasingly evident that investigations into these cases could not be conducted in the same way as ordinary criminal investigations or mass disasters. The investigation into cases of human-rights violations demanded a new approach both from the legal and scientific perspectives. The perpetrator was not an individual but the state and its apparatus (Bernardi and Fondebrider 2007; Doretti and Snow 2003).

The multiplicity of traces left by the administrative bureaucracy of the state, particularly in societies with a predominantly urban structure as Argentina, facilitated the use of a scientific methodology that aimed at reconstructing the logic behind the state's operation. This involved not only collecting but also analyzing data in the proper and meaningful context.

The secrecy of the perpetrators, the difficulties in accessing relevant documents (especially those held by intelligence agencies), the widely scattered information sources, and the relatively short passage of time made the search for the disappeared as well as the confirmation and rejection of hypotheses extremely slow.

The exhumation and analysis of the bodies found were mere steps in a broader and more complex process, the explanation of which is the purpose of this paper. An open mind and a deep knowledge of how the repressive regime operated do not guarantee a proper investigation but are a solid foundation on which to proceed. In this regard, each identification enabled forensic anthropologists to ratify or rectify a course of action adopted; therefore, every time a victim was successfully identified, the achievement went far beyond the humanitarian or legal field since it had a multiplicative effect upon the entire investigation.

Furthermore, it is worth mentioning that strong suspicion had fallen on other auxiliary investigative bodies, such as the police, for which reason they had to be excluded from these investigations, at least in the early years. As a result of this context, which was inherent not only in Argentina's early democracy but also in all Latin America, the traditional role of a forensic anthropologist was expanding to a field that usually is in hands of the police or the lawyers; that is, to collect and interpret information coming from oral and written sources.

1983/1984, First Exhumations: Dr Clyde Snow Arrives in Argentina

In early 1984, shortly after democracy returned to Argentina, judges began to order exhumations of graves in cemeteries thought to contain the remains of disappeared persons. Relatives of the disappeared, desperate to find out what had happened to their loved ones and hoping to recover their remains, often attended the exhumations. However, these exhumations were problematic in several ways. First, official medical doctors in charge of the work had little experience in the exhumation and analysis of skeletal remains; in daily professional experience they worked mainly with cadavers. Further, exhumations were carried out in a nonscientific manner by cemetery workers and the bones were frequently broken, lost, mixed up, or left behind in the graves. At that time, in the mid-1980s, discussions about archaeology or anthropology in a forensic context were very unusual. Only forensic physicians were deemed experts in the field, supported by odontologists and radiologists; in fact, there was scarcely room left for any other scientific discipline. The consequence of this was that hundreds of bodies were damaged, thus losing valuable evidence that was necessary to identify the remains found and to sustain a legal case against those responsible for the crimes.

Second, some of the forensic doctors involved in these efforts were complicit, either by omission or commission, with the crimes of the previous regime. In Argentina, as in most Latin American countries, forensic experts are part of the police and/or the judicial system, and their independence is often severely hindered during nondemocratic periods. Thus, there is a conflict of interest when state bodies have to investigate the state without external oversight. This accounts for the fact that forensic experts from official institutions had very little credibility with victims' families. For these reasons, it was necessary to find an independent, scientific alternative for such procedures.

Early in 1984, CONADEP and the Grandmothers of Plaza de Mayo, a nongovernmental human-rights organization searching for children who disappeared with their parents or who were born in captivity, requested assistance from the Science and Human Rights Program at the American Association for the Advancement of Science (AAAS; Washington DC, USA) with the intention of creating a means by which to (i) establish a biological link between disappeared babies and their family members and (ii) scientifically carry out exhumations and identify the remains found. E. Stover, then Director of the Science and Human Rights Program at AAAS, organized a delegation of American forensic experts to travel to Argentina to assist with the determination of a methodology and its implementation. The delegation found several exhumed, unidentified skeletons stored in plastic bags in dusty storerooms at numerous forensic medicine institutes. Many bags held the bones of more than one individual. The delegation called for an immediate halt to the exhumations due to improper excavation, storage, and analysis.

Among the AAAS delegation members was Dr Clyde Snow, one of the world's foremost experts in forensic anthropology. Snow called on local archaeologists, anthropologists, and physicians to begin exhumations and analysis of skeletal remains using traditional archaeological and forensic anthropology techniques. Snow returned to Argentina repeatedly over the next five years, trained the founding EAAF members, and helped form the team. In addition, Snow has since helped to start similar teams in Chile, Guatemala, and Peru.

THE CREATION OF EAAF

When Snow arrived in Argentina, he immediately called on professional archaeologists and physical anthropologists. As he did not get the desired response, he invited students of these disciplines and from the field of medicine, who would then create the EAAF in mid-1984 as a nongovernmental, not-for-profit scientific organization.

In other words, its origin did not stem from an academic or judicial decision, but from an initiative taken by a group of students who had the support and guidance of Snow, as well as the initial trust of human-rights organizations and of some judges concerned with investigations into cases of disappeared people.

Since the beginning, it was clear for all EAAF members that finding and identifying remains was a highly sensitive issue, since it touched upon many painful and distressing topics for the victims' families: the uncertainty as to whether their loved ones were dead or alive, whether it would be possible to bury them and carry out customary funeral rites, and whether justice could be pursued. Unfortunately, these basic and crucial questions could not always be answered, either because the political power was not fully committed to the search and the judicial power was not willing or able to investigate, or simply because the criminals involved would not provide any information and there was lack of power for, or interest in, getting their declarations.

It is in this general context, with little specific variations, that the search for the remains of people detained, disappeared, or killed has been carried out in many parts of the world over the last 30 years. The organizations of civil society, such as those made up of victims' relatives, have been the first to take to the streets to inquire about their loved ones, and they are the ones who persist in the search, far beyond truth commissions or the wish of governments to close the investigation proceedings.

Thus, Snow and EAAF's first application was strongly influenced from the beginning by the awareness that, beyond any consideration as to the best technical procedures to exhume and analyze human remains, the relationship with the victims' families and their communities was absolutely vital before, during, and after the investigation. This does not mean that families influence the scientific work performed, but when they are encouraged to participate in decision-making and information-sharing processes, scientific work gains credibility and transparency, which are indispensable for cases of this kind.

How to Investigate Cases of Disappeared People

It is often assumed that an investigation into cases of political, ethnic, and/or religious violence may be conducted in a way similar to an ordinary criminal case: for example, an individual kills somebody and runs away. However, cases of disappeared people are much more complex crimes because the responsibility lies mainly with the state and its apparatus. This is not exclusive to the Argentine case described, but common to all contexts where crimes of this kind are perpetrated. Furthermore, the opening of graves and the recovery of remains without a well-informed strategy is a mistaken approach that accounts for the great number of unidentified bodies kept in storage.

Therefore, before recovering and analyzing the remains, it is recommended that what may be called a *preliminary* investigation be conducted for the purpose of determining the identity of the person for whom the search is being made, what happened to him/her, what circumstances surround the case, etc. To sum up, the goal is to reconstruct the victim's life, but also to learn what he/she was like from a biological viewpoint, and to put forth hypotheses as to where he/she might be buried.

In order to conduct an investigation of this kind, written and oral sources have to be used. The former involve, first and foremost, documents issued by state bodies in relation to the facts concerned, such as intelligence documents from the security and armed forces, judicial investigations, autopsy reports, cemetery and morgue records, photographs of corpses, and death certificates. It is also necessary to survey mass media archives (many times, journalists are the first to arrive at the site, or conduct investigations on their own initiative) as well as academic research works. Oral sources involve, first of all, the victims' relatives, but also witnesses, political activists, and police and military officers involved in the events. All the information collected should be entered in databases specifically designed to store the information related to the searches and help propose specific working hypotheses.

This preliminary investigation, an unusual task among forensic anthropologists engaged in more traditional contexts, is then followed by the (archaeological) field-work and laboratory (analysis of the remains) stages.

From the archaeological point of view, the different burial sites pose new challenges. Mass graves containing possibly 30, 50, or 200 sets of remains in different states of preservation, mined areas, 20 m-deep wells full of remains, graves under new buildings, graves in damp soil, or graves at the bottom of a cliff are only some of the burial sites likely to be found when dealing with human-rights violation cases. As a result, the strategy for the archaeological excavation should be adapted to the type of environment as well as to the general context. For example, security factors are usually complex, since these sites can be located in rural areas where perpetrators may still be around. Another issue is the logistic challenges posed by road and communication deficiencies. It is important to bear in mind that these cases do not occur in developed countries (the so-called first-world countries) but in developing or very poor countries, which have considerable deficiencies in terms of infrastructure and security. Indeed, all of these factors affect and condition work strategies.

An analogous situation holds for the analysis of remains. In cases like these, conditioning factors do not result from infrastructure issues but from the population under analysis. On the one hand, the use of standards developed for other population groups, such as those in the USA, usually raises doubts as to the accuracy of, for example, age estimation in a different population group than that from which the standard was developed. This certainly impacts the identification process, which is already intrinsically complex where antemortem information is lacking. The target victims were usually people who had no access to dentists or physicians, accounting for a lack of records that makes matching attempts more difficult.

FORENSIC ANTHROPOLOGY AND INVESTIGATIONS INTO HUMAN-RIGHTS VIOLATIONS

Snow's initial tasks in Argentina were a source of inspiration for groups of forensic anthropologists who were later trained in Chile, Guatemala, and Peru. Thanks to his work, generosity and capacity, forensic anthropology has evolved in parts of Latin America differently from its development in other regions of the world. These differences can be summarized as follows:

- constant interaction with the victims' families and their communities before, during, and after the investigations, ensuring the credibility of the results obtained and transparency in the procedures adopted;
- the involvement of forensic anthropologists during preliminary investigations makes it possible to better integrate the contextual and historical information of a case with the scientific data gathered in the archaeological and laboratory (anthropological, medical, dental, and genetic) analysis stages;
- collection of antemortem data includes direct involvement in the interviews with the victims' family members, taking an approach closer to that used by social, rather than strictly forensic, anthropology. This creates a climate of trust with the

victims' relatives, allowing attainment of greater detail, thus ensuring better and more accurate data collection;

- the use of both the forensic archaeologist and the forensic anthropologist, in an integrated way, ignoring the often-sharp division between the professionals from such disciplines and their roles;

- finally, a broad experience in the search, excavation, and interpretation of complex graves located in different environments. Similarly, a substantial case load of perimortem injury analysis, which make forensic pathologists to appreciate more the role that forensic anthropologists could play in the interpretation of trauma in bone.

CREATION OF THE LATIN AMERICAN ASSOCIATION OF FORENSIC ANTHROPOLOGY

The *Asociación Latinoamericana de Antropología Forense* (or ALAF, the Latin American Association of Forensic Anthropology) was founded on February 28, 2003, in Sherman, Texas, USA, at a meeting attended by all the Latin American forensic teams along with others who actively work in the field of forensic anthropology in the region. The Argentine Forensic Anthropology Team invited a total of 17 people from seven countries: Guatemala, Peru, Colombia, Mexico, Argentina, Chile, and Venezuela.

The ALAF is a non-profit association pursuing the following objectives:

- establishing ethical and professional criteria for the application of forensic anthropology in order to guarantee its quality;
- promoting the official use of forensic anthropology and archaeology in judicial investigations in Latin America;
- promoting the accreditation of professionals in forensic anthropology by creating an independent board that would guarantee compliance with high professional standards;
- in keeping with the recommendations of international treaties and agreements, promoting the mechanisms that guarantee the deceased person's family access to the information on the procedures and to the results of forensic investigations;
- promoting the protection of the physical integrity of those associated with the investigation and their families due to the risks involved in working as a forensic anthropologist in some Latin American countries;
- defending the scientific and technical autonomy of forensic anthropological investigations in Latin America and the Caribbean.

ALAF's first public activity was the organization of a scientific session, "Forensic Anthropology in Latin America," at the 56th Annual Meeting of the American Academy of Forensic Sciences in Dallas, Texas. Representatives of forensic teams from Argentina, Colombia, Guatemala, and Peru participated in this event. In its first year, ALAF created a fully functional website. To date, ALAF has 67 active members plus numerous students, and has already organized five annual meetings in Guatemala, Peru, Colombia (two times), and Argentina.

The International Evolution of the EAAF

After 27 years of work in approximately 40 countries, the EAAF has become a reference in terms of the application of forensic anthropology to cases of political violence. In addition, the EAAF has been instrumental in delivering training and providing assistance to other similar teams. The initial collaboration with Snow, which helped train teams in Chile and Guatemala, has continued with the training of teams in Cyprus and South Africa.

Furthermore, this method of work has proved that relationships with the victims' families, the respect for their rights, doubts, and uncertainties, and the constant dialogue with them produce reliable results that become acceptable to the community. Contrary to what was previously thought, the relationship with the victims' families does not condition the scientific task but adds meaning to the job performed by the judicial authorities and forensic scientists.

In conclusion, the application of forensic anthropology to the kind of investigations described in this paper had expanded his traditional role, incorporating subjects and roles that enrich the identification process and the establishment of cause of death.

REFERENCES

Bernardi, P. and Fondebrider, L. (2007). Forensic archaeology and the scientific documentation of human rights violations: an Argentinean example from the early 1980s. In R.Ferllini (ed.), *Forensic Archaeology and Human Rights Violations.* Charles C. Thomas, Springfield,. IL.

Blau, S. and Skinner, M. (2005). The use of forensic archaeology in the investigation of human rights abuse: unearthing the past in East Timor. *International Journal of Human Rights* 9(4): 449–463.

Blau, S., Fondebrider, L., and Saldanha, G. (2010). Working with families of the missing: a case study from East Timor. In K. Lauritsch (ed.), *'We Need the Truth' – Enforced Disappearances in Asia* (pp. 136–144). ECAP, Guatemala.

Doretti, M. and Snow, C. (2003). Forensic anthropology and human rights: the Argentine experience. In D.W. Steadman (ed.), *Hard Evidence: Case Studies in Forensic Anthropology* (pp. 290–310). Prentice Hall, Upper Saddle River, NJ.

Fondebrider, L. (2008). The application of forensic anthropology to the investigation of cases of political violence: perspectives from South America. In S. Blau and D. Ubelaker (eds), *Handbook of Forensic Archaeology and Anthropology* (pp. 67–75). Left Coast Press, Walnut Creek, CA.

Simmons, T. and Haglund, W.D. (2005). Anthropology in a forensic context. In J.R. Hunter and M. Cox (eds), *Advances in Forensic Archaeology* (pp. 159–176). Routledge, New York.

Skinner, M. and Sterenberg, J. (2005). Turf wars: authority and responsibility for the investigation of mass graves. *Forensic Science International* 151 (2–3): 221–223.

Skinner, M., Alempijevic, D., and Djuric-Srejic, M. (2003). Guidelines for international forensic bio-archaeology monitors of mass grave exhumations. *Forensic Science International* 134 (2–3): 81–92.

Snow, C.C., Tedeschi, L.G., Levine, L., Lukash, L., and Stover, E.C. (1984). The investigation of the human remains of the "disappeared" in Argentina. *American Journal of Forensic Medicine and Pathology* 5(4): 297–299.

Sterenberg, J. (2008). Dealing with the remains of conflict: an international response to crimes against humanity, forensic recovery, identification, and repatriation in the former Yugoslavia. In S. Blau and D. Ubelaker (eds), *Handbook of Forensic Archaeology and Anthropology* (pp. 416–425). Left Coast Press, Walnut Creek, CA.

Stover, E. and Shigekane, R. (2002). The missing in the aftermath of war: when do the needs of victim's families and international war crimes tribunals clash? *International Review of the Red Cross* 84(848): 845–866.

Tidball-Binz, M. (2006). Forensic investigations into the missing. In A. Schmitt, E. Cunha, and J. Pinheiro (eds), *Forensic Anthropology and Medicine: Complementary Sciences from Recovery to Cause of Death* (pp. 387–407). Humana Press, Totowa, NJ.

CHAPTER 32 The Pervasiveness of Daubert

*Stephen D. Ousley and
R. Eric Hollinger*

INTRODUCTION

There have been many recent publications (e.g., Christensen and Crowder 2009; Foster and Huber 1999) about the meaning and impact of the Daubert decision (*Daubert v. Merrell Dow Pharmaceuticals* 1993) and subsequent rulings on forensic science. The role of science in legal proceedings has become preeminent and the "CSI effect," in which many juries expect to see cutting-edge scientific evidence, is undeniable: one jury recently refused to convict, despite enough circumstantial evidence, because a DNA test was not run on a half-eaten hamburger (Stockwell 2005). At the same time, in response to the Daubert ruling, previously used forensic methods (such as bite-mark analysis) have been largely discontinued and even fingerprint and DNA analyses are being reevaluated. The Daubert decision also played a role in the recent National Academy of Sciences findings (Holden 2009) calling for an independent governmental body to review scientific methods in the forensic sciences. We maintain that Daubert was necessary, forensic anthropology has improved since Daubert, and all of the forensic sciences are becoming more scientific because of it.

While technically applying only to scientific testimony, the main thrust of the Daubert decision is that scientists doing forensic work must do good science. Most importantly, scientific results do not stand alone; there are no results, *prima facie*, that prove any disputed fact: scientific conclusions must be qualified, or, ideally, quantified, as to the probability that the conclusions are correct, through estimated error rates. Daubert diminished the authority of the expert witness based solely on experience and reputation, and instead emphasized the quality of the expert's methods. The Frye standard for scientific evidence of "general acceptance" was in part superseded by emphasizing the more neutral concepts of reliability and validity as criteria to evaluate

A Companion to Forensic Anthropology, First Edition. Edited by Dennis C. Dirkmaat.
© 2012 John Wiley & Sons Ltd. Published 2015 by John Wiley & Sons Ltd.

the correctness of scientific conclusions. Daubert thus allows more innovation in scientific methods. The Daubert decision resulted in significant changes in the Federal Rules of Evidence rule 702, which is in force for federal civil law and incorporated into many state criminal law proceedings (Foster and Huber 1999). Additionally, a recent supreme court case (*Melendez-Diaz v. Massachusetts* 2009) established that scientists and technicians who evaluate evidence in any way must be prepared to testify about any methods and techniques used to establish a fact, even the methods for determining the weight of drugs recovered from a crime scene.

In this chapter we will define the terms "reliability" and "validity" and explain why they are important in any scientific investigation, especially after Daubert, and how they can be estimated, especially in forensic anthropology. Following Carmines and Zeller (1979), we will emphasize that validity and reliability are not absolute, but are matters of degree. We will explore an additional factor in evaluating scientific work, which we refer to as "stupidity" that incorporates practitioner error. Stupidity is important because it is unavoidable. In contrast to validity and reliability, stupidity is inherently unquantifiable; it can however be intuitively qualified. We also will clarify the terms "error," "precision," and "accuracy" in scientific investigations, especially statistical analyses. Understanding all of these concepts is necessary to help choose the best methods in scientific investigations, to understand why incorrect conclusions will be reached at times even using the best methods, and to judge which conclusions are more likely to be true when conclusions derived from different methods disagree.

ERROR

The word "error" is ambiguous because it has been applied to a mistake, to an incorrect conclusion, or, in statistics, to normal variability, or to "noise." In this chapter we will only use "error" to mean a mistake, and as such we will use it in measuring reliability and qualifying stupidity. We also advocate using error in this restricted sense in discussing scientific concerns, as the following explanations illustrate.

Mistakes

If observations (measurements or trait states) are recorded incorrectly, then the use of those observations will be worthless or misleading. W.W. Howells (1973) delineated seven sources of mistakes when recording osteological measurements: (i) application of technique: how things are measured, defined, or observed; (ii) interobserver error: inconsistencies among individuals measuring; (iii) intraobserver error: inconsistencies by an individual measurer; (iv) instrument errors: defective or uncalibrated instruments; (v) instrument reading errors: misreading numbers; (vi) recording error: writing down an observation incorrectly; and (vii) computer data entry error: final stage error. All of these kinds of mistakes influence reliability in that they will tend to produce inconsistent recorded measurements from the same bone. The magnitude of intra- and interobserver errors (or, more neutrally, differences) can be calculated using various statistics. For instance, differences of 1–2 mm are not uncommon when measuring a large measurement such as glabello-occipital length, with a mean of about 175 mm, and is often at most 1.5% of the mean measurement. However, 1–2 mm represents a relatively larger error when measuring something like interorbital breadth, with a mean measurement of 23 mm. Miscalibrated instruments, most often digital calipers that were not zeroed before measurement, or a spreading caliper with bent arms, will likewise lead to inconsistent measurements compared to properly calibrated instruments. Naturally, in order to help minimize interobserver differences, practitioners must study measurement definitions and must practice measuring. Practice is even more important in scoring nonmetric traits because the consequences of interobserver differences are much greater: for a three-state trait (small, medium, large), a difference of one grade in scoring results in 50% disagreement between observers; for a two-state trait (presence/absence), a difference of one grade in scoring results in 100% disagreement between observers. Additionally, qualifying a nasal breadth as small, medium, or large will likely be influenced by the last 100 nasal apertures seen by the subjective observer. In contrast, measuring a nasal breadth will produce an objective quantity that contains more information. Also, measured nasal breadths can be converted later to small, medium, and large if desired, but not vice versa. Therefore, explicit trait definitions and standardization are essential for replicating the observation methods used in scientific publications and for high reliability (consistency or repeatability), and measures of repeatability, such as intra- and interobserver differences, are important and currently are published more often. Observational methods with higher repeatability are inherently better than those with lower repeatability. If there is inconsistency in recording observations, the observations will show low reliability and will be of little practical use in analyses.

Incorrect conclusions

Daubert emphasizes a method's known or potential "error rate." It is advised to use techniques that have the lowest error rates, but what is meant really is a lower probability of drawing an incorrect conclusion; in other words, we should use methods with the highest validity. Because the *potential* error rate of *any* method is 100%, we should choose the method that most likely provides correct answers, assuming no mistakes in observations, analysis, interpretations, and conclusions using the data and method. Following Daubert, the expert witness should be able to estimate the probability of being correct objectively, using the specified data and methods, rather than subjectively. There are no techniques that are 100% correct because of a persistent nonzero probability of stupidity, but certain techniques are more likely to produce correct answers than other techniques. Partial fingerprints are not nearly as informative for individual identification as a complete 10-fingerprint set. A single bone is not as informative as a complete skeleton. A sex classification function that classifies male and female reference samples 95% correctly has, in the long run and under the best conditions, a 5% chance of incorrectly classifying all future remains analyzed, which is sometimes termed a 5% classification error rate. When the expert has little evidence to work with, the expert must simply accept that only weak conclusions can be made with confidence. Depending on the condition and completeness of the remains present, estimating sex may be no better than guessing. Incomplete remains having a 50% probability of being male only involves "error" if one draws a firm conclusion about a particular sex, because there is a 50% probability that the conclusion is incorrect. A higher probability of correctness means that the probability of making an incorrect conclusion is lower. A conclusion can end up being correct for the right reasons, or for the wrong reasons, such as estimating sex by flipping a coin. We maintain that a conclusion with a higher probability of being correct that later is shown to be incorrect is more defensible than a conclusion with a lower probability of being correct that later is shown to be correct. We maintain that coming to a conclusion that has greater support is more defensible than coming to a conclusion with less support, no matter which one eventually is shown to be correct. It would indeed be an error, a mistake, to draw a conclusion with less support rather than drawing a different conclusion with greater support.

Of course, the value of reaching a correct conclusion depends on the background, or prior, probabilities. Predicting heads or tails correctly 50% of the time when flipping a coin is not very impressive. As mentioned in Chapter 16 in this volume, on stature estimation, our conclusions should be more accurate and specific, when possible, than those that can be made using no information from the present case and merely based on the prior probabilities. Importantly, although randomness is often emphasized, it is most often thought of as equal probabilities, a 50/50 chance when there are two choices, as in the case of coin flipping. But that is not always the case. In the classroom one of us (SDO) demonstrates an amazing computer program that predicts handedness of an individual based on mathematical manipulations of his or her date of birth. In every undergraduate class tested so far, it has consistently been at least 85% correct, and in many classes it is 100% correct. The students are impressed by the demonstration, in which every prediction is that the student is right-handed. Once informed that in the general population, 90% are right handed, the students are not so impressed. The high prior probability of being right handed is the reason that methods for determining handedness from the skeleton are difficult to justify, because any proposed method would need to be much better than 90% correct.

Noise or variability

"Error," the "error term," and the "standard error" are all related to statistical analysis but can have very different underlying meanings depending on the analysis. Early statistical methods were employed to estimate constants such as the circumference of the Earth. Any deviations from the constants were, by definition, errors in estimation or calculation, artificial, and therefore statistical "noise." When biological measures first were studied, such as human statures, people from different countries showed different mean measurements, and a normal curve was often observed within groups. Unfortunately, "error" was also used to describe these normal deviations from the mean, normal variations, which are seen in nearly all linear measurements in virtually all animals (Stigler 1999). Likewise, in linear regression, the "standard error" represents deviations in actual values from predicted values represented by the regression line. In forensic stature estimation, there is not a perfect relationship between bone length and stature, because a bone length is one small component, along with many other components, that make up stature. Accounting for normal variation is exceedingly important in stature estimation, with an important trade-off between precision and correctness: the more precise the stature estimation is, the less certain we can be that the actual stature is contained within the prediction interval. Stature estimates are often given in prediction intervals, which define how often we should be correct in the long run; this is an estimate of validity. We will be incorrect less often when we use a 99% prediction interval for stature given a bone length than when we use a 90% prediction interval. The 99% prediction interval is rather wide and may not help narrow down possible identifications, but we must accept the variability in stature given specific bone lengths. As long as the stature prediction is accurate (unbiased), we can be correct nearly 100% of the time if we use a plus/minus 15 cm prediction interval, but we would sacrifice precision for correctness.

RELIABILITY

"Reliability" in a scientific and legal sense is frequently misunderstood due to its ordinary language meaning, similar to dependability, which is actually closer in meaning to validity. Reliability is necessarily a component of validity but, in a forensic sense, using reliable methods – that is, measuring something correctly and consistently – may have little or nothing to do with reaching the correct conclusion, which reflects validity. We follow Hand (2004), in that reliability is best thought of as consistency in measurement, or how closely different observers measure or score the same phenomenon. Highly reliable measurements show very low or no interobserver errors and high repeatability. A clock that shows the correct time consistently is a reliable and valid clock; a clock that is consistently 1 hour ahead is reliable but invalid; in each case the clock presents a consistent representation of time. Thus, reliability ideally should be used only for observations and measurement, the beginnings of any analysis that will lead to meaningful conclusions. Inter- as well as intraobserver differences are naturally of paramount concern when measuring reliability. Collecting reliable bone measurements involves using an instrument, and for small bones a digital sliding caliper accurate to 0.01 mm should be more reliable than using an osteometric board, which should be more reliable than holding a bone next to a meter stick and "eyeballing" a measurement. When measuring a long bone such as a femur, using an osteometric board is likely more reliable than using multiple measurements from a 200 mm digital sliding caliper. Of course, the digital caliper and osteometric board should be calibrated. If digital calipers are not properly zeroed, they can nonetheless produce repeatable and reliable results for many observers, at least until they are zeroed. In this limited case it could be said that the measurements written down are reliable but invalid (in the strict sense of "face validity") representations of the bone length (Carmines and Zeller 1979). In other words, the measurements are consistent, but consistently wrong. Also, some measurements are inherently more reliable than others, no matter what instrument is used. Measurements involving maximum lengths of long bones show higher reliability than a bone measurement such as the condylomalleolar length of the tibia, which requires the correct orientation of the tibia and a specialized osteometric board (Moore-Jansen et al. 1994). On the cranium, the measurement of frontal chord (nasion–bregma) shows lower interobserver differences than measuring orbital breadth, and especially mastoid height, because the frontal chord involves measuring the distance between two well-defined points. Importantly, sources of observational error occur with DNA sequencing, and extracting an incorrect DNA sequence due to DNA contamination may be more likely than due to computer or instrument error. It should be clear that reliability is not absolute, in that no data-extraction method can be said to be perfectly reliable. Rather, the degree of reliability represents consistency and repeatability of measurements or observations using specific instruments and procedures.

VALIDITY

Validity is the strength of the connection between a hypothesis and the real world application or conclusion (Carmines and Zeller 1979). Validity can be subdivided but as a whole it covers the entire process of data collection, data entry, data analysis, reaching general conclusions, drawing inferences, and then coming to specific conclusions about the case at hand (Footer and Huber 1999). For example, measurements are collected from the bones of an unidentified person; statistical analyses are employed; overall, the measurements are more similar to those in a reference sample of males as opposed to females; due to the greater similarity with males, we infer that the remains come from a male. Validity is estimated through the process of validation, in which the same techniques and methods are used under similar circumstances and any discrepancies in conclusions are noted. Reliability is a part of validity because reliability, or consistency in measurements or observations, is required during data collection. Unreliable data cannot lead to valid conclusions. Validity is more often applied to methods, rather than data, except when an instrument measures something of direct importance, such as a clock, which measures time. As mentioned, a clock that is consistently 1 hour ahead is reliable but invalid, because it never shows the correct time (although with the knowledge that it consistently runs ahead, making it reliable, we can calculate the correct time). A stopped clock is neither reliable nor correct because it is inconsistent and never (well, okay, twice a day) tells the correct time. In analyzing a reliable mitochondrial DNA sequence, for example, heteroplasmy can cause false exclusions or false matches, and is due to the imperfect validity of matching an unidentified mitochondrial DNA sample to a known sample.

Estimating validity, specifically what is known as predictive validity, is related to a method's known or potential error rate (i.e., how often it provides a correct conclusion) referenced in Daubert. Predictive validity can be estimated and quantified most easily using statistical methods such as discriminant function analysis (DFA) or other classification procedures, of course, assuming that the measurements are taken correctly. DFA provides classification accuracies for specified measurements and groups as part of the procedure using a validation sample that is very similar to the reference samples (Hastie et al. 2001). In DFA, bone measurements of the reference groups are statistically manipulated to separate them as much as possible. Then, individuals in the validation sample as well as the unknown individual simply are classified into the group to which they are most morphologically similar. Some of the classifications will be incorrect, but the classifications are tabulated and the classification percentage correct is calculated. As you may imagine, with more measurements, the accuracy increases; there is more information and greater intergroup differences when using 10 measurements than when using two. Higher classification accuracies represent higher validity for the method of estimating sex from multiple bone measurements. Naturally, the inference must be made that the individual comes from one of the groups, the groups are most relevant to the questions at hand, and the estimated percentages correct will hold up in the long run. If improper samples are used, the conclusions will be erroneous, which involves another aspect of validity, termed external validity, because the inference that the group samples are appropriate may not be correct. For example, Giles and Elliot (1962) published a novel way of using DFA to classify skeletal remains into three groups using cranial measurements, but their samples were from nineteenth-century American blacks and whites and a limited sample of American Indians. As Ayres and Jantz (1990) illustrated, secular changes in cranial morphology cause more incorrect classifications of modern crania using the Giles and Elliot formulas. Likewise, İşcan and Cotton's (1990) DFA using postcranial measurements performed very poorly when tested against modern Americans (Ousley and Jantz 1997).

STUPIDITY

Stupidity is a term we use largely for practitioner mistakes, any errors in execution that can affect reliability and validity, or can be largely independent of them. This category may be necessary because Christensen and Crowder (2009: 5) remarked that "Practitioner mistakes, especially those that result in misidentification, challenge the view of *method reliability* regardless of the validity of the method" (emphasis added). As mentioned, we agree with other authors in that analytical methods are best understood in terms of validity rather than reliability. Otherwise, there is a virtually unlimited number of analytical methods depending on which specific tools and procedures are used at what time. For example, working on a case while fatigued will probably result in more errors, and if these errors are more probable because of the method, then this analysis, in the presence of fatigue, is technically a different method. However, practitioner mistakes, or stupidity, can affect the validation process, or can be independent of reliability and validity. Elements of stupidity would include incorrectly writing down a measurement even though the measurement was taken correctly (possibly affecting reliability and validity), inadvertently entering the wrong number into a computer or calculator when calculating something such as a discriminant function score (possibly affecting validity), or incorrectly writing down the result of a test that results in an incorrect conclusion (affecting validity). More specific examples of the role of stupidity involve analyzing craniometrics to estimate sex or ancestry. For instance, to analyze a case, I could take measurements using a tape measure, enter those measurements into a calculator, and use the Giles and Elliot (1962) formulas; or I could take measurements using a spreading caliper, enter those measurements into a calculator, and use the Giles and Elliot (1962) or other published formulas; or I could measure using a vernier caliper, calculate the DFA coefficients using Fordisc 3 (Jantz and Ousley 2005), and use a calculator to analyze the measurements; or I could measure using a digital caliper and analyze the measurements using any number of methods available in commercial statistical packages such as R (R Development Core Team 2009), SAS (SAS Institute 2001), or SYSTAT (Systat Software 2004); or I could measure using a digital caliper and enter the measurements directly into Fordisc 3; or I could use a three-dimensional digitizer and software to register landmarks, have software calculate the standard measurements, then import the measurements directly into Fordisc 3 for analysis. In each of these analyses, there are different reliability and validity concerns, and the method may change only slightly, but the opportunity for stupidity to affect results diminishes with each successive approach. From these examples we can see that the same analytical method can be used in a variety of ways, and not only are some methods better than others, some procedures for employing the same method are more likely to produce errors than others. In general, automated methods are more easily replicable, but certain kinds of mistakes are harder to detect using automated procedures. Most importantly, unlike measures of reliability and validity, the measurement of stupidity cannot be quantified, but can only be qualified because we know that certain methods (such as manual data recording and entry) are more prone to mistakes than others.

Two Cautionary Tales

Estimating sex using the mandible

Loth and Henneberg (1996) proposed a method for estimating sex from the adult mandible by examining the flexure of the posterior border of the ramus, which was a trait they discovered and defined, and supposedly found only in males. They reported an overall accuracy in sex estimation of 94% for healthy mandibles with teeth present. Koski (1996) concluded that the method did not perform as well as claimed, but his validation study was limited to radiographs from females only, many of whom were under 10 years of age; it was a poor test of the method. Later, Donnelly et al. (1998) tested the method more appropriately using dry bones from adults. In their blind tests, they found that 63% of the mandibles were sexed correctly on average, with a male bias in estimating sex, the mandibular flexure trait was difficult to recognize and score as present or absent, and interobserver differences were high. Estimating sex using mandibular flexure, then, is an extreme case: it is unreliable due to interobserver differences and invalid because classifications using it are no better than by chance. There have been several other validation studies that echo the conclusions of Donnelly et al. (1998). The importance of validation studies through independent and blind tests of methods, which incorporate the concerns of reliability, validity, and stupidity, cannot be overemphasized.

The bones of Everett Ruess?

In May 2008, skeletal remains were found in a remote part of Utah, USA, after following up on a Navajo family's story of the murder of a white man in 1934. An initial DNA test from a molar indicated that the DNA came from a European, rather than a Native American. Dennis Van Gerven, at the University of Colorado at Boulder, excavated the burial site further, analyzed the remains, and decided that the remains came from a white male about 5 feet 8 inches (172.7 cm) tall, aged between 19 and 22 years old. Everett Ruess, a 5 foot 8 inch, 20-year-old writer and artist, exploring isolated parts of Utah, had disappeared in 1934, so a possible identification was obvious. Van Gerven and his assistant reassembled the cranium and then used skull–photo superimposition to compare the remains and photographs of Everett Ruess. Van Gerven was quite certain in his conclusions: "Everett had unique facial features, including a really large, jutting chin. This guy had the same features. And the bones match the photos in every last detail, even down to the spacing between the teeth.

The odds are astronomically small that this could be a coincidence.... I'd take it to court. This is Everett Ruess" (Roberts 2009). Later, a more precise DNA test was conducted comparing the DNA of the remains to DNA from saliva of four of Ruess' nephews and nieces. Dr Kenneth Krauter, a molecular biologist at the University of Colorado at Boulder, analyzed 600 000 DNA markers (much more than the usual 18 or so used in forensic comparisons) and found that the individual shared 25% of the markers with Ruess' nieces and nephews, the expected percentage in such relatives, and did not share nearly as many of the same markers with a random sample of other individuals. Krauter concluded "This is a textbook case.... The evidence is irrefutable that the bones are from a close relative of the four, ... Combined with the facial reconstruction, that makes this an irrefutable case," and "The combination of the forensic and genetic analyses makes it an open and shut case.... I believe it would hold up in any court in the country" (National Geographic 2009; Roberts 2010). Van Gerven later added "If this were going before a court of law, you'd want to build a case.... That's what we've done here, with Navajo oral tradition, the forensic analysis and now the DNA test. We can be certain that this is Ruess" (National Geographic 2009).

Ruess' relatives received the remains and were planning to cremate them but concerns were raised by the Utah state archaeologist and a physical anthropologist (Jones and Kopp 2009): most importantly, the lack of peer review, with results only available in magazine articles and press releases; haphazard recovery techniques; inconsistencies in the Navajo family story about the murder; a much greater amount of tooth wear seen in photos of the mandible than found in contemporary European Americans and more typical for American Indians; a probable large untreated tooth cavity, which Ruess would have had treated; questions about the skull–photo super-imposition methods; and finally, they suggested the DNA results should be tested by a laboratory with experience analyzing ancient DNA. The relatives of Ruess agreed to a second DNA test, done this time by the Armed Forces DNA Identification Laboratory, which has extensive experience in analyzing DNA from American soldiers and testing against possible relatives. In October 2009, the shocking results came back: not only were the remains not from Everett Ruess, they were from a Native American.

What went wrong? How could different analyses, especially of the same DNA, give different conclusions? Apparently, there was no contamination of the individual's DNA. Also, the DNA sequences were read correctly, which indicates that reliability is not the issue. The analysis of the DNA was at fault. Krauter used sophisticated software to compare the 600 000 DNA markers that is used in medical research in the living. When using the software on ancient DNA, which in many cases is degraded, for some reason the software treated the greater noise associated with degraded DNA as DNA marker matches with the relatives. Krauter paid far more attention to the results of the method than to the process of arriving at those results. He later said "We screwed up by relying on the technology too much," and "Fortunately, the error uncovered how the extreme sensitivity can be misleading if a researcher takes its output at face value" (Roberts 2010). Thus, while DNA evidence is often compelling, it can be misleading. The methods of analyzing DNA evidence are evolving still, and each new method must be evaluated as to its validity in providing answers to forensic questions.

What about Van Gerven's inexorable conclusions from the independent skull–photo superimposition? Van Gerven expressed disappointment that his conclusions were shown to be incorrect and offered no further comments on the "unique" skeletal and dental features that identified the remains as Ruess. Despite the impression of unique features, it may well simply be possible to manipulate antemortem photographs and images of a cranium to produce a minimum of discrepancies. The Ruess case illustrates why skull–photo superimposition may involve as much art as science, because we do not know the random probability of matching a number of features in the cranium to photos of different people. Uncanny resemblances and similarities, using skull–photo superimposition or other methods, may be mere coincidence. In other words, skull–photo superimposition has uncertain validity for uniquely identifying an individual.

We suspect that the Ruess case is an example of confirmation bias; namely, when you look for something, you will probably find it, and you rarely will rethink your own results when they meet your expectations. We are all susceptible to confirmation bias, which has been demonstrated in a blind test of fingerprint experts, who were told the opinion (which disagreed with the experts' previous assessment) of a fictitious FBI fingerprint expert who examined the same prints. When the experts reexamined the prints, the experts often unknowingly reversed their former conclusions (Dror et al. 2006). Van Gerven likely was influenced by the initial DNA finding indicating European ancestry and possible identification, and his results in turn likely influenced Krauter's DNA conclusions, though technically they all should be independent: results from one independent test should not be considered as supporting the results of another test.

CONCLUSIONS

The Daubert ruling clarified what good scientific methods involve: accurate, consistent, and repeatable measurement techniques (reflecting reliability), making assumptions and inferences explicit, and using the most accurate methods available for reaching conclusions (reflecting validity). Methods with greater reliability and validity are preferred over those with lower reliability and validity. At the same time, stupidity should be minimized, and precision maximized, as much as is practical. When different methods produce different conclusions we will choose the method that most likely produces the correct answer given the concerns of reliability, validity, and stupidity. Nevertheless, we will at times come to an incorrect conclusion even when we use the best methods; this is only a mistake if we made mistakes during data collection or analysis. Validity and reliability are probabilistic, and no method is 100% correct at producing results or conclusions.

REFERENCES

Ayres, H.G., Jantz, R.L., and Moore-Jansen, P.H. (1990). Giles and Elliott race discriminant functions revisited: a test using recent forensic cases. In Gill, G.W. and Rhine, J.S. (eds), *Skeletal Attribution of Race* (pp. 65–71). Anthropological paper no. 4. Maxwell Museum of Anthropology, Albuquerque, NM.

Carmines, E.G. and Zeller, R.A. (1979). *Reliability and Validity Assessment*. Sage Publications, Newbury Park, CA.

Christensen, A.M. and Crowder, C.M. (2009). Evidentiary standards for forensic anthropology. *Journal of Forensic Sciences* 54: 1211–1216.

Daubert v. Merrell Dow Pharmaceuticals, 509 US 579, 113 S.Ct. 2786, 125 L.Ed. 2d 469 (US June 28, 1993) (No. 92–102).

Donnelly, S.M., Hens, S.M., Rogers, N.L., and Schneider, K.L. (1998). Technical note: a blind test of mandibular ramus flexure as a morphologic indicator of sexual dimorphism in the human skeleton. *American Journal of Physical Anthropology* 107: 363–366.

Dror, I.E., Charlton, D., and Peron, A.E. (2006). Contextual information renders experts vulnerable to making erroneous identifications. *Forensic Science International* 156: 74–78.

Foster, K.R. and Huber, P.W. (1999). *Judging Science: Scientific Knowledge and the Federal Courts*. MIT Press, Cambridge, MA.

Giles, E. and Elliott, O. (1962). Race identification from cranial measurements. *Journal of Forensic Sciences* 7: 147–157.

Hand, D.J. (2004). *Measurement Theory and Practice: The World Through Quantification*. Arnold, London.

Hastie, T., Tibshirani, R., and Friedman, J. (2001). *The Elements of Statistical Learning: Data Mining, Inference, and Prediction*. Springer, New York.

Holden, C. (2009). Forensic science needs a major overhaul, panel says. *Science* 323: 1155.

Howells, W.W. (1973). *Cranial Variation in Man: A Study by Multivariate Analysis of Patterns of Difference Among Recent Human Populations*. Papers of the Peabody Museum, vol. 67. Peabody Museum of Archeology and Ethnology, Harvard University, Cambridge, MA.

İşcan, M.Y. and Cotton, T.S. (1990). Osteometric assessment of racial affinity from multiple sites in the postcranial skeleton. In Gill, G.W. and Rhine, J.S. (eds), *Skeletal Attribution of Race* (pp. 83–90). Anthropological paper no. 4. Maxwell Museum of Anthropology, Albuquerque, NM.

Jones, K. and Kopp, D. (2009). *Everett Ruess – A Suggestion to Take Another Look*. http://history.utah.gov/archaeology/ruess.html.

Koski, K. (1996). Mandibular ramus flexure–indicator of sexual dimorphism? *American Journal of Physical Anthropology* 101: 545–546.

Jantz, R.L. and Ousley, S.D. (2005). *FORDISC 3: Computerized Forensic Discriminant Functions*. Version 3.1. University of Tennessee, Knoxville, TN.

Loth, S.R. and Henneberg, M. (1996). Mandibular ramus flexure: a new morphologic indicator of sexual dimorphism in the human skeleton. *American Journal of Physical Anthropology*, 99:473–85.

Melendez-Diaz v. Massachusetts, 129 S.Ct. 2527, 69 Mass. App. 1114, 870 N.E. 2d 676, reversed and remanded (2009) (No. 07–591).

Moore-Jansen, P.M., Ousley, S.D., and Jantz, R.L. (1994). *Data Collection Procedures for Forensic Skeletal Material*, 3rd edn. University of Tennessee, Knoxville, TN.

National Geographic. (2009). *Press release: after 75 years, National Geographic ADVENTURE solves mystery of lost explorer*. April 30. http://adventure.nationalgeographic.com/2009/04/everett-ruess/dna-test-text.

Ousley, S.D. and Jantz, R.L. (1997). The forensic data bank: documenting skeletal trends in the United States. In K. Reichs (ed.), *Forensic Osteology*, 2nd edn (pp. 297–315). Charles C. Thomas, Springfield, IL.

R Development Core Team. (2009). *R: a Language and Environment for Statistical Computing*. R Foundation for Statistical Computing, Vienna. www.R-project.org.

Roberts, D. (2009). *Finding Everett Ruess*. April/May. http://adventure.nationalgeographic.com/print/2009/04/everett-ruess/david-roberts-text.

Roberts, D. (2010). *Everett Ruess Update: How the DNA Test Went Wrong*. February 2. http://ngadventure.typepad.com/blog/2010/02/everett-ruess-how-the-dna-test-went-wrong.html.

SAS Institute (2001). *SAS/SHARE 9 User's Guide*. SAS Institute Inc., Cary, NC.

Stigler, S.M. (1999). *Statistics on the Table: The History of Statistical Concepts and Methods.* Harvard University Press, Cambridge, MA.

Stockwell, J. (2005). Defense, prosecution play to new 'CSI' savvy. *Washington Post* May 22. www.washingtonpost.com/wp-dyn/content/article/2005/05/21/AR2005052100831.html.

Systat Software (2004). *Systat Version 11.* Systat Software Inc., Point Richmond, CA.

PART IX Ethics, Overview, and the Future of Forensic Anthropology

Introduction
to Part IX

Dennis C. Dirkmaat

Throughout this book, forensic anthropology has been portrayed as a field experiencing a state of evolution. Previously considered as a laboratory-based field conducted by physical anthropologists, with the addition of better analytical methods – applicable to modern human skeletal samples, forensic archaeology, and forensic taphonomy – the field of forensic anthropology has been reconfigured in the last 20–30 years to include new definitions, perspectives, goals, activities, roles, and duties. Not only are forensic anthropologists invited more often (and earlier) into the medicolegal investigation to work with forensic pathologists on a diversity of cases, including those involving fatal-fire victims and complex skeletal trauma, they are called out by law enforcement and coroner/medical examiner's offices to assist in the recovery of a wide variety of outdoor scenes from surface scatters to mass disaster scenes. In fact, top students interested in forensic anthropology can even aspire to have a career as a forensic anthropologist as there are now full-time jobs (with benefits) as professional forensic anthropologists.

Within this reconfigured field the forensic anthropologist is portrayed not just as a technician, applying skeletal biology techniques to the interpretation of human remains from forensic cases, but as a scientist and researcher in the field of human skeletal biology and physical anthropology, validating old, tried-and-true analytical methods applicable to human bones, and conducting research in the relatively new field of forensic taphonomy. As this new role becomes rapidly clarified, perhaps it is time to step back and examine where the field is now, and where it might be 20–30 years in the future. One important issue to consider is that, unlike other anthropology disciplines, forensic anthropologists handle evidence, and the analysis and interpretation of that evidence has serious and lifelong effects on victims, families of victims, and even perpetrators and suspects. It is critical that the work completed by the forensic anthropologist be held to the highest of standards relative to conducting good science;

A Companion to Forensic Anthropology, First Edition. Edited by Dennis C. Dirkmaat.
© 2012 Blackwell Publishing Ltd. Published 2012 by Blackwell Publishing Ltd.

conducting the work ethically, and conducting it to standards familiar to other forensic scientists, including other anthropologists.

With respect to *conducting good science*, the recently reconfigured expectations that (i) expert witnesses present their results and interpretations of analyses to the highest of scientific standards (i.e., Daubert) and (ii) all scientific disciplines reconsider the scientific basis and merit of their field through validation of methods, as well as certification of practioners and accreditation of laboratories conducting the analyses (i.e., National Academy of Sciences report), will produce an immediate and far-reaching impact on all of the forensic sciences (except DNA, of course, which has been operating at those levels). The historic location of forensic anthropology within academia, museums, and laboratories suggests that forensic anthropology might already be well positioned to successfully address these expectations.

Ousley and Hollinger in Chapter 32 use the Daubert ruling as a springboard to discuss the concept of "doing good science." Their basic premise is that the renewed emphasis on science in the forensic sciences is a positive sign. However, doing good science is not as easy as it seems. What may appear at first glance to be good valid scientific results may, in fact, have resulted from poor scientific design, improper or inappropriate study samples, or the application of inappropriate statistical tests, and, thus, may be subject to misinterpretation. Ousley and Hollinger provide a very informative discussion in which they attempt to clarify the key common terms and concepts in scientific investigation by providing very precise definitions of the terms reliability, variability, and validity, while introducing a new term: stupidity.

France in Chapter 33 provides a discussion relevant to conducting forensic anthropology work *ethically*. She examines the ethical basis of the field of forensic anthropology and focuses not only on what should be done, but also on what should not be done. Although most scientific organizations provide a well-worded description of the ethics of the group, it is the boundary areas that provide the most interesting discussions. A consideration of how individuals who cross to the dark side of the boundary should be disciplined is also presented in the chapter.

Within the field of forensic science and medicolegal death investigation and research, a controversial issue is the culling of biological material at autopsy: Who owns it?, Can it be used for research?, What to do with the material after conducting analyses and obtaining results?, When is permission from next of kin required?, and Can it be used for teaching purposes? Although no definitive answers can be given to these questions, a foundation and framework for further discussion is provided. France includes appendices of the codes of ethics of the major organizations that forensic anthropologists work with and belong to: the American Academy of Forensic Sciences, the American Board of Forensic Anthropologists, and the Scientific Working Group for Forensic Anthropology.

James Adovasio, a world-renowned archaeologist, provides an overview of the field of forensic anthropology in Chapter 34 from the perspective of an "outsider" who is not directly involved in forensic anthropology cases, though he does conduct forensic geoarchaeological investigations. He is, therefore, familiar with the proper handling of evidence and the primary goals of forensic science investigation and thus provides an interesting perspective regarding how forensic anthropology is conducted and perceived. His primary critique of forensic anthropology is that many (most?) forensic anthropologists attempt to analyze evidence, including the human remains, from crime scenes with little or no understanding of the context and association of that evidence to the scene and to other evidence. This is a situation that is incomprehensible in archaeological investigation or, for that matter, any science. Adovasio reaffirms the contention that the best way to recover outdoor crime scene evidence is through the employment of archaeological principles and methods. He indicates that through the melding of the laboratory analysis of the remains, with the thorough understanding of the contextual setting, forensic anthropology then becomes a scientific *pursuit*, although he does not see it as a stand-alone *science*, for it lacks a "corpus of theory." The challenge for forensic anthropology in the future, then, is to find that body.

CHAPTER 33 Ethics in Forensic Anthropology

Diane L. France

INTRODUCTION

Writing a chapter concerning ethics in forensic anthropology appears at first glance to be a straightforward task: just discuss all of the principles that define an ethical professional practice (don't lie, don't be biased, conduct yourself in accordance with all laws, etc.), and advise the audience not to do anything that you wouldn't want your mother to know about. In fact, writing a chapter about ethics brings up many questions about exactly where to draw the line between ethical and unethical behavior in practice and in research.

The overwhelming majority of qualified forensic anthropologists practices ethically, and recognizes and reports unethical behavior in someone else. Most belong to at least one professional organization – the American Academy of Forensic Sciences (AAFS), the American Board of Forensic Anthropology (ABFA), the American Association of Physical Anthropology, and others – and comply with the codes of ethics and conduct of those organizations. That relatively few complaints of unethical conduct are brought before any of these professional organizations is testimony to the fact that most forensic anthropologists endeavor to practice in a responsible, ethical manner.

The primary concern about developing codes of ethics and conduct is only in part about setting guidelines for the majority of professionals who already practice ethically. It is equally important to set parameters for effectively disciplining those who practice unethically. It is important to give "teeth" to the codes to protect the profession and all those who are affected by forensic anthropologists from unethical behavior. The codes inform our clients that it is reasonable for them to expect ethical behavior from us. However, it is equally important that the codes protect those who

A Companion to Forensic Anthropology, First Edition. Edited by Dennis C. Dirkmaat.
© 2012 John Wiley & Sons Ltd. Published 2015 by John Wiley & Sons Ltd.

are wrongly accused from inaccurate and unfair complaints of unethical behavior. The primary difficulty in creating codes of ethics and conduct is always to make them specific enough to give direction to the professional in practice and to be general enough so that the disciplinary body will find appropriate wording that will allow it to discipline individuals who conduct themselves unethically.

This chapter is divided into three sections. The first section outlines and discusses standard codes of conduct for the practitioner. The second discusses a more recent question concerning the ethics of conducting research on human remains (a subject that does not commonly appear in codes of ethics), and the third section outlines the ways in which disciplinary actions should and do resemble court cases.

CODES OF ETHICS AND CONDUCT IN THE PRACTICE OF FORENSIC ANTHROPOLOGY

Codes of ethics and conduct of the AAFS, the ABFA, and those developed by the Scientific Working Group for Forensic Anthropology (SWGANTH) have some basic concepts in common. Some of these concepts are also presented in publications such as Barnett (2001) and Bowen (2010). Although the wording or exact content of the sections relating to the same subject matter may differ between groups, and neither SWGANTH nor the American Association of Physical Anthropologists have a disciplinary role, all provide guidelines for proper conduct. The codes of ethics and conduct for AAFS, ABFA, and SWGANTH are given as Appendices A, B, and C in this chapter. The following are generalized compilations of the codes. All of the organizations have the first three rules in common.

1. Do not *materially*[1] misrepresent your education, training, experience, or expertise. Do not claim membership in any organization or any *level* of that organization (associate member, member, fellow, etc.) if it is not true.
2. Do not *materially* misrepresent data or evidence. Base your opinions and conclusions on the evidence and only to the extent justified by the evidence. Your opinions shall be technically correct and scientifically based.

This rule is supported by the Daubert guidelines for expert testimony in that conclusions must be based on scientific evidence. Conclusions must be testable and falsifiable, have a known or potential error rate, and be subject to peer review. The forensic community has largely embraced the concept of reliability and validity in evidence, particularly after the publication in 2009 of *Strengthening Forensic Science in the United States: A Path Forward* (Committee on Identifying the Needs of the Forensic Sciences Community, National Research Council 2009), which was critical of some forensic evidence that has not, in the opinion of the panel, been based on the scientific method.

It is tempting to go beyond the evidence to reach stunning conclusions more appropriate to an episode of the television crime series *CSI* than to an actual case. You need not apologize for not being able to answer questions about a case when the evidence is not present.

3. Do not conduct yourself in ways that are adverse to the best interests and objectives of the profession or to the organization(s) of which you are member. Do not

make statements that appear to represent the position of the organization to which you belong without having the authority to do so.

4. Remain intellectually independent and impartial. Do not be biased in favor of the prosecution or defense. Answer questions with answers based upon the evidence and not based upon which side is asking the questions. Disclose *all* findings to the agency entitled to receive those findings.

An agency may ask you to limit your investigation or report to only those factors that are beneficial to their side of the case. You may decide to present a report with the full analysis or to decline further involvement with the case, but you must not be biased in your report. Provide a report with impartial conclusions and leave it to the agency to decide what to do with the information.

This rule applies to scientists but it does not apply to attorneys. It is interesting that some forensic scientists, including forensic anthropologists, complain that defense attorneys, for example, behave unethically when they are providing a vigorous defense for their client, particularly if the scientist "knows" that the defendant is the perpetrator. In fact, the defense attorney would be guilty of unethical behavior if he did not provide his client with the most vigorous defense possible within legal limits.

5. Set a reasonable fee or fee structure and accept no work based on a contingency fee.

A contingency fee is one in which payment will not be demanded unless a favorable conclusion is reached. Obviously this implies, and creates the impression of, bias in a case.

Do not be guilty of "double-dipping" in which you *inappropriately* charge more than one agency for work. Your total income from a case should not be over 100% of your fee schedule.

Be aware that your fee may become an issue in the courtroom, so be prepared to justify what you were paid for a case and by whom. As an example, some attorneys will try to show that you were paid to produce results for one side or the other. Avoid falling into that trap, and, as long as it is true, make it clear that you formed conclusions based only on the evidence, not on the paycheck.

6. Maintain the confidentiality required in a case. Do not, without permission, discuss aspects of a case that cannot be obtained from public records.

Be very careful about using e-mail because of potential problems with confidentiality. Be aware that all records, including e-mails, can be subpoenaed. If you do exchange e-mails about a case, be sure that you do not include extra comments unrelated to the case, including the description of the great date you had the night before. Anything in that e-mail could be used against you in court.

7. Maintain the integrity of the evidence in a case.

Maintain a chain of custody so that you can account for the location of the evidence in the case at all times. Keep a log of who has access to evidence when they had access, and state their role in the case.

Make sure that anyone who has access to evidence is qualified to have that access, and make it clear to that individual that he or she is subject to subpoena. If the individual is uncomfortable with that possibility, that person should not have access to the evidence. To allow an unqualified person unsupervised access to evidence is unethical.

8. Do not insinuate yourself into cases in which you have not been invited.

It is perfectly reasonable to introduce yourself to local authorities and to inform them that you are available to assist in your professional capacity. It is also reasonable to advertise your services on, for example, a website. However, it is unethical to appear uninvited at a crime scene, a mass fatality incident, or at the medical examiner's office for the purpose of insinuating yourself into a case. Also, do not contact family members for the purpose of insinuating yourself into a case.

Even if you are an expert with years of experience, you don't know everything about the case if you have not been invited to assist. You may be just as likely to damage a case as to be advantageous, and you could be arrested for interference! If you believe that the medicolegal personnel are investigating the case improperly or inadequately, there are ways to advise the appropriate authorities without inserting yourself into the case directly.

9. Do not engage in practices in which there is a conflict of interest or an appearance of a conflict of interest.
10. Treat all remains with respect and dignity.
11. Report ethical violations to the appropriate authorities.

Usually an organization such as the AAFS or the ABFA cannot discipline nonmembers, but ethical violations can still be reported, for example, to the judge in a court case. Illegal activities must be reported to the appropriate authorities.

Probably the most important rules are to remain *unbiased*, *truthful*, and *transparent*.

CODES OF ETHICS AND CONDUCT IN RESEARCH

- Who "owns" human remains?
- Is it ethical to remove portions of remains for research if that research will result in future identifications of unknowns? Is research justified if it is for the common good?

The codes listed in the previous section are straightforward and, as stated, are followed by the vast majority of practitioners in forensic anthropology. Most codes do not, however, cover research involving human remains. Everyone recognizes that in order for a forensic anthropologist to practice forensic anthropology based on current standards, extensive research on modern skeletons must be conducted. Historically, the rules about research on human remains or portions of those remains were relatively relaxed, but as the laws are changing, the determination of what constitutes ethical behavior in research is changing as well.

As part of a postmortem examination in the USA and in many other countries, a medical examiner or coroner has the authority to remove and study any organs or other body parts that will assist in the identification of the remains and in the determination of the cause and manner of death. They also have the right to retain them for as long as questions of medicolegal significance exist. But does this confer ownership of the remains to the medical examiner, or does the medical examiner merely have possession of the remains? Does the next of kin have ownership of the remains or of parts of remains? Can the next of kin prohibit research on remains?

Rohan Hardcastle (2009: 203) argues "…common law and statutory regimes in England, the USA and Australia fail to provide clear or coherent legal principles for determining the legal status of, and rights pertaining to, biological materials separated from dead bodies or living persons." He cites many interesting cases that discuss ownership of remains, but in fact there has been no clear universal consensus about rights to the body of a deceased person.

The Human Tissue Act of 2004 (www.legislation.gov.uk/ukpga/2004/30/contents) in the UK is commonly cited when the discussion arises concerning the removal of tissues from either living or deceased individuals, as well as the disposal of tissues no longer needed for investigation or research. The Act arose in part as a reaction to a discovery that Professor Dick van Velzen, at the Alder Hey Children's Hospital in Liverpool, routinely removed whole organs from children and fetuses at autopsy but did not use them to develop his conclusions, nor did he even properly catalogue and store the remains. Further, he was found to have lied to parents who thought that they were burying their children intact, including all organs. The investigation of this case resulted in a report, commonly called the Redfern Report, that prompted the change in laws in the UK regarding retention of organs and other body parts (Royal Liverpool Children's Inquiry 2001).

The Human Tissue Authority in the UK now states that a postmortem "…examination and removal and storage of relevant material to determine the cause of death do not require consent from the relatives if these activities have been authorised by the coroner. However, following the cessation of the coroner's authority, it is unlawful to use the retained material for a scheduled purpose set out in the [Human Tissue] Act, or to continue to store it with the intention of using it for a scheduled purpose, without appropriate consent" (Human Tissue Authority, nd). In general, taking tissue that may include bone samples from the deceased for purposes other than the coroner's investigation must be based upon consent from someone authorized to give that consent (as defined in the Act), starting with the next of kin.

In the USA as well, laws are becoming more prohibitive when addressing the removal of tissue from the deceased for research. For example, a state statute recently passed by Minnesota (similar laws are in place in some other states) asserts:

> Autopsies performed pursuant to this subdivision may include the removal, retention, testing, or use of organs, parts of organs, fluids or tissues, at the discretion of the coroner or medical examiner, when removal, retention, testing, or use may be useful in determining or confirming the cause of death, mechanism of death, manner of death, identification of the deceased, presence of disease or injury, or preservation of evidence. Such tissue retained by the coroner or medical examiner pursuant to this subdivision shall be disposed of in accordance with standard biohazardous hospital or surgical material and does not require specific consent or notification of the legal next of kin. When removal, retention, testing, and use of organs, parts of organs, fluids, or tissues is deemed beneficial, and is done only for research or the advancement of medical knowledge and progress, written consent or documented oral consent shall be obtained from the legal next of kin, if any, of the deceased person prior to the removal, retention, testing, or use. (Minnesota Statutes 2011).

The overwhelming impetus for the adoption of these laws appears to be the history of removing tissues for any reason other than autopsy without notifying the next of kin. Such cases have arisen in the USA and in the UK (and elsewhere not discussed in this chapter) because family members have discovered that whole organs were removed from their loved one and that the body was buried without those organs. In the minds of many cultures and individuals, the body cannot truly be laid to rest unless the entire body is present. But even in situations in which there is no religious or cultural basis for this belief, the assertion that the next of kin has a keen interest in the disposition of the body and its elements is becoming more prevalent.

Should long-term retention of remains, or portions of remains, for research without consent from the next of kin, or any other legitimate source, be considered unethical? It may soon be a moot question as laws continue to be enacted that would prohibit that practice. At this point, the removal of body parts for anthropological research without consent should be considered unethical. Should we extend that consideration to situations in which we are, for example, exhuming a mass grave in another country? Of course we should. We must not treat individuals from other countries differently than they would be treated in the USA.

What if the next of kin of the remains cannot be identified because the remains are not identified? Mass fatality incidents, mass grave excavations, and badly decomposed remains present this situation. Laws concerning the removal of remains, or portions of remains, for the purpose of identification and determining the circumstances surrounding death are the same (in most countries) as in any other medicolegal context. The coroner or medical examiner is permitted to remove the elements for those purposes. If the next of kin cannot be attributed to specific remains, permission from other legitimate sources (again, variously defined in different situations) *might* be obtained for the purpose of research and long-term retention. Again, transparency in your actions is important.

What about gathering information for research by noninvasive means? The UK Human Tissues Act of 2004 does not address the use of photographs or electronic images, but it subscribes to the guidance by the 2011 UK General Medical Council's guidelines (General Medical Council 2011). Those guidelines state that permission is needed from the subject of images of internal organs, images of pathology slides, ultrasound images, X-rays, and a few other images and recordings (see the list at www.gmc-uk.org/guidance/ethical_guidance/7840.asp) *unless* those images can be anonymised "…by removal or coding of any identifying marks such as writing in the margins of an X-ray…" (General Medical Council 2011).

In the USA, the Basic Health and Human Services Policy for Protection of Human Research Subjects (Subpart A, Title 45, Section 46.101; see US Department of Health and Human Services 2009) regulates research on a *living individual* "conducted, supported, or otherwise subject to regulation by any federal department or agency which takes appropriate administrative action to make the policy applicable to such research…. It also includes research conducted, supported, or otherwise subject to regulation by the federal government outside the United States…." The policy goes on to state that research involving the collection or study of existing data, documents, records, pathological specimens, or diagnostic specimens *is acceptable* if the information is publicly available or if the information is recorded in such a manner that the subjects cannot be identified.

Photographs at autopsy and as a part of research on remains are under less scrutiny in the courts, with at least one notable exception. Florida Statute 406.135, commonly referred to as the Earnhardt Family Protection Act or the Earnhardt Law (Florida Statute 2010), states that only the next of kin may view autopsy photographs without a court order under penalty of a felony in the third degree. After Dale Earnhardt was killed in an accident during a NACSCAR race, the family feared that the photographs would be shown or distributed as part of public record and, therefore, would be identifiable as Dale Earnhardt. They filed suit to have photographs removed from the public record, resulting in the current, relatively aggressive and limiting legislation.

Therefore, creating photographs, radiographs, and other imaging that cannot be linked to the identity of the deceased is most likely legal in most circumstances and in most states (with Florida as an exception). Other noninvasive techniques, such as measurements, certainly should be performed if they will assist in the identification of the victim(s) or to assist in the determination of the circumstances surrounding death. In most circumstances, it is likely not illegal to use those measurements in future research as long as the identity of the victim cannot be linked to the data. This same threshold should be considered for photographs or radiographs. However, before photographing or X-raying subjects, one should be familiar with the laws regarding that practice with the very general rule that if it is illegal, it is unethical.

DISCIPLINARY ACTIONS

After an allegation of unethical behavior is reported to the appropriate agency or organization, the case usually proceeds much like a court case to protect both the accused and the organization. The accused retains rights to a vigorous defense against the allegations and the agency must prove the unethical behavior beyond a reasonable doubt. Keep in mind that any allegations and resulting disciplinary action could potentially end a career. The specific courses of action at each step may differ between organizations, but overall there are strong similarities. As noted above, these organizations do not have the jurisdiction to discipline nonmembers, even if the accused leaves the organization during the investigation or disciplinary process. Of course, illegal actions on the part of the accused must always be reported to legal authorities. In the AAFS and ABFA, the Ethics Committee and the Board of Directors maintain confidentiality about the case until the final stages.

Before a complaint to the ABFA can be investigated, two Diplomates must file ethics complaints against an individual (Appendix B). In the AAFS, the Ethics Committee may use any evidence that may come to its attention from any source to initiate an investigation, although a formal investigation will proceed only if at least two committee members determine that probable cause exists that the allegation is well founded (Appendix A). This threshold protects the accused from spurious allegations. Apart from this difference in the initiation of an investigation, the disciplinary actions of the AAFS and the ABFA are similar as the code for disciplinary actions for the ABFA were based, in part, on the codes for the AAFS.

If the committee decides to proceed with the formal investigation, the accused and the complainant (and the Chair or Secretary of the section of the AAFS of which the accused is a member) are notified of the complaint. In each organization, the accused has a limited time to respond but will be given access to all relevant information about the complaint, an opportunity to respond to the accusation (and to present evidence in defense) and to be represented by counsel.

During the course of the investigation, a hearing may be scheduled in which the accused and the accuser are given an opportunity to respond to the investigation and to each other (although the exact nature of the hearing is different in the AAFS and the ABFA).

The nature of the vote to discipline the accused is different in the AAFA and the ABFA, but in both organizations the accused is allowed to appeal the decision to censure, suspend, or expel the individual found guilty of a violation of the rules of ethics.

FINAL CONSIDERATIONS

This chapter has outlined many of the situations in which an ethical decision is fairly easily prescribed. Naturally the rules presented here cannot cover every situation, so use at least these two tools to determine whether your decision is ethical.

1. Be transparent. Do nothing that you would be ashamed or embarrassed for the world to judge. Discuss your decisions with others who may not agree with you.
2. As commonly, but not quite accurately, attributed to the Hippocratic Oath in medicine: "First do no harm":

   - to the victim, who deserves an honest and unbiased voice. This is likely the most complicated of these guidelines because it is easier to convince yourself that the ends justify the means. "If I say in my report that these insults to the skeleton are caused by the knife that the police removed from the crime scene, it will tie the suspect to the crime and bring justice to this victim." Perhaps you will help the case and take a dangerous criminal off of the streets, but perhaps you will help tie the wrong suspect to the crime. Losing sight of the science by acting as judge and jury does not help the victim, and the act is dishonest and unethical;
   - to the next of kin who deserve respect and truthful information, even if the next of kin are not immediately identified. Remember that the next of kin may have religious and spiritual beliefs about death that are different from yours, and they must be respected;
   - to the forensic anthropology profession. Remember that you represent the profession and other forensic anthropologists with every act;
   - to the justice system and to society. Evidence obtained in an unethical manner may destroy a case in court with unforeseeable consequences.

APPENDICES

Appendix A: codes of ethics and conduct for the American Academy of Forensic Sciences

Article II: Code of Ethics and Conduct
SECTION 1 - THE CODE: As a means to promote the highest quality of professional and personal conduct of its members and affiliates, the following constitutes the Code of Ethics and Conduct which is endorsed by all members and affiliates of

the American Academy of Forensic Sciences: a. Every member and affiliate of the Academy shall refrain from exercising professional or personal conduct adverse to the best interests and objectives of the Academy. The objectives stated in the Preamble to these bylaws include: promoting education for and research in the forensic sciences, encouraging the study, improving the practice, elevating the standards and advancing the cause of the forensic sciences. b. No member or affiliate of the Academy shall materially misrepresent his or her education, training, experience, area of expertise, or membership status within the Academy. c. No member or affiliate of the Academy shall materially misrepresent data or scientific principles upon which his or her conclusion or professional opinion is based. d. No member or affiliate of the Academy shall issue public statements that appear to represent the position of the Academy without specific authority first obtained from the Board of Directors.

SECTION 2 - MEMBER AND AFFILIATE LIABILITY: Any member or affiliate of the Academy who has violated any of the provisions of the Code of Ethics (Article II, Section 1) may be liable to censure, suspension or expulsion by action of the Board of Directors, as provided in Section 5 h below.

SECTION 3 - INVESTIGATIVE BODY: There shall be constituted a standing Ethics Committee (see Article V for composition), the primary function of which shall be: a. To order and/or conduct investigations and, as necessary, to serve as a hearing body concerning conduct of individual members or affiliates which may constitute a violation of the provisions of Article II, Section 1. b. To act as an advisory body, rendering opinions on contemplated actions by individual members or affiliates in terms of the provisions of Article II, Section 1.

SECTION 4 - INVESTIGATION INITIATING ACTION: The following are the principal forms by which the Ethics Committee may initiate investigative action: a. A member or affiliate of the Academy may submit a written complaint alleging violation(s) of Article II, Section 1 by a member or affiliate to the Academy Office (see Article II, Section 5, Rules and Procedures, below) or to the Chair of the Ethics Committee. Such a complaint should be made in a timely manner. b. The Ethics Committee may institute an inquiry based on any evidence that may come to its attention from any source which in its opinion indicates the need for further query or action under the provisions of these bylaws.

SECTION 5 - RULES AND PROCEDURES: The following procedures shall apply to any allegation of unethical conduct against a member or affiliate of the Academy: a. Allegations of unethical conduct against a member or affiliate received by the Academy shall be transmitted promptly to the Chair of the Ethics Committee. b. The Ethics Committee shall determine whether the alleged unethical conduct falls within its jurisdiction and whether there is probable cause to believe that the allegation is well founded. c. If the Ethics Committee, in its preliminary determination, finds that it does not have jurisdiction or that there is a lack of probable cause to believe that the allegation is well founded, it shall close the case. It shall issue a report of such determination to the Board of Directors, setting forth the basic facts but omitting the names of the parties, and stating the reasons for its decision to close the case. Notice of the allegation, including the source, and its disposition, shall be given to the accused. Notice of the disposition shall also be given to the complainant(s). d. If the Ethics Committee finds that it has jurisdiction and that there is probable cause to believe that the allegation is well founded, it shall give notice of the filing of the allegation and its sources to the accused. In accordance with Rules and Procedures formulated by the Ethics Committee and approved by the Board of Directors, the Committee shall assemble such information from both the accused and the complainant(s) which shall permit it to determine whether the allegation requires further action. e. The Ethics Committee may appoint an Academy Fellow or Fellows to investigate the allegation and/or to present the evidence to the Committee. f. If, based on the results of an investigation, the Ethics Committee decides to dismiss the allegation without a formal hearing, it may do so. It shall notify the accused and the complainant(s) of its decision and shall issue a report to the Board of Directors setting forth the basic facts and stating the reason(s) for its decision, but omitting the names of the accused and complainant(s). g. If the Ethics Committee decides to formally hear the case, it shall give the accused a reasonable opportunity to attend and be heard. The complainant(s) shall also be given a reasonable opportunity to be heard. Following the hearing, the Committee shall notify the accused and the complainant(s) of its decision. The Ethics Committee shall also submit a report on its decision to the Board of Directors. If the Committee finds unethical conduct, the report shall include the reasons for its decision, and any recommendations for further action by the Board. The accused may also submit to the Board of Directors a written statement regarding what sanctions, if any, should be imposed. h. If the Ethics Committee's decision is that unethical conduct on the part of the accused member or affiliate has occurred, the Board of Directors shall review the report, and ratify or overturn the decision, or remand the case to the Committee for further action. If the Board of Directors ratifies the Committee's decision, it shall also review any written submission provided by the member or affiliate found to be in violation of the Code. The member or affiliate may then, upon a vote of three-fourths of the members of the Board present and voting, be censured, suspended or expelled. The nature and conditions of any sanction must be provided to the member or affiliate. A suspended member or affiliate may only be reinstated by the procedure set forth in Article II, Section 6. i. A member or affiliate who has been found in violation of the Code of Ethics has the right to appeal the actions of the Board of Directors to the membership of the Academy. To initiate an appeal, the member or affiliate must file a brief written notice of the appeal, together with a written statement, with the Academy Secretary not less than one hundred twenty days prior to the next Annual Business Meeting of the Academy. j. The Board of Directors shall then prepare a written statement of the reasons for its actions and file the same with the Academy Secretary not less than forty days prior to the next Annual Business Meeting of the Academy. k. Within twenty days thereafter, the Academy Secretary shall mail to each voting member of the Academy a copy of the appellant's notice of appeal and supporting statement, and a copy of the Board of Directors' statement. l. A vote of three-fourths of the members present and voting at the Academy's Annual Business Meeting shall be required to overrule the action of the Board of Directors in regard to censure, suspension or expulsion of a member or affiliate. m. No member of the Board of Directors who is the subject of an Ethics Committee investigation, or who has any other conflict of interest, shall participate in any matter before the Board concerning ethics. n. The Ethics Committee shall formulate internal Rules and Procedures designed to facilitate the expeditious, fair, discreet, and impartial handling of all matters it considers. The Rules and Procedures shall be subject to the approval of the Board of Directors.

SECTION 6 - SUSPENSION OF MEMBERS AND AFFILIATES: Members or affiliates who have been suspended may apply to the Board of Directors for reinstatement once the period of suspension is completed. A suspended member or affiliate shall not be required to pay dues during the period of suspension. If reinstated, the required dues payment shall be the annual dues less the pro-rated amount for the period of suspension.

American Academy of Forensic Sciences (2011)

Appendix B: code of ethics and conduct for the American Board of Forensic Anthropology

Proposed February 16, 1999; Approved by the Board February 20, 2001; Changes proposed January 16, 2008; Approved by the Board January 23, 2008.

SECTION I - THE CODE: As a means to promote the highest quality of professional and personal conduct of its members, the following constitutes the Code of Ethics and Conduct which is endorsed and adhered to by all Diplomates of the American Board of Forensic Anthropology (ABFA).

Diplomates of the American Board of Forensic Anthropology shall:

a) Refrain from exercising professional or personal conduct adverse to the best interests and purposes of the ABFA.
b) Refrain from providing any material misrepresentation of education, training, experience, or area of expertise. Misrepresentation of one or more criteria for certification shall constitute a violation of this section of the code.
c) Refrain from providing any material misrepresentation of data upon which an expert opinion or conclusion is based. Diplomates shall render opinions and conclusions strictly in accordance with the evidence in the case (hypothetical or real) and only to the extent justified by the evidence.
d) Not make statements in his/her written reports, public addresses, or testimony that are not technically correct and scientifically based.
e) Act at all times in a completely impartial manner by employing scientific methodology to reach logical, unbiased conclusions and by reporting all findings in a clear, concise manner.
f) Set a reasonable fee for services if it is appropriate to do so; however, no services shall be rendered on a contingency fee basis.
g) Treat all information from an agency or client with the confidentiality required.
h) Refrain from issuing public statements which appear to represent the position of the ABFA without specific authority first obtained from the Board of Directors.

SECTION II - GROUNDS FOR DISCIPLINE: Any Diplomate whose professional conduct becomes adverse to the best interests and purposes of the ABFA shall be liable to censure, suspension, or expulsion with revocation and recall of certification granted in Article XIII, Section 4 of the Bylaws. The Diplomate shall be censured, suspended, or expelled by action of the ABFA Board of Directors acting on the findings and recommendations of the Ethics Committee, following the appeal period or any other actions required. Investigative action may be initiated due to alleged violations of any of the following provisions:

a) An intentional misstatement or misrepresentation or concealment or omission of a material fact or facts in an application or any other communication to the Board or its representative(s).
b) Conviction of an applicant for certification or holder of a certificate by this Board by a court of competent jurisdiction of a felony or of any crime involving moral turpitude.
c) Issuance of a certificate contrary to or in violation of any of the laws, standard rules, or regulations governing the Board and its certification programs at the time of its issuance; or determination that the person certified was not in fact eligible to receive such certificate at the time of its issuance.
d) Violations of the Rules of Professional Conduct of the ABFA by an applicant or holder of a certificate of this Board.
e) Non-payment of annual renewal fees after the delinquent notice by the Treasurer. Reinstatement may be granted in that fiscal year upon payment of the outstanding fees plus a $50.00 reinstatement charge.
f) Failure to complete recertification documentation as required.
g) Upon the recommendation of any two Diplomates of the American Board of Forensic Anthropology, the qualifications of any Diplomate may be reviewed by the Board to determine whether the Certificate of Qualification issued by the Board should be revoked. The candidate shall have the right to present his or her case to the Board, but the final decision rests with the Board.
h) Upon recommendation of the Ethics Committee and approval by two-thirds (2/3) of the Board, action to suspend or revoke certification may only be taken after at least thirty (30) days advance notice of the nature of the charges or reasons for such action has been given to the individual concerned and a reasonable opportunity for such person to be heard has been provided by the Board.

SECTION III - INVESTIGATIVE BODY: There shall be constituted a standing Ethics Committee, the primary composition and function of which will be:

a) The standing Ethics Committee shall serve as the investigative body to which the chairperson of the Ethics Committee shall refer all cases for consideration.
b) The members of the Ethics Committee shall be appointed by the President of the Board with the advice and consent of the Board of Directors. Each member, with the exception of the non-voting attorney member, if present, will serve a one-year term. The Ethics Committee shall elect a chairperson from its membership annually.
c) The President of the ABFA may chair the Ethics Committee in the absence of the Ethics Committee chairperson, if the chairperson is under investigation, has a conflict of interest in that particular case, or for other valid reasons, is unable to participate.

d) The Ethics Committee can order investigations and serve as a hearing agency concerning past or present conduct of individual members of the ABFA which may constitute a violation of the provisions of the Code of Ethics and Conduct.

SECTION IV - INVESTIGATION INITIATING ACTION: The following are the principal forms by which the Ethics Committee may initiate investigative proceedings:

a) Any two Diplomates of the ABFA may submit formal written allegations of violations concerning a Diplomate to the Secretary of the ABFA (see Judiciary Process below) or to the chairperson of the Ethics Committee.

b) The Ethics Committee may institute an inquiry based on any evidence brought to its attention which indicates the need for further query or positive action under the provisions of the Bylaws. Appropriate to this form of action, ABFA officers, upon receipt of a complaint concerning the professional or personal conduct of a Diplomate, may refer said complaint to the Ethics Committee in writing, accompanied by a recommendation, if any, concerning need for further investigation. Such recommendations, however, shall not be binding on the Ethics Committee.

SECTION V - JUDICIARY PROCESS

a) Written allegations against a Diplomate, if delivered to the ABFA Secretary, shall immediately be transmitted to the chairperson of the Ethics Committee.

b) The Ethics Committee shall immediately give notice of the filing of a complaint to the accused, and in accordance with the Rules and Regulations formulated by the Ethics Committee and approved by the Board of Directors, assemble such written data from both the accused and the accuser(s) which will permit the Ethics Committee to arrive at a preliminary determination as to whether the complaint is well founded and requires further investigation. The accused will be advised of the nature of the complaint and of the identity of the complainant(s).

c) The Ethics Committee may contact any individual or entity (whether an ABFA Diplomate or not) in its investigations.

d) If the Ethics Committee, in its preliminary determination, finds that the complaint is not well founded, it shall dismiss the complaint. It shall issue a report of such determination to the Board of Directors, setting forth the basic facts, but omitting the names of the parties if possible and stating the reasons for its decision to dismiss.

e) If the Ethics Committee determines the complaint is well founded, the Ethics Committee will investigate the allegations. The Ethics Committee shall then formally hear the charges and shall give both the accused and the accuser(s) a reasonable opportunity to be heard and confront each other.

  1. Notice shall be sent by certified mail, return receipt requested, to both the accused and the accuser(s) for the purpose of setting up a formal hearing.

  2. The accused shall receive a copy of the written complaint. He/she is entitled to see the document in its entire form.

  3. After receipt of the return notice (by certified mail, return receipt requested) a formal hearing date will be mutually agreed to by both parties and the Ethics Committee. This date will be at least ninety (90) days from said receipt of official notice in order to give both parties adequate time to prepare for the hearing. If

agreeable to both parties, the hearing shall be held at or about the time of the annual meeting of the ABFA in order to keep costs to a minimum. If one or both parties request a hearing date at a time other than the annual meeting, the costs of said hearing shall be the responsibility of the party/parties requesting the hearing and not the ABFA.

4. At this hearing no legal counsel for either party may be present. The non-voting attorney member of the ABFA, as a non-voting member of the Ethics Committee, will be present for the purpose of assuring that propriety, protocol, and adherence to proper procedures is maintained during the hearing. The attorney Board member shall act in an advisory position to the committee only and shall not be involved in the presentation of the case for either party.

5. The Ethics Committee shall make a report, which will include a recommendation to the ABFA Board of Directors at the conclusion of the hearing(s).

f) Upon a vote of three-fourths (3/4) of the members of the Board of Directors present and voting, the party accused of unethical or wrongful conduct may be censured, suspended, or expelled. No Board of Director member or member of the Ethics Committee who is the subject of a pending accusation under the provisions of the ABFA Code of Ethics and Conduct shall sit in deliberation on any manner concerning ethics.

g) The accused has the right to appeal the action of the Board of Directors to the Diplomates of the ABFA. In effecting an appeal, the appellant must file a brief typewritten notice of the appeal, together with any typewritten statement he/she may wish to submit in his/her behalf, with the ABFA Secretary not more than thirty (30) days after receiving notice of the action of the Board. Punitive actions of the Board against the accused shall not commence until after thirty (30) days from the date of notification of the accused to allow time for the appeal. The Secretary shall immediately advise each member of the Board of Directors of the appeal and shall forward to each a copy of the supporting papers submitted by the appellant. If no appeal is received by the Secretary within thirty (30) days, the actions of the Board shall be implemented, and may no longer be appealed.

h) If an appeal is received within thirty (30) days, the Executive Committee shall prepare a written statement of the reasons for the Board of Directors' actions and file the same with the ABFA Secretary not more than thirty (30) days from receipt of the appeal.

i) Within ten (10) days of receipt of the statement from the Executive Committee, the Board of Directors shall choose seven (7) Diplomates-at-Large (not on the Board of Directors), and the appellant shall choose eight (8) Diplomates-at-Large (not on the Board of Directors) to hear the appeal.

j) The Secretary of the Board shall send the statements from the appellant and the Executive Committee to each of the fifteen (15) members of this Appeal Committee within ten (10) days after the Board is convened.

k) The fifteen (15) members of this Appeal Committee shall arrange a closed meeting either in person or by conference call(s). The meeting(s) shall be strictly confidential. Neither the appellant, the accuser(s), nor any member of the Board of Directors shall be part of the meeting. The non-voting attorney member shall be part of the meeting(s) to assure propriety, protocol, and adherence to procedures. The attorney shall not represent either party involved in the hearing.

l) Decisions of the Diplomates in the closed hearing will be based upon the written information provided by the appellant and the Board of Directors. A written vote of three-fourths (3/4) of the Appeal Committee present and voting at the closed meeting shall be required to overrule the action of the Board of Directors in regard to censure, suspension, or expulsion of a Diplomate.

SECTION VI - CONFIDENTIALITY, RULES, AND AAFS ETHICS

a) Any member of the Ethics Committee, Board of Directors, or Appeals Committee divulging information on matters previously considered or being considered could be in violation of the Code of Ethics and Conduct and is subject to charges of same being filed. This does not apply to written statements made and worded by the Board of Directors concerning ethical matters or about the case(s) being considered which may be distributed to the Appeals Committee by said person after Executive Committee approval.
b) The Ethics Committee shall formulate internal Rules and Procedures, and from time to time propose changes to such Rules and Procedures, designed to facilitate the expeditious, fair, discreet, and impartial handling of all complaints or matters brought before the Ethics Committee. The Rules and Procedures, and any subsequent deletions, additions, or amendments thereto, shall be subject to the approval of the Board of Directors.
c) In order to prevent a conflict of interest between the ABFA and the American Academy of Forensic Sciences (AAFS), a written report of the action of the Board of Directors of the ABFA concerning censure, suspension, or expulsion of the Diplomate will be forwarded to the chairperson of the Ethics Committee of the AAFS. The AAFS Ethics Committee will be notified if a notice of appeal is filed and ultimately the results of said appeal. It is assumed that if an ethical problem occurs with the Ethics Committee of the AAFS with an ABFA Diplomate who is a member of the AAFS, a report would be given to the President or Secretary or chairperson of the Ethics Committee of the ABFA for any consideration or action.
American Board of Forensic Anthropology (2008).

Appendix C: Scientific Working Group for Forensic Anthropology code of ethics and conduct

1.0 Principle, spirit and intent
The profession of Forensic Anthropology demands a high level of scientific competence and an ethical practice that is beyond reproach. Forensic anthropologists shall maintain a high level of personal and professional integrity in accordance with the trust placed upon them by the government, the legal system and the public.

2.0 Purpose and scope
Because the recovery, analysis, and identification of human remains are complex procedures, often performed for diverse jurisdictional entities, this document outlines general principles to assist Forensic Anthropologists in performing their work in an

ethical manner. In the absence of specific guidelines or procedures, the principle, spirit and intent of this code should be met.

3.0 Code of ethics and conduct

Forensic Anthropologists shall not misrepresent themselves or their work, misappropriate tangible or intellectual property, evade the truth, conspire to deceive, demonstrate disrespect to the dead or their family members, or otherwise betray the confidence placed in them by the public. Specifically, Forensic Anthropologists shall:

1. Refrain from professional or personal conduct adverse to the best interests of the profession of Forensic Anthropology, including illegal or unethical conduct and the use of their name and credentials in support of illegal or unethical activity.
2. Refrain from misrepresenting their education, training, experience, or expertise.
3. Refrain from knowingly engaging in misrepresenting data upon which expert opinion or conclusion is based. This includes plagiarism, failure to appropriately credit work done by others, falsification of data, falsification of the conditions under which data were obtained, and falsification of the results derived from the data.
4. Avoid actual or apparent undue internal and external political, commercial, financial, and other pressures, influences, and conflicts of interest adversely affecting the quality and integrity of their work.
5. Avoid or disclose involvement in any activities diminishing confidence in their competence, impartiality, independence of judgment, or operational integrity.
6. Treat all human remains with respect and dignity.
7. Treat all evidence with the care and control necessary to ensure its integrity.
8. Comply with all pertinent jurisdictional laws and guidelines regarding the collection, use, and disposition of evidence, particularly those related to human remains.
9. Ensure that all analytical techniques and methods used are appropriate, accurate and reliable to current scientific standards.
10. Refrain from unauthorized data collection.
11. Treat all information received from an investigating agency or client with the confidentiality required.
12. Refrain from extrajudicial statements relating to on-going investigations except those that a reasonable person would expect to be disseminated by means of public communication. Refrain from unauthorized statements relating to on-going investigations that do more than state without elaboration the factual and analytical information contained in a public record.
13. Fully and completely disclose all findings to the submitting agency.
14. Testify in a clear, straightforward manner and refrain from speculation beyond their scope of expertise and professional competence.
15. Avoid solicitation of work when doing so runs counter to the public interest.
16. Not render services on a contingency fee basis.
17. Maintain intellectual independence and impartiality during the analysis of the evidence.
18. Ensure that personnel under their direction are appropriately trained and are vetted prior to assisting with anthropological casework.
19. Report violations of the Code of Ethics using the appropriate mechanism.
20. Carry out the duties of their profession in such a manner so as to instill the confidence of the public, the forensic science community, and the medicolegal system.
Scientific Working Group for Forensic Anthropology (2010).

NOTE

1   A material fact is one that is important or crucial to an issue, and is sometimes subject to interpretation. For example, if a forensic anthropologist said that she had been practicing for "about 30 years" when she had actually practiced for 27 years, is that a material misrepresentation?

REFERENCES

American Academy of Forensic Sciences (2011). American Academy of Forensic Sciences Bylaws. Article II. Codes of Ethics and Conduct. www.aafs.org/aafs-bylaws.

American Board of Forensic Anthropology (2008). ABFA Manual. Code of Ethics and Conduct. www.theabfa.org/American%20 Board%20of%20Forensic%20 Anthropology%20Diplomate%20 Reference%20Manual%20060311.pdf .

Barnett, P.D. (2001). *Ethics in Forensic Science: Professional Standards for the Practice of Criminalistics*. CRC Press, Boca Raton, FL.

Bowen, R.T. (2010). *Ethics and the Practice of Forensic Science*. CRC Press, Boca Raton, FL.

Committee on Identifying the Needs of the Forensic Sciences Community, National Research Council. (2009). *Strengthening Forensic Science in the United States: A Path Forward*. National Academies Press, Washington DC.

Florida Statute (2010). 406.135, last modified 2010, signed May 7, 2010. Title XXIX Public Health Section revised House Bill No. 7115, Chapter 2006–263 (filed June 20, 2006).

General Medical Council (2011). Recordings for Which Separate Consent is not Required. www.gmc-uk.org/guidance/ ethical_guidance/7840.asp.

Hardcastle, R. (2009). *Law and the Human Body: Property Rights, Ownership and Control*. Hart Publishing, Portland, OR.

Human Tissue Authority. (nd). Code of Practice 3. Post-Mortem Examination. www.hta.gov.uk/legislationpoliciesand-codesofpractice/codesofpractice/code3 post-mortem.cfm?FaArea1=customwid gets.content_view_1&cit_id=687&cit_ parent_cit_id=680)

Minnesota Statutes (2011). 390.11 Subd.2. Autopsies, revised 2006, in House Legislative Session 84; Bill HF2656 (filed June 2, 2006).

Royal Liverpool Children's Inquiry. (2001). Summary and Recommendations. www. legislation.gov.uk/ukpga/2004/30/ contents.

Scientific Working Group for Forensic Anthropology. (2010). Code of Ethics and Conduct, issue date June 1, 2010. http://swganth.startlogic.com/ Ethics%20Rev0.pdf.

US Department of Health and Human Services. (2009). Policy for Protection of Human Research Subjects (Subpart A, Title 45, Section 46.101). United States Basic Health and Human Services. www. hhs.gov/ohrp/humansubjects/ guidance/45cfr46.html#46.101.

CHAPTER 34 An "Outsider" Look at Forensic Anthropology

James M. Adovasio

INTRODUCTION

When I initially was asked to prepare an "outsider's" view of forensic anthropology, I was somewhat hesitant for several reasons. First, and perhaps most importantly, I am not exactly a bona fide stranger to this arena. Indeed, since the late 1980s, I and several associates have been involved in a series of Archaeological Resource Protection Act (ARPA) prosecutions, the successes of which pivoted around a somewhat arcane applied specialty of geoarchaeology known as forensic sedimentology. In fact, I have been involved heavily in every ARPA prosecution requiring sedimentological recovery and analysis and participated in the first ARPA case ever argued before a jury. Moreover, I am currently involved in several such investigations.

My second reservation about preparing this commentary is rather more complicated. In addition to my forays into the rarefied world of forensic sedimentology, I have lectured in Dennis Dirkmaat's highly successful Short Courses in Forensic Anthropology series since their initiation. In these courses, I normally address some of the same topics I mention below, namely the critical importance of hyper-rigorous data collection, documentation, and analytical protocols in *one* aspect of forensic anthropology: the investigation of outdoor crime scenes and/or mass disaster loci. While I feel both competent and confident in offering comments on this subset of forensic anthropological activities as well as my own generally obscure forensic specialization, my exposure to the broader field of forensic anthropology is rather more circumscribed. Put most simply, I have *no* first-hand familiarity with most of what

A Companion to Forensic Anthropology, First Edition. Edited by Dennis C. Dirkmaat.
© 2012 John Wiley & Sons Ltd. Published 2015 by John Wiley & Sons Ltd.

forensic anthropologists *actually do, especially in regard to laboratory analyses.* While I understand that these investigations involve both traditional and innovative new ways to age, sex, characterize, and otherwise interpret human remains, I have little personal experience with these activities. Furthermore, outside of the narrow forensic applications noted previously, I have not read extensively the literature in the field. Given these not inconsiderable caveats, the following personal observations are offered.

FORENSIC ANTHROPOLOGY AS A DISCIPLINE

Even a casual perusal of the literature of forensic anthropology from the past several decades reveals that there is little general agreement, and perhaps even none, concerning the fundamental nature of this field. As is well known to the readers of this volume, there is a school of thought which sees forensic anthropology as simply an applied version of physical anthropology. In this perspective, forensic anthropology is mainly about the analysis of materials – principally osteological remains – in a laboratory context. As such, from that perspective it is not viewed as a discipline or even a subdiscipline, but rather as the practical application of a minor component of the parent field of study.

To others, notably the volume's editor and most of its contributors, forensic anthropology is a far broader academic and applied pursuit. In this larger perspective, it includes not only the "traditional" mainly laboratory-based activities but also a substantial and critical field aspect. This field-based component of forensic anthropology has been labeled "forensic archaeology" (Connor and Scott 2001; Dirkmaat and Adovasio 1997; Hochrein 2002; Lovis 1992; Skinner and Lazenby 1983; Stoutamire 1983; see also Chapter 2 in this volume). In this guise, forensic anthropology adds the rigorous collection and documentation of field data to the equally rigorous treatment of recovered materials in the laboratory.

In my own view, *neither* the narrower nor broader perspective on forensic anthropology characterizes or describes a stand-alone discipline. Whether one restricts this area of study to the laboratory or includes a field component, forensic anthropology conspicuously lacks a body of accumulated theory upon which it is based or grounded. Historical archaeology, for example, is a recognized subdiscipline of archaeology which has such a theoretical grounding derived in part from archaeology and history. However, no matter how forensic anthropology is defined in terms of the scope of its activities, it does not seem to have a comparable theoretical foundation. If there is a corpus of theory in any or all of it iterations, it remains unclear.

FORENSIC ANTHROPOLOGY AS A SCIENCE

Already having offended some of its practitioners by denying its status as a stand-alone field, I might as well offend most of the remainder of its players by suggesting it is not a science either. Before the cries of "blasphemy" from the offended or disconsolate rise to a crescendo, I also should state for the record (and with due respect to my colleague and friend Lew Binford and his many disciples) that nor is anthropological archaeology a science. Rather, as I have attempted to tell generations of my students, most archaeologists use scientific methods or the protocols of scientific inquiry as tools to examine, illuminate, and explain (with an intentionally small "e") the often very remote post. For a given field of inquiry to use science, however, does not necessarily mean that that field is a science.

This is precisely the same way I see forensic anthropology. Forensic anthropology in either the field or the laboratory *uses* scientific procedures to examine past human activities and the actors and actresses who performed them. The only difference is that, unlike archaeology, the past for a forensic anthropologist is often defined in years, weeks, days, hours, or even minutes: not centuries or millennia.

As a final note on this particular matter, I offer the observation that while there is a vocal cadre of archaeologists who believe that archaeology is not and never can be a science on the level of mathematics, chemistry, or physics, there is also a subset who subscribe to the notion that archaeologists should not employ scientific protocols as investigative methods. For those only too well-known postmodernists, anthropological archaeology can exist quite apart from the rigorous application of scientific methods. Suffice to say, forensic anthropology without scientific protocols could not exist at all.

SCIENCE IN FORENSIC ANTHROPOLOGY

If the premise is accepted that the rigorous application of scientific methods is the *sine qua non* of forensic anthropology, then how is this to be accomplished? I submit that much of this volume as well as a substantial portion of the available literature (or, minimally, that segment which I have read), adequately addresses scientific methods in laboratory contexts. Hence, there is no need to reiterate that information here. On the other hand, with a few notable exceptions (e.g., Dirkmaat and Adovasio 1997; Dirkmaat and Cabo 2006; Dirkmaat et al. 2008; Morse et al. 1983; Skinner and Lazenby 1983), the application of suitably resolute recovery and documentary protocols in field situations often is addressed minimally or ignored entirely.

I find this lacuna to be both curious and ultimately totally unacceptable. It is curious because some of the forensic anthropologists I have known who insist on the most stringent analytical and documentary methodologies in the laboratory have no similar concerns with data collection and documentation in the field.

Jesse D. Jennings, the late Dark Lord of the Desert, once referred to the products of a certain theoretically elegant but methodologically impoverished anthropology department by stating "they can't dig their way out of a sand box!" (Jennings, personal communication, 1970). In my view, the same admittedly harsh condemnation may be applied to many exclusively laboratory-focused forensic anthropologists. Why does this matter and why is this unacceptable?

Reduced to its simplest rendering, the quality and ultimate utility of data analyzed and processed in the laboratory is only as good as its recovery and documentation in the field. This obviously applies to archaeological and geoarchaeological materials, and even more so to forensic evidence employed in courtroom situations. Archaeological data are used to support or rebut some position about events in the often distant past; forensic evidence is used to exonerate or incarcerate a living person in the present. Again, rendered most succinctly, the establishment and precise documentation of context and the demonstration of association for most archaeologists serves to reinforce or demolish some paradigm or hypothesis about long-dead folk. For forensic anthropologists, context and association literally can be life and death issues. If the officers who arrested O.J. Simpson or investigated the scene of his alleged crime did not appreciate this fully, his defense attorneys certainly did. Paradoxically, many forensic anthropologists – as well as the agencies who hire them or with whom they collaborate – often do not.

As I have frequently stressed to the attendees of Dr Dennis Dirkmaat's forensic anthropology summer programs, the rigorous documentation of an outdoor or field crime scene or disaster setting is accomplished best using methods drawn from contemporary archaeology. Again, why? Archaeologists have developed exquisite protocols to document and recover data from scenes of human interaction with the landscape which are, in some cases, hundreds or thousands of years old. Moreover, these methods can address the relative integrity or degree of disturbance of the landscape before, during, and after the episode of human activity. If these methods work with sites as old as the dawn of the cultural record, they certainly should work with or at loci of human activity occupied or visited a few hours ago, yesterday, or last week.

This is not to say, however, that the employment of scientific techniques in forensic field contexts is simply about borrowing. Instead, and much more importantly, it implies using these techniques within a mindset that understands and appreciates fully *why* these protocols are being employed. In my view, the use of high-resolution field protocols *does not* mean that the practitioner has the ability to use a total station or submeter GPS unit like a professional surveyor; to take field photographs with the very latest Nikon or Canon camera with dedicated lenses; or to wield a trowel like a Jedi light saber. It *does* mean that the individuals in charge of forensic field operations, like their archaeological counterparts, should grasp the fundamental realization that he or she has three elemental responsibilities: (i) the delineation of stratigraphy from observable stratification, (ii) the establishment of context, and (iii) the demonstration of association.

Though often used synonymously, stratigraphy and stratification are not the same thing. Stratification refers to the physical presence of strata; that is, bodies of sediment with a constellation of actual, measurable, quantifiable, physical attributes. These notably include grain size and its corollary, texture; mineralogical, geochemical, and elemental composition; and color. These attributes, in turn, can reflect the source area(s) of the sediment, mode(s) of transportation, mechanism(s) of deposition, or the kind(s) and degree of disturbance. Strata have tops and bottoms (contacts and interfaces respectively) which bracket the enclosed sediments and whose successful and precise recognition is critical to interpretation. In short, strata are products; how they came to be and their arrangement as they are found is a process, the study of and the substance of which is stratigraphy. While all outdoor forensic sites, like their archaeological counterparts, are not vertically stratified, many are. Furthermore, those lacking vertical stratification often exhibit overlapping evidence of sequential use of the same surface in the form of horizontal stratification. In any case, ignorance of stratification precludes the determination of stratigraphy, which, in turn, prohibits the establishment of context or the demonstration of association.

Context is simply the place of an artifact, ecofact, or other object in time and space. If the field operative effectively delineates stratification and excavates and documents accordingly, context follows logically therefrom.

Though also derived from the delineation of stratification, association is among the most difficult things to demonstrate for the field archaeologist or the field forensic anthropologist. It should be stressed that association is not simply co-occurrence in the same stratigraphic level or on the same depositional surface. Rather, and put simply, association implies that two or more items entered or were emplaced in the depositional record of a site penecontemporaneously as a result of the same process. In graphic terms, it is the projectile point lodged *in* the bone not simply *with* the bone. In forensic terms, it is primary, direct, and probative: not circumstantial evidence.

Consistent and effective implementation of the foregoing requires not simply a working appreciation of which specific tools and techniques may be employed in various situations but also and, far more basically, an operative knowledge of the fundamental principles of stratigraphy as defined by the seventeenth century Danish scientist Nicolas Steno in his seminal work *dissertationis prodromus* (1669). This understanding of stratigraphy, in turn, requires a grasp not simply of Steno's so-called First Law – superposition – but also of the seldom-cited or appreciated other laws: original horizontality, lateral continuity, and (especially) intersecting relationships.

I could continue in this vein, but only at the risk of fundamentally altering the purpose of this commentary. Suffice to note that without regular adherence to scientifically rigorous data recovery and documentation protocols in the field – as in the laboratory – there is *no* science in or unique to forensic anthropology.

Sadly, the usual reactions to all of this are that the use of such field methods requires highly trained personnel or that such procedures take too much time and/or cost too much! Unfortunately, this is also the response of many cultural resource management (CRM) archaeologists, not a few academic archaeologists and especially the agencies, institutions, or organizations who employ them. With this myopic view in mind, I can state only the obvious. When an outdoor forensic site or, for that matter, an indoor crime scene is processed, its integrity simultaneously is compromised or destroyed. If that examination and concurrent documentation are not properly done at the highest order of resolution, there is *no* corrective recourse. While this situation may be "only" career-damaging or ego-bruising in an archaeological context, it literally can be fatal in a forensic case.

Forensic Anthropology: *Quo Vadis?*

While I may have some notions of what I think forensic anthropology is and is not, I feel quite unsure as to the assessment of where the field may be heading. As an academician within anthropology for the past four decades, I have witnessed forensic anthropology evolve from what was once primarily a laboratory enterprise to something larger and broader. However, as noted above, for many it remains and always will be a laboratory-focused, applied area of study. In this iteration, forensic anthropology was, is, and will continue to be the poor stepsister of physical anthropology just as CRM is viewed by some as the lesser sibling of archaeology. As for those like Dr Dennis Dirkmaat and his colleagues in this volume, who embrace and promote the larger vision that forensic anthropology must include a rigorous field component, I am uncertain how many others subscribe or will ever subscribe in the future to that enlarged view.

As an academic administrator for the better part of the last three decades, I also have witnessed the number of students attracted to a specialization in forensic anthropology steadily increase, at least in the institutions where I have worked. This growth has been almost exponential in the very recent past. However, I fear that at least in the case of the contemporary "spike" that the interest in forensic anthropology may be due to the halo effect of the contemporary fascination with forensic investigative crime drama television programs. When that glow fades, perhaps the allure will diminish as well.

On a more abstract level, I presently do not see any tendency for forensic anthropology, however narrowly or broadly it is defined, to achieve the disciplinary status of archaeology or physical anthropology, let alone that of anthropology at large. The absence of a grounding body of theory remains a fatal flaw and shows no sign of correction. This may explain why so few institutions offer a specialization in forensic anthropology at the doctoral level.

Whatever its academic status, I believe forensic anthropology can remain a viable area of research and practice *if* certain conditions are met. First, it would be potentially disastrous if the laboratory-focused side of the field split away from its more field-oriented brethren. Similarly, I believe that forensic anthropology will remain vibrant *only* as long as it retains relevancy and credibility not only to, or within, its parent field(s) but also within the umbrella of forensic science at large. To do this, of course, will necessitate the maintenance of rigorous scientific standards across the board and the ability to convince the field's "clients" of the utility of the information it provides.

Fortunately, from my vantage point as an occasional contributor to an esoteric forensic specialization – as well as from my view as a dean and, now, provost – I foresee neither a fatal bifurcation in forensic anthropology nor any abandonment of scientific rigor in the laboratory or in field situations.

Finally, while not exactly analogous, I ultimately view the future potential of forensic anthropology in much the same way as I consider the history and trajectory of geoarchaeology. When this interface field first emerged three decades ago, it struggled for self-definition. It was not highly regarded by most geologists who found it "too soft" and many archaeologists who didn't know even what geoarchaeologists were trying to do. Now, however, geoarchaeology is modestly thriving and is widely recognized as an academically legitimate area of inquiry. Based on this admittedly imperfect analogue, I am confident that the next "outsider" who comments on forensic anthropology several decades hence will find this area of scholarly inquiry and practice alive and well.

REFERENCES

Connor, M. and Scott, D.D. (2001). Paradigms and perpetrators. *Historical Archaeology* 35(1): 1–6.

Dirkmaat, D.C. and Adovasio. J.M. (1997). The role of archaeology in the recovery and interpretation of human remains from an outdoor forensic setting. In W.D. Haglund and M.H. Sorg (eds), *Forensic Taphonomy: The Postmortem Fate of Human Remains* (pp. 39–64). CRC Press.

Dirkmaat, D.C. and Cabo, L.L. (2006). The shallow grave as an option for disposing of the recently deceased: goals and consequences [abstract]. *Proceedings of American Academy of Forensic Sciences* 12: 299.

Dirkmaat, D.C., Cabo, L.L., Ousley, S.D., and Symes, S.A. (2008). New perspectives in forensic anthropology. *Yearbook of Physical Anthropology* 51: 33–52.

Hochrein, M.J. (2002). An autopsy of the grave: recognizing, collecting, and processing forensic geotaphonomic evidence. In W.D. Haglund and M.H. Sorg (eds), *Advances in Forensic Taphonomy: Method, Theory, and Anthropological Perspectives* (pp. 45–70). CRC Press, Boca Raton, FL.

Lovis, W.A. (1992). Forensic archaeology as mortuary anthropology. *Social Sciences and Medicine* 34(2): 113–117.

Morse, M., Duncan, J., and Stoutamire, J. (eds) (1983). *Handbook of Forensic Archaeology and Anthropology.* Rose Printing Co., Tallahassee, FL.

Skinner, M. and Lazenby, R.A. (1983). *Found! Human Remains: A Field Manual for the Recovery of the Recent Human Skeleton.* Archaeology Press, Burnaby, BC.

Stoutamire, J. (1983). Excavation and recovery. In D. Morse, J. Duncan, and J. Stoutamire (eds), *Handbook of Forensic Archaeology and Anthropology* (pp. 20–47). Rose Printing Co., Tallahassee, FL.

Index

abrasions, 350
  see also trauma
academic research, 644
accidental death, 538
accumulated degree-days (ADD),
    484, 492
  see also taphonomic processes [agents]
Adams, Bradley, 542–3
ADBOU, 217–18
adipocere, 484, 487–8, 490, 513–14
Adovasio, James, 53
  see also Meadowcroft Rockshelter
aerial photographs, 86, 105
Africa
  Rwanda, 640
  Sierra Leone, 640
  South Africa, 627–40, 647
  sub-Saharan, 627
age estimation, 175, 202–19, 224–38, 333,
    351, 466, 553–4, 556–8, 565
  accuracy, 203, 208, 230
  adult, 202–19
  age indicators, 225–8
  age intervals, 230
  ancestry and, 215–16
  asymmetry and, 216
  auricular surface, 226, 228
  bias, 232
  biological age, 202–3, 225, 230

Buckberry and Chamberlain method,
    206–8
chronological age estimation, 183, 202–3,
    225–32, 629–32
computed tomography and, 228
cranial suture closure, 6, 183, 204,
    208–11, 217, 226, 228, 630
dental maturation sequences, 630
environmental influences and, 203
epiphyseal-diaphyseal fusion patterns, 143,
    183, 203–4, 226
history of, 9
Igarashi et al. method, 208
juvenile, 203–4, 226
logit or probit models, 17
Lovejoy et al. method, 206, 208, 217
morphological features, 225–6
multifactorial approach, 216–19
Osborne et al. method, 208
osteoarthritis, 215
phase analysis, 16
problems with, 230–3
pubic symphysis, 204–6, 209, 211,
    216–17, 226, 228, 232
quantitative methods, 227, 234
reference samples, 228
retroauricular area, 206
sex and, 215–16
spheno-occipital synchondrosis, 204

A Companion to Forensic Anthropology, First Edition. Edited by Dennis C. Dirkmaat.
© 2012 John Wiley & Sons Ltd. Published 2015 by John Wiley & Sons Ltd.

Printed and bound by CPI Group (UK) Ltd, Croydon, CR0 4YY

16/04/2025

14658833-0004